Genetic Improvement of Solanaceous Crops

Volume 2: Tomato

Genetic Improvement of Solanaceous Crops

Volume 2: Tomato

Editors

Maharaj K. Razdan
University of Delhi, Delhi
India

Autar K. Mattoo
USDA-ARS
Sustainable Agricultural Systems Laboratory
Beltsville, MD
USA

CRC Press
Taylor & Francis Group
Boca Raton London New York

CRC Press is an imprint of the
Taylor & Francis Group, an **informa** business
A SCIENCE PUBLISHERS BOOK

First published 2007 by Science Publishers Inc.

Published 2019 by CRC Press
Taylor & Francis Group
6000 Broken Sound Parkway NW, Suite 300
Boca Raton, FL 33487-2742

First issued in paperback 2019

No claim to original U.S. Government works

ISBN 13: 978-0-367-45390-9 (pbk)
ISBN 13: 978-1-57808-179-0 (hbk)

Visit the Taylor & Francis Web site at
http://www.taylorandfrancis.com

and the CRC Press Web site at
http://www.crcpress.com

CIP data will be provided on request.

Cover Illustration

The photograph of *Transgenic Tomato Fruit Bunch* is reproduced from the paper **Engineered polyamine Accumulation in Tomato enhances phytonutrient Content, Juice Quality, and Vine Life** (*Roshni A. Mehta, Tatiana Cassol, Ning Li Nasreen Ali, Avtar K. Handa, Autar K. Mattoo*) Published in *Nature Biotechnology 20, 613-618 (01 June 2002)*. The photograph was taken by an USDA photographer, Peggy Greb. The Editors and Publisher express their gratitude to her.

Preface

Tomato is one of the most consumed vegetables in the world and is the dietary source of vitamins, minerals and fiber, which are important for human nutrition and health. Fresh fruits are used in salads, various culinary preparations, juices, or processed in the form of purees, concentrates, condiments and sauces. Tomato plants are grown worldwide in the field, or in greenhouses. Genetic improvement of this Solanaceous crop has been an on-going process with the objective of gaining high fruit yield, enhanced fruit nutritive value, controlled fruit maturation and ripening, and developing resistance to phytophagous insects, microbial pathogens, and various abiotic stresses. More importantly, with the increase in the world population, the quantum of tomato consumption has considerably increased and farmers, agronomists and horticulturists have had to walk a tight rope to enhance yield without losing sight of the production quality to meet the demands of the fresh market and the processing industry. Of the nearly 3 million hectares under vegetable cultivation the tomato crop occupied one-third of this global area with total tomato production in 1994 reported as 77.5 Mt, averaging 27 t ha^{-1}. Most of the production increases hitherto have been achieved using conventional methods of selection and breeding coupled with improved growth practices: use of fertilizer, improved irrigation, and pest management. Other advancements have been possible through the application of molecular markers to ease selection process and technological innovations such as development of genetically enhanced tomatoes engineered for high quality and resistance to disease and extreme environments. This book presents a critical appraisal of the state-of-the-art findings on this crop in the form of overviews, emphasizing various approaches and strategies used for its improvement through research conducted at various research institutes, organizations and universities world over.

Improvement of a particular crop can best be envisaged when comprehensive information is known of its origin and available genetic resources. The controversy over the taxonomic status of the cultivated tomato *(Lycopersicon esculentum* Mill.) has been resolved. Application of molecular breeding techniques (RAPDs, RFLPs) and genomics research has now convinced the research community to place tomato under the genus *Solanum,* namely, *Solanum lycopersicum* L.

(http://www.sgn.cornell.edu/about/solanum_nomenclature.pl). This implies that one could explore the gene pool among all *Solanum* species for improvement of this and other Solanaceous crops. Conservation of all tomato genetic resources is, therefore, all the more necessary. Chapter 1 starts with history, origin and early cultivation of tomato. Interestingly, Peru is considered to be a likely place of domestication of tomato and yet another hypothesis sees its first domestication in Mexico. Tomato gene banks have been established in USA and other countries where currently more than 75000 accessions of tomato are preserved. These gene banks maintain data concerning the reproductive biology of conserved accessions, world production scenario of fresh-market as well as processed tomatoes, and descriptive list of characteristics of various collections. Information dealing with these aspects is given in Chapter 2. Role of cytogenetics in evolution and selection of tomato variants with improved traits is elaborated in Chapter 3. This chapter further provides an overview of tomato genomics through genetic maps constructed by applying conventional and molecular breeding techniques in order to assess variability among various tomato genetic resources such as mutants, wild species, intra- and inter-specific populations as well as introgressed lines derived from recombination experiments. By integrating classical gene linkage maps with the high-resolution molecular maps it is now possible to evaluate the degree of similarity in basic genomic structure of tomato sp. Some plants maintain superior traits vigorously only in hybrid form. This phenomenon called heterosis is recognized as one of the primary factors contributing to manifestation of superiority in respect of some quantitative traits of tomato. Chapter 4 describes the strategies for using heterosis for developing tomatoes with certain quantitative traits, whereas Chapter 5 elaborates on improvement of quality traits using traditional and enhanced breeding methods. Tomato fruits are major dietary sources of antioxidant lycopene, and vitamins A and C, besides other micronutrients/antioxidants, which largely contribute to tomato fruit quality. Tomato breeders have been examining a wealth of genetic variability available in the present day heirloom cultivars, land races, and related wild tomato species in respect of various dietary sources. Approaches to genetically enhance tomato fruit's nutritive value are highlighted in Chapter 6. Molecular markers have proven very useful in selection of elite tomato germplasm, and efforts to best utilize various molecular genetic approaches have led to an understanding of the physiological basis of drought resistance response in tomatoes. An overview of research done on these aspects is provided in Chapters 7 and 8. Recent advances in plant genetic engineering have made it possible to produce transgenic tomato plants with characteristics for a number of improved traits. A general account of genetic engineering technology applied for production of tomatoes transformed for various traits is given in Chapter 9. Hormonal control of fruit maturation, its molecular basis, and future prospects of applying microarray analysis, as well as proteomics, essentially for producing designer-tomato fruits with enhanced shelf-life are discussed in Chapter 10. Biochemical and molecular mechanisms in fruit ripening have projected insights on existence of molecular links between distinct fruit ripening types in tomato. A number of genes involved in ethylene biosynthesis as well as light signaling are implicated, which reportedly provide targets for manipulation of fruit color, nutrient content, and cell-wall breakdown during the process of ripening. Details of these, based on model systems proposed for fruit ripening of both climacteric and nonclimacteric fruits, are summarized in Chapter 11. An inherent problem encountered in crops, including tomato, is the huge annual yield losses incurred due to diseases caused by pathogens and pests. Molecular breeding coupled with application of transgenic technology has to a greater extent the potential to circumvent this problem for tomato by producing cultivars resistant to bacteria, fungi, viruses, and insects as well as mite pests. Recent findings on these aspects of tomato resistance are reviewed in Chapters 12-15. Finally, the considerable progress made toward understanding the physiological bases of plant tolerance to different abiotic stresses and characterization of tolerant tomato genotypes to stresses, such as salinity, cold and heat, are discussed in Chapter 16.

There has been a long-felt need to have documented, comprehensive information on the improvement of tomato in one place. This book attempts to accomplish this goal. It should be useful not only to breeders, or other specialists, but equally benefit teachers as well as students seeking information on aspects of tomato biology, genetics and biotechnology. It will be apparent to the readers that some authors have used the revised classification of tomato as *Solanum lycopersicum* while others have kept to the older nomenclature, viz., *Lycopersicon esculentum*. We have let both these usages in the book till the transition over the next few years is completed.

This compendium has become a reality only through the expert contributions of the 33 authors from 6 countries and generous support and encouragement from the publishers. We sincerely thank them all.

October 2006   **Maharaj K. Razdan**
                                                                                    Delhi, India

                                                                              Autar K. Mattoo
                                                                              Beltsville, USA

List of Contributors

Atanassova, Bistra, Department of Applied Genetics, (formerly Department of Heterosis), Institute of Genetics "Prof. D. Kostov", Bulgarian Academy of Sciences, Sofia 1113, Bulgaria, *e-mail: bistra_a@yahoo.com.*

Causse, Mathilde, INRA, Fruit and Vegetable Genetics and Breeding Research Station, BP 94 - 84143 Avignon, France. *e-mail: mathilde.causse@avignon.inra.fr*

Chetelat, Roger T., Department of Plant Sciences, University of California, One Shields Avenue, Davis, CA 95616, USA. *e-mail: trchetelat@ucdavis.edu*

Damidaux, René, INRA, Fruit and Vegetable Genetics and Breeding Research Station, BP 94 - 84143, Avignon, France.

Foolad, Majid R., Department of Horticulture, 217 Tyson Building, The Pennsylvania State University, University Park, PA 16802, USA. e-mail: mrf5@psu.edu

Fox, Elizabeth, Department of Plant Biology, Cornell University, Ithaca, NY 14853, USA.

Francis., David M., Department of Horticulture and Crop Science, The Ohio State University, OARDC, 1680 Madison Ave, Wooster, OH 44691, USA. e-mail: francis. 77@osu.edu

Gardner, R.G., Mountain Horticultural Crops Research and Extension Center, North Carolina State University, 455 Research Drive, Fletcher, NC 28732, USA. e-mail: randy-gardner@mcsu.edu

Georgiev, Hristo, Department of Applied Genetics, (former Department of Heterosis) Institute of Genetics "Prof. D. Kostov", Bulgarian Academy of Sciences, Sofia 1113, Bulgaria.

Giovannoni, Jim, USDA-ARS Plant, Soil and Nutrition Lab, Boyce Thompson Institute for Plant Research, Tower Rd., Ithaca, NY 14853, USA. *e-mail: jjg33@cornell.edu*

Handa, Avtar K., Department of Horticulture and Landscape Architecture, 625 Agricultural Mall Drive, Purdue University, West Lafayette, IN 47907-2010, USA. *e-mail: ahanda@purdue.edu*

Hazarika, P.J., Department of Genetics, University of Delhi–South Campus, Benito Juarez Road, New Delhi 110021, India.

Ji, Yuanfu, Gulf Coast Research and Education Center, University of Florida, 14625 CR 672, Wimauma, Fl. 33598, USA.

Kennedy, George G., Department of Entomology, Box 7630, North Carolina State University, Raleigh, NC 27695-7630, USA. *e-mail: george_kennedy@ncsu.edu.*

Labate, Joanne A., Plant Genetic Resources Unit, United States Department of Agriculture, Agricultural Research Service, 630 West North Street, Geneva, New York 14456-0462, USA.

Madhulatha, P., Department of Genetics, University of Delhi–South Campus, Benito Juarez Road, New Delhi 110021, India.

Medina, Andrea L., Department of Plant and Environmental Sciences, New Mexico State University, P.O. Box 3003, MSC3Q Las Cruces, NM 88003, USA.

O'Connell, Mary A., Department of Plant and Environmental Sciences, New Mexico State University, PO Box 30003, MSC 3Q, Las Cruces, NM 88003, USA. *e-mail: moconnel@nmsu.edu*

Pandey, R., Department of Genetics, University of Delhi–South Campus, Benito Juarez Road, New Delhi 110021, India.

Peralta, Iris E., Facultad de Ciencias Agrarias, Universidad Nacional de Cuyo, C.C. 7, Chacras de Coria 5505, Lujan, Mendoza, Argentina, and CONICET-IADIZA, C.C. 507, Mendoza 5500, Argentina.

Perla, Venu, Department of Horticulture and Landscape Architecture, 625 Agricultural Mall Drive, Purdue University, West Lafayette, IN 47907-2010, USA.

Rajam, M.V., Department of Genetics, University of Delhi–South Campus, Benito Juarez Road, New Delhi 110021, India. *e-mail: mv_rajam@hotmail.com*

Razdan, M.K., Department of Botany, Ramjas College, University of Delhi (Main Campus), Delhi 110007, India. Present Address: Principal, Shyam Lal College (University of Delhi), Delhi 110032, India.

Robbins, Matthew D., Department of Horticulture, University of Wisconsin Madison, 1575 Linden Drive, Madison, WI 53706, USA.

Robertson, Larry D., Plant Genetic Resources Unit, United States Department of Agriculture, Agricultural Research Service, 630 West Street, Geneva, New York 14456-0462, USA. e-mail: lrobertson@pgru.ars.usda.gov

Rousselle, Patrick, INRA, Fruit and Vegetable Genetics and Breeding Research Station, BP 94 - 84143 Avignon, France.

Sánchez Peña, Pedro, Department of Plant and Environmental Sciences, New Mexico State University, P.O. Box 3003, MSC 3Q Las Cruces, NM, 88003, USA.

Scott, J.W., Gulf Coast Research and Education Center, University of Florida, Institute of Food and Agricultural Sciences, 14625 CR 672, Wimauma, FL 33598, USA. *e-mail: jwsc@ufl.edu*

Spooner, David M., USDA, Agricultural Research Service, Vegetable Crops

Research Unit, 'Department of Horticulture, University of Wisconsin, 1575 Linden Drive, Madison, Wisconsin 53706-1590, USA. *e-mail: dspooner@wisc.edu*

Srivastava, Alka, Department of Horticulture and Landscape Architecture, 625 Agricultural Mall Drive, Purdue University, West Lafayette, IN 47907-2010, USA.

Stevens, Mikel R., Department of Plant and Animal Sciences, 287 Widstoe Building, Brigham Young University, Provo, Utah 84602, USA. *e-mail: mikel_stevens@byu.edu*

Stommel, John R., United States Department of Agriculture, Agricultural Research Service, Vegetable Laboratory, Building 010A, BARC-West, 10300 Baltimore Avenue, Beltsville, MD 20705, USA. *e-mail: stommelj@ba.ars.usda.gov*

Treviño, Marcela B., Department of Plant and Environmental Sciences, New Mexico State University, P.O. Box 3003, MSC 3Q, Las Cruces, NM, 88003, USA.

Yang Wencai, Department of Vegetable Science, College of Agronomy and Biotechnology, China Agricultural University, No.2 Yuanmingyuan Xi Lu, Haidian District, Beijing 100094, The People's Republic of China.

Abbreviations Used Throughout the Book

2-ip	6-γ-γ-dimethylamino purine
ABA	Abscisic acid
AB-QTL	Advanced backcross QTL analysis
ACC	1-Aminocyclopropane-1-carboxylic acid
ACO	1-Aminocyclopropane-1-carboxylic acid oxidase
ACS	1-Aminocyclopropane-1-carboxylic acid synthase
ADC	Arginine decarboxylase
ADP	Adenosine diphosphate
AFLP	Amplified fragment length polymorphism
AIS	Alcohol insoluble acids
AMOVA	Analysis of Molecular Variance
AMV	Alfa mosaic virus
ARDC	Asian Research Development Centre
ARFs	Auxin response factors
ARS	Agriculture Research Service
AUX	Auxins
AVDRC	Asian Vegetable Research and Development Centre, Taiwan
AVG	Aminoethoxyvinylglycine
aw	Anthocyanin without marker
BAC	Bacterial artificial chromosome
BADH	Betaine aldehyde dehydrogenase
BAP	6-benzylaminopurine
BC	Back cross
BCTV	Beet curly top virus
BP	Before present
BRs	Brassinosteroids
CAAS	Institute of Crop Germplasm Resources, China
CaMV	Cauliflower mosaic virus
CAPS	Cleaved amplified polymorphic sequence
CATIE	Centro Agronomico Tropical de Investigacion y Ensenanza, Coata Rica
CATIE	Centro Agronomico Tropical de Investigacion y Ensenanza, Costa Rica

CBFI	C-repeat/dehydration responsive element biding factor Gene
CBFI	C-repeat/dehydration responsive element binding gene 1
CDKs	Cyclin dependant kinase
cDNA	Complementary DNA
CGC	Crop Germplasm Committee
CGN	Centre for Genetic Resources, The Netherlands
CHI	Chalcone isomerase
CHS	Chalcone synthase
CIP	International Potato Center, Lima, Peru
CKs	Cytokinins
cM	Centimorgan
CMS	Cytoplasmic male sterility
CMV	Cucumber mosaic virus
CN	Controlled nutrient experiment
CORPOICA	Corporacion Columbiana de Investigacion Agropecuaria, Colombia
COS	Conserved ortholog set
CP	Coat Protein
cpDNA	Chloroplast DNA
CS	Cold stress
CT	Cold tolerance
CTV	Curly Top Virus
DArT	Diversity Array Technology
DMW	Dry matter weight
DNA	Deoxyribonucleic acid
DS	Drought stress
DT	Drought tolerance
DW	Dry weight
EBDC	Ethylene-bis-dithiocarbamate
EC	Electrical conductivity
ELISA	Enzyme-linked immunosorbent-assay
EMBPRA	European Molecular Biology and Plant Research Association
EMS	Ethylmethane sulphonate
ER	Extreme resistance
EST	Expressed sequence tag
ETC	Electricity transport chain
ex	Exerted stigma
FAO STAT	FAO statistics base
FAO	Food and Agriculture Organisation
FISH	Fluorescent *in situ* hybridization
FS	Flavone synthase
FW	Fresh weight

G:F	Glucose and fructose ratio
GAs	Gibberrellins
GBSST	Granule bound starch synthase gene 1 waxy
GCA	General combining ability
gDW	Gram dry weight
GISH	Genomic *in situ* hybridization
GM	Genetically modified
GMA	Generation Mean Analysis
GRIN	Germplasm Resources Information Network
GRSV	Ground nut ring spot virus
GTP	Guanosine triphosphate
GUS	β-glucronidase marker
h2	heritability
ha	Hectare
HQT	Hydoxycinnamoyl transferase
HS	Heat stress
HT	Heat tolerance
IAA	Indole acetic acid
IAC	Institut Agronomico de Campines, Brazil
IBC	Inbred backcross
IL	Isogenic line/Introgression line
INRA	Institut Nationale des Research Agronomiques
IPGRI	International Plant Genetic Resources Institute
ITS	Internal transcribed spacer
JA	Jasmonic acid
Kg	Kilogram
LEA	Late embryogenesis abundance
LIS	Linalool synthase gene
LMGV	Lab. De Melhoramento Genetico Vegetale, Brazil
LTPs	Lipid transfer proteins
LTs	Low temperatures
M	Mild
MA	Monosomic addition
MABC	Marker assisted backcross
MAS	Marker assisted selection
Mbp	Mega base pairs
MCP	1-methylcyclopropene
MeJA	Methyl jasmonic acid
Meq	Milliequivalent
MG	Mature green
MS	Male sterile
mt DNA	Mitochondrial DNA
Mt	Million ton

MW	Molecular weight
N	Necrotic
NAA	1-naphthalene-acetic acid
NAFTA	Northern American Free Trade Agreement
NBPGR	National Bureau of Plant Genetic Resources India
NCED	9-cis-epoxycarotenoid dioxygenase
NCGRP	National Center for Genetic Resources Preservation
Ne	No. of individuals in a theoretically ideal population having the same magnitude of drift as the actual population
NIL	Near isogenic line
NJ	Neighbor joining analysis
NOR	Nucleolar Organizing region
NPGS	National plant germplasm system
NPTII	Neomycin phosphotransferase
NS	Nonstress conditions
NSF	National Science Foundation
NSL	National seed laboratory
nt	Nucleotide region
ODC	Ornithine decarbocylase
PABA	p-aminobenzoate
PAs	Polyamines
PAT	Polar auxin transport
PCA	Principal components analysis
PCR	Polymerase chain reaction
PDS	Phytoene desaturase
PepMV	Pepino mosaic virus
PG	Polygalacturonase
PGRs	Plant growth regulators
PGRU	Principal genetic resources unit
PI	Plant introduction
PLD	Phospholipase D
PLRV	Potato leaf roll virus
PME	Pectinmethyl esterase
PMI	Phosphomannose isomerase
PN	Photosynthetic rate
PPP	Pectinase phosphate pathway
ps	positional sterile
PTGS	Post-transcriptional silencing
PVE	Phenotypic variation explained
PVX	Potato virus X
PVY	Potato virus Y
QTL	Quantitative trait loci
RAPD	Randomly amplified polymorphic DNA

rDNA	Ribosomal DNA
REP-PCR	Repetitive element sequence PCR
RFLP	Restriction length fragment polymorphism
RGA	Resistance gene analog
RIL	Recombinant inbred line
RN	Recombination nodules
RNA	Ribose nucleic acid
RNAi	RNA interference
RS	Ringspot
RSV	Respiratory syncytial virus
SAAT	Strawberry alcohol acyltransferase
SAM	S-adenosyl methionine/S-adenosine-L-methionine
SARS	Severe acute respiratory syndrome
SC	Synaptonemal complex/self-compatible
SCA	Specific combining ability
SDM	Seedling decapitation method
SG	Seed germination
SGN	Solanaceae genomic network
SI	Self-incompatible
sl	Stamen less
SL	Substitution line
SNP Wave	SNP multiplexed genotyping technology
SNP	Single nucleotide polymorphism
SnRK1	Sucrose non-fermenting related kinase 1
SS	Salt stress
SSD	Single seed descent
SSR	Single sequence repeat
ST	Salt tolerance
STMS	Selected microsatellite markers
STS	Silver thiosulphate
SuSy	Sucrose synthase
T50	Time to 50% germination
TB	Tip blight
TCSV	Tospovirus chloritic spot virus
T-DNA	Ti-plasmid DNA segment
TEV	Tobacco etch virus
TGRC	Tomato genetic research cooperative
TI	Tolerance index
TILLING	Targeting induced local lesions in genomes
TLCV	Tomato leaf curl virus
ToMoV	Tomato mottle virus
ToMV	Tomato mosaic virus
TSWV	Tomato spotted wilt virus

TYLC	Tomato yellow leaf curl virus
TYTV	Tomato yellow top virus
UPGMA	Unweighted pair group method of arithmetic averages
USDA	United States Department of Agriculture
USDA-ERS	USDA Economic Research Service
USDA-NASS	USDA National Agricultural Statistics Center
UTR	Untranslated region
VIGS	Virus induced gene silencing
VIP	Vegetative insecticidal proteins
VM	Very mild
VS	Vegetative stage
WIS	Water insoluble solids
WUE	Water use efficiency
WVC	World vegetable center
XET	Xyluglucan endotransglycolase
XGH	Xyluglucan hydrolase
XTH	Xyluglucan transglycosylation hydrolysis
YAC	Yeast artificial chromosome

Contents

History, Origin and Early Cultivation of Tomato (Solanaceae)

Iris E. Peralta[1] and **David M. Spooner**[2*]

[1]*Facultad de Ciencias Agrarias, Universidad Nacional de Cuyo, C.C. 7, Chacras de Coria 5505, Lujan, Mendoza, Argentina, and CONICET-IADIZA, C.C. 507, Mendoza 5500, Argentina email: iperalta@lab.cricyt.edu.ar.*

[2]*USDA, Agricultural Research Service, Vegetable Crops Research Unit, Department of Horticulture, University of Wisconsin, 1575 Linden Drive, Madison, Wisconsin 53706-1590, USA email: dspooner@wisc.edu*

INTRODUCTION

Tomatoes rank fourth among the leading world vegetables. In 2001, over 100 million metric tons were produced, with the 15 leading countries being (in descending order) China, US, India, Turkey, Egypt, Italy, Spain, Brazil, Islamic Republic of Iran, Mexico, Greece, Russian Federation, Ukraine, Chile, and Uzbekistan (FAO 2002; Fig. 1.1). There has been a general upward trend in tomato production during the period 1992-2002 (Fig. 1.2). Interestingly, the countries that produce higher yields (Fig. 1.3) do not possess the ideal climate for the tomato crop and have less land area devoted to tomato production (Fig. 1.4). Northern European countries, as well as Canada and New Zealand, produce most of their tomatoes under controlled greenhouse conditions. Tomato consumption has also shown a general increased trend of consumption over a period of time (FAO 2002). Tomatoes supply a mean of 12.1 kg/cap/yr, and tomato consumption is higher in Mediterranean and Arab countries (usually between 40 and 60 kg/cap/yr). Tomatoes are highly popular in Egypt, Italy, Israel, Lebanon, Turkey and United Arab Emirates (60-70 kg/cap/yr), whereas people from Greece and Libya have the highest preference consuming more than 100 kg of tomatoes per capita and year. Tomatoes are also a popular food in Latin and North America.

Tomatoes rank second among the leading vegetables of the US (Ensminger et al. 1995), with a production of 10.25 million metric tons in

*Corresponding author: David M. Spooner

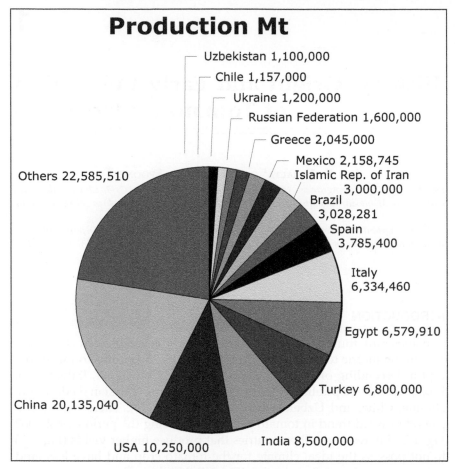

Fig. 1.1 Tomato production worldwide, 2001.

2001. Much of the US production is processed, with major products being canned tomatoes, ketchup, chilli sauce, juice, paste, powder, puree, salad dressings, sauces, soups, and vegetable and juice cocktails.

The US farmgate (point of first sale) value of tomatoes in 2001 was 1.12 billion dollars for fresh tomatoes and 0.54 billion dollars for processed tomatoes ($1.66 billion dollars total) (USDA, National Statistics Service 2002a). California and Florida clearly dominate the US market, with Florida accounting for 40.3% of the fresh US market, and California accounting for 24.1% of the fresh market and 90.7% of the processed market (USDA, National Statistics Service 2002b). Tomato consumption has substantially increased in the US since the beginning of the last century. In 1920, the per capita consumption was only 8.2 kg/yr, which in 1978 increased to

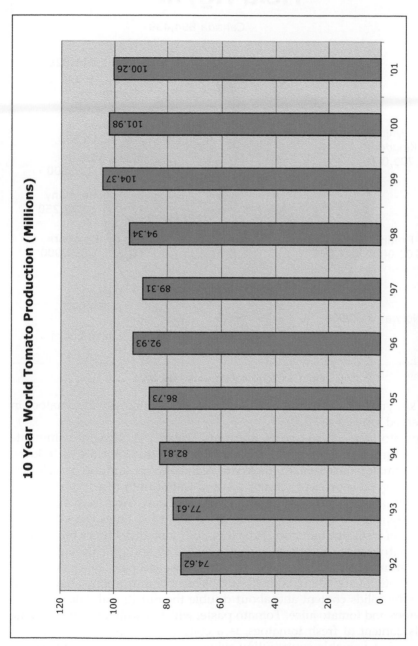

Fig. 1.2 Tomato production worldwide, 1992-2001.

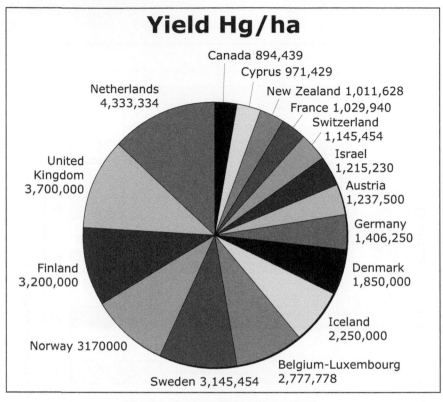

Yield Hg/ha

Canada 894,439

Cyprus 971,429

Netherlands
4,333,334

New Zealand 1,011,628

France 1,029,940

Switzerland
1,145,454

United
Kingdom
3,700,000

Israel
1,215,230

Austria
1,237,500

Germany
1,406,250

Finland
3,200,000

Denmark
1,850,000

Iceland
2,250,000

Norway 3170000

Belgium-Luxembourg
2,777,778

Sweden 3,145,454

Fig. 1.3 Tomato yield worldwide, 2001.

25.5 kg/yr (Rick 1978), and now is 40.5 kg/yr, but mostly of tomatoes in preserved forms (FAO 2002).

Tomato is a rich source of nutrients (Table 1.1). General comments (Ensminger et al. 1995) made in particular from this table are as follows: Fresh tomatoes and tomato juices are high in water and low in calories. Both are good sources of vitamins A and C, but unfortified tomato juice has only about 2-3 the vitamin C content of raw, ripe (red) tomatoes. Similarly, canned tomatoes contain only about 3-4 times the vitamin C content of fresh ripe tomatoes. Ripe tomatoes contain 3-4 times the vitamin A as mature green tomatoes, but otherwise red and green tomatoes are about equal in nutritional value. Tomato puree and plain types of tomato sauce (without added ingredients such as meat or mushrooms) have about twice the solids content and about double the nutritional value of fresh tomatoes and tomato juice. Tomato paste, which has about four times the solids content of fresh tomatoes, is a concentrated source of nutrients, making it a valuable contribution when used in preparation of pastas, pizzas, and other foods. Ketchup and chilli sauce are about equal in

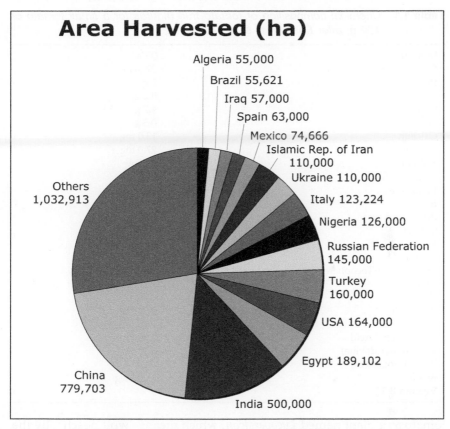

Fig. 1.4   Area of tomato harvest worldwide, 2001.

nutritional value, since each item is made with similar ingredients and contains about 32% solids (about 5 times the content of fresh tomatoes and tomato juice). However, the nutrients per calorie of these products are significantly less than those furnished by tomato paste, because the solids content and caloric values are boosted by added salt and sugar.

Tomato popularity and its high level of consumption make this vegetable one of the major sources of vitamins and minerals in human diet, and provides healthy benefits that will be discussed further in the following chapters.

TAXONOMY

Since the tomato was introduced to Europe in the sixteenth century, early botanists recognized the close relationships of tomatoes with the genus *Solanum*, and commonly identified them as *S. pomiferum* (Sabine 1820, Luckwill 1943a). Anguillara (1561) identified the newly introduced

Table 1.1 *Chemical composition of tomato fruit (figures for a small tomato of 100 g; after Ensminger et al. 1995).*

Moisture	95%
Food energy	22 kcal
Protein	1 g
Fats	0.2 g
Carbohydrates	4.7 g
Fiber	0.5 g
Calcium	13.0 mg
Phosphorus	27.0 mg
Sodium	3.0 mg
Magnesium	17.7 mg
Potassium	244.0 mg
Iron	0.50 mg
Zinc	0.20 mg
Copper	0.01 mg
Vitamin A	900.0 IU
Vitamin D	0
Vitamin E (α-Tocopherol)	0.40 mg
Vitamin C	23 mg
Thiamin	0.06 mg
Riboflavin	0.04 mg
Niacin	0.70 mg
Panthothenic Acid	0.33 mg
Vit. B-6 (pyridoxine)	0.10 mg
Folacin (folic acid)	39.00 mcg
Biotin	4.00 mcg
Vitamin B-12	0

tomato as a plant named *Lycopersicon*, which means "wolf peach", by the Greek naturalist Galen fourteen centuries earlier. However, the actual plant described by Galen is unknown, and it certainly did not refer to any form of tomato because all tomato species are not native of the Old World. Tournefort (1694) was the first to consider cultivated tomatoes within a distinct genus under the early name *Lycopersicon*. He used the multilocular character of the fruit as a criterion to differentiate *Lycopersicon* from *Solanum*. Tournefort listed nine taxa but only seven of them correspond to fasciated-fruited varieties that differed in the color and size of their fruits, and the other two described taxa belong to different Solanaceae (Luckwill 1943a).

Linnaeus (1753) classified tomatoes in the genus *Solanum*, and under the specific name of *Solanum lycopersicum* grouped all the cultivated multilocular forms that Tournefort described as separate species. He also described a second wild species from Peru, *S. peruvianum*. Jussieu (1789), in his natural classification, also included tomatoes in *Solanum*. On the other hand, Miller (1754) reconsidered Tournefort's classification and formally described the genus *Lycopersicon*. This classification of tomatoes under *Lycopersicon* continued as the prevailing treatment by several classical and modern authors (e.g., Dunal 1813, 1852, Bentham and Hooker 1873, Müller 1940a, Luckwill 1943a, Correll 1958, D'Arcy 1972, 1987, 1991, Hunziker 1979, Rick 1979, 1988, Symon 1981, 1985, Taylor 1986, Warnock 1988, Hawkes 1990, Rick et al. 1990b).

More recently, the phylogenetic relationships within the Solanaceae have been examined with molecular data. Spooner et al. (1993) examined outgroup relationships of tomato to potato and other members of the Solanaceae based on chloroplast DNA restriction site data (Fig. 1.5). Subsequent molecular studies unequivocally supported tomato to be firmly internested in the genus *Solanum* L., then this tomato-potato sister group relationship is now clearly established (Olmstead and Palmer 1997, Bohs and Olmstead 1997, 1999, Peralta and Spooner 2001). Based on these results, a new phylogenetic classification has assigned tomato to the genus *Solanum* (Spooner 2005). This classification of tomatoes in *Solanum* matches the original treatment of Linnaeus (1753), as well as prior taxonomists who insightfully foresaw this generic relationship based on morphological data (Wettstein 1895, MacBride 1962, Seithe 1962, Heine 1976, Fosberg 1987, Child 1990). Börner (1912) also recognized the close affinities among tomatoes and potatoes, and proposed a new genus *Solanopsis* to segregate them. Although most taxonomists today place tomato in *Solanum*, most agronomists and horticulturists do not use this name (see Doco et al. 1997, Shichijo et al. 2001, Van der Heuvel 2001, Weller et al. 2001). Most users of the classification in *Lycopersicon* clearly base their reluctance to use the *Solanum* names on tradition or the practical goal of maintaining familiar names rather than adherence to any particular classification philosophy. In this chapter tomato species are classified in the genus *Solanum* and their comparative *Lycopersicon* synonyms are given in Table 1.2.

Hypotheses of ingroup relationships within tomato also have varied greatly. Müller (1940a), Luckwill (1943a), and Child (1990) classified to-mato based on morphological criteria, while Rick (1963, 1979) and Rick et al. (1990b) classified tomato quite differently based on biological (inter-breeding) criteria. Peralta and Spooner (2001) produced a phylogeny of tomato based on DNA sequences of the single-copy GBSSI (*waxy*) gene, and Spooner et al. (2005) based on Amplified Fragment Length Polymor-phisms. The results support allogamy, self-incompatibility, and green fruits as primitive in tomatoes, and most closely match the classification of Child (1990). One of the self-incompatible species, the highly polymorphic *Solanum peruvianum* L., was supported to consist of one group of popula-tions from northern Peru and another group of populations from central to southern Peru. A phenetic morphological study by Peralta and Spooner (2003) supported all species, including the "northern" and "southern" group of populations of *S. peruvianum* as distinct taxa. Peralta et al. (2005) used these results, and morphological data, to divide the former *S.*

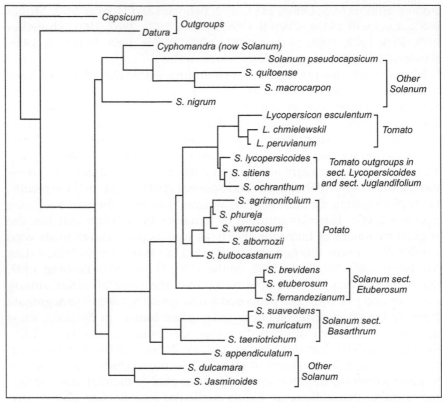

Fig. 1.5 One of two-most parsimonious cladograms (as a phylogram) of chloroplast DNA restriction site data examining wild tomatoes (here labeled *Lycopersicon*), their sister groups (*Solanum* sect. *Lycopersicoides*, sect. *Juglandifolium*), wild potatoes (*Solanum* sect. *Petota*), and further outgroups in *Solanum* sect. *Etuberosum*, sect. *Basarthrum*, and other *Solanum* (modified from Spooner et al., 1993).

peruvianum into four species. A taxonomic monograph of tomato, based partly on these new molecular and morphological data, is in preparation by the present authors and Sandra Knapp (Natural History Museum, London).

DISTRIBUTION, HABITATS, MORPHOLOGY, MATING SYSTEMS AND GENETIC RESOURCES OF WILD TOMATOES

The wild relatives of the cultivated tomato are native of western South America along the coast and high Andes from central Ecuador, through Peru, to northern Chile, and in the Galápagos Islands (Table 1.2). The most likely ancestor of cultivated tomatoes is the wild cherry tomato (usually identified as *S. lycopersicum* var. *cerasiforme*), which is more widespread,

Table 1.2 *Comparison of wild tomato species (*Solanum *L. section* Lycopersicon *(data comapiled from Müller, 1940a; Luckwill 1943a; Esquinas Alcazar 1981; Rick 1982b, 1986b; Taylor 1986; Peralta et al., 2005). The* Lycopersicon *synonyms follow the* Solanum *names.*

Species	Lycopersicon synonyms	Fruit color	Breeding system	Distribution and Habitat	Comments and interesting features for breeding purposes
S. lycopersicum L.	L. esculentum Miller	Red	SC, facultative allogamous	Native from Ecuador and Peru. Wide range of habitats, weed in newly open areas	Moisture-tolerance, resistance to wilt, root-rotting, and leaf-spotting fungi
S. cheesmaniae (Riley) Fosberg	L. cheesmaniae Riley	Yellow, yellow green, orange, purple	SC, exclusively autogamous	Endemic of the Galápagos Archipelago. From low elevations in the saline seashore up to 500 m in volcanic areas	Closely related to S. galapagense. Salt tolerance, lepidoptera and virus resistances, and genes involved in the retention of fruits and thick pericarp
S. galapagense S. Darwin and Peralta	Part of L. cheesmaniae L. Riley (previously known as forma or var. minor)	Pale to deep orange	SC, exclusively autogamous	Endemic of the Galápagos Archipelago. Mostly occurring on coastal lava to within 1 m of high tide mark within range of salt spray, but occasionally inland up to 50 m	Closely related to S. cheesmaniae. Salt tolerance.
S. pimpinellifolium B. Juss.	L. pimpinellifolium (B. Juss.) Miller	Red	SC, autogamous, facultative allogamous	Central Peru to central chile, dry coastal habitats, 0 – 500 m, but exceptionally upto 1400 m.	Closely related to S. lycopersicum (some natural introgression with it). Contributed to improve color and fruit quality. Insect, nematode, and disease resistances.

(Contd.)

(Contd.)

Taxon	Fruit	Breeding system	Distribution	Notes
S. arcanum Peralta	Part of *L. peruvianum* (L.) Miller	Green	SI	100-2800 m; N Peru, lomas, dry valleys, and dry rocky slopes	
S. chilense (Dunal) Reiche	*L. chilense* Dunal	Small green, with purple stripe	SI, alleallogamous	Sea level-3250 m; S Peru to N Chile, grows in dry river beds, survives by deep roots	Typically erect becoming decumbent; post-syngamic barriers with *S. peruvianum*. Drought resistance
S. chmielewskii (C. M. Rick, Kesicki, Fobes & M. Holle), D. M. Spooner, G. J. Anderson & R. K. Jansen	*L. chmielewskii* C. M. Rick, Kesicki, Fobes & M. Holle	Green	SC, facultatively allogamous	1600-3200 m, Pacific side, South-Central Peru to N Bolivia; moist habitats; slightly better-drained sites that *S. neorickii*	Sympatric with *S. neorickii*. Contributed to improve high sugar content in the crop
S. corneliomuelleri J. F. Macbr.	Part of *L. peruvianum* (L.) Miller; also known as *Lycopersicon glandulosum* C. F. Mull.	Green	SI	Landslides and rocky slopes, (40)200-3300 m, Central to S Peru	
S. habrochaites S. Knapp & D. M. Spooner	*L. hirsutum* Dunal	Green	Typically SI, 1-2 collections SC, but with later inbreeding depression	Typically high elevations, (40) 200-3300 m, in moist well drained soils; Central Ecuador to Central Peru	Cold and frost tolerance. Insect resistance (glandular hairs), and other resistances
S. huaylasense Peralta	Part of *L. peruvianum* (L.) Miller	Green	SI	Rocky slopes, (940) 1700-3000 m, N Peru, Ancash along Río Santa.	

(Contd.)

(Contd.)

Species	Color	Breeding system	Distribution and habitat	Characteristics
S. neorickii C.M. Rick, Kesicki, Fobes & M. Holle, D.M. Spooner, G.J. Anderson, & R.K. Jansen / L. parviflorum C. M. Rick, Kesicki, Fobes & M. Holle	Pale green	SC, highly autogamous	(920)1950-2600 m, Pacific side, South Ecuador to South-central Peru; moist and well-drained rocky environments; more common than S. chmielewskii.	Sympatric with S. chmielewskii; probably evolved from S. chmielewskii; yet no natural introgression reported with S. neorickii.
S. pennellii Correll / L. pennellii (Correll) D'Arcy	Green	Usually SI, some SC in Southern range	sea level to 2300 m; N cent to S cent Peru (8-16 °S); hot dry habitats but subject to dew and fog; (many stomata adaxially, poor root system).	Drought resistence; covered with glandular hairs imparts insect resistance; hybridizes unilaterally (as male) with many other species except S. chilense or S. peruvianum.
S. peruvianum L. / L. peruvianum (L.) Miller	Green	Typically SI, allogamous,	Sea level-600 m; Central Peru to N Chile. Coastal lomas formations and occasionally as a weed at fields edges.	Virus, bacteria, fungi, aphid, and nematode resistances.

and perhaps more recently distributed into Mexico, Colombia, Bolivia, and other South American countries (Rick and Holle 1990). The prior taxonomies recognized the cherry tomato as *L. esculentum* var. *cerasiforme* or *S. lycopersicon* var. *cerasiforme* but we do not recognize this variety and combine all variants of this species (cultivated and wild) into *S. lycopersicon*. The wild cherry tomato grows spontaneously in tropical and subtropical areas worldwide, where it might have been accidentally introduced or escaped from cultivation.

Wild tomatoes grow in a variety of western South American habitats, from near sea level to over 3,300 m in elevation (Rick 1973, Taylor 1986). These habitats include the arid Pacific coastal lowlands and adjacent lower valleys to mesic uplands in the high Andes. Numerous valleys, formed by rivers draining into the Pacific, characterize the western side of the Andes. Wild tomato populations grow at different altitudes in these narrow and geographically isolated valleys, and are adapted to particular microclimatic and soil conditions. Certainly, the Andean geography, diverse ecological habitats, and different climates contributed to wild tomato diversity (Warnock 1988).

Wild tomatoes are perennial herbaceous plants, although in their natural habitat tomatoes most probably behave as annuals and might die after the first growing season due to frost or drought. They have an erect or prostrate growth habit, and possess taxonomically useful differences in leaf, inflorescence, flower, fruit, and seed characters. Leaves are pinnately dissected with 2-6 opposite or sub-opposite, sessile, subsessile or petiolate pairs of leaflets. There is great interspecific variation in leaf dissection with primary, secondary, tertiary, and interjected leaflets. The basic inflorescence is a cyme with different branching patterns (monochasial, dichotomous, and polychotomous), and with or without axial bracts. Flowers are typically yellow; the anthers are united laterally to form a flask-shaped cone with an elongated sterile tip at the apex (except in *S. pennellii*). Flowers are buzz pollinated. Fruit size, color, and pubescence are variable (Table 1.2), as are seed size, color and development of radial walls of the seed coat cells (Müller 1940a, Luckwill 1943a). Fruits are usually bilocular in the wild species, and bilocular or multilocular in the cultivated varieties.

Mating systems have played an important role in the evolution of wild tomato species, varying from allogamous self-incompatible, to facultative allogamous, and self-compatible, to autogamous and self-compatible (Rick 1963, 1979, 1986a; Table 1.2). The self-incompatibility system in tomatoes is gametophytic and controlled by a single, multiallelic S locus (Rick 1982a). Large flowers and greater stigma exsertion from the anther tube have been associated with self-incompatibility resulting in greater outcrossing and genetic variation in wild tomatoes (Rick 1982a). Similarly, in the self-compatible species *S. pimpinellifolium*, greater outcrossing and genetic variation is related to large flowers and greater stigma exsertion; marginal populations of this species are highly autogamous with little or no genetic variation, bearing small flowers, with little or no stigma exsertion (Rick et al. 1977). Self-incompatibility is most probably regulated by different unlinked genes or gene complexes (Rick 1982a), and changes in mating systems in wild tomatoes occurred from self-incompatibility, as the ancestral condition, to self-compatibility, which probably never reversed to self-incompatibility. Change from self-incompatibility to self-compatibility is expected to have arisen infrequently and independently (Rick 1982a). Hybridization is another possible source of genetic variation. Evidence of natural interspecific hybridization and gene flow among wild self-compatible tomato species have been documented in native sympatric populations of *S. pimpinellifolium* and *S. lycopersicum*, and cultivated tomatoes in Ecuador and Peru (Rick 1958). The reciprocal introgression of traits into both taxa generates complex morphological gradation between them that makes their taxonomic identification difficult (Rick 1958).

The traditional breeding for pure lines in the cultivated tomato has narrowed its genetic base (Stevens and Rick 1986). Fortunately, genetic resources from the primary center of diversity provide a wealth of useful genetic traits to improve the crop (Rick 1982b, 1995). All wild tomato species are diploid ($2n = 2x = 24$) and can be crossed (but sometimes with difficulty) to the cultivated tomato (Rick 1979). They are of great use in breeding programs as sources of disease resistances and agronomic traits (Esquinas Alcazar 1981, Rick 1982b, 1986b, Rick et al. 1987, Stevens and Rick 1986, Laterrot 1989). The International Plant Genetic Resources Institute (IPGRI) recognized the need for maintaining valuable vegetable genetic resources and nominated tomatoes for priority conservation status. Ross (1998) considered that the diversity of tomato is likely to be well conserved, and cited 62,832 accessions maintained in gene banks around the world, although most of these accessions are *S. lycopersicum*. The genetic variation among *S. lycopersicum* accessions at the Asian Research and Development Center (ARDC–one of the largest collections of cultivated tomato germplasm) was evaluated with Random Amplified Polymorphic DNA (RAPDs) by Villand et al. (1998). RAPD diversity was greater in accessions from the primary center (Ecuador, Peru, Chile), and for breeding purposes variation can be obtained at a faster rate by sampling accessions from this area than from other geographic regions. The largest and most important collection of wild species genetic resources exists at the Tomato Genetics Resources Center (TGRC, University of California, Davis).

Tomato also serves as a model organism to understand the basic genetics of diploid plants. Features that enhance the usefulness of tomatoes for genetic studies are: the naturally occurring variability in the species, self-pollination that lead to the expression of recessive mutations, the possibility of controlled hybridization within and among species, the lack of gene duplication, and the possibility to easily identify the 12 chromosomes (Rick 1978). In recent years there have been great advances in tomato genetics. New methodological approaches like molecular mapping of important agronomical characters have provided powerful tools for the improvement of the tomato crop (Tanksley and McCouch 1997).

DOMESTICATION OF CULTIVATED TOMATOES— PERU OR MEXICO?

Methods for Inferring Location of Crop Origins

Two competing hypotheses have been advanced to ascertain the place of domestication of the cultivated tomato, one from Peru, and another from Mexico. How does one search for origins of crops? The first systematic attempt was outlined by DeCandolle (1886). He used an eclectic approach based on the following four criteria: 1) "Botany", or observing natural spontaneous geographic distributions of the crop or its putative wild relatives. These data could be gathered from floras or herbaria, but this could be complicated by recent adventive introductions; 2) "Archaeology and paleontology", gathered from fossil evidence of plant remains in caves, burial sites, or other preserved deposits; 3) "History", searching for evidence in early accounts of peoples; 4) "Philology", or linguistic evidence, or comparison of native names of plants to prior languages. DeCandolle, however, placed the least credence on the linguistic evidence. Since DeCandolle's time, additional techniques have been used in determining the origin of crops which include radiocarbon dating, scanning electron microscopy, palynology, refined archaeological methods as flotation techniques, and genetic and molecular evidence (Smith 1995).

Peruvian Hypothesis

DeCandolle (1886) advanced the Peruvian hypothesis for the site of do-mestication of tomato. He reviewed botanical (Bauhin 1623, Ruiz and Pavón 1797), linguistic (Roxburgh 1832), and historical (Hernández 1651) evidence and concluded: 1) there were no unambiguous natural records of tomato outside of the Americas before its European discovery there; 2) Bauhin (1623) referred to tomato as "mala peruviana" and "pomi del Peru," which suggested initial domestication and transport of tomato from Peru to Europe; 3) its origin was from the wild cherry tomato (*S. lycopersicum*) that by DeCandolle's time was known to occur from coastal Peru, Mexico, to southwestern US (California); 4) the distribution of cultivated tomato and its progenitor outside of Peru originated by garden escapes; and 5) the plant was domesticated before the discovery of America but not very long before that. This Peruvian origin was later supported by other authors (Moore 1935, Müller 1940a,b, Luckwill 1943a,b).

Mexican Hypothesis

Jenkins (1948) developed the Mexican hypothesis. He pointed out that the first reference to tomato in Europe was made by Matthiolus (1544) who provided a short description of tomato. Matthiolus (1554) amplified the description where he first provided the Italian name "pomi d'oro", and Latin name "mala aurea." A later edition of his work (Matthiolus 1586) provided an illustration showing an unambiguously identified tomato plant that made his concept of tomato clear, but there was no reference to its geographical origin. Only later, Anguillara (1561) first used the name "pomi del Peru," along with the name "pomi d'oro," but his reference is ambiguous as to whether he was referring to the same plant. Jenkins argued that "pomi del Peru" was used by early botanists to refer to other solanaceous plants such as *Datura stramonium* L. and had nothing to do with tomato, weakening DeCandolle's linguistic evidence.

Jenkins's second argument was that there was no evidence for pre-Colombian domestication of tomatoes in South America, yet good evidence for early domestication in Mexico. This comes from a reference from Guilandini (1572) who referred to tomato as "tumatle ex Themistitan," using an indigenous Mexican name for tomato. Jenkins interpreted the name "Themistitan" as a variant spelling of "Temixtitan" which in turn is a corruption of "Tenochtitlan", the native name for Mexico City. He therefore concluded that tomatoes came from Mexico. During the seventeenth century the Nahua name "tomatl" was often mentioned by botanists, and variants of this name are used in different languages at present (tomate in Spanish, tomato in English, etc.). Interestingly, the early name "Pomi d'oro" is still used in Italy. According to Jenkins (1948), evidence for early Mexican domestication also came from Hernández (1651) who documented early cultivation of tomato in Mexico at least before 1578 (the year of his death) and possibly from Acosta (1590); although Acosta could be referring to Mexico or Peru. Nevertheless, Yakovleff and Herrera (1935) considered that Acosta documented the uses of tomatoes in ancient Peru.

Jenkins's third argument was that there was considerably more variation of the landrace cultivars in Mexico than in Peru. Following ideas of Vavilov (1926), Jenkins argued that var. *cerasiform*, the small bilocular fruit form of *S. lycopersicum*, was introduced into Mexico in pre-Columbian times and it was domesticated in the central area that he considered as a secondary center of diversity. Jenkins agreed with DeCandolle (1886) that *S. lycopersicum* was the progenitor of the domesticated cultivars, but disagreed with the place of domestication in Peru.

Our Conclusion

We consider the question of the original site of domestication of cultivated tomato to be unsolved, and likely to forever be so. Like DeCandolle (1886), we consider linguistic evidence to be a weak source of data, and the existing linguistic sources for tomato are scant, ambiguous, and subject to various interpretations. Contrary to Jenkins's (1948) statements that there are no indigenous Peruvian names for tomato, Horkheimer (1973) documented a Quechua name for tomato (pirca), and Yakovleff and Herrera (1935) cited another Quechua name (pescco-tomate) possibly referring to the small bilocular fruit form of *S. lycopersicum*. The historical evidence also is sparse and ambiguous in their references to tomatoes. From the analysis of the original description by Hernández (1651), it is not clear that the plant cited as "tomatl" referred to the true tomatoes or a native *Physalis* species. Unless some new document is uncovered that clearly identifies introductions of tomato to Europe from a certain area (see McCue 1952, for a comprehensive summary of historical references), the first European site of introduction will forever remain unknown. However, even such a clear reference would not determine a first site of domestication, viz. Mexico vs. Peru.

Jenkins's (1948) Vavilovian argument of more diversity of cultivars in Mexico is not supported by comparative data (Villand et al. 1998) from South America (Ecuador, Peru and Chile). Tomatoes from Europe and North America share similar isozymes with those from Mexico and Central America, suggesting the tomato was introduced to Europe and North America from Mexico or Central America (Rick and Fobes 1975). Nevertheless, comparisons among genetic variability of primitive tomato cultivars found in Mexico, Central America and Peru, and modern varieties have neither substantiated nor disproved the hypothesis that Mexico was the centre of domestication (Rick et al. 1974, Rick and Fobes 1975, Rick and Holle 1990). Rick and Holle (1990) provided an isozyme study of different accessions of var. *cerasiforme* of the wild cherry tomato *(S. lycopersicum)* from South America, but they did not include cultivars or landraces from Mexico. The only comparative molecular studies (RAPDs and/or nuclear RFLPs) of diversity of landrace cultivars (Williams and St. Clair 1993, Villand et al. 1998) of tomato do not address the Peruvian/Mexican hypothesis.

A molecular study may be useful to elucidate the origin of tomato domestication by comparing a large number of accessions. However, it would be complicated by relative lack of variation within *S. lycopersicum* (including its landraces), and by the difficulty to identify existing landraces from Mexico and Peru as truly native today. The only putative archaeological evidence of tomato is decorated functional ceramics "spindle whorls" produced by the native Quimbaya culture (500-1000 AD) of Colombia (McMeekin 1992). However, our examination of the figures in this publication do not convince us that these are unequivocally tomato flowers, and could be other *Solanum* flowers (possibly potato). Like Rick and Holle (1990), we conclude that none of the evidence is conclusive regarding either a Mexican or a Peruvian initial site of domestication, and that tomatoes may have been domesticated independently in both areas.

EARLY HISTORY OF THE CULTIVATED TOMATO IN EUROPE

What were the first morphotypes of cultivated tomatoes exported from the Americas and where did they come from? McCue (1952) examined these questions through an extensive search from the literature, herbarium specimens, and early drawings. Despite this extensive search, we still know very little. The first European contact with Mexico was in 1519 (taking of Mexico City), and with Peru in 1531 (completion of the Peruvian conquest). Botanists at that time were mainly interested in the medicinal and culinary properties of plants and had little interest or knowledge of distribution or origin of cultivars. The first tomato references mentioned above were from sixteenth century herbalists, who were mainly interested with the medicinal values or "virtues" of plants, but they knew them only from exchange among botanical gardens.

These early botanists classified new plants by comparison with plants already known in Europe and from classical Greek references. *Lycopersicon*, the ancient Greek name for the tomato attributed to Galen is a clear example. By this method, Matthiolus (1544) described tomato by comparison to mandrake, a solanaceous plant known to the classical Greek botanist Dioscorides as: "Another species (of mandrake) has been brought to Italy in our time, flattened like the melerose (variety of apple) and segmented, green at first and when ripe of a golden color, which is eaten in the same manner as the egg plant, fried in oil with salt and pepper, like mushrooms." From this we glean that early introduced tomatoes had yellow fruits. In a later edition of his work, Matthiolus (1554) cited both yellow and red fruits, and mentioned the Italian name for the tomato "pomi d'oro" and its Latin equivalent "mala aurea" or golden apple. Another early common name for tomato is "poma amoris", or "love apples," because at that time it was believed that fruits had aphrodisiacal properties. All these ancient names persisted well into the nineteenth century (Moore 1935).

The earliest tomato herbarium specimens also came from this period (McCue 1952). Jerna (1947) reported specimens labeled as "Malus insana, Mandragorae species Poma amoris" attributed to Francesso Petrolini dated between 1550 and 1560. Mattirolo (1899) mentioned another tomato specimen found in the sixteen-volume herbarium of Ulisse Aldrovandi, which was most probably cultivated in Bologna, and is the oldest extant herbarium specimen of tomato, and is now preserved at the Botanical Garden of Bologna.

Georg Oelinger was a Nürnberg aphotecarian, an avid plant collector. He cultivated tomatoes in his garden probably as curiosity or as medicinal plants. The complete edition of Oelinger's (1553) work had a picture of red- and yellow-fruited tomatoes, and all the fruits are deeply furrowed (fasciated fruits). It is clear from both illustrations that the flowers had duplications of sepals and petals (6-7-parted).

Dodoens (1553) listed the Latin, German, and French names for the tomatoes along with an illustration of the plant, but he did not mention uses. His later publications (Dodoens 1574, 1583) illustrated round fruits with furrows, flowers with 7-8 petals, and two of the flowers with exserted styles.

The illustrations in L'Obel (1576) and Tabernaemontanus (1591) were similar to Dodens's (1553). Gesner (1561) mentioned that the tomato fruit was easily grown in Germany, matured early, and had fruits varying in color from gold, red, and white; one illustration showed a plant with round fruits without furrows, and flowers with six sepals.

Teppner (1993) discussed the descriptions and illustrations of early tomatoes cultivated in Europe. He included a copy of Dodens's (1583) drawing that clearly showed a plant with large, horizontally compressed, furrowed fruits, characteristic of early tomato cultivars. According to Sabine (1820) the cherry tomato must have been introduced at the same time as a large-fruited cultivar. Nevertheless, Aymonin (in Besler 1613) considered that the cherry tomato appeared in Europe around 1625. In Europe, tomatoes initially were cultivated mainly as ornamental plants in gardens, and they were considered inedible or poisonous because they were similar to the poisonous mandrake or belladonna.

Tomatoes were first accepted for culinary purposes in southern Europe (Ray 1673, Miller 1752) during the seventeenth and eighteenth century. Filippo (1811) reported three varieties in Italy and gave instructions for their cultivation. Sabine (1820) reported four varieties of red tomatoes and two of yellow tomatoes that were cultivated in Europe; he also discussed the condition for cultivation in England based on the experience of native gardeners. Alefeld (1866) mentioned seven varieties in Germany. According to McCue (1952), Salmon (1710) mentioned tomatoes for the first time in North America. Although tomato cultivation was not difficult, the crop gained economic importance only by the end of the nineteenth century or beginning of the twentieth century when tomato breeding programs were established (Lehmann 1955, Rick 1978, 1995).

According to Rick (1995), domestication and improvement of tomato fruit production have been accompanied by changes in the position of stigma from the anther tube. The closely related wild species, and older Latin American cultivars (and their wild species progenitor), tend to have well exserted stigmas. Rick (1995) emphasized that in the absence of appropriate pollinators, flowers with exserted stigmas diminished the percentage of fruit set. Strong artificial selection for less exserted stigmas must have occurred after the tomato was first introduced to Europe, and even more selection under greenhouse culture. As a result, the stigma of most cultivars is shortened and now positioned at the mouth of the anther tube or even completely included in the anther tube. This shortening eliminated outcrossing and increased fruit yield in the modern varieties, but reduced the genetic variation of the crop.

The successful improvement of tomato agronomical traits is based on the understanding of basic genetics, continuous advancement of molecular genetic studies and breeding methods, which will be developed in the following chapters.

SUMMARY

Tomato is a major crop of world economy and supplies essential nutrients in human diets. There have long existed controversies regarding the place of domestication, early history, and taxonomy of tomato. The wild tomato species are native to western South America, from Ecuador south to northern Chile, and the Galapagos Islands. The putative progenitor of the cultivated species (*Solanum lycopersicum* = *Lycopersicon esculentum* var. *cerasiforme*) currently is widespread throughout warm regions of the world, but many of these are recent introductions. There are two competing hypotheses of the place of domestication of tomato, one supporting Peru, another in Mexico. While the Mexican origin is reasonable, we cannot discount a Peruvian origin, or even parallel domestication in both areas. Tomatoes were first recorded outside the Americas in Italy in 1544. They were cultivated first as ornamental or curiosity plants and thought by many to be poisonous. It was first accepted as a vegetable crop in southern Europe during the late sixteenth century. The first European cultivars had yellow to red flattened fruits, with deep furrows, and flowers with stigmas exserted from the anther tube. Derived cultivars had a wider range of fruit colors and shapes, smoother fruits, and stigmas included in the anther tube that led to increased fruit set but reduced the genetic variation of the crop. The taxonomy of tomato always has been controversial. This controversy involves not only generic placement in *Lycopersicum* or *Solanum*, but also hypotheses on interspecific relationships. Recent molecular data support treatment of tomato in *Solanum* (as we treat it here), and support allogamy, self-incompatibility, and green fruits as primitive of tomatoes. These studies support at least two distinct taxa in the formerly recognized *S. peruvianum*.

Acknowledgments

The authors thank John Stommel and Claudio Galmarini for critical comments and review of this chapter.

REFERENCES

*Acosta, J. de. 1590. Historia natural y moral de las Indias. Sevilla.

Alefeld, F.G.C. 1866. Landwirthschaftliche Flora. Wiegandt and Hempel. Berlin, 363 pp.

*Anguillara, L. 1561. Semplici dell' eccellente M.L.A., Liquali in piu Pareri a diversi nobili huomini scritti appaiono, et. Nouvamente da M. Giovanni Martinello mandati in luce.

Bauhin, C. 1623. Pinax theatri botanici. Ludovici Regis. 522 pp.

Bentham, G. and J.D. Hooker. 1873. Solanaceae. Genera Plantarum 2: 882-913.

Besler, B. 1613. Hortus Eystettensis. Broadsheet, Nürnberg, 367 plates.

Bohs, L. and R.G. Olmstead. 1997. Phylogenetic relationships in *Solanum* (Solanaceae) base on ndhF sequences. Syst Bot 22: 5-17.

Bohs, L. and R.G. Olmstead. 1999. *Solanum* phylogeny inferred from chloroplast DNA sequence phylogeny. *In:* M. Nee, D. E. Symon, R. N. Lester, and J. P. Jessop [eds.], Solanaceae IV, Advances in Biology and Utilization. Royal Botanic Gardens, Kew, UK, pp. 97-110.

Börner, C. 1912. *Solanum* L. und *Solanopsis* gen. nov. Abh. Naturwiss. Vereine Bremen 21: 282.

Child, A. 1990. A synopsis of *Solanum* subgenus *Potatoe* (G. Don) (D'Arcy) (*Tuberarium* (Dun.) Bitter (s.l.). Feddes Repert Specierum Nov Regni Veg 101: 209-235.

Correll, D.S. 1958. A new species and some nomenclatural changes in *Solanum*, section *Tuberarium*. Madroño 14: 232-236.

D'Arcy, W.G. 1972. Solanaceae studies II: typification of subdivisions of *Solanum*. Ann Missouri Bot Gard 59: 262-278.

D'Arcy, W.G. 1987. The circumscription of *Lycopersicon*. Solanaceae Newslett 2: 60-61.

D'Arcy, W.G. 1991. The Solanaceae since 1976, with a review of its biogeography. *In:* J. G. Hawkes, R. N. Lester, M. Nee, and N. Estrada [eds.], Solanaceae III: Taxonomy, Chemistry, Evolution. Royal Botanic Gardens, Kew pp. 75-137.

DeCandolle, A. 1886 (reprint 1959). Origin of cultivated plants. Hafner Publishing Company, New York, 468 pp.

Doco, T., P. Williams, S. Vidal, and P. Pellerin. 1997. Rhamnogalacturon II, a dominant polysaccharide in juices produced by enzymic liquefaction of fruits and vegetables. Carbohydr Res 297: 181-186.

Dodoens, R. 1553. Trivm priorvm de stirpium historis commentariorum imaginesad viuum expressae. Christophori Plantini, Antwerp.

Dodoens, R. 1574. Purgantium aliarumque eo facientium, tum et radicum, conuoluulorum ac deletriarum herbarum historiae, libri iiii. Christophori Plantini, Antwerp. 505 pp.

*Dodoens, R. 1583. Stirpium historiae pemptades sex, sive libri XXX Ed. 1. Christophori Plantini, Antwerp, 860 pp.

Dunal, M.F. 1813. Histoire naturelle, medicinale et economique des *Solanum*. Amand Koenig, Paris, 298 pp.

Dunal, M.F. 1852. Solanaceae. *In:* A.L.P.P. de Candolle [ed.], 1852. Prodromus systematis naturalis regni vegetabilis 13. Treuttel and Würtz, London, pp. 1-690.

Ensminger, A.H., M.E. Ensminger, J.E. Konlande and J.R.K. Robson. 1995. The concise encyclopedia of foods and nutrition. CRC Press, Boca Raton, Florida, 1178 pp.

Esquinas Alcazar, J. T. 1981. Genetic resources of tomatoes and wild relatives. Rep. IBPGR No. AGP. IBPGR, Rome, 80-103: 1-65.

FAO. 2002. On-line crop production statistics. http://apps.fao.org/. FAO, Rome, Italy.

*Filippo, R. 1811. L'Ortolano. Milan.

Fosberg, F.R. 1987. New nomenclatural combination for Galapagos plant species. Phytologia 62: 181-183.

*Gesner, C. 1561. Horti Germaniae. Argentorati, Strasbourg.

*Guilandini, M. 1572. Papyrus hoc est comentarius in tria C. Plinii maioris de papryo capita. Venice.

Hawkes, J.G. 1990. The Potato: Evolution, Biodiversity and Genetic resources. Belhaven Press, London.

Heine, H. 1976. Flora de la Nouvelle Caledonie, vol. 7. Musee Nacional D'Histoire Naturelle, Paris, 212 pp.

Hernández, F. 1651. Nova plantarum animalium et mineralium Mexicanorum historia. Vitalis Mascardi, Rome.

Horkheimer, H. 1973. Alimentación y obtención de alimentos en el Perú prehispánico. Universidad Nacional Mayor de San Marcos, Lima. 190 pp.

Hunziker, A.T. 1979. South American Solanaceae: a synoptic survey. *In:* J. G. Hawkes, R. N. Lester, and A. D. Skelding [eds.], The Biology and Taxonomy of Solanaceae, Academic Press, London, pp. 49-85.

Jenkins, J.A. 1948. The origin of the cultivated tomato. Econ Bot 2: 379-392.

Jerna, G. 1947. Qualche Cenno di Storia sul Pomodoro in Italia. Humus. Volume III, no. 9.

Jussieu, A.L. de. 1789. Genera Plantarum. Viduam Herissant et Theophilium Barrios, Paris, 498 pp.

Lehmann, C. O. 1955. Das morpholische System der Kulturtomaten (*Lycopersicum esculentum* Miller). Züchter, 3. Sonderheft, 64 Seiten.

Laterrot, H. 1989. The tomato. Advantages and use of wild varieties for varietal creation. Rev Hort 295: 13-17.

Linnaeus, C. 1753. Species Plantarum, 1st ed. Holmiae, Stockholm, 2 volumes, 1200 pp.

L'Obel M. de. 1576. Plantarum seu stirpium historia. Christophori Plantini, Antwerp, 676 pp.

Luckwill, L.C. 1943a. The genus *Lycopersicon*: an historical, biological, and taxonomical survey of the wild and cultivated tomatoes. Aberdeen Univ Stud 120: 1-44.

Luckwill, L.C. 1943b. The evolution of the cultivated tomato. J R Hortic Soc 68: 19-25.

MacBride, J.F. 1962. Flora of Peru: Solanaceae. Field Mus Nat Hist Publ Bot Ser 13: 3-267.

Mattirolo, O. 1899. Illustrazione del Primo Volume dell'Erbario di Ulisse Aldrovandi. Genova.

Matthiolus, P.A. 1544. Di Pedacio Dioscoride Anazarbeo libri cinque della historia, et materia medicinale trodotti in lingua volgare Italiana. Venice.

Matthiolus, P.A. 1554. [another edition] Commentarii, in libros sex Pedacci Dioscoricis Anazarbei, de medica materia. Venice.

Matthiolus, P.A. [ed. I. Camerarius]. 1586. [another edition] De plantas epitome ultissima, Petri Andreae Matthioli Frankfort.

McCue, G.A. 1952. The history and use of the tomato: an annotated bibliography. Ann Missouri Bot Gard 39: 289-348.

McMeekin, D. 1992. Representations of pre-Columbian spindle whorls of the floral and fruit structure of economic plants. Econ Bot 46: 171-180.

Miller, P. 1752. The gardeners dictionary. 6th ed. John and Francis Rivington, London.

Miller, P. 1754. The gardeners dictionary, Abridged 4th ed. John and James Rivington, London.

Moore, J.A. 1935. The early history of the tomato or loveapple. Missouri Botanical Garden Bulletin 23:134-138.

Müller, C.H. 1940a. A revision of the genus *Lycopersicon*. U S Dep Agric Misc Publ 382: 1-28 + 10pl.

Müller, C.H. 1940b. The taxonomy and distribution of the genus *Lycopersicon*. Nat Hort Mag 19: 157-160.

Oelinger, G. 1553. Herbarium des Georg Oellinger anno 1553 zu Nürnberg. Edited by Eberhard Lutze and Hans Retzlaff. Salzburg, Akademischer Gemeinschaftsverlag 1949.

Olmstead, R.G. and J.D. Palmer. 1997. Implications for phylogeny, classification, and biogeography of *Solanum* from cpDNA restriction site variation. Syst Bot 22: 19-29.

Peralta, E. and D.M. Spooner. 2001. GBSSI gene phylogeny of wild tomatoes (*Solanum* L. section *Lycopersicon* [Mill.] Wettst. subsection *Lycopersicon*). Am J Bot 88: 1888-1902.

Peralta, I.E. and D.M. Spooner. 2003. Relationships and morphological characterization and relationships of wild tomatoes (*Solanum* L. section *Lycopersicon*) Monogr Syst Bot Missouri Bot Gard 104: 227-257.

Peralta, I., S. Knapp, and D.M. Spooner. 2005. New species of wild tomatoes (*Solanum* section *Lycopersicon*: Solanaceae) from northern Peru. Syst Bot 30: 424-434.

Ray, J. 1673. Observations topographical, moral and physiological, made in a journey through part of the Low-Countries, Germany, Italy, and France. John Martyn, London, 499 pp.

Rick, C.M. 1958. The role of natural hybridization in the derivation of cultivated tomato of western South America. Eco Bot 12: 346-367.

Rick, C.M. 1963. Barriers to interbreeding in *Lycopersicon peruvianum*. Evolution 17: 216-232.

Rick, C.M. 1973. Potential genetic resources in tomato species: clues from observations in native habitats. *In:* A.M. Srb [ed.], Genes, Enzymes, and Populations. Plenum Press, New York, pp. 255-269.

Rick, C.M. 1978. The Tomato. Sci Amer 239 (2): 76-87.

Rick, C.M. 1979. Biosystematic studies in *Lycopersicon* and closely related species of *Solanum*. *In:* J.G. Hawkes, R.N. Lester and A.D. Skelding [eds.], The Biology and Taxonomy of Solanaceae. Academic Press, New York, pp. 667-677.

Rick, C.M. 1982a. Genetic relationships between self-incompatibility and floral traits in the tomato species. Biol Zentralb 101: 185-198.

Rick, C.M. 1982b. The potential of exotic germplasm for tomato improvement. *In:* I.K. Vasil, W.R Scowcroft and K.J. Frey [eds.], Plant Improvement and Somatic Cell Genetics, Academy Press, New York, pp. 1-28.

Rick, C.M. 1986a. Reproductive isolation in the *Lycopersicon peruvianum* complex. *In:* W.G. D'Arcy [eds.], Solanaceae, Biology and Systematics. Colombia University Press, New York, pp. 477-495.

Rick, C.M. 1986b. Germplasm resources in the wild tomato species. Acta Hort 190: 39-47.

Rick, C.M. 1988. Tomato-like nightshades: affinities, autoecology, and breeders opportunities. Eco Bot 42: 145-154.

Rick, C.M. 1995. Tomato. *Lycopersicum esculentum* (Solanaceae). *In:* J. Smartt and N.W. Simmonds [eds.], Evolution of Crop Plants, Second Edition. Longman Scientific & Technical, Essex, UK, pp. 452-457.

Rick, C.M., R.W. Zobel and J.F. Fobes. 1974. Four peroxidase loci in red-fruited tomato species: genetics and geographic distribution. Proc Natl Acad Sci USA 71: 835-839.

Rick, C.M. and J.F. Fobes.1975. Allozyme variation in the cultivated tomato and closely related species. Bull Torrey Bot Club 102:376-384.

Rick, C.M., J.F. Fobes and M. Holle. 1977. Genetic variation in *Lycopersicon pimpinellifolium*: evidence of evolutionary change in mating systems. Pl Syst Evol 127: 139-170.

Rick, C.M., J.W. Deverna, R.T. Chetelat , and M. A. Stevens. 1987. Potential contributions of wide crosses to improvement of processing tomatoes. Acta Hort 200: 45-55.

Rick, C.M. and M. Holle. 1990. Andean *Lycopersicon esculentum* var. *cerasiforme*: genetic variation and its evolutionary significance. Econ Bot 43(3 Suppl.): 69-78.

Rick, C.M, H. Laterrot, and J. Philouze. 1990. A revised key for the *Lycopersicon* species. TGC Rep 40: 31.

Ross, R.J. 1998. Review paper: global genetic resourees of vegetables. Plant var and seeds 11: 39-60.

Roxburgh, W. 1832. Flora Indica, or Descriptions of Indian Plants. Vol. 2. W. Thacker and Co., Calcutta, 691 pp.

Ruiz, H. and J.A. Pavon. 1797. Flora peruvianae, et chilensis Prodromus, sive novorum generum plantarum peruvianarum, et chilensium descriptions, et icons. Ed. 2. De órden del Rey, Madrid, 152 pp.

Sabine, J. 1820. On the love apple or tomato. Transantions, Royal Hort Soc London 3: 342-354.

Salmon, W. 1710. Botanologia, The English Herbal, or History of Plants. H. Rhodes and J. Taylor, London.

Shichijo, C., K. Katada, O. Tanaka, and T. Hashimoto. 2001. Phytochrome A-mediated inhibition of seed germination in tomato. Planta 213: 764-769.

Seithe, A. 1962. Die Haararten der Gatun *Solanum* L. und ihre taxonomische Verwertung. Bot Jahrb Syst 81: 261-336.

Smith, B.D. 1995. The emergence of agriculture. Sci Am Lib, New York, 230 pp.

Spooner, D. 2005. New species of wild tomatoes (*Solanum* section *Lycopersicon:* Solanaceae). Systematic Botany 30(2) : 424-434.

Spooner, D.M., G.J. Anderson and R.K. Jansen. 1993. Chloroplast DNA evidence for the interrelationships of tomatoes, potatoes, and pepinos (Solanaceae). Am J Bot 80: 676-688.

Spooner, D.M., I. Peralta and S. Knapp. 2005. Comparison of AFLPs with other markers for phylogenetic inference in wild tomatoes [*Solanum* L. section *Lycopersicon* (Mill.) Wettst.]. Taxon 54: 43-61.

Stevens, M.A. and C.M. Rick. 1986. Genetics and breeding. *In:* J. G. Atherton and J. Rudich [eds.], The Tomato Crop: A Scientific Basis for Improvement. Chapman and Hall, London, pp. 35-109.

Symon, D.E. 1981. The solanaceous genera *Browallia, Capsicum, Cestrum, Cyphomandra, Hyoscyamus, Lycopersicon, Nierembergia, Physalis, Petunia, Salpichroa, Withania,* naturalized in Australia. J Adelaide Bot Gard 3: 133-166.

Symon, D.E. 1985. The Solanaceae of New Guinea. J Adelaide Bot Gard 8: 1-177.

Tabernaemontanus, J. T. 1591. Neuw Kreuterbuch. Franckfurt am Mayn, Germany Volumen II...digerirt vnd vollbracht durch Nicholaum Braun (2 vols.).

Tanksley, S.D. and S.R. McCouch. 1997. Seed banks and molecular maps: unlocking genetic potential from the wild. Science 277: 1063-1066.

Taylor, I.B. 1986. Biosystematics of the tomato. *In:* J.G. Atherton and J. Rudich [eds.], The Tomato Crop: a Scientific Basis for Improvement. Chapman and Hall, London, pp. 1-34.

Teppner, H. 1993. Die Tomate, Verwandtschaft, Geschichte, Blütenökologie. Neue Folge 61 (S): 189-212.

Tournefort, J.P. de. 1694. Elemens de Botanique. l'Imprimerie royale, Paris, (3 Vols.)

USDA National Agricultural Statistics Service. 2002a. Crop Values. U. S. Dep. Agric. Nat. Agric. Stat. Serv., Publ. PR2(02).

USDA National Agricultural Statistics Service. 2002b. Vegetables, 2001 Summary. U. S. Dep. Agric. Nat. Agric. Stat. Serv., Publ. VG1(02).

Van der Heuvel, K.J.P.P., D.W.M. Barendse and G.J. Wullems 2001. Effects of giberellic acid on cell division in anthers of the giberellin deficient gib-1 mutant of potato. Plant Biol. 3: 124-131.

Vavilov, N.I. 1926. Studies on the origin of cultivated plants. Tr. Prikl. Bot. Genek. Sel. 26: 135-199.

Villand, J., P.W. Skroch, T. Lai, P. Hanson, C.G. Kuo and J. Nienhuis. 1998. Genetic variation among tomato accessions from primary and secondary centers of diversity. Crop Sci 38: 1339-1347.

Warnock, S.J. 1988. A review of taxonomy and phylogeny of the genus *Lycopersicon*. HortScience 23: 669-673.

Weller, J.L., G. Perrotta, M. E. L. Schreuder, A. van Tuinen, M. Koorneef, G. Guiliano, and R. E. Kendrick. 2001. Genetic dissection of blue-light sensing in tomato using mutants deficient in cryptochrome 1 and phytochromes A, B1 and B2. Plant J 25: 427-440.

Wettstein, R. von. 1895. Solanaceae, In: A. Engler and K. Prantl (eds.), Die Natürlichen Pflanzenfamilien 4(3b): 4-38.

Williams, C.E. and D.A. St. Clair. 1993. Phenetic relationships and levels of variability detected by restriction fragment length polymorphism and random amplified polymorphic DNA analysis of cultivated and wild accessions of *Lycopersicon esculentum*. Genome 36: 619-630.

Yakovleff, E and F.L. Herrera. 1935. El mundo vegetal de los antiguos peruanos. Revista del Museo Nacional 3. Botánica etnológica,Lima tomo 3, n°3:55.

---

*Not seen in original.

2

Genetic Resources of Tomato (*Lycopersicon esculentum* Mill.) and Wild Relatives

LARRY D. ROBERTSON AND JOANNE A. LABATE

Plant Genetic Resources Unit, United States Department of Agriculture, Agricultural Research Service, 630 West North Street, Geneva, New York 14456-0462, USA
email: lrobertson@pgru.ars.usda.gov

CLASSIFICATION AND TAXONOMY

The genus and species designation of tomato has been the subject of much debate, consequently, many synonyms exist. The designation most frequently used today for the cultigen is *Lycopersicon esculentum* Mill., with *Solanum lycopersicum* L. preferred by some authors. Arguments supporting the transfer of *Lycopersicon* species into *Solanum* L. have been made while recognizing the convenience of maintaining the generic designation *Lycopersicon* for the sake of nomenclatural stability (Peralta and Spooner 2001; see also chapter 1).

The common names used for tomato are numerous. Tomati is the word used by the Indians of Mexico, who have grown the plant for food since prehistoric times. Other names reported by early European explorers were tomatl, tumatle, and tomatas, probably variants of Indian words. Most common names have a root of tomat (used in Danish and Swedish). Tomate is used in French, German, and Portuguese. Tomato is used in English and Spanish. Tomast is used in Dutch and Pomodoro is used in Italian.

Traditional classification (reviewed by Taylor 1986) places tomato within the family Solanaceae (nightshade), sub-family *Solanoideae* (chromosome number x = 12), and tribe *Solaneae*. *Lycopersicon* is one of the smallest genera within this tribe, containing cultivated tomato (*Lycopersicon esculentum*) and its eight wild relatives (*L. pimpinellifolium* (L.) Mill., *L. cheesmanii* L.

Riley, *L. parviflorum* C.M. Rick et al., *L. chmielewskii* C.M. Rick et al., *L. hirsutum* Dunal, *L. chilense* Dunal, *L. pennellii* (Correll) D'Arcy, and *L. peruvianum* (L.) Mill.). Under this classification system *Solanum* is closely related to but distinct from the genus *Lycopersicon*, the two genera being separated on the basis of anther morphology. A comprehensive list of synonyms for *Lycopersicon* species is published on the Germplasm Resources Information Network (GRIN, http://www.ars-grin.gov/). GRIN recognizes *L. glandulosum* C.H. Mull. as a distinct species, whereas many authors treat this taxa as *L. peruvianum var. glandulosum*.

The family Solanaceae contains many plants of economic importance including tomato, potato, eggplant, petunia, tobacco, pepper (*Capsicum*), and *Physalis*. This has promoted detailed studies of many groups but the vast number of species (upwards of 1,000) and their extensive morphological complexity have lent an intractability to comprehensive taxonomic treatments. Consequently, taxonomic studies have focused on subgenera or regional floras (Bohs and Olmstead 1997). *Solanum* contains seven subgenera and 60 to 70 sections (D'Arcy 1991).

During the past decade, molecular genetic techniques have greatly impacted plant taxonomic studies at all levels, including the elucidation of the closest relatives of many crop species (reviewed by Soltis and Soltis 2000). Results of such studies can inform plant breeders, and those interested in comparative genomics, as to which species are potential sources of new alleles or model genetic systems. Recent molecular evidence has placed potato (*S. tuberosum*) and tomato as sister taxa deeply nested within *Solanum*. Under this scheme cultivated tomato is designated as *S. lycopersicum* within section *Lycopersicum*, subgenus *Potatoe*. Molecular evidence supporting this classification is briefly reviewed here.

Restriction site analysis of chloroplast DNA (cpDNA) of *Solanum* subgenus *Potatoe* (including potatoes and pepinos), *Cyphomandra* (tree tomatoes), and *Lycopersicon* (tomatoes), using *Capsicum* and *Datura* as outgroups supported two main clades among the studied taxa: 1) *Solanum* subgenus *Potatoe* and *Lycopersicon*; and 2) other *Solanum* and *Cyphomandra* (Spooner et al. 1993; see also Chapter 1). The authors argued that cpDNA and morphological data supported the transfer of *Lycopersicon* into *Solanum* subgenus *Potatoe*, section *Lycopersicum*, and recognized cultivated tomato as *Solanum lycopersicum* L. The approximately 2 kb chloroplast gene *ndhF* has been found to be useful for phylogenetic studies at inter- and intrafamilial levels of plants (Bohs and Olmstead 1997 and references therein). This gene was sequenced in a broad sample of 18 *Solanum* species, species representing five genera from subfamily Solanoideae, and one outgroup (*Nicotiana tabacum* L. from subfamily Cestroideae). A strict consensus tree of the 12 most parsimonious trees from unweighted parsimony analysis showed 100% bootstrap support (500 bootstrap replicates) for potato and tomato as sister taxa (Bohs and Olmstead 1997). Presence versus absence of restriction sites for ten restriction enzymes were surveyed for the entire chloroplast genome for 36 broadly sampled *Solanum* species and 13 outgroup species (Olmstead and Palmer 1997). Main findings of cladistic analysis of 567 variable restriction sites included: 1) monophyly of *Solanum*, including *Lycopersicon* and *Cyphomandra*, supported by 25 restriction site synapomorphies and 100% bootstrap value (100 bootstrap replicates), and 2) three primary clades within *Solanum*, designated I, II, and III. Clade II consisted of *S. muricatum*, *S. tuberosum*, and *S. lycopersicum*, all within the subgenus *Potatoe*. This clade was supported by 19 synapomorphies and 100% bootstrap value.

The cpDNA studies of Bohs and Olmstead (1997) and Olmstead and Palmer (1997) discerned three or four primary DNA lineages within broadly sampled *Solanum*. The latter authors argue that these data are expected to accurately reflect evolutionary relationships within *Solanum* rather than processes such as lineage sorting or hybridization and introgression. Lineage sorting refers to the differential fixation of shared polymorphisms among descendent species in such a way that obscures phylogenetic relationships. This would require intraspecific polymorphism of cpDNA in a progenitor species, with different haplotypes subsequently becoming fixed in different descendent species. This is purported to be unlikely in these studies because cpDNA evolves slowly, and estimates of cpDNA divergence among closely related species of *Solanum* are often low (Spooner et al. 1993). Hybridization and introgression are more likely to be problematic in obscuring evolutionary relationships among closely related species than among the distantly related species that were sampled for these studies.

Although tomato and potato are near-relatives and have the same basic chromosome number (x=12), multiple rearrangements prevent them from cross-hybridizing (Bonierbale et al. 1988, Tanksley et al. 1992). Recently, tomato, potato, and capsicum have been extensively studied via molecular mapping in order to understand the evolution of genome structure in the Solanaceae (Livingstone et al. 1999 and references therein).

Phylogenetic Relationships among *Lycopersicon* Species

Early taxonomic studies subdivided the genus into two groups, Eulycopersicon, which are color-fruited and Eriolycopersicon, which are green-fruited species (Muller 1940). Rick (1976) designated species into two groups, the esculentum and the peruvianum complexes, based on their reproductive compatibility with cultivated tomato. Key distinguishing taxonomic characters in the following species descriptions were taken from "A revised key for the *Lycopersicon* species" (Rick et al. 1990).

Esculentum Complex

i) *Lycopersicon esculentum* (Mill.). Cultivated tomato (*L. esculentum* var. *esculentum*) is distributed world-wide. Its precise origin in Mexico and/or Peru is uncertain (see below). Fruit interior is red when ripe with seeds 1.5 mm or longer. Leaf margins are typically serrate. Fruit diameter (3 cm or larger) and number of locules (2-to-many) distinguish it from the cherry tomato (*Lycopersicon esculentum* var. *cerasiforme* (Dunal) A. Gray), another cultivated form of tomato that is derived from *L. esculentum* var. *esculentum* crosses with *L. pimpinellifolium* with a fruit diameter 1.5 to 2.5 cm with 2 locules. Wild and weedy forms of *Lycopersicon esculentum* var. *cerasiforme* are the only wild species found outside of South America (Taylor 1986). It has traditionally been regarded as the most likely direct wild ancestor of the cultigen but molecular genetic evidence has challenged this view (Nesbitt and Tanksley 2002, see below).

ii) *Lycopersicon pimpinellifolium* can reciprocally hybridize with the cultigen and displays natural introgression. It may be a direct ancestor of *L. esculentum* or the two species may have evolved in parallel from a green-fruited ancestor (Rick 1976). Fruit interior is red when ripe with seeds 1.5 mm or longer. Relatively smaller fruit diameter (less than 1.5 cm) and generally undulate or entire leaf margins distinguish this species from *L. esculentum*. In the wild its typical habitat is at relatively low elevations (less than 1000 m) in Peru although there are known exceptions at altitudes of 1200 to 1400 m (Taylor 1986).

iii) *Lycopersicon cheesmanii* can reciprocally hybridize with the cultigen but does not do so in the wild because it is geographically restricted to the Galapagos Islands. Fruit interior is yellow or orange when ripe with seeds 1.0 mm or shorter. Subspecies *L. cheesmanii* f. *minor* (Hook. f.) C.H. Mull. is found in relatively lower altitude xeric habitats (Taylor 1986) and is characterized by highly ornate and elaborately subdivided leaflets (Muller 1940).

iv) *Lycopersicon parviflorum* is easily reciprocally hybridized with the cultigen although there may be exceptions for some accessions (Taylor 1986). Fruit inside is green or whitish when ripe with seeds 1.0 mm or shorter. Sympodia have two leaves, inflorescences with small or no bracts, and flowers small (corolla diameter 1.5 cm or less). Its center of diversity is inter-Andean Peru, where it prefers moist habitats along rocky banks of small streams (Taylor 1986). *L. parviflorum* and its sister taxa *Lycopersicon chmielewskii* comprise the *minutum*-complex. The highly autogamous inbreeding *L. parviflorum* is thought to have been derived from the primarily outcrossing *L. chmielewskii*.

v) *Lycopersicon chmielewskii* easily hybridizes with the cultigen (Taylor 1986). Fruit interior is green or whitish when ripe. Taxonomic traits distinguishing this species from *L. parviflorum* are larger seeds (1.5 mm or longer) and larger flowers (corolla diameter 2.0 cm or more). It is sympatric with *L. parviflorum* but more limited in its distribution and prefers slightly better-drained habitats (Taylor 1986).

vi) *Lycopersicon hirsutum* can act successfully as the pollen parent when crossed with cultivated tomato but the reciprocal cross does not set fruit. *Lycopersicon hirsutum* f. *glabratum* C.H. Mull. and the cultigen are reciprocally compatible (Taylor 1986). The two forms of *L. hirsutum* are not fully compatible with each other. Fruit interior is green or whitish when ripe and sympodia have three leaves. The *glabratum* biotypes are characterized by less hairy leaves, stems, and fruit, a smaller corolla, less showy flowers, and a tendency to inbreed. *L. hirsutum* typically grows in moist river valleys at the relatively highest elevations for *Lycopersicon* species (500 to 3300 m), with *glabratum* occupying the northern extremes of the distribution. Both are distributed in Ecuador and Peru (Taylor 1986).

vii) *Lycopersicon pennellii* easily hybridizes with cultivated tomato. Fruit interior is green or whitish when ripe and sympodia have two leaves. Inflorescences have large bracts and anthers are free and poricidal. Distribution is relatively restricted along coastal Peru with some populations found in extreme xeric habitats experiencing temperatures of 25° to 30° C (Taylor 1986).

Peruvianum Complex

i) *Lycopersicon chilense* can act as the pollen parent when crossed with the cultigen but viable seeds are rare and embryos must be cultured. In the reciprocal cross, *L. chilense* will not accept *L. esculentum* pollen. Fruit interior is green or whitish when ripe and sympodia have two leaves. Inflorescences have large bracts, anthers are attached in a tube and are dehiscent by lateral apertures. Flowers are congested and anther tubes are straight. Peduncles are longer than 15 cm and plants are erect. *L. chilense* is the most southerly of *Lycopersicon* species distributed in Chile and southern Peru and prefers extremely arid habitats (Taylor 1986).

ii) *Lycopersicon peruvianum* is genetically and morphologically the most diverse of the *Lycopersicon* species and several varieties have been described (*dentatum, humifusum, peruvianum*). In general, the species exhibits severe barriers to crossing with *L. esculentum*. Fruit interior is green or whitish when ripe and sympodia have two leaves. Inflorescences have large bracts, anthers are attached in a tube and are dehiscent by lateral apertures. Flowers are loosely arrayed with anther tubes generally bending distally. Peduncles are shorter than 15 cm and plants are spreading. Distribution ranges from northern Peru to northern Chile, encompassing a broad range of habitats and including many mountain races that are geographically isolated from each other (Taylor 1986).

iii) *Lycopersicon glandulosum* is synonymous with *Lycopersicon peruvianum* var. *glandulosum*. This is a mountain race common to central Peru. Although it is reproductively compatible with coastal populations of *L. peruvianum* it is distinguished from them by its very thin stems, short dense glandular hairs, and narrow leaflets. It grows at elevations as high as 3000 m where temperatures reach as low as 4° to 8° C (Taylor 1986).

Within the genus *Lycopersicon* closely related interspecies and intraspecies heterogeneity have made the resolution of precise interspecific relationships difficult. Molecular genetic evidence examining relationships has been accumulating during the previous two decades. Speciation within the genus based on these data has been relatively recent. Estimated divergence times for the genus on the basis of pooled silent sites and a rate of 6.03×10^{-9} silent substitutions per site year showed that the genus began its initial radiation ~ 7 million years ago. *L. esculentum* and its nearest relatives *L. cheesmanii* and *L. pimpinellifolium* shared a recent common ancestor ~1 million years BP (before present) (Nesbitt and Tanksley 2002).

Molecular markers support three to four major groupings within the genus, reflecting mating system and fruit color, with autogamy and red-orange fruit being true synapomorphs. These results are generally consistent with morphological and crossability data (Rick 1979). Relationships among species within the major groups are not well-resolved. The traditional distinction between the esculentum and peruvianum complexes is somewhat obscured by molecular phylogenetic evidence. This is because of relatively large genetic distances between *L. hirsutum* and *L. pennellii* and the remaining *Lycopersicon* species, although both species can hybridize with the highly-derived *L. esculentum*, and so are considered to be part of the esculentum complex. The main findings from molecular genetic studies are reviewed here.

The pionering studies examined organellar genomes. Restriction fragment length polymorphism (RFLP) of chloroplast (cpDNA) and hybridization of mitochondrial (mtDNA) DNA studies supported the transfer of *L. pennellii* from the genus *Solanum* into the genus *Lycopersicon* (Palmer and Zamir 1982, McClean and Hanson 1986). At the subgeneric level parsimony analysis supported red fruit color to be monophyletic in *L. esculentum*, *L. pimpinellifolium*, and *L. cheesmanii* (Palmer and Zamir 1982). cpDNA results also gave evidence for a close relationship of *L. chilense* and *L. chmielewskii* to *L. peruvianum*. This apparent anomaly may be explained by the observation that *L. peruvianum* is a heterogeneous taxon with northern populations being somewhat closely related to the esculentum complex.

Eight *Lycopersicon* species were surveyed for RFLPs using 40 single-copy nuclear probes (Miller and Tanksley 1990). The sample of 156 plants represented two to five plants per accession and with the exception of L. *parviflorum*, multiple accessions per species. Unweighted pair group method using arithmetic averages (UPGMA) analysis using genetic distances based on proportion of shared bands gave two major groupings of species. The first group distinguished the self-incompatible (SI) *L. hirsutum*, *L. pennellii*, and *L. peruvianum* from the self-compatible (SC) *L. esculentum*, *L. pimpinellifolium*, *L. cheesmanii*, *L. parviflorum*, and *L. chmielewskii* (Table 2.1). The second grouping delineated green versus red-fruited species. An exception was a northern Peruvian representative of *L. peruvianum* var. *humifusum* (LA2150) which grouped with the SC species albeit most distantly. The authors postulated that this accession may be a modern representative of the SI gene pool that gave rise to the SC *L. parviflorum* and *L. chmielewskii*.

Three main groupings resulted from cluster analysis of allozyme data from eight *Lycopersicon* species: i) SC and red-fruited *L. esculentum*, *L. pimpinellifolium*, *L. cheesmanii*, and *L. peruvianum*, ii) the SC and green-fruited *L. chmielewskii* and *L. parviflorum*, and iii) the SI and green fruited *L. pennellii* and *L. chilense* (Breto et al. 1993).

Random amplified polymorphic DNA (RAPD) markers were used to estimate genetic distances among 22 accessions of *L. peruvianum*; 12 accessions of *L. chilense*; and two accessions from each of *L. esculentum* var. *esculentum*, *L. esculentum* var. *cerasiforme*, *L. pimpinellifolium*, *L. cheesmanii*,

L. chmielewskii, *L. parviflorum*, *L. hirsutum*, and *L. pennellii* (Egashira et al. 2000). Neighbor-joining (NJ) analysis of genetic distances based on 435 RAPD fragments distinguished four groups consisting of three clusters and two outlier taxa; i) *L. esculentum*, *L. pimpinellifolium*, *L. cheesmanii*, and *L. cheesmanii* var. *minor*, ii) *L. chmielewskii* and *L. parviflorum*, iii) the peruvianum complex *L. peruvianum* and *L. chilense*, and iv) *L. hirsutum* and *L. pennellii* which were distinct from all other taxa, including each other. Bootstrap support using 1000 bootstrap replicates was generally low (less than 40%) for most of these groups.

Sixteen polymorphic simple sequence repeat (SSR) markers were used to genotype five to ten plants per accession from one to three accessions each of *L. esculentum* var *esculentum*, *L. esculentum* var *cerasiforme*, *L. pimpinellifolium*, *L. cheesmanii*, *L. parviflorum*, *L. chmielewskii*, *L. chilense*, *L. hirsutum*, *L. pennelli*; and from 11 accessions of *L. peruvianum* (Alvarez et al. 2001). A NJ tree constructed using eight low-diversity loci (gene diversity less than 0.245) was found to be more reliable than trees constructed using all loci. The authors inferred that relatively more diverse SSR loci were more mutable and exhibited higher degrees of homoplasy among species. SSR results supported genetic relationships based on RFLP (Miller and Tanksley 1990) and morphological (Rick 1979) data. With the low-diversity SSR data the northern accessions of *L. peruvianum* (LA2334, LA2172, LA1708, LA2157) clustered closer to *L. esculentum*, *L. pimpinellifolium*, *L. cheesmanii*, *L. parviflorum*, and *L. chmielewskii* than to the southern *L. peruvianum* accessions (LA372, LA462, LA1333, LA1373, LA1274, LA1945, LA1955) and *L. chilense*.

DNA sequence variation of the 5′ portion of granule-bound starch synthase gene (GBSS1 *waxy*) was examined for three accessions from each of *L esculentum* var. *cerasiforme*. *L. cheesmanii*, *L. chmielewskii*, *L. parviflorum*, *L. hirsutum*, *L. pennellii*, *L. chilense*; five accessions of *L. pimpinellifolium*; 39 accessions of *L. peruvianum* one cultigen; plus one to two accessions from each of nine closely related outgroups (Peralta and Spooner 2001). The approximately 1300 nucleotide (nt) region included eight exons and seven introns. A strict consensus tree of 15000 most parsimonious trees did not give good resolution of closely related species. It did support allogamy, self-incompatibility and green fruit as primitive and further supported the splitting of *L. peruvianum* into two groups with the northern populations close to the self-compatible taxa *L. chmielewskii*, *L. parviflorum*, *L. cheesmanii*, *L. pimpinellifolium*, and *L. esculentum* and the southern populations close to *L. chilense*. *L. hirsutum* and *L. pennellii* were distantly related to the remaining esculentum complex species. *Solanum jungandifolium* and *Solanum ochranthum* were the closest outgroup to *Lycopersicon* with *Solanum lycopersicoides* and *Solanum sitiens* basal to these.

DNA sequences of the approximately 700 nt internal transcribed spacer (ITS) region of nuclear ribosomal DNA (rDNA) from one accession each of *L. esculentum* var. *esculentum*, *L. esculentum* var. *cerasiforme*, *L. cheesmanii*, *L. pimpinellifolium*, *L. parviflorum*; three accessions each from *L. peruvianum*, *L. hirsutum*, and *L. pennellii*; plus one accession from each of seven outgroups were studied using parsimony analysis (Marshall et al. 2001). The single most-parsimonious tree displayed a fully resolved topology but with low bootstrap support on some of the branches. Results showed three clades: i) esculentum containing *L. pimpinellifolium*, *L. esculentum* var. *esculentum*, *L. esculentum* var. *cerasiforme*, *L. cheesmanii*, *L. chmielewskii*, and *L. parviflorum*, ii) peruvianum containing *L. peruvianum* and *L. chilense*, and iii) hirsutum containing *L. hirsutum* and *L. pennellii*. These results reflected mating system (SC versus SI) and fruit color supporting the synapomorphy of both SC and red fruit. The northerly *L. peruvianum* var. *humifusum* (LA2150) was nested within the peruvianum clade rather than close to *L. chmielewskii* and *L. parviflorum* (inconsistent with Miller and Tanksley 1990) but with low bootstrap support.

In the first study of *Lycopersicon* relationships using nuclear DNA sequences at multiple loci, four regions were included for a total of approximately 7 kb (Nesbitt and Tanksley 2002). Loci consisted of a fruit weight quantitative trait locus (QTL) *fw2*, the 5′ untranslated region (UTR) of *fw2.2*, *orf44* (an open reading frame of unknown function immediately adjacent to *fw2.2*), three loci on chromosomes other than *fw2.2*: alcohol dehydrogenase 2 (*Adh2*), and two anonymous single-copy loci TG10 and TG11. Species were represented by four accessions of *L. esculentum* var. *esculentum*; three accessions of *L. pimpinellifolium*; and one accession each of *L. cheesmanii*, *L. parviflorum*, *L. hirsutum*, *L. pennellii*, and *L. peruvianum*. The single most-parsimonious tree from data pooled across loci and using *L. pennellii* as the outgroup gave strong support (100% bootstrap values for 100 replicates) for three major groupings: i) *L. esculentum*, *L. cheesmanii*, and *L. pimpinellifolium*, ii) *L. parviflorum* and *L. peruvianum*, and iii) *L. hirsutum* and *L. pennellii*. The *L. peruvianum* accession in this study (LA1708) was a northern type that also clustered close to *L. parviflorum* based on SSR data (Alvarez et al. 2001).

L. esculentum var. *cerasiforme* accessions were also sequenced in Nesbitt and Tanksley's (2002) study but were not used in estimating interspecific relationships because they introduced many incongruities into the trees. Gene trees of *fw2.2* 5' UTR, *Adh2*, TG10, and TG11 show ten *L. esculentum var. cerasiforme* alleles to be interspersed among four *L esculentum var. esculentum* and three *L. pimpinellifolium* alleles. The authors suggested that this apparent admixture of alleles may represent hybridizations between *L esculentum var. esculentum* and *L. pimpinellifolium* giving rise to *L. esculentum var. cerasiforme*.

To summarize the molecular phylogenetic studies of *Lycopersicon*, there is general support for the following relationships (from derived to ancestral groups): i) red fruit SC *L. esculentum, L. cheesmanii, and L. pimpinellifolium*, ii) green fruit SC species *L. chmielewskii* and *L. parviflorum*, iii) green fruit SI northern *L. peruvianum*, iv) green fruit SI southern *L. peruvianum* and *L. chilense*, and v) green fruit SI *L. hirsutum* and *L. pennellii*.

ORIGIN AND MOLECULAR DIVERSITY OF CULTIVATED TOMATO

Cultivated tomato has been documented to have been in existence for only about 400 years (Boswell 1937). There are detailed written accounts of its presumed origin, migration, selection, and introgression (Boswell et al. 1933, Boswell 1937, Jenkins 1948, Stevens and Rick 1986). Peru is the center of origin for the genus and while *L. esculentum var. cerasiforme* was believed to be the direct progenitor of large-fruited cultigens (Bailey 1896), it may more likely be an admixture of cultivated and wild species (Nesbitt and Tanksley 2002). *L. esculentum* is thought to have originally been domesticated and planted in maize fields by ancient Mexicans, although this remains uncertain (Jenkins 1948). Tomato spread to Europe in the early 1500s, initially in Italy and Spain, and thereafter became widely distributed. It was not grown and consumed in large quantities until the late 1700s. Since then it has been selected and bred within a broad range of climates from cool-temperate to tropical. The first cultigens grown in the U.S. came from England and France and the first U.S. improved cultigens were Tilden, released in 1865, and Trophy, released in 1870. Trophy ushered in a new epoch of tomato popularity in the U.S. and is believed to have contributed to the parentage of most of the cultigens developed during the next few decades (Bailey 1896).

Small companies became important sources of seed in late 1800s to 1900s and around 1910 public breeders started introducing disease resistant cultigens, e.g., Tennessee Red and Louisiana Wilt Resistant. These early resistant types dominated the U.S. market in the 1920s and 1930s. In the early 1940s closely related wild species within the genus *Lycopersicon* began to be screened for additional disease resistance, and wild sources provided much of the breeding germplasm during subsequent decades (Stevens and Rick 1986). Wild germplasm continues to play a major role in tomato breeding. Recent work has demonstrated that favorable alleles in wild relatives can remain cryptic until expressed in an improved background (Tanksley et al. 1996) and techniques with which to incorporate wild alleles into modern cultigens continue to be refined (Monforte et al. 2001).

Although migration of the cultigen from Latin America to Europe and subsequently throughout the world has been documented, pedigrees for the majority of *L. esculentum* accessions held in germplasm collections are largely unknown. Molecular markers have been used to estimate genetic diversity of sets of cultigens. Two trends have been observed: i) a narrowing of the germplasm base caused by genetic bottlenecks and selection, and ii) an increase in molecular genetic variation in and around regions introgressed from wild relatives.

Genetic variation among accessions originating from primary and secondary centers of diversity was surveyed using 41 RAPD primers (Villand et al. 1998). This study included 21 accessions from the primary center of diversity for the genus (Chile, Peru, and Ecuador), 37 accessions from secondary centers contiguous with the primary center (South America, Central America, and Mexico), and 38 accessions from secondary centers other than South America, Central America, and Mexico (including Asia, Africa, Cuba, Europe, and U.S.). Average genetic distance for all possible 4,560 pairwise comparisons of accessions (estimated as the complement to the simple matching coefficient; Gower 1972) was 0.164 ± 0.084. Genetic diversity based on allele frequencies (Nei 1987) for groups of accessions was estimated. Average genetic diversity was relatively greater in the primary center (0.219) compared to contiguous (0.175) and other (0.137) secondary centers. This supported allozyme data that showed greater diversity in *L. esculentum* from Peru and Ecuador compared to material from secondary centers (Rick and Fobes 1975). There was also more genetic variation in a set of 20 processing tomato cultigens (Villand 1995) compared to random samples of 20 cultigens from this study. The authors suggested that this could be explained by recent introgression of favorable alleles from wild *Lycopersicon* species into processing cultivars. This was consistent with an observed increase in RAPD variation among modern cultigens compared to cultigens released prior to around 1960 (Williams and St. Clair 1993).

Genetic relationships among 19 tomato cultigens from a geographically isolated regions accessions originating from outside the center of origin (1 South Africa, 5 Europe, 1 Russia, 1 China, 1 Australia, 9 from Canada and U.S., and 1 unknown) were studied using 65 polymorphic SSR markers (He et al. 2003). UPGMA clustering based on genetic distances (Nei and Li 1979) did not reveal a clear pattern reflecting geographical origins of accessions. European accessions were widely dispersed throughout the dendrogram, implying that most non-European germplasm is closely related to germplasm of European origin. However, equal numbers of accessions from various geographical regions would need to be compared in order to test this hypothesis.

RAPD markers showed a narrowing of the germplasm base between tomato cultigens bred and released in India during the 1970s compared to the 1990s (Archak et al. 2002). Average gene diversity (Nei 1987) based on 174 RAPD fragments was 0.265 (n = 5 accessions) in pre-1979 material versus 0.068 (n = 7) and 0.118 (n = 8) in two sets of modern material originating from different breeding programs. This was interpreted as hav-

Table 2.1 *Reproductive biology of* Lycopersicon *species.*

Taxon	Self-compatible	Mating System	Cross-compatibility with cultigen
L. cheesmanii	yes	autogamous	reciprocal
L. chilense	no	allogamous	can act as male with embryo rescue
L. chmielewskii	yes	facultative	reciprocal
L. esculentum var. esculentum	yes	autogamous	reciprocal
L. esculentum var. cerasiforme	yes	autogamous	reciprocal
L. glandulosum	no	allogamous	no
L. hirsutum f. typicum	no, except for some biotypes	allogamous or facultative	can act as male
L. hirsutum f. glabratum	yes	facultative	reciprocal
L. parviflorum	yes	autogamous	reciprocal
L. pennellii	no, except for some biotypes	allogamous or facultative	reciprocal
L. peruvianum	no, except for some biotypes	allogamous or facultative	no, can occasionally be overcome with technical difficulty
L. pimpinellifolium	yes	autogamous	reciprocal

ing resulted from a trend towards breeding for a specific type, i.e., determinate plants bearing uniform fruit.

Although the genetic base of cultivated tomato is narrow, researchers have found it possible to distinguish cultivars using small numbers (four to five) of polymorphic loci (Bredemeijer et al. 1998, four SSR loci distinguishing 16 cultivars; Suliman-Pollatschek et al. 2002, four SSR loci distinguishing 10 cultivars; He et al. 2003, five SSR loci distinguishing 19 cultivars).

ECONOMIC IMPORTANCE

Tomato is a major vegetable crop in the United States and worldwide. The crop is used both fresh and in processed products. Fresh tomato is eaten by itself, in salads, and is used in many recipes as an ingredient. Processed products include paste, canned tomatoes (diced, crushed, and whole), salsa, ketchup, and as an ingredient in many condiments. Tomatoes are also dried and used for cooking.

Worldwide, tomato is produced for the fresh market and processing on approximately 4 million hectares, with an average yield of 27.2 ton ha^{-1} and a yearly production of 108.5 million tons (FAOSTAT 2002; Table 2.2). The top five leading tomato producing countries are China, the United States, Turkey, India, and Egypt. Statistics of regions and for leading tomato producers is given in Table 2.2. Tomato area has increased by 38% and production has increased by 45% in the past ten years. Most of this worldwide increase in production has come from China, where the area has more than tripled from 0.30 million ha to 0.97 million ha, with an increase in production from 8.5 million tons to 25.5 million tons, propelling China from number two to the number one producer of tomato over the United States. China now accounts for 23.5% of the world's tomato production. Leading producers of tomato are listed in Table 2.2.

Production of tomato for fresh market and processing in the United States is given in Table 2.3 (USDA-NASS 1995, 2003). Area of tomato cultivation in the United States for fresh market production has decreased from 53.4 thousand ha to 50.6 thousand ha; however, the area of cultivation for processed products has increased from 110.9 thousand ha to 126.2 thousand ha. Therefore, tomato production for fresh market has decreased in the past ten years from 1.4 million tons to 1.2 million tons; whereas, production of processed tomato has increased in the past ten years from 5.1 million tons to 6.8 million tons.

Table 2.2 *World production of tomatoes in 1992 and 2002 (FAOSTAT 2002).*

	1992			2002		
Location	Area (ha × 10³)	Yield (ton ha⁻¹)	Production (ton × 10³)	Area (ha × 10³)	Yield (ton ha⁻¹)	Production (ton × 10³)
World	2,883	25.9	74,757	3,989	27.2	108,499
Africa	442	20.2	8,918	622	20.0	12,428
North and Central America	329	37.9	12,476	300	52.8	15,838
South America	150	30.0	4,496	149	43.4	6,481
Asia	1,254	23.7	29,768	2,238	23.8	53,290
Europe	698	26.8	18,688	670	29.8	19,969
Oceania	10	40.8	410	10	50.0	492
Leading Countries						
China	304	27.9	8,501	974	26.1	25,466
United States of America	164	59.2	9,730	177	69.4	12,267
Turkey	172	37.5	6,450	225	40.0	9,000
India	309	15.7	4,850	520	14.3	7,420
Egypt	152	30.8	4,694	181	35.0	6,329
Italy	118	46.6	5,483	123	49.3	6,055
Spain	56	47.4	2,647	60	65.2	3,878
Brazil	52	41.0	2,141	62	56.5	3,518
Iran, Islamic Rep. of	92	25.8	2,371	110	27.3	3,000
Mexico	102	16.5	1,677	70	30.0	2,084
Russian Federation	130	12.3	1,600	142	12.8	1,820
Greece	52	35.6	1,850	38	45.1	1,700
Chile	19	40.0	780	20	66.0	1,287
Ukraine	116	11.2	1,303	105	10.5	1,100
Uzbekistan	47	29.1	1,370	28	35.7	1,000
Portugal	20	35.0	700	18	55.2	994
Morocco	25	35.4	894	19	52.0	991
Nigeria	40	10.0	400	126	7.0	879

The value of fresh market tomato produced in the United States is 1.17 billion dollars and the value of processed tomato production is 683 million dollars (Table 2.3). Tomato accounts for approximately 14.5% of the United States, fresh vegetable production value, while it accounts for approximately 50.7% of the processed vegetable production value. Significant production of fresh market tomato is found in 17 states. The two leading producers are Florida and California, which account for 67% of the total value of fresh market tomato production value. There are four primary producers of tomato for the processing market, but California accounts for

Table 2.3   *United States fresh market production of tomatoes 1992 and 2002 (USDA-NASS, USDA 1995 and 2003)*

Location	1992			2002		
	Area (ha × 10³)	Production (ton × 10³)	Value ($ × 10⁶)	Area (ha × 10³)	Production (ton × 10³)	Value ($ × 10⁶)
Fresh Market						
Total Vegetables	611.8	15,087.1	5,889.3	679.1	17,659.1	8,087.0
Tomato	53.4	1,770.5	1,397.0	50.6	1,692.0	1,171.0
AL	1.3	19.0	6.5	0.5	14.6	7.4
AR	0.5	16.0	9.1	0.5	15.2	14.8
CA	14.6	457.2	343.7	15.6	523.9	293.4
FL	20.8	946.1	821.8	18.2	653.2	508.3
GA	1.2	48.0	36.1	1.0	34.0	15.0
IN	0.5	7.1	5.5	0.7	11.2	15.9
MD	0.8	9.1	7.5	0.8	12.2	12.6
MI	1.0	15.9	11.3	0.7	16.2	10.9
NJ	1.9	27.2	21.4	1.3	34.4	27.3
NY	0.9	8.0	6.8	1.1	17.1	24.0
NC	0.6	10.9	4.8	1.1	40.4	24.1
OH	1.2	20.4	14.0	2.7	112.4	89.2
PA	1.7	32.4	15.1	1.7	31.8	16.7
SC	1.5	48.3	22.5	1.3	38.0	20.8
TN	1.8	32.7	15.1	1.6	60.1	43.8
TX	1.3	3.8	2.2	0.4	8.2	5.8
VA	1.3	59.5	40.7	1.5	68.9	41.0
Processed						
Total Vegetables	585.3	12,915.0	1,126.4	545.9	15,528.1	1,346.9
Tomato	110.9	7,962.8	509.4	126.2	10,574.8	683.1
CA	97.1	7,195.8	447.4	117.8	10,029.8	632.4
IN	2.7	148.4	13.0	3.3	232.6	22.1
MI	2.3	165.5	12.6	1.3	101.6	9.3
OH	5.9	339.9	26.0	2.5	135.7	12.1
Other states	2.9	113.2	10.5	1.3	75.1	7.2

approximately 93% of the production value. The second producer, Indiana, accounts for only 3% of the production value.

While Florida and California are the number 1 and 2 producers of fresh market tomato, respectively; the value of their production has decreased by approximately 31% in the past ten years with a decrease of the total production value from 83.5% to 50.7%. However, 12 of the other 15 leading production states have increased their production value of fresh market tomato in the past ten years. The largest increases have been in New York, North Carolina, Ohio, and Tennessee. This shows a trend of consumer preference for locally produced tomatoes because of the perception of higher quality.

The reduction in fresh market tomato can mostly be explained by the increase of imports and the decrease of exports of fresh market tomato by the United States (Table 2.4). In the past ten years fresh market production of tomato in the United States has declined by 78,500 tons. However, the trade deficit of the United States in fresh market tomato production has increased by 634,800 tons. In 2002, most of the United States import of fresh market tomato has come from Mexico, with an increase of imports by 573,400 tons in the past ten years. Imports accounted for approximately 11% of fresh market tomato consumption in the United States in 1992, but by 2002 imports accounted for 32% of the fresh market tomato consumption. Cantliffe (1997) has extensively reviewed the impact of the North American Free Trade Agreement (NAFTA) on Mexican fresh market tomato production and its importation into the United States for the period of 1982 to 1997. The United States has maintained a trade surplus in processed tomato products which has slightly increased in the past ten years (Table 2.4).

Although the majority of fresh market fruit is field grown, there has been an expansion of greenhouse grown tomato. Greenhouse production of fresh market tomato is significant in Europe, especially the Netherlands (Snyder 1996, Table 2.5). Greenhouse production of tomato is predominantly produced with rockwool (Logendra and Janes 1997), moving towards hydroponics (Jensen 1997).

Table 2.4 *United States exports and imports of fresh market and processed tomato (USDA-ERS 2003)*

Exports (ton $\times 10^3$)			Imports (ton $\times 10^3$)		
Product	*1992*	*2002*	*Product*	*1992*	*2002*
Fresh	166.7	150.6	Fresh	196.0	860.1
Paste	73.2	99.1	Paste	19.8	17.8
Canned (whole and pieces	14.0	34.5	Prepared (excluding sauce)	43.3	14.7
Sauce	59.9	116.1	Sauce (including pulp and puree)	7.9	124.1
Catsup and chili sauce	24.5	38.7	Dried	7.8	12.9
Canned pulp (puree)	6.5	16.1			

Table 2.5   *Greenhouse tomato area in selected countries*

Country	Hectares
Canada	287
England/Wales	1214
The Netherlands	4613
Spain	12,140
United States	182

Source: Snyder (1996)

GERMPLASM CONSERVATION

Germplasm Collections

Major germplasm collections of tomato are maintained in the United States at the Plant Genetic Resources Unit, United States Department of Agriculture (USDA) at Geneva, NY, and at the Tomato Genetic Resources Center (TGRC) located in the Department of Vegetable Crops at the University of California Davis. The Asian Vegetable Research and Development Center (AVDRC),now referred to as the World Vegetable Center (WVC), located at Tainan, Taiwan is an international center affiliated with the Consultative Group of International Centers, which maintains the major international collection of tomato germplasm.

Worldwide, there are more than 75,000 accessions of *Lycopersicon* germplasm accessions maintained in more than 120 countries in a number of national institutions (Battencourt and Konopka 1990, updated on web). Table 2.6 lists the largest collections (except the USDA, AVDRC and TGRC which are listed in Table 2.7). Most data reported in the IPGRI listing has been updated since the mid-1990s, with many updates, especially for the larger collections since 2000. This report is available on the Internet at: http://www.ipgri.cgiar.org/ germplasm/dbintro.htm. There are URLS available within this report for many of the institutions where large collections of tomato are reported. The countries with the greatest number of germplasm accessions of *Lycopersicon*, besides the USA, are Brazil, Bulgaria, Canada, China, Colombia, Germany, Hungary, the Philippines, and Spain. These countries have two thousand or more accessions of *Lycopersicon* species conserved, mostly *L. esculentum*.

The collections of USDA and AVDRC are mainly cultivated *Lycopersicon*, though each has significant wild *Lycopersicon* collections (Table 2.7) with AVDRC conserving 659 accessions of wild *Lycopersicon* species and the USDA conserving 458 accessions of wild *Lycopersicon* species. Both of these genebanks have large collections of *L. peruvianum* and *L. pimpinellifolium*

Table 2.6   Worldwide *Lycopersicon germplasm conservation (For USDA, AVDRC, TGRC; see Table 2.7).*

Country	Institution	No. accessions
Australia	Australian Tropical Crops & Forages Genetic Resources Centre	1116
Brazil	Centro Nacional de Pesquisa de Hortali‚as, EMBPRA	2070
	Lab. de Melhoramento Genetico Vegetal (LMGV)-CCTA-VENF	508
	Instituto Agronomico de Campinas (IAC)	500
	Departamento de Fitotechnia-Universidad Federal de Vicosa	600
Bulgaria	Institut de resources phytogénétiques 'K Malkov'	580
Canada	Saskatoon Research Centre, Agriculture and AgriFood Canada	1897
	Horticultural Experiment Station, Simcoe, Ontario	1070
China	Institute of Crop Germplasm Resources (CAAS)	1942
Colombia	Corporacion Columbiana de Investigacion Agropecuaria-CORPOICA	2018
Costa Rica	Estacion Experimental AGricola Fabio Baudrit, Univ. de Costa Rica	700
	Centro Agronomico Tropical de Investigacion y Ensenanza (CATIE)	457
Cuba	Banco de Germoplasma, Inst. de Invest. Fund. en Agricultura	630
Czech Rep.	Genebank Department-Vegtable Secion Olomouc	1613
France	Unité Expérimentale d'Angers Geves	1254
	Station d'Amélioratoin des Plantes, INRA Avignon	1246
Germany	Genebank, Inst. for Plant Genetics and Crop Plant Research (IPK)	3262
Hungary	Institute for Agrobotany	2043
India	National Bureau of Plant Genetic Resources (NBPGR)	940
Italy	Dip. di Agronomia & Genetica Veg. Unversita degli Studi de Napoli	804
Japan	Department of Genetic Resources, Nation. Inst. of Agrobiol. Resour.	452
The Netherlands	Centre for Genetic Resources, The Netherlands (CGN)	1159
Nigeria	National Centre for Genetic Resources and Biotechnology, (FMST)	451
Peru	Universidad Nacional Agraria La Molina	936
Philippines	National Plant Genetic Resources Laboratory, IPB/UPLB	4793
Poland	Plant Breeding and Acclimatization Institute (IHAR)	419
	Plant Genetic Resources Laboratory Research Inst. of Vegt. Crops	427
Spain	Centro de Recursos Fitogeneticos, INIA	1267
	Experimental Station La Mayora CSIC	801
	Univ.Politécde Valencia, Escuela Té Sup. de Ing. Agron. Banco de Germ plasmo	1405
	Banco de Germoplasma de Horticolas-Diputacion General de Aragon	1381
Turkey	Plant Genetic Resources Dept. Aegean Agricultural Research Inst.	544
Vietnam	Centre for Introduced Crops Vietnam Inst. Agric. Sci. & Tech.	487
Yugoslavia	Institute of Field and Vegetable Crops	1030

Table 2.7   Lycopersicon *holdings of the USDA at Geneva, NY (USDA), the Tomato Genetic Resources Center (TGRC), and the Asian Vegetable Development and Research Center (AVDRC)*

Species	Subtaxa	USDA	TGRC	AVDRC
Lycopersicon				
cheesmanii		7	39	18
cheesmanii	f. minor	5	30	9
chilense		1	83	31
chmielewskii		1	27	11
esculentum		4913	427	5311
esculentum	var. cerasiforme	267	275	109
esculentum x esculentum var. cerasiforme		-	-	15
esculentum hybrids		158	*	123
glandulosum		12	13	12
hirsutum		39	76	49
hirsutum	f. glabratum	21	41	17
parviflorum		6	53	12
pennellii		10	40	61
pennellii	var. puberulum	-	8	-
peruvianum		124	155	120
peruvianum	var. humifusum	2	11	4
pimpinellifolium		230	247	315
sp.		13	-	1014
Solanum				
juglandifolium		-	7	-
lycopersicoides		-	16	-
ochranthum		-	4	-
sitiens			5	
Total		5809	1557	7231

*See Table 2.8

(Table 2.7). The collection at TGRC has an emphasis on wild species and various genetic stocks though it also has over 700 accessions of *L. esculentum*. Both the USDA and AVDRC have hybrid populations of *L. esculentum* with other *Lycopersicon* species (Table 2.7), while TGRC maintains a significant number of *Lycopersicon* introgression populations (Table 2.8). A small collection of *Solanum* species (Section *Petota*, sub-section *Potatoe*, Series Juglandfolium) is also maintained by the TGRC with germplasm conserved for *S. juglandifolium*, *S. lycopersicoides*, *S. ochranthum*, and *S. sitiens* (Table 2.7).

Geographical distributions for *Lycopersicon esculentum* accessions maintained by AVDRC and the USDA are given in Table 2.9. The distributions of accessions by country for both collections are similar; with the exception

Table 2.8 *Introgression and special purpose populations of TGRC*

Material	Parental Material	Number
Lycopersicon pennelli introgression lines	LA0716; M-82	76
L. hirsutum introgression lines	LA1777; E-6203 (LA4024)	98
Solanum lycopersicoides introgression lines	LA2951; VF36 (LA0490)	80
L. pennellii alien substitution lines	LA0716	7
L. pimpinellifolium backcross recombinant inbreds	LA1589, E6203	99
S. lycopersicoides monsomic addition lines	LA1964; Vendor	10
High soluble solids derivatives of *L. chmielewskii*	LA1208 deriatives	3
Mutants derived from *L. cheesmanii*	*L. esculentum* derivates of *L. cheesmanii*	5
L. pimpinellifolium exserted stigma derivative	LA1585	1
S. lycopersicoides hybrid	LA2951; VF36	1
Total		380

that AVDRC has a larger representation from Asian countries, such as India, Korea, Malaysia, the Philippines, Sri Lanka, and Taiwan. Both institutions have large number of accessions from Brazil, Canada, China, El Salvador, the former USSR, Guatemala, Honduras, Hungary, India, Mexico, Peru, Turkey, the USA, and Yugoslavia. There is probably much unintentional duplication between these collections and also for those of Table 2.6. Unwanted redundancy within the USDA collection and among different germplasm collections will be discussed in a later section.

Large genetic resources collections are usually duplicated at a second backup location. Backup germplasm collections are highly desirable since they provide a valuable resource to replace germplasm collections that are lost or destroyed by natural disasters (such as fires, floods etc.), political disturbances or mechanical breakdowns. Both the USDA and the TGRC collections are backed up at the National Center for Genetic Resources Preservation (NCGRP) located at Ft. Collins, Colorado. The backup status of the USDA tomato collection at Ft. Collins is given in Table 2.10. Overall, the USDA collection has an 89% backup, though some of the wild taxa have lower backup rates. The TGRC collection is also almost entirely backed up (95%) at NCGRP (Anonymous 2003).

The majority of *Lycopersicon* germplasm used for improvement of tomato has been wild species germplasm. The TGRC maintains a series of special purpose collections of wild species germplasm (Table 2.11). A number of accessions are available with tolerance to drought, flooding, high

Table 2.9 Number of accessions and Country of origin for Lycopersicon esculentum at the Asian Vegetable Development and Research Center (AVDRC), and the USDA germplasm collection maintained at Geneva, NY (USDA)

Country	AVDRC	USDA	Country	AVDRC	USDA	Country	AVDRC	USDA	Country	AVDRC	USDA
Afghanistan	11	13	England	29	-	Korea	27	5	Spain	16	434
Albania	-	20	Ethiopia	16	17	Lao Pdr	3	-	Sri Lanka	23	1
Argentina	62	79	Former USSR	90	104	Lebanon	2	2	Surinam	-	2
Armenia	-	3	France	22	30	Lithuania	-	1	Sweden	7	10
Australia	54	33	French Guiana	10	10	Malawi	2	-	Switzerland	1	2
Bangladesh	5	-	Gambia	2	-	Malaysia	38	1	Syria	26	6
Belgium	-	1	Georgia	-	1	Mauritius	-	1	Taiwan	217	11
Belize	-	2	Germany	17	32	Mexico	86	137	Tanzania	1	1
Bhutan	1	-	Ghana	8	10	Moldova	-	4	Thailand	91	3
Bolivia	79	85	Greece	5	6	Morocco	15	7	Trinidad	2	-
Borneo	1	-	Guadeloupe	18	14	Nepal	7	2	Turkey	193	199
Brazil	119	119	Guam	2	1	Netherlands	44	30	Uganda	1	-
Bulgaria	31	65	Guatemala	220	223	New Caledonia	1	1	Uruguay	1	1
Burkina Faso	-	4	Guyana	1	1	New Zealand	3	2	USA	1061	783
Canada	145	223	Honduras	98	94	Nicaragua	32	32	Venezuela	18	16
Chile	46	49	Hong Kong	10	-	Nigeria	16	46	Vietnam	6	-
China	449	415	Hungary	144	163	Norway	1	4	Yugoslavia	116	159
Colombia	36	142	India	125	91	Pakistan	9	6	Zaire	1	1
Cook Islands	2	2	Indonesia	13	3	Panama	50	39	Zambia	1	-
Costa Rica	46	44	Iran	61	62	Papua N. Guinea	4	-	Unknown	91	9
Croatia	-	1	Iraq	2	2	Peru	221	170	Total	5420	5180
Cuba	7	7	Israel	29	19	Philippines	160	20			
Czech Republic	-	13	Italy	63	94	Poland	46	62			
Czechoslovakia	70	10	Jamaica	-	1	Puerto Rico	17	28			
Denmark	-	3	Japan	60	17	South Africa	19	24			
Ecuador	112	84	Jordan	2	-	Romania	5	7			
Egypt	4	3	Kazakhstan	-	13	Russian Federation	-	23			
El Salvador	410	419	Kenya	1	-	South Yemen	1	-			

Table 2.10 *Backup of the USDA Lycopersicon collection at Geneva, NY*

Species	Subtaxa	No. Acc.	Backup	%Backup
L. cheesmanii		7	5	100.0
L. cheesmanii	f. minor	5	3	42.86
L. chilense		1	1	100.0
L. chmielewskii		1	1	100.0
L. esculentum		4913	4362	89.4
L. esculentum	var. cerasiforme	267	264	68.9
L. esculentum x esculentum var. L. cerasiforme		-	-	-
L. esculentum hybrids		158	158	100.0
L. glandulosum		12	10	83.3
L. hirsutum		39	23	76.2
L. hirsutum	f. glabratum	21	16	59.0
L. parviflorum		6	6	100.0
L. pennellii		10	3	30.0
L. pennellii	var. puberulum	-	-	-
L. peruvianum		124	95	76.6
L. peruvianum	var. humifusum	2	2	100.0
L. pimpinellifolium		230	219	95.2
L sp.		13	1	7.7
Total		5809	5169	89.0

temperatures, aluminum toxicity, chilling injury, salinity-alkalinity, and arthropod damage. Accessions with tolerance are listed in Table 2.10. Lists for these are available on the Internet at: http://tgrc.ucdavis.edu/. These stress tolerant wild species stocks have been extensively utilized in tomato crop improvement.

As mentioned previously, the TGRC maintains a number of introgression populations and special purpose populations, e.g., *Lycopersicon esculentum* X *L. pennellii* and *L. hirsutum* introgression populations are available (Table 2.8). Additionally, TGRC maintains *Solanum lycopersicoides* X *L. esculentum* populations. Various other substitution, backcross recombinant, alien and monosomic addition lines, various mutant populations, and high soluble solid derivatives are also maintained. The majority of the germplasm of *Lycopersicon* maintained at the TGRC is of various genetic stocks (Table 2.12). Monogenic stocks account for approximately $2/3^{rd}$ of these genetic stocks. There are also large numbers of chromosome markers and miscellaneous marker combinations.

Table 2.11 *TGRC stress tolerant wild species stocks*

Stress	Taxon	Accessions
Drought	L. pennellii	LA0716 [a]
	L. chilense	LA1958, LA1959, LA1972[a]
	S. sitiens	LA1974, LA2876, LA2877, LA2878, LA2885
Flooding	L. esculentum var. cerasiforme	LA1421 [a]
	S. juglandifolium [b]	LA2120
	S. ochranthum [b]	LA2682
High temperature	L. esculentum	LA2661, LA2662, LA3120, LA3320
Aluminum	L. esculentum var. cerasiforme	LA2710 [c]
Chilling	L. hirsutum	LA1363, LA1393, LA1777
	L. chilense	LA1969, LA1971, LA4117A
	S. lycopersicoides	LA1964, LA2408, LA2781
Salinity-alkalinity	L. cheesmanii	LA1401, LA108, LA3124, LA3909
	L. chilense	LA1930, LA1932, LA1958, LA2747, LA2748, LA2880, LA2931
	L. esculentum	LA2711
	L. esculentum var. cerasiforme	LA1310, LA2079, LA2081, LA4133
	L. pennellii	LA0716, LA1809, LA1926, LA1940, LA2656
	L. peruvianum	LA0462, LA1278, LA2744
	L. pimpinellifolium	LA1579
Arthropod	L. hirsutum	LA0407 [a]
	L. pennellii	LA0716 [a]

[a] others available
[b] Listed as general characteristic of taxon
[c] Suspected

Conservation

Conservation of genetic resources of tomato in the broad sense encompasses germplasm collection, maintenance, distribution, characterization and evaluation. Collection of tomato germplasm is influenced by the breeding system of the species to be collected and the ease in transfer of traits to the cultigen (Table 2.1). Maintenance of genetic resources of seed crops such as tomatoes involves two separate but inter-related activities; a) long-term maintenance of seed, and b) regeneration of seed as required to maintain sufficient quantities of high-quality seed for storage and distribution. Both of these processes require sufficiently large numbers of plants and or

Table 2.12   *Various genetic stocks available at TGRC*

Type	Number
Translocations	37
Trisomics	31
Autotetroploids	20
Chromosome markers	194
Linkage screening testers	13
Miscellaneous marker combinations	377
Monogenic stocks	994
Total	1666

seeds to avoid the loss of genetic diversity within the collection and to maintain the genetic identity of accessions conserved.

Germplasm characterization and evaluation greatly increases the usefulness of tomato germplasm. Major traits of interest in tomato include quality traits such as soluble solids, yield improvement, and increase of resistance/tolerance to major biotic and abiotic stresses. Conservation of germplasm has the primary goal of providing germplasm for crop improvement and research, and the success of germplasm conservation efforts is measured by the distribution and utilization of the germplasm conserved.

Germplasm Collection

As can be seen from Tables 2.6 and 2.7, the cultivated *Lycopersicon* is well represented in many genebanks around the world. Germplasm acquisition through collections also requires proper sampling procedures to obtain a representative sample for conservation. Outcrossing wild *Lycopersicon* taxa require adequate sampling to obtain a representative sample for conservation while cultivated taxa require less seed per sample.

In the past 20 years there have been a number of collections of the wild *Lycopersicon* species, mostly through the efforts of the TGRC located at Davis, California. Plant collection expeditions have been sponsored by IBPGR in 1980, 1984, 1986, and 1987 (Anonymous 2003) and, additional wild/primitive germplasm collected in 1985, 1995, 1996, and 2001 on trips sponsored by the TGRC and the USDA Plant Exchange Office. The 2001 trip resulted in collection of *L. peruvianum*, *L. chilense*, and two *Solanum* species; *S. lycopersicoides*, and *S. sitiens* from Chile (Chetelat 2001). These expeditions covered areas that were previously poorly represented in existing collections. Collections have also been conducted by researchers from Spain in the Galapagos Islands and in Peru (Nuez and Cuartero 1984).

The Crop Germplasm Committee (CGC) for tomato in the United States National Plant Germplasm System (NPGS) has concluded in its 2003 report (Anonymous 2003) that the status of wild tomato germplasm is in good shape and vastly superior than that of many other crop plants. Recommendations were made for germplasm collection in certain remote areas of Peru, especially in the Rio Maranon watershed because of the presence of *L. peruvianum* and the importance of the genetic diversity found in other populations of *L. peruvianum*. Other acquisition priorities established by this committee include acquisition of germplasm from research projects which have terminated or that are expected to be terminated in the near future. Germplasm from such collections include cultigens, breeding lines, genic and chromosomal variants, and other stocks.

Regeneration

To avoid the loss of diversity and genetic identity through genetic drift, mutation, and selection during the regeneration, standards must be established and used for minimal numbers of plants and seeds. The procedures used for regeneration and storage of tomato are dependent on the breeding system of the plant, with cross-pollinated species requiring larger samples during collection, regeneration, and conservation. For self-pollinated species the requirements are determined by needs to maintain a sufficiently large sample for storage to reduce the number of regenerations required and is determined by the crop husbandry of the plant. Cross-pollinated taxa require a sufficient number of plants to adequately represent the accession and to prevent genetic drift during regeneration.

Specifically the USDA collection of *L. esculentum* is maintained by regenerants from approximately 24 plants, with accessions planted in the field without pollination control. Most genebanks also maintain cultivated *Lycopersicon* with similar numbers of plants, usually in the field with no pollination control. With the self-pollinated cultivated tomato, number of plants used for regeneration is mostly determined by numbers of selfings in field operations and by the desired number of seeds obtained for regeneration. Since tomato seed stores well, production of a large amount of seed can significantly reduce the chances of genetic drift, by increasing the time between regenerations. Wild taxa of *Lycopersicon* are both cross and self-pollinated and both self-compatible and self-incompatible (Table 2.1). Some species are facultative with accessions from some areas self-incompatible and others highly self-compatible. Ideally, during regeneration of the cross-pollinated species upwards of 50 plants are used for regeneration to obtain a representative sample by reducing the effects of genetic drift and selection during the regeneration process.

Regeneration of cultivated tomato is usually conducted by producing transplants that are taken to the field once the danger of frost has passed. At the USDA in Geneva, these are planted using a transplanter into plastic mulch. Seed production requires constant monitoring for diseases with timely application of pesticides to allow sufficient production of quality, disease-free seed for maintenance and distribution. Small scale processing equipment is used for processing cultivated tomato seed with custom designed equipment such as in Fig. 2.1, which is used to separate the seed and gel from the skin and pulp. This material is then fermented for two to three days to ease the processing. After this the seed is washed and dried. Once dried the 'hair' is separated from the seed by tumbling in mesh bags and dried in a clothes-dryer or air only. To reduce incidence of seed borne viruses (such as TMV) seed is treated with a 20% solution of bleach and dried. Seed processing for wild taxa is done by hand. Final processing and storage of seed is discussed in the next section.

Seed Storage Conditions

Methods used for storage are aimed to increase the duration seed remains viable and useful for distribution. Requirements for long-term storage of species with orthodox seeds such as tomato have been well established and are related to moisture content of the seed stored, the type of storage container and the temperature of the environment used for seed storage (Harrington 1970, Roberts 1975, IBPGR 1976, Ellis et al. 1991, 1994).

Equations to predict seed longevity in storage have been developed and refined by Ellis and Roberts (1980) to take into account the variation in initial seed quality along with seed moisture content and temperature of storage. The suggested moisture content for storage of orthodox seeds for genetic conservation is at a moisture content of 5 ± 1%. This moisture content is achieved at Geneva, NY by drying the seed in a room that is maintained at 20% relative humidity at a temperature of 4 to 5°C. The suggested temperature for long-term storage is –20°C. Storage containers should be airtight but not vacuum-sealed because of damage to seed from the evacuation process.

Based on the results of seed vigor studies, Zheng et al. (1998) suggested use of 'ultra-dry' (less than 5% moisture content) seed storage for long-term genetic conservation of plants. They suggested that ultra-dry seed could be stored at higher temperatures compared to –20°C for long-term storage in genebanks. Ellis and Roberts (1998), and Walters et al. (1998),

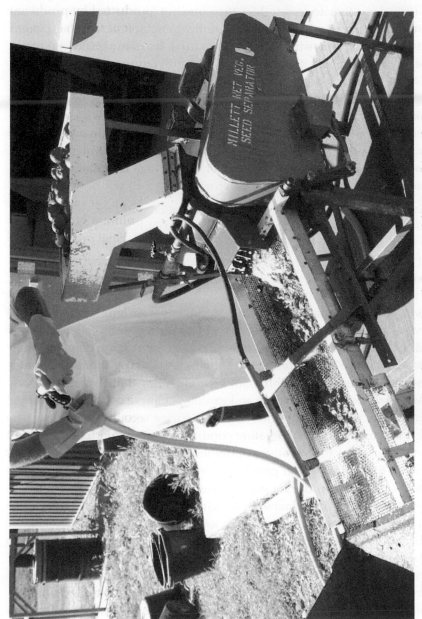

Fig. 2.1 Experimental wet seed processor for separation of tomato seed from small lots of tomato genotypes

both question the usefulness of ultra-dry seed storage at ambient temperatures as a substitute for long-term cold storage in genebanks. XiangHui et al. (1998) dried tomato seed to 1.5% moisture but found that seed could not be stored for long periods at ambient temperatures. Pandey (1995) found that storage of 2% moisture seeds of tomato in hexylene glycol improved short- or medium term storage at ambient temperatures.

Others (Stanwood and Sova 1995, Iriondo et al. 1992) have made suggestions for use of cryopreservation of seed for long-term genetic conservation (conservation using liquid nitrogen) but there is need for more research before a definitive answer of whether there is any advantage to this for orthodox seeds. Sacks and St. Clair (1996) found that pollen cryopreservation can be used successfully for storage of tomato pollen and for tomato breeding and germplasm storage.

Maintaining Genetic Integrity during Regeneration and Storage

Because seed stocks are depleted through distribution and seeds eventually die during storage, periodic regenerations are necessary. Regeneration and maintenance procedures must minimize genetic changes within accessions. Maintaining genetic integrity involves maintaining genetic identity and genetic diversity. Loss of genetic identity occurs through contamination during regeneration by foreign pollen, seed adulteration during harvesting, threshing, and packaging, and through gene mutations (Steiner et al. 1997). Genetic erosion (or loss of genetic diversity) occurs through genetic drift due to random loss of alleles particularly in small populations and through genetic shifts due to unintentional natural selection.

Most new mutations are rare and random. Some studies indicated that major chromosomal aberrations occurred during seed storage and senescence, especially under desiccating conditions, but these gross mutations did not persist in regenerated populations (Wu et al. 1998). Inadvertent selection for more adapted genotypes or genotypes with relatively more viable seed can occur during regeneration (Wu et al. 1998). Selection acts on specific loci. Without knowing which loci these might be or even if we do know, selection is always difficult to distinguish from genetic drift.

Genetic drift and gene flow are relatively easier to detect than mutation and selection. Several studies have found evidence for one or both of these during regeneration in genebanks (del Rio et al. 1997, Wu et al. 1998, Börner et al. 2000, Parzies et al. 2000). Genetic drift refers to random changes in allele frequencies in an accession caused by random sampling of gametes during sexual reproduction. The rate of drift depends on effective population size (*Ne*), which is defined as the number of individuals in a theoretically ideal population having the same magnitude of drift as the actual population (Hartl and Clark 1989). *Ne* is usually smaller than census size and will be substantially so if an accession undergoes a genetic bottleneck.

Theoretical studies indicate that carefully sampling equal numbers of seeds from as many seed parents as possible can effectively prevent drift during regeneration (Crossa and Vencovsky 1997, Vencovsky and Crossa 1999) but this can become so labor-intensive as to make it impractical (Johnson 1998). An alternative strategy is to sample a single inflorescence per plant, rather than the whole plant, when bulking seed of outcrossing plants, such as outcrossing *Lycopersicon* taxa. This will improve *Ne* providing the relative variation in seeds per spike as less than that of seeds per plant (Johnson 1998). In annual ryegrass (*Lolium multiflorum*) balanced (equal numbers of seeds per plant combined), spike (one inflorescence per plant combined), and bulk (seeds combined proportionally according to seeds per plant) regeneration samples were compared. The first method was clearly superior at maintaining genetic integrity of accessions. The latter two methods maintained diversity of eight isozymes as estimated by heterozygosity and mean numbers of alleles per locus, but allele frequencies shifted using those two methods.

Prevention of genetic drift and contamination in outcrossing species such as most wild *Lycopersicon* spp. requires more resources to be invested in regeneration per accession compared to the self-pollinated *L. esculentum*. Molecular markers showed evidence of genetic drift during regeneration in one of eight wheat (*Triticum aestivum*; Börner et al. 2000) and one of six wild potato (*Solanum jamesii*; del Rio et al. 1997) accessions. In two barley (*Hordeum vulgare*) landraces that had been maintained for over 70 years, *Ne* was estimated to be 4.7 using morphological and isozyme markers, even though census sizes in regeneration plots were routinely 600 plants (Parzies et al. 2000). The authors stated that either intense directional selection or a single bottleneck event could explain the extreme loss of genetic diversity.

The mating system of *L. esculentum* results in homozygosity within accessions. Census sizes of regenerated accessions are typically around 25 plants. This is considered to be large enough to prevent sudden extinction of an accession through rapid fixation of rare, deleterious mutations or an accidentally stressful environmental condition leading to inadvertent selection (Treuren and Hintum 2001). Decline of genetic integrity in a tomato accession would most likely occur from gene flow through contaminating pollen or mishandling. For seven oat (*Avena sativa*) lines maintained for 124 years, electrophoresis of storage proteins showed results ranging from no contamination of a line to complete replacement of a line by a foreign phenotype (Steiner et al. 1997). Both mishandling of seed and pollen contamination during maintenance were implicated.

Reducing Redundancy in Collections

Many empirical studies have addressed the question of unintentional duplication of conserved germplasm by examining subsets of collections. Various types of duplications have been defined - identical duplication refers to genetically identical accessions, common duplication denotes accessions derived from a common parental population, partial and compound duplication implies that not all alleles are duplicated, and parental duplication refers to the relationship between a particular cross and the resultant offspring (Hintum and Visser 1995). Although precise language such as this can lead to refined studies of redundancy the use of terms such as these has not become widespread.

A commonly applied experimental approach has been to compare identically or similarly named accessions of a particular species using passport, phenotypic, and/or molecular marker data. The first conclusion that can be reached is that substantial amounts of duplication have been found whenever it was looked for. A study of three European lettuce (*Lactuca*) collections estimated a mean duplication of 12% within and 37% among the collections based on passport data (Hintum 2000). The author pointed out that this may be an underestimate for two reasons: i) because poorly documented accessions were considered to be distinct, and ii) there may be overrepresentation of certain small fractions of the gene pool due to recent shared ancestry or over-collecting in certain geographical areas. We have evidence for both of these problems in our tomato collection and will likely encounter them in our other vegetable collections.

Studies on duplication often address the pooling of duplicate material and estimating how much genetic variation will be lost by doing so. One strategy for pooling is to maximize the ratio of the similarity within groups to the dissimilarity between groups (Hintum et al. 1996). Many studies have applied this model and examples include all accessions of sorghum (*Sorghum bicolor*) named "Orange" in the NPGS (Dean et al. 1999), Dutch landraces of *B. oleracea* accessions at CGN (Hintum et al. 1996), flax (*Linum usitatissimum*) accessions designated as "breeder's lines" at CGN (Treuren et al. 2001), all Peruvian sweet potato accessions at CIP (Huamán et al. 1999), and all accessions of cabbage named Golden Acre (*B. oleracea* var.

capitata L.) in the NPGS (Phippen et al. 1997). In general these studies recommended pooling duplicate accessions based on AMOVA (Analysis of Molecular Variance) in a way that would retain greater than 90% of the total molecular genetic variation. In a pilot study, the USDA at Geneva (unpublished data) applied eight microsatellite markers to six identical San Marzano and six Globe types of tomato and found one of six accessions to be clearly genetically distinct in both cases.

Hintum (2000) has developed methods for quantification of duplications to the level of duplications both within and between germplasm collections along with standard errors of estimates by use of set theory. The definition used for duplicates was accessions with passport data that implies that they are genetically similar or the same. Four lettuce (*Lactuca sativa* L.) accessions were used to apply the equations developed for estimation of duplications within and between germplasm collections. This study points out that: a) most of the probable duplicates were from the exchange of material between collections, b) most of the duplicate materials consisted of older named varieties, and c) accessions unique to one collection were; i) recently added varieties, ii) recently collected material, and iii) poorly documented duplicates which made their identification difficult. The collections reported in Tables 2.6 and 2.7 probably have a high level of redundancy both within and between genebanks.

The USDA germplasm collection has been surveyed for duplications recently, and 1333 accessions have been found to form 455 sets of putative duplicate accessions. An empirical approach using accession identifiers was used to identify these duplicate accessions. An approach of using passport data to identify the known original source of these accessions to keep as representing the cultigen has been used to reduce duplication for approximately 90% of these accessions. The other 10% of accessions will be grown out at several locations for identification of duplications.

Development of Core Subsets
Core subsets are tools for users to efficiently work with a large fraction of the total genetic diversity in a collection. Random sampling methods are applied to develop core subsets. In order to broadly capture diversity, sampling may be from subsets of accessions initially grouped according to phenotype or ecogeographic origin (Li et al. 2002). Phenotypic traits are not always good indicators of genetic variation because they can be influenced by environment or result from independent genetic bases. Molecular markers can help overcome these limitations.

Molecular markers have been used: i) to compare various techniques applied to generate a core (e.g., *Sorghum bicolor*, Grenier et al. 2000), ii) as criteria to decide which accessions to include in a core [e.g., cashew (*Anacardium occidentale*), Dhanaraj et al. 2002, Andean potato (*Solanum phureja*), Ghislain et al. 1999], and iii) to validate that a core is representative of a given base collection [e.g., Mexican common bean (*Phaseolus vulgaris*), Skroch et al. 1998]. Computer simulations, multivariate statistical techniques such as Principal Components Analysis (PCA), and genetic variation measures such as numbers of alleles, genetic diversity, and percentage of rare alleles, are frequently used to establish and validate core subsets.

Molecular markers reflect pedigrees, which may not be significantly correlated with gross morphology. In the Brazilian cassava (*Manihot esculentum*) collection cultigens with similar agronomic traits were very heterogeneous at the molecular level (Carvalho and Schaal 2001). For four major *Sorghum* races, grouping of accessions based on agronomic descriptors did not correlate with groupings produced from RAPD markers (Dahlberg et al. 2002). The authors stated that such correlations should not be expected because in most cases genes underlying an agronomic trait represent a very small fraction of the genome.

Core subsets of *Lycopersicon esculentum* and the wild *Lycopersicon* taxa have been assembled by the USDA-Geneva and TGRC, respectively. The cultivated core subset is a dynamic collection that is being refined. Presently, approximately 200 accessions of cultivated tomato, mostly from the United States, comprise the core collection. The tomato CGC is making efforts to modify this collection to have a balanced representation globally and to have better representation of tomato usage and of plant types. The TGRC at Davis has also established a core collection of wild *Lycopersicon* taxa.

GENETIC DIVERSITY IN *Lycopersicon* SPECIES

Mating system, life history traits (e.g., annual versus perennial, longevity, etc.), and ecological factors such as those causing frequent extinction and recolonization, all shape genetic variation within and among populations of a species. For the genus *Lycopersicon*, mating system is the most-extensively studied of these factors. The genus has been viewed as a model system among plants for studying the effect of mating system within-species variation (Stephan and Langley 1998, Baudry et al. 2001), because it consists of closely-related species with a range of mating systems from selfing (*L. esculentum, L. pimpinellifolium, L. cheesmanii,* and *L. parviflorum*) to facultative outcrossing (*L. chmielewskii*) to obligate-outcrossing (*L. pennellii, L. hirsutum, L. chilense,* and *L. peruvianum*). In addition, some biotypes of *L. pennellii, L. hirsutum,* and *L. peruvianum* have been found to be self-compatible (Rick 1982, Taylor 1986).

Neutral theory predicts that polymorphism within a species is a function of mutation rate and effective population size (Kimura 1983). Effective population size is defined as the number of individuals in a theoretically ideal population having the same magnitude of genetic drift as the actual population (Hartl and Clark 1989). Compared to random-mating, selfing is expected to decrease effective population size by one-half and hence the genetic variation by the same amount (Pollak 1987). Additional reasons why selfing (self-pollinated) species are predicted to have reduced variation are related to frequent bottlenecks and reduced recombination. A single self-fertilizing plant can act as a founder for a new population; selfing species may frequently undergo such bottlenecks. A deficiency of double heterozygotes commonly characterizing selfing populations decreases effective recombination rate, thereby increasing linkage disequilibrium. Extensive linkage disequilibrium is associated with reduced variation because selection will effect more of the genome, i.e., the selected locus plus tightly linked, unselected loci (Charlesworth et al. 1993). In the plant genera *Arabidopsis* (Savolainen et al. 1999) and *Leavenworthia* (Liu et al. 1999) selfing species were found to have greater than two-fold reduction in genetic variation relative to outcrossing species.

In the genus *Lycopersicon,* levels of polymorphism for RFLP (Miller and Tanksley 1990), RAPD (Egashira et al. 2000), SSR (Alvarez et al. 2001), and DNA sequences (Baudry et al. 2001) have all been shown to be correlated with mating system.

In an RFLP study, the number of unique restriction fragments and the number of unique restriction patterns were used to estimate genetic diversity of accessions (Miller and Tanksley 1990). SI species *L. hirsutum, L. pennellii,* and *L. peruvianum* were found to be much more diverse than the SC species *L. esculentum, L. pimpinellifolium, L. cheesmanii, L. parviflorum,* and *L. chmielewskii.* Genetic distances among accessions were calculated based on proportion of shared restriction fragments (Nei 1987, equations 5.53 to 5.55). Average genetic distances among accessions were approximately ten-fold greater for *L. peruvianum,* and five-fold greater for *L. hirsutum* and *L. pennelli* than any other SC species. In addition, most of the diversity was distributed among rather than within accessions for the SC species.

Genetic distances based on proportion of shared bands (Nei and Li 1979) between species, between accessions within species, and between plants within an accession were estimated for *L. esculentum* var. *esculentum* and its wild *Lycopersicon* relatives using RAPD markers (Egashira et al. 2000). The facultative outcrosser *L. chmielewskii* contained the highest average within accession variation among the SC species (0.045). This was almost six-fold lower than the highest estimate found among the SI species (0.252 in *L. chilense*). Average genetic distances between accessions within species ranged from 0.036 for *L. esculentum* var. *esculentum* to 0.677 for *L. chilense*. Northern Peruvian accessions of *L. peruvianum* and southern Peruvian accessions of *L. chilense* showed the greatest within-species genetic diversity.

L. hirsutum, *L. pennellii*, and *L. peruvianum* were found to be more diverse than *L. esculentum*, *L. pimpinellifolium*, *L. cheesmanii*, *L. parviflorum*, and *L. chmielewskii* in a study of 16 polymorphic SSR loci (Alvarez et al. 2001). Two measures of variation were used: i) numbers of unique alleles (species-specific), and ii) gene diversity based on allele frequencies (Weir 1996). All species except *L. esculentum* harbored at least one unique allele. Sixty-six of 144 alleles (46%) were found to be unique. *L. chilense* contained the relatively highest proportion of unique alleles (0.80) when corrected for differences in numbers of plants sampled per species. Gene diversity within SC species was lowest in *L. esculentum* (0.03) and highest in *L. pimpinellifolium* (0.20). In *L. pimpinellifolium* all gene diversity was partitioned among the three sampled accessions, i.e., alleles were fixed at all loci within accessions of this species. For the SI species, gene diversity ranged from 0.24 in *L. pennellii* to 0.57 in southern representatives of *L. peruvianum*. In contrast to estimates based on RAPDs, SSR markers showed northerly accessions of *L. peruvianum* contain less diversity than southerly accessions (Egashira et al. 2000).

Effects of mating system and recombination on intraspecific DNA sequence polymorphism were studied by comparing *L. chilense*, *L. hirsutum*, *L. peruvianum*, *L. chmielewskii*, and *L. pimpinellifolium* (Baudry et al. 2001). Five plants per species were sampled, except for *L. hirsutum* (three plants), and sequenced at five single-copy genes in chromosomal regions with either high (2.33×10^{-8} - 2.73×10^{-8} per site per generation) or low (0.00 - 0.46 $\times 10^{-8}$ per site per generation) rates of recombination. More than 8 kb of DNA was sequenced per plant, including four anonymous, single-copy cDNA markers (Tanksley et al. 1992) CT208, CT251, CT268, CT143, and the sucrose accumulator gene *sucr*. Intraspecific polymorphism was estimated as θ for non-coding nucleotide sites (Nei 1987). Mating system was found to have a significant effect on polymorphism. *L. pimpinellifolium* and *L. chmielewskii* had on an average approximately 4 and 40-fold less polymorphism than *L. hirsutum*, the least polymorphic of the SI species. *L. chmielewskii* was the least variable species, with all estimates of θ equalling zero with the exception of CT143, where θ was less than 0.01. *L. peruvianum* was the most polymorphic species, with θ values ranging from approximately 0.01 to greater than 0.03 across loci. In addition, a high proportion (14% - 40% across loci) of fixed differences among the other four species were observed as polymorphisms within *L. peruvianum*. This may represent lineage sorting of alleles among species and imply that *L. peruvianum* is representative of the ancestral species from which other species were derived. Although reduced recombination rate has been found to be significantly correlated with lower polymorphism in other species (e.g., *Drosophila*, Aquadro et al. 1994), in this study recombination and polymorphism were only weakly positively correlated.

Theta estimates for *L. esculentum* across four loci (fw2.2 5' UTR, Adh2, TG10, and TG11) ranged from 0.0016 - 0.0048 (Nesbitt and Tanksley 2002). These estimates largely reflected polymorphism within *L. esculentum* var. *cerasiforme*. Four modern cultigens of *L. esculentum* var. *esculentum* included in this study contained only one polymorphic site in more than 7 kb of total sequence. The high degree of monomorphism within *L. esculentum* var. *esculentum* has likely resulted from severe bottlenecks. However, polymorphism should be higher within regions of the genome containing introgressed loci from wild relatives (Miller and Tanksley 1990).

To summarize, selfing species of *Lycopersicon* contain significantly less genetic variation relative to outcrossing species. The reduction in variation exceeds the predicted 50% reduction that could be explained by mating system alone. Similar observations have been made in other plant genera (Savolainen et al. 1999, Liu et al. 1999). Additional factors such as founder events, fluctuating population size, population substructure, selection, and linkage must be better understood in order to explain relative levels of genetic variation observed within *Lycopersicon* species (Baudry et al. 2001).

CHARACTERIZATION AND EVALUATION

Characterization and evaluation of germplasm of tomato is essential to promote its utilization for crop improvement. The CGC for tomato has identified a number of biotic and abiotic stresses, as well as quality traits as having high priority for evaluation to provide sources for use in improvement programs on tomato (Table 2.13). High priorities have been established for screening disease resistance for verticillium wilt race 2, bacterial canker, geminiviruses, pepino mosaic virus, spotted wilt, and bacterial spot. High priorities for screening for insect resistances have been established for silverleaf whitefly and nematodes. Priority has also been established for abiotic stresses such as heat and cold. A number of quality traits have been prioritized, especially for soluble solids, flavor, and color.

An international descriptor list has been established for *Lycopersicon* spp. (IPGRI 1996). This list has three major types of descriptors; passport,

Table 2.13 *Tomato problems where genetic improvements would benefit U.S. production.*

Type	Priority	Description
DISEASES		
	High	Verticillium wilt race 2
		Bacterial canker
		TYLCV & other geminiviruses
		Pepino mosaic virus
		Spotted wilt
		Bacterial spot
	Medium	Late blight
		Corky root
		Phytophthora root rot
		Fruit rots
		CMV
		Beet curly top virus
	Low	Bacterial speck race 2
		PVY
		Target spot
		Powdery mildew
INSECTS & PESTS (new screening protocols important)		
	High	Silverleaf whitefly
		Nematodes, heat stable
	Medium	Aphids
ABIOTIC STRESSES		
	High	Cold tolerance
	Medium	Heat tolerance
		Salinity tolerance
		Color disorders
HORTICULTURAL		
	High	Soluble solids
		Flavor (need to define components)
	Medium	Antioxidants/nutritional content
		Color
		Sugar type
		Peelability/dicing
	Low	Pectin chemistry
		Blossom-end smoothness

characterization and evaluation descriptors. Also included are management descriptors and environment and site descriptors.

The passport descriptors provide the basic identification of the accession which includes the genebank's accession number along with the other identifiers associated with the accession, such as collection number and/or other institution identifier numbers. If the accession is a variety or cultigen, name and if available, pedigree and breeding method are included in passport descriptors. The taxonomic classification of the accession is also included in the passport descriptors. The other passport descriptors provide information on where an accession was collected, including political and eco-geographic information about the collection site. Eco-geographic information often is very useful in selecting accessions, especially when little or no characterization or evaluation data are available. Management descriptors associated with an accession provide information about seed availability, viability, etc., but are usually not publicly available.

Characterization and evaluation descriptors are often publicly available and provide information to aid users in selecting accessions for use in crop improvement programs. Characterization descriptors have high heritability and are usually only recorded in one environment and include descriptors such as plant type, leaf type, fruit shape, fruit color, among many others. Evaluation descriptors are more detailed and are often replicated over environments. Important evaluation descriptors include biotic and abiotic stress resistance and/or tolerance, quality descriptors such as soluble solids and fruit pH. Other descriptors, especially with evaluation traits, are provided which describe the environment and site where the evaluation was conducted, methodologies used, and person(s) who conducted the research. Examples of these would be the geographical coordinates of the evaluation site, the soil type, weather conditions during the season of the evaluation (temperatures, rainfall), and dates of planting and harvest. This type of data aids in interpretation of many evaluation descriptors.

The United States tomato CGC has established a minimal descriptor list for tomato (Table 2.14). This list was decided in order to provide a guideline for the genebank of which descriptors in the overall list were felt to be of high importance in making selections of germplasm accessions. This was to facilitate provision of a manageable number of descriptors to record during regeneration, since the IPGRI list includes a number of morphological descriptors, many of which are species specific. The minimal descriptor list includes plant descriptors, fruit descriptors, and chemical composition.

Table 2.14 *Minimal descriptor list for tomato.*

I. Plant characteristics

A. Plant growth type
  1. Miniature Dwarf
  2. Dwarf
  3. Determinate
  4. Semi-determinate
  5. Indeterminate
B. Canopy size
  3. Small   [Red Rock]
  5. Intermediate   [Florida MH-1]
  7. Large   [Mountain Pride-determinate; Tropic-indeterminate]
C. Leaf type
  1. Rugose
  2. Potato leaf
  3. Standard
  4. Curled
  5. Others [Specify in Notes]
D. Flowers per inflorescences
  3. Low
  5. Medium
  7. High
E. Type of inflorescence
  1. Simple
  2. Forked
  3. Compound
F. Number of fruit set
  Recorded on second truss
G. Number of days to maturity
  From sowing until 50% of the plants have at least one fruit ripened

---

II. Fruit Descriptors

A. Exterior color of immature fruit
  1. Light green
  2. Medium green (Apple)
  3. Dark green (hp, dg)
B. Exterior color of mature fruit
  1. White
  2. Green
  3. Yellow
  4. Gold
  5. Orange
  6. Pink
  7. Red
  8. Other [Specify in Notes]
C. Exterior mature fruit appearance
  1. Dull
  2. Medium
  3. Glossy

(Contd.)

(Contd.)

D. Shoulder Color
    1. Green
    2. Gray Green (light green)
    3. Uniform
E. Mature fruit interior flesh color
    1. White
    2. Green
    3. Yellow
    4. Orange
    5. Pink
    6. Red
    7. Other (Specify)
F. Mature fruit interior flesh color intensity
    1. Pale
    2. Intermediate
    3. Deep
G. Fruit shape
    1. Flattened (oblate)
    2. Slightly flattened (deep oblate)
    3. Globe
    4. Deep Globe
    5. Blocky (square round)
    6. Heart-shaped
    7. Ellipsoid (plum-shaped)
    8. Cylindrical (long oblong)
    9. Pyriform (pear-shaped)
H. Pistil scar
    1. Dot
    2. Stellate
    3. Linear
    4. Irregular
I. Fruit weight (g, average of 10 fruits)
    Checks   [Sweet 100]
                 [Red Cherry Large]
                 [Roma or New Yorker]
                 [Flora-Dade]
                 [Tropic]
                 [Florida 7060]
J. Uniformity of fruit size
    3. Low
    5. Intermediate
    7. High
K. Fruit firmness
    3. Soft
    5. Medium
    7. Hard
L. Nippled fruit [mature fruits]
    1. Absent

(Contd.)

(Contd.)

    2. Present
    3. Inverted
M. Presence/absence of jointless pedicel
    1. Absent
    2. Present
    3. Arthritic (not complete)
N. Radial cracking
    0. Absent
    3. Slight
    5. Intermediate
    7. Severe
P. Concentric cracking
    0. Absent
    3. Slight
    5. Intermediate
    7. Severe
Q. Cuticle cracking
    0. Absent
    3. Slight
    5. Intermediate
    7. Severe

III. Chemical composition

A. Soluble solids
   Measured with refractometer from 4 fruits

UTILIZATION

The major resource of tomato germplasm for crop improvement in the past 20 years has been the use of the wild species as sources of disease and insect resistance, and for improvement of quality traits (Rick and Chetelat 1995). This is because tomato being a self-pollinated crop has germplasm strongly reduced in variability for domestication and breeding. A thorough summarization of the use of related wild tomato species for crop improvement of tomato through 1995 (with literature citations) has been provided by Rick and Chetelat (1995). Introgression of many disease resistant genes into cultivars has been accomplished through the identification of linked molecular markers (Table 2.15; Causse et al. 2000, Grube et al. 2000). Resistance and/or tolerance has been transferred from wild species of *Lycopersicon* for bacterial, fungal, nematode, viral diseases, and for resistance to parasitic plants (broomrape and dodder). Wild species have also been used as sources of tolerances of abiotic stresses and for improvement of quality traits. Resistance and/or tolerance has also been transferred for insect pests namely, *Coleoptera, Diptera, Homoptera, Lepidoptera,* and *Acarina* arthropods. *L. pennellii* has been found to be a

Table 2.15 Sources of common disease resistance alleles transferred to cultivated tomato

Chr	Locus	Pathogen	Source	Reference
1	Cf-4	*Cladosporium fulvum*	*L. hirsutum* 3833 (Univ. of Toronto)[a]	Haanstra et al. (2000), Kerr and Bailey (1964)
1	Cf-9	*Cladosporium fulvum*	*L. pimpinellifolium* PI 126915	Haanstra et al. (1999)
2	Tm-1	TMV	*L. pimpinellifolium, L. hirsutum, L. peruvianum, L. chilense*	Levesque et al. (1990), Pelham (1966), Holmes (1954), Frazier and Dennett (1949), Kikuta and Frazier (1947)
3	py-1	*Pyrenochaeta lycopersici*	*L. peruvianum* var. *glandulosum* Pannevis 02126 K	Doganlar et al. (1998), Laterrot (1993), Laterrot (1983)
4	Hero	*Globodera rostochiensis*	*L. pimpinellifolium* B6173 (LA0121)	Sobczak et al. (2005), Ellis and Maxon-Smith (1971)
5	Pto	*Pseudomonas syringae* pv. *tomato*	*L. pimpinellifolium* PI 370093	Martin et al. (1991), Pitblado et al. (1984), Kerr and Cook (1983), Pitbaldo and Kerr (1980)
6	Mi (Meu-1)	*Meloidogyne* spp., *Macrosiphum euphorbiae*	*L. peruvianum* PI 128657	Yaghoobi et al. (1995)
6	Ty-1	TYLCV	*L. chilense* LA 1969	Zamir et al. (1994)
6	Cf-2	*Cladosporium fulvum*	*L. pimpinellifolium* PI 370093	Jones et al. (1992), Kerr et al. (1980), Pitabaldo and Kerr (1980), Langford (1937)
6	Cf-5	*Cladosporium fulvum*	*L. esculentum* x *L. pimpinellifolium* PI 187002	Dixon et al. (1998), Dickinson et al. (1993), Kerr et al. (1971)
7	I-1	*Fusarium oxysporum* f. sp. *lycopersici*	*L. pennellii* PI 414773	Scott et al. (2004)

(Contd.)

(Contd.)

7	I-3	Fusarium oxysporum f. sp. lycopersici	L. pennellii PI 414773	Sarfatti et al. (1989)
7	Ph-1	Phytopthera infestans	L. esculentum var. cerasiforme PI 108245	Chunwongse et al. (1998), Clayberg et al. (1965), Clayberg et al. (1959), Gallegly and Marvel (1955)
9	Ve	Verticillium dahliae	Peru Wild (synonymous to Utah 665) PI 303801	Kawchuk et al. (1998), Schaible et al. (1951)
9	Ph-3	Phytopthera infestans	L. pimpinellifolium L3708 (AVDRC)	Chunwongse et al. (1998)
9	Tm-2[a]	TMV	L. peruvianum PI 128650	Young et al. (1988)
9	Frl	Fusarium oxysporum	L. peruvianum PI 126944	Vakalounakis et al. (1997), Vakalounakis (1988), Scott and Farley (1983)
9	Sw-5	TSWV	L. peruvianum[b]	Roselló et al. (1998), Stevens et al. (1992), Stevens (1964)
10	Ph-2	Phytopthera infestans	L. pimpinellifolium WVa 700	Moreau et al. (1998)
11	Sm	Stemphylium spp.	L. pimpinellifolium PI 79532	Behare et al. (1991)
11	I	Fusarium oxysporum f. sp. lycopersici	L. pimpinellifolium PI 79532	Sarfatti et al. (1989)
11	I-2	Fusarium oxysporum f. sp. lycopersici	L. pimpinellifolium × L. esculentum PI 126915	Sarfatti et al. (1989)
12	Mi-3	Meloidogyne incognita, M. javanica	L. peruvianum PI 126443	Yaghoobi et al. (1995)
12	Lv	Leveillula taurica	L. chilense LA 1969	Chunwongse et al. (1994)

[a] Kerr and Bailey (1964) reported that this accession was lost.

[b] Stevens et al. (1992) reported that the identity of the original PI number was lost after the death of J.M. Stevens but list PI 126928, PI 126929, PI 126944, PI 128645, PI 128654, PI 129109 and two L. peruvianum var. dentatum accessions with unknown PI numbers as potential sources based on his breeding records.

promising source for drought tolerance and salt tolerance. The work on fruit quality has concentrated on increasing soluble solids content. Higher levels of soluble solids content have been discovered in *L. cheesmanii, L. chmielewskii,* and *L. hirsutum.*

A number of studies have identified QTLs in wild species of *Lycopersicon* that provide improvement in the cultivated tomato for horticultural traits. QTLs associated with horticultural yield in *L. pennellii* and *L. hirsutum* were identified by Eshed et al. (1996) and Bernacchi et al. (1998a). *L. hirsutum* alleles were found that gave a 16% increase in total yield and a 20% improvement was achieved by combining introgressions from *L. pennellii.* Introgression lines have been used for mapping of QTL for improved fruit characteristics in *L. chmielewskii* (Frary et al. 2003) and for yield associated QTL using *L. pennellii* (Eshed and Zamir 1995). Materials developed showed promise for improvement of cultivated tomato. Genetic gains from introgressions for desirable wild QTL-alleles from *L. hirsutum* and *L. pimpinellifolium* for quality traits such as fruit firmness, soluble solids content, and brix *X* red yield have been reported (Bernacchi et al. 1998b).

SUMMARY

Cultivated tomato (*Lycopersicon esculentum* Mill.) is an important vegetable, with a worldwide area of 4 million hectares and a production of 108.5 million tons. Tomato cultivation area has increased by 38% and production has increased by 45% in the past ten years, with most of this increase in China, which has increased production from 8.5 to 25.5 million tons, propelling it to the number one tomato producer in the world. Other leading tomato producers are the United States, Turkey, India, and Egypt. In the United States, tomato accounts for 14.5% of the economic value of fresh market vegetable production and 50.7% of the economic value of processed production of vegetables.

Domestication of tomato is relatively recent, within the past 400 years. *L. esculentum* is thought to have originally been domesticated in maize fields by ancient Mexicans, with Peru as the center of diversity for the genus. Tomato spread to Europe in the early 1500s and thereafter became widely dispersed. Tomato has eight (nine by some authorities) related wild species relatives which are extensively utilized for crop improvement.

There are more than 75,000 accessions of tomato conserved in genebanks around the world, with the largest of these at AVDRC, TGRC, and the USDA genebank at Geneva, NY. These genebanks maintain large collections of the wild relatives in addition to the cultigen. Several collections are also available of related *Solanum* taxa and *Lycopersicon* introgression populations. While cultivated tomato is self-pollinated, the other taxa provide a mixture from self-pollinated to obligate cross-pollinated, with self compatibility and self-incompatibility. This has led to development of methodologies and standards for maintenance of tomato germplasm to minimize the effects of genetic drift, mutation, and selection.

The cultivated tomato has undergone a narrowing of the germplasm base caused by genetic bottlenecks and selection. The major utilization of tomato germplasm for crop improvement in the past 20 years has been the use of wild species as sources of genetic variation. This has led to a major utilization of wild species introgressions which have resulted in an increase in molecular genetic variation in and around regions that have been introgressed from wild species. Wild species have been used as sources of variation for disease and insect resistances and/or tolerances, abiotic stress tolerances, and for fruit quality.

REFERENCES

A.E., C.C.M. van de Wiel, M.J.M. Smulders, and B. Vosman. 2001. Use of microsatellites to evalute genetic diversity and species relationships in the genus *Lycopersicon*. Theor Appl Genet 103: 1283-1292.

Anonymous. 2003. Tomato Crop Germplasm Committee (CGC) Report. http://www.ars-grin.gov/npgs/cgclist.html#Tomato.

Aquadro, C.F., D.J. Begun, and E.C. Kindahl. 1994. Selection, recombination, and DNA polymorphism in *Drosophila*. *In*: G.B. Golding. [ed.], Non-neutral evolution: Theories and molecular data. Chapman and Hall, New York.

Archak, S., J.L. Karihaloo, and A. Jain. 2002. RAPD markers reveal narrowing genetic base of Indian tomato cultivars. Curr Sci 82: 1139-1143.

Bailey, L.H. 1896. The amelioration of the garden tomato 1896. The survival of the unlike: A collection of evolution essays suggested by the study of domestic plants. The Macmillan Co., London, England.

Battencourt, E. and J. Konopka. 1990. Vegetables: *Abelmoschus*, *Allium*, *Amaranthus*, Brassicaceae, *Capsicum*, Cucurbitaceae, *Lycopersicon*, *Solanum* and other vegetables. Updated at http://www.ipgri.cgiar.org/germplasm/dbintro.htm. International Plant Genetic Resources Institute (IPGRI), Maccarese, Italy.

Behare, J., H. Laterrot, M. Sarfatti, and D. Zamir. 1991. Restriction fragment length polylmorphism mapping of the *Stemphylium* resistance gene in tomato. Mol Plant Microbe In 4:489-492.

Bernacchi, D., T. Beck-Bunn, D. Emmatty, Y. Eshed, S. Inai, J. Lopez, V. Petiard, H. Sayama, J. Uhlig, D. Zamir, and S. Tanksley. 1998b. Advanced backcross QTL analysis of tomato. II. Evaluation of near-isogenic lines carrying single-donor introgressions for desirable wild QTL-alleles derived from *Lycopersicon hirsutum* and *L. pimpinellifolium*. Theor Appl Genet 97:170-180.

Bernacchi, D., T. Beck-Bunn, Y. Eshed, J. Lopez, V. Petiard, J. Uhlig, D. Zamir, and S. Tanksley. 1998a. Advanced backcross QTL analysis in tomato. I. Identification of QTLs for traits of agronomic importance from *Lycopersicon hirsutum*. Theor Appl Genet 97:381-397.

Bohs, L., and R.G. Olmstead. 1997. Phylogenetic relationships in *Solanum* (Solanaceae) based on *ndhF* sequences. Syst Bot 22:5-17.

Bonierbale, M.W., R.L. Plaisted, and S.D. Tanksley. 1988. RFLP maps based on a common set of clones reveal modes of chromosomal evolution in potato and tomato. Genetics 120:1095-1103.

Börner, A., S. Chebotar, and V. Korzun. 2000. Molecular characterization of the genetic integrity of wheat (*Triticum aestivum* L.) germplasm after long-term maintenance. Theor Appl Genet 100:494-497.

Cantliffe, D. J. 1997. Impression of west Mexican fresh market vegetable agriculture, Pre and post NAFTA (1982-1997) *In:* C. Vavrina, P. Gilreath and J. Noling [eds.], Citrus and Vegetable Magazine. University of Florida, Gainesville, FL , pp. 5-6.

Carvalho, L.J.C.B., and B.A. Schaal. 2000. Assessing genetic diversity in the cassava (*Manihot esculenta* Crantz) germplasm collection in Brazil using PCR-based markers. Euphytica 120:133-142.

Causse, M., C. Caranta, V. Saliba-Colombani, A. Moretti, R. Damidaux, and P. Rousselle. 2000. Enhancement of tomato genetic resources via molecular markers. Cahiers d'études et de recherches francophones/Agricultures 9:197-210.

Chetelat, R. 2001. Plant exploration in northern Chile to collect wild tomato species, with emphasis on *Solanum lycopersicoides* and *S. sitiens*. Final Report 6/28/2001, TGRC. http://tgrc.ucdavis.edu/chile.html.

Chunwongse, J., Chunwongse, C., Black, L., and P. Hanson. 1998. Mapping of the *Ph-3* gene for late blight from *L. pimpinellifolium* L 3708. TGC Reports 48:13-16.

Chunwongse, J., T.B. Bunn, C. Crossman, J. Jiang, and S.D. Tanksley. 1994. Chromosomal localization and molecular-marker tagging of the powdery mildew resistance gene (*Lv*) in tomato. Theor Appl Genet 89:76-79.

Clayberg, C.D., L. Butler, C.M. Rick, and R.W. Robinson. 1965. List of tomato genes as of January, 1965. TGC Reports 15:7-21.

Clayberg, C.D., L. Butler, P.A. Young, and C.M. Rick. 1959. List of tomato genes as of January 1959. TGC Reports 9:6-18.

Crossa, J., and R. Vencovsky. 1997. Variance effective population size for two-stage sampling of monoecious species. Crop Sci 37:14-26.

Dahlberg, J., X. Zhang, G. Hart, and J. Mullet. 2002. Comparative assessment of variation among sorghum germplasm accessions using seed morphology and RAPD measurements. Crop Sci 42:291-296.

Dean, R.E., J.A. Dahlberg, M.S. Hopkins, S.E. Mitchell, and S. Kresovich. 1999. Genetic redundancy and diversity among 'Orange' accessions in the U.S. National Sorghum collection as assessed with simple sequence repeat (SSR) markers. Crop Sci 39:1215-1221.

del Rio, A., J. Bamberg, and Z. Huaman. 1997. Assessing changes in genetic diversity of potato gene banks. 1. Effects of seed increase. Theor Appl Genet 95.

Dhanaraj, A.L., E.V.V.B. Rao, K.R.M. Swamy, M.G. Bhat, D.T. Prasad, and S.N. Sondur. 2002. Using RAPDs to assess the diversity in Indian cashew (*Anacardium occidentale* L.) germplasm. J Hort Sci Biotech 77:41-47.

Dickinson, M.J., D.A. Jones, and J.D.G. Jones. 1993. Close linkage between the *Cf-2/Cf-5* and *Mi* resistance loci in tomato. Mol Plant Microbe In 6:341-347.

Dixon, M.S., K. Hatzixanthis, D.A. Jones, K. Harrison, and J.D.G. Jones. 1998. The tomato *Cf*-5 disease resistance gene and six homologs show pronounced allelic variation in leucinerich repeat copy number. Plant Cell 10:1915-1925.

Doganlar, S., J. Dodson, B. Gabor, T. Beck-Bunn, C. Crossman, and S.D. Tanksley. 1998. Molecular mapping of the *py-1* gene for resistance to corky root rot (*Pyrenochaeta lycopersici*) in tomato. Theor Appl Genet 97:784-788.

Ellis, P.R. and J.W. Maxon-Smith. 1971. Inheritance of resistance to potato cyst-eelworm (*Heterodera rostochiensis* Woll.) in the genus *Lycopersicon*. Euphytica 20:93-101.

Ellis, R.H., and E.H. Roberts. 1980. Improved equations for the prediction of seed longevity. Ann Bot 45:13-30.

Ellis, R.H., and E.H. Roberts. 1998. How to store seeds to conserve biodiversity. Nature 395:758.

Ellis, R.H., T.D. Hong, and E.H. Roberts. 1991. Seed moisture content, storage, viability and vigour. Seed Sci Res 1:275-277.

Ellis, R.H., T.D. Hong, D. Astley, and H.L. Kraak. 1994. Medium-term storage of dry and ultra-dry seeds of onion at ambient and sub-zero temperatures. Onion Newsletter for the Tropics 6:65-68.

Eshed, Y. and D. Zamir. 1995. An introgression line population of *Lycopersicon pennellii* in the cultivated tomato enables the identification and fine mapping of yield-associated QTL. Genetics 141:1147-1162.

Eshed, Y., G. Gera and D. Zamir. 1996. A genome-wide search for wild-species alleles that increase horticultural yield of processing tomatoes. Theor Appl Genet 93:877-886.

FAOSTAT. 2002. Food and agriculture organization of the United Nations statistical databases. http://apps.fao.org/default.htm.

Frary, A., S. Doganlar, A. Frampton, T. Fulton, J. Uhlig, H. Yates, and S. Tanksley. 2003. Fine mapping of quantitative trait loci for improved fruit characteristics from *Lycopersicon chmielewskii* chromosome 1. Genome 46:235-243.

Frazier, W.A. and R.K. Dennett. 1949. Tomato lines of *Lycopersicon esculentum* type resistant to tobacco mosaic virus. Proc Amer Soc Hort Sci 54:265-71.

Gallegly, M.E. and M.E. Marvel. 1955. Inheritance of resistance to tomato race 0 of *Phytophthora infestans*. Phytopathology 45:103-09.

Ghislain, M., D. Zhang, D. Fajardo, Z. Huaman, and R. Hijmans. 1999. Marker-assisted sampling of the cultivated Andean potato *Solanum phureja* collection using RAPD markers. Genetic Resources and Crop Evolution 46:547-555.

Gower, J. C. 1972. Measures of taxamonic distances and their analysis: In J. S. Weiner and J. Huizinga [eds.], The Assessment Of Population Affinties In Man. Clarendon Press, Oxford, UK.

Grenier, C., M. Deu, S. Kresovich, P.J. Bramel-Cox, and P. Hamon. 2000. Assessment of genetic diversity in three subsets constituted from the ICRISAT sorghum collection using random vs. non-random sampling procedures B. Using molecular markers. Theor Appl Genet 101:197-202.

Grube, R., E. Radwanski, and M. Jahn. 2000. Comparative genetics of disease resistance within the Solanaceae. Genetics 155:873-887.

Haanstra, J.P.W., C.M. Thomas, J.D.G. Jones, and P. Lindhout. 2000. Dispersion of the *Cf-4* disease resistance gene in *Lycopersicon* germplasm. Heredity 85:266-270.

Haanstra, J.P.W., R. Lauge, F. Meijer-Dekens, G. Bonnema, P.J.G.M. de Wit, and P. Lindhout 1999. The *CF-ECP2* gene is linked to, but not part of, the *Cf-4/Cf-9* cluster on the short arm of chromosome 1 in tomato. Mol Gen Genet 262:839-845.

Harrington, J.F. 1970. Seed pollen storage for conservation of plant gene resources, *In:* O. Frankel and E. Bennett [eds.], Genetic Resources In Plants. Blackwell Scientific Publishers, Oxford, UK, pp. 501-520.

Hartl, D. and A. Clark. 1989. Principles of Population Genetics. Sinauer Associates, Inc., Sunderland, MA.

He, C., V. Poysa, and K. Yu. 2003. Development and characterization of simple sequence repeat (SSR) markers and their use in determining relationships among *Lycopersicon esculentum* cultivars. Theor Appl Genet 106:363-373.

Holmes, F.O. 1954. Inheritance of resistance to infection by tobacco-mosaic virus in tomato. Phytopathology 44:640-42.

Huamán, Z., C. Aguilar, and R. Ortiz. 1999. Selecting a Peruvian sweet potato core collection on the basis of morphological, eco-geographical, and disease and pest reaction data. Theor Appl Genet 98:840-844.

IBPGR. 1976. Report of IBPGR working group on engineering, design and cost aspects of long-term seed storage facilities. International Board for Plant Genetic Resources, Rome, Italy.

IPGRI. 1996. Descriptors for tomato (*Lycopersicon* spp.) : http://www.ipgri.cgiar.org/germplasm/dbintro.htm. International Plant Genetic Resources Institute, Maccarese, Italy.

Iriondo, J., C. Perez, and F.G. Perez. 1992. Effect of seed storage in liquid nitrogen on germination of several crop and wild species. Seed Science and Technology 20:165-171.

Jenkins, J.A. 1948. The origin of cultivated tomato. Eco Bot 2:379-392.

Jensen, M.H. 1997. Protected Agriculture: A Global Review. World Bank, Washington D.C, USA.

Johnson, R.C. 1998. Genetic structure of regeneration populations of annual ryegrass. Crop Sci 28:851-857.

Jones, D.A., Balint-Kurti, P.J., Dickinson, M.J., Dixon, M.X., and Jones, J.D.G. 1992. Locations of genes for resistance to *Cladosporium fulvum* on the classical and RFLP maps of tomato. TGC Reports 42:19-22.

Kawchuk, L.M., J. Hachey, and D.R. Lynch. 1998. Development of sequence characterized DNA markers linked to a dominant verticillium wilt resistance gene in tomato. Genome 41:91-95.

Kerr, E.A. and D.L. Bailey. 1964. Resistance to *Cladosporium fulvum* cke. obtained from wild species of tomato. Can J Bot 42:1541-54.

Kerr, E.A. and F.I. Cook. 1983. Ontario 7710 – a tomato breeding line with resistance to bacterial speck, *Pseudomonas syringae* pv *tomato* (okabe). Can J Plant Sci 63:1107-1109.

Kerr, E.A., E. Kerr, Z.A. Patrick, and J.W. Potter. 1980. Linkage relation of resistance to *Cladosporium* leaf mold (*Cf*-2) and root-knot nematodes (*Mi*) in tomato and a new gene for leaf mold resistance (*Cf*-11). Can J Genet Cytol 22:183-186.

Kerr, E.A., Z.A. Patrick, and D.L. Bailey. 1971. Resistance in tomato species to new races of leaf mold (*Cladosporium fulvum* cke). Hort Res 11:84-92.

Kikuta, K. and W.A. Frazier. 1947. Preliminary report on breeding tomatoes for resistance to tobacco mosaic virus. Proc Amer Soc Hort Sci 49:256-62.

Kimura, M. 1983. The neutral theory of molecular evolution. Cambridge University Press, Cambridge, UK.

Langford, A.N. 1937. The parasitism of *Cladosporium fulvum* cooke and the genetics of resistance to it. Can J Res 15:108-28.

Laterrot, H. 1983. La lutte genetique contre la maladie des racines liegeuses de la tomate. Revue Horticole 238:23-25.

Laterrot, H. 1993. Revised list of near isogenic tomato lines in Moneymaker type with different genes for disease resistances. TGC Reports 43:79.

Levesque, H., F. Vedel, C. Mathieu, and A.G.L. de Courcel. 1990. Identification of a short rDNA spacer sequence highly specific of a tomato line containing *Tm*-1 gene introgressed from *Lycopersicon hirsutum*. Theor Appl Genet 80:602-608.

Li, Z. H. Zhang, Y. Zeng, Z. Yang, S. Shen, C. Sun, and X. Wang. 2002. Studies on sampling schemes for the establishment of core collection of rice landraces in Yunnan, China. Genetic Resources and Crop Evolution 49:67-74.

Liu, F., D. Charlesworth and M. Kreitman. 1990. The effect of mating system differences on nucleotide diversity at the phosphoglucose isomerase locus in the plant genus Leavenworthia. Genetics 151:343-357.

Livingstone, K.D., V.K. Lackney, J.R. Blauth, and R. van Wijk. 1999. Genome mapping in Capsicum and the evolution of genome structure in the Solanaceae. Genetics 152:1183-1202.

Logendara, L.S., H.W. Janes, and A.P. Papadopoulos. 1997. Hydroponic tomato production: growing media requirements. *In:* K. Wignarajah [ed.], Proceedings 18th Annual Conference on Hydroponics. Hydroponic Society of America, San Ramon, CA, USA, pp. 119-123.

Marshall, J.A., S. Knapp, M.R. Davey, J.B. Power, E.C. Cocking, M.D. Bennett, and A.V. Cox. 2001. Molecular systematics of *Solanum* section *Lycopersicon* (*Lycopersicon*) using the nuclear ITS rDNA region. Theor Appl Genet 103:1216-1222.

Martin, G.B., J.G.K. Williams, and S.D. Tanksley. 1991. Rapid identification of markers linked to a *Pseudomonas* resistance gene in tomato by using random primers and near-isogenic lines. Proc Nat Acad Sci USA 88:2336-2340.

McClean, P.E. and M.R. Hanson. 1986. Mitochondrial DNA sequence divergence among *Lycopersicon* and related *Solanum* species. Genetics 112:649-667.

Miller, J.C. and S.D. Tanksley. 1990. RFLP analysis of phylogenetic relationships and genetic variation in the genus *Lycopersicon*. Theor Appl Genet 80:437-448.

Monforte, A.J., E. Friedman, D. Zamir, and S.D. Tanksley. 2001. Comparison of a set of allelic QTL-NILs for chromosome 4 of tomato: Deductions about natural variation and implications for germplasm utilization. Theor Appl Genet 102: 572-590.

Moreau, P., P. Thoquet, J. Olivier, H. Laterrot, and N. Grimsley. 1998. Genetic mapping of *Ph*-2, a single locus controlling partial resistance to *Phytophthora infestans* in tomato. Mol Plant Microbe In 11:259-269.

Muller, C.H. 1940. A revision of the genus *Lycopersicon*. USDA Misc. Publ. 328:29.

Nei, M. 1987. Molecular evolutionary genetics. Columbia Univ. Press, New York.

Nei, M. and W.H.Li. 1979. Mathematical model for studying genetic variation in terms of restriction endonucleases. Proc Natl Acad Sci 76:5269- 5273.

Nesbitt, T. C. and S.D. Tanksley. 2002. Comparative seqencing in the genus *Lycopersicon*: Implications for the evolution of fruit size in the domestication of cultivaed tomatoes. Genetics 162:365-379.

Nuez, F. and J. Cuartero. 1984. Collections of *Lycopersicon* and *Solanum pennellii* in Peru. Plant-Genet-Resour-Newsl. Rome:42-45.

Olmstead, R.G. and J.D. Palmer. 1997. Implication for the phylogeny, classifications and biogeography of *Solanum* from cpDNA restriction site variation. Syst Bot 22:19-29.

Palmer, J.D. and D. Zamir. 1982. Chloroplast DNA evolution and phylogenetic relationships in *Lycopersicon*. Proc Natl Acad Sci 79:5006-5010.

Pandey, D.K. 1995. Liquid preservatives to improve longevity of tomato (*Lycopersicon esculentum* L.) seeds. Sci Hort 62:54-62.

Parzies, H., W. Spoor, and R. Ennos. 2000. Genetic diversity of barley landrace accessions (*Hordeum vulgare* ssp. *vulgare*) conserved for different lengths of time in ex situ gene banks. Heredity 84:476-486.

Pelham, J. 1966. Resistance in tomato to tobacco mosaic virus. Euphytica 15:258-67.

Phippen, W.B., S. Kresovich, F.G. Candelas, and J.R. McFerson. 1997. Molecular characterization can quantify and partition variation among genebank holdings: a case study with phenotypically similar accessions of *Brassica oleracea* var. *capitata* L. (cabbage) 'Golden Acre'. Theor Appl Genet 94:227-234.

Pitblado, R.E. and E.A. Kerr. 1980. Resistance to bacterial speck (*Pseudomonas tomato*) in tomato. Acta Hortic 100:379-382.

Pitblado, R.E., B.H. MacNeill, and E.A. Kerr. 1984. Chromosomal identity and linkage relationships of *Pto*, a gene for resistance to *Pseudomonas syringae* pv. *tomato* in tomato. Can J Plant Pathol 6:48-53.

Pollack, E. 1987. On the theory of partially inbreeding finite populations. I. Partial selfing. Genetics 117:353-360.

Rick, C.M. 1976. Tomato (family *Solanaceae*). *In:* N. W. Simmonds [ed.], Evolution of Crop Plants. Longman Publications, UK, pp. 268-273.

Rick, C.M. 1979. Biosystematic studies in *Lycopersicon* and closely related species of *Solanum*. *In:* J.G. Hawkes, R.N. Lester and A.D. Skelding [eds.], The Biology and Taxonomy of the *Solanaceae*. Academic Press, London, UK, pp. 667-677.

Rick, C.M. 1982. A new self-compatible wild population of *L. peruvianum*. Report of the Tomato Genetics Cooperative 32:43-44.

Rick, C.M. and J.F. Fobes. 1975. Allozyme variation in the cultivated tomato and closely related species. Bull Torrey Bot Club 102:376-386.

Rick, C.M. and R.T. Chetelat. 1995. Utilization of related wild species for tomato improvement. Acta Hort 412:21-38.

Rick, C.M., H. Laterrot, and J. Philouze. 1990. A revised key for the *Lycopersicon* species. Report of the Tomato Genetics Cooperative 40:31.

Roberts., E. 1975. Problems of long-term storage of seed and pollen for genetic resources conservation, *In:* O.H. Frankel and J.G. Hawkes [eds.], Crop Genetic Resources for Today and Tomorrow. Cambridge University Press, New York, pp. 269-295.

Roselló, S., M.J. Diez, and F. Nuez. 1998. Genetics of tomato spotted wilt virus resistance coming from *Lycopersicon peruvianum*. Eur J Plant Pathol 104:499-509.

Sacks, E. and D. St. Clair. 1996. Cryogenic storage of tomato pollen: effect on fecundity. HortScience 31:447-448.

Sarfatti, M., J. Katan, R. Fluhr, and D. Zamir. 1989. An RFLP marker in tomato linked to the *Fusarium oxysporum* resistance gene *I2*. Theor Appl Genet 78:755-759.

Savolainen, O., C.H. Langley, B.P. Lazzaro, and H. Freville. 2000. Contrasting patterns of nucleotide polymorphism at the alcohol dehydrogenase locus in the outcrossing *Arabidopsis lyrata* and the selfing *Arabidopsis thalania*. Mol Biol Evol 17:645-655.

Schaible, L., O.S. Cannon, and V. Waddoups. 1951. Inheritance of resistance to verticillium wilt in a tomato cross. Phytopathology 41:986-90.

Scott, J.W. and J.D. Farley. 1983. 'Ohio CR-6' tomato. HortScience 18:114-115.

Scott, J.W., H.A. Agrama, and J.P. Jones. 2004. RFLP-based analysis of recombination among resistance genes to fusarium wilt races 1, 2, and 3 in tomato. J Am Soc Hortic Sci 129:394-400.

Skroch, P.W., J. Nienhuis, S. Bebee, J. Tohme, and F. Pedraza. 1998. Comparison of Mexican common bean (*Phaseolus vulgaris* L.) core and reserve germplasm collections. Crop Sci 38:488-496.

Snyder, R.G. 1996. Greenhouse tomatoes - The basics of successful production. Greenhouse Tomato Seminar. ASHS Press, American Society of Horticultural Science, Alexandria, VA, USA, pp. 3-6.

Sobczak, M., A. Avrova, J. Jupowicz, M.S. Phillips, K. Ernst, and K. Amar. 2005. Characterization of susceptibility and resistance responses to potato cyst nematode (*Globodera* spp.) infection of tomato lines in the absence and presence of the broad-spectrum nematode resistance *Hero* gene. Mol Plant Microbe In 18:158-168.

Soltis, E.D. and P.S. Soltis. 2000. Contributions of plant molecular systematics to studies of molecular evolution. Plant Mol Biol 42:45-75.

Spooner, D.M., G.J. Anderson, and R.K. Jansen. 1993. Chloroplast DNA evidence for the interrelationships of tomatoes, potatoes, and pepinos (Solanaceae). Am J Bot 80:676-688.

Stanwood, P.C. and S. Sowa. 1995. Evaluation of onion (*Allium cepa* L.) seed after 10 years of storage at 5, -18, and -196 °C. Crop Sci 35:852-856.

Steiner, A.M., P. Ruckenbauer, and E. Goecke. 1997. Maintenance in genebanks, a case study:

contaminations observed in the Nürnberg oats of 1831. Genetic Resources and Crop Evolution 44:533-538.

Stephan, W. and C.H. Langley. 1998. DNA polymorphism in *Lycopersicon* and crossing-over per physical length. Genetics 150:1585-1593.

Stevens, J.M. 1964. Tomato breeding. Acta Hortic 194:69-75.

Stevens, M.A., and C.M. Rick. 1986. Chapter 2. Genetics and breeding, *In:* J. Atherton and J. Rudich. [eds.], The Tomato Crop. Chapman and Hall, NY, USA, pp. 35-109.

Stevens, M.R., S.J. Scott, and. R.C. Gergerich. 1992. Inheritance of a gene for resistance to tomato spotted wilt virus (TSWV) from *Lycopersicon peruvianum Mill.* Euphytica 59:9-17.

Suliman-Pollatschek, S., K. Kashkush, H. Shats, J. Hillel, and U. Lavi. 2002. Generation and mapping of AFLP, SSRs, and SNPs in *Lycopersicon esculentum.* Cellular and Molecular Biology Letters 7:583-597.

Tanksley, S.D., M.W. Ganal, J.P. Prince, M.C. de Vicente, M.W. Bonierbale, P. Broun, T.M. Fulton, J.J. Giovannoni, S. Grandillo, and G.B. Martin. 1992. High density molecular linkage maps of the tomato and potato genomes. Genetics 132:1141-1160.

Tanksley, S.D., S. Grandillo, T.M. Fulton, D. Zamir, Y. Eshed, V. Petiard, J. Lopez, and T. Beck-Bunn. 1996. Advanced backcross QTL analysis in a cross between an elite processing line of tomato and its wild relative *L. pimpinellifolium.* Theor Appl Genet 92:213-224.

Taylor, I.B. 1986. Chapter 1. Biosystematics of the tomato. *In:* J. Atherton and J. Rudich [eds.], The Tomato Crop. Chapman and Hall, NY, USA, pp. 1-34.

USDA-ERS. 2003. U.S. tomato statistics, 1960-2002. USDA-Economic Research Service. http://usda.mannlib.cornell.edu/reports/nassr/fruit/pvg-bban/vgan0195.txt

USDA-NASS. 1995. Vegetables 1994 summary. USDA-National Agricultural Statistics Service. http://usda.mannlib.cornell.edu/reports/nassr/fruit/pvg-bban. Vegetables Annual Summary, 01.18.95.

USDA-NASS. 2003. Vegetables 2002 summary. USDA-National Agricultural Statistics Service. http://usda.mannlib.cornell.edu/reports/nassr/fruit/pvg-bban. Vegetables Annual Summary, 01.29.03.

Vakalounakis, D.J. 1988. The genetic analysis of resistance to fusarium crown and root rot of tomato. Plant Pathol 37:71-73.

Vakalounakis, D.J., H. Laterrot, A. Moretti, E.K. Ligoxigakis, and K. Smardas. 1997. Linkage between *Frl* (*Fusarium oxysporum* f.sp. *radicislycopersici* resistance) and *Tm-2* (tobacco mosaic virus resistance-2) loci in tomato (*Lycopersicon esculentum*). Ann Appl Biol 130:319-323.

van Hintum, T.J.L. 2000. Duplication within and between germplasm collections. III. A quantitative mode. Genetic Resources and Crop Evolution 47:65-68.

van Hintum, T.J.L., and D.L. Visser. 1995. Duplication within and between germplasm collections. Genetic Resources and Crop Evolution 42:1235-145.

van Hintum, T.J.L., I.W. Boukema, and D. Visser. 1996. Reduction of duplication in a *Brassica oleracea* germplasm collection. Genetic Resources and Crop Evolution 43:343-349.

van Treuren, R., L. J.M. van Soest, and T. J.L. van Hintum. 2001. Marker-assisted rationalisation of genetic resource collections: a case study in flax using AFLPs. Theor. Appl. Genet. 103:144-152.

Vencovsky, R. and J. Crossa. 1999. Variance effective population size under mixed self and random mating with applications to genetic conservation of species. Crop Sci 39:1282-1284.

Villand, J., P.W. Skroch, T. Lai, P. Hanson, C.G. Kuo, and J. Nienhuis. 1998. Genetic variation among tomato accessions from primary and secondary centers of diversity. Crop Sci 38:1339-1347.

Villand, J.M. 1995. Comparison of molecular marker and morphological data to determine genetic distance among tomato cultivars. M.S., Univ. of Wisconsin, Madison.

Walters, C., E.E. Roos, D.H. Touchnell, P. Stanwood, L. Towill, L. Wiesner, and S. Eberhart. 1998. Refrigeration can save seeds economically. Nature 395:758.

Weir, B. 1996. Genetic Data Analysis II. Sinauer Associates, Inc., Sunderland, Massachusetts.

Williams, C.E. and D.A. St Clair. 1993. Phenetic relationships and levels of variability detected by restriction fragment length polymorphism and random amplified polymorphic DNA analysis of cultivated and wild accessions of *Lycopersicon esculentum*. Genome 36:619-630.

Wu, X., N. Wu, X. Qian, R. Li, F. Huang, and L. Zhu. 1998. Phenotypic and genotypic changes in rapeseed after 18 years of storage and regeneration. Seed Sci Res 8:55-64.

XiangHui, K., Z. HaiYing, K. Kong, and H. Zhang. 1998. The effect of ultradry methods and storage on vegetable seeds. Seed Sci Res 8:41-45.

Yaghoobi, J., I. Kaloshian, Y. Wen, and V.M. Williamson. 1995. Mapping a new nematode resistance locus in *Lycopersicon peruvianum*. Theor Appl Genet 91:457-464.

Young, N.D., D. Zamir, M.W. Ganal, and S.D. Tanksley. 1988. Use of isogenic lines and simultaneous probing to identify DNA markers tightly linked to the *Tm-2a* gene in tomato. Genetics 120:579-585.

Zamir, D., I. Ekstein-Michelson, Y. Zakay, N. Navot, M. Zeidan, M. Sarfatti, Y. Eshed, E. Harel, T. Pleban, H. van Oss, N. Kedar, H.D. Rabinowitch, and H. Czosnek. 1994. Mapping and introgression of a tomato yellow leaf curl virus tolerance gene, *Ty-1*. Theor Appl Genet 88:141-146

Zheng, G.M., X.M. Jing, and K.L. Tao. 1998. Ultradry seed storage cuts cost of gene bank. Nature 393:223-224.

3

Cytogenetics and Evolution

Roger T. Chetelat and Yuanfu Ji[1]

*Department of Plant Sciences, University of California, One Shields Avenue,
Davis, CA 95616, USA,
e-mail: trchetelat@ucdavis.edu*
[1]*Present address: Gulf Coast Research and Education Center, University of Florida, 14625
CR 672, Wimauma, FL 33598, USA*

INTRODUCTION

As an experimental organism, tomato presents a number of genetic, biological, and economic advantages that have led to its development as a
model for cytogenetic and evolutionary studies. The cultivated tomato,
Lycopersicon esculentum (= *Solanum lycopersicum*), and related wild species
traditionally classified as genus *Lycopersicon*, more recently as *Solanum* sect.
Lycopersicon (Spooner et al. 2005) are diploids, with a chromosome number
of 2n=2x=24. Eleven of the 12 chromosomes in the haploid tomato nucleus
are metacentric or submetacentric (Lapitan et al. 1989). The exception, chromosome 2, is acrocentric with a heterochromatic short arm consisting
primarily of the nucleolus organizing region (NOR). During late prophase
of meiosis (diakinesis), only chromosome 2 can be distinguished from the
others, by virtue of its association with the nucleolus. However, in early
prophase (pachytene), each of the 12 chromosomes can be identified by the
position of the centromere, the length of chromatic and achromatic segments, and the pattern of heterochromatic knobs (chromomeres) (Khush
1963). These features are illustrated in corresponding cytological maps for
each chromosome in the set (Rick and Butler 1956, Khush and Rick 1968).

The tomato genome is also well defined by genetic maps based on morphological and molecular markers. High density molecular marker maps
based on RFLP, SSR, or AFLP markers are available (Tanksley et al. 1992,
Pillen et al. 1996, Haanstra et al. 1999, Frary et al. 2005). Estimates for total
map length are ~1200-1300 cM. The relatively low haploid DNA content of

---

Corresponding author: Roger T. Chetelat

tomato, ca. 950 Mbp or 0.95 pg per C (Arumuganathan and Earle 1991, Michaelson et al. 1991), makes it well suited for molecular studies. Though larger than *Arabidopsis* or rice (~125 and 425 Mbp, respectively), the tomato genome is smaller than many other model plant species, such as maize or wheat (~2,500 and 16,000 Mbp, respectively). The average ratio of physical to genetic distance is ~750 kb/cM, a value low enough to make positional cloning of genes practical in most genomic regions. Furthermore, recombination is essentially limited to the euchromatic regions, which constitute less than one fourth of the total DNA content (see below). The tomato genome encodes a total of ~35,000 genes, which are located primarily in euchromatin (Van der Hoeven et al. 2002). Therefore, recombination rates per unit physical distance are substantially higher within the genetically active fraction than in the genome as a whole.

In comparison to other crop plants, many aspects of growth and development in tomato have been beneficial to cytogenetic studies. It is naturally self-pollinated, which simplifies the maintenance of stocks, yet hybridizations are easy to perform and yield large quantities of seed of controlled parentage. Tomato can be grown under a wide range of environmental conditions and propagated through seed or asexually via rooted cuttings. Its photoperiodic insensitivity and relatively short generation time permit the propagation of up to 3 generations per year. The structure of the tomato plant, particularly its compound leaves and sympodial growth habit, allows detection of an enormous array of hereditary variations, such as altered growth habit, leaf shape, texture and color, flower morphology, color and function, and fruit size, shape and color, among others. Tomato also provides a popular model for physiological and biochemical studies of fruit development, quality, and ripening.

As a result of its economic importance as a crop, industry-sponsored research on tomato contributed to early advances in genetics and wide hybridizations. A large number of mutants, aneuploids, and various other spontaneous genetic defects were discovered in growers' fields (Rick 1945). The relative ease with which it can be transformed by *Agrobacterium tumefaciens* (Fillatti et al. 1987) has made tomato a popular organism for testing biotechnological approaches to enhancing fruit quality and other traits. As a result of these and other advantages, the first genetically engineered (GMO) food plant marketed in the USA was a tomato (Bruening and Lyons 2000).

Research on tomato has depended to a large extent on genetic resources such as mutants, cytogenetic stocks, and wild species populations (Chetelat 2005). Tomato germplasm is preserved at genebanks such as the C.M. Rick

Tomato Genetics Resource Center (TGRC) and the USDA's Plant Genetic Resources Unit (PGRU), which provide seed samples to interested researchers worldwide. The TGRC, located at Univ. of California, Davis, maintains over 1,000 monogenic stocks, consisting of spontaneous or induced mutations at 600+ loci affecting most aspects of plant development and morphology (http://tgrc.ucdavis.edu). Over 1,400 other genetic stocks are also available through the TGRC, including several types of trisomics (primary, secondary, tertiary, and compensating), as well as autotetraploids, and derivatives of wild species such as alien additions, substitutions, and introgression lines. Lastly, the TGRC maintains over 1,100 wild species accessions, representing 13-20 species of *Lycopersicon* and related *Solanum* taxa collected in their native regions. The PGRU, located in Geneva, New York, maintains a large collection of open-pollinated cultivars, as well as accessions of several wild species (http://www.ars-grin.gov/). The Hebrew University of Jerusalem maintains a large population of mutants useful for the analysis of gene function (http://www.zamir.sgn.cornell.edu/mutants/). Together, these genetic resources provide important tools for analysis of the tomato genome.

The present chapter summarizes and integrates recent advances in the cytogenetics and evolution of tomato, with emphasis on applications in genetics and breeding. The following sections describe novel genetic resources and their uses, genetic maps of the tomato genome, meiotic behavior of wide hybrids, and chromosome structure and evolution. Related topics covered by 'previous' reviews include: genome mapping (Pillen et al. 1996), classical and molecular genetics (Rick and Yoder 1988), genetics and breeding (Stevens and Rick 1986, DeVerna and Paterson 1991), and other aspects of tomato cytogenetics (Gill 1983, Quiros 1991). The early literature in this field was summarized in comprehensive reviews by Rick and Khush (1966) and Rick and Butler (1956).

SOURCES OF VARIATION

Hyper- and Hypoploidy

A rich assortment of hyper- and hypoploid stocks were identified in early work on tomato cytogenetics (Khush 1973). Haploids, monosomics, and segmental deficiencies were induced by irradiation of mature pollen (i.e. post-gametogenesis) and identified by pollination of recessive marker stocks using the pseudo-dominant method (Khush and Rick 1968). However, none could be reliably propagated, sexually or vegetatively, and all are now extinct. Haploids of tomato do not set seed because they rarely produce viable gametes. Segmental deletions and monosomics cannot be maintained because the gametes with deficiencies are inviable and thus do not transmit to the next generation. Monosomics produce nullisomic gametes (e.g. n-1, n-2, etc), which are eliminated during development of the male or female gametophytes. The only primary monosomic recovered was for chromosome 11, which is one of the shortest chromosomes of the set (Rick and Khush 1961). Although several tertiary monosomics were recovered, none were transmissible to the next generation (Khush and Rick 1966). These results are consistent with the presumed diploid nature of the tomato genome, which, lacking any large scale duplications, does not tolerate deficiencies at the gametophytic stage.

In contrast, hyperploidy is tolerated to a much greater extent, and both triploids and tetraploids are relatively fertile. Spontaneous autotriploids were the most common type of unfruitful ('bull') tomatoes in commercial fields in California (Rick 1945). Pollination of triploids provided useful variants, particularly the trisomics (Khush 1973). However, like the hypoploids, the triploids are genetically unstable since they produce unbalanced gametes, and cannot be maintained through seed. In contrast, autotetraploids are stable, though partially sterile, and 4x stocks of several cultivars and wild species are maintained by the TGRC. Nearly all currently available tomato tetraploids were induced by colchicine treatment or were recovered as spontaneous variants in field cultures. Cultivated tomato and its related wild species are virtually all diploids, in contrast to potato, where a range of ploidy levels (e.g. 2x, 3x, 4x) are found amongst native and domesticated forms. However, there are two known instances of natural tetraploid populations, both in *L. chilense* (= *S. chilense*), one of which also happens to be the northernmost population of this species (Rick 1990).

A complete set of primary trisomics, as well as many secondary, tertiary, and compensating trisomics have been produced in tomato (Khush 1973). The first linkage maps, based on morphological markers, were associated with individual pachytene chromosomes by trisomic segregation analysis. Though no longer the most efficient method for placement of mutant loci on chromosomes—linkage tester stocks are more informative—the trisomics are still useful for assigning molecular markers to their respective chromosomes. For example, chromosomal assignment of RFLP markers was accomplished with the primary trisomics by dosage analysis of hybridization intensity (Young et al. 1987). The primary trisomics were also used to identify individual chromosomes in synaptonemal complex spreads (Sherman and Stack 1992). Secondary and tertiary trisomics have been useful for determining the positions of centromeres on the genetic map (Pillen et al. 1996).

Allopolyploid hybrids have been created in tomato as vehicles for transferring genes from certain related wild species. This has been a particularly useful strategy for the tomato-like nightshades, *S. lycopersicoides* and *S. sitiens*. In case of *S. lycopersicoides*, the first diploid hybrids were readily obtained by embryo culture, but were highly sterile, due in part to low rates of pairing between homeologous chromosomes (Rick 1951); in contrast, allotetraploid (amphidiploid) hybrids produced by colchicine treatment showed relatively normal meiotic behavior, with preferential pairing among homologues, and much improved fertility. Despite these initial, promising results, little if any progress was made in breeding *S. lycopersicoides* with tomato until the first allotriploid (sesquidiploid) hybrids were reported some 35 years later (Rick et al. 1986). The sesquidiploids eventually yielded a complete series of monosomic alien additions (2n+1), each containing a single extra *S. lycopersicoides* chromosome in the background of *L. esculentum* (Chetelat et al. 1998). Like the primary trisomics, transmission rates and fertility vary widely among the individual monosomic additions. Nonetheless, they are generally fertile enough to be sexually propagated. A limited series of diploid substitution lines, heterozygous for a single *S. lycopersicoides* chromosome, were also derived (Ji and Chetelat 2002).

Allopolyploid hybrids representing the genome of *S. sitiens* (syn. *S. rickii*) have been derived in similar, though more circuitous, fashion. Although *S. sitiens* has not been successfully hybridized with cultivated tomato, it does cross readily with *S. lycopersicoides*, reflecting a close genetic affinity between these species (Rick 1979, 1988b). As a result, *S. sitiens* is also cross compatible with the previously synthesized *L. esculentum*—*S. lycopersicoides* sesquidiploids (DeVerna et al. 1990). Since the extra *S. lycopersicoides* chromosomes tend to be eliminated in the progeny of the sesquidiploid, it serves as a convenient donor of the *L. esculentum* genome (i.e. acts as a bridging genotype). The resulting diploid F_1 *L. esculentum* x *S. sitiens* hybrids are highly sterile, but chromosome doubling with colchicine produces more fertile amphidiploids, from which a few monosomic additions, substitutions, and recombinant diploids have been obtained (Pertuze et al. 2003).

Introgression

The wild tomato species are potentially rich sources of allelic variation for genetic studies and for cultivar development following introgression.

Crosses between the cultivated tomato and all but two of its wild relatives are feasible, although the ease of hybridizations varies greatly. In contrast to the cultigen, whose early history of domestication and breeding led to severe depletion of its genetic variation, the wild species are far more diverse. Populations of the self-incompatible species, such as *L. peruvianum* (= *S. peruvianum*), are especially heterogeneous, containing more variation within a single accession than all accessions of any one of the self-compatible species, including *L. esculentum* (Miller and Tanksley 1990). In addition, genetic variation within populations of *L. hirsutum* (= *S. habrochaites*) and *L. pimpinellifolium* (= *S. pimpinellifolium*) varies according to geographic location, with populations from the center of each species' range containing more diversity than those collected at the northern or southern limits (Rick et al. 1977, 1979). Considering that the TGRC alone maintains over 1,100 wild species populations, nearly all of which are cross-compatible with tomato, they represent an amazingly rich and accessible source of genetic variation.

At the phenotypic level, variation is sometimes apparent only in backcross derivatives of interspecific hybrids. In the study of quantitative characters, for example, backcross populations sometimes display 'transgressive variation', in which individual genotypic combinations produce phenotypic values exceeding either parental extreme (for example, de Vicente and Tanksley 1993). A similar phenomenon has been observed for qualitative characters, in which mutant phenotypes not expressed by either parent may appear in progeny of wide hybrids (Rick 1967). Such 'novel variation' may arise from a number of sources, including genic or plasmatic interactions, latent variation in the wild species, and de novo mutation. For example, the *B* gene for high β-carotene is present in all of the green-fruited species but expressed only in the genetic background of *L. esculentum*, suggesting an interaction between genes determines carotenoid accumulation. Another type of interaction, between the nuclear and cytoplasmic genomes, appears to control expression of cytoplasmic male sterility (CMS) in tomato, as it is only observed when cytoplasm of *L. esculentum* is transferred into the nuclear genome of wild species such as *L. pennellii* (= *S. pennellii*). Latent variation (i.e. residual heterozygosity) in the wild species can also account for novel traits, such as the *old gold* mutant (B^{og}) from *L. chilense*.

The wild species have also been the main sources of marker polymorphisms required for development of high density genetic maps. Using DNA-based markers such as RFLPs and RAPDs, there is generally very little variation detectable within or between *L. esculentum* varieties or landraces

(Miller and Tanksley 1990, Villand et al. 1998, Williams and St. Clair 1993). In contrast, the wild species can be highly variable. For example, a single inbred accession of the wild species *L. pennellii* was polymorphic relative to cultivated tomato for ~71% of AFLP markers (Haanstra et al. 1999). Approximately 81% of RFLPs (probe x RE combinations) were informative in *S. lycopersicoides* vs. tomato (Chetelat et al. 2000).

Mapping populations used for the construction of molecular marker maps in tomato include conventional interspecific F_2 or BC populations. Such segregating populations are usually difficult to propagate indefinitely, unless they can be immortalized by tissue culture or other means. Since each unique progeny array is ephemeral, they do not provide an optimal, long term mapping resource. A recombinant inbred line (RIL) population, such as one derived from *L. esculentum* x *L. cheesmanii* f. *minor* (= *S. galapagense*), is a more permanent resource and provides greater map resolution than the corresponding F_2 (Paran et al. 1995). However, some of the *L. cheesmanii* RILs had a higher than expected degree of residual heterozygosity and/or reduced fertility. Another type of permanent mapping resource, and one that has been pioneered in tomato, is the introgression line (IL) population. ILs consist of overlapping homozygous chromosome segments introgressed from a wild donor genome into a constant genetic background, in this case, of cultivated tomato. A set of 50 ILs contains an entire *L. pennellii* genome bred into *L. esculentum*, the first such population synthesized for tomato (Eshed and Zamir 1995), with an additional 26 sublines providing increased map resolution (Pan et al. 2000). Similar types of prebred lines have been created for *L. hirsutum* (Monforte and Tanksley 2000a), *L. pimpinellifolium* (Doganlar et al. 2002), and *S. lycopersicoides* (Canady et al. 2005). ILs have a number of advantages for fine mapping of QTLs, gene identification, and related breeding applications (see reviews by Zamir and Eshed 1998; Zamir 2001). Furthermore, ILs tend to have greater viability and fertility than corresponding RILs from the same interspecific hybrids, due to the more limited contribution of wild species genome in each IL. As a result, they provide a more realistic genetic background for evaluation of vegetative and reproductive characters. However, a permanent 'reference' mapping population that offers a high level of linkage resolution is not yet available in tomato. Towards this goal, Vision et al. (2001) developed a backcross recombinant inbred population from the cross *L. esculentum* x *L. pennellii*. Using marker assisted selection on a large population, a subset of individual genotypes were chosen to provide maximum map resolution for the population as a whole.

Mutation

Spontaneous and induced mutations affecting development and morphology were essential ingredients of early genetic research on tomato. Mutations provided markers for the first classical maps, for studies of segregation and recombination in wide crosses, and for integration of cytological and genetic maps. The characterization of spontaneous unfruitful ('bull') plants occurring in field plantings led to the identification of a large number of useful variants, including male-sterile mutants, trisomics, tetraploids, triploids, haploids, and meiotic defects (Rick 1945). The number of mutants described in tomato increased steadily as a result of these and other studies. The Tomato Genetics Cooperative (TGC) was established by C.M. Rick and associates in 1950 to promote exchange of information and germplasm amongst geneticists, and to coordinate linkage studies (Robinson 1982). First published in 1951, the TGC Report evidenced the accelerating pace of research on tomato genetics: new mutants were described, segregation, allelism and linkage tests reported, and lists of available stocks issued. Large-scale mutagenesis studies, in particular those of Hans Stubbe from Gatersleben, Germany (summarized in Stubbe 1972a, 1972b), greatly expanded the available collections of mutants. At the time of the first comprehensive review (Rick and Butler 1956), only 118 mutant loci were known (of which 56 had been mapped). Today, ~1,200 mutations at 1,000 loci have been described, of which ~400 have been mapped or assigned to a chromosome, and a small but growing number cloned and sequenced. The TGRC currently maintains and distributes over 1,000 monogenic stocks with mutations at over 600 loci, affecting most aspects of development and morphology (Chetelat 2005).

Mutagenesis studies in tomato have employed a variety of artificial means for generating new mutations. The standard methods, widely used in other model organisms, include treatment of seeds or pollen with alkylating agents (mainly EMS) or radiation (primarily X-rays and fast neutrons). Additional mutagenesis strategies that have been used to a limited extent include somaclonal variation (Evans and Sharp 1983, Gavazzi et al. 1987, van den Bulk et al. 1990), and transposon tagging using the maize *Ac/Ds* elements (Meissner et al. 1997, 2000). Transposon tagging has several attractive features. First, the gene responsible for a mutant phenotype is readily identified by sequencing DNA on either side of the insertion site (e.g. by inverse PCR). Second, libraries of *Ds* insertions at different positions in the genome have been established (Knapp et al. 1994, Thomas et al. 1994), since transposition of *Ds* occurs preferentially to linked sites, the probability of tagging a gene in the same region is thereby increased. Thirdly, the chance of identifying *Ds* insertions into target genes can be improved by 'site selected insertional mutagenesis', a combination of DNA pooling and nested PCR (Cooley et al. 1996).

Despite these improvements, there is still a need for a high throughput mutagenesis system in tomato that will allow gene isolation for functional genomics. Large-scale insertional mutagenesis by T-DNA tagging is not practical in tomato due to limitations in current *Agrobacterium* transformation methods. Recently, EMS mutagenesis has been revived by the development of technology for identifying point mutations. The TILLING (Targeting Induced Local Lesions In Genomes) method screens pooled DNA samples from segregating populations to identify individuals with point mutations in a gene of interest (McCallum et al 2000). EMS mutagenesis, which causes primarily C/G to T/A transitions, has been a highly effective method of generating mutants in tomato. A population of 13,000 M2 families was generated by EMS and fast neutron treatments in cv. M-82, and includes 3,417 catalogued mutant phenotypes (Menda et al. 2004). Examples of allelism with existing mutations suggest the M-82 mutant population is nearly saturated (i.e. likely to contain at least one mutation in each gene). Phenotypes and images of these mutants are available online (http://www. zamir.sgn.cornell.edu/mutants/). The tomato genome – euchromatic regions only – is currently being sequenced by the international 'SOL' initiative (http://www.sgn.cornell.edu). Combined with existing EST databases (Van der Hoeven et al. 2002), the increased sequence information will allow more genes to be identified as potential targets based on their sequence alone. Candidate genes can also be identified by comparison of gene or QTL locations in tomato to sequence information from orthologous regions in model species such as tomato and *Arabidopsis* (Ku et al. 2000, 2001).

LINKAGE MAPS

Classical Maps

The 'classical' linkage maps of tomato are based on the simultaneous segregation of multiple morphological markers, almost always in intraspecific (*L. esculentum*) crosses. Initially limited to existing mutants of spontaneous origin, establishment of linkage groups was greatly facilitated by a large influx of induced mutations. Cooperation among TGC members in mapping the increasing number of mutant loci was also a key element; for a period of time, each chromosome was 'assigned' to a particular investigator to work out its linkage relations and develop new marker combinations (Robinson 1982). Efficient detection of linkage was possible with the development of sets of chromosome-specific linkage tester stocks, which combined multiple markers on a single chromosome (see Chetelat and Petersen 2003). Additional linkage tester stocks combined two strategically situated markers on each of two chromosomes, so that in theory a maximum of six segregation tests would be required to detect linkage anywhere in the genome. The number of morphological markers that can be simultaneously and independently genotyped in this fashion is in many cases severely limited by their phenotypic effects. Problems frequently encountered are epistasis among genes controlling related traits, sterility or inviability of multiple marker combinations, and lack of 'good' markers (e.g. seedling stage expression) for certain genomic regions. In addition, linkage tests of new morphological markers typically segregate in repulsion phase with respect to the tester combination, which limits precision of recombination fraction estimates for recessive genes. While coupling phase linkage tests, particularly testcrosses, provided greater precision, they require the prior synthesis of new marker combinations (i.e. recombinant genotypes), which becomes limiting with more than just a few markers per chromosome. These factors limited saturation and resolution of the classical linkage map of tomato, as in other organisms. Isozyme markers were integrated with the mutant loci beginning in the 1970's (Tanksley and Rick 1980). Due to their neutral phenotypes and codominant expression, allozymes overcame many of the limitations of morphological mutants as genetic markers. However, available enzyme staining technology restricted the number of protein markers, and only a few have been added to the map recently (Bernatzky and Tanksley 1986, Chetelat et al. 2000).

At the present time, approximately 400 morphological and isozyme markers have been at least assigned to their respective chromosomes, and the majority have also been positioned within their linkage groups. The last comprehensive linkage summaries are now 19+ years old (Mutschler et al. 1987, Stevens and Rick 1986), and sorely in need of revision. Fortunately, the classical maps for several chromosomes have been updated, usually as a result of integration with molecular markers. These include chromosome 1 (Balint-Kurti et al. 1995, van Tuinen et al. 1997), chromosome 3 (Koorneef et al. 1993, van der Biezen et al. 1994), chromosome 6 (van Wordragen et al. 1996, Weide et al. 1993), chromosome 7 (Burbidge et al. 2001), chromosome 10 (van Tuinen et al. 1997), and chromosome 11 (van Tuinen et al. 1998).

Molecular Marker Maps

The development of molecular linkage maps of the tomato genome based on DNA markers provided many advantages over the existing classical maps. Due to the low level of polymorphism detectable within *L. esculentum* using DNA markers, the molecular linkage maps are based on segregation and recombination in interspecific crosses. F_2 progeny from the cross *L. esculentum* x *L. pennellii* have been favored for this purpose because of their relative ease of hybridization, the fertility and normal meiotic behavior of F_1 hybrids, and the high polymorphism rate that the distinguishes these two species. Because the number of DNA marker loci is not limiting, the molecular linkage maps have high marker density and good saturation of linkage groups. The framework map, based on RFLP markers, contains over 1000 loci, with an average distance between markers of only 1.2 cM (Pillen et al. 1996, Tanksley et al. 1992). This map is also populated with many genes of known function or phenotype, including morphological markers, isozyme loci, and cloned genes. Additionally, the approximate locations of centromeres have been determined for each linkage group (Pillen et al. 1996), providing anchor points to the cytological maps.

In addition to RFLP markers, a number of other DNA marker systems have been used to create linkage maps in tomato, with varying success. AFLPs provide thousands of polymorphic bands, and were used to generate an ultra-dense genetic map of tomato, consisting of over 1200 markers (Haanstra et al. 1999, Spooner et al. 2005). Marker distribution along the chromosomes was decidedly nonrandom, with the majority of AFLPs (particularly the *Eco*RI-*Mse*I derived markers) tightly clustered in the centromeric regions. AFLPs may thereby provide markers for genomic regions less readily detected by RFLPs (Bonnema et al. 2002). RAPD and SSR markers also map primarily to proximal regions of chromosomes, although SSRs identified in EST sequences are more randomly dispersed (Grandillo and Tanksley 1996, Areshchenkova and Ganal 1999, 2002). Despite the greater time and expense of applying RFLP markers, they have the significant advantage of providing multiallelic genetic probes that can be compared across species and populations. This makes RFLPs ideally suited for mapping in tomato, with its rich collection of wild relatives, which provide not only abundant marker variation, but also many traits of interest to breeders. The availability of a set of reference maps and corresponding RFLP markers have provided the genetic tools to expedite a vast array of genetic studies and plant breeding applications in tomato too numerous to summarize here.

An RFLP linkage map has been developed for tomato using conserved ortholog set (COS) markers (Fulton et al. 2002). These represent ESTs that are single or low copy in the tomato genome, and have a high degree of homology to a single ortholog in *Arabidopsis*, as determined by sequence comparisons. Over 1000 COS markers were identified, of which ~550 have been mapped. The COS map is anchored to previous maps with a large number of conventional RFLPs, and includes SSR loci identified within the ESTs sequences. Map resolution has been improved by increasing population size, and total map length is substantially increased over the original molecular map based on RFLPs. Many RFLPS have been converted to CAPS (cleaved amplified polymorphic sequence) markers (Frary et al. 2005). Together with SSRs, these provide a set of framework PCR-based markers. Current versions of these maps, as well as sequence databases, are available through the Solanaceae Genomics Network (SGN) (http://www.sgn.cornell.edu).

Rates of recombination within a given marker interval may vary greatly in tomato due to the influence of several factors. Recombination is generally higher in female than in male gametes (de Vicente and Tanksley 1991, van Oijen et al. 1994, Ganal and Tanksley 1996). Recombination is also elevated in progeny of F_1 interspecific hybrids relative to advanced backcross generations (Rick 1969, 1971), and higher in whole chromosomes than in introgressed segments (Paterson et al. 1990, van Wordragen et al. 1996, Chetelat and Meglic 2000, Monforte and Tanksley 2000b, Ji and Chetelat 2002). In addition, recombination rates vary according to species divergence. For example, recombination in intraspecific *L. peruvianum* crosses was higher (10% on average) than interspecific *L. esculentum* x *L. pennellii* (van Oijen et al. 1994). Similarly, recombination in *L. esculentum* x *S. lycopersicoides* was ca. 30% lower on average than in *L. esculentum* x *L. pennellii* (Chetelat et al. 2000). Finally, reduced recombination is observed between markers in the centromeric regions, as a result of which marker density (in genetic terms) is higher (Tanksley et al. 1992, see below).

Comparative Maps of *Lycopersicon* and Related *Solanum* Species

Comparisons of genetic maps from inter- and intraspecific crosses involving *Lycopersicon* spp. indicate nearly complete colinearity between them. Comparative maps of the following species have been developed from interspecific crosses to *L. esculentum*: *L. pimpinellifolium* (Grandillo and Tanksley 1996), *L. cheesmanii* f. *minor* (Paran et al. 1995), *L. chmielewskii* (= *S. chmielewskii*) (Paterson et al. 1990), *L. pennellii* (Tanksley et al. 1992), *L. hirsutum* (Bernacchi and Tanksley 1997), *L. parviflorum* (= *S. neorickii*) (Fulton et al. 2000), and *L. peruvianum* (van Oijen et al. 1994, Fulton et al. 1997). Although there were significant differences among these maps for total genetic length and/or recombination rates in specific marker intervals, they were all essentially consistent with the framework map in terms of marker order along each chromosome. One noteworthy exception is a region on chromosome 7 which is inverted in *L. pennellii* relative to *L. esculentum* (Van der Knaap et al. 2004). The otherwise strong conservation of gene order indicated by these comparative maps is consistent with observations of normal chromosome pairing and fertility in most F_1 interspecific hybrids between *Lycopersicon* species. The genomes of all species in the *Lycopersicon* clade can therefore be considered essentially colinear and homologous.

In contrast, genetic maps of the *S. lycopersicoides* and *S. sitiens* genomes indicate these species have a different genome structure. A genetic map based on BC_1 *L. esculentum* x *S. lycopersicoides* showed a genome-wide reduction in recombination of about 30% compared to interspecific *Lycopersicon* maps (Chetelat et al. 2000). No recombination could be detected on the long arm of chromosome 10, suggesting the intergeneric F_1 was heterozygous for a structural rearrangement in this region. Following introgression of individual chromosomes into *L. esculentum*, recombination between markers on *S. lycopersicoides* 10L remained undetectable (Ji and Chetelat 2002). To determine the gene order of *S. lycopersicoides* chromosome 10, recombination between homologous chromosomes would be required. A map based on *S. sitiens* x *S. lycopersicoides* is ideal for this purpose, since the two nightshades are closely related and cross-compatible (i.e. their genomes are homologous), yet present a higher rate of marker polymorphism than intraspecific populations of either species. Results of this map showed colinearity with *Lycopersicon* for all regions of the genome, except 10L, where a paracentric inversion distinguishes the two groups (Pertuze et al. 2002). The location of this inversion explains the absence of recombination between *L. esculentum* and *S. lycopersicoides* chromosomes in this region.

Furthermore, the breakpoint of this inversion appears to be identical to the one described for chromosome 10L in cultivated potato (*S. tuberosum*), one of five such rearrangements that distinguish it from tomato (Bonierbale et al. 1988, Tanksley et al. 1992). *S. lycopersicoides* and *S. sitiens* have the same marker order on this chromosome as potato, a configuration that is also observed in pepper (*Capsicum*) and eggplant (*S. melongena*) (Livingstone et al. 1999, Doganlar et al. 2002). Given the close relationship between *Lycopersicon* and the much larger *Solanum* genus (Spooner et al. 2005), the potato/eggplant/pepper arrangement must be ancestral and the tomato inversion derived. The presence of the potato arrangement in *S. sitiens* and *S. lycopersicoides*, which are among the closest relatives of tomato outside the *Lycopersicon* group, supports this hypothesis. Furthermore, the observed colinearity between tomato, *S. lycopersicoides* and *S. sitiens*, in the regions of the other four potato rearrangements suggests that the 10L inversion evolved most recently, presumably coinciding with divergence of *Lycopersicon* from a *Solanum* ancestor. As such, the 10L inversion is a cyto-taxonomic marker for the *Lycopersicon* genome. These two basic genomes, designated L and S (Fig. 3.1), appear to be the only large scale differences in chromosome structure separating tomato from any of the species with which it is cross-compatible. In comparison, five basic genomes have been postulated for the cultivated potato (A genome) and related *Solanum* species (B, C, D, and E) on the basis of chromosome pairing and fertility in hybrids between them (Matsubayashi 1991).

MEIOSIS IN WIDE HYBRIDS

Hybrids Between Species of *Lycopersicon*

As mentioned above, sexual crosses are possible between cultivated tomato and any of the wild *Lycopersicon* species, although ease of hybridization and fertility of the resulting F_1's varies greatly (Rick 1979). The red- or orange-fruited species – *L. esculentum*, *L. pimpinellifolium*, and *L. cheesmanii* (= *S. cheesmaniae or S. galapagense*) – can be freely intercrossed to form highly fertile hybrids. Crosses between the red/orange and the green-fruited species generally succeed only when the former are used as female parent (unilateral incompatibility), and hybrids are less fertile. Obtaining hybrids with *L. peruvianum* or *L. chilense* can be more problematic: embryo/ ovule rescue or other techniques are usually required, and F_1 hybrids between the groups are less fertile.

Despite differences in crossability and hybrid fertility, meiosis is relatively normal in all interspecific combinations examined to date (Rick 1979). In each case, parental chromosomes synapse along their entire length at pachytene, form 12 bivalents with chiasmata at diakinesis, leading to proper alignment at metaphase and regular anaphase of the first division (Afify 1933, Lesley and Lesley 1943, McGuire and Rick 1954, Sawant 1958, Chmielewski 1962, Khush and Rick 1963). Minor differences in chromosome morphology were observed between chromosomes of *L. esculentum* and *L. pennellii*, primarily in the lengths of heterochromatic regions and

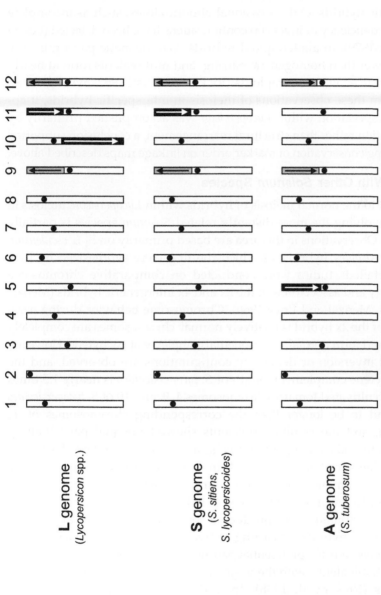

Fig. 3.1 Comparative idiograms of the L genome of tomato (*Lycopersicon* spp.), S genome of the tomato-like nightshades *S. lycopersicoides* and *S. sitiens*, and A genome of cultivated potato (*Solanum tuberosum*). The locations of five paracentric inversions that distinguish these genomes are indicated by arrows. Ancestral chromosome configurations are shown in white, derived inversions in black, and regions of uncertain ancestry in gray, based on results of Livingstone et al. (1999), and Pertuze et al. (2002).

their pattern of chromomeres (Khush and Rick 1963). However, little evidence for structural differentiation has been observed at meiosis in interspecific hybrids. Only occasional abnormalities, such as incomplete pairing, or deficiency or inversion configurations have been detected (Lesley 1950, Rick 1979). In allotetraploid hybrids, heterogenetic pairing is only slightly lower than homogenetic pairing, and multivalents form at nearly the same rate as in autotetraploids (Rick and Khush 1962, Sybenga et al. 1994). From these observations of meiosis in interspecific hybrids, it appears that speciation within *Lycopersicon* was accomplished primarily by gene mutation rather than structural rearrangement, a conclusion supported by the strong conservation of marker order on linkage maps described above.

Hybrids With Other *Solanum* Species

In contrast to the normal meiosis of hybrids within *Lycopersicon*, meiosis in hybrids involving the more distantly related *Solanum* species is partially disrupted. Observations in this area are based primarily on F_1 *L. esculentum* x *S. lycopersicoides* hybrids, of which the first were synthesized by Rick (1951). Detailed studies were conducted on comparative chromosome morphology and associations in the 2x and 4x intergeneric hybrids (Menzel 1962, 1964, Menzel and Price 1966). Chromosome behavior during early prophase of the 2x hybrid is relatively normal: chromosomes are completely synapsed and form normal synaptonemal complexes at pachytene. However, occasional inversion or deficiency configurations are observed, and the total pachytene complement length of *S. lycopersicoides* is nearly 1.5 times that of the cultivated tomato. Chromosomes 4, 9, and 10 of *S. lycopersicoides* were found to be longer than the corresponding chromosomes of *L. esculentum,* and the resultant bivalents showed unequal pairs. During meiosis, 2x hybrids undergo reduced chiasma formation and produce about four univalents/cell at metaphase I. In 4x allotetraploids, pairing occurs preferentially among homologous chromosomes, resulting in mostly bivalents and a few multivalents. The 4x hybrids also exhibit greater pollen fertility than diploids, and produce a few viable seeds. In allotriploid (sesquidiploid) hybrids, consisting of two genomes of *L. esculentum* and one of *S. lycopersicoides*, preferential pairing of the *L. esculentum* homologues produces 12 bivalents, with the *S. lycopersicoides* chromosomes forming 12 univalents (Rick et al. 1986). In addition, the condensation of *S. lycopersicoides* chromosomes during early diakinesis of the sesquidiploids is significantly delayed with respect to their *L. esculentum* counterparts. These observations of incomplete chromosome pairing and lack of synchronization during meiosis indicate the chromosomes of *L. esculentum* and *S. lycopersicoides* are homeologous.

A similar relationship exists between the genomes of tomato and *S. sitiens*. Though sexually incompatible with *L. esculentum*, *S. sitiens* crosses easily with its sister taxon *S. lycopersicoides* to form fully fertile hybrids with normal meiotic behavior (Rick 1979, DeVerna et al. 1990, Pertuze et al. 2002). Taking advantage of this chain relationship, sesquidiploid *L. esculentum - S. lycopersicoides* hybrids were used as donors of the *L. esculentum* genome in crosses to *S. sitiens*, as described above. Chromosome pairing at diakinesis of the resulting diploid F_1 *L. esculentum* x *S. sitiens* hybrids was disrupted, with an average of 5.7 univalents observed per cell. Amphidiploids showed a strong preference for homologous pairing, with a great majority of cells containing 24 bivalents (DeVerna et al. 1990). The results indicate that chromosomes of *S. sitiens* are homeologous with those of *L. esculentum*, and homologous with *S. lycopersicoides*.

While attempts to cross tomato with more distantly related *Solanum* species have failed, somatic hybrids have been produced for some combinations. For example, *L. esculentum* (+) *S. ochranthum* cell fusions resulted in allotetraploid and allohexaploid hybrids (Stommel 2001). Though many were aneuploid, and highly sterile, a few of the 4x hybrids had moderate fertility. Multivalent formation during meiosis in the these hybrids provides evidence of pairing between homeologous chromosomes, a prerequisite for eventual recombination and introgression. Prospects in this area are bolstered by recent success in introducing tomato chromosomes into potato (see next section).

Genomic *In Situ* Hybridization

The field of molecular cytogenetics has been revolutionized by advances in techniques such as fluorescence *in situ* hybridization (FISH), which provides powerful tools for investigating chromosome structure and function of complex genomes. A specific application of FISH is genomic *in situ* hybridization (GISH), which utilizes total genomic DNA of one species as a probe to distinguish parental genomes or chromosomes in sexual and somatic hybrids. GISH analysis of hybrids between *Lycopersicon* species has been limited because the taxa involved are closely related, and therefore not easily differentiated by hybridization. GISH experiments using standard stringency conditions usually result in complete hybridization of probe DNA to both genomes if they share a high degree of sequence homology. However, hybridization and/or post-hybridization stringencies can be increased to effectively differentiate more closely related genomes. For example, GISH of hexaploid *L. esculentum* (+) *L. peruvianum* somatic hybrids and their diploid derivatives showed extensive pairing and chiasmata formation between chromosomes of the two species (Parokonny et al. 1997). This is not unexpected, based on genetic evidence of recombination in progeny of sexual hybrids (Fulton et al. 1997). Meiosis has been examined by GISH in other hybrids between closely related species, including *L. esculentum* x *L. pennellii* (Haider Ali 2001) and *S. lycopersicoides* x *S. sitiens* (see Fig. 3.2B) sexual hybrids.

GISH cytology has been more widely used to analyse meiosis in hybrids between *Lycopersicon* and more distantly related species. For example, *S. lycopersicoides* has been hybridized with *L. esculentum* by somatic cell fusion, in addition to the earlier conventional crosses (Handley et al. 1986, Hossain et al. 1994, Matsumoto et al. 1997). There was little cross hybridization between their genomes, and GISH was a useful tool for determining the genetic constitution of 4x and 6x hybrids and their progeny (Escalante et al. 1998). Pairing between *S. lycopersicoides* and *L. esculentum* chromosomes was studied in monosomic addition and substitution lines, which contain individual alien chromosomes introgressed into cultivated tomato (Ji and Chetelat 2002, Fig. 3.2). In the monosomic additions (2n+1=25), the extra *S. lycopersicoides* chromosome forms a univalent and its tomato counterparts form bivalents in ~90% of meiocytes, indicating a strong preference for homologous pairing. In the corresponding substitution lines (2n=24), pairing between the *S. lycopersicoides* chromosome and its *L. esculentum* homeologue occurs at a higher frequency (up to 90% of cells) due to the absence of homologous partners. However, homeologous pairing was greatly reduced in the substitution for *S. lycopersicoides* chromosome 10, due to a lack of crossing over within the paracentric inversion on 10L that differentiates these genomes.

GISH has also been a useful tool for dissecting the genetic composition of complex hybrids involving more than two parental species (Ji et al. 2004). For example, the amphidiploid *L. esculentum* x *S. sitiens* hybrid described previously was backcrossed to sesquidiploid *L. esculentum* x *S. lycopersicoides*, resulting in a complex trigenomic hybrid. A combination of GISH and RFLP analysis revealed the following genetic composition: two genomes of *L. esculentum*, one genome of *S. sitiens*, and two residual chromosomes from *S. lycopersicoides* (Pertuze et al. 2003, Ji et al. 2004). During meiosis, chromosomes of the two *Solanum* parents pair regularly, as do the *L. esculentum* set, while the remaining *S. sitiens* chromosomes are usually unpaired. Though less fertile than a true sesquidiploid, this trigenomic

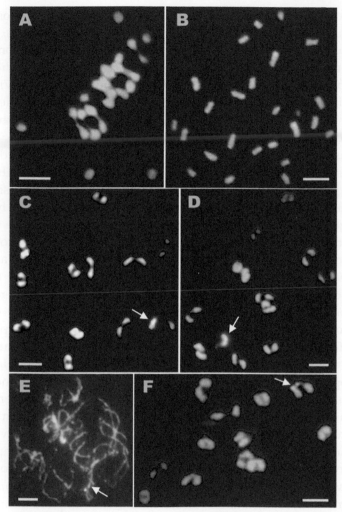

Fig. 3.2 Examples of the use of genomic in situ hybridization (GISH) for genome analysis of tomato interspecific/intergeneric hybrids, and monosomic addition (MA), substitution (SL) and introgression (IL) lines. (A) disrupted pairing at metaphase I of F$_1$ *L. esculentum* x *S. lycopersicoides*; (B) partial differentiation at mitotic metaphase of homologous chromosomes in F$_1$ *S. lycopersicoides* x *S. sitiens*; (C) *S. lycopersicoides* SL-8 at diakinesis with pairing between homeologous chromosomes (bivalent; arrow); (D) *S. sitiens* MA-8 at diakinesis showing pairing between homeologues (bivalent; arrow) and an unpaired *L. esculentum* chromosome (univalent; arrowhead); (E) *S. lycopersicoides* SL-7 at pachytene showing the unpaired *S. lycopersicoides* chromosome (arrow); (F) heterozygous introgression line containing a segment from *S. lycopersicoides* chromosome 7 of ~42 cM (TG499 - TG128) in length (arrow). (A, C, E-F) Red = *S. lycopersicoides*, Blue = *L. esculentum*; (B) Blue = *S. lycopersicoides*, Red = *S. sitiens*; (D) Blue = *L. esculentum*, Red = *S. sitiens*. Bars represent 5 µm.

hybrid nonetheless yielded a few monosomic additions, substitutions, and recombinant diploids containing individual *S. sitiens* chromosomes, or segments thereof, in *L. esculentum*. Despite representing only a portion of the *S. sitiens* genome, the introgressions so far obtained demonstrate the feasibility of breeding specific traits from this nightshade into tomato. Virtually ignored in previous searches for disease resistance or other desiderata of interest to breeders, *S. sitiens* is most remarkable for its adaptation to the hyperaridity of its native habitat, the Atacama desert of Chile (Rick 1988b).

The successful regeneration of potato (+) tomato fusion hybrids from protoplasts, first reported by Melchers et al. (1978), raised the possibility of eventual gene transfer between these economically important, yet sexually incompatible, solanaceous crop species. Unfortunately, the initial hybrids were highly sterile, and produced neither fruit nor tubers. Since then, a hexaploid potato (+) tomato fusion hybrid was successfully backcrossed to tetraploid potato, from which a single BC_1 plant was generated (Jacobsen et al. 1994). GISH cytology indicated this plant possessed nine tomato chromosomes: at meiosis, they formed three homologous bivalents and three univalents, hence represented only six of the 12 possible tomato chromosomes (Jacobsen et al. 1995). Additional crosses yielded BC_1 progeny with different numbers of extra chromosomes, and in which an entire haploid tomato genome was represented (Garriga-Caldere et al. 1997). Following additional backcrosses, a complete set of 12 tomato monosomic additions in a potato background were identified (Garriga-Caldere et al. 1998, Haider Ali et al. 2001). Pairing and recombination between potato and tomato chromosomes was observed, albeit at very low rates, demonstrating the potential for gene transfer between these important solanaceous crops (Garriga-Caldere et al. 1999). A low level of heterogenetic pairing was also observed by Gavrilenko et al. (2001) in 4x *L. esculentum* (+) *S. etuberosum* hybrids; anther culture of the amphidiploid resulted in regeneration of 2x hybrids, suggesting a route for possible transfer of *S. etuberosum* chromosomes into tomato via allohexaploid fusions and androgenic sesquidiploid derivatives.

CHROMOSOME STRUCTURE AND VARIATION

Pachytene Chromosome Structure

The first studies of tomato pachytene chromosomes revealed the general picture of chromosome morphology (Lesley and Lesley 1935) and identified two chromosomes of the set (Brown 1949). On the basis of centromere positions, and the relative lengths of heterochromatic and euchromatic regions and heterochromatic knobs, Barton (1950) was able to distinguish each of the 12 tomato chromosomes at pachytene. The chromosomes were numbered from 1 to 12 according to their length, with no. 1 being the longest. The main nucleolar organizing region (NOR) was associated with chromosome 2. A cytological map of the tomato genome was thereby constructed, which incorporated the distinctive features of each chromosome.

The pachytene maps were later integrated with linkage maps by determining the location of genetic markers relative to chromosome landmarks, such as centromeres, telomeres, and breakpoints of various origins. Radiation-induced deletions were used to locate genetic markers on the cytogenetic map of each chromosome (Khush and Rick 1968). Deletions were identified by the pseudo-dominant technique, in which recessive marker stocks were pollinated with irradiated wild type pollen. The appearance of recessive phenotypes in the progeny indicated the loss of the wild type (dominant) alleles. By this method, the positions and lengths of 74 deletions were determined on pachytene chromosomes, and 35 genetically mapped morphological mutations were located on 18 of the 24 chromosome arms. Marker genes were observed to be non-randomly distributed within chromosomes, being located almost exclusively in euchromatin. Genetic data and the distribution of cytological chiasmata pointed to much less recombination in heterochromatin than in euchromatin.

Ratio of Physical to Genetic Distance

The ratio of physical to genetic distance is not constant within the tomato genome. Departures from the genome-wide average value (~750 kb/cM) include local recombination hotspots: for example, ~10 kb/cM observed within *Lin5* on chromosome 9 (Fridman et al. 2000), ~24 kb/cM around *ms-14* on chromosome 11 (Gorman et al. 1996), and 55-110 kb/cM near *Cf-4/Cf-9* on chromosome 1 (Bonnema et al. 1997). At the other extreme, lower than average recombination rates are observed near centromeres: for example, ~4-16 Mb/cM around *Tm-2* on chromosome 9 (Ganal et al. 1989) and ~60 Mb/cM around *Mi* on chromosome 6 (Kaloshian et al. 1998). Another manifestation of this recombination suppression is the clustering of random cDNA and genomic clones in proximal regions of most chromosomes on the molecular linkage map (Tanksley et al. 1992, Pillen et al. 1996). This suppressive effect is most pronounced within alien introgressions, such as those comprising the resistance genes *Mi* and *Tm-2* bred into *L. esculentum* from *L. peruvianum*. Substantially higher recombination rates were observed in the wild species background for the same marker intervals. Prospects for map-based cloning of desirable genes or QTLs identified in wild relatives will therefore depend to a large extent on their location within the genome.

Synaptonemal Complex

A more accurate estimate of the size and chromatin composition of the tomato genome has been obtained from studies of the synaptonemal complex (SC) from pachytene stage nuclei. Tomato is well suited to cytological visualization of SCs using electron microscopy (Stack 1982). The SC karyotype (like the conventional pachytene version) indicates the pericentromeric regions are composed of large blocks of heterochromatin (Sherman and Stack 1992). Approximately 36% of total SC length is composed of DNA packaged as heterochromatin (Peterson et al. 1996). However, because DNA in heterochromatin is more dense (on a per unit SC length basis), it represents approx. 77% of the total DNA in the tomato genome. These heterochromatic regions of the chromosomes are considered genetically inactive. For example, few mutant loci were mapped to heterochromatin by deletion analysis (Khush and Rick 1968). This means that the effective genome size of tomato (considering only euchromatin) could be as little as 0.22 pg DNA/ C, slightly larger than that of *Arabidopsis* (Peterson et al. 1996). However, the tomato genome is composed of mostly single or low copy number sequences: from independent estimates based on hybridization or reassociation studies, this fraction represents ~70% of nuclear DNA (Zamir and Tanksley 1988, Peterson et al. 1998). FISH localization to pachytene chromosomes showed that while euchromatin contains primarily single-copy DNA, a majority of this fraction, as well as most of the repetitive DNA, is located in pericentromeric heterochromatin (Peterson et al. 1999).

The physical basis for centromeric recombination suppression has been examined using recombination nodules (RNs), which are the manifestation of crossing over events on SCs. In a detailed study of the location and frequency of RNs in tomato, Sherman and Stack (1995) elucidated several significant trends. First, RNs are nonrandomly distributed, being located primarily in euchromatin rather than heterochromatin, and absent from the telomere ends of SCs and centromeres. The relatively low frequency of RNs within pericentromeric heterochromatin and near the telomeres explains the clustering of RFLP markers on linkage maps in these regions. Second, RN frequency per unit length of euchromatin is not constant, which might explain the observed gaps and recombination hotspots on genetic maps. Thirdly, a 1:1 relationship between RNs and chiasmata was observed. Assuming each RN is a crossover event, the *L. esculentum* genome would contain a total of ca. 1095 map units, a value similar to those of the classical molecular linkage maps.

Effects of Recombination on Natural Variation

Natural genetic variation at the population level is strongly influenced by mating systems, which can vary among tomato species from obligate outcrossing to complete inbreeding (Rick 1988a). As expected, the strictly allogamous (self-incompatible) species or accessions show vastly more within-population variation than the autogamous or facultative (both self-compatible) groups (Rick et al. 1979, Miller and Tanksley 1990). In addition to the effects of reproducitve biology, genetic variation is influenced by gene position along the chromosome. Levels of DNA polymorphism at a locus are positively correlated with the rate of crossing over per unit physical distance (Stephan and Langley 1998). For example, genes close to the centromere (low recombination) tend to have less within species diversity than genes farther away from the centromere. The strongest association between crossing-over and heterozygosity is found in species with intermediate levels of diversity. Relatively speaking, the effect of recombination levels on DNA polymorphism in tomato is weaker than in some species, such as *Drosophila*, and is far less than the influence of mating system (Baudry et al. 2001).

Centromere Mapping

The approximate locations of centromeres on the chromosomes have been determined by several methods (summarized by Pillen et al. 1996). A combination of RFLP mapping and localization of rDNA loci were used for chromosomes 1 and 2, deletion mapping for chromosomes 3 and 6, analysis of the tomato-potato inversion breakpoints for chromosomes 5, 9, 10, 11, and 12, or comparison of cytological, classical and molecular maps for chromosomes 4 and 8. The centromeres of chromosomes 7 and 9 were more precisely localized on the molecular linkage map through dosage analysis in trisomic stocks, including complementary telo-, secondary and tertiary trisomics (Frary et al. 1996). Both centromeres were localized within a cluster of tightly linked markers. To order markers within these clusters, high resolution maps for both centromeric regions were constructed from F_2 *L. esculentum* x *L. pennellii* and F_2 *L. esculentum* x *L. pimpinellifolium* populations. Similar approaches would be feasible for other chromosomes, for which the pertinent trisomic stocks are available from the TGRC.

Analysis of Genome Structure by FISH

Repetitive Elements

In situ hybridization (ISH) techniques have been used in tomato to analyse genome organization and to determine the physical localization of DNA sequences on the chromosomes. ISH involves use of biotin-labeled probes and detection of hybridization signals with colored immuno-chemical precipitates. Fluorescence *in situ* hybridization (FISH) employs a system of coupled fluorochromes to provide higher resolution in detecting target DNA molecules. Both detection methods have been used to map repetitive and single-copy DNA in tomato. Ganal et al. (1989) mapped four classes of repetitive sequences to meiotic metaphase chromosomes using ISH. One of these, TGRI, is a subtelomeric satellite DNA repeat of 162bp. On somatic chromosomes, TGRI is located at 20 of the 24 telomeres, as well as at centromeres and interstitial sites on some chromosomes (Lapitan et al. 1989). TGRI is separated from the telomeric repeat by a few hundred kilobases, and both are found in the heterochromatic terminal knobs observed on pachytene chromosomes (Ganal et al. 1991).

Another class of repeats, the rDNA genes, are located on several tomato chromosomes. The 5S rDNA sequence was localized by FISH on pachytene spreads to the first heterochromatic knob adjacent to the centromere on the short arm of chromosome 1 (Xu and Earle 1996a). The 45S rDNA sequence was estimated to be present in approximately 2300 copies in the tomato genome and was mapped to a distal position on the short arm of chromosome 2 by linkage analysis (Vallejos et al. 1986). Using ISH, hybridization of the 45S sequence was observed at the end of an acrocentric chromosome, presumably chromosome 2 (Ganal et al. 1988). This result agrees with the known location of the main NOR in this region of chromosome 2 (Brown 1949, Barton 1950). Additional rDNA loci were found on chromosomes 6, 9 and 11 (Xu and Earle 1994, 1996b).

Single and Low Copy Sequences
Unlike repetitive elements, detection of short, single-copy DNA sequences on plant chromosomes using FISH is technically difficult. However, efficient detection of single or low copy sequences is possible using large inserts, such as yeast artificial chromosomes (YACs) or bacterial artificial chromosomes (BACs) (Fuchs et al. 1996). For example, the Colorless nonripening (*Cnr*) locus was mapped to a small interval on chromosome 2 using FISH of BAC clones to pachytene spreads (Tor et al. 2002).

Peterson et al. (1999) demonstrated the usefulness of synaptonemal complex spreads for detecting very short single copy sequences by FISH. SC spreads have several advantages over mitotic metaphase chromosome preparations for FISH applications: SC spreads are relatively free of debris that can interfere with probe penetration, have relatively decondensed chromatin that is highly accessible to probes, and are about ten times longer than their metaphase counterparts, which permits FISH mapping at higher resolution. Genomic clones of ~14 kb, containing RFLP probes previously mapped to chromosome 11, were localized by FISH on SC 11. Marker order and the physical distances between them could be ascertained in this fashion.

The resolution of target DNA sequences on chromosomes can be further enhanced by hybridization to extended DNA fibers from interphase nuclei. Initially developed for human DNA, extended fiber FISH has been adapted to several plant species, including tomato (Fransz et al. 1996, Zhong et al. 1996). In this method, genomic DNA fibers from lysed interphase leaf nuclei are uniformly stretched on a microscope slide. DNA molecules thus linearized are stretched to ~3 kb/μm, allowing DNA targets to be mapped by FISH at the level of a few kilobases and a detection sensitivity of only a few hundred base pairs. This can be helpful for ordering markers that co-segregate on genetic maps due to low recombination rates; for example, fiber FISH of BACs spanning the nematode resistance gene *Mi* showed it is located near the junction of euchromatin and heterochromatin on chromosome 6S (Zhong et al. 1999). Extended fiber FISH was also used to study the molecular and chromosomal organization of individual telomere domains (Zhong et al. 1998).

CONCLUSIONS AND OUTLOOK

Continued genetic improvement of cultivated tomato is certain, given its advantageous cytogenetic and biological features, rich germplasm resources, and well-developed genomics infrastructure. The genetic basis of important economic traits, such as disease resistance, yield, and fruit quality, have been thoroughly investigated, resulting in abundant genetic markers—and in many cases the underlying gene sequences themselves—to facilitate their transfer into new varieties. Enlargement of the genetic base of tomato through introgression from related nightshade taxa is expected to provide novel traits. With further development of genomics tools and expansion of sequence databases, tomato will continue to be used as a genetic model for other solanaceous crops.

Tomato and its wild relatives also provide excellent material with which to study the evolution of adaptive traits and reproductive barriers. Speciation within *Lycopersicon* was accomplished primarily by mutation rather than chromosomal rearrangement, and as a result experimental hybridization and introgression between most species are readily accomplished. Another advantage is that genetic variation within and among populations of wild *Lycopersicon* species displays a geographic pattern of distribution (for example, Rick et al. 1977, 1979; Caicedo and Schaal 2004). This must be due in large part to the geologic history and climate of the native Andean region, which produced habitat differentiation and geographic isolation of populations. Locally adapted populations of wild tomatoes differ in their responses to abiotic stresses, including extremes of moisture, temperature, and salinity, and their susceptibility to diseases and insect pests. In addition, mating systems range from complete autogamy to strict allogamy, and are accompanied by various types of reproductive barriers between species or populations. Tomato is therefore an attractive plant in which to study the molecular and genetic basis of these traits, and their evolutionary significance particularly in native plant populations.

SUMMARY

This chapter summarizes recent advances in the cytogenetics and evolution of tomato, with emphasis on applications in genetics and breeding. As an experimental plant material, the tomato presents a number of genetic, biological, and commercial advantages. Research on tomato has depended to a large extent on genetic resources such as mutants, wild species populations, and other genetic stocks. Recently synthesized introgression line populations representing the genomes of related wild species provide powerful tools for genome analysis and breeding. The genetic base of tomato has been expanded by hybridization and recombination with previously inaccessible tomato-like nightshades. The relatively small genome of tomato is now well delineated with genetic maps of various types. The early classical maps have been superceded by, and in some cases integrated with, high resolution molecular linkage maps based on RFLPs and other types of markers. These maps provide a framework for comparative genetic analysis of the *Lycopersicon,* and related *Solanum* species, as well as abundant markers useful in breeding programs. Linkage maps from different tomato species indicate all share the same basic genome structure, consistent with evidence from chromosome pairing in interspecific hybrids, and that speciation must therefore have been accomplished by gene mutation rather than genome rearrangement. Two tomato-like nightshades, *Solanum lycopersicoides* Dun. and *S. sitiens* Johnst., differ from the *Lycopersicon* clade by a paracentric inversion on chromosome 10L, an arrangement which appears to be ancestral to that of tomato. Maps based on conserved ortholog set (COS) markers identify regions of microcolinearity between tomato and *Arabidopsis*. Advances in cytological methods provide new tools to study genome structure. Fluorescence *in situ* hybridization (FISH) techniques have been used for physical mapping of DNA elements, determination of marker order within regions of suppressed recombination, and analysis of chromosome pairing in wide hybrids. Examination of the ultrastructure of synaptonemal complexes (SCs) reveal that recombination nodules, the sites of crossing-over, occur preferentially in euchromatin, and are rare in the pericentromeric heterochromatin. Up to 77% of the tomato genome is heterochromatic and underrepresented on linkage maps. Natural variation within populations of wild species is correlated with rates of crossing over per physical distance along chromosomes; as a result, genes near the centromere tend to be less variable than those in more distal positions. With recent improvements in genomics infrastructure, tomato will continue to serve as a useful model for the genomes of related solanaceous crops.

Acknowledgements

The authors gratefully acknowledge Dr. Gurdev Khush and Dr. Carlos Quiros for their helpful comments and suggestions during review of the manuscript.

REFERENCES

Afify, A. 1933. The cytology of the hybrid between *Lycopersicon esculentum* and *L. racemigerum* in relation to its parents. Genetica 15: 225-240.

Areshchenkova, T. and M.W. Ganal. 1999. Long tomato microsatellites are predominantly associated with centromeric regions. Genome 42: 536-544.

Areshchenkova, T. and M.W. Ganal. 2002. Comparative analysis of polymorphism and chromosomal location of tomato microsatellite markers isolated from different sources. Theor Appl Genet 104: 229-235.

Arumuganathan, K. and E.D. Earle. 1991. Nuclear DNA content of some important plant species. Plant Mol Biol Rep 9: 208-218.

Balint-Kurti, P.J., D.A. Jones, and J.D.G. Jones. 1995. Integration of the classical and RFLP linkage maps of the short arm of tomato chromosome 1. Theor Appl Genet 90: 17-26.

Barton, D.W. 1950. Pachytene morphology of the tomato chromosome complement. Am J Bot. 37: 639-643.

Baudry, E., C. Kerdelhue, H. Innan, and W. Stephan. 2001. Species and recombination effects on DNA variability in the tomato genus. Genetics 158: 1725-1735.

Bernacchi, D. and S.D. Tanksley. 1997. An interspecific backcross of *Lycopersicon esculentum* x *L. hirsutum*: linkage analysis and a QTL study of sexual compatibility factors and floral traits. Genetics 147: 861-877.

Bernatzky, R. and S.D. Tanksley. 1986. Toward a saturated linkage map in tomato based on isozymes and random cDNA sequences. Genetics 112: 887-898.

Bonierbale, M.W., R.L. Plaisted, and S.D. Tanksley. 1988. RFLP maps based on a common set of clones reveal modes of chromosomal evolution in potato and tomato. Genetics 120: 1095-1103.

Bonnema, G., D. Schipper, S. Van Heusden, P. Zabel, and P. Lindhout. 1997. Tomato chromosome 1: high resolution genetic and physical mapping of the short arm in an interspecific *Lycopersicon esculentum* x *L. peruvianum* cross. Mol Gen Genet 253: 455-462.

Bonnema, G., P. van den Berg, and P. Lindhout. 2002. AFLPs mark different genomic regions compared with RFLPs: a case study in tomato. Genome 45: 217-221.

Breuning, G. and J.M. Lyons. 2000. The case of the FLAVR SAVR tomato. Calif Agric 54: 6-7.

Brown, S.W. 1949. The structure and meiotic behavior of the differentiated chromosomes of tomato. Genetics 34: 437-461.

Burbidge, A., P. Lindhout, T.M. Grieve, K. Schumacher, K. Theres, A.W. van Heusden, A.B. Bonnema, K.J. Woodman, and I.B. Taylor. 2001. Re-orientation and integration of the classical and interspecific linkage maps of the long arm of tomato chromosome 7. Theor Appl Genet 103: 443-454.

Caicedo, A.L., and B.A. Schaal. 2004. Population structure and phylogeography of *Solanum pimpinellifolium* inferred from a nuclear gene. Mol Ecol 13: 1871-1882.

Canady, M.A., V. Meglic, and R.T. Chetelat. 2005. A library of *Solanum lycopersicoides* introgression lines in cultivated tomato. Genome 48: 685-697.

Chetelat, R.T. 2005. Revised list of monogenic stocks. Tomato Genetics Coop. Report 55: 48-69.

Chetelat, R.T. and V. Meglic. 2000. Molecular mapping of chromosome segments introgressed from *Solanum lycopersicoides* into cultivated tomato (*Lycopersicon esculentum*). Theor Appl Genet 100: 232-341.

Chetelat, R.T., V. Meglic, and P. Cisneros. 2000. A genetic map of tomato based on BC$_1$ *Lycopersicon esculentum* x *Solanum lycopersicoides* reveals overall synteny but suppressed recombination between these homeologous genomes. Genetics 154: 857-867.

Chetelat, R.T. and J. Petersen. 2003. Revised list of miscellaneous stocks. Tomato Genet Coop Rep 53: 44-61.

Chetelat, R.T., C.M. Rick, P. Cisneros, K.B. Alpert, and J.W. DeVerna. 1998. Identification, transmission, and cytological behavior of *Solanum lycopersicoides* Dun. monosomic alien addition lines in tomato (*Lycopersicon esculentum* Mill.). Genome 41: 40-50.

Chmielewski, T. 1962. Cytogenetical and taxonomical studies of a new tomato form, Part I. Genetica Polonica 3: 253-264.

Cooley, M.B., A.P. Goldsbrough, D.W. Still, and J.I. Yoder. 1996. Site-selected insertional mutagenesis of tomato with maize *Ac* and *Ds* elements. Mol Gen Genet 252: 184-194.

de Vicente, M.C. and S.D. Tanksley. 1991. Genome-wide reduction in recombination of backcross progeny derived from male versus female gametes in an interspecific backcross of tomato. Theor Appl Genet 83: 173-178.

de Vicente, M.C. and S.D. Tanksley. 1993. QTL analysis of transgressive segregation in an interspecific tomato cross. Genetics 134: 585-596.

DeVerna, J.W. and A.H. Paterson. 1991. Genetics of *Lycopersicon*. In: G. Kaltoo [ed.], Genetic Improvement of Tomato. Springer-Verlag, Berlin/Heidelberg, Germany, pp. 21-38.

DeVerna, J.W., C.M. Rick, R.T. Chetelat, B.J. Lanini, and K.B. Alpert. 1990. Sexual hybridization of *Lycopersicon esculentum* and *Solanum rickii* by means of a sesquidiploid bridging hybrid. Proc Natl Acad Sci (USA) 87: 9496-9490.

Doganlar, S., A. Frary, M.C. Daunay, R.N. Lester, and S.D. Tanksley. 2002. A comparative genetic linkage map of eggplant (*Solanum melongena*) and its implications for genome evolution in the Solanaceae. Genetics 161: 1697-1711.

Doganlar, S., A. Frary, H-M. Ku, and S.D. Tanksley. 2002. Mapping quantitative trait loci in inbred backcross lines of *Lycopersicon pimpinellifolium* (LA1589). Genome 45: 1189-1202.

Escalante, A., S. Imanishi, M. Hossain, N. Ohmido, and K. Fukui. 1998. RFLP analysis and genomic *in situ* hybridization (GISH) in somatic hybrids and their progeny between *Lycopersicon esculentum* and *Solanum lycopersicoides*. Theor Appl Genet 96: 719-726.

Eshed, Y. and D. Zamir. 1995. An introgression line population of *Lycopersicon pennellii* in the cultivated tomato enables the identification and fine mapping of yield-associated QTL. Genetics 141: 1147-1162.

Evans, D.A. and W.R. Sharp. 1983. Single gene mutations in tomato plants regenerated from tissue culture. Science 221: 949-951.

Fillatti, J.J., J. Kiser, R. Rose, and L. Comai. 1987. Efficient transfer of a glyphosate tolerance gene into tomato using a binary *Agrobacterium tumefaciens* vector. Bio/Technology 5: 726-730.

Fransz, P.F., C. Alonso-Blanco, T.B. Liharska, A.J.M. Peeters, P. Zabel, and J. Hans De Jong. 1996. High-resolution physical mapping in *Arabidopsis thaliana* and tomato by fluorescence *in situ* hybridization to extended DNA fibres. Plant J 9: 421-430.

Frary, A., G.G. Presting, and S.D. Tanksley. 1996. Molecular mapping of the centromeres of tomato chromosomes 7 and 9. Mol Gen Genet 250: 295-304.

Frary, A., Y. Xu, J. Liu, S. Mitchell, E. Tedeschi, and S. Tanksley. 2005. Development of a set of PCR-based anchor markers encompassing the tomato genome and evaluation of their usefulness for genetics and breeding experiments. Theor Appl Genet III: 291-312.

Fridman, E., T. Pleban, and D. Zamir . 2000. A recombination hotspot delimits a wild-species quantitative trait locus for tomato sugar content to 484 bp within an invertase gene. Proc Nat Acad Sci (USA) 97: 4718-4723.

Fuchs, J., D.U. Kloos, M.W. Ganal, and I. Schubert. 1996. *In situ* localization of yeast artificial chromosome sequences on tomato and potato metaphase chromosomes. Chrom Res 4: 277-281.

Fulton, T., R. van der Hoeven, N. Eannetta, and S. Tanksley. 2002. Identification, analysis and utilization of conserved ortholog set (COS) markers for comparative genomics in higher plants. Plant Cell 14: 1457-1467.

Fulton, T.M., S. Grandillo, T. Beck-Bunn, E. Fridman, A. Frampton, J. Lopez, V. Petiard, J. Uhlig, D. Zamir, and S.D. Tanksley. 2000. Advanced backcross QTL analysis of a *Lycopersicon esculentum* x *Lycopersicon parviflorum* cross. Theor Appl Genet 100: 1025-1042.

Fulton, T.M., J.C. Nelson, and S.D. Tanksley. 1997. Introgression and DNA marker analysis of *Lycopersicon peruvianum*, a wild relative of the cultivated tomato, into *Lycopersicon esculentum*, followed through three successive backcross generations. Theor Appl Genet 95: 895-902.

Ganal, M.W., N.L.V. Lapitan, and S.D. Tanksley. 1988. A molecular and cytogenetic survey of major repeated DNA sequences in tomato (*Lycopersicon esculentum*). Molec Gen Genet 213: 262-268.

Ganal, M.W., N.L.V. Lapitan, and S.D. Tanksley. 1991. Macrostructure of the tomato telomeres. Plant Cell 3: 87-94.

Ganal, M.W. and S.D. Tanksley. 1996. Recombination around the *Tm2a* and *Mi* resistance genes in different crosses of *Lycopersicon peruvianum*. Theor Appl Genet 92: 101-108.

Ganal, M.W., N.D. Young, and S.D. Tanksley. 1989. Pulsed field gel electrophoresis and physical mapping of large DNA fragments in the $Tm\text{-}2^a$ region of chromosome 9. Mol Gen Genet 215: 395-400.

Garriga-Caldere, F., D.J. Huigen, A. Angrisano, E. Jacobsen, and M.S. Ramanna. 1998. Transmission of alien tomato chromosomes from BC_1 to BC_2 progenies derived from backcrossing potato(+)tomato fusion hybrids to potato: the selection of single additions for seven different chromosomes. Theor Appl Genet 96: 155-163.

Garriga-Caldere, F., D.J. Huigen, E. Jacobsen, and M.S. Ramanna. 1999. Prospects for introgressing tomato chromosomes into the potato genome: an assessment through GISH analysis. Genome 42: 282-288.

Garriga-Caldere, F., D.J.F.F. Huigen, E. Jacobsen, and M.S. Ramanna. 1997. Identification of alien chromosomes through GISH and RFLP analysis and the potential for establishing potato lines with monosomic additions of tomato chromosomes. Genome 40: 666-673.

Gavazzi, G., C. Tonelli, G. Todesco, E. Arreghini, F. Raffaldi, F. Vecchio, G. Barbuzzi, M.G. Biasini, and F. Sala. 1987. Somaclonal variation versus chemically induced mutagenesis in tomato (*Lycopersicon esculentum* L.). Theor Appl Genet 74: 733-738.

Gavrilenko, T., R. Thieme, and V.M. Rokka. 2001. Cytogenetic analysis of *Lycopersicon esculentum* (+) *Solanum etuberosum* somatic hybrids and their androgenetic regenerants. Theor Appl Genet 103: 231-239.

Gill, B.S. 1983. Tomato cytogenetics - a search for new frontiers, In: M.S. Swaminathan, P.K. Gupta and U. Sinha [eds.], Cytogenetics of Crop Plants. Macmillan India, India, pp. 456-480.

Gorman, S.W., D. Banasiak, C. Fairley, and S. McCormick. 1996. A 610 kb YAC clone harbors 7 cM of tomato (*Lycopersicon esculentum*) DNA that includes the *male sterile 14* gene and a hotspot for recombination. Mol Gen Genet 251: 52-59.

Grandillo, S. and S.D. Tanksley. 1996. Genetic analysis of RFLPs, GATA microsatellites and RAPDs in a cross between *L. esculentum* and *L. pimpinellifolium*. Theor Appl Genet 92: 957-965.

Haanstra, J.P.W., C. Wye, H. Verbakel, F. Meijer-Dekens, P. Van Den Berg, P. Odinot, A.W. Van Heusden, S. Tanksley, P. Lindhout, and J. Peleman. 1999. An integrated high-density RFLP-AFLP map of tomato based on two *Lycopersicon esculentum* x *L. pennellii* F$_2$ populations. Theor Appl Genet 99: 254-271.

Haider Ali, S.N., M.S. Ramanna, E. Jacobsen, and R.G.F. Visser. 2001. Establishment of a complete series of a monosomic tomato chromosome addition lines in the cultivated potato using RFLP and GISH analysis. Theor Appl Genet 103: 687-695.

Haider Ali, S.N., M.S. Ramanna, E. Jacobsen, and R.G.F. Visser. 2002. Genome differentiation between *Lycopersicon esculentum* and *L. pennellii* as revealed by genomic in situ hybridization. Euphytica 127: 227-234.

Handley, L.W., R.L. Nickels, M.W. Cameron, P.P. Moore, and K.C. Sink. 1986. Somatic hybrid plants between *Lycopersicon esculentum* and *Solanum lycopersicoides*. Theor Appl Genet 71: 691-697.

Hossain, M., S. Imanishi, and A. Matsumoto. 1994. Production of somatic hybrids between tomato (*Lycopersicon esculentum*) and nightshade (*Solanum lycopersicoides*) by electrofusion. Breeding Sci 44: 405-412.

Jacobsen, E., M.K. Daniel, J.E.M. Bergervoet-van Deelen, D.J. Huigen, and M.S. Ramanna. 1994. The first and second backcross progeny of the intergeneric fusion hybrids of potato and tomato after crossing with potato . Theor Appl Genet 88: 181-186.

Jacobsen, E., J.H. de Jong, S.A. Kamstra, P.M.M.M. van den Berg, and M.S. Ramanna. 1995. Genomic in situ hybridization (GISH) and RFLP analysis for the identification of alien chromosomes in the backcross progeny of potato (+) tomato fusion hybrids. Heredity 74: 250-257.

Ji, Y. and R.T. Chetelat. 2003. Homeologous pairing and recombination in *Solanum lycopersicoides* monosomic addition and substitution lines of tomato. Theor Appl Genet 106: 979-989.

Ji, Y., R.A. Pertuzé, and R.T. Chetelat. 2004. Genome differentiation by GISH in interspecific and intergeneric hybrids of tomato and related nightshades. Chromosome Res 12: 107-116.

Kaloshian, I., J. Yaghoobi, T. Liharska, J. Hontelez, D. Hanson, P. Hogan, T. Jesse, J. Wijbrandi, G. Simons, P. Vos, P. Zabel, and V.M. Williamson. 1998. Genetic and physical localization of the root-knot nematode resistance locus *Mi* in tomato. Mol Gen Genet 257: 376-385.

Khush, G.S. 1963. Identification key for pachytene chromosomes of *L. esculentum*. Tomato Genet Coop Rep 13: 12-13.

Khush, G.S. 1973. Cytogenetics of aneuploids. Academic Press, New York, USA.

Khush, G.S. and C.M. Rick. 1963. Meiosis in hybrids between *Lycopersicon esculentum* and *Solanum pennellii*. Genetica 33: 167-183.

Khush, G.S. and C.M. Rick. 1966. The origin, identification, and cytogenetic behavior of tomato monosomics. Chromosoma 18: 407-420.

Khush, G.S. and C.M. Rick. 1968. Cytogenetic analysis of the tomato genome by means of induced deficiencies. Chromosoma 23: 452-484.

Knapp, S., Y. Larondelle, M. Rossberg, D. Furtek, and K. Theres. 1994. Transgenic tomato lines containing *Ds* elements at defined genomic positions as tools for targeted transposon tagging. Mol Gen Genet 243: 666-673.

Koornneef, M., J. Bade, C. Hanhart, K. Horsman, J. Schel, W. Soppe, R. Verkerk, and P. Zabel. 1993. Characterization and mapping of a gene controlling shoot regeneration in tomato. Plant J 3: 131-141.

Ku, H.-M., J. Liu, S. Doganlar, and S.D. Tanksley. 2001. Exploitation of *Arabidopsis*-tomato synteny to construct a high-resolution map of the *ovate*-containing region in tomato chromosome 2. Genome 44: 470-475.

Ku, H.-M., T. Vision, J. Liu, and S.D. Tanksley. 2000. Comparing sequenced segments of the tomato and arabidopsis genomes: large-scale duplication followed by selective gene loss creates a network of synteny. Proc Nat Acad Sci (USA) 97: 9121-9126.

Lapitan, N.L.V., M.W. Ganal, and S.D. Tanksley. 1989. Somatic chromosome karyotype of tomato based on *in situ* hybridization of the TGRI satellite repeat. Genome 32: 992-998.

Lesley, M.M. 1950. A cytological basis for sterility in tomato hybrids: evidence for an inversion in one chromosome of the F_1 between *Lycopersicon esculentum* var. Pearson and *L. peruvianum* P.I. 126946. J Hered 26-28.

Lesley, M.M. and J.W. Lesley. 1935. Heteromorphic A chromosomes of the tomato differing in satellite size. Genetics 23: 485-493.

Lesley, M.M. and J.W. Lesley. 1943. Hybrids of the Chilean tomato: sterile and fertile plants of *Lycopersicon peruvianum* var. *dentatum* Dun. (*L. chilense* Dun.) and diploid and tetraploid hybrids with cultivated tomatoes. J Hered 34: 199-205.

Livingstone, K.D., V.K. Lackney, J.R. Blauth, R.I.K. Van Wijk, and M. Kyle Jahn. 1999. Genome mapping in *Capsicum* and the evolution of genome structure in the Solanaceae. Genetics 152: 1183-1202.

Matsubayashi, M. 1991. Phylogenetic relationships in potato and its related species. In: T. Tsuchiya and P.K. Gupta [eds], Chromosome Engineering in Plants: Genetics, Breeding, Evolution, Part B. Elsevier Science Publishers, Amsterdam, pp. 93-118.

Matsumoto, A., S. Imanishi, M. Hossain, A. Escalante, and H. Egashira. 1997. Fertile somatic hybrids between F_1 (*Lycopersicon esculentum* x *L. peruvianum* var. *humifusum*) and *Solanum lycopersicoides*. Breeding Sci 47: 327-333.

McCallum, C.M., L. Comai, E.A. Greene, and S. Henikoff. 2000. Targeting induced local lesions in genomes (TILLING) for plant functional genomics. Plant Physiol 123: 439-442.

McGuire, D.C. and C.M. Rick. 1954. Self-incompatibility in species of *Lycopersicon* Sect. Eriopersicon and hybrids with *L. esculentum*. Hilgardia 23: 101-123.

Meissner, R., V. Chague, Q. Zhu, E. Emmanuel, Y. Elkind, and A.A. Levy. 2000. A high throughput system for transposon tagging and promoter trapping in tomato. Plant J 22: 265-274.

Meissner, R., Y. Jacobson, S. Melamed, S. Levyatuv, G. Shalev, A. Ashri, Y. Elkind, and A. Levy. 1997. A new model system for tomato genetics. Plant J 12: 1465-1472.

Melchers, G., M.D. Sacristan, and A.A. Holder. 1978. Somatic hybrid plants of potato and tomato regenerated from fused protoplasts. Carlsberg Res Commun 43: 203-218.

Menda, N., Y. Semel, D. Peled, Y. Eshed, and D. Zamir. 2004. *In-silico* screening of a saturate mutation library of tomato. Plant J 38: 861-872.

Menzel, M.Y. 1962. Pachytene chromosomes of the intergeneric hybrid *Lycopersicon esculentum* x *Solanum lycopersicoides*. Am J Bot 49: 605-615.

Menzel, M.Y. 1964. Differential chromosome pairing in allotetraploid *Lycopersicon esculentum* - *Solanum lycopersicoides*. Genetics 50: 855-862.

Menzel, M.Y. and J.M. Price. 1966. Fine structure of synapsed chromosomes in F_1 *Lycopersicon esculentum* - *Solanum lycopersicoides* and its parents. Am J Bot 53: 1079-1086.

Michaelson, M.J., H.J. Price, J.R. Ellison, and J.S. Johnston. 1991. Comparison of plant DNA contents determined by Fuelgen microspectrophotometry and laser flow cytometry. Am J Bot 78: 183-188.

Miller, J.C. and S.D. Tanksley. 1990. RFLP analysis of phylogenetic relationships and genetic variation in the genus *Lycopersicon*. Theor Appl Genet 80: 437-448.

Monforte, A.J. and S.D. Tanksley. 2000a. Development of a set of near isogenic and backcross recombinant inbred lines containing most of the *Lycopersicon hirsutum* genome in a *L. esculentum* genetic background: a tool for gene mapping and gene discovery. Genome 43: 803-813.

Monforte, A.J. and S.D. Tanksley. 2000b. Fine mapping of a quantitative trait locus (QTL) from *Lycopersicon hirsutum* chromosome 1 affecting fruit characteristics and agronomic traits: breaking linkage among QTLs affecting different traits and dissection of heterosis for yield. Theor Appl Genet 100: 471-479 .

Mutschler, M.A., S.D. Tanksley, and C.M. Rick. 1987. Linkage maps of the tomato (*Lycopersicon esculentum*). Tomato Genet Coop Rep 37: 5-34.

Pan, Q., Y.-S. Liu, O. Budai-Hadrian , M. Sela, L. Carmel-Goren, D. Zamir, and R. Fluhr. 2000. Comparative genetics of nucleotide binding site-leucine rich repeat resistance gene homologues in the genomes of two dicotyledons: tomato and *Arabidopsis*. Genetics 155: 309-322.

Paran, I., I. Goldman, S.D. Tanksley, and D. Zamir. 1995. Recombinant inbred lines for genetic mapping in tomato. Theor Appl Genet 90: 542-548.

Parokonny, A.S., J.A. Marshall, M.D. Bennett, E.C. Cocking, M.R. Davey, and J.B. Power. 1997. Homeologous pairing and recombination in backcross derivatives of tomato somatic hybrids (*Lycopersicon esculentum* (+) *L. peruvianum*). Theor Appl Genet 94: 713-723.

Paterson, A.H., J.W. DeVerna, B. Lanini, and S.D. Tanksley. 1990. Fine mapping of quantitative trait loci using selected overlapping recombinant chromosomes, in an interspecies cross of tomato. Genetics 124: 735-742.

Pertuzé, R.A., Y. Ji, and R.T. Chetelat. 2002. Comparative linkage map of the *Solanum lycopersicoides* and *S. sitiens* genomes and their differentiation from tomato. Genome 45: 1003-1012.

Pertuzé, R.A., Y. Ji, and R.T. Chetelat. 2003. Transmission and recombination of homeologous *Solanum sitiens* chromosomes in tomato. Theor Appl Genet 107: 1391-1401.

Peterson, D.G., N.L.V. Lapitan, and S.M. Stack. 1999. Localization of single- and low-copy sequences on tomato synaptonemal complex spreads using fluorescence *in situ* hybridization (FISH). Genetics 152: 427-439.

Peterson, D.G., W.R. Pearson, and S.M. Stack. 1998. Characterization of the tomato (*Lycopersicon esculentum*) genome using *in vitro* and *in situ* DNA reassociation. Genome 41: 346-356.

Peterson, D.G., H.J. Price, J.S. Johnston, and S.M. Stack. 1996. DNA content of heterochromatin and euchromatin in tomato (*Lycopersicon esculentum*) pachytene chromosomes. Genome 39: 77-82.

Pillen, K., O. Pineda, C.B. Lewis, and S.D. Tanksley. 1996. Status of genome mapping tools in the taxon Solanaceae. *In*: A.H. Paterson [ed.], Genome Mapping in Plants. R.G. Landes Co., pp. 282-308.

Quiros, C.F. 1991. *Lycopersicon* cytogenetics. *In*: T. Tsuchiya and P.K. Gupta, [eds.], Chromosome Engineering in Plants: Genetics, Breeding, Evolution, Part B. Elsevier, Amsterdam, pp. 119-138.

Rick, C.M. 1945. A survey of cytogenetic causes of unfruitfulness in the tomato. Genetics 30: 347-362.

Rick, C.M. 1951. Hybrids between *Lycopersicon esculentum* Mill. and *Solanum lycopersicoides* Dun. Proc Natl Acad Sci (USA) 37: 741-744 .

Rick, C.M. 1967. Exploiting species hybrids for vegetable improvement. Proc XVII Int'l Hort Congress III: 217-229.

Rick, C.M. 1969. Controlled introgression of chromosomes of *Solanum pennellii* into *Lycopersicon esculentum*: segregation and recombination. Genetics 62: 753-768.

Rick, C.M. 1971. Further studies on segregation and recombination in backcross derivatives of a tomato species hybrid. Biol Zentralbl 90: 209-220.

Rick, C.M. 1979. Biosystematic studies in *Lycopersicon* and closely related species of *Solanum*. *In*: J.G. Hawkes, R.N. Lester and A.D. Skelding [eds.], The Biology and Taxonomy of the Solanaceae, Academic Press, New York, pp. 667-678.

Rick, C.M. 1988a. Evolution of mating systems in cultivated plants. *In*: L.D. Gottlieb, and S.K. Jain. [eds.], Plant Evolutionary Biology. Chapman and Hall, London, pp. 133-147.

Rick, C.M. 1988b. Tomato-like nightshades: affinities, autecology, and breeders opportunities. Econ Bot 42: 145-154 .

Rick, C.M. 1990. New or otherwise noteworthy accessions of wild tomato species. Tomato Genet Coop Rep 40: 30.

Rick, C.M. and L. Butler. 1956. Cytogenetics of the tomato. Adv Genet 8: 267-382.

Rick, C.M., J.W. DeVerna, R.T. Chetelat, and M.A. Stevens. 1986. Meiosis in sesquidiploid hybrids of *Lycopersicon esculentum* and *Solanum lycopersicoides*. Proc Natl Acad Sci (USA) 83: 3580-3583.

Rick, C.M., J.F. Fobes, and M. Holle. 1977. Genetic variation in *Lycopersicon pimpinellifolium*: evidence of evolutionary change in mating systems. Plant Syst Evol 127: 139-170.

Rick, C.M., J.F. Fobes, and S.D. Tanksley. 1979. Evolution of mating systems in *Lycopersicon hirsutum* as deduced from genetic variation in electrophoretic and morphological characters. Plant Syst Evol 132: 279-298.

Rick, C.M. and G.S. Khush. 1961. X-ray-induced deficiencies of chromosome 11 in the tomato. Genetics 46: 1389-1393.

Rick, C.M. and G.S. Khush. 1962. Preferential pairing in tetraploid tomato species hybrids. Genetics 47: 979-980.

Rick, C.M. and G.S. Khush. 1966. Chromosome engineering in *Lycopersicon*. *In*: R. Riley and K.R. Lewis. [eds.], Chromosome Manipulation and Plant Genetics. Oliver & Boyd, Edinburgh, pp. 8-20.

Rick, C.M. and J.I. Yoder. 1988. Classical and molecular genetics of tomato: highlights and perspectives. Ann Rev Genet 22: 281-300.

Robinson, R.W. 1982. A history of the Tomato Genetics Cooperative. Tomato Genet Coop Rep 32: 1-2.

Sawant, A.C. 1958. Cytogenetics of interspecific hybrids, *Lycopersicon esculentum* Mill. x *L. hirsutum* Humb. and Bonpl. Genetics 43: 502-514.

Sherman, J.D. and S.M. Stack. 1992. Two-dimensional spreads of synaptonemal complexes from solanaceous plants. V. Tomato (*Lycopersicon esculentum*) karyotype and idiogram. Genome 35: 354-359.

Sherman, J.D. and S.M. Stack. 1995. Two-dimensional spreads of synaptonemal complexes from solanaceous plants. VI. High resolution recombination nodule map for tomato (*Lycopersicon esculentum*). Genetics 141: 683-708.

Spooner, D.M., I.E. Peralta, S. Knaap. 2005. Comparison of AFLPs with other markers for phylogenetic inference in wild tomatoes [*Solanum* L. section *Lycopersicon* (Mill.) Wettst.] Taxon 54: 43-61.

Stack, S. 1982. Two-dimensional spreads of synaptonemal complexes from Solanaceous plants. I. The technique. Stain Technol 57: 265-272.

Stephan, W. and C.H. Langley. 1998. DNA Polymorphism in Lycopersicon and crossing-over per physical length. Genetics 150: 1585-1593.

Stevens, M.A. and C.M. Rick. 1986. Genetics and breeding. *In:* J.G. Atherton and J. Rudich [eds.], The Tomato Crop: A Scientific Basis for Improvement. Chapman and Hall, London, pp. 35-109.

Stommel, J.R. 2001. Barriers for introgression of *Solanum ochranthum* into tomato via somatic hybrids. J Am Soc Hort Sci 126: 387-388.

Stubbe, H. 1972a. Mutanten der Kulturtomate *Lycopersicon esculentum* Miller VI. Kulturpflanze 19: 185-230.

Stubbe, H. 1972b. Mutanten der Wildtomate *Lycopersicon pimpinellifolium* (Jusl.) Mill. IV. Kulturpflanze 19: 231-263.

Sybenga, J., E. Schabbink, J. van Eden, and J.H. de Long. 1994. Pachytene pairing and metaphase I configurations in a tetraploid somatic *Lycopersicon esculentum* x *L. peruvianum* hybrid. Genome 37: 54-60.

Tanksley, S.D., M.W. Ganal, J.P. Prince, M.C. de Vicente, M.W. Bonierbale, P. Broun, T.M. Fulton, J.J. Giovanonni, S. Grandillo, G.B. Martin, R. Messeguer, J.C. Miller, L. Miller, A.H. Paterson, O. Pineda, M. Roder, R.A. Wing, W. Wu, and N.D. Young. 1992. High density molecular linkage maps of the tomato and potato genomes. Genetics 132: 1141-1160.

Tanksley, S.D. and C.M. Rick. 1980. Isozymic gene linkage map of the tomato: applications in genetics and breeding. Theor Appl Genet 57: 161-170.

Thomas, C.M., D.A. Jones, J.J. English, B.J. Carroll, J.L. Bennetzen, K. Harrison, A. Burbidge, G.J. Bishop, and J.D.G. Jones. 1994. Analysis of the chromosomal distribution of transposon-carrying T-DNAs in tomato using the inverse polymerase chain reaction. Mol Gen Genet 242: 573-585.

Tor, M., K. Manning, G.J. King, A.J. Thompson, G.H. Jones, G.B. Seymour, and S.J. Armstrong. 2002. Genetic analysis and FISH mapping of the Colourless non-ripening locus of tomato. Theor Appl Genet 104: 165-170.

Vallejos, C.E., S.D. Tanksley, and R. Bernatzky. 1986. Localization in the tomato genome of DNA restriction fragments containing sequences homologous to the ribosomal RNA (45S), the major chlorophyll A/B binding polypeptide and the ribulose bisphosphate carboxylase genes. Genetics 112: 93-106.

van den Bulk, R.W., H.J.M. Loffler, W.H. Lindhout, and M. Koorneef. 1990. Somaclonal variation in tomato: effect of explant source and a comparison with chemical mutagenesis. Theor Appl Genet 80: 817-825.

van der Biezen, E.A., B. Overduin, H.J.J. Nijkamp, and J. Hille. 1994. Integrated genetic map of tomato chromosome 3. Rpt Tomato Genet Coop 44: 8-10.

Van Der Hoeven, R., C. Ronning, J. Giovannoni, G. Martin, and S. Tanksley. 2002. Deductions about the number, organization, and evolution of genes in the tomato genome based on analysis of a large expressed sequence tag collection and selective genomic sequencing. Plant Cell 14: 1441-1456.

van der Knaap, E., A. Sanyal, S.A. Jackson, and S.D. Tanksley. 2004. High-resolution fine mapping and fluorescence *in situ* hybridization analysis of *sun*, a locus controlling tomato fruit shape, reveals a region of the tomato genome prone to DNA rearrangements. Genetics 168: 2127-2140.

van Ooijen, J.M., J.M. Sandbrink, M. Vrielink, R. Verkerk, P. Zabel, and P. Lindhout. 1994. An RFLP linkage map of *Lycopersicon peruvianum*. Theor Appl Genet 89: 1007-1013.

van Tuinen, A., M.-M. Cordonnier-Pratt, L. Pratt, R. Verkerk, P. Zabel, and M. Koornneef. 1997. The mapping of phytochrome genes and photomorphogenic mutants of tomato. Theor Appl Genet 94: 115-122.

van Tuinen, A., A.H.L.J. Peters, and M. Koornneef. 1998. Mapping of the *pro* gene and revision of the classical map of chromosome 11. Tomato Genet Coop Rep 48: 62-70.

van Wordragen, M.F., R.L. Weide, E. Coppoolse, M. Koornneef, and P. Zabel. 1996. Tomato chromosome 6: a high resolution map of the long arm and construction of a composite integrated marker-order map. Theor Appl Genet 92: 1065-1072.

Villand, J., P.W. Skroch, T. Lai, P. Hanson, C.G. Kuo, and J. Nienhuis . 1998. Genetic variation among tomato accessions from primary and secondary centers of diversity. Crop Sci 38: 1339-1347.

Vision, T.J., Y. Xu, N. Van Eck, D.G. Brown, and S.D. Tanksley. 2001. A *L. esculentum* x *L. pennellii* backcross recombinant inbred population. Tomato Genet Coop Rep 51: 36.

Weide, R., M.F. van Wordragen, R.K. Lankhorst, R. Verkerk, C. Hanhart, T. Liharska, E. Pap, P. Stam, P. Zabel, and M. Koorneef. 1993. Integration of the classical and molecular linkage maps of tomato chromosome 6. Genetics 135: 1175-1186.

Williams, C.E. and D.A. St. Clair. 1993. Phenetic relationships and levels of variability detected by restriction fragment length polymorphism and random amplified polymorphic DNA analysis of cultivated and wild accessions of *Lycopersicon esculentum*. Genome 36: 619-630.

Wolters, A.M.A., H.C.H. Schoenmakers, S. Kamstra, J. van Eden, M. Koorneef, and J.H. de Jong. 1994. Mitotic and meiotic irregularities in somatic hybrids of *Lycopersicon esculentum* and *Solanum tuberosum*. Genome 37: 726-735.

Xu, J. and E.D. Earle. 1994. Direct and sensitive fluorescence *in situ* hybridization of 45S rDNA on tomato chromosomes. Genome 37: 1062-1065.

Xu, J. and E.D. Earle. 1996a. Direct FISH of 5S rDNA on tomato pachytene chromosomes places the gene at the heterochromatic knob immediately adjacent to the centromere of chromosome 1. Genome 39: 216-221.

Xu, J. and E.D. Earle. 1996b. High resolution physical mapping of 45S (5.8S, 18S and 25S) rDNA gene loci in the tomato genome using a combination of karyotyping and FISH of pachytene chromosomes. Chromosoma 104: 545-550.

Young, N.D., J.C. Miller, and S.D. Tanksley. 1987. Rapid chromosomal assignment of multiple genomic clones in tomato using primary trisomics. Nucleic Acids Res 15: 9339-9348.

Zamir, D. 2001. Improving plant breeding with exotic genetic libraries. Nat Rev Genet 2: 983-989.

Zamir, D. and Y. Eshed. 1998. Tomato genetics and breeding using nearly isogenic introgression lines derived from wild species. *In*: A.H. Paterson [ed.], Molecular Dissection of Complex Traits. CRC Press Inc., FL, pp. 207-217.

Zamir, D. and S.D. Tanksley. 1988. Tomato genome is composed largely of fast evolving low copy number sequences. Mol Gen Genet 213: 254-261.

Zhong, X., P.F. Fransz, J. Wennekes-van Eden, P. Zabel, A. van Kammen, and J.H. de Jong. 1996. High-resolution mapping of pachytene chromosomes and extended DNA fibres by fluorescence in-situ hybridisation. Plant Mol Biol Rep 14: 232-242.

Zhong, X.-B., J. Bodeau, P.F. Fransz , V.M. Williamson, A. Van Kammen, J.H. De Jong, and
  P. Zabel. 1999. FISH to meiotic pachytene chromosomes of tomato locates the root-
  knot nematode resistance gene *Mi-1* and the acid phosphatase gene *Aps-1* near the
  junction of euchromatin and pericentromeric heterochromatin of chromosome arms
  6S and 6L, respectively. Theor Appl Genet 98: 365-370.

Zhong, X.-B., P.F. Fransz, J. Wennekes-van Eden, M.S. Ramanna, A. van Kammen, P. Zabel,
  and J.H. de Jong . 1998. FISH studies reveal the molecular and chromosomal
  organization of individual telomere domains in tomato. Plant J 13: 507-517.

4

Expression of Heterosis by Hybridization

BISTRA ATANASSOVA AND HRISTO GEORGIEV

Department of Applied Genetics (former Department of Heterosis),
Institute of Genetics "Prof. D. Kostov",
Bulgarian Academy of Sciences, Sofia 1113, Bulgaria
email: bistra_a@yahoo.com

INTRODUCTION

The phenomenon of heterosis is defined by Shull (1911) as "the superiority of heterozygous genotypes with respect to one or more characters in comparison with the corresponding homozygotes. Heterosis is the phenotypic result of gene interaction in heterozygotes and thus confined (at least in maximal amount) to that state." Hence, to observe heterosis, the F_1 must be superior to the two breeding lines that are its parents. Nevertheless, the heterozygote may also be inferior to both homozygotes. This performance is defined as negative heterosis (Jinks 1983).

The expression of hybrid vigor in plant crosses has been recognized for nearly 250 years (Zirkle 1952). The phenomenon received higher degree of attention since the publications of Shull (1908, 1909, 1914), East (1908, 1909) and Jones (1918), which resulted in the development of the first commercial maize hybrids and their introduction in practice at about 1930 (Sprague 1983). Within the years that followed, heterosis was observed in several crops.

Heterosis is now recognized as one of the primary factors that contributed to the success of plant breeding in many crops. Duvick (1997a) reported that maize, sorghum, and sunflower were produced as hybrids in all the industrialized world; hybrid rice was grown extensively in China and was recently introduced in India; many commercial vegetables and flowers were grown almost entirely as hybrids. The application of heterosis was

Corresponding author: Bistra Atanassova

determined as one of the greatest achievements in the twentieth century (Barabas 1992) and as the single greatest applied achievement of the discipline of genetics (Griffing 1990).

Despite the numerous studies on heterosis, however, and although this phenomenon is exploited wherever possible, its biological basis remains unknown and its elucidation is still a major challenge for scientists. Sprague (1983) mentioned two separate courses that investigations of the phenomenon of heterosis have followed. The first has been descriptive and concerned with properties, frequency and the magnitude of the observed effects. The second course centered primarily on biochemical-physiological studies and during the recent years on molecular studies.

The genetic theories advanced on this subject differ in the relative importance of overdominance, epistasis and linkage and how they contribute to hybrid performance. Up to now, none of them is acceptable to all. Since the rediscovery of Mendelism in 1900 two principal hypotheses were suggested as the genetic basis of heterosis: dominance and overdominance. Tsaftaris (1995) summarized these hypotheses as follows: The dominance hypothesis attributes the increased vigor of heterozygosity to dominant alleles mainly because of the observed correlation between recessiveness and detrimental effects. The overdominance hypothesis assumes that heterozygosity *per se* is important; i.e. there exist loci, albeit relatively rare, at which the heterozygote is superior to either homozygote. Jinks (1954) implicated epistasis in the expression of heterosis and recently multilocus epistatic interactions are recognized as the third theory of heterosis (Allard 1996, Stuber 1999, Monforte and Tanksley 2000).

The last decade has witnessed a period of renewed interest toward resolving some of the long standing issues related to heterosis. New interest has been sparked by the application of molecular genetics. The knowledge and the experience accumulated throughout the years as well as advances in molecular genetics and technologies provided new tools and stimulated scientists' endeavors to shed additional light on heterosis. This resulted in a significant number of studies in which new aspects of the phenomenon have been evaluated, new approaches to investigate its mechanisms have been developed and new theories and hypotheses have been advanced (McDaniel 1986, Griffing 1990, Georgiev 1991, Nienhuis and Sills 1992, Khanna-Chopra et al. 1993, Verma and Chahal 1993, Houle 1994, Stuber 1994, 1999, Pooni 1994, Pooni and Treharne 1994, Tsaftaris 1995, David 1997, Lee 1997, Kang 1997, Deng et al. 1998, Tsaftaris and Kafka 1998, Milborrow 1998, Melchinger 1999, Lamkey and Edwards 1999, Goodnight 1999, Wang and Yin 1999, Omholt et al. 2000). Despite the significant contribution in this respect, however, the opinion of the scientists summarized by Stuber (1999) was that "the causal factors for heterosis are today as obscure as they were 50 years ago". According to Hallauer (1999) "The exact genetic basis of heterosis may never be known and understood because of allelic interactions: interactions of alleles at a locus, interactions of alleles at different loci, interactions of the nucleus and cytoplasm, and interactions of the genotype and environment. But heterosis will continue to have a major role in the future plant improvement even though our knowledge on its genetic basis is limited."

Given the present wide use of heterosis in practice on the one hand, and a lack of real understanding of its nature, on the other, recent studies on heterosis center on: a) Increasing knowledge on the genetic mechanisms that includes efforts on physiological, biochemical, and molecular levels. A better understanding of these mechanisms would enhance the ability of the breeders not only to predict the performance of a given hybrid, but also to form new genotypes that might be used directly, without preliminary tests, as F_1 hybrids; b) based on the available knowledge of this phenomenon, increasing the efficiency in the strategies for developing parental lines whose F_1 hybrids would be promising ones; and c) improving the process of hybrid seed production by developing easier and more reliable and efficient technologies.

MANIFESTATION OF HETEROSIS IN TOMATO

Studies on heterosis in tomato were initiated at the beginning of the twentieth century, almost at the same time as those in maize (Hedrick and Booth 1907, East and Hayes 1912). Tomatoes are a self-pollinated inbred crop and there was a view that high level of heterosis could not be manifested in tomato F_1 hybrids. However, heterosis in tomato, as well as in other self-pollinated crops, was observed in several crosses. It provided evidence that this phenomenon was not limited to cross-pollinated crops. One of the theories of this performance was that natural cross-pollination predominated within the wild forms of tomatoes in the centers of their origin. Therefore, crosses between tomato lines might be considered as F_1 between inbred lines of a species which is naturally cross-pollinating (Rick 1950, Rick and Butler 1956).

Heterosis is a widely documented phenomenon in tomato. More than 50-60% of the studies on heterotic performance refer to heterosis for yield and yield components. This percentage was relatively stable even throughout the last 10 years when efforts of tomato breeders strongly emphasized nutritional value, safety and sensory quality of a food product, tolerance to abiotic stress, etc. Based on these studies, as well as on earlier ones, it might be concluded that the frequency and the level of heterosis for yield and yield components in tomato are relatively high. Kravchenko (1990), for example, reported manifestation of heterosis for yield in 80% and for earliness in 88% of the hybrids studied all over 15 years period. According to Wehner (1997), the level of heterosis for yield in tomato was significant (estimates average 60%). Suresh Kumar et al. (1995) reported 193.55 % of heterosis (over superior parent) for one of the yield components - fruit number.

According to Yordanov (1983), besides yield, enhanced plant vigor, earliness, higher adaptability to unfavorable environment and uniformity, were the manifestations of heterosis most often encountered in the tomato. These traits are of significant economic importance. Khanna-Chopra et al. (1993), for example, reported that vigor helped in the efficient utilization of environmental factors. Uniformity is one of the principal benefits of hybrids as this trait in plant structure and maturation permits efficient mechanical harvest. Earliness, generally defined as the number of days from sowing to the appearance of the first ripe fruit (Kemble and Gardner 1992), was characterized by Doganlar et al. (2000) as crucial for regions with short growing season, as desirable for taking advantage of high prices during the early season and as a prerequisite for reducing the heating and lighting expenses of greenhouse-grown tomato. Based on experiments aiming at the development of tomato cultivars adapted to low growing temperature, Nieuwhof (1990) found no important genotype x temperature interactions. This finding suggested that for satisfying the need to reduce energy consumption in greenhouses it was preferable to breed early varieties characterized by rapid growth at normal temperature instead of selecting for adaptation to low growing temperature. In some crosses the magnitude of heterosis for earliness is significant and certainly has, and will be exploited in developing early hybrids. For example, in a study of 92 F_1 tomato hybrids between 19 ultra early cultivars used as seed parents and 5 late cultivars. Boe (1988) found that early yield in the F_1 hybrids ranged from 52% to 307% when compared to that of the early seed parent.

Genetic studies on tolerance to abiotic stress, nutritive and market quality provide evidence that manifestation of heterosis in tomato is not limited to traits related to plant vigor or yield. Bhatt et al. (1998) reported relatively high levels of heterosis (60.4% and 52.4%) for vitamin C content detected in a number of F_1 hybrids. Heterosis for vitamin C was also reported by Dod and Kale (1992), and Mageswari and Natarajan (1999). Heterosis was observed for total soluble solids and dry matter content (Patil and Patil 1988, Yadav et al. 1988, Shrivastava 1998 a, Daskloff et al. 1990, Dod and Kale 1992, Mageswari and Natarajan 1999) and lycopene and β-carotene content in fruits (Chen and Zhao 1990, Amaral Junior et al. 1997). It must be noted that there is also information reporting low level or negative heterosis for the content of compounds related to the nutritive value of tomato such as ascorbic acid, lycopene, total soluble solids, etc. (Chen and Zhao 1990, Wang et al. 1998, Amaral Junior et al. 1999). In a study of a number of tomato lines and hybrids for their suitability to prepare ketchup and paste, manifestation of heterosis for ketchup recovery and paste yield was detected (Jawaharlal et al. 1999). Heterosis for pericarp thickness was described by Patil and Patil (1988), Daskaloff et al. (1990) and Dod and Kale (1992), and for fruit firmness by Wang et al. (1995), Resende et al (1999) and Atanassova et al. (2005).

Besides traits related to fruit market and nutritive quality, heterosis was reported for characteristics related to plant tolerance to stress, plant physiology etc. Heterotic performance for these traits is very important as it enhances the ability of plants to cope better with the environment, that usually results in higher yield. Experiments demonstrated, for example, that yield gains in hybrid maize were due primarily to improvements in tolerance to abiotic and biotic stress and that the improvement occurred in parental inbreds as well as in their F_1 hybrids (Duvick 1997a). In a study of viability of pollen from intra- and inter-specific crosses of tomato, produced at low temperature, Fernàndez-Muñoz et al. (1995) observed positive heterosis for pollen viability in the cross between *L. esculentum* x *L. pimpinellifolium*. Philouze (1997) reported high level of heterosis for yield in the hybrid Monfavet n° 63-5 when grown in non-heated greenhouse. It was found that heterosis was mainly due to higher percentage of fruit set in the hybrid. Under low temperature the fertilizing ability of the hybrid plants pollen was higher than that of the parent lines pollen. Hassan et al. (1999) reported high level of heterosis for tolerance to salinity at seedling stage, that ranged from 24.9% to 100%(based on better parent values). Zhacote and Kharti (1990) observed heterosis for net photosynthetic production in hybrids between cultivated and wild forms of tomato and Titok et al. (1994) reported manifestation of heterosis for chlorophyll content both in leaves and stems, a higher level of heterosis being observed in stems.

Based on these reports, as well as on numerous other studies [some of them reviewed by Yordanov (1983), Kalloo (1988) and Georgiev (1991)], it might be concluded that heterosis in tomato was observed for a large scale of quantitative traits, almost all of them being of breeding interest.

It is commonly assumed that genetic stability (homeostasis) in hybrids refers to reduced genotype - environment interactions. A number of studies provided, however, evidence that the majority of quantitative traits were significantly affected by environmental factors and that heterosis was also dependent on the environment (Yordanov 1983, Russel et al. 1993, Cooper and Podlich 1999, Temperini et al. 2001). Therefore, the evaluation of a given tomato hybrid when grown in different locations is necessary for getting reliable information on its performance. This kind of information is of great importance for tomato growers as it can help them to make intelligent cultivar decisions (Murray et al. 1999).

GENETIC BASIS OF HETEROSIS IN TOMATO

The design and efficiency of breeding programs depends on the relative importance of different types of gene action. Therefore, manipulating heterosis in breeding programs requires knowledge of its quantitative genetic basis. Gene action in tomato, as well as in many other species has been approached by studying the various types of genetic variance in populations and by generation means analysis. Diallel sets of F_1 crosses between collections of tomato lines are usually used for obtaining a preliminary impression of the genetic variation for characters of economic importance. As a result, a relatively large amount of information on the nature of gene action for several quantitative traits in tomato is available, the predominant part of this information concerns total yield, yield components and earliness. Additive and non-additive gene effects have been reported to be important for yield and its main components. It was found that their magnitude varied depending on the genotype and on the environmental conditions (Dod et al. 1992, Natarajan 1992, Sherif and Hussein 1992, Vallejo Cabrera and Estrada 1993, Ramos et al. 1993, Rai et al. 1997, Singh et al. 1998, Surjan et al. 1999). Studies on the genetics of earliness have indicated that dominance plays an important role in this trait (Banerjee and Kalloo 1989, Kemble and Gardner 1992). Information concerning genetics of quantitative characteristics related to seed, cotyledons, leaf, stem, fruit, growth stages, early and total yield etc. was summarized by Georgiev (1991).

As already mentioned, during the last two decades a significant number of genetic studies and breeding programs in tomato emphasized enhancing plant tolerance to abiotic stresses and fruit nutritional value, sensory and market quality. It resulted in determining the genetic basis of characteristics related to these traits. Data of some of these studies are summarized in Table 4.1 and Table 4.2.

Table 4.1 *Gene expression of some characters related to nutritive and market quality in tomato*

Character	Gene action	Reference
Lycopene content	Dominant	Daskaloff et al. (1990)
β-carotene content	Additive and non-additive	Daskaloff et al. (1990)
Vitamin C content	Dominant Non-additive	Daskaloff et al. (1990)
Percentage of reducing sugar	Additive and dominant	Stommel and Haynes (1993)
Glucose/fructose ratio	Additive	Stommel and Haynes (1993)
Sugar content	Dominant	Daskaloff et al. (1990)
Fruit firmness and longevity	Additive	Al-Falluji et al. (1982), Dobhal et al. (1999), Atanassova et al. (2005).
Pericarp thickness	Additive	Dobhal et al.(1999), Rai et al. (1997)
	Additive and non-additive	Singh et al.(1998), Dod et al. (1995)
	Non-additive	Daskaloff et al. (1990)
TSS	Additive and non-additive	Dobhal et al. (1999), Singh et al. (1998), Dod et al. (1995).
	Non-additive	Dhaliwal et al., (2000), Wang et al., (1998)
	Additive	Shrivastava (1998 b)
Reducing sugars content	Additive	Shrivastava (1998 b)
Dry matter content	Additive	Shrivastava (1998 b)
Resistance to cuticle cracking	Additive and dominant	Emmons and Scott (1998)

The numerical value recorded for a complex trait (such as total or early yield) is known to be a function of its components. In tomato, for example, earliness is usually divided into four different components and number of fruits per plant and mean fruit weight are considered as the two primary ones determining yield. Hence, the expression of heterosis for a complex trait is also studied and explained on the basis of component interactions. Bos and Sparnaaij (1993) showed that component analysis provided the necessary data for the exploitation of recombinative heterosis in plant breeding. Recombinative heterosis was defined as the phenomenon that the progeny value of complex character exceeded the mid-parent value as a result of the multiplicative relationship between the complex character

Table 4.2   *Gene expression of some characters related to tolerance to abiotic stress in tomato*

Character	Gene action	Reference
Calcium use efficiency (based on total plant dry weight) in tomato grown under low-calcium stress	Additive and dominant	Li and Gabelman (1990)
Low temperature tolerance during germination	Additive	Foolad and Lin (1998)
Salt tolerance during vegetative growth	Additive and dominant	Foolad (1996)
Absolute and relative growth under salt stress and Na^+ and Ca^2 accumulations in the leaf	Additive	Foolad (1997)
Endosperm effect on germination performance under salt stress	Additive	Fooland and Jones (1991, 1992)
Testa effect on germination performance under salt stress	Dominant	Fooland and Jones (1991, 1992)
Pollen fertility and fruit set under high field temperature	Additive	Dane et al. (1991)

and its components traits. It was suggested that this form of heterosis may be an important cause of Specific Combining Ability (SCA).

Besides the investigations that focused on acquiring and increasing knowledge on genetic variation for characters of economic importance, a number of studies aiming at getting better understanding of heterosis in tomato have been also carried out during the last decade. These studies might contribute to developing new approaches for more efficient exploitation of heterosis in tomato breeding.

Griffing (1990) tested three heterosis hypotheses in a controlled-nutrient (CN) experiment, with reference to tomato yield and its components for a set of two inbred lines and their hybrid that had previously exhibited heterosis under field conditions. Heterosis was not exhibited by yield or yield components at any of the four nutrient levels. Hence, the total heterosis phenomenon was classified as nutrient-dependent heterosis occurring under field conditions, but not under the nutritional restrictions of the CN experiment. Such a performance fitted the hypothesis suggesting that heterosis was a consequence of a faster hybrid growth rate. Under this hypothesis lack of heterosis at all four levels of CN experiments was due to the CN procedure which forced all three genotypes to have the same growth rate. Under the differential growth rate hypothesis, the F_1 would utilize nutrients from a given allocation most quickly while the parent would utilize the nutrients more slowly. It was speculated that the indeterminate pattern of plant development responsible for yield and its components was due to two major gene systems: genes that determined morphogenetic responses and genes that determined growth rate manifestations.

Studies on relationship between assimilatory surface, growth rates and net photosynthetic rate (P_N) of tomato hybrid FMHy1 and its parents provided evidence that the hybrid was characterized by the greatest total leaf area and biomass and that heterosis greatly affected the early development of the hybrid (Rao et al. 1992). Greater net assimilation rate of the hybrid during early growth suggested a greater P_N per unit leaf area. It was concluded that early growth of the hybrid and higher growth rates were responsible for higher dry matter production and yield per plant.

In a study of glycolysis, the pentose phosphate pathway (PPP) for oxidation of carbohydrates and the electron transport chain (ETC) in the mitochondria of 5 tomato genotypes and their F_1 hybrids differing in yield, Titok et al. (1998) found that glycolysis and PPP, but not ETC, were inhibited in green leaves of the hybrids while in the parental genotypes both these processes were not inhibited. It suggested higher energy potential in the hybrids which produced favorable metabolic conditions for growth and was supposed to be the main reason for improved yield.

Studies on F_1 hybrids between line B 317 and a set of isogenic/near isogenic lines (IL/NIL) of tomato cv. Ailsa Craig differing in genes *baby lea syndrome (bls)*, *high pigment (hp)*, *sunny (sy)*, *venosa (ven)*, *curly mottled (cm)*, *entire (e)*, *ripening inhibitor (rin)*, *relaxata (rela)*, *lutea (lut)* and *clausa (clau)* showed that heterosis for productivity, early yield and mean fruit weight occurred only in three of the 10 hybrids (Atanassova et al. 2002). The pollen parents of two of these three hybrids (Ailsa Craig *ven* and Ailsa Craig *rela*) were characterized as possessing genetically controlled low vitality, i.e., the occurrence of heterosis was more or less limited to F_1 hybrids of a given group of mutants. Such results might be consistent with those reported by Strunnikov (1983), who observed high level of heterosis in F_1 hybrids of silkworm (*Bombyx mori*) where one of the populations exhibited genetically controlled low vitality. A hypothesis was drawn out that such populations might possess the so called "compensatory gene complex" (CGC) that might contribute not only to the survival and reproduction of these populations but also to heterosis for some traits in their F_1 hybrids.

An attempt of heterosis dissection was made using near-isogenic line TA 523 (*L. esculentum*) containing a 40-cM introgression at the bottom of chromosome 1 from *L. hirsutum* (Monforte and Tanksley 2000). A set of recombinant lines (sub NILs) derived from the original NIL TA 523 were developed in order to fine-map the genetic factors included within the original introgression. Analysis of the subNILs revealed that the gene action of the QTL for yield was dominant (d/a=0.7) which eliminated the possibility that yield increase was due to true overdominance at a single gene locus. On the other hand, negative yield effects in other regions of the introgressed segment that would be predicted by the dominance complementation model, were not detected. Epistatic interactions among genetic factors along the introgressed segment were suggested as the cause of yield heterosis.

STRATEGIES FOR DEVELOPING TOMATO HYBRIDS

The plant characters chosen as selection criteria depend on the goals of the breeding program. When the trait is characterized by high heritability, a direct selection is possible. Selection for polygenic characters (such as early yield total yield, flavor), influenced also by environment, is complicated. Hence, the development and the selection of parents in hybrid breeding programs can be difficult. One way to facilitate breeding for complex characters is to make them more amenable to improvement by determining and analyzing their components. Most of these components are also of a quantitative nature and may influence each other. According to Bos and Sparnaaij (1993), this not only causes SCA effects, it also causes problems in the identification of markers (phenotypic or molecular) for complex characters. In most cases there is no question of a single marker for complex character: in one genotype a high value for the complex character may be due to high level for component x_1, in a second genotype it may be due to a high level for x_2. Thus, one should look for marker genotypes corresponding with favorable levels of the important component traits, rather than for marker genotypes corresponding with high levels of the complex character.

Components that contribute to the performance of a given complex character in tomato, such as total and early yield, were investigated since the early studies on heterosis in this species (Powers 1945, Burdick 1954, Williams 1959, etc.) and are still largely evaluated till date as well as exploited in breeding programs (Szwadiak and Kordus 1992, Vallejo Cabrera and Estrada 1993, Rai et al. 1997, Wang et al. 1998, Doganlar et al. 2000 etc.). As already mentioned, number of fruits/plant, mean value of fruit weight, plant height etc., were determinant main components for yield.

Based on a number of studies, Doganlar et al. (2000) divided the determinants of earliness into four different components, each one being a heritable trait: 1) days from sowing/transplanting to the first flowering (anthesis); 2) days from anthesis to the first fruit set; 3) days from the first fruit set to the first ripe fruit; 4) days from the first ripe fruit to the end of ripening.

Besides determining the components of a given complex trait, knowledge on the relationships between them is necessary. One way to get such information is to establish correlation between characters. Correlation might be evaluated between components of a complex character: between complex character and one of its components; between components of different complex characters; between a given character and environment; and between physiological character and yield, etc. Based on a significant number of references, Andruchtchenko (1987) summarized data on correlation between: a) early and total yield, or yield components and biochemical characteristics of fruit quality; b) fruit weight/ firmness and biochemical characteristics of fruit quality; c) environmental factors (e.g. soil and atmosphere) and biochemical characteristics of fruit quality; d) content of fruit compounds and the taste; e) market quality (fruit longevity, firmness, cracking) and characteristics of fruit pericarp, weight, form, etc. Evaluation of relationships between productivity/growth rate and morphological/physiological characteristics of plants grown under unfavorable conditions might be useful in breeding genotypes tolerant to stress (Nieuwhof et al. 1993, Nkansah and Ito 1994, Matsunaga and Monma 2000). Knowledge on the significance of correlation (positive or negative) between two characters might not only help in indirect selection of a trait that is difficult to be controlled all over the breeding cycle, but can also be useful in foreseeing (at least partially) results in some breeding programs. A number of studies, for example, provided evidence that the gain in earliness might cause a reduction in fruit weight (Boe 1988, Banerjee and Kalloo 1989, Lindhout et al. 1994).

A significant number of studies on correlation between different traits in tomato also suggest that the knowledge on the relationship between characters of breeding interest could be useful and exploited in improving parental lines and hybrids. It must be noted, however that the reliability of the genetic correlation established between different traits has been found to vary and is influenced by genotype – environment interactions (Aastveit and Aastveit 1993). This peculiarity needs to be taken into consideration while designing selection strategies.

The real value of the lines designated for developing hybrids, however, is not based on their own performance *per se* but the performance of their F_1 hybrids. Therefore, the last and the most important step in breeding hybrids is the evaluation of the lines for their combining ability. Diallel sets of F_1 crosses between collections of tomato lines have long been used, and still are used, for selecting the best parents for F_1, and for identifying the best crosses from which to extract improved inbred lines. The biometrical analysis of Griffing (1956) provides information on the combining ability of the parents and the magnitude of additive and non-additive gene action. There is a significant number of studies that refer to general and specific combining ability (GCA and SCA) for a large scale of traits of breeding interest in tomato, such as yield and yield components (Rai et al. 1997, Wang et al. 1998), total soluble solids, dry matter (Dod et al. 1995, Wang et al. 1998, Shrivastava 1998), resistance to cuticle-cracking (Emmons and Scott 1998), etc. GCA and SCA can interact with the environment that would result in changes in parental combining abilities over the environments (Singh 1973). For this reason evaluation of parents in more than one environment is recommended. According to Shattuck et al. (1993), if conducting the diallel in only one environment, plant geneticists should attempt to match the diallel with the environment of interest.

The strategies above mentioned, although widely used in the development of tomato hybrids and their parental lines, are far from being perfect and obligatory for leading to the results expected. Moreover, they are costly because results relate to many years of tests and evaluations not only of the parental lines developed but also of their hybrids. For this reason, enhancing the efficiency of the process of developing parental lines, and hybrids, is and will be of primary importance in the future.

The advent of molecular markers provided tools for mapping genes involved in quantitative trait loci (QTL) and the possibility for plant genetic improvement based on molecular marker assisted selection (MAS). Molecular markers might be used both for trait identification and trait introgression. Therefore, their utility extends throughout all phases of breeding programs. In theory, MAS was shown to produce greater selection gains than phenotypic selection for normally distributed quantitative traits. Even without direct effect of the marker on the quantitative trait locus, detecting linkage between such traits is of interest for increasing the response in selection. Based on a model for estimating the probability of selecting one or more superior genotypes, using MAS, Knapp (1993) found that a breeder for phenotypic selection without applying MAS had to test 1.0 - 16.7 times more progenies than a breeder using MAS. It was concluded that MAS might substantially decrease the resources needed for a selection goal of a low to moderate heritability trait when both the selection goal and the selection intensity were high.

Tomato has been a model plant for QTL mapping. Several qualitative and quantitative trait loci in tomato have been mapped over the past decade, such as earliness and fruit ripening time (Lindhout et al. 1994, Monforte et al. 1999, Doganlar et al. 2000), total yield (Bernacchi et al. 1998), plant height and fresh mass (Paran et al. 1997), fruit size and shape (Grandillo et al. 1999, van der Knaap and Tanksley 2001), fruit firmness (Bernacchi et al. 1998), lycopene content (Chen et al. 1999), β-carotene accumulation in fruits (Zhang and Stommel 2000), organoleptic quality (Causse et al. 2001, Saliba Colombani et al. 2001), cold and salt tolerance during different developmental stages (Bretó et al. 1994, Foolad et al. 1998a, Foolad et al. 1998b), etc.

Molecular markers (such as restriction fragment length polymorphism (RFLP), amplified fragment length polymorphism (AFLP), random amplified polymorphic DNA (RAPD) and microsatellites (SSRs) are also expected to contribute to easier and more efficient identification of the lines whose crosses would result in promising hybrids. It was commonly assumed that hybrids produced from lines having different origin (i.e. developed from different cultivars) tended to have greater, consistent yield levels than hybrids of inbred lines originating from the same source population. The concept of heterotic groups gradually evolved from empirical evidence of crosses from inbred lines. Theoretically, the more distant the parents, the greater is the number of genes they differ, thereby the greater the potential interaction of the genes in the form of dominance and epistasis, and the greater will be potential for heterosis (Falconer 1989). In this sense genetic diversity might be an important issue in predicting F_1 performance. The assignment of tomato lines to heterotic groups before field testing may allow the breeder to avoid crosses within groups that would result in lower costs.

Genetic diversity can be measured by several means including pedigree data, genotypic origin from contrasting geographic regions, etc. although molecular methodologies are considered as the most reliable ones. The latter might contribute to increasing the accuracy in determining the divergence between the genotypes of interest, that could ultimately lead to improved classification. Comparison of molecular markers with pedigree data of related genotypes, based on theoretical and experimental results is, however, recommended (Melchinger 1993). The idea of exploiting the genetic distance of the parents as an indicator in the pursuit of heterosis in tomato hybrids is not a new one. Daskaloff (1942) reported high level of heterosis for yield and earliness in F_1 between tomato lines developed on the basis of hybridization between the tomato cultivar Sarya and *L. racemigerum* (Lange) (accession unknown), lately classified by Muller (Zhutchenko 1973) as *L. pimpinellifolium* Mill. It was also reported that lines developed on the basis of interspecific hybridizations were character-ized by high combining ability of economically important traits. Based on these findings it was concluded that for acquiring heterosis for early and total yield, the hybrids should include lines of different origins (Daskaloff 1955, 1967, Yordanov 1983).

Recently, a number of studies aimed at cultivar identification, determining relatedness, and comparing the magnitude and structure of genetic variation among different tomato accessions have been carried out. RFLP analysis of phylogenetic relationship and genetic variation in the genus *Lycopersicon* show that the ratio of within vs. between accession diversity was much lower for self-compatible species. It indicated that most of the diversity within these species existed between populations, rather than within populations. Overall the amount of genetic variation in the self-incompatible species far exceeded that found in self-compatible species (Miller and Tanksley 1990). Villand et al. (1998) used RAPD in estimating relationships between accessions collected from Old and New World regions. Differences in RAPD marker frequencies indicated uniqueness of accessions from the Old and New World collections. Accessions from Ecuador, Peru and Chile had a larger magnitude of marker diversity than Old World ones. Comparison of subpopulations of *L. esculentum* and its subspecies *L. esculentum* var. *cerasiforme* indicated that the two were distinct but had similar levels of diversity. Noli et al. (1999) reported results from RAPD analyses of modern and vintage cultivated tomato accessions and eight accessions of wild *Lycopersicon* species (*L. esculentum* var. *cerasiforme*, *L. pimpinellifolium* and *L. peruvianum*). Cluster analysis allowed *L. esculentum* to be clearly distinguished from the wild species. Within *L. esculentum* two major groups were identified, the first including all the fresh market varieties and vintage processing varieties and the second including most of the modern processing varieties. RAPD analysis did not distinguish cultivars indicated as synonyms or selected from the same standard variety. Bredemeijer et al. (1998) reported that sixteen tomato cultivars were DNA-typed for 20 selected microsatellite markers (STMS) using the fluorescent approach. Length polymorphism among the PCR products was detected with 18 of these markers, yielding gene diversity valued from 0.06 to 0.74. As few as four STMSs were sufficient to differentiate between the 16 cultivars, indicating that these markers were especially suitable for a species like tomato which had low levels of variation as detected by other types of markers. In tomato, a high-density AFLP map has been constructed using an interspecific population (Haanstra et al. 1999). Studies using AFLPs to assess genetic diversity among tomato cultivars demonstrated that AFLP markers were effective for obtaining unique fingerprints of, and assessing genetic diversity among, tomato cultivars (Park et al. 2004).

It has to be noted, however, that studies on the relationship based on marker assisted genetic distance of the parents and heterosis in maize, wheat, soybean, chickpea, oilseed rape and other crops show non-obligatory linearity to heterosis. It might be due to insufficient genome coverage because of the low number of marker loci. It might also suggest that genetic distance at the molecular level as determined by RAPD, RFLPs etc. in some cases could have a limited utility as indicator or predictor of heterotic performance. Several studies on the applicability of molecular markers provide evidence, however, that their application in plant breeding holds promise for increasing the accuracy of prediction of genotypic values (Tsaftaris 1995, Melchinger 1999, Hallauer 1999, Alvarez et al. 2001, Archak et al. 2002, Bredemeijer et al. 2002, He et al. 2003 etc.). According to Melchinger (1999), groupings of germplasm based on molecular marker information can provide the basis for establishing new heterotic groups or broadening the genetic base of existing ones. This must be supplemented by evaluating the performance of crosses among these subgroups to asses their heterotic response, which is essential for identifying heterotic patterns. By using examples from different crops Melchinger (1999) demonstrated that genetic distances based on molecular markers can be used for: a) revealing genetic relationship among different germplasm; b) assessing germplasm to groups and subgroups of similar materials, and c) detecting pedigree relatedness between germplasm.

The recent advent of molecular linkage maps might also make it possible to detect and individually analyze the loci underlying heterosis. The use of molecular markers to identify QTLs responsible for heterosis may help in improving the genetic gain in some selection breeding schemes.

IMPROVING THE PROCESS OF HYBRID SEED PRODUCTION

Tomato is a self-pollinated inbred crop, its flower is bisexual and hand emasculation and pollination are used for producing hybrid seed. Biological bases of hybrid seed production including flower morphology characteristics, anther emasculation, hybridization techniques etc. are reviewed and presented in detail by Yordanov (1983) and Georgiev (1991).

Significant quantities of hybrid seed are produced using a fertile seed parent. Tomato is considered a high value crop grown for either fresh market or processing. According to Duvick (1997b), seeding rates are low compared to the fruit value of this commercial crop. One noteworthy reason is that crossing is performed in countries where labor costs are very low. The quantum of research being carried out to improve the process of hybrid seed production suggests that the present technology is not perfect and does not give itself to easy adaptation to economic needs.

The benefit of incorporating male sterility into hybrid breeding programs was recognized not long after the appreciation of the advantages of heterosis and the detection of male sterile genotypes in tomato. For the first time male sterility was used in tomato hybrid seed production by Rick (1945) and till date this phenomenon is still discussed as the most promising way for facilitating the process of hybrid seed production (Sawhney 1994, 1997, 2004, Gorman and McCormick 1997, Atanassova 2000).

Genetic male sterility occurs widely in *Lycopersicon esculentum,* since cytoplasmic male sterility (CMS) does not occur naturally in the genus *Lycopersicon.* Georgiev (1991) defined the male sterility in tomato as autosterility and classified the sterile mutants into two groups: *male sterile* [including the male sterile (*ms*) and *stamenless* (*sl*) series] and functional sterile [including *positional sterile* (*ps*), *positional sterile 2* (*ps 2*), and exserted stigma (*ex*)]. Based on anther development and the phenotype, the male sterile mutants in higher plants were classified into structural, sporogenous and functional types (Kaul 1988; Fig. 4.1, Fig. 4.2, Fig. 4.3, Fig. 4.4 and Fig. 5). The majority of the mutants belong to the *male sterile* (*ms*) or *stamenless* (*sl*) series, while the frequency of the mutations controlling functional sterility is lower (Kaul 1988, Sawhney 1994, Gorman and McCormick 1997, Atanassova et al. 2001).

Over the years it was established that several requirements such as complete male sterility controlled by recessive gene, normal female fertility with no defects in morphology, and stability of sterility expression etc., must be met by designated male sterile plants to be used in breeding programs (Gorman and McCormick 1997, Atanassova et al. 2001). Contrarily, it has been found that each type of male sterility in tomato exhibited not only significant advantages but also significant disadvantages when used in hybrid seed production, (Table 4.3). This table elaborates on several approaches for correcting the disadvantages of using male sterility

Fig. 4.1 Sporogenous (*ms 10*) sterility in tomato. Flowers with exserted and non-exserted stigma

Fig. 4.2 Structural (*sl*) sterility in tomato.

Fig. 4.3 Functional (*ps*) sterility in tomato.

Fig. 4.4 Functional (*ps 2*) sterility in tomato. Longitudinal section of tomato anther cones; on the left: *ps 2*-indehiscent anthers; on the right: fertile anthers

Fig. 4.5 Functional (*ex*) sterility in tomato. On the left: fertile flower (P_2); on the right: *exserted stigma* sterile flower (P_1); in the middle: flower with exserted stigma (F_1)

in tomato hybrid seed production. Despite the numerous attempts of testing almost all these approaches the number of cultivars possessing male sterile seed parent remains rather limited. In our knowledge, (although our

Table 4.3 *Advantages and disadvantages in using different types of male sterility in tomato hybrid seed production*

Sterility	Advantages	Disadvantages	Approaches for correcting or escaping disadvantages
Sporogenous	1. Elimination of the process of emasculation (stigma accessible because of anther deformations) 2. Complete sterility. 3. Stable expression of sterility independent of the environmental conditions. 4. Complete restoration of fertility in F_1 5. Easy transfer of sterility genes to any genotype	1. Maintenance of the line as a population of sterile and fertile plants. Sterile plants could be assessed at anthesis 2. Occurrence at some periods of high percentage of flowers with non exserted stigma that would require hand emasculation	1. Exploitation of closely linked markers expressed before flowering (Philouze 1974, Durand 1981, Tanksley and Zamir 1988). 2. Temporary restoration of fertility by chemical (Shmidt and Shmidt 1981, Ma et al. 1999), or environmental (Masuda et al. 2000) treatment. 3. *In vitro* micropropagation of male sterile lines
Structural	The same as in the sporogenous mutants	1. Maintenance of the line as a population of sterile and fertile plants. Sterile plants could be assessed at anthesis	1. Restoration of male fertility to produce pure male sterile seed by hormonal or environmental treatments (Sawhney 1997)
Functional	1. Easy maintenance by artificial selfing 2. Elimination of the process of emasculation (valid for *exserted stigma* genotypes)	1. Necessity of stamen emasculation (valid for *ps* and *ps 2* sterility in tomato) 2. Undesirable selfing	1. Combining *ps* or *ps 2* with exserted stigma 2. Combining *ps* or *ps 2* with short style (Georgiev and Atanassova 1981) 3. Use of *ps 2 ful* recombinants for easier breeding of *ps 2* lines, (Atanassova 1991) 4. Use of RAPD markers linked to *ps* gene for easier breeding of *ps* lines (Staniaszek et al. 2000)

information might be incomplete) only *male sterile 10^{35} anthocyanin absent* *(ms10^{35} aa), positional sterile (ps)* and *positional sterile 2 (ps 2)* seed parents have found application in release of a number of commercial hybrids during the last two decades (Atanassova, 1999, Staniaszek et al., 2000).

The *male sterile (ms)* and *stamenless (sl)* mutants seem to be the most applicable in breeding programs aimed at the facilitation of hybrid seed production mainly because of their complete male sterility and accessible stigma (Stevens and Rick 1986, Sawhney 1994, Gorman and McCormick 1997). Because of anther deformation, some *ms* mutants such as *ms 10, ms 15, ms 32* exhibit exserted stigma, and are therefore accessible for pollination without emasculation. By developing *ms10^{35}aa* genotypes, the main disadvantage in using *ms*-sterility in hybrid seed production (i.e., assessment of sterile plants at anthesis) was eliminated (Philouze 1974). The anthocyaninless sterile plants are easy to be distinguished since early developmental stages. Moreover, no effect of genes *ms10^{35}aa* on plant and fruit characteristics was established (Gardner 2000). This technology might be applied also if using *ms 15 anthocyanin without (aw)* genotypes as the two genes are closely linked (Clayberg 1965). According to Jorgensen (1987), tight linkages may be synthesized also through genetic transformation, which would allow introduction of an appropriate marker gene to a random location in each of a large number of plants carrying a suitable *ms* gene.

Studies on *ms* mutants and *exserted stigma* manifestation provided evidence, that the latter, usually a beneficial character in hybrid seed production, might sometimes be harmful. First of all, it is important to note that these mutants were characterized by lower hybrid seed yield (Atanassova 1999), which was suggested probably due to the fast drying of the stigma (Georgiev 1991). Secondly, observations on stigma exsertion variability within a number of *ms* sterile lines developed at the Station des Plantes Maraîchères, INRA, Monfavet, Avignon, France, showed that the percentage of flowers with accessible stigma strongly varied depending on the environment and on the genotype (Table 4.4, Fig. 4.1). In some genotypes (and this during the period of hybrid seed production), the percentage of flowers without exserted stigma was so high, that for using them in commercial hybrid seed production, anther emasculation would become necessary. These results are consistent with those of Levin et al. (1994). Based on a study of the effects of the *ms 10* gene, polygenes, and their interaction on pistil and anther-cone length, it was concluded that emasculation of the *ms 10* male sterile parent appeared to be unavoidable for the efficient production of hybrid seeds. Removal of the shrunken *ms* sterile anthers was found, however, to be more difficult and required a

Table 4.4   *Manifestation of non-exserted stigma in male sterile tomato lines observed in the experimental fields of the Institute of Genetics, Sofia, during the period of hybrid seed production (May 25 - July 5, 2001)*

Genotype	*Percentage of flowers with non-exserted stigma ± SE**				
	May, 25th	*June, 5th*	*June, 15th*	*June, 25th*	*July, 5th*
Monalbo *ms 10*	3.2 ± 0.2	27.4 ± 5.7	64.7 ± 8.2	87.5 ± 11.0	16.6 ± 2.1
Monalbo *ms 32*	10.0 ± 1.3	33.2 ± 4.9	63.0 ± 9.4	69.6 ± 7.4	21.5 ± 3.4
Porphyre *ms 10*	11.4 ± 3.0	44.0 ± 4.1	59.6 ± 7.6	92.4 ± 6.8	90.2 ± 5.5
Porphyre *ms 32*	9.5 ± 1.7	41.4 ± 7.6	87.1 ± 8.3	74.2 ± 5.9	56.4 ± 6.7
Monfavet 167 *ms 10*	1.5 ± 0.2	13.7 ± 2.8	1.8 ± 0.1	14.0 ± 0.0	9.2 ± 2.2
Monfavet 167 *ms 32*	6.6 ± 1.3	3.9 ± 0.8	19.6 ± 4.1	3.4 ± 0.5	5.8 ± 1.8

SE* - standard error between means of replicates

longer time than emasculation of fertile flower buds (Atanassova 1999). Recent technologies have made it possible to identify and isolate *ms* genes and to engineer transgenic male sterile plants (Gorman et al. 1996) as well to chemically induce male sterility (Sakaki and Yamazaki 1990, Cross and Schultz 1997), this particularity should be outlined and taken into consideration. Chen and Tanksley (2001) reported that the fine mapping of stigma exsertion QTLs se2.1 revealed that se2.1 was located in the chromosomal interval between RFLP markers T1301 and T 662 and that the change of style length might be the function of se2.1 genes. This finding might act as tool for more efficacious manifestation of stigma exsertion.

According to Bar and Frankel (1993), some *ms* mutants (*ms 14, ms 17, ms 18, ms 31, ms 33, ms 47*) were found to exercise pleiotropic effect on a number of economically important traits such as percentage of early marketable yield, average fruit weight, and total marketable yield. This suggests that detailed studies on *ms* sterile lines are necessary before including them in breeding programs.

Functional male sterile mutants offer the advantage of reproduction by artificial selfing in order to produce 100% sterile progeny. Despite this advantage they are considered as less promising for use in hybrid seed production because they display two significant disadvantages - occasional lapses in their expressivity resulting in undesirable selfing and except in *exserted stigma*-sterility, necessity of stamen emasculation. Observations on a large number of *ps 2* sterile lines throughout their manipulation in breeding and hybrid seed production have shown, however, that it was possible to deal with the first disadvantage (lapses in *ps 2* expressivity) by taking into consideration some specific characteristics related to their performance (Table 4.5; Atanassova 1999, Atanassova et al. 2001).

Table 4.5 *Ways of dealing with the lapses in* ps 2 *gene expressivity and its consequences in breeding and hybrid seed production*

Specific characteristics in the performance of ps 2 lines to be taken into consideration when using ps 2 sterility in breeding and hybrid seed production.	Approaches for dealing with specific characteristics in the performance of ps 2
1. The *ps* 2 gene expressivity varies with the genotypes	This enables the breeding of *ps* 2 lines that exhibit very low percentage of selfing.
2. The percentage of selfing in the *ps* 2 lines varies within and between the years of growing, being forever lowest during the period of hybrid seed production.	Strict control of this characteristic is necessary throughout the entire breeding process. The percentage of selfing recorded at the end of the growing season, instead at the end of the period of producing hybrid seed, might give a wrong idea on the use fulness of the *ps* 2 line.
3. If occurring, anther dehiscence in the *ps* 2 lines occurs usually after the second day of flowers opening.	Regular emasculation and pollination of the plants at least each two days is necessary. If missing this term, all flowers at anthesis have to be eliminated.
4. The percentage of selfed seeds when using a *ps* 2 line as seed parent in producing hybrid seed is significantly lower than the percentage of selfing, observed on the same line.	The usefulness of a *ps* 2 line has to be evaluated on the basis of both, percentage of selfing and hybridity of the seeds obtained by artificial pollination.

The need of stamen emasculation is the second disadvantage that usually prevents breeders from using *ps* 2 sterility in breeding programs. Comparative study on the time necessary for the emasculation of floral buds in the fertile plants, and flowers at anthesis in the sterile lines, made it clear that this disadvantage, in terms of limiting factor, was exaggerated. Emasculation during anthesis (as practiced when using a *ps* 2 line as seed parent) was easier and almost twice as rapid as emasculation applied on the fertile floral buds (Atanassova 1999). Anther emasculation could be made even easier if *ps* 2 was combined with short style (Georgiev and Atanassova 1981; Fig. 4.6). Such flowers can be emasculated without using forceps: the anther cone and the petals can be easily separated manually from the flower by plucking out with two fingers the tip of the cone (or the petals) (Fig.4.7). Certainly, the idea of developing genotypes combining *ps* 2 sterility with *exserted stigma* is a tempting one as it would result in efficacious technology: easy maintenance of the seed parent by artificial selfing and no need of anther emasculation. Such a technology, however, could be acceptable only by finding a way to deal with the strong variability in the rate of stigma exsertion. While testing *ex*-lines for their usefulness in

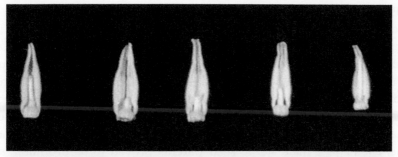

Fig. 4.6 Longitudinal section of tomato flowers with normal and short style.

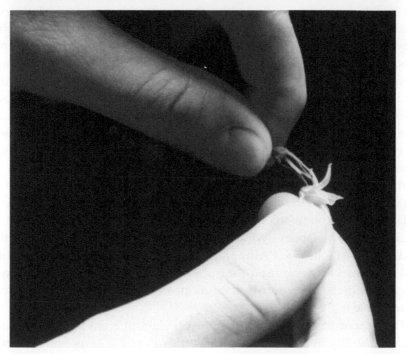

Fig. 4.7 Emasculation in tomato *ps 2-* line possessing relatively low level stigma without using forceps

hybrid seed production it was established that it was really difficult to determine and fix the right rate of stigma exsertion. On the one hand, lines possessing 1.0-1.5 mm stigma exsertion were found occasionally to be like normal ones requiring stamen emasculation. On the other hand, F_1 hybrids of the lines possessing steadily manifested exserted stigma (2.0 mm or more above the anther cone), performed sometimes as longuistylic, that resulted in lower percentage of fruit setting (Fig. 4.5).

Easier and more rapid anther emasculation is not the only criterion for an efficient hybrid seed production. Hybrid seed yield is also of great importance. Comparative studies on hybrid seed yield obtained from *ps 2* sterile lines, depending on the developmental stage of stamen emasculation and pollination, showed that a significantly higher (1.5 to 3 times) hybrid seed yield resulted from pollination at anthesis. This is the stage in which the *ps 2*-lines are usually manipulated for producing hybrid seed (Atanassova and Georgiev 2002). Therefore, in terms of expenditure and time, use of *ps 2* male sterile seed parents proved to be profitable and economically justifiable (Atanassova 1999, Atanassova and Georgiev 2002). These data were also confirmed by the production figures. In Bulgaria, until 1990, about 1 ton of hybrid seed per year (30% of the total quantity of hybrid seed produced) was produced using *ms 10 aa* and *ps 2* seed parents (Georgiev 1991), while presently about 80% of the hybrids released and spread are practically from *ps 2* seed parent. Hybrids using *ps 2* sterile seed parent were released also in Czech Republic and Moldova (Atanassova 1999, 2000). It suggests that the functional, and more precisely *ps 2* male sterility, is not to be underestimated. It should be taken into consideration for elaborating systems or breeding programs aimed at the reduction of the time and costs associated with hybrid seed production.

According to Potaczek and Kubicki (1986) and Staniaszek et al. (2000) the use of *positional sterility (ps)* in tomato hybrid seed production under Polish climatic conditions was advantageous. The process was found to be cheaper due to labour-saving procedures up to 30% as compared to the traditional method. Two RAPD markers linked to the *ps* gene were identified. The markers were used for purity determination of maternal lines carrying gene *ps* gene and F_1 hybrids possessing *ps* sterile seed parent (Staniaszek et al., 2000).

The experience accumulated while applying different types and genes of male sterility in the practice (at least three - $ms10^{35}aa$, *ps* and *ps 2*) has shown that they should not be categorized as "more or less promising". Sometimes a given type of sterility might initilly look promising but during its manipulation in breeding and hybrid seed production unexpected difficulties may crop up (e.g. the occurrence at times of a high percentage of flowers with non-exserted stigma in some *ms* tomato lines). Converselly, some mutants or sources of sterility might display undesirable traits, but while using some approaches they could be corrected or eliminated, (e.g. the *ps 2* sterility in tomato).

Thus, the application of male sterility in breeding and hybrid seed production in tomato is not merely a theory, but is being practised in a number of countries. Nevertheless, it is difficult to determine how widely cultivars possessing male sterile seed parent are spread, or what is their percentage based on the total number of hybrid varieties, as it is difficult to track data on the release of varieties possessing a male sterile seed parent. The availability of such hybrids is, however, a fact not to be neglected. It confirms that induction of male sterility is one of the right ways for facilitating hybrid seed production in tomato and a start has been made in this direction.

Economizing hybrid seed production includes not only facilitating its process *per se*, but also increasing its efficiency by improving the quality of the final product, i.e. of the hybrid seed. Besides the high germination ability which is an obligatory characteristic for each kind of commercial seed, the high percentage of hybridity is of primary importance in hybrid seeds. For this reason, breeders are tempted to introduce male sterile seed parents to facilitate the production of hybrid seed. Using male sterile seed parent would be a warranty for production of 100% hybrid seed and would eliminate the necessity of testing the seed for hybridity. In view of the fact that male sterile seed parents could not be widely used in developing tomato hybrids, different morphological markers such as *potato leaf* (*c*), *anthocyaninless of Hoffmann* (*ah*), *anthocyanin without* (*aw*) etc. were introduced in fertile or *ps 2* seed parents of a number of commercial hybrids (Farkas 1993, Xue 1994, Atanassova et al. 2001). It permitted testing the hybridity of the seed at germination or seedling stage. Recently, some molecular techniques made it possible to rapidly evaluate the purity of the hybrid seed produced (Rom et al. 1995, Paran et al. 1995, Chuang et al. 1999) and probably in the near future these techniques will be widely used for this purpose.

TOMATO HYBRID ADOPTION IN PRACTICE—HISTORY AND PRESENT SITUATION

As already mentioned, studies on heterosis in tomato were initiated almost simultaneously with such studies on maize. The introduction of tomato hybrids, however, came into practice 30-35 years later than maize hybrids. The first commercial tomato hybrids were developed much earlier in some countries. In USA, for example, the first hybrid "Burpee Hybrid" was developed by Dr. Oved Shifriss in the early 1940's, probably 1942, and marketed by the W. Atlee Burpee Co. (Prof. Rick C.M., personal communication). The first tomato hybrid in Bulgaria (Saria x Komet) was developed in 1932 by Prof. Daskaloff, (Daskaloff, 1937) at the Agricultural

Experimental Station (now Institute of Vegetable Crops "Maritza"), Plovdiv, where for the first time large quantities of hybrid seed were produced. Since 1949 and up to the late 1960's, for example, the number one cultivar for early field production was the hybrid No. 10 x Bizon, developed also by Daskaloff. In Japan one of the first hybrids "Fukuju" was developed by Prof. Fujii (Fujii 1948, 1952) and released in 1938 by Osaka Agricultural Station, (now Osaka Prefectural Agricultural and Forestry Research Station). The first Dutch tomato hybrid Single Cross (Vetomold x Ailsa Craig), combining traits of resistance to *Cladosporium fulvum*, was developed by Bruinsma in 1939 and introduced in 1946. The hybrid exceeded the other varieties by taste and earliness (Anonymous 1992). In France, the first hybrid Fournaise F_1 was developed by Vilmorin and released in 1956 almost simultaneously with the hybrids Monfavet n° 63-5, Monfavet n° 63-4, and Monfavet n° 63-18 developed by INRA, (Philouze 1986, 1997). In the early 60's these hybrids enjoyed a significant success and were rapidly adopted by the producers. In Israel, the first tomato hybrid "Urit" was developed at Volcani Center and released in 1971 (Pilowski et al. 1971). In China, the first tomato hybrid "Beijing Zhaohoug x Aonong No 2" was developed in 1969 at the Zheijiang Agricultural University (Wang et al. 1988).

Tomato hybrids began to take over the market towards the end of sixties or the beginning of seventies. Their use increased dramatically throughout the following decades. In 1997, Duvick (1997 b) reported that 100% of fresh market and 80% of processed tomatoes in USA were F_1 hybrids. At present, there is a similar situation in several countries in Europe, Asia and Australia.

FUTURE PROSPECTS AND CHALLENGES IN EXPLOITATION OF HETEROSIS IN TOMATO

The goal of the breeders is to develop hybrids superior in one or a number of traits to the standard hybrid or hybrids previously released and used in practice, and most of them do not develop heterotic hybrids *per se*. In a way each hybrid might be considered as a challenge for developing new, superior hybrids. Meanwhile, consumers and growers demand as well as handling requirements have dramatically changed during the past two decades. Increasing yield, for example, is not any more the major focus of commercial tomato breeding programs. It has shifted to breeding for a complex of traits, such as improved quality, flavor, more efficacious plants coping with abiotic and biotic stresses for at least a substantial part of their life, etc. The longevity of fruit, once considered as a trait of primary importance, is not now appreciated if not combined with good flavor and/or texture. Hence, the new hybrids developed have to be superior to the previously released ones in terms of traits that the latter already possess plus permanently cumulate new valuable agronomic traits in order to satisfy the demand of consumers for high nutritive quality and food safety, and of growers - for economic profitability and handling requirements. To achieve these objectives plant breeding research should focus on improving fruit flavor, texture, composition and studying the potential of plants to synthesize desirable components by developing new methods that lead to accelerated screening, adjusting the precision of selection stages, minimizing costs etc. The achievement of such ambitious, complex and perhaps even challenging future breeding goals would be difficult (if not impossible) without the extensive use of the phenomenon of heterosis. As already mentioned, heterosis in tomato may be expressed at any developmental stage and observed for a large scale of quantitative traits related to plant productivity, adaptability, physiology and fruit nutritive and market quality. It is commonly known that the phenomenon of heterosis was widely used in developing early and high yield commercial hybrids. It is rather doubtful (as it would be difficult to trace down such information) that heterosis was widely pursued in breeding programs for traits, other than earliness, total yield and yield components.

Hence, given that tomatoes are one of the most important crops in the world because of their volume of consumption, and overall contribution to nutrition, and bearing in mind that the new developed varieties have to satisfy complex consumer and grower demands as well as handling requirements, the manifestation of heterosis for traits related to plant tolerance to biotic and abiotic stresses, photosynthetic efficiency, nutritive and market quality, efficient root system, etc., may be considered as a reserve to be exploited in future breeding programs.

The recent advances in the molecular genetics (tagging and isolation of genes, QTL controlling a given trait, expression of desirable alien genes in transgenic plants, improving the efficiency of breeding via marker assisted selection, etc.) is expected to contribute to the more efficient and extensive exploitation of heterosis in developing tomato hybrids. Combining conventional and molecular breeding techniques might offer help in improving screening efficiency for many traits of agronomic value, estimating genetic diversity, reducing the time for new line development, assessing heterotic groups, and detecting as well as individually analyzing the loci underlying heterosis. The rapidly increasing number of investigations aiming at identification and isolation of male sterile genes in tomato, engineering transgenic male sterile plants, synthesizing tight linkages between an appropriate marker gene and male sterile gene through genetic transformation might also generate useful approaches contributing to the creation of better systems of hybrid seed production. The fact that the application of male sterility in breeding and hybrid seed production in tomato is no longer a theoretical one, gives ground to believe that such systems would be easily accepted and applied in practice.

The prospects and expectations outlined above need to be taken cautiously. According to Young (2000), although DNA markers hold great promise, realizing this promise remains elusive, as most markers associations are not significantly successful in MAS. At the same time, the effectiveness of MAS in breeding programs would depend also on the genetic determinism of the traits of interest (Hospital et al. 1997) and on its cost (Young 2000). Kearsey and Farquhar (1998) consider that unreliability of QTL location may suggest its false candidacy. Therefore, the new knowledge gained creates new problems to solve, such as developing reliable molecular markers, refining the techniques, decreasing the cost associated with molecular markers assays etc. The capabilities of recently developed molecular techniques suggest their wide application in breeding programs and in the genetic improvement of cultivated plants, including exploitation of heterosis.

In 1908, Shull noted that the efficiency of maize breeding programs would considerably improve by finding a suitable method of predicting hybrid performance before field evaluation. The numerous studies on heterosis and the large experience of exploiting this phenomenon over an entire century, complemented by the rapid advent of molecular techniques during the last decade, make it possible to believe that this endeavor might be on the way to come true.

SUMMARY

Studies on heterosis in tomato were initiated at the beginning of the twentieth century. Although its biological basis remains unknown, this phenomenon is now recognized as one of the primary factors contributing to the success of plant breeding in tomato and many other crops.

This subject is surveyed under the headings: **1.** Introduction; **2.** Manifestation of heterosis in tomato—Heterosis in tomato is observed for a large scale of economic quantitative traits such as total and early yield and yield components, tolerance to stress, dry matter, vitamin C, lycopene content etc. More than 50-60% of the studies on heterotic performance in tomato refer, however, to heterosis for yield and yield components; **3.** Genetic basis of heterosis in tomato–Manipulating heterosis in breeding programs requires knowledge on its quantitative genetic basis. Data concerning gene action for tolerance to abiotic stresses and fruit nutritional value, sensory and market quality is summarized, and recent studies of heterosis in tomato are reviewed; **4.** Strategies for developing tomato hybrids–Making complex characters more amenable to improvement by determining their components as well as relationship between them is analyzed as one way to facilitate breeding for complex characters. Possibilities for tomato genetic improvement based on mapping genes involved in quantitative trait loci (QTL) and more efficient identification of the lines whose crosses would result in promising hybrids, based on molecular markers, are reviewed and discussed; **5.** Improving the process of hybrid seed production–Advantages and disadvantages in incorporating different types of genic male sterility into hybrid breeding programs are discussed; **6.** Tomato hybrid adoption in ptactice–history, present situation and information concerning the development of the first tomato hybrids in USA, France, Holland, Israel, Japan, Bulgaria, China is presented; and **7.** Future prospects and challenges in exploitation of heterosis in tomato discussed. Achievement of the ambitious and complex breeding goals in tomato improvement will be difficult without the extensive use of heterosis. Combining conventional and molecular breeding techniques might offer help in improving screening efficiency for many traits of agronomic value, estimating genetic diversity, reducing the time for line development, assessing heterotic groups, detecting and individually analyzing the loci underlying heterosis.

Acknowledgments

Many thanks to R. Chetelat, University of California, Davis and S. Daskalov, Institute of Genetics, Sofia for their critical reading and helpful comments on the manuscript. Thanks to H. Laterrot, (INRA, Monfavet, France), Li Junming (IVF, Beijing, China), N. V. Marrewijk (PRI Wageningen, Netherland), M. Friedmann (Volcani Center, Israel), H. Egashira (Faculty of Agriculture, Yamagata University, Tsuroka, Japan) for their assistance in providing information included in the paper.

REFERENCES

Aastveit, A.H. and K. Aastveit. 1993. Effects of genotype-environment interactions on genetic correlation. Theor Appl Genet 86, 8:1007-1013.
Al-Falluji R.A., D.H. Trinklein, and V.N. Lambeth. 1982. Inheritance of pericarp firmness in tomato by generation mean analysis. HortScience 17, 5:763-764.

Allard, R.W. 1996. Genetic basis of the evolution of adaptedness in plants. Euphytica 92:1-11.

Amaral Jùnior, A.T. Do, V.W.D. Casali, F.L. Finger, and R. F. Daher. 1997. Heterosis in tomato for content of carotenoids with medicinal end-use. SOB Informa 15/16, 1:20.

Amaral Jùnior, A.T. Do, V.W.D. Casali, C.D. Cruz and F.L. Finger. 1999. Genetic inferences on yield and quality of tomato in a diallel cross. Pesquisa Agropecuària Brasileira. 34, 8:1407-1416.

Andruchtchenko, V.K. 1987. Genetic approaches for quality improvement of vegetables. (In Russian: Selektzionno-geneticheskie metodi uluchenya kachestva ovochtey). Chtinza, Kicinev, Moldova.

Anonymous, 1992. In: Dutch Horticultural Seeds Association. [eds.], Twee eeuwen tuinbouwzaden (Two centuries horticultural seeds), p. 76.

Alvarez, A.E., C.C.M. van de Viel, M.J.M. Smulders and B. Vosman. 2001. Use of microsatellites to evaluate genetic diversity and species relationship in the genus *Lycopersicon*. Theor Appl Genet 103:1283-1292.

Archak, S., J.L. Karihaloo and A. Jain. 2002. RAPD markers reveal narrow genetic base of Indian tomato cultivars. Curr Sci 82:1139-1143.

Atanassova, B. 1991. Linkage studies of the *"positional sterility-2"* mutant in tomato. J Genet. & Breed 45:293-296.

Atanassova, B. 1999. Functional male sterility (*ps 2*) in tomato (*Lycopersicon esculentum* Mill.) and its application in breeding and hybrid seed production. Euphytica 107:13-21.

Atanassova, B. 2000. Functional male sterility in tomato (*Lycopersicon esculentum* Mill) and its application in hybrid seed production. Acta Physiol Plantarum 22, 3:221-225.

Atanassova B., E. Balacheva, E. Molle, and Hr. Georgiev. 2005. Genetic study on the prolonged fruits longevity in tomato (*Lycopersicon esculentum* Mill.). Proc. of the XV Meeting of the Eucarpia Tomato Working Group, September 20-24, Bari, Italy.

Atanassova, B., S. Daskalov, and V. Nikova. 2001. Male sterility in three Solanaceae genera (*Capsicum, Lycopersicon, Nicotiana*) and its application in breeding and hybrid seed production. In: R.G. Berg van den, G.W.M. Barendse, G.M. van der Weerden, and C. Mariani [eds.], Solanaceae V. Advances in Taxonomy and Utilization. Nijmegen University Press, Netherlands, pp. 349-361.

Atanassova, B., S. Daskalov, L. Shtereva, and E. Balatcheva. 2001 Anthocyaninless mutations improving tomato and pepper tolerance to adverse climatic conditions. Euphytica 120:357-365.

Atanassova, B. and Hr. Georgiev. 2002. Using genic male sterility in improving hybrid seed production in tomato(*Lycopersicon esculentum* Mill). Acta Hort 579:185-188.

Atanassova, B., L. Shtereva, and E. Balatcheva E. 2002. Estimation of heterosis for productivity and early yield in F_1 hybrids of tomato (*Lycopersicon esculentum* Mill) mutants differing in their vitality. Acta Hort 579:45-48.

Banerjee, M.K. and G. Kalloo. 1989. The inheritance of earliness and fruit weight in crosses between cultivated tomatoes and two wild species of *Lycopersicon*. Plant Breed 102:140-152.

Bar, M. and R. Frankel 1993. Pleiotropic effects of male sterility genes in hybrid tomatoes (*Lycopersicon esculentum* Mill). Euphytica 69:149-154.

Barabàs, Z. 1992. A new era in the production of hybrid varieties? Hungarian Agric Res 1:17-21.

Bernacchi, D., T. Beck-Bunn, D. Emmatty, Y. Eshed, S. Inai, J. Lopez, V. Petiard, H. Sayama, J. Uhlig, D. Zamir D. and S. Tanksley. 1998. Advanced backcross QTL analysis of tomato. II. Evaluation of near-isogenic lines carrying single-donor introgressions for desirable wild QTL-alleles derived from *Lycopersicon hirsutum* and *L. pimpinellifolium*. Theor Appl Genet 97, 1/ 2:170-180.

Bhatt, R.P., V.R. Biswas, H.K. Pandey, G.S Verma, and N. Kumar. 1998. Heterosis for vitamin C in tomato (*Lycopersicon esculentum*). Indian J Agric Sci 68, 3:176-178.

Boe, A.A. 1988. Effect of using ultra early tomato lines as seed parents on the earliness of F_1 hybrid lines. HortScience 23, 3-1:452.

Bos, I. and L.D Sparnaaij. 1993. Component analysis of complex characters in plant breeding. II. The pursuit of heterosis. Euphytica 70, 3:237-245.

Bredemeijer, G.M.M., P. Arens, D. Wouters, D. Visser, and B. Vosman. 1998. The use of semi-automated fluorescent microsatellite analysis for tomato cultivar identification. Theor Appl Genet 97, 4:584-590.

Bredemeijer, G.M.M., R.J. Cooke, M.W. Ganal, R. Peters, P. Isaak et al. 2002. Construction and testing of microsatellite database containing more than 500 tomato varieties. Theor Appl Genet 105:1019-1026.

Bretó, M.P., M.J Assíns, and E.A. Carbonell. 1994. Salt tolerance in *Lycopersicon* species. III. Detection of quantitative trait loci by means of molecular markers. Theor Appl Genet 88, 3/ 4:395-401.

Burdick, A. !954. Genetics of heterosis for earliness in the tomato. Genetics 39: 488-505.

Causse, M., V. Saliba-Colombani, I. Lesschaeve, and M Buret. 2001. Genetic analysis of organoleptic quality in fresh market tomato. 2. Mapping QTLs for sensory attributes. Theor Appl Genet 102:273-283.

Chen, F.Q., M.R. Foolad, J. Hyman, D.A. St.Clair, and R.B. Beelaman. 1999. Mapping of QTLs for lycopene and other fruit traits in a *Lycopersicon esculentum* x *L. pimpinellifolium* cross and comparison of QTLs across tomato species. Molec Breed 5, 3:283-299.

Chen, K.-Y. and S.D. Tanksley. 2001. Fine mapping of stigma exsertion QTLs se2.1. Report Tomato Genet Coop 51:15-16.

Chen, Q.S. and Y.W. Zhao. 1990. Study on genetical effects on four characteristics of tomato. J Jiangsu Agric College 11, 4:33-38.

Chuang, S.J., H.H. Wang, and T.K. Hu. 1999. Purity identification of F_1-hybrid tomato (*Lycopersicon esculentum* Mill.) cultivars by RAPD markers. J Agriculture and Forestry 48, 2:103-112.

Clayberg, C.D. 1965. A linked seedling marker for *ms 15*. Report Tomato Genet. Coop. 15:29.

Cooper, M. and D.W. Podlich. 1999. Genotype x environment interactions, selection response and heterosis. *In:* J.G Coors and Shivaji Pandey [eds.], The Proceedings of an International Symposium "The Genetics and Exploitation of Heterosis in Crops", CIMMYT, Mexico City, Mexico, 17-22 August, 1997. Madison, USA, American Society of Agronomy, pp. 81-92.

Cross, J.W. and P.J. Schultz. 1997. Chemical induction of male sterility. *In:* K.R. Shivanna and V.K. Sawhney [eds], Pollen Biotechnology for Crop Production and Improvement. Cambridge University Press, Cambridge, UK, pp. 218-236.

Dane, F., A.G. Hunter, and O.L. Chambliss. 1991. Fruit set, pollen fertility and combining ability of selected tomato genotypes under high-temperature field conditions. J Am Soc Hort Sci 116, 5:906-910.

Daskaloff, Ch. 1937. Beitrag zum studium der heterosis bei den tomaten in bezug auf die herstellung von heterosis sorten fur die praxis. Die Gartenbauwissenschaft, XI, 2:129-143.

Daskaloff, Ch. 1942. Ergebnisse aus Kreuzungen: Sol. Racemigerum x Sarya und Plowdiwer. Züchter SIV/5:105-111.

Daskaloff, Ch. 1955. Die Heterosis und ihre Ausnutzung in Gemüsebau. Dtsch Landwirtsch. 8:2-6.

Daskaloff, Ch., M. Konstantinova, E. Molle, and D. Baralieva.1990. Genetic studies on tomato quality. Bulg. Academy of Sciences Press, Sofia, Bulgaria.

Daskaloff, Ch., M. Yordanov, and A. Ognianova. 1967. Heterosis in tomatoes. Bulg. Academy of Sciences Press, Sofia, Bulgaria.

David, P. 1997. Modeling the genetic basis of heterosis: tests of alternative hypotheses. Evolution 51, 4:1049-1057.

Deng H.W., Y.X. Fu, and M. Lynch. 1998. Inferring the major genomic mode of dominance and overdominance. Genetica 102/103:559-567.

Dhaliwal, M.S., S. Singh, and D.S. Cheema. 2000. Estimating combining ability effects of the genetic male sterile lines of tomato for their use in hybrid breeding. J Genet Breed 54 (3): 199-205.

Dobhal, V.K., U.K. Kohli, and D. Mehta. 1999. Genetic analysis of fruit firmness and related traits in tomato. J Hill Res 12, 1:31-33.

Dod, V.N. and P.B. Kale. 1992. Heterosis for certain quality traits in tomato *Lycopersicon esculentum* Mill. Crop Res. (Hisar) 5, (2):302-308.

Dod, V.N., P.B. Kale, and R.V. Wankhade. 1992. Genetic analysis of fruit yield of tomato. Crop Res. (Hisar), 5, 2:319-325.

Dod, V.N., P.B. Kale, and R.V. Wankhade. 1995. Combining ability for certain quality traits in tomato. Crop Res. (Hisar) 9, (3):407-412.

Doganlar, S., S.D. Tanksley and M.A. Mutschler. 2000. Identification and molecular mapping of loci controlling fruit ripening time in tomato. Theor Appl Genet 100, (2):249-255.

Durand, Y. 1981. Relationship between marker genes *aa* and *Wo* and male sterility gene *ms* 35. *In:* J. Philouze [ed.], The proceedings of the Meeting of the Eucarpia Tomato Working Group, 18-21 May, 1981, Avignon, INRA, France, pp. 225-228.

Duvick, D.N. 1997 a. Heterosis: Feeding people and protecting natural resources. *In:* Book of abstracts of the International Symposium "The Genetics and Exploitation of Heterosis in Crops", 17-22 August, Mexico City, Mexico, pp. 6-7.

Duvick, D.N. 1997 b. Commercial strategies for exploiting heterosis. *In:* Book of abstracts of the International Symposium "The Genetics and Exploitation of Heterosis in Crops", 17-22 August, Mexico City, Mexico, pp. 206-207.

East, E.M. 1908. Inbreeding in corn. Rept. Connecticut Agric Expt Sta for 1907, pp. 419-428.

East, E.M. 1909. The distinction between development and heredity in inbreeding. Am Nat 43:173-181.

East, E.M. and H.K. Hayes. 1912. Heterozygosis in evolution and in plant breeding. USDA Bur Plant Ind Bull 243:58.

Emmons, C.L.W. and J.W. Scott. 1998. Diallel analysis of resistance to cuticle cracking in tomato. J Am Soc Hort Sci 123, 1:67-72.

Falconer, D.S. 1989. Introduction to quantitative genetics (2nd ed). John Wiley & Sons, New York.

Farkas, J. 1993. Current problems in heterosis breeding of tomato. Zöldségtermesztési Kutató Intézet Bulletinje 25:23-36.

Fernàndez-Muñoz, R., J.J. Gonzalez-Fernandez, and J. Cuartero 1995. Genetics of the viability of pollen grain produced at low temperatures in *Lycopersicon* Mill. Euphytica 84:139-144.

Foolad, M.R. 1996. Genetic analysis of salt tolerance during vegetative growth in tomato *Lycopersicon esculentum* Mill. Plant Breeding 115, 4: 245-250.

Foolad, M.R. 1997. Genetic basis of physiological traits related to salt tolerance in tomato *Lycopersicon esculentum* Mill. Plant breeding 116, 1:53-58.

Foolad, M.R., F.Q. Chen, and G.Y. Lin. 1998a. RFLP mapping of QTLs conferring salt tolerance during germination in an interspecific cross of tomato. Theor Appl Genet 97, 7:1133-1144.

Foolad, M.R., F.Q. Chen, and G.Y. Lin. 1998b. RFLP mapping of QTLs conferring cold tolerance during seed germination in an interspecific cross of tomato. Molec Biol 4, 6:519-529.

Foolad, M.R. and R.A. Jones. 1991. Genetic analysis of salt tolerance during germination in *Lycopersicon*. Theor. App. Genet. 81:321-326.

Foolad, M.R. and R.A. Jones. 1992. Models to estimate maternally controlled genetic variation in quantitative seed characters. Theor App Genet. 83:360-366.

Foolad, M.R. and G.Y. Lin. 1998. Genetic analysis of low-temperature tolerance during germination in tomato *Lycopersicon esculentum* Mill. Plant breeding. 117 (2):171-176.

Fujii, T. 1948. Tomato (Series of encycropedia of horticulture of Vegetable Crops). Sangyo-Tosyo Inc. (in Japanese).

Fujii, T. 1952. Comments on vegetable cultivars - Review of new vegetable cultivars adaptable to each districts in Japan. Asakura publisher Inc. Tokyo (in Japanese).

Gardner, R.G. 2000. A male-sterile cherry tomato breeding line, NC 2C *ms-10 aa*. HortScience. 35, 5: 964-965.

Georgiev, Hr. 1991. Heterosis in tomato breeding. *In:* G. Kalloo [ed.], Genetic Improvement of Tomato. Springer-Verlag, Berlin, pp. 83-98.

Georgiev, Hr. and B. Atanassova. 1981. Positional male sterile line tomatoes *ps 2* with a low level stigma . Compt Rend Acad Bulg Sci 34, 3:423-424.

Goodnight, C.J. 1999. Epistasis and heterosis. *In:* J.G Coors and Shivaji Pandey [eds.], The Proceedings of an International Symposium "The Genetics and Exploitation of Heterosis in Crops", CIMMYT, Mexico City, Mexico, 17-22 August, 1997, Madison, USA, American Society of Agronomy, pp. 59-68.

Gorman, S.W., D. Banasiak, C. Fairley, and S. McCormick. 1996. A 610 kb YAC clone harbors 7 cM of tomato (*Lycopersicon esculentum*) DNA that includes the *male sterile 14* gene and a hotspot for recombination. Mol Gen Genet 251 (1):52-59.

Gorman, S.W. and S. McCormick.1997. Male sterility in tomato. Crit Rev Plant Sci16, 1:31-53.

Grandillo, S., H.M. Ku, and S.D. Tanksley. 1999. Identifying the loci responsible for natural variation in fruit size and shape in tomato. Theor Appl Genet 99 (6):978-987.

Griffing, B. 1956. Concept of general and specific combining ability in relation to diallel crossing system. Aust J Biol Sci 9:463-493.

Griffing, B. 1990. Use of a controlled - nutrient experiment to test heterosis hypotheses. Genetics. 126 (3):53-767.

Haanstra, J.P.W., C. Wye, H. Verbakel, F. Meijer-Dekens, P. van den Berg. 1999. An integral high density RFLP-AFLP map of tomato based on two *Lycopersicon esculentum* x *L. pennellii* F$_2$ populations. Theor Appl Genet 106:363-373.

Hallauer, A.R 1999. Heterosis. What have we learned? What have we done? Where are we headed? *In:* J.G Coors and Shivaji Pandey [eds.], The Proceedings of an International Symposium "The Genetics and Exploitation of Heterosis in Crops", CIMMYT, Mexico City, Mexico, 17-22 August, 1997, Madison, USA, American Society of Agronomy, pp. 483-492.

Hassan, A.A., H.H. Nassar, M.A. Barakat, and M.S. Tolba. 1999. Tomato breeding for salinity tolerance. III. Genetics of tolerance. Egyptian J Hort 26, 3:391-403.

He, C., V. Poisa and K. Yu. 2003. Development and characterization of simple sequence repeat (SSR) markers and their use in determining relationship among *Lycopersicon esculentum* cultivars. Theor Appl Genet 106:363-373.

Hedrick, U.P. and N.O. Booth. 1907. Mendelian characters in tomatoes. Proc Am Soc Hortic Sci 5:19-24.

Hospital, F., L. Moreau, F. Lacoudre, A. Charcosset, and A. Gallais. 1997. More on the efficiency of marked assisted selection. Theor Appl Genet 95 (8):1181-1189.

Houle, D. 1994. Adaptive distance and the genetic basis of heterosis. Evolution 48, 4:1410-1417.

Jawaharlal, M., Seemanthini Ramadas, and F. Veerara-Gavathatham. 1999. Evaluation of parents and hybrids of tomato (*Lycopersicon esculentum Mill.*) for their suitability to prepare ketchup and paste. South Indian Horticulture. 47, 1/6:38-41.

Jinks, J.L. 1954. The analysis of continuous variation in a diallel cross of *Nicotiana rustica* varieties. Genetics 39:767-788.

Jinks, J.L. 1983. Biometrical genetics of heterosis. *In:* R. Frankel [ed.], Heterosis: Reappraisal of Theory and Practice. Springer-Verlag, Berlin, pp. 1-46.

Jones, D.F. 1918. The effects of inbreeding and crossbreeding upon development. Connecticut Agric Expt Sta Bul 207:100.

Jorgensen, R.A. 1987. A hybrid seed production method based on synthesis of novel linkages between marker and male-sterile genes. Crop Sci, 27:806-810.

Kalloo, Dr. 1988. Vegetable breeding. v. III., CRC Press, Inc., Boca Raton, Florida, USA.

Kang, M.S. 1997. Phenotypic plasticity, heterosis and environmental stress: a concise review. *In:* Book of abstracts of the International Symposium "The Genetics and Exploitation of Heterosis in Crops". 17-22 August, Mexico City, Mexico, pp. 140-143.

Kaul, M.L.H. 1988. Male sterility in higher plants. Monograph of Theor Appl Genet, Springer-Verlag, Berlin, (vol. 10): 1005 pp.

Kearsey, M.J. and A.G.L. Farquhar. 1998. QTL analysis in plants: where are we now? Heredity 80: 137-142.

Kemble, J.M. and R.G. Gardner. 1992. Inheritance of shortened fruit maturation in the cherry tomato Cornell 871213-1 and its relation to fruit size and other components of earliness. J Am Soc Hort Sci 117:646-650.

Khanna-Chopra, R., M. Maheswari, D.G. Rao, and S.K. Sinha. 1993. Expression of heterosis - a physiological analysis, *In:* M.M Verma, D.S. Virk, G.S. Chahal, and Dhillon B.S. [eds.], The proceedings of a Symposium "Heterosis breeding in crop plants - theory and application" Ludhiana, India, 23-24 February 1993, pp. 20-36.

Knaap, van der E. and S.D. Tanksley. 2001. Identification and characterization of a novel locus controlling early fruit development in tomato. Theor Appl Genet 103:353-358.

Knapp, S.J. 1993. Marker assisted selection as a strategy for increasing the probability of selecting superior genotypes. Crop Sci 38, 5:1164-1174.

Kravchenko, V. A. 1990. Complex hybridization in breeding tomato for earliness. Selektsiya i semenovodstvo. 2:14-16.

Lamkey, K.R. and J.W. Edwards 1999. Quantitative genetics of heterosis, *In:* J.G. Coors and Shivaji Pandey [eds.], The Proceedings of an International Symposium "The Genetics and Exploitation of Heterosis in Crops", CIMMYT, Mexico City, Mexico, 17-22 August, 1997. Madison, USA, American Society of Agronomy, pp. 31-48.

Lee, M. 1997. Towards understanding and manipulating heterosis in crops - can molecular genetics help? *In:* Book of abstracts of the International Symposium "The Genetics and Exploitation of Heterosis in Crops". Mexico City, Mexico. 17-22 August, 1997, pp. 110-111.

Levin, I., A. Cahaner, H.D. Rabinowitch, and Y. Elkind. 1994. Effects of the *ms 10* gene, polygenes and their interaction on pistil and anther-cone lengths in tomato flowers. Heredity 73, 1:72-77.

Li, Y.M. and W.H. Gabelman. 1990. Inheritance of calcium use efficiency in tomatoes grown under low calcium stress. J Am Soc Hort Sci 115:835-838.

Lindhout, P., S. van Heusden, G. Pet, J.W. van Ooijen, H. Sandbrink, R. Verkerk, R. Vrielink, and P. Zabel. 1994. Perspectives of molecular marker assisted breeding for earliness in tomato. Euphytica 79:279-286.

Ma, Y., K. Sakata, and M. Masuda. 1999. Partial inhibition of pollen degradation by gibberellic acid in male sterile tomato mutants derived from cv. First (*Lycopersicon esculentum* Mill.). Scientific Reports of the Faculty of Agriculture, Okayama University 88:57-63.

Mageswari, K. and S. Natarajan. 1999. Studies on heterosis for yield and quality in tomato (*Lycopersicon esculentum* Mill.). South Ind Hort 47, 1/6, 216-217.

Masuda, M., K. Uchida, K. Kato, and S.G.Agong. 2000. Restoration of male fertility in seasonally dependent male sterile mutant tomato *Lycopersicon esculentum* cv. First. J Jap. Soc Hort Sci 69, 5:557-562.

Matsunaga, H. and S. Monma. 2000. Fruit productivity and related plant characteristics of tomato cultivars grown under low nitrogen application. Bulletin of the National Res. Institute of Vegetables, Ornamental Plants and Tea, 15: 107-114

McDaniel, R.G. 1986. Biochemical and physiological basis of heterosis. Crit Rev Plant Sci 4, 3.227-246.

Melchinger, A.E. 1993. Use of RFLP markers for analysis of genetic relationship among breeding materials and prediction of hybrid performance. *In:* D.R. Buxton, R. Shibles, R.A. Forsberg, B.L. Blad, K.H Asay, G.M. Paulsen, and R.F. Wilson. [eds.], International Crop Science I. International Crop Science Congress, Ames, Iowa, USA, 14-22 July, 1992, Madison, WI, USA, Crop Science Society of America, pp. 621-628.

Melchinger, A.E. 1999. Genetic diversity and heterosis. *In:* J.G Coors and Shivaji Pandey [eds.], The Proceedings of an International Symposium "The Genetics and Exploitation of Heterosis in Crops", CIMMYT, Mexico City, Mexico, 17-22 August, 1997. Madison, USA, American Society of Agronomy, pp. 99-118.

Milborrow, B.V. 1998. A biochemical mechanism for hybrid vigor. J Exp Bot 49, 324:1063-1071.

Miller, J.C. and S.D. Tanksley. 1990. RFLP analysis of phylogenetic relationships and genetic variation in the genus *Lycopersicon.* Theor Appl Genet 80, 4, 437-448

Monforte, A.J., M.J. Asins, and E.A. Carbonell. 1999. Salt tolerance in *Lycopersicon* spp. VII. Pleiotropic action of genes controlling earliness on fruit yield. Theor Appl Genet 98, 3/4:593-601.

Monforte, A.J. and S.D. Tanksley. 2000. Fine mapping of a quantitative trait loci (QTL) from *Lycopersicon hirsutum* chromosome 1 affecting fruit characteristics and agronomic traits: breaking linkage among QTLs affecting different traits and dissection of heterosis for yield. Theor Appl Genet 100, 3/4:471-479.

Murray, M., M. Cahn, J. Caprile, D. May, G. Miyao, B. Mullen, J. Valencia, and B. Weir. 1999. University of California Cooperative Extension processing tomato cultivar evaluation program. HortTechnol. 9, 1:36-39.

Natarajan, S. 1992. Inheritance of yield and its components in tomato under moisture stress. Madras Agric. J. 79 (12):705-710.

Nienhuis, J. and G. Sills. 1992. The potential of hybrid varieties in self-pollinating vegetables. *In:* Y. Dattée, C. Dumas, and A. Gallais [eds.], Reproductive Biology and Plant Breeding. Springer-Verlag, Berlin, Germany, pp. 387-396.

Nieuwhof , M. 1990. Adaptation of tomatoes to low growing temperatures through breeding, a realistic goal? Prophyta, 44, 5:115-117.

Nieuwhof, M., J. Jansen, and J.C. van Oeveren . 1993. Genotypic variation for relative growth rate and other growth parameters in tomato (*Lycopersicon esculentum* Mill.) under low energy conditions. J Genet Breed 47, 1:35-44.

Nkansah, G.O. and T. Ito. 1994. Relationship between some physiological characters and yield of heat-tolerant, non-tolerant and tropical tomato cultivars grown at high temperature. J Japan Soc Hort Sci 62, 4:781-788.

Noli, E., S. Conti, M. Maccaferri, and M.C. Sanguineti. 1999. Molecular characterization of tomato cultivars. Seed Sci Technol 27, 1:1-10.

Omholt, S.W., E. Plahte, L. Oyehaug, and Xian KeFang. 2000. Gene regulatory networks generating the phenomena of additivity, dominance and epistasis. Genetics 155, 2:969-980.

Paran, I., I. Goldman, and D. Zamir. 1997. QTL analysis of morphological traits in a tomato recombinant inbred line population. Genome 40, 2:242-248.

Paran, I., M. Horowitz, D. Zamir, and S. Wolf. 1995. Random amplified polymorphic DNA markers are useful for purity determination of tomato hybrids. HortScience 30, 2:377.

Park, Y.H., M.A.L. West, and D.A.St. Clair 2004. Evaluation of AFLPs for germplasm fingerprinting and assessment of genetic diversity in cultivars of tomato (*L. esculentum* Ll.). Genome 47:510-518.

Patil, A.A. and S.S. Patil. 1988. Heterosis for some quality attributes in tomato. J. Maharashtra Agric Univ 13, 2:241.

Philouze, J. 1974. Gènes marqueur liés aux gènes de sterilité mâle *ms-32* et *ms-35* chez la tomate. Ann Amelior Plantes 24, 1:77-82.

Philouze, J. 1986. Evolution et situation varietale actuelle chez la tomate. A.I.C.P.C./ A.C.F.E.V./B.R.G. La diversité des plantes legumières, pp. 33-38.

Philouze, J. 1997. Tomate: des variétés fixées aux hybrides F_1. Fruits & Légumes 151:8-9.

Pilowski, M., D. Lapushner, and R. Frankel. 1971. URIT (hybrid 1589). Rep Tomato Genet Coop 23:53.

Potaczek, H. and B. Kubicki. 1986. Functional male sterility and heterostyly of cultivated tomato (*Lycopersicon esculentum* Mill). Genetica polonica 27 : 309-314.

Pooni, H.S. 1994. Genetics of heterosis and its implications for crop improvement. Ann. Biol. 9:323-332.

Pooni, H.S. and A.J. Treharne. 1994. The role of epistasis and background genotype in the expression of heterosis. Heredity 72:628-635.

Powers, L. 1945. Relative yields of inbred lines and F_1 hybrids of tomato. Bot Gaz 106:247-268.

Rai, N., M.M. Syamal, A.K. Joshi, and P.K. Ghosh. 1997. Diallel analysis for pericarp thickness and storability in tomato *Lycopersicon esculentum* Mill.). Ann Agric Res 18, 1:71-75.

Rai, N., M.M. Syamal, A.K. Joshi, and C.B.S. Rajput. 1997. Genetics of yield and yield components in tomato (*Lycopersicon esculentum* Mill.). Indian J Agric Res 31, 1:46-50.

Ramos, B.F., F.A. Vallejo Cabrera, and P.C. Tavares de Melo. 1993. Genetic analysis of the character mean fruit weight and its components in a diallel cross between cultivars of tomato, *Lycopersicon esculentum* Mill. Acta Agronómica, Universidad Nacional de Colombia, 43, 1/ 4:15-29.

Rao, N.K.S., R.M. Bhatt, and N. Anand. 1992. Leaf area, growth and photosynthesis in relation to heterosis in tomato. Photosynthetica. 26, 3:449-453.

Reiger, R., A. Michaelis and M.M. Green. 1976. Glossary of Genetics and Cytogenetics. VEB Gustav Fisher Verlag Jena, pp. 276.

Resende, L.V., W.R. Maluf, L.A.A. Gomes, F.M.F. Da Mota, and J.T.V. Resende. 1999. Diallel analysis of fruit firmness of cultivars and lines of tomatoes (*Lycopersicon esculentum* Mill.). Ciência e Agrotecnologia 23:12-18.

Rick, C.M. 1945. Field identification of genetically male-sterile tomato plants to use in producing F_1 hybrid seed. Proc Am Soc Hort Sci 46:277-283.

Rick, C.M. 1950. Pollination relations of *Lycopersicon esculentum* in native and foreign regions. Evolution 4:110-122.

Rick, C.M. and L. Butler. 1956. Cytogenetics of the tomato. Adv Genet 8:267-382.

Rom, M., M. Bar, A. Rom, M. Pilowsky, and D. Gidoni. 1995. Purity control of F_1-hybrid tomato cultivars by RAPD markers. Plant Breeding 114, 2:188-190.

Russell, G., P. van Gardingen, and G.W. Wilson. 1993. Using physiological information about varieties: the way forward? Aspects of Appl. Biol. 34:47-56.

Sakaki, M.T. and H. Yamazaki H. 1990. Plant male sterilant. US Patent no 4,976,775, Dec., 11, 1990.

Saliba-Colombani, V., M. Causse, D. Langlois, J. Philouze, and M. Buret. 2001. Genetic analysis of organoleptic quality in fresh market tomato. I. Mapping QTLs for physical and chemical traits. Theor Appl Genet 102:259-272.

Sawhney, V.K. 1994. Genetic male sterility in tomato and its manipulation in breeding, In: E. G. Williams, A.E. Clarke, and R.B. Knox [eds.], Genetic Control of Self-Incompatibility and Reproductive Development in Flowering Plants. Kluwer Acad. Publ., Dordrecht, Netherlands. pp. 443-458.

Sawhney, V.K. 1997. Genic male sterility. In: K.R. Shivanna and V.K. Sawhney [eds]. Pollen biotechnology for crop production and improvement. Cambridge University Press, Cambridge, UK, pp. 183-198.

Sawhney, V.K. 2004. Photoperiod-sensitive male-sterile mutant in tomato and its potential use in hybrid seed production. J Hort Sci Biotech 79:138-141.

Shattuck, V.I., B. Christie, and C. Corso. 1993. Principles for Griffing's combining ability analysis. Genetica 90: 73-77.

Sherif, T.H.I. and H.A. Hussein. 1992. A genetic analysis of growth and yield characters in the tomato (Lycopersicon esculentum Mill.) under the heat stress of late summer in Upper Egypt. Assiut J Agric Sci 23, 2:3-28.

Shmidt, von H. and V. Shmidt. 1981. Untersuchungen an pollensterilen, stamenless-ähnlichen Mutanten von Lycopersicon esculentum Mill. II. Normalisierung von ms-15 and ms-33 mit Gibberellinsäure (GA$_3$). Biol Zentralbl 100:691-696.

Shrivastava, A.K. 1998 a. Heterosis and inbreeding depression for acidity, total soluble solids, reducing sugar and dry matter content in tomato (Lycopersicon esculentum Mill.). Advances in Plant Sci 11, (2):105-110.

Shrivastava, A.K. 1998 b. Combining ability analysis for total soluble solids, reducing sugars, dry matter content and seeds weight in tomato. (Lycopersicon esculentum Mill.). Advances in Plant Sci 11:17-22.

Shull, G.H. 1908. The composition of a field of maize. Rept Am Breeders' Assoc 4:296-301.

Shull, G.H. 1909. A pure line method in corn breeding. Rept Am Breeders' Assoc 5:51-59.

Shull, G.H. 1911. Experiments with maize. Bot Gaz 52: 480-483.

Shull, G.H. 1914. Duplicate genes for capsule form in Bursa bursa Bastoris. J Ind Abst Vererb 12: 97-149.

Singh, D. 1973. Diallel analysis for combining ability over several environments. II. Indian J. Genet. Plant Breed 33:469-481.

Singh, D.N., A. Sahu, and A.K. Parida. 1998. Stability of fruit yield and its attributing traits in tomato (Lycopersicon esculentum). Indian J Agric Sci 68, 7:373-374.

Singh, S., M.S. Dhaliwal, D.S. Cheema, and G.S. Brar. 1998. Diallel analysis of some processing attributes in tomato. J Genet Breed 52, 3:265-269.

Sprague, G. F.1983. Heterosis in maize: Theory and practice. In: R. Frankel [ed.], Heterosis: Reappraisal of Theory and Practice. Springer-Verlag, pp.47-70.

Staniaszek, M., W. Marczewski, H. Habdas, and H. Potaczek. 2000. Identification of RAPD markers linked to the ps gene and their usefulness for purity determination of breeding lines and F$_1$ tomato hybrids. Acta Physiol Plant 22:303-306.

Stevens, M.A. and C.M. Rick. 1986. Genetics and breeding. In: J.G. Atherton and J. Rudich [eds.], The Tomato Crop. Chapman and Hall, London, pp. 35-109.

Stommel, J.R. and K.G. Hayness. 1993. Genetic control of fruit sugar accumulation in a Lycopersicon esculentum x L. hirsutum cross. J Am Soc Hort Sci 118:859-863.

Strunnikov, V.A. 1983. New hypothesis for heterosis: theoretical and practical usefulness. Vestnik Selskohoz. Nauki. 1:34-40.

Stuber, C.W. 1994. Heterosis in plant breeding. In: J. Janick [ed.], Plant Breeding Reviews. V. 12. John Wiley & Sons Inc., New York, pp. 227-251.

Stuber, C.W. 1999. Biochemistry, molecular biology and physiology of heterosis. In: J.G Coors and Shivaji Pandey [eds.], The Proceedings of an International Symposium

"The Genetics and Exploitation of Heterosis in Crops", CIMMYT, Mexico City, Mexico, 17-22 August, 1997. Madison, USA, American Society of Agronomy, pp. 173-183.

Suresh Kumar, M.K. Banerjee, and P.S Partap. 1995. Studies on heterosis for various characters in tomato. Haryana J Hort Sci 24, 1:54-60.

Surjan, S., M.S. Dhaliwal, D.S. Cheema, and G.S. Brar. 1999. Breeding tomato for high productivity. Advances in Hort Science 13:95-98.

Szwadiak, J. and R. Kordus. 1992. A diallel analysis of yield-contributing traits in tomato (*Lycopersicon esculentum* Mill.). Genetica Polonica 33:219-225.

Tanksley, S.D. and D. Zamir. 1988. Double tagging of a male sterile gene in tomato using a morphological marker gene. HortScience 23:387-388.

Temperini, O., G. Colla, R. Campinelli, C. Piccioni, and R. Martellucci. 2001. Agronomic and commercial evaluation of some processing tomato hybrids. Informatore Agrario. 57:75-79.

Titok, V.V., V.A. Lemesh, O.V. Rusinova, and V.L. Podlisskikh. 1994. Leaf area, chlorophyll content and biomass of tomato plants and their heterotic hybrids under *in vitro* culture. Photosynthetica 30:255-260.

Titok, V.V., S.L. Yurenkova, and L.V. Khotyleva. 1998. Energy metabolism in the leaves of F_1 tomato hybrids. Vestsi Akademii Navuk Belarusi, Seriya Biyalagichnikh Navuk. 2:37-41.

Tsaftaris, A.S. 1995. Molecular aspects of heterosis in plants. Physiol Plant 94:362-370.

Tsaftaris, A.S. and M. Kafka. 1998. Mechanisms of heterosis in crop plants. J Crop Production 1:95-111.

Vallejo Cabrera, F.A. and S.E.I Estrada. 1993. Estimation of genetic parameters for the character yield and its primary components in a diallel cross between different lines of tomato *Lycopersicon esculentum* Mill. Acta Agronómica Universidad Nacional de Colombia 43:30-43.

Verma, M.M. and G.S. Chahal. 1993. Genetic basis of heterosis. In: M.M Verma, D.S. Virk, G.S. Chahal, and B.S. Dhillon. [eds.], The Proceedings of Symposium "Heterosis breeding in Crop plants - Theory and Application", Ludhiana, 23-24 February 1993. Crop Improvement Society of India, pp.1-18.

Villand, J., P.W. Skroch, T. Lai, P. Hanson, C.G. Kuo, and J. Nienhuis. 1998. Genetic variation among tomato accessions from primary and secondary centers of diversity. Crop Sci. 38:1339-1347.

Wang, D.Y. and Q.M.Yin. 1999. A study on crop heterosis mechanism: a theory of resultant genetic vibration. Acta Agrculturae Universitatis Jiangxiensis 21(3):314-319.

Wang, F., J.F. Li, and G.Y. Li. 1995. A study on inheritance and correlation of fruit firmness in tomato. Acta Hort 402:253-258.

Wang, H., M. Wang and Ch. Li. 1988. The use of heterosis, In: Tomato Breeding. Shanghai Technology and Science Publishing Company (In Chinese), pp. 153-213.

Wang, L., M. Wang, Y. Shi, S.P. Tian, and Q.H.Yu. 1998. Genetic and correlation studies on quantitative characters in processing tomato. Adv Horti 2:378-383.

Wang, Y.F., M. Wang, D.Y. Wang, and L. Wang. 1998. Studies on heterosis in some processing tomato (*Lycopersicon esculentum* Mill.) lines. Acta Agric Shanghai 14 (3):29-34.

Wehner, T.C. 1997. Heterosis in important US vegetable crops. In: Abstracts of the International Symposium "The Genetics and Exploitation of Heterosis in Crops" 17-22 August, Mexico City, Mexico, pp. 527-530.

Williams, W. 1959. Heterosis and the genetics of complex characters. Nature 184:527-530.

Xue, B.Y. 1994. Breeding tomato green-stemmed (*aw*) lines and the utilization of heterosis. China Vegetables 4:32-33.

Yadav, E.D., P.N Kale, and K.N. Wavhal.1988. Genetic analysis of fruit dry matter in tomato. Vegetable Sci 15:49-54.

Yordanov, M. 1983. Heterosis in tomato. *In:* R. Frankel [ed.], Heterosis: Reappraisal of Theory and Practice. Springer-Verlag, pp. 189-219.

Young, N.D. 2000. A cautiously optimistic vision for marked-assisted breeding. Mol breed 5:505-510.

Zhakote, A.G. and V.G. Kharti. 1990. Features of photosynthesis in wild, semicultivated and cultivated genotypes of tomato and their F_1 hybrids in relation to breeding for yield. Sel'skohozyaistvennaya Biologiya, 5:82-88.

Zhang, Y. and J.R. Stommel. 2000. RAPD and AFLP tagging and mapping of *Beta* (B) and *Beta* modifier (Mo_B), two genes which influence β-carotene accumulation in fruit of tomato (*Lycopersicon esculentum* Mill.). Theor Appl Genet 100, 3:368-375.

Zhutchenko, A.A. 1973. Tomato genetics. Shtintza, Kichinev, Moldova.

Zirkle, C. 1952. Early ideas on inbreeding and crossbreeding. *In:* Gowen [ed.], Heterosis. Iowa State College Press, Ames, IA, pp.1-13.

Traditional and Enhanced Breeding for Quality Traits in Tomato

MATHILDE CAUSSE, RENÉ DAMIDAUX AND PATRICK ROUSSELLE

INRA, Fruit and Vegetable Genetics and Breeding Research Station
BP 94 - 84143 Avignon, France
email:mathilde.causse@avignon.inra.fr

INTRODUCTION

Tomato breeding initially focused on yield and ancillary traits (adaptation, disease resistance, earliness). Presently a large number of cultivars adapted to specific growing conditions (glasshouse, plastic tunnel or field) and for commercial use (fresh-market or processing tomatoes) are released each year. Tomato breeding objectives evolved with the performance of released cultivars and the modification in growing conditions, with increasing emphasis on fruit quality.

Tomato breeders use classical methods devoted to inbred crops such as pedigree or backcross methods. They also exploit several monogenic mutations, which are particularly useful. To accomplish any crop improvement goal, a breeding program requires clearly defined objectives and efficient breeding strategy. Whereas breeding objectives depend on the environmental conditions, mode of culture, mode of harvest, and quality of the end product (whether the product is to be used fresh or processed), breeding efficiency mainly depends on the available genetic diversity and trait inheritance. Selection for quality attributes is difficult for three reasons: (i) many traits must be simultaneously improved, (ii) most quality traits inherited quantitatively are controlled by several genes and strongly influenced by the environment, and (iii) organoleptic quality (not being first breeding objective) is evaluated by sensory analysis, which cannot be applied to the large number of genotypes usually screened during the

Corresponding author: Mathilde Causse

breeding process; thus indirect selection is usually performed on fruit composition traits, which are only partly linked to the consumer perception.

Plant molecular biology offers breeders a new prospective of genetic improvement based on molecular markers and genetic engineering. These new tools considerably modified the paradigm of plant breeding allowing a new access to remote variability. Today we are only at the beginning of the integration of these tools into breeding strategies.

This chapter mainly focuses on quality traits, as specific chapters are devoted to adaptation to abiotic stress (chapters 16) and to disease resistances (chapters 12, 13, 14, 15). The first part of this chapter presents the major quality attributes, their inheritance and the available genetic variability, and the second part is devoted to traditional tomato breeding methods. Applications and limitations of molecular markers and genetic engineering for the improvement of tomato quality are also analyzed.

GENETIC VARIABILITY AND INHERITANCE OF QUALITY ATTRIBUTES

The major quality attributes depend on the end uses of tomato (processing or fresh market). Quality of fresh market is determined by appearance, firmness and flavor, whereas processing tomato quality is mainly determined by total soluble solids content, color, pH and firmness. However, the parameters that influence these attributes are often the same: sugars and organic acids are the major components of dry matter weight, but also contribute to the flavor of the product (Table 5.1). Events occurring during ripening (ethylene biosynthesis, cell wall modifications) also control texture traits, which are important for both organoleptic quality and processed tomato quality. Inheritance of fruit quality components were extensively reviewed by Davies and Hobson (1981) and Stevens (1986). Dorais et al (2001) reviewed the genetic and environmental basis of greenhouse tomato fruit quality.

Commercial Quality Attributes

Yield, Fruit Weight and Dry Matter Weight

Yield, fruit weight and dry matter weight are strongly dependent on cultivars and growing conditions. Yield results from fruit number and fruit weight. Cultivars are classified based on their fruit size and shape from the cherry tomato (less than 20g) to beef tomato (fruit weight higher than 200g). The potential size depends on cell number established in pre-anthesis stage, but final fruit size depends on the rate and duration of cell

Table 5.1   *Range of composition of ripe tomato fruit and related quality traits (adapted from Davies and Hobson 1981, Stevens 1986, and Hobson and Grierson 1993)*

Component	Typical composition (% fresh weight)	Range of content	Related quality trait
Dry matter	6.5		
Total soluble solids	4.5 *	45-65 mg/g	Flavor
Reducing sugars	3.0	20-37 mg/g	"
Glucose	1.4	9-17 mg/g	"
Fructose	1.6	11-20 mg/g	"
Sucrose	0.1	0-1 mg/g	"
Acids	0.3	70-130 meq/liter	Flavor ; product safety
Malic acid	0.1	6-21 meq/liter	"
Citric acid	0.2	65-120 meq/liter	"
Insoluble solid content	0.5	7-25 mg/g	Firmness (texture) and shelf life ; consistency (viscosity)
Fat	0.15	1-3 mg/g	
Proteins	0.4	6-20 mg/g	
Ash	0.7	5-7 mg/g	Health value
Potassium	0.25	1-3 mg/g	Buffering system
Calcium	0.01	0.04-0.21 mg/g	Firmness - Blossom end rot
Carotenoids	0.03	40-65 µg/g	Color
Lycopene	0.03	35-60 µg/g	"
Carotene	0.005	3-8 µg/g	"
Vitamin C	0.02	15-30 µg/g	Health value
Volatils	0.01	10 µg/g	Aroma

*determined by refractometer (° Brix)

enlargement (Ho 1996). Seed number and competition among fruits also affect the final fruit size (Bertin et al. 1998). A number of quantitative trait loci (QTLs) control the variation in fruit weight, but small fruit size is usually partly dominant.

Fruit shape varies from flat to long fruit and is determined at the carpel development stage. A mutation (*o* for ovate) controls the ovate fruit shape (Ku et al. 1999), together with minor QTLs. Environmental conditions may also influence the final fruit shape.

Tomato dry matter weight varies from 5 to 9% (Davies and Hobson 1981) depending on its contents. Soluble and total solids relate to yield of concentrated tomato product. The total soluble solid content is usually measured by the refractive index (° Brix), and is strongly influenced by cultivars and environmental conditions (Gautier et al. 2005a, b). Organic acids (malic and citric acids) and reducing sugars (glucose and fructose) represent the main part of total soluble solids. A large genetic variation for the contenst in organic acids has been shown (Stevens 1986). Acidity influences the storability of processed tomatoes and tomato products by inhibiting the germination of thermophilic organisms. A pH above 4.5 is thus commercially undesirable as it would require increased temperature and processing time to avoid spoilage. Titratable acidity and pH are often negatively correlated, but not systematically. Citrate contributes to a greater extent to sourness because of its higher content. An inverse relationship between yield and fruit solids is frequently observed (Grandillo et al. 1999). The physiological basis of fruit solid content are to be searched in sink strength and photosynthetic efficiency (Hewitt and Stevens 1981).

Insoluble solids, the second important component of total solids, are made of cell wall components, proteins, pectins, cellulose and polysaccharides. They are important both for the fruit firmness and for the viscosity of tomato juice, ketchup, soup sauce and paste, and thus for the end-product quality.

External Properties of the Fruit

Fruit Color   Fruit color depends on the content of the carotenoid pigments, the red pigment lycopene, and to a lesser extent β-carotene. Carotenoids are not homogeneously spread in the fruit and outer pericarp shows the highest concentration in the fruit (Thakur et al. 1996). Several monogenic mutations alter the biosynthesis of carotenoids (Stevens 1986). Some reduce the synthesis of lycopene, leading to yellow or orange fruits (*r, B, t, Del, at* mutations). On the contrary, plants carrying the mutations *og^c*, *dg* or *hp* show more intense red fruits. Table 5.2 shows the pigment composition in these mutants. These mutants allowed the elucidation of the carotenoid biosynthetic pathway (Ronen et al. 1999, 2000). Another mutation (coded *y*) provides tomatoes with transparent epidermis and leads to pink fruits, particularly appreciated in Asia. The absence of dark green coloration of the collar before ripening is due to a single mutation (*u*, for uniform ripening), which is now widely used in modern varieties. In addition to these mutations, quantitative variation for color intensity and pigment content has been observed (Chen et al. 1999, Saliba-Colombani et al. 2001).

External Defaults   Several physiological disorders, such as fruit cracking, gold speck, blossom-end rot or puffiness, affect fruit quality. Additional to the loss in yield caused by defaults, this external quality in fruits is of prime concern in the grading of fresh-market tomato. All these disorders are induced by the climate and cultural practices, but genetic variation may also influence their apparition (Dorais et al. 2001). The sensitivity to these defaults is quantitatively inherited and linked to several factors such

Table 5.2 Modification in the composition in pigments (µg/g) of several mutants (after Tomes 1963)

Mutation	Phytoene	Phytofluene	Neuros-porene	Lycopene	β-carotene	ζ-carotene	γ-carotene	δ-carotene	α-carotene
Normal red	29	8	-	44	5	-	1	-	-
r	10	trace	-	-	trace	-	-	-	-
hp	15	-	-	56	7	-	2	-	-
t	51	17	3	-	1	45	-	-	-
Del	13	-	-	17	6	1	6	33	2
og^c	25	-	-	50	3	-	1	-	-

as the fruit sugar content, water stress tolerance (Taylor et al. 2004), plant architecture and foliage density.

Fruit Firmness and Shelf Life   Answering to the demand of producers and retailers of fresh-market tomatoes, breeders have considerably improved yield, firmness and shelf life of tomato fruit. The improvement in shelf life was obtained either by the use of the ripening mutations or by the cumulative effect of several genes improving fruit firmness.

Several mutations affecting fruit ripening are known. The most widely used in tomato breeding is *rin* (ripening inhibitor). Described in 1968, it corresponds to a spontaneous mutation appeared in breeding material of *L. esculentum*. Two other mutations interest tomato breeders, (a) *nor* (non-ripening) found in a Canadian variety, which was first used for winter consumption, and (b) *alc* (alcobaca) first observed in Portugal. In plants carrying these mutations at the homozygous state all the ripening processes are blocked and fruits remain yellow to light orange. Fruits may be kept several months in a room without any alteration. When the mutations are in the heterozygous state, fruits become red and ripen more slowly than non-mutants and may thus be kept for a longer time on the shelf (Davies and Hobson 1981). The genes controlling the *rin* and *nor* mutations were recently cloned (chapter 11).

Undoubtedly, long shelf cultivars have invaded the tomato market, but their quality (particularly their color and flavor) has not been liked by consumers (Jones 1986, McGlasson et al. 1987). Recently it has been shown that the genetic background in which the non-ripening genes were introduced strongly influenced the expression of these defaults (Fig. 5.1) which could thus be overcome by selecting good hybrid combinations (Causse et al. 2003). The impact of the enzymes involved in cell wall modifications during ripening and fruit firmness including shelf life have been extensively studied and modifications of polygalacturonase or pectin methyl esterase activity were proposed to increase fruit shelf life and texture properties (Hobson and Grierson 1993).

Organoleptic Quality Attributes

For fresh-market tomato, breeders have successfully improved yield, disease resistances, adaptation to greenhouse conditions, extrinsic fruit characteristics, etc. but have lacked clear targets for improving organoleptic fruit quality. Consumers have complained about tomato taste for years (Bruhn et al. 1991, Hobson 1988). Recent studies on consumer habits in the US as well as in Europe showed that consumers would accept a higher

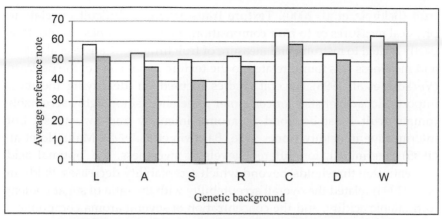

Fig. 5.1 Comparison of the preference notes attributed by consumers to several pairs of isogenic hybrids differing by the *rin* genotype (*rin rin+* in grey, *rin+rin+* in white). Two isogenic lines differing by the presence of the *rin* gene were crossed to 7 lines of different origins and fruit size (C, D, W are cherry tomato lines, L, A, S, and R, large-fruited lines), and fruits of the hybrids were assessed by consumers (about 45 repetitions per genotype); Causse et al. unpublished.

price for better flavor and higher nutritional value of fresh vegetables and fruits (Grasselly et al. 2000). Breeders thus need efficient selection criteria with potential for improvement, i.e. the range of genetic variability available, the mode of inheritance and the influence of growing conditions and genetic and environmental factors on quality traits.

Important Components of Organoleptic Quality

Tomato fruit quality for fresh consumption is determined by a set of attributes: extrinsic (size, color, firmness) and intrinsic (flavor, aroma, texture) properties. Relationships between tomato taste and fruit characteristics have been investigated. Flavor is dependent on the ratio of reducing sugars and organic acids (Stevens et al. 1977, 1979; Bucheli et al. 1999), and the composition of volatile aromas. Sweetness and acidity are related to sugars and acids content (Malundo et al. 1995, Janse and Schols 1995). Sweetness seems more influenced by the content in fructose than in glucose, while acidity is mostly due to the citric acid, present in higher content than malic acid in mature fruits (Stevens et al. 1977). Depending on the studies, acidity is more related to the fruit pH or to the titratable acidity (Baldwin et al. 1998, Auerswald et al. 1999). Both sugars and acids contribute to the sweetness and to the overall aroma intensity (Baldwin et al. 1998). More than 400 volatiles have been identified (reviewed by Petro-Turza 1987), a few of them contributing to the particular aroma of tomato fruit (Baldwin et al. 2000). Texture traits are more difficult to relate to physical measures or to fruit composition, although firmness in mouth is partly related to instrumental measure of fruit firmness (Causse et al. 2002), and mealiness was found related to the texture parameters of the pericarp (Verkeke et al. 1998). Several studies focused on identifying the most important characteristics for consumer preferences since highly acceptable tomato fruit must be good in aroma intensity and sweetness, but intermediate in acidity (Jones 1986, Baldwin et al. 1998). Malundo et al. (1995) determined, for different levels of sweetness, the 'optimal acid concentration thresholds' beyond which acceptability decreases. Baldwin et al. (1998) related the overall acceptability with the ratio of sugar content to titratable acidity, and the concentration of several aroma compounds. Verkeke et al. (1998) underlined the role of texture traits in the preference of consumers. In processed tomato juice samples, fruitiness intensity is a good descriptor of the overall tomato flavor, correlated positively with the contents of glucose and reducing sugars and negatively with glutamic acid content (Bucheli et al. 1999). Fruit color is also important in the perception of fruit quality (Stommel et al. 2005).

Genotypic Variation in Organoleptic Quality Traits of Fresh-market Tomato
Although production of high quality fruits is dependent on environmental factors (light and climate) and cultural practices, a large range of genetic variation in quality traits could be used for breeding tomato quality. Genetic variability in quality traits has been reviewed by Davies and Hobson (1981), Stevens (1986), and Dorais et al. (2001). A few analyses revealed genetic variation for sugars (Stevens 1972), acids, volatile compounds (Langlois et al. 1996) and secondary metabolites (Davies and Hobson 1981, Grolier and Rock 1998). Most of the studies on genetic variation in fruit quality are limited to only a few cultivars or comparable groups of cultivars. Preferences of consumers to exploit genetic variability have rarely been considered. Causse et al. (2003) showed the importance of flavor and secondarily of texture traits in consumer appreciation. Cherry tomatoes are identified as having the best flavor (Hobson and Bedford 1989), with fruits rich in acids and sugars. Long shelf life cultivars have been described as generally less tasty (Jones 1986), because of lower volatile content (Baldwin et al. 1991). The studies on trait inheritance, through classical analysis of cross designs, were devoted to a few traits, particularly the solids and acids content (Stevens 1986) revealed their polygenic control.

Wild relatives of *L. esculentum* may be interesting for improving fruit content. Mutations in enzymes involved in carbon metabolism were found in

L. chmielewskii and in *L. hirsutum*, leading to particular sugar compositions. For example, the *sucr* mutation in an invertase gene identified in *L. chmielewskii* provides fruits with sucrose instead of glucose and fructose (Chetelat et al. 1995). In *L. hirsutum*, an allele of the ADP glucose pyrophosphorylase enzyme was identified as much more efficient than the allele of the cultivated species since it leads to an increase in the final sugar content of fruit (Schaffer et al. 2000). Another locus *Fgr* modulates the fructose-glucose ratio in mature fruit of *L. hirsutum* (Levin et al. 2000). A gene *Lin5* encoding an apoplastic invertase has been shown to be a QTL modulating sugar partitioning, an allele of *L. pennellii* expressing higher sugar concentrations than in the *L. esculentum* (Fridman et al. 2000). Wild Lycopersicon species may also provide original aromas, favorable to tomato quality, as found in a *L. peruvianum* accession (Kamal et al. 2001) or unfavorable as found in a *L. pennellii* accession (Tadmor et al. 2002).

Quality Attributes Specific to Processing Tomato

The self-pruning mutation (*sp*), characteristic of all the processing varieties, controls the determinate growth habit of tomato plants. This recessive mutation spontaneously appeared in Florida in 1914. In processing cultivars the *sp* mutation is associated with concentrated flowering, fruit firmness and resistance of mature fruits to over-ripening, thus allowing a unique mechanical harvest. The sp gene was recently cloned (Pnueli et al. 1998). This mutation does not only affect the plant architecture, but also modulates the expression of genes controlling fruit weight and content (Stevens 1986, Fridman et al. 2002). The jointless mutations, provided by the *j* and *j2* genes, are also useful to processing tomato production. The *j2* mutation discovered in a *L. cheesmanii* accession is characteristic of no abscission zone in fruit pedicel enabling harvest without calyx and pedicel during vine pick-up.

Viscosity is among the major quality attributes specific to processing tomato. It is influenced by alcohol-insoluble solids content under a polygenic control (Stevens 1986). Modification of the polygalacturonase activity through antisense transformation increased the viscosity of processed tomato (Hobson and Grierson 1993).

Nutritional Value

Tomato consumption has been shown to reduce the risks of certain cancers and cardiovascular diseases (Giovannucci 1999). Its health value is mostly due to its content in lycopene and carotene (Bramley 2000), but it also constitutes an important source of vitamin C (Offord 1998). Genetic variation for the content in carotenoids and other micronutrients has been shown in *L. esculentum* cultivars and to a larger extent in the wild relatives (Stevens 1986). In spite of considerable efforts in developing cultivars with higher content in carotenoids (Stommel 2001), or in vitamin C, none has reached a commercial importance, in part because of a negative relation between yield and these traits.

Numerous other potentially beneficial compounds are found in tomato fruits (folate, potassium, phenylpropanoids, and flavonoids) and a wide range of variation is shown among cultivars. Aspects of tomato fruit-nutritive value and the progress made for its enhancement are described in chapter 6.

Relationship Among Quality Traits

Many efforts for improving fruit quality have failed because of the complex correlations between the various components or between yield or fruit weight and fruit components. Negative correlations limit the potential improvement derived from a specific cross. Furthermore sensory assessment is time consuming and expensive, thus breeders need instrumental measures which could replace trained panels. Correlations among traits may vary from one progeny to the other, but some general trends may be observed: (a) instrumental measures of fruit color are related to pigment content, as are pH and titratable acidity (Causse et al. 2002, Table 5.3); (b) sugar content is positively correlated with dry matter weight and with soluble solid content. In several studies involving sensory evaluation and fruit content analyses, sweetness was positively correlated with reducing sugar content and sourness with titratable acidity (Baldwin et al. 1998, Causse et al. 2002). Firm texture is positively correlated with the instrumental firmness (Lee et al. 1999, Causse et al. 2002). The correlation between fruit weight and sugar content is frequently negative (Causse et al. 2001a), but may be positive in other samples (Grandillo and Tanksley 1996a). Correlations were also detected between fruit size and antioxidant composition (Hanson et al. 2004). High throughput metabolic profiling is a new approach to get insight on the whole metabolic changes in tomato fruits during fruit development or in various genotypes (Baxter et al. 2005, Overy et al. 2005, Schauer et al. 2005).

TRADITIONAL BREEDING METHODS

Several reviews describe in detail the breeding methods used for tomato improvement (Peirce 1991, Tigchelaar 1986).

Table 5.3 Phenotypic correlations between sensory attributes and major physical and chemical traits in a RIL population derived from a cross between a cherry tomato line and a large-fruited line (from Causse et al. 2002).

Sensory attributes	Sweetness	Sourness	Tomato aroma	Lemon aroma	Candy aroma	Citrus aroma	Pharmaceutical aroma	Firm texture	Melting texture	Juicy texture	Mealy texture	Skin difficult to swallow
Physical traits												
Fruit weight	-	-0.45	-0.53	-0.39	-0.44	-0.24	0.25	-0.21	0.33	-0.20	0.19	-
Firmness	-	-	-	-	-	-	-	0.29	-	-	-	-
Elasticity	-	-0.47	-0.48	-0.40	-0.33	-	-	-	-	-0.32	-	-
Color (L)	-	-0.37	-0.36	-0.23	-0.31	-	-	-0.20	0.32	-	-	-
Chemical traits												
Sugar content	0.64	-	0.65	-	0.67	0.55	-0.23	-	-	-	-0.24	-0.20
Dry matter weight	0.38	0.48	0.76	0.35	0.62	0.42	-0.23	0.24	-0.32	-	-	-0.27
pH	-	-0.26	-	-0.28	-	-	-	-	-	-	-0.24	0.20
Titratable acidity	-0.22	0.82	0.56	0.70	0.20	-	-	0.21	-0.30	0.20	-	-
Lycopene content	-	-	-	-	0.25	-	-	-	-	-	-	-
Carotene content	-	-	0.24	-	0.25	-	-	-	-	0.20	-0.25	-

Only significant correlations are shown (P<0.05).

Inbred Lines and F_1 Hybrids

The choice of a breeding method is mainly determined by the mating system of the crop. The cultivated tomato is a self-pollinated species. In the past, cultivars were inbred and lines produced after several selfing generations. Nevertheless, most of the modern cultivars are F_1 hybrids, produced by crossing two complementary inbred lines. F_1 hybrids are chosen for several reasons: they allow the easy combination of several dominant disease resistance genes, they drive benefit from heterosis for several traits (details in chapter 4), and they protect breeder's innovations.

For the production of F_1 seeds, a set of nuclear recessive male sterility genes have been described, but are not used for a commercial purpose (Kaul 1991). The use of a functional male sterility gene, controlled by the positional sterile mutation (*ps2*) whose anthers open only after a contact, has been proposed (Atanassova 1999). Due to the difficulty of carrying sterility genes along the selection schemes and to the rapid turnover of tomato cultivars, F_1 hybrids are more frequently produced by hand pollination in countries with low labor cost.

The Choice of Parental Lines

To achieve breeding goals, the choice of parental lines is crucial. The choice of parents requires a good knowledge of the available germplasm. Traditional breeding is usually performed starting from the cross of two elite lines found among adapted cultivars, rather than unadapted exotic germplasm. Indeed, the production of a new cultivar from the cross between two elite lines takes 5 to 7 years, whereas the incorporation of new genes from wild relatives takes about 20 years. The complexity of a plant breeding program increases with the number of genes contrasting the two parents. Nevertheless, a few examples of the use of wild relatives for the improvement of tomato can be mentioned: (a) the use of *L. hirsutum* for improving cold tolerance, (b) *L. chilense* for drought tolerance, and (c) *L. cheesmanii* for soluble solids and salt tolerance (Hobson and Grierson 1993). Firmness was selected in the United States in a progeny descending from of *L. pimpinellifolium*, carrying very small fruits, whose firmness is difficult to assess (Fig. 5.2). The first crosses, performed around the 40s, led to the cultivars Florida MH1 and Flora Dade, which were cultivated worldwide and latter used as genitors of firmness. Indeed, a wide range of variability is shown in wild relatives of cultivated tomato for most of the fruit components (Davies and Hobson 1981). For example, some accessions were identified with lycopene content or ascorbic acid content twice as observed in traditional cultivars.

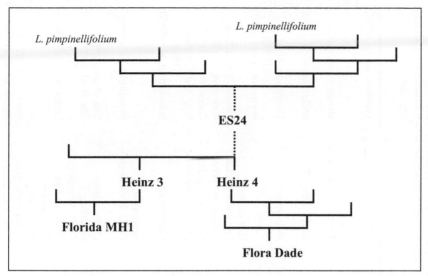

Fig. 5.2   Pedigree of two firm fruit varieties, Florida MH1 and Flora Dade widely used in tomato breeding (after Scott, 1983)

It also seems possible to strongly increase carotene content using *L. hirsutum*. Increase in sugar content was also obtained by crosses with *L. pimpinellifolium*, *L. cheesmanii* and *L. chmielewskii* (Stevens 1986).

Breeding Methods

Following hybridization, the segregated populations are generally selected for improved traits by backcross or pedigree method. Whereas selfing or backcrossing will progressively lead to homozygozity, some other methods (single seed descent, recurrent selection) have been proposed but are rarely used in tomato breeding.

Backcrossing for Monogenic Traits
Tomato is one of the cultivated plant species with the highest number of monogenic traits useful in breeding process. The mutations in these traits either appeared spontaneously in cultivated varieties, or were detected in a relative species, or were induced by a mutagenesis treatment (Chetelat and Rick 1996). The majority of disease resistance genes, however, are derived from wild species, most of the fruit mutations originated from *L. esculentum*. Some mutations are known since years (hp, og^c, pat2), but were rarely used because of unfavorable pleiotropic effects or linkage to unfavorable genes. Table 5.4 shows the major mutations used in tomato cultivars or potentially useful for quality improvement. They concern plant

Table 5.4 *Major mutations used for breeding fruit quality and plant architecture in tomato*

Mutation	Phenotype	Origin	Chromosome	Activity	Reference
rin (ripening inhibitor)	inhibited ripening (semi-dominant)	*L. esculentum*	5	MADS-box gene	Vrebalov et al. (2002)
alc (alcobaca)	inhibited ripening (semi-dominant)	*L. esculentum*	10	Not cloned	-
nor (non-ripening)	inhibited ripening (semi-dominant)	*L. esculentum*	10	Transcription factor	Moore et al. (2002)
Nr (Never-ripe)	inhibited ripening (dominant)	*L. esculentum*	9	C2H4 receptor	Wilkinson et al. (1995)
Cnr (Colorless non-ripening)	inhibited ripening (dominant)	*L. esculentum*	2	Not cloned	Thompson et al. (1999), Seymour et al. (2002)
at (apricot)	orange fruits	*L. esculentum*	5	Not cloned	-
B (Beta)	yellow fruits	*L. hirsutum*	6	Lycopene cyclase	Ronen et al. (2000)
og^c (old gold-crimson)	higher lycopene content	*L. esculentum*	6	Lycopene cyclase	Ronen et al. (2000)
Del (Delta)	orange fruits	*L. hirsutum*	12	Lycopene cyclase	Ronen et al. (1999)
r (yellow flesh)	yellow fruits	*L. esculentum*	3	Phytoene synthase	Fray and Grierson (1993),
t (tangerine)	orange fruits	*L. esculentum*	10	Carotenoid isomerase	Isaacson et al. (2002)
hp-2 (high pigment)	higher lycopene content	*L. esculentum*	12	DET1 homologue	Mustilli et al. (1999)
hp-1 (high pigment)	higher lycopene content	*L. esculentum*	2	DDB1 Light signalling	Yen et al. (1997), Liu et al. (2004)
dg (dark green)	higher lycopene content	*L. esculentum*	12	DET1 homolog; allelic to hp1	Levin et al. (2003)
u (uniform ripening)	absence of green collar	*L. esculentum*	10	Not cloned	-
y	uncolored epidermis	*L. esculentum*	1	Not cloned	-
j (jointless)	absence of abscission zone	*L. esculentum*	11	MADS-box gene	Mao et al. (2000)
sp (self-pruning)	determinate growth	*L. esculentum*	6	CEN homologue	Pnueli et al. (1998)

architecture, color, fruit content or shelf life. Several genes responsible for these mutations were recently cloned. More than 1000 mutations are identified (Rick and Chetelat 1996; http://tgrc.ucdavis.edu) and new ones are being produced and characterized (http://www.sgn.cornell.edu/mutants/mutants_web).

The easiest way to transfer a mutation of interest into an elite cultivar is a succession of backcrosses to the desirable parent. If the gene is recessive, its expression is masked when heterozygous and selfing generations must be added to identify plants carrying the gene. The transfer thus takes more than 10 generations. In spite of an important number of backcrosses, a restriction of recombination in the introgressed region is frequent, leading to fragments which may cover a large portion of the donor chromosome. For instance, it has been shown that after more than 10 backcrosses, the region carrying the *Tm-2a* gene of resistance to the tobacco mosaic virus still existed on a full arm of chromosome, carrying a number of unfavorable genes of *L. peruvianum* (Young and Tanksley 1989).

Pedigree Method
Generally, several traits are to be improved simultaneously, thus pedigree selection is preferred. In this method, seeds form each single plant selection are maintained separately. This method allows a close monitoring of parentage and a maximum selection pressure in each generation. Usually selection is performed in the first generations for highly heritable traits, while the selection for yield is made in the latter generations, which have high homozygosity and can be used for repeated trials. Many programs combine backcross and pedigree selection. A gene of interest is transferred in the first generations by backcross, followed by selection for other traits by pedigree selection.

Single Seed Descent
Single seed descent (SSD) was proposed to enable breeders to preserve a wide genetic base until genotypes are sufficiently fixed to select effectively for traits with low heritability. In the F_2 and subsequent generations only a low selection pressure is applied. Application of this technique to tomato was investigated by Casali and Tigchelaar (1975). They showed that family performance was better evaluated than in pedigree selection, but the efficiency of SSD was not higher than pedigree selection.

Recurrent Selection
The previous methods are devoted to the progeny of a cross between two parents. Population improvement, based on a broader initial variability has been developed for allogamous crops, such as maize (Hallauer and

Miranda 1981). It has been shown to be particularly effective for improving combining ability in hybrids. There have been few reports of such selection system in tomato, because a number of controlled crosses are involved, first for creating an initial population and then for selection of the improved population in each cycle.

A combination of several methods is sometimes useful. For instance, in order to increase soluble solid content in processing tomato, a selection scheme combining backcrosses and recurrent selection was used (Damidaux, unpublished data). Following two backcross generations, from a cross involving *L. cheesmanii* and two *L. esculentum* varieties, selection alternated selfing with intercrosses of the best plants. At each generation, selection was performed among 15 to 20 families, each represented by 20 plants. Selection was based on an index including plant architecture, vigor, yield, fruit weight and brix. Table 5.5 shows the progress obtained at each generation for fruit weight and brix. After several generations, fruit weight equivalent to that of the control variety was recovered but with an improved soluble solid content (Brix >6°).

Selection of F_1 Combinations

The production of F_1 hybrids instead of pure lines did not considerably change the breeding methods. However, test crosses for combining ability must be included in the selection schemes. In the simplest approach, superior inbreds are crossed to elite lines arising from a selection scheme and the best hybrids are kept for evaluation in replicated and multi-site trials.

Genetic Parameters Affecting Selection for Genetic Gain

Heritability

The selection efficiency will depend not only on the selection pressure during a breeding scheme, but also on the inheritance of the trait and the way it will be influenced by the environment. The broad sense heritability, which represents the ratio of genetic variability over the overall phenotypic variability, may vary from one trait to another and from one sample to the other. For example, Stevens (1986) mentioned values of heritability for soluble solid content ranging from 0.13 to 0.70.

Genotype × Environment Interaction

Genotype by environment interaction (G × E), which describes the inconsistent response of a genotype in a series of environments, is a major impediment in selection. Genotypes with broad adaptation are thus preferred. G × E interaction is strongly significant for marketable yield

(Peirce 1991). Growing conditions are known to influence quality traits at the composition level (reviewed by Dorais et al. 2001 for glasshouse conditions), and at the sensory level (Hobson and Bedford 1989). In a study of the inheritance of quality traits, Causse et al. (2003) showed that the influence of environmental conditions varied from one trait to another, G × E being strong for fruit weight and aroma intensity but not too much significant for firmness and fruit content. Auerswald et al. (1999) and Johansson et al. (1999) observed G × E interactions were usually not strongly significant in comparison to the main effects. They found that variety differences affected more fruit quality than growing conditions.

Multitrait Selection

Selection is straightforward when applied to a single trait. When several traits are to be selected, negative genetic correlations may limit the overall progress. For instance, fruit size and earliness are negatively correlated. Several strategies have been proposed for multitrait selection, namely tandem selection, independent culling levels or index selection (detailed in Peirce 1991). Molecular markers enable a new approach of genetic correlations.

ENHANCED BREEDING USING MOLECULAR MARKERS

Molecular Markers for New Genotype Propagation

Mapping genes useful for breeding purposes has always been of prime importance for tomato breeders (Buttler 1952). Molecular markers have enabled biologists to construct saturated maps of the genome and systematically localize genes of interest on these maps (chapter 7). The first markers, based on the detection of Restriction Fragment Length Polymorphisms, allowed the construction of a reference map of the tomato genome (Tanksley et al. 1992). With more than 1000 loci, spread on the 12 chromosomes, it has been possible to localize several mutations and genes of interest (Table 5.4). Genes of interest were first mapped thanks to pairs of near isogenic lines differing only in the region of the interesting gene (Laterrot 1996, Philouze 1991). Bulks of individuals were later used (following the Bulk Segregant Analysis method), together with markers based on PCR amplification of the DNA (RAPD or AFLP markers). Following the identification of PCR markers linked to the gene of interest, specific PCR markers are set up, simplifying the genotyping step for breeders. Nevertheless, PCR markers such as RAPD or AFLP map in majority genes close to the centromeres, reducing their potential efficiency for gene mapping

Table 5.5 Evolution of the average fruit weight and soluble solid content of families along a selection scheme combining backcrosses (BC), selfing (S) and intercrosses (I) of the best plants. Progress expressed as percentage of the control (Cannery Row).

Generation	Number of families	Fruit weight (g)			Soluble solid Content (° Brix)		
		Average of the generation	Control	Percentage of Cannery Row	Average of the generation	Control	Percentage of Cannery Row
BC2	24	61.5	62.4	-1.3	5.1	5.3	-2.3
BC2 S1	13	47.9	58.1	-17.6	5.9	5.3	11.4
BC2 S2	29	54.7	67.5	-19.0	6.4	5.7	13.3
I1 (BC2 S2)	21	60.1	75.1	-19.9	6.1	5.0	21.2
I2 (BC2 S2)	19	69.0	86.5	-20.2	6.4	4.7	37.6
BC3	10	66.8	70.0	-4.6	6.5	4.9	32.2
BC3 S1	17	70.5	88.3	-20.2	6.2	4.9	27.8
I1 (BC3 S1)	16	70.1	76.3	-8.1	6.3	4.9	27.8
I2 (BC3 S1)	19	91.4	82.4	11.0	5.4	4.6	17.8

in tomato (Grandillo and Tanksley 1996b, Haanstra et al. 1999, Saliba-Colombani et al. 2000). Recent advances in using molecular markers for genetic enhancement of tomato are comprehensively detailed in chapter 12.

Applications of QTL Maps in Tomato

Thanks to molecular markers, genes controlling quantitative traits (QTLs) can be detected and genetic maps constructed. QTL mapping strategies allowed screening wild species related to cultivated tomato for alleles (at QTL) of interest from a horticultural point of view (Tanksley and McCouch 1997). The information based on these new approaches have thus been proposed to be introduced in breeding schemes depending on the objective. QTLs have now found application in the improvement of various traits in tomato (see other chapters in the book).

Several studies were performed to detect QTLs for adaptations and quality traits in tomato. Due to the low molecular polymorphism in tomato cultivars, most of these studies concerned interspecific progenies (Table 5.6). They not only address to tomato processing quality attributes, but also to quality attributes for fresh market (Saliba-Colombani et al. 2001, Causse et al. 2001a), horticultural traits, and adaptation to abiotic stresses, such as salinity (Breto et al. 1994), drought (Martin et al. 1989) or heat (see chapter 16). Several population types were used, from F2 or BC1 in the first studies, to populations of more homozygous populations such as recombinant inbreds (Goldman et al. 1995, Saliba-Colombani et al. 2001), advanced backcross generations (BC2 or BC3, Tanksley and Nelson 1996, Fulton et al. 1997, Bernacchi et al. 1998a) or introgression lines (Eshed and Zamir 1995). Most of the wild species were studied, but often only one accession per species has been analyzed, due to the difficulty to obtain the first generations. The salient features of results obtained are emphasized below:

- QTLs have been detected in all cases, sometimes with strong effects. A few QTLs depicting a large part (20 to 50%) of the phenotypic variation, acting together with minor QTLs, frequently act in an additive manner so that they are sometimes dominant and even over-dominant QTLs (Paterson et al. 1988, 1991, de Vicente et al. 1993).
- QTLs can be separated into two types: QTLs stable over the various parameters, such as the different environments, years or types of progeny, and QTLs more specific to one condition (Paterson et al. 1991).
- Some regions involved in the variation of a trait are found in progenies of different accessions of a species, or from different species (Fulton et al. 1997, Bernacchi et al. 1998b, Chen et al. 1999, Grandillo

Table 5.6 QTL detection studies in tomato

Species Progeny	progeny size	Traits -Objectives	Reference
L. cheesmanii			
L. esc. (UC204B) x *L. cheesmanii* (LA483)	350 F2, F3	QTLs of fruit weight, SSC, pH; comparison of locations and tests; comparison to other cross	Paterson et al. (1991)
L. esc. (UC204B) x *L. cheesmanii* (LA483)	97 F8 RIL	QTLs of fruit weight, SSC, seed weight, color	Goldman et al. (1995)
L. chmielewskii			
L. esc. (VF36) x *L. chmielewskii* (LA1028)	165 F2	QTLs of SSC, verification of the association with 2 introgressed markers	Osborn et al. (1987)
L. esc. (VF36, VF145) x *L. chmielewskii* (LA1028)	200 F2	QTLs of SSC, search for introgressed segments and verification in F2	Tanksley et al. (1988)
L. esc. (UC82B) x *L. chmielewskii* (LA1028)	237 BC1	QTLs of fruit weight, SSC, pH; 1st presentation of the interval mapping methodology	Paterson et al. (1988)
L. esc. (UC82B) x *L. chmielewski* (LA1028)	200 BC2F1	QTLs of fruit weight, SSC, pH; fine mapping of QTLs from the previous experiment	Paterson et al. (1990)
L. esc. (VF145) x *L. chmielewski* (LA1028)	64 BC2F5	QTL of SSC - follows Tanksley and Hewitt (1998), physiological bases of SSC QTLs	Azanza et al. (1994)
L. esculentum var. cerasiforme			
L. esc. (Levovil) x *L. esculentum var. cerasiforme* (Cervil)	144 RIL	QTLs of organoleptic quality traits	Causse et al. (2001a), Saliba-Colombani et al. (2001)
L. hirsutum			
L. esc. (E6203) x *L. hirsutum* (LA1777)	149 BC1	major self-incompatibility locus and linked QTLs	Bernacchi et al. (1997)
L. esc. (E6203) x *L. hirsutum* (LA1777)	315 BC2	Advanced Backcross for 19 agronomical traits - 3 locations	Bernacchi et al. (1998a)
L. esc. x *L. hirsutum* (LA407)	BC1F5	Fruit color	Kabelka et al, (2004)
L. pennellii			
L. esc. (M82) x *L. pennellii* (LA716)	20 F3 et 30 BC1S1	drought tolerance QTLs	Martin et al. (1989)
L. esc. (Vendor Tm2a) x *L. pennellii* (LA716)	432 F2 + BC	QTLs of 11 growth traits ; transgression analysis	De Vicente et al. (1993)

(Contd.)

(Contd.)

Cross	Population	Description	Reference
L. esc. (M82) x L. pennellii (LA716)	20 F$_3$ and 30 Bc ISl	QTLs of fruit weight, SSC, pH; transgressions; introgression lines, fine mapping	Eshed et al. (1995, 1996)
L. esc. (UCT5) x L. pennellii (LA716)	100 F2	salt tolerance QTLs; extreme genotype selection	Foolad et al. (1997)
L. esc. (E6203) x L. pennellii (LA1657)	165 BC2	25 horticultural traits	Frary et al. (2004)
L. esc. (M82) x L. pennellii (LA716)	75 IL	Fruit color and fruit composition and candidate genes	Lu et al. (2003), Gur et al. (2004), Causse et al. (2004)
L. pimpinellifolium			
L. esc. (Madrigal) x L. pimpinellifolium (LA1589) // E6203	206 F2	salt tolerance QTLs	Breto et al. (1994)
L. esc. (M82) x L. pimpinellifolium (LA1589) // E6203	257 BC1	QTLs of 19 horticultural traits	Grandillo and Tanksley (1996a)
L. esc. (M82) x L. pimpinellifolium (LA1589)	170 BC2F1	QTLs of several traits ; Advanced Backcross	Tanksley et al. (1996b)
L. esc. (M82) x L. pimpinellifolium (LA1589)	249 BC4F3	fine mapping of a fruit shape QTL (fs8.1)	Grandillo et al. (1996)
L. esc. (NC84173) x L. pimpinellifolium (LA722)	119 BC1	QTLs of lycopene content and other traits; comparison of QTLs among species	Chen et al. (1999)
L. esc (Giant Heirloom) x L. pimpinellifolium (LA1589)	200 F2	QTLs of fruit size and locule number	Lippman and Tanksley (2001)
L. esc (E6203) x L. pimpinellifolium (LA1589)	170 BC2F6	22 quality traits	Doganlar et al. (2002)
L. esc (Yellow Stuffer) x L. pimpinellifolium (LA1589)	160 F2	Fruit size and shape	Van der Knaap and Tanksley (2003)
L. peruvianum			
L. esc. (E6203) x L. peruvianum (LA1706)	200 BC4	35 horticultural traits - 4 locations Advanced Backcross	Fulton et al. (1997)
Several species			
L. esc. x L. pennellii (LA716); L. esc. (M82) x L. pimpinellifolium (LA1589)	86 F2 260 BC1	fine mapping of QTL for fw2.2 cloning; Identification of a YAC carrying the QTL	Alpert et al. (1995,1996)
L. esc. (M82) x L. pimpinellifolium (LA1589); L. esc. (E6203) x L. hirsutum (LA1777)	23 NIL	Advanced Backcross; comparison of QTLs among species; identification of transgressive QTLs in NILs	Bernacchi et al. (1998b)

et al. 1999, Fulton et al. 2002). For example, Figure 5.3 summarizes the chromosome regions carrying several QTLs of fruit weight and soluble solids, based on ten populations involving seven different species. Fruit weight and soluble solids are negatively correlated and QTLs for these traits are often in the same regions, with opposite allele effects, thereby suggesting a pleiotropic effect as a basis of this correlation. Such information is important to identify the limits in introgressing QTLs for solid content in large-fruit varieties.

- The dissection of complex traits in relevant components together with the QTL mapping of these dissected components allowed the genetic basis of the variability of complex traits to be understood. For example, a map of QTLs controlling several attributes of organoleptic quality in fresh-market tomato revealed relations between QTLs for sensory attributes and chemical components of the fruit (Causse et al. 2002). The analysis of biochemical composition of a trait is also important. Indeed, the fine study of two soluble solid QTLs, introgressed from *L. chmielewskii*, showed their different roles in the accumulation of metabolites during fruit development (Azanza et al. 1995).

- Fine mapping experiments allow to precisely map the QTLs in a chromosome region and to verify the existence of several QTLs linked in the same region (Paterson et al. 1990, Frary et al. 2003, Lecomte et al. 2004a). For example, by reducing the size of an introgressed fragment from *L. pennellii*, Eshed and Zamir (1995) identified three linked QTLs controlling fruit weight on a single chromosome arm. Fine mapping is also an important step for cloning QTLs, as shown by the recent successes in cloning QTLs controlling fruit weight (Alpert et al. 1996, Frary et al. 2000), fruit shape (Tanksley 2004) and soluble solid content (Fridman et al. 2000, 2004).

- Wild species, in spite of their commercially low characteristics in comparison to cultivars, may carry alleles contributing to the improvement of agronomic traits in cultivated tomato (de Vicente and Tanksley 1993). For example, in two progenies involving two different species, 10 - 50% of the QTLs had positive allelic effects of the wild species from a breeding point of view, involving 20 agronomical traits (Fig. 5.4). Some of these results, were difficult to foresee, for example, QTL enhancing fruit color in BC3 population introgressed from *L. hirsutum*, a green-fruited species (Bernacchi et al. 1998b). Gur and Zamir (2004) obtained a large progress by pyramiding independent yield-promoting regions introduced from the wild species *L. pennellii*.

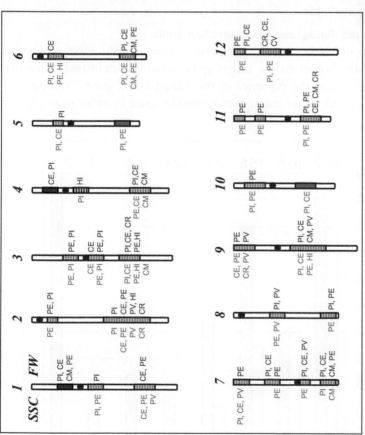

Fig. 5.3 Genetic map of the tomato genome showing the regions where QTLs were detected in at least two different progeny for fruit weight (right of the chromosome) or soluble solid content (left). The map is based on data involving *L. pimpinellifolium* **P I** (Grandillo and Tanksley 1996a, Tanksley et al. 1996, Chen et al. 1999) *L. cheesmanii* **C E** (Paterson et al. 1991, Goldman et al. 1995), *L. chmielewskii* **C M** (Paterson et al. 1988), *L. hirsutum* **H I** (Bernacchi et al. 1998a), *L. peruvianum* **P V** (Fulton et al. 1997), *L. pennellii* **P E** (Eshed and Zamir 1996), *L. esculentum var cerasiforme* **C R** (Saliba-Colombani et al. 2001).

Marker-assisted Selection: Applications and Limitations

Many agricultural important loci have been mapped and tagged with molecular markers. Marker-assisted selection (MAS) allows breeders to select genomic regions involved in the expression of traits of interest. The efficiency and complexity of MAS depend on the genetic nature of the trait (monogenic or polygenic). For monogenic traits, marker-assisted backcross (MABC) is the most straightforward strategy, whereas for polygenic traits various strategies are available.

Marker-Assisted Backcross for Mendelian traits

The principle of MABC for a single gene is quite simple. First, molecular markers tightly linked to the target gene must be identified, allowing efficient detection of the presence of the introgressed gene ("foreground selection"). Second, other markers may also be used in order to accelerate

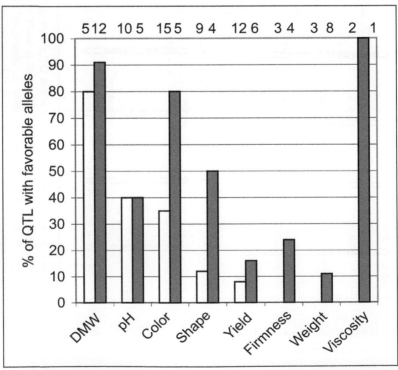

Fig. 5.4 Percentage of QTLs with favorable alleles (from a horticultural and technological perspective) detected in *L. hirsutum* (white) and *L. pimpinellifolium* (black). The total number of QTL per trait is indicated above the bars (adapted from Bernacchi et al. 1998b).

its return to the recipient parent genotype at other loci ("background selection"). Background selection is based not only on markers located on the chromosomes carrying the gene to introgress (carrier chromosome), but also on other chromosomes. Markers devoted to background selection on a carrier chromosome allow the identification of individuals for which recombination events took place on one or both sides of the gene, in order to reduce the length of the donor type segment of genome dragged along with the gene (Young and Tanksley 1989). Background selection on non carrier chromosomes was investigated by Hospital et al. (1992). In three generations of MABC, isogenicity is higher than that obtained by classical methods. Today, tomato breeders frequently use molecular markers for the introgression of monogenic traits such as disease resistance or fruit specific traits.

Marker-assisted Selection for QTLs

Traits showing a quantitative variation are usually controlled by several QTLs, each with different individual effects. Due to the genetic complexity of such traits, several QTLs with limited effects must be simultaneously manipulated. Depending on their number, the nature and range of their effect, and the origin of favorable alleles, different MAS strategies are proposed.

Marker-Assisted Backcross for QTLs   As for monogenic traits, MABC is the most effective strategy when a small number of QTLs, all coming from the same parent, must be transferred into an elite line. Hospital and Charcosset (1997) determined the optimal number and positions of the markers needed to control the QTLs during the foreground selection step and the maximum possible number of QTLs that could be simultaneously monitored with realistic population sizes (a few hundred individuals). They also investigated the use of markers for background selection. In practice, the position of the QTL is not precisely estimated and the true position of the QTL is unknown, but is supposed to be within a confidence interval. From this confidence interval length, Hospital and Charcosset (1997) deduced the number of markers and their position relative to the estimated position of the QTL, in order to ensure an optimal control of the QTL. On average, using at least three markers per QTL allows a good control over several generations by providing a low risk in possessing the donor type alleles without having the desired genotype at the QTL. However, the minimum number of individuals that should be genotyped at each generation depends on (i) the confidence interval length, (ii) the number of markers and (iii) the number of QTLs. It seems illusive to transfer more than four or five QTLs with the simultaneous design unless a very large population is to be considered, or the precision of the QTL location achieved is very high.

Transfer of Large Number of QTLs through Marker-assisted Backcross   A program of QTL detection for fruit quality traits has been achieved using a population of recombinant inbred lines derived from a cross between a cherry tomato line selected for its good taste and intense aroma with a large-fruited line with unremarkable taste. An almost saturated map was constructed (Saliba-Colombani et al. 2000). Each line was evaluated for its physical (fruit weight, color and firmness) and chemical (dry matter weight, titratable acidity, pH, and the contents of soluble solids, sugars, lycopene, carotene and 12 aroma volatiles) traits. Trained panels formed a descriptive profile of each line, assessing sweetness, sourness, aroma (overall aroma intensity, candy, lemon, citrus fruit and pharmaceutical aromas) and texture. Firmness, meltiness, mealiness, juiciness and difficulty to swallow the skin characterized the texture. For all traits, molecular markers were used to map QTLs (Saliba-Colombani et al. 2001, Causse et al. 2001a). Several clusters of QTLs were identified: A total of 86 QTLs over 130 (66%) mapped to about 14% of the map length. QTL co-localizations were observed for related sensory and instrumental traits (Causse et al. 2002). Most of the favorable alleles for tomato organoleptic quality improvement came from the cherry tomato line (coded C). Consecutively, a MABC scheme was set up in order to transfer the five regions of the cherry tomato genome (with the largest effects on fruit quality) into three recurrent lines (coded L, B and D) that had large fruits and different levels of fruit firmness. The QTL regions were chosen according to the QTL effects and their involvement in complementary quality traits (Table 5.7). A single RIL with favorable alleles at all the five QTL regions was crossed to the three recurrent lines. Marker-assisted selection was performed during three successive backcrosses, followed by two selfing generations necessary to fix the five QTLs and recover the genome of the recurrent parents for non-carrier chromosomes (Lecomte et al. 2004b).

The population size as well as the number and location of markers to be used at each generation were optimized, based on the analytical formulas proposed by Hospital and Charcosset (1997). A sequential elimination of the plants carrying the unwanted alleles was performed on BC2 and BC3. The population size allowed a successful transfer of the five segments into each recurrent line. MAS scheme allowed reducing the proportion of donor genome on the non carrier chromosomes under the level expected without

Table 5.7 *Characteristics of the five QTL regions controlling the variation of organoleptic quality traits chosen to be transferred into elite tomato lines (from Causse et al. 2001b).*

Chromosome region	Main QTLs (sensory trait)	R^2 (%)	Size of the region (cM)	Other QTLs
1	Sourness (+)	12.4	31	(-) : Elasticity
2	Sweetness (+)	24.8	31	(+) : Candy, lemon, citrus fruit aromas, firmness, sugar,
	Overall aroma			carotene content, color
	intensity (+)	24.1		(-) : Elasticity, fruit weight
4	Mealiness (+)	13.4	19	(+) : Content in various volatile
	Firmness (+)	31.9		compounds, embarrassing skin, color, firmness
9a	Sourness (+)	16.8	15	(+) : Lemon aroma, juiciness, dry matter weight, firmness, acidity, color
9b	Pharmaceutical aroma (+)	28.5	17	(+) eugenol content

(+) or (-) sign indicates that the cherry tomato line carries favorable (or unfavorable) alleles to the quality value.

selection. Background selection was not applied to markers linked to the QTL segments, and large pieces of donor genome were often fixed around these segments.

Plants carrying from one to five QTLs were selected in order to study their individual or combined effects. Most of the QTLs were recovered in lines carrying one introgression region and new QTLs were detected (Chaib et al. 2006). The line carrying the five segments in L progeny was crossed to several other lines and the fruit quality of the hybrids was assessed by fruit composition and sensory evaluation. It appeared that although fruit size was reduced, hybrids had improved fruit quality, in comparison to parental lines, promising a potential improvement for the pleasure of consumers. Fig. 5.5 illustrates the improvement obtained for several fruit composition traits in the line carrying the favorable alleles at 5 QTLs at the homozygous or heterozygous state. The same trends were observed in the other genetic backgrounds. Nevertheless fruit weight in these genotypes was always lower than expected due to the effect of unexpected QTLs, and this constitutes a limit to marker-assisted breeeding whose effect was masked in the RIL population.

Advanced Backcross, a Strategy for the Simultaneous Discovery and Transfer of New Alleles The advanced backcross QTL analysis is another strategy tailored for the discovery and transfer of valuable QTL alleles

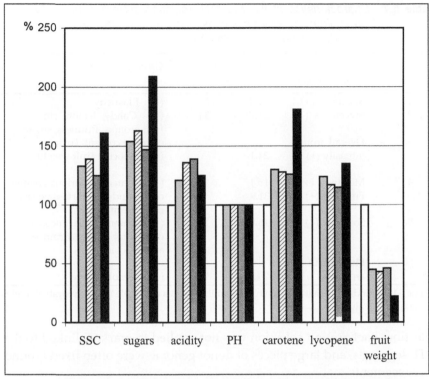

Fig. 5.5   Progress in fruit composition following marker-assisted backcross. Percentages of improvement of genotypes in which five QTLs segments were introgressed, at the homozygous level (black) or in three hybrid combinations (grey and hatched), compared to the recurrent line (white). (Adapted from Lecomte et al. 2004b and unpublished data).

from unadapted donor lines into established elite inbred lines (Tanksley and Nelson 1996). The QTL analysis is delayed until an advanced generation (BC_3 or BC_4), while negative selection is performed to reduce the frequency of deleterious donor alleles during the preliminary steps. BC_3 / BC_4 populations reduce linkage drag by (a) reducing the size of introgressed fragments, and (b) limiting epistatic effects and decreasing the amount of time later needed to develop near isogenic lines carrying the QTL (Fulton et al. 1997). Tanksley and colleagues have applied this strategy for screening positive alleles in 5 wild species, *L. pimpinellifolium* (Tanksley et al. 1996 a,b), *L. hirsutum* (Bernacchi et al. 1998a), *L. peruvianum* (Fulton et al. 1997), *L. pennellii* (Eshed et al. 1996) and *L. parviflorum* (Fulton et al. 2000). They showed a number of important transgressions potentially useful for processing tomato and demonstrated that beneficial alleles could be identified in unadapted germplasm and simultaneously transferred into elite cultivars, thus exploiting the hidden value of exotic germplasm (Tanksley and Nelson 1996, Bernacchi et al. 1998b).

Pyramidal Design When the number of QTLs to introgress becomes important, Hospital and Charcosset (1997) proposed to use a pyramidal design. QTLs are first monitored one by one by MABC, to benefit from higher background selection intensity, and then the selected individuals are intercrossed to cumulate favorable alleles at the QTLs in the same genotype. When favorable alleles come from different sources, van Berloo and Stam (1998) proposed an index method to select among recombinant inbred lines to be crossed in order to obtain a single genotype containing as many favorable quantitative trait alleles as possible. Plants showing the optimal index are crossed together. This strategy was shown to be effective for obtaining transgression in offspring populations of *Arabidopsis* (van Berloo and Stam 1999).

Population Improvement The use of genetic markers to improve populations was proposed using a statistical approach based on an index combining phenotypic and marker information (Lande and Thompson 1990). The efficiency of this MAS was investigated either analytically (Lande and Thompson 1990, Moreau et al. 1998) or by computer simulation (Hospital et al. 1997). MAS could be more effective in comparision to pure line phenotypic selection particularly in quite large populations and for traits showing relatively low heritabilities. Nevertheless, this strategy is more devoted to allogamous crops such as maize than to autogamous tomato (Moreau et al. 1998).

Towards Mixed Breeding Strategies

MAS as explained above is particularly effective for traits difficult and expensive to evaluate such as fruit quality. Once molecular markers closely linked to the desirable alleles are identified, marker-assisted selection can be performed in segregating populations even at early stage of plant development. It is thus possible to conduct several rounds of selection in a year. MABC quickly cumulates up to five QTLs in a single genotype and is very effective. The availability of reliable PCR-based markers proved crucial for the success of such selection scheme. It seems also important to re-evaluate QTL effect in advanced generation, as unexpected results may limit the success of MABC. In the future, it will be important to explore the complementarity between marker-assisted selection and conventional breeding and to develop overall strategies that tightly and interactively integrate the two approaches.

Biotechnology for Quality Enhancement

Use of Genetic Engineering for Tomato Breeding

Several genes, related to quality traits, have been used for transformation experiments (chapters 9-11). Transgenic approaches to modify tomato fruit quality concern all the quality attributes. The modification of sink/source relations has been proposed, screening several target enzymes (sucrose synthase, sucrose phosphate synthase, invertase, ADPG pyrophosphorylase; reviewed by Herbers and Sonnewald 1998). Fruit shelf life and texture showed improvement through down regulation of key enzymes of the ethylene biosynthesis or of cell wall enzymes (Hobson and Grierson 1993, Brummell et al. 2002, Powell et al. 2003). Taste was improved in transgenic plants carrying a thaumatin gene from *Thaumatococcus daniellii* (Bartoszewski et al. 2003). Carotenoid synthesis pathway was dissected using transgenic approach, and it has also been proposed to use this approach to increase the content in carotenoids (Fraser et al. 2001). Aroma composition was modified based on the expression of alcohol dehydrogenase (Speirs et al. 1998), or introducing new enzymes such as yeast Δ9 desaturase (Wang et al. 1996), borage Δ6 desaturase (Cook et al. 2002) or S-linalool synthase from a flower *Clarkia brewrii* (Lewinsohn et al. 2001). Improvement of nutritional value was proposed for several targets such as quercetin (Bovy et al. 2002) or polyamines (Mehta et al. 2002). Numerous examples of experiments can be reviewed but the only case developed until a commercial use was the antisense polygalacturonase gene for the modification of fruit texture (Kramer and Redenbaugh 1994).

Towards the Knowledge of the Whole Genome

Tomato is a model crop for the study of fleshy fruit development and a number of genes expressed during fruit development and ripening have been cloned. Recently a program of systematic EST sequencing provided about 150,000 tags corresponding to more than 30,000 unique genes (van der Hoeven et al. 2002). The identification of a set of conserved ortholog genes between *L. esculentum* and *A. thaliana* will also facilitate synteny studies and fasten gene and QTL characterization (Fulton et al. 2002). All the genomic tools available will allow a better knowledge of gene function and regulation as well as the development of precise and more efficient gene-assisted selection, avoiding large segments to be introgressed. Once an important gene is characterized, it will be important to find some new allelic variants. Some new techniques have been proposed for screening mutant populations (McCallum et al. 2000), which in the future may be helpful in the discovery of such new alleles (Emmanuel and Levy 2002).

CONCLUSION: ENHANCED TOMATO BREEDING THANKS TO THE BETTER KNOWLEDGE OF THE GENOME

Molecular markers have permitted the dissection of the genetic basis of complex traits into individual components and the location of genes/QTLs on chromosomes has become individually accessible to selection. These genetic markers have also allowed breeders to access the wild species in a more efficient way than in the past. Exotic libraries, which consist of marker-defined genomic regions taken from wild species and introgressed onto the background of elite crop lines, provide plant breeders with an important opportunity to improve the agricultural performance of modern crop varieties. These libraries are also powerful tools for the discovery and characterization of genes that underlie traits of agricultural value (Zamir 2001).

Nevertheless, the use of molecular markers has some limitations. Polymorphic markers are needed, but the level of polymorphism revealed within *L. esculentum* accessions by the traditional markers is restricted. However, the study of single nucleotide polymorphisms (SNP) may provide in the future a new source of polymorphic markers. Furthermore, epistatic effects together with the effect of genetic background on the expression of QTLs have been shown to hamper the use of molecular markers in marker-assisted selection. There is thus a need to optimize the breeding schemes by combining the use of molecular markers with phenotypic evaluations.

Biotechnology has provided considerable change in the breeding paradigm. Genetic transformation widens the realm of possibility, by the use of genes of other species or the modulation of tomato genes. Until now, the examples of commercial use of GMO in tomato have been rare, but genetic transformation is very useful for fundamental research. The method needs to be perfected to commercialize GMO plants. The use of specific promoters, together with the absence of reporter genes, will make these tools more efficient and acceptable to consumers, from technical point of view, and should be combined in the future to other breeding techniques. On the philosophical point of view, the use of GMO for fruit quality improvement can be accepted in some societies and not in others. It is hoped that in the future technologies available would allow to design and optimize breeding methods without frontiers by application of all techniques (GMO, markers and traditional breeding). Thus, it would be important to combine the empirical approach of breeders based on an intimate knowledge of the tomato crop with the power of biotechnology. Integration of related disciplines are more and more important in order to (1) develop more efficient methods to evaluate quality, (2) enhance knowledge of the biochemical and molecular bases of the traits, and (3) better understand G x E to increase the adaptation of new varieties to new conditions.

REFERENCES

Alpert, K.B., S. Grandillo, and S.D. Tanksley. 1995. fw2.2: a major QTL controlling fruit weight is common to both red- and green-fruited tomato species. Theor Appl Genet 91: 994-1000.

Alpert, KB. and S.D. Tanksley. 1996. High-resolution mapping and isolation of an yeast artificial chromosome contig containing fw2.2: a major fruit weight quantitative trait locus in tomato. Proc Natl Acad Sci USA 93: 15503-15507.

Auerswald, H., P. Peters, B. Bruckner, A. Krumbein, and R. Kuchenbuch. 1999. Sensory analysis and instrumental measurements of short-term stored tomatoes (*Lycopersicon esculentum* Mill.). Postharvest Biology and Technology 15:323-334.

Atanassova, B. 1999. Functional male sterility (ps2) in tomato (*Lycopersicon esculentum* Mill.) and its application in breeding and seed production. Euphytica 107: 1, 13-21.

Azanza, F., D. Kim, S.D. Tanksley , and J.A. Juvik. 1995. Genes from *Lycopersicon chmielewskii* affecting tomato quality during fruit ripening. Theor Appl Genet 91: 495-504.

Azanza, F., T.E. Young, D. Kim, S.D. Tanksley, and J.A. Juvik. 1994. Characterization of the effects of introgressed segments of chromosome 7 and 10 from *Lycopersicon chmielewskii* on tomato soluble solids, pH and yield. Theor Appl Genet 87: 965-972.

Baldwin, E.A., M.O. Nisperos-Carriedo, R. Baker, and J.W. Scott. 1991. Quantitative analysis of flavor parameters in six Florida tomato cultivars. J Agric Food Chem 39: 1135-1140.

Baldwin, E.A, J.W. Scott, M.A. Einstein, T.M.M. Malundo, B.T. Carr, R.L. Shewfelt, and K.S. Tandon. 1998. Relationship between sensory and instrumental analysis for tomato flavor. J Am Soc Hort Sci 123:906-915.

Baldwin, E. A., J. W. Scott, C. K. Shewmaker, and W. Schuch. 2000. *Flavor trivia* and tomato aroma: biochemistry and possible mechanisms for control of important aroma components. HortScience 35: 6, 1013-1022.

Bartoszewski, G., A. Niedziela, M. Szwacka, and K. Niemirowicz-Szczytt. 2003. Modification of tomato taste in transgenic plants carrying a thaumatin gene from *Thaumatococcus daniellii* Benth. Plant Breeding 122: 347- 351.

Baxter, C.J., M. Sabar, W.P. Quick, and L.J. Sweetlove. 2005. Comparison of changes in fruit gene expression in tomato introgression lines provides evidence of genome-wide transcriptional changes and reveals links to mapped QTLs and described traits. J Exp Bot 56 :1591-1604.

Bernacchi, D. and S.D. Tanksley. 1997. An interspecific backcross of *Lycopersicon esculentum* x *L hirsutum*: linkage analysis and a QTL study of sexual compatibility factors and floral traits. Genetics 147: 861-877.

Bernacchi, D., T. Beck-Bunn, Y. Eshed, J. Lopez, V. Petiard, J. Uhlig, D. Zamir, and S. Tanksley. 1998a. Advanced backcross QTL analysis in tomato. I. Identification of QTLs for traits of agronomic importance from *Lycopersicon hirsutum*. Theor Appl Genet 97 (3): 381-397.

Bernacchi, D., T. Beck-Bunn, D. Emmatty, Y. Eshed, S. Inai, J. Lopez , V. Petiard, H. Sayama, J. Uhlig, D. Zamir, and S. Tanksley 1998b. Advanced backcross QTL analysis in tomato. II. Evaluation of near-isogenic lines carrying single-donor introgressions for desirable wild QTL-alleles derived from *Lycopersicon hirsutum* and *L. pimpinellifolium*. Theor Appl Genet 97 (1/2): 170-180; erratum 97(7): 1191-1196.

Bertin, N., C. Gary, M. Tchamitchian, and B. E. Vaissiere. 1998. Influence of cultivar, fruit position and seed content on tomato fruit weight during a crop cycle under low and high competition for assimilates. J Hort Sci Biotech 73 : 541-548.

Bovy, A., R. de Vos, M. Kemper, E. Schijlen, M. Almenar Pertejo, S. Muir, G. Collins, S. Robinson, M. Verhoeyen, S. Hughes, C. Santos-Buelga, A. van Tunen. 2002. High-flavonol tomatoes resulting from the heterologous expression of the maize transcription factor genes LC and C1 . The Plant Cell 14: 2509-2526.

Bramley, P.M. 2000. Is lycopene beneficial to human health? Phytochemistry 54:233-236.

Breto, M.P., M.J. Asins, and E.A. Carbonell. 1994. Salt tolerance in *Lycopersicon* species. III. Detection of QTLs by means of molecular markers. Theor Appl Genet 88: 395-401.

Bruhn, C.M., N. Feldman, C. Garlitz, J. Harwood , E. Ivans, M. Marshall , A. Riley, D. Thurber, and E. Williamson. 1991. Consumer perceptions of quality: Apricots, cantaloupes, peaches, pears, strawberries, and tomatoes. J Food Qual 14:187-195.

Brummell, D.A., W.J. Howie, C. Ma, and P. Dunsmuir. 2002. Postharvest fruit quality of transgenic tomatoes suppressed in expression of a ripening-related expansin. Postharvest Biology and Technology 25: 209-220.

Bucheli, P., E. Voirol, R. Delatorre, J. Lopez, A. Rytz, S.D. Tanksley, and V. Petiard. 1999. Definition of nonvolatile markers for flavor of tomato (*Lycopersicon esculentum* Mill.) as tools in selection and breeding. J Agric Food Chem 47:659-664.

Butler, L. 1952. The linkage map of the tomato. J. Hered 43: 25-35.

Casali, V.W.D. and E.C. Tigchelaar. 1975. Breeding progress in tomato with pedigree selection and single seed descent. J Am Soc Hort Sci 100: 362-364.

Causse, M, V. Saliba-Colombani, M. Buret, I. Lesschaeve, and S. Issanchou. 2001a. Genetic analysis of organoleptic quality in fresh market tomato: 2. Mapping QTLs for sensory attributes. Theor Appl Genet 102:273-283.

Causse, M, L. Lecomte, N. Baffert, P. Duffe, and F. Hospital. 2001b. Marker-Assisted Selection for the transfer of QTLs controlling fruit quality traits into tomato elite lines. Acta Hort 546:557-564.

Causse, M, V. Saliba-Colombani, L. Lecomte, P. Duffé, P. Rousselle, and M. Buret. 2002. Genetic analysis of fruit quality attributes in fresh market tomato. J Exp Bot 53: 2090-2098.

Causse, M., M. Buret, K Robini, and P. Verschave. 2003. Inheritance of nutritional and sensory quality traits in fresh market tomato and relation to consumer preferences. J Food Sci 68: 2342-2350.

Causse, M., P. Duffe, M.C. Gomez, M. Buret, R. Damidaux, D. Zamir, A. Gur, C. Chevalier, M. Lemaire-Chamley, and C. Rothan. 2004. A genetic map of candidate genes and QTLs involved in tomato fruit size and composition. J Exp Bot 55 : 1671-85.

Chaib, J., L. Lecomte , M. Buset, M. Causse. 2006 stability over genetic backgrounds, generations and years of quantitative trait locus (QTLS) for organoleptic quality in tomato. Theor Appl Genet in press.

Chen, F.Q., M.R. Foolad, J. Hyman, D.A. St Clair and R.B. Beeleman. 1999. Mapping QTLs for lycopene and other fruit traits in a *Lycopersicon esculentum* x *L pimpinellifolium* cross and comparison of QTLs across tomato species. Mol Breeding 5: 283-299.

Chetelat, R.T., J.W. De Verna, and A.B. Bennett. 1995. Introgression into tomato (*Lycopersicon esculentum*) of the *L. chmielewskii* sucrose accumulator gene (*sucr*) controlling fruit sugar composition. Theor Appl Genet 91 : 327-333.

Cook, D., D. Grierson, C. Jones, A. Wallace, G. West, and G. Tucker. 2002. Modification of fatty acid composition in tomato (*Lycopersicon esculentum*) by expression of a borage Delta(6)-desaturase. Mol Biotech 21: 123-128.

Davies, J.N. and G.E. Hobson. 1981. The constituents of tomato fruit - The influence of environment, nutrition and genotype. Crit Rev Food Sci Nutr 15:205-280.

De Vicente, MC, and S.D. Tanksley. 1993. QTL analysis of transgressive segregation in an interspecific tomato cross. Genetics 134: 585-596.

Doganlar, S, A. Frary, H.M. Ku, and S.D. Tanksley. 2002. Mapping quantitative trait loci in inbred backcross lines of *Lycopersicon pimpinellifolium* (LA1589). Genome 45: 1189-1202.

Dorais, M., AP. Papadopoulos, A. Gosselin. 2001. Greenhouse tomato fruit quality. Hort Rev 26:239-319.

Emmanuel, E. and A. Levy 2002. Tomato mutants as tools for functional genomics. Curr Opin in Plant Biol 5(2): 112-117.

Eshed, Y. and D. Zamir. 1995. An introgression line population of *Lycopersicon pennellii* in the cultivated tomato enables the identification and fine mapping of yield-associated QTLs. Genetics 141: 1147-1162.

Eshed, Y. and D. Zamir. 1996. Less-than-additive epistatic interactions of quantitative trait loci in tomato. Genetics 143: 1807-1817.

Eshed, Y., G. Gera, and D. Zamir. 1996. A genome-wide search for wild-species alleles that increase horticultural yield of processing tomato. Theor Appl Genet 93: 877-886.

Foolad, M.R., T. Stoltz, C. Dervinis, R.L. Rodriguez, and R.A. Jones. 1997. Mapping QTLs conferring salt tolerance during germination in tomato by selective genotyping. Mol Breed 3: 269-277.

Frary, A., T.C. Nesbitt, S. Grandillo, E. Knaap, B. Cong, Liu J, J. Meller, R. Elber, K.B. Alpert, and S.D. Tanksley. 2000. fw2.2: a quantitative trait locus key to the evolution of tomato fruit size. Science 289: 85-88.

Frary, A., S. Doganlar, A. Frampton, T. Fulton, J. Uhlig, H. Yates, and S. Tanksley. 2003. Fine mapping of quantitative trait loci for improved fruit characteristics from *Lycopersicon chmielewskii* chromosome 1. Genome 46 : 235-243.

Frary, A., T.M. Fulton, D. Zamir, and S.D. Tanksley. 2004. Advanced backcross QTL analysis of a *Lycopersicon esculentum x L. pennellii* cross and identification of possible orthologs in the Solanaceae. Theor Appl Genet 108: 485-496.

Fraser, P. D., S. Romer, J. W. Kiano, C. A. Shipton, P. B. Mills, R. Drake, W. Schuch, and P. M. Bramley, 2001: Elevation of carotenoids in tomato by genetic manipulation. J Sci Food Agric 81. 822-827.

Fray, R.G. and D. Grierson 1993. Identification and genetic analysis of normal and mutant phytoene synthase genes of tomato by sequencing, complementation, and co-suppression. Plant Mol Biol 22: 589–602.

Fridman, E., T. Pleban, and D. Zamir. 2000. A recombination hotspot delimits a wild-species quantitative trait locus for tomato sugar content to 484 bp within an invertase gene. Proc Nat Acad Sci USA 97(9):4718-23.

Fridman, E., Y.S. Liu, L. Carmel-Goren, A. Gur, M. Shoresh, T. Pleban, Y. Eshed, and D. Zamir. 2002. Two tightly linked QTLs modify tomato sugar content via different physiological pathways. Mol Gen Genet 266 (5): 821-826.

Fridman, E., F. Carrari, Y.S. Liu, A.R. Fernie, and D. Zamir. 2004. Zooming in on a quantitative trait for tomato yield using interspecific introgressions. Science 305 : 1786-1789.

Fulton, T.M., T. Beck-Bunn, D. Emmatty, Y. Eshed, J. Lopez, V. Petiard, J. Uhlig, D. Zamir, and S.D. Tanksley. 1997. QTL analysis of an advanced backcross of *Lycopersicon peruvianum* to the cultivated tomato and comparisons with QTLs found in other wild species. Theor Appl Genet 95: 881-894.

Fulton, T.M., S. Grandillo, T. Beck-Bunn, E. Fridman, A. Frampton, J. Lopez, V. Petiard, J. Uhlig, D. Zamir, and S.D. Tanksley. 2000. Advanced backcross QTL analysis of a *Lycopersicon esculentum x Lycopersicon parviflorum* cross. Theor Appl Genet 100: 1025-1042.

Fulton, T. M., R. van der Hoeven, N.T. Eanetta, and S.D. Tanksley. 2002. Identification, analysis, and utilization of conserved ortholog set markers for comparative genomics in higher plants. The Plant Cell 14: 1457-1467.

Fulton, T.M., P. Bucheli, E. Voirol, J. Lopez, V. Petiard, and S.D. Tanksley. 2002. Quantitative trait loci (QTL) affecting sugars, organic acids and other biochemical properties possibly contributing to flavor, identified in four advanced backcross populations of tomato. Euphytica 127: 163-177.

Gautier, H., A. Rocci, M. Buret, D. Grasselly, Y. Dumas, and M. Causse. 2005a. Effect of photoselective filters on the physical and chemical traits of vine-ripened tomato fruits. Can J Plant Sci 85 (2): 439-446.

Gautier, H., A. Rocci, M. Buret, D. Grasselly, and M. Causse. 2005b. Fruit load or fruit position alters response to temperature and subsequently cherry tomato quality. J Sci Food Agri 85 (6): 1009-1016.

Giovannucci, E. 1999. Tomatoes, tomato-based products, lycopene, and cancer : review of epidemiologic literature. J Nat Canc Inst 91: 317-331.

Goldman, J.L., I. Paran, and D. Zamir. 1995. Quantitative trait locus analysis of a recombinant inbred line population derived from a*Lycopersicon esculentum* x *Lycopersicon cheesmanii* cross. Theor Appl Genet 90: 925-932.

Grandillo, S., H.M. Ku, and S.D. Tanksley. 1996. Characterization of fs8.1, a major QTL influencing fruit shape in tomato. Mol Breed 2: 251-260.

Grandillo, S. and S.D. Tanksley. 1996a. QTL analysis of horticultural traits differentiating the cultivated tomato from the closely related species *Lycopersicon pimpinellifolium*. Theor Appl Genet 95:935-951.

Grandillo, S. and S.D. Tanksley. 1996b. Genetic analysis of RFLPs, GATA microsatellites and RAPDs in a cross between *L. esculentum* and *L. pimpinellifolium*. Theor Appl Genet 92: 957-965.

Grandillo, S. D. Zamir, and S.D. Tanksley. 1999. Genetic improvement of processing tomatoes: a 20 years perspective. Euphytica 110: 85-97.

Grasselly, D., B. Navez, and M. Letard. 2000. Tomate : Pour un produit de qualité, Ed CTIL, pp. 222

Grolier, P. and E. Rock. 1998. Composition of tomato in antioxidants : variations and methodology. *In* Proc of the Tomato and Health Seminar, Pamplona, Spain, 25-28 May 1998, pp. 23-25.

Gur, A., Y. Semel, A. Cahaner, and D. Zamir. 2004. Real time QTL of complex phenotypes in tomato interspecific introgression lines. Trends In Plant Sci 9: 107-109.

Gur, A. and D. Zamir, 2004. Unused natural variation can lift yield barriers in plant breeding PLOS. Biology 2 : 1610-1615.

Hanson, P.M., R.Y. Yang, J. Wu, J.T. Chen, D. Ledesma, S.C.S. Tsou, and T.C. Lee. 2004. Variation for antioxidant activity and antioxidants in tomato. J Am Soc Hort Sci 129: 704-711.

Haanstra, J.P.W., C. Wye, and H. Verbakel. 1999. An integrated high-density RFLP-AFLP map of tomato based on two *Lycopersicon esculentum* x *L. pennellii* F2 populations . Theor Appl Genet 99: 254-271.

Hallauer, A.R. and J.B. Miranda. 1981 Quantitative genetics in maize breeding. Iowa State Univ Press, pp. 468.

Herbers, K. and U. Sonnewald. 1998. Molecular determinants of sink strength. Curr Opin in Plant Biol 1: (3): 207-216.

Hewitt, J.D. and M.A. Stevens. 1981. Growth analysis of two tomato genotypes differing in total fruit solids content. J Am Soc Hort Sci 106: 723-727.

Ho, L.C. 1996. The mechanism of assimilate partitioning and carbohydrate compartmentation in fruit in relation to the quality and yield of tomato. J Exp Bot 47: 1239-1243.

Hobson, G.E. and L. Bedford. 1989. The composition of cherry tomatoes and its relation to consumer acceptability. J Hort Sci 64:321-329.

Hobson, G.E. and D. Grierson. 1993. Tomato. *In:* J. Taylor and G Tucker (eds), Biochemistry of Fruit Ripening. G. Seymour, Ced.3 pp. 405-442.

Hobson, G.E. 1988. How the tomato lost its taste. New Sci 19:46-50.

Hospital, F. and A. Charcosset. 1997. Marker-assisted introgression of Quantitative Trait Loci. Genetics 147: 1469-1485.

Hospital, F., C. Chevalet, and P. Mulsant. 1992. Using markers in gene introgression breeding programs. Genetics 132: 1199-1210.

Hospital, F., L. Moreau, F. Lacoudre, A. Charcosset, and A. Gallais. 1997. More on the efficiency of marker-assisted selection. Theor Appl Genet 95: 1181-1189.

Isaacson, T., G. Ronen, D. Zamir, and J. Hirschberg. 2002. Cloning of tangerine from tomato reveals a carotenoid isomerase essential for the production of beta-carotene and xanthophylls in plants. Plant Cell 14(2): 333-342.

Janse, J. and M. Schols. 1995. Une préférence pour un goût sucré et non farineux. Groenten + Fruit 26:16-17.

Johansson, L., A. Haglund, L. Berglund, P. Lea, and E. Risvik. 1999. Preference for tomatoes affected by sensory attributes and information about growth conditions. Food Quality and Preference 10:289-298.

Jones, R.A. 1986. Breeding for improved post-harvest tomato quality : genetical aspects. Acta Hort 190:77-78.

Kabelka, E., W.C. Yang, and D.M. Francis. 2004. Improved tomato fruit color within an inbred backcross line derived from *Lycopersicon esculentum* and *L. hirsutum* involves the interaction of loci. J Amer Soc Hort Sci 129: 250-257.

Kamal, A. H. M., T. Takashina, H. Egashira, H. Satoh, and S. Imanishi. 2001. Introduction of aromatic fragrance into cultivated tomato from the *'peruvianum* complex'. Plant Breed 120: 179-181.

Kaul, M.L.H. 1991. Reproductive biology in tomato. *In:* G. Kalloo [ed], Genetic Improvement of Tomato. Springer Verlag, pp. 39-50.

Kramer, M. G. and K. Redenbaugh. 1994: Commercialization of a tomato with an antisense polygalacturonase gene: the FLAVR SAVR™ tomato story. Euphytica 79, 293-297.

Ku, H.M., S. Doganlar, K.Y. Chen, and SD. Tanksley. 1999. The genetic basis of pear-shaped tomato fruit. Theor Appl Genet 99: 844-850.

Lande, R. and R. Thompson. 1990. Efficiency of marker-assisted selection in the improvement of quantitative traits. Genet 124: 743-756.

Langlois, D., P.X. Etievant, P. Pierron, and A. Jorrot. 1996. Sensory and instrumental characterisation of commercial tomato varieties. Zeitsch Lebens Unters Forsch 203:534-540.

Laterrot, H. 1996. Twenty near isogenic lines in Moneymaker type with different genes for disease resistance. Rept Tom Genet Coop 46: 34.

Lecomte, L., V. Saliba-Colombani, A. Gautier, M.C. Gomez-Jimenez, P. Duffé, M. Buret, and M. Causse. 2004a. Fine mapping of QTLs for the fruit architecture and composition in fresh market tomato, on the distal region of the long arm of chromosome 2. Mol Breed 13: 1-14.

Lecomte, L., P. Duffé, M. Buret, B. Servin, F. Hospital, and M. Causse. 2004b. Marker-assisted introgression of 5, QTLs controlling fruit quality traits into three tomato lines revealed interactions between QTLs and genetic backgrounds. Theor Appl Genet 109: 658-668.

Levin, I., N. Gilboa, E. Yeselson, S. Shen, and AA. Schaffer. 2000. Fgr, a major locus that modulates the fructose to glucose ratio in mature tomato fruits. Theor Appl Genet 100: 2, 256-262.

Levin, I., P. Frankel, N. Gilboa, S. Tanny, and A. Lalazar, 2003. The tomato dark green mutation is a novel allele of the tomato homolog of the DEETIOLATED1 gene. Theor Appl Genet 106: 454-460.

Lee, S.Y., I. Luna-Guzman, S. Chang, D.M. Barrett, and J.X. Guinard. 1999. Relating descriptive analysis and instrumental texture data of processed diced tomatoes. Food Quality and Preference 10:447-455.

Lewinsohn, E., F. Schalechet, J. Wilkinson, K. Matsui, Y. Tadmor, K.H. Nam, O. Amar, E. Lastochkin, O. Larkov, U. Ravid, W. Hiatt, S. Gepstein, and E. Pichersky. 2001. Enhanced levels of the aroma and flavor compound S-Linalool by metabolic engineering of the terpenoid pathway in tomato fruits. Plant Physiol. 127: 1256-1265.

Lippman, Z. and S.D. Tanksley. 2001. Dissecting the genetic pathway to extreme fruit size in tomato using a cross between the small-fruited wild species *Lycopersicon pimpinellifolium* and *L.esculentum* var Giant Heirloom. Genet 158: 413-422.

Liu, Y.S., A. Gur, G. Ronen, M. Causse, R. Damidaux, M. Buret, J. Hirschberg, and D. Zamir. 2003. There is more to fruit colour than candidate carotenoid genes. Plant Biotech J 1: 195-207.

Liu, Y.S., S. Roof, Z.B. Ye, C. Barry, A. van Tuinen, J. Vrebalov, C. Bowler, and J. Giovannoni. 2004. Manipulation of light signal transduction as a means of modifying fruit nutritional quality in tomato. Proc Nat Acad Sci USA 101: 9897-9902.

Malundo, T.M.M., R.L. Shewfelt, and J.W. Scott. 1995. Flavor quality of fresh market tomato (*Lycopersicon esculentum* Mill.) as affected by sugar and acid levels. Postharvest Biol Technol 6:103-110.

Mao, L., D. Begum, H.W. Chuang, M.A. Budiman, E.J. Szymkowiak E.E. Irish, and R.A. Wing. 2000. JOINTLESS is a MADS-box gene controlling tomato flower abscission zone development. Nature 406: 910-913.

Martin, B., J. Nienhuis, G. King, and A. Schaeffer. 1989. Restriction fragment length polymorphisms associated with water use efficiency in tomato. Science 243: 1725-1726.

McCallum, C.M., L Comai, EA Greene, and S. Henikoff. 2000. Targeted screening for induced mutations. Nature Biotechnol 18: 455-457.

McGlasson, W.B., J.H. Last, K.J. Shaw, and S.K. Meldrum. 1987. Influence of the non-ripening mutants rin and nor on the aroma of tomato fruit. HortScience 22:632-634.

Mehta, R. A., T. Cassol, N. Li, N. Ali, A. K. Handa, and A. K. Mattoo. 2002: Engineered polyamine accumulation in tomato enhances phytonutrient content, juice quality, and vine life. Nature Biotechnol 20: 613-618.

Moore, S., J.P. Vrebalov, J. Paxton, J. Giovannoni. 2002. Use of genomics tools to isolate key ripening genes and analyse fruit maturation in tomato. J Exp Bot 53: 2023-2030.

Moreau, L., A. Charcosset, F. Hospital, and A. Gallais. 1998. Marker-assisted selection efficiency in populations of finite size. Genetics 148: 1353-1365.

Mustilli, A.C., F. Fenzi, R. Ciliento, F. Alfano, and C. Bowler. 1999. Phenotype of the tomato high pigment-2 mutant is caused by a mutation in the tomato homolog of DEETIOLATED1. The Plant Cell 11: 145-157.

Offord, E.A. 1998. Nutritional and health benefits of tomato products. *In:* Proc of the Tomato and Health Seminar, Pamplona, Spain, 25-28 May 1998, pp. 5-11.

Osborn, T.C. and D.C. Alexander, F. Fobes. 1987. Identification of restriction fragment length polymorphisms linked to genes controlling soluble solids content in tomato fruit. Theor Appl Genet 73: 350-356.

Overy, S.A., H.J. Walker, S. Malone, T.P. Howard, C.J. Baxter, L.J. Sweetlove, S.A. Hill, and W.P. Quick. 2005. Application of metabolite profiling to the identification of traits in a population of tomato introgression lines. J Exp Bot 56: 287-296.

Paterson, A.H., S. Damon, J.D. Hewitt, D. Zamir, H.D. Rabinowitch, S.E. Lincoln, E.S. Lander, and S.D. Tanksley. 1991. Mendelian factors underlying quantitative traits in tomato: comparison across species, generations, and environments. Genetics 127: 181-197.

Paterson, A.H., J.W. Deverna, B. Lanini, and S.D. Tanksley. 1990. Fine mapping of quantitative trait loci using selected overlapping recombinant chromosomes in an interspecific cross of tomato. Genetics 124: 735-742.

Paterson, A.H., E.S. Lander, J.D. Hewitt, S. Peterson , S.E. Lincoln, and S.D. Tanksley. 1988. Resolution of quantitative traits into Mendelian factors by using a complete linkage map of restriction fragment length polymorphisms. Nature 335: 721-726.

Peirce, L.C. 1991. Selection systems for tomato improvement. *In:* G. Kalloo [ed.]., Genetic Improvement of Tomato. Springer Verlag, pp. 59-72.

Petro-Turza, M. 1987. Flavor of tomato and tomato products. Food Rev Int 2:309-351.

Philouze, J. 1991. Description of isogenic lines, except for one, or two, monogenically controlled morphological traits in tomato, *Lycopersicon esculentum* Mill. Euphytica 56: 121-131.

Pnueli, L., L. Carmel-Goren, D. Hareven, T. Gutfinger, J. Alvarez, M. Ganal, D. Zamir, and E. Lifschitz. 1998. The SELF-PRUNING gene of tomato regulates vegetative to reproductive switching of sympodial meristems and is the ortholog of CEN and TFL1. *Development* 125(11):1979-89.

Powell, A.L.T., M.S. Kalamaki, P.A. Kurien, Gurrieri,. and S. Bennett. 2003. Simultaneous transgenic suppression of LePG and LeExp1 influences fruit texture and juice viscosity in a fresh market tomato variety. J Agric Food Chem 51: 7450-7455.

Rick, C.M. and R.T. Chetelat. 1993. Revised list of monogenic stocks. TGC Report 43: 53-77.

Ronen, G., M. Cohen , D. Zamir, and J. Hirschberg. 1999. Regulation of carotenoid biosynthesis during tomato fruit development: expression of the gene for lycopene epsilon-cyclase is down-regulated during ripening and is elevated in the mutant Delta. Plant J 17(4):341-51.

Ronen, G., G.L. Carmel, D. Zamir and J. Hirschberg. 2000. An alternative pathway to beta-carotene formation in plant chromoplasts discovered by map-based cloning of Beta and old-gold color mutations in tomato. Proc Nat Acad Sci USA 97: 11102–11107.

Saliba-Colombani, V., M. Causse, L. Gervais, and J. Philouze. 2000. Efficiency of AFLP, RAPD and RFLP markers for the construction of an intraspecific map of the tomato genome. Genome 43: 29-40.

Saliba-Colombani, V., M. Causse, D. Langlois, J. Philouze, and M. Buret. 2001. Genetic analysis of organoleptic quality in fresh market tomato: 1. Mapping QTLs for physical and chemical traits. Theor Appl Genet 102: 259-272.

Schaffer, A.A., I. Levin, I. Oguz, M. Petreikov, F. Cincarevsky, Y. Yeselson, S. Shen, N. Gilboa, and M. Bar. 2000. ADPglucose pyrophosphorylase activity and starch accumulation in immature tomato fruit: the effect of a *Lycopersicon hirsutum*-derived introgression encoding for the large subunit. Plant Sci 152: 135-144.

Schauer, N., D. Zamir, and A.R. Fernie. 2005. Metabolic profiling of leaves and fruit of wild species tomato: a survey of the *Solanum lycopersicum* complex. J Exp Bot 56 : 297-307.

Scott, J.W. 1983. Genetic sources of tomato firmness. In Proc 4th Tomato Quality Workshop. Fla Exp Stn Res Rept 60-67.

Seymour, G.B., K. Manning, E.M. Eriksson, A.H. Popovich, and G.J. King. 2002. Genetic identification and genomic organization of factors affecting fruit texture. J Exp Bot 53 : 2065-2071.

Speirs, J., E. Lee, K. Holt, K. Yong-Duk, N.S. Scott, B. Loveys, and W. Schuch. 1998. Genetic manipulation of alcohol dehydrogenase levels in ripening tomato fruit affects the balance of some flavor aldehydes and alcohols. Plant Physiol 117: 1047-1058.

Stevens, M.A. 1972. Citrate and malate concentrations in tomato fruits: genetic control and maturational effects. J Am Soc Hort 97:655-658.

Stevens, M.A. 1986. Inheritance of tomato fruit quality components. Plant Breed Rev 4:273-311.

Stevens, M.A., A.A. Kader, M. Albright-Holton, and M. Algazi. 1977. Genotypic variation for flavor and composition in fresh market tomatoes. J Am Hort Sci 102:680-689.

Stevens, M.A., A.A. Kader, and M. Albright. 1979. Potential for increasing tomato flavor via increased sugar and acid content. J Am Soc Hort Sci 104:40-52.

Stommel, J.R. 2001. USDA 97L63, 97L66 and 97L97: tomato breeding lines with high fruit beta-carotene content. HortScience 36: 2, 387-388.

Stommel, J., J.A. Abbott, R.A. Saftner, and M.J. Camp. 2005. Sensory and objective quality attributes of beta-carotene and lycopene-rich tomato fruit. J Am Soc Hort Sci 130 : 244-251.

Tadmor, Y., E. Fridman, A. Gur, O. Larkov, E. Lastochkin, U. Ravid, D. Zamir, and E. Lewinsohn. 2002. Identification of malodorous, a wild species allele affecting tomato aroma that was aelected against during domestication. J Agric Food Chem 50(7): 2005-2009.

Tanksley, S.D. and J.H. Hewitt. 1988. Use of molecular markers in breeding for soluble solids content in tomato - a reexamination. Theor Appl Genet 75: 811-823.

Tanksley, S.D., M.W. Ganal, J.P. Prince, M.C. de Vicente, M.W. Bonierbale, P. Broun, T.M. Fulton, J.J. Giovannoni, S. Grandillo, G.B. Martin, R. Messeguer, J.C. Miller, L. Miller, A.H. Paterson, O. Pinedo, M.S. Roder , R.A. Wing , W. Wu, and N.D. Young. 1992. High density molecular linkage maps of the tomato and potato genomes. Genetics 132: 1141-1160.

Tanksley, S.D. and J.C. Nelson. 1996a. Advanced backcross QTL analysis: a method for the simultaneous discovery and transfer of valuable QTLs from unadapted germplasm into elite breeding lines. Theor Appl Genet 92: 191-203.

Tanksley, S.D., S. Grandillo, T.M. Fulton , D. Zamir, T. Eshed, V. Pétiard, J. Lopez, and T. Beck-Bunn. 1996b. Advanced backcross QTL analysis in a cross between an elite processing line of tomato and its wild relative *L. pimpinnellifolium*. Theor Appl Genet 92 (2): 213-224.

Tanksley, S.D. and S.R. McCouch. 1997. Seed banks and molecular maps: Unlocking genetic potential from the wild. Science 277: 1063-1066.

Tanksley, S.D. 2004. The genetic, developmental, and molecular bases of fruit size and shape variation in tomato. Plant Cell 16 Suppl. S181-S189.

Taylor, M.D., S.J. Locascio, and M.R. Alligood. 2004. Blossom-end rot incidence of tomato as affected by irrigation quantity, calcium source, and reduced potassium. HortScience 39: 1110-1115.

Thakur, B. R., R.K. Singh, and P.E. Nelson. 1996. Quality attributes of processed tomato products: a review. Food Rev Int 12: 375-401.

Thompson, A.J., M. Tor, C.S. Barry, J. Vrebalov, C. Orfila, M.C. Jarvis, J.J. Giovannoni, D. Grierson, and G.B. Seymour. 1999. Molecular and genetic characterization of a novel pleiotropic tomato-ripening mutant. Plant Physiol 120: 383–389.

Tigchelaar, E.C. 1986. Tomato breeding. *In:* M.J. Bassett [Ed.], Breeding Vegetable Crops. 135-172.

Tomes, M.L. 1963. Temperature inhibition of carotene synthesis in tomato. Bot. Gaz 121: 180-185.

van Berloo, R. and P. Stam. 1998. Marker-assisted selection in autogamous RIL populations: a simulation study. Theor Appl Genet 96: 147-154.

van Berloo, R. and P. Stam. 1999. Comparison between marker-assisted selection and phenotypical selection in a set of *Arabidopsis thaliana* recombinant inbred lines. Theor Appl Genet 98: 113-118.

Van der Hoeven, R., C. Ronning, J. Giovannoni, G. Martin, and S. Tanksley. 2002. Deductions about the number, organization and evolution of genes in the tomato genome base don analysis of a large expressed sequence tag collection and selective genomic sequencing. The Plant Cell 14: 1441-1456.

Van der Knaap, E. and S.D. Tanksley. 2003. The making of a bell pepper-shaped tomato fruit: identification of loci controlling fruit morphology in Yellow Stuffer tomato. Theor Appl Genet 107: 139-147.

Verkeke, W., J. Janse, and M. Kersten. 1998. Instrumental measurement and modelling of tomato fruit taste. Acta Hort 456:199-205.

Vrebalov, J., D. Ruezinsky, V. Padmanabhan, R. White, D. Medrano, R. Drake , W. Schuch, and J. Giovannoni. 2002. A MADS-box gene necessary for ripening at the tomato ripening-inhibitor (rin) locus. Science 296, 343–346.

Wang, C., C.K. Chin , C.T. Ho, C.F. Hwang, J.J. Polashock, and C.E. Martin (1996). Changes of fatty acids and fatty acid derived flavor compounds by expressing the yeast -9 desaturase gene in tomato. J Agric Food Chem 44: 3399-3402.

Wilkinson, J., M. Lanahan, H. Yen, J.J. Giovannoni, and H.J. Klee. 1995. An ethylene-inducible component of signal transduction encoded by Never-ripe. Science 270: 1807–1809.

Yen, H., A. Shelton, L. Howard, J. Vrebalov, and J.J. Giovannoni. 1997. The tomato high pigment (hp) locus maps to chromosome 2 and influences plastome copy number and fruit quality. Theor Appl Genet 95, 1069–1079.

Young, N.D. and S.D. Tanksley. 1989. RFLP analysis of the size of chromosomal segments retained around the Tm-2 locus of tomato during backcross breeding. Theor Appl Genet 77: 353-359.

Zamir, D. 2001. Improving plant breeding with exotic genetic libraries. Nature Rev Genet 2(12): 983-989.

6

Genetic Enhancement of Tomato Fruit Nutritive Value

JOHN R. STOMMEL

United States Department of Agriculture, Agricultural Research Service
Vegetable Laboratory, Building 010A, BARC-West
Beltsville, MD 20705, USA
email: stommelj@ba.ars.usda.gov

INTRODUCTION

During the past century, plant breeding and genetics have contributed significantly to improve nutritive value of horticultural and agronomic crops. Crop nutritive value is influenced by available cultivars, as well as production and postharvest environments. Significant strides have been made, especially in altering content of macronutrients such as proteins, carbohydrates, and oils in crop plants. Current research on improving phytonutrient content includes greater focus on study and assignment of health-promoting properties to micronutrients as well. A vast number of phytonutrients have been identified which are believed to impart health benefits. Relatively few of these phytonutrients, however, are well proven to impart this function. Of this large group of compounds, the genetics of biosynthesis and accumulation is well characterized in relatively few examples. The knowledge required by plant breeders and geneticists to improve plant nutritive value requires input from many disciplines to contribute to the improvement of health benefits. These disciplines include nutrition, food science, medicine, postharvest physiology, molecular biology, and biochemistry.

Tigchelaar (1987) identified four alternatives to enhance the contribution of horticultural crops to human nutrition. These include: 1) genetic enhancement of significant nutrients, 2) improvement of fruit quality to encourage greater per capita consumption, 3) facilitate culture to encourage greater production and availability, and 4) reduce the seasonality of supply to encourage consumption. This review focuses on genetic enhancement of tomato fruit nutritive quality. Varietal development programs have long focused on selection for yield and disease resistance. Enhancement of constituents that are not major yield components take on new importance today in crop improvement programs as market demands support these efforts. A survey conducted in 1990 identified 11 crucial issues affecting horticulture and ranked food quality and safety as the number one issue in terms of present importance (Mitchell 1990). Survey respondents encouraged the development and production of crops having higher nutritional value. Current high levels of consumer health awareness suggest that a survey conducted today would yield similar priorities.

The cultivated tomato, *Solanum lycopersicum**, is an extremely popular and versatile vegetable crop. The tomato fruit is utilized fresh and as a processed product. It boasts a United States commercial crop value of nearly 2 billion dollars (United States Department of Agriculture 2005). Worldwide, over 115 million metric tons of tomatoes were produced in 2004 for fresh and processed consumption on approximately 4.4 million hectares (Food and Agriculture Organization of the United Nations 2004). The United States is one of the leading world tomato producers, second only to China. In 2004, China and United States produced 30.1 million and 12.4 million metric tons of tomatoes, respectively (Food and Agriculture Organization of the United Nations 2004). Tomato ranks second in importance ·
among commercial vegetable crops in the United States in terms of yield and consumption, and is the favorite homegrown crop of four out of five backyard gardeners. Yearly per capita consumption of fresh tomatoes increased 15% between the early 1990s and the early 2000s to nearly 18 pounds per person. Conversely, use in processed products declined 9% to 68 pounds per person (United States Department of Agriculture 2005).

By virtue of the volume of tomato products that are consumed, the tomato crop makes a significant dietary contribution to human health (Table 6.1). A typical tomato fruit, for example, contains only intermediate levels of provitamin A, carotenoids and vitamin C, yet makes important contributions to the dietary intake of vitamin A (9.5%) and vitamin C (11.5%) (Senti and Rizek 1975, United States Department of Agriculture 2002). An average carrot, by comparison, contains provitamin A levels 75 to 85 times higher than that of a typical red pigmented tomato, and supplies 14% of the total U.S. vitamin A consumption. Vitamin A deficiency is a major cause of childhood mortality and is also a likely contributing factor to maternal

---

*Author has adopted new taxonomic nomenclature *Solanum lycopersicum* of cultivated tomato (in lieu of *Lycopersicon esculentum*) in this chapter and also treated other *Lycopersicon* species under the genus *Solanum* based on recent publication of Spooner (2005)

Table 6.1 *Nutrient content of fresh tomatoes: red, ripe, raw, year-round average.*

Nutrient	Value per 100g of edible portion	Nutrient	Value per 100g of edible portion
Proximates		*Vitamins*	
Water	94.5 g	Vitamin C,	
Protein	0.88 g	total ascorbic acid	12.7 mg
Total fat (lipid)	0.20 g	Thiamin	0.04 mg
Carbohydrate	3.92 g	Riboflavin	0.02 mg
Fiber, total dietary	1.2 g	Niacin	0.59 mg
Sugars, total	2.63 g	Pantothenic acid	0.09 mg
Sucrose	0.00 g	Vitamin B_6	0.08 mg
Glucose	1.25 g	Folate, total	15 mcg
Fructose	1.37 g	Vitamin A	833 IU
		Vitamin E (α-tocopherol)	0.54 mg
Minerals		γ-Tocopherol	0.12 mg
Calcium	10 mg	Vitamin K	7.9 mcg
Iron	0.27 mg		
Magnesium	11 mg	*Lipids*	
Phosphorous	24 mg	Fatty acids,	
Potassium	237 mg	total saturated	0.05 g
Sodium	5 mg	total monounsaturated	0.05 g
Zinc	0.17 mg	total polyunsaturated	0.14 g
Copper	0.06 mg	Phytosterols	7 mg
Manganese	0.11 mg		
Selenium	0 mcg	*Carotenoids*	
		β-carotene	449 mcg
		α-carotene	101 mcg
		Lycopene	2573 mcg
		Lutein + zeaxanthin	123 mcg

Adapted from: United States Department of Agriculture, National Nutrient Database for Standard Reference, Release 17 (2004).

deaths in developing countries (World Health Organization 2005). Tomatoes are a major dietary source of lycopene, the red pigment in tomato fruit that is associated with dietary health benefits (Gerster 1997). In addition to these well-known phytonutrients, other compounds in tomato fruit with antioxidant properties include chlorogenic acid, rutin, plastoquinones, tocopherol, and xanthophylls (Beecher 1998, Leonardi et al. 2000). Tomatoes also contribute carbohydrates, fiber, flavor compounds, minerals, proteins, and glycoalkaloids to the diet (Davies and Hobson 1981, Gundersen et al. 2001).

Various aspects of the inheritance of tomato fruit nutritive constituents have been condensed in larger reviews of tomato genetics and fruit quality (e.g., Stevens 1986, Stevens and Rick.1986, Tigchelaar 1987, Berry and Uddin 1991). The present chapter reviews new developments relevant to prior treatments of the subject and highlights new applications of molecular genetics to enhancement of tomato nutritive value.

MICRONUTRIENTS

Considerable genetic variation exists in tomato for micronutrients with antioxidant activity (Hanson et al. 2004, Schauer et al. 2004). A number of these micronutrients, particularly carotenoids, have been the subject of considerable research due to their contribution to the quality of fresh and processed tomato products. Increased recognition of their health promoting properties has stimulated new research and provided added justification for ongoing investigations in this area.

Carotenoids

Horticultural crops are the main source of dietary carotenoids and much of the world dietary vitamin A is derived from vegetable carotenes. Tomato fruit color is determined by carotenoid content. Enhanced fruit pigmentation has long been a goal in tomato cultivar development programs due to the positive association between the intensity of pigmentation and perceived product quality. Recognition of the health benefits associated with carotenoid consumption provides added justification for improving tomato fruit pigmentation (Paiva and Russell 1999). In addition to β-carotene, which is a provitamin A carotenoid, lycopene has received increasing attention for its potential health benefits due to recent research that demonstrated positive associations between tomato products and improved human health (Giovannucci et al. 1995). Consumption of tomato products is associated with decreased risk of cardiovascular disease and certain cancers, including prostate and cervical cancer (Clinton 1998, Gerster 1997, Giovannucci 1999). Recent discussions note the positive association between lycopene and health, but also emphasize that there are a family of compounds in tomato that are of benefit (Laquatra et al. 2005). Relative to carotenoids, studies on the colorless carotenoid phytofluene demonstrated that it is more bioavailable than expected, based on its concentrations in tomatoes (Paetau et al. 1998). These studies suggest that phytofluene may play a synergistic role with other carotenoids in protecting human health.

Total carotenoid content of a typical tomato fruit varies between 70 and 190 µg/g fresh weight (Gross 1991). Carotenes constitute 70 to 95% of the total carotenoids. Lycopene, the red pigment in tomato fruit, is the predominant carotene, comprising up to 90% of the total carotenoids. Its colorless polyene precursors, phytoene and phytofluene account for 15 to 30% of total carotenoids. Lesser amounts of the provitamin A carotenoid, β-carotene (2 to 15%), plus γ-carotene, δ-carotene, ζ-carotene, lutein, and neurosporene can be measured. Within the tomato fruit, the outer pericarp tissue contains the highest concentration of carotenoids. Lycopene levels in the outer pericarp may be six-fold higher than that in the locular tissue (Gross 1991). Levels of carotenoids in tomato products are dependent on the variety and on growing conditions. Summer- and winter-greenhouse grown fruits are lower in carotene content than fruits produced outdoors during summer months. Fruit picked green and ripened in storage may be much lower in carotene than vine-ripened fruit (Gould 1992).

The identification and cloning of genes (e.g. cDNA) that code for nearly all of the enzymes required for carotenoid biosynthesis in plants has stimulated considerable interest in engineering plants with altered carotenoid content (Cunningham and Gantt 1998). Numerous successful and unsuccessful efforts to alter fruit carotenoid composition by manipulating expression of carotenogenesis transgenes have been reported. Early studies employing constitutively expressed phytoene synthase in transgenic tomatoes caused ectopic production of carotenoids and dwarfism (Fray et al. 1995). Dwarfism in these plants was inversely related to expression of phytoene synthase, a likely diversion of geranyl geranyl pyro phosphate away from the gibberellin biosynthetic pathway. Subsequent studies have focused on expression of the bacterial carotenoid genes, *crtI* encoding phytoene desaturase and *crtB* encoding phytoene synthase, in transgenic plants (Romer et al. 2000, Fraser et al. 2002). This strategy resulted in successful modifications of fruit carotenoid content without the adverse effects associated with constitutive expression of the tomato phytoene synthase *Psy1* cDNA. Expression of *crtI* resulted in a three-fold increase in β-carotene content (Romer et al. 2000). Total carotenoid content was not increased and transgene expression did not affect plant growth and development. Expression of *crtB* in transgenic plants resulted in 2 to 4-fold increases in total carotenoid content. Levels of phytoene, lycopene, β-carotene, and lutein were increased 2.4-, 1.8-, and 2.2-fold, respectively (Fraser et al. 2002). Rosati et al. (2000) successfully altered carotenoid levels in transgenic plants with constructs intended to up-regulate or down-regulate the expression of the lycopene β-cyclase gene. Expression of an *Arabidopsis* lycopene β-cyclase cDNA resulted in increased β-carotene content. Conversely, transformation with a tomato antisense lycopene β-cyclase cDNA construct reduced lycopene β-cyclase activity and was accompanied by a slight increase in lycopene content. In a related study, overexpression of the genes encoding the *Arabidopsis* lycopene β-cyclase and pepper β-carotene hydroxylase, under the control of a phytoene desaturase (*Pds*) tomato fruit specific promoter, produced transformants with significantly increased levels of β-carotene and the xanthophylls β-cryptoxanthin, and zeaxanthin (Dharmapuri et al. 2002). The xanthophyll overproduction trait was inherited as a single dominant character. Overall, these studies demonstrate that expression of foreign transgenes that influence carotenoid biosynthesis can successfully overcome the co-suppression reported to occur when using tomato coding sequences.

Utilizing a novel approach, Mehta et al. (2002) demonstrated that fruit-specific expression of an yeast S-adenosylmethionine decarboxylase gene in transgenic plants resulted in increased ripening-specific accumulation of spermidine and spermine, levels of which normally decline in ripening fruit, and ethylene levels that were higher than those observed in non-transformants. Increased polyamine accumulation was associated with a 2- to 3-fold increase in lycopene content. Elucidation of the mechanism by which higher polyamines influence carotenoid accumulation in an ethylene-independent manner should reveal new strategies to improve fruit quality and nutritive value.

Mutants

A large genetic database exists for simply inherited genes that influence carotenoid content in tomato. The elucidation of carotenoid biochemistry was aided in large part through the study of single gene mutants of tomato. Based upon the results of inheritance studies in tomato and on the information available at that time on the structures of proposed intermediates, Porter and Lincoln (1950) detailed the first biochemical pathway of carotenoid biosynthesis in 1950. Subsequent modifications and extensions of that pathway have occurred (Porter and Anderson 1962, 1967, Goodwin 1971, Jones and Porter 1986, Hirschberg 2001). Carotene biosynthesis in higher plants has been reviewed extensively in recent years (Bramley 1992, Bartley and Scolnik 1994, Hirschberg 2001). The molecular genetics of carotenoid biosynthesis in plants has advanced rapidly due to the wealth of information available from bacterial and fungal models and extension of this knowledge to plant genomes (Piechulla et al. 1985).

More than 20 genes have been characterized in tomato that influence the type, amount, or distribution of fruit carotenoids depicted in Table 6.2 (Khudairi 1972, Darby 1978, Stommel 1992a). Carotenoid profiles in these variant genotypes have allowed speculation on points of gene action in the carotenoid biosynthetic pathway. Many of the available color variants were first identified as mutants that arose spontaneously in cultivars of *S. lycopersicum*. Color mutants also occur in the wild tomato species, and in the case of the green-fruited species, are often manifested transgressively upon introduction into *S. lycopersicum*.

Table 6.2 *Selected loci that influence tomato fruit carotenoid, anthocyanin, and carbohydrate content.*

Mutant (symbol)	Fruit Description	Reference
Carotenoid		
Apricot (*at*)	Fruit yellow with pinkish blush	Jenkins and McKinney (1955)
Beta (*B*)	Fruit flesh orange; increased β-carotene, reduced lycopene	Lincoln and Porter (1950), Tomes et al. (1954), Ronen et al. (2000)
Beta modifier (*Mo$_B$*)	>90% of total carotenes as β-carotene (*Mo$_B$*), ca. 50% of total carotenes as β-carotene (*Mo$_B^+$*); action specific to *B*.	Tomes et al. (1954), Zhang and Stommel (2000)
Colorless skin (*y*)	Unpigmented fruit epidermis; Colorless skin over red flesh results in pink fruit	Rick and Butler (1956)
Dark green (*dg*)	Immature fruit dark green; increased levels of carotenoids in mature fruit	Konsler (1973), Levin et al. (2003)
Delta (*Del*)	Orange-red flesh; enhanced δ-carotene and α-carotene fractions	Tomes (1963), Ronen et al. (1999)
Diospyros (*dps*)	Fruit tissue is dusky orange	Rick (1967)
Ghost (*gh*)	Phytoene synthesis normal, no colored carotenoids	Rick et al. (1959), Scolnik et al. (1987)
Green flesh (*gf*)	Chlorophyll retained in ripe fruit, normal lycopene synthesis; fruit reddish-brown	Kerr (1958a)
Green ripe (*Gr*)	Green-pigmented flesh in ripe fruit	Kerr (1958b), Barry et al. (2005)
High pigment 1 (*hp1*)	Immature fruit dark green; increased levels of carotenoids in mature fruit	Clayberg et al. (1960), Van Tuinen et al. (1997), liu et al. (2004)
High pigment 2 (*hp2*)	Similar to *hp1*	Yen et al. (1997), Mustilli et al. (1999)
Intensified pigmentation (*Ip*)	Enhanced lycopene synthesis	Rick (1974)
Old gold crimson (*og^c*)	Enhanced red color; increased lycopene, reduced β-carotene	Thompson et al. (1967), Ronen et al. (2000)
Red yellow (*ry*)	Modifier for red color in yellow fruit	Young (1956)
Sherry (*sh*)	Fruit flesh yellow with reddish tinge	Zscheile and Lesley (1967)
Tangerine (*t*)	Orange fruit; colored carotenoids principally prolycopene	MacArthur (1934), Isaacson et al. (2002)
Yellow *flesh* (*r*)	Reduced polyenes, very low levels of carotenes; fruit flesh yellow	Rick & Butler (1956), Fray and Grierson (1993)

(Contd.)

(Contd.)

Anthocyanin		
Anthocyanin fruit (*Aft*)	Variable purple pigmentation; anthocyanin in skin and outer pericarp	Giorgiev (1972), Jones et al. (2003)
Atroviolacium (*atv*)	Excess anthocyanin on fruit	Rick (1964)
Aubergine (*Abg*)	Fruit epidermis purple, particularly on shoulder	Rick et al. (1994)
Carbohydrate		
Apoplastic invertase (*LIN5*)	Increased hexose sugars	Fridman et al. (2000)
Fructose glucose ratio (*Fgr*)	Increased fructose to glucose ratio	Schaffer et al. (1999)
Fructokinase 2 (*FK2*)	Modifier of *Fgr*	Levin et al. (2000)
Sucrose accumulator (*sucr*)	Mature fruit accumulate predominantly sucrose	Stommel and Haynes (1993) Chetelat et al. (1993)
Transient starch accumulation (*AGPaseL1*)	Increased transient starch accumulation	Schaffer et al. (2000)

An example of such a transgressive segregant is the *Beta* (*B*) allele, first characterized in progeny descended from a cross between *S. lycopersicum* and the green-fruited species *S. habrochaites** (Lincoln et al. 1943, Kohler et al. 1947). Introgression of the *B* allele into *S. lycopersicum* increases β-carotene content at the expense of lycopene. Inheritance studies suggested that high concentrations of β-carotene found in orange-pigmented fruit were controlled by a single gene exhibiting incomplete dominance (Lincoln and Porter 1950). Subsequent studies by Tomes et al. (1954) determined that *B* was dominant but subject to influence by a modifier gene, Mo_B , that segregated independently of *B*. Utilizing molecular markers linked to *B* and Mo_B, genotypic evaluations discounted incomplete dominance to explain inheritance of fruit carotene content, but revealed that *B* and Mo_B were linked on chromosome 6 and did not segregate as independent genes as originally proposed (Zhang and Stommel 2000). In the presence of the homozygous recessive form of the allele, Mo_B Mo_B, β-carotene represents more than 90% of the total carotene content and fruit are orange pigmented (Tomes et al. 1954). Expression of the dominant Mo_B+ form of the allele, however, reduces β-carotene content to more than 50% of the total carotenoids and increases lycopene to less than 50% of total carotenoids resulting in red-orange pigmented fruit. Orange-fruited accessions of *S. cheesmaniae, S. galapagense, S. pimpinelli folium, S. chilense,* and *S. chmielewskii* containing high concentrations of β-carotene have also been described (Rick 1956, Manuelyan et al. 1975, Chalukova 1988, Stommel and Haynes 1994).

*Name changed from *L. hirsutum*

The inheritance of β-carotene content in these wild tomato species is consistent with that described for the dominant *B* gene from *S. habrochaites*.

Molecular analysis revealed that *B* encodes a novel lycopene β-cyclase that converts lycopene to β-carotene (Ronen et al. 2000). Increased transcription of *B* in the *Beta* mutant results in conversion of lycopene to β-carotene and orange pigmented fruit. Based upon variable expression of lycopene-β-cyclase transgenes in fruit tissue of transformed plants, Ronen et al. (2000) suggested that variation observed in tissue-specific β-carotene accumulation could be attributed to differences in the rate of gene expression as a result of additional sequence elements upstream to the promoter.

Use of *B* resulted in the development of the fresh market tomato varieties Caro-Red (Tomes and Quackenbush 1958) and Caro-Rich (Tigchelaar and Tomes 1974) that contain nearly ten times the β-carotene of normal red fruited tomato varieties. Consumer preference for red fruit pigmentation has unfortunately, limited widespread use of these cultivars. Increased consumer education may be required to ensure consumption of high β-carotene containing varieties. Breakage of the close genetic linkage between *B* and *sp*, the gene for indeterminate growth habit, has eliminated this limitation of *B* for use in cultivars intended for commercial production (Stommel 2001, Stommel et al. 2005).

In contrast with *B*, the recessive *old gold crimson* mutant (og^c) enhances lycopene content at the expense of β-carotene (Thompson et al. 1967, Lee and Robinson 1980). Ronen et al. (2000) demonstrated that og^c is an allele of *B* and that null mutations in the *B* gene are responsible for the *old gold crimson* phenotype. Despite up to 40% reductions in β-carotene levels in crimson plants, crimson cultivars have been developed for the market for their desirable dark red pigmentation. The relatively recent recognition of health benefits attributable to lycopene in the diet has negated any negative consequences of the loss in nutrients from reduced levels of β-carotene. Under low temperature conditions, plants expressing og^c exhibit orange flower pigmentation, thus facilitating screening in segregating populations.

Introgression of the monogenic recessive *high pigment* (non-allelic *hp1* and *hp2*, mapping to chromosome 2 and 1, respectively; Van Tuinen et al. 1997, Yen et al. 1997) mutant alleles enhance total fruit carotenoid content without significantly altering the relative percentage of different carotenoid constituents (Cookson et al. 2003). As a result, levels of a number of carotenoids can be increased without altering desirable red fruit pigmentation. These mutants are light hypersensitive and characterized by seedlings with higher anthocyanin levels and short hypocotyls as well as higher fruit and foliage pigmentation (Mochizuki and Kamimura 1984). Within respective *hp1* and *hp2* loci, two mutant alleles have been identified that exhibit varying photoresponsiveness: *hp1* and *hp1^w* and *hp2* and *hp2^j* (Kerckhoffs and Kendrick 1997). A third mutant exhibiting exaggerated photoresponsiveness is the *dark green* (*dg*) mutant (Konsler 1973). Recent evidence indicates that *dg* is allelic to *hp2* (Levin et al. 2003). Incorporation of these alleles boosts total carotene content 30 to 50%. β-carotene content of *dg* lines is approximately 50% greater than that found in *hp* lines and 250% greater than that typical of normal red-fruited tomatoes. All three mutants produce dark green immature fruit due to increased chlorophyll content, but mature green fruit of *dg* mutants is characteristically much darker green in comparison to *hp1* and *hp2* fruit (Baker and Tomes 1964, Palmieri et al. 1978). Both *dg* and *hp* mutations also increase fruit firmness and ascorbic acid levels (Jarret et al. 1984). Undesirable pleiotropic effects that include slow seed germination, increased seedling mortality, brittle stems and premature defoliation is associated with these mutants and has thus far limited their practical use. Plants expressing both *crimson* and *high pigment* alleles produce fruit with lycopene levels three to four times that of conventional red-fruited tomatoes. If utilized as a source of natural lycopene, the negative effect of *hp* on yield can be offset by the industries high dollar value of the extracted pigments.

The *hp2* mutant allele encodes the tomato homolog of the nuclear protein *DEETIOLATED1* (*DET1*) from *Arabidopsis* that is involved in light signal transduction (Mustilli et al. 1999). Similarly, *dg* encodes a novel allele of the *DET1* gene (Levin et al. 2003). Recent studies demonstrated that *hp1* is a mutation in a tomato *UV-DAMAGED DNA-BINDING PROTEIN 1* (*DDB1*) homolog whose *Arabidopsis* counterpart interacts with *DET1* (Liu et al. 2004). Studies on these high pigment mutants reveal that light signal transduction regulates the carotenoid pathway in a manner that affects total fruit carotenoid content and that genes encoding components of light signal transduction may provide new genetic tools for manipulating fruit nutritional value (Liu et al. 2004, Yen et al. 1997). Liu et al. (2004) demonstrated that two tomato light signal transduction genes, *LeHY5* and *LeCOP1LIKE*, are positive and negative regulators of fruit pigmentation, respectively. Further studies (Levin et al. 2004) report additional putative light responsive genes that modulate carotenoid profiles in fruit of these light hypersensitive tomato mutants.

The gene *Intensified Pigmentation*, which has effects similar to that of *dg* and *hp*, was described in progeny descended from an *S. lycopersicum* x *S. chmielewskii* cross (Rick 1974). Fruit expressing *Ip* also exhibit dark green immature fruit and intensified carotenoid pigmentation in ripe fruit.

Unlike *dg* and *hp*, *Ip* behaves as a dominant gene and appears to have reduced detrimental effects on seed germination and plant vigor.

A variety of mutants affecting tomato fruit carotenoid content distinct from the high pigment and high β-carotene types have been characterized and may have beneficial affects in enhancing or evaluating fruit nutritional composition. The recessive gene *r* (*Yellow Flesh*) is responsible for yellow fruit flesh and results in greatly reduced levels of polyenes and very low levels of colored carotenoids (Rick and Butler 1956). Many yellow-fleshed home garden type tomatoes have been developed. The *r* locus corresponds to a null mutation for a chromoplast-specific phytoene synthase, *Psy1* (Fray and Grierson 1993), thus accounting for the lack of carotenoid accumulation in yellow-fruited tomatoes and green-fruited species. A second phytoene synthase gene, *Psy2*, has been identified which is also expressed in ripening tomato fruit (Bartley and Scolnik 1993). Its transcripts are relatively more abundant, however, in mature leaves. Phytoene synthase has been described as the "pacemaker" in carotenoid synthesis in ripening fruit and is regulated at the level of transcription (Fraser et al., 1994). Lois et al. (2000) have proposed that a second gene, *DXS*, encoding the first enzyme of isoprenoid synthesis in the plastids, works in concert with *Psy1* to control fruit carotenoid synthesis.

The dominant *delta* (*Del*) allele conditions increased δ-carotene and re-duced lycopene content, resulting in reddish-orange colored fruit (Tomes 1963). Ronen et al. (1999) demonstrated cosegregation of the *Crtl-e* locus encoding ε-cyclase with the *Del* mutation located on chromosome 12. ε-cyclase converts lycopene to δ-carotene. Transcript for *Crtl-e* was shown to increase 30-fold in ripening fruit of the *Del* mutant. *Del* does not alter carotenoid composition or *Crtl-e* mRNA in leaves or flowers, thus indicat-ing that *Del* is likely an allele of the gene for ε-cyclase (Ronen et al. 1999).

The recessive *tangerine* (*t*) mutant conditions orange fruit color due to the accumulation of poly-cis-lycopene, also referred to as prolycopene (MacArthur 1934, Tomes 1963). Fruit of the *tangerine* mutant also exhibit elevated phytoene and phytofluene. Trans-lycopene is the principal form of lycopene in red tomato fruit. Located on chromosome 10, a clone of the *tangerine* gene, designated *CRTISO*, was shown to encode an authentic carotenoid isomerase that is required during carotenoid desaturation (Isaacson et al. 2002). CRTISO is a redox-type enzyme structurally related to the bacterial-type phytoene desaturase CRTI. Analysis of two alleles of *tangerine* demonstrated that in one case, loss of function in CRTISO was attributable to a deletion mutation in *CRTISO*, and in the second, expression of this gene was impaired. *CRTISO* from tomato is normally expressed in all green tissues but is upregulated during fruit ripening and in flowers. Unlike orange-pigmented fruit that accumulate β-carotene due to the *Beta* allele, *tangerine* mutants do not have increased pro-vitamin A carotenoids. Evidence that cis-lycopene is more bioavailable, than trans-lycopene (Unlu et al. 2003, Boileau et al. 1999), has focused considerable interest on this mutant in human nutrition-related studies. Pigment compositions of nearly all orange-fleshed cultivars available for commercial and home use attribute their color to the *tangerine* allele.

Incorporation of the recessive *y* allele results in a colorless epidermis lacking the normal yellow pigmentation (Rick and Butler 1956). Presence of *y* in red-fleshed genotypes results in a pink fruit phenotype. Pink cultivars are popular in home garden and specialty markets. The combination of *y* with *r* (*yellow flesh*) results in pale yellow or "white" fruit. Several novel cultivars, such as White Queen, owe their pigmentation to this gene combination. Plants with the recessive *apricot* (*at*) allele bear fruit that are characteristically yellow with a pinkish blush at maturity (Jenkins and Mackinney 1955). Variations on this include a modifier gene for red color in yellow fruit (*ry*; Young 1956) and *sherry* (*sh*; Zscheile and Lesley 1967). Additional color variants include the *diospyros* (*dps*) mutant with dusky orange fruit.

Presence of the recessive *green flesh* (*gf*) allele prevents breakdown of chlorophyll that normally occurs in maturing fruit (Kerr 1958a). Retention of green chlorophylls and the red pigment, lycopene, results in reddish-brown colored fruit. Anthers of *gf* plants exhibit a characteristic lemon-green pigmentation. *Green ripe* (*Gr*) results in green-pigmented flesh in ripe fruit (Kerr 1958b). Barry et al. (2005) mapped the dominant *Gr* locus to the long arm of chromosome 1 and determined that *Gr* is an ethylene insensitive mutant that may encode a novel ethylene signaling component. Novel heirloom cultivars have been described with unusual pigmentation patterns. 'Green Grape', for example, exhibits yellow/green exterior and green interior attributed to a lesion in up-regulation of the carotenoid biosynthetic pathway. Expression of the *ghost* (*gh*) allele on chromosome 11 results in fruit that contain only phytoene and no colored carotenoids due to a block in the desaturation of phytoene (Rick et al. 1959, Scolnik et al. 1987). Expression of *gh* may be quite variable. Similar to fruit color, leaf tissue is generally white or light yellow, although mosaics of white and green are not uncommon.

Quantitative Trait Loci

The described monogenic mutants have a dramatic effect on fruit pigmentation. Nonetheless, they have not contributed widely to carotenoid pigmentation in commercial cultivars. Extensive genetic and molecular characterization of simply inherited tomato pigment mutants has not established a molecular genetic basis for quantitatively inherited variation in fruit pigmentation.

Analogous to fruit firmness, soluble solids, and other fruit quality traits, quantitative trait loci (QTL) associated with variation in fruit pigmentation have been described that begin to explain dissimilarity in intensity of red pigmentation in modern tomato cultivars. Numerous QTL introgressed from *S. pimpinellifolium* (Tanksley and Nelson 1996, Chen et al. 1999), *S. habrochaites* (Bernacchi et al. 1998, Kabelka et al. 2004, Yates et al. 2004), *S. peruvianum* (Fulton et al. 1997, Yates et al. 2004), and *S. neorickii* (Fulton et al. 2000) have been described that influence fruit color. Analysis of QTL identified in a *S. lycopersicum* cross also revealed loci associated with enhanced fruit color (Saliba-Colombani et al. 2001, Causse et al. 2002). Not surprisingly, these QTL may have negative or positive effects on ripe fruit color and epistasis as well as pleiotropy may occur. Whereas QTL studies often focus upon the positive effect of loci introgressed from wild relatives of tomato, Kabelka et al. (2004) identified loci from *S. lycopersicum* that contributed to improved tomato color and noted that the trend for introgressions from *S. habrochaites* was to shift the population toward undesirable color. Encouragingly, all of these studies identify some QTL associated with fruit qualities that have also been identified for quality attributes by others in different interspecific tomato crosses. These conserved major loci and minor loci with positive epistatic effects will be of great interest in marker-assisted breeding strategies to improve tomato quality and nutritive value.

Vitamins

Vitamin A and vitamin C are the principal vitamins in tomato fruit. The tomato makes important contributions to the dietary intake of these vitamins. Tomatoes also provide moderate levels of folate and potassium in the diet and lesser amounts of vitamin E and several water-soluble vitamins.

Vitamin A

β-carotene is the principal provitamin A carotenoid and is an essential nutrient in the human diet because of its retinoid activity (Tee 1992, Omenn et al. 1994). Epidemiological evidence suggests that increased intake of high β-carotene containing fruits and vegetables may be associated with a reduced risk of heart disease and certain cancers (Ziegler 1989, Doll 1990, Block et al. 1992, Omenn et al. 1994). Vitamin A deficiency has been described as one of the most serious nutritional disorders of children in the world, especially in developing countries (Sommer 1997, World Health Organization 2005). Traditional breeding and transgene strategies to improve β-carotene content have yielded germplasm with increased levels of this provitamin A carotenoid (see Carotenoids).

Vitamin C

Fruits and vegetables supply approximately 91% of the vitamin C in the U.S. food supply. In tomato, considerable variation exists among cultivars and wild species accessions for ascorbic acid content, thereby providing good opportunities for genetic enhancement of fruit vitamin C potential. Within the *Solanum* section *Lycopersicon*, ascorbic acid levels are reported to range from 10-120 mg/100 g fresh weight (Lambeth et al. 1966, Hobson and Davies 1971). Depending upon cultivar, environment, fruit maturity and postharvest treatment, ascorbic acid comprises 40-90% of the organic acids (Bradley 1946, Carangal et al. 1954, Davies 1965, McClendon et al. 1959). Exposure of fruit to sunlight is an important factor affecting the level of ascorbic acid (Hassan and McCollum 1954, Dumas et al. 2003). Malic acid is the other principal organic acid and ranges from 10-60% of that of ascorbate (Davies 1965). Ascorbic acid content typically increases as fruits mature, reaching a maximum just before full red color, and declining with full ripeness and overmaturity (Malewski and Markakis 1971). A comparison of fresh versus processed purees found that processing caused a 3% to 17% loss in ascorbic acid content (Warnock 1983). Warnock (1983) suggested that genotypic differences accounted for the degree of ascorbate loss upon processing.

Stevens (1972) evaluated the inheritance of citrate and malate in two tomato accessions. The inheritance of citrate and malate concentration was controlled by a single gene for each compound. The genes are linked and exhibit 18% recombination. The dominant form of the alleles condition a high concentration of citrate and a low concentration of malate. Since high malate concentrations are often associated with sour fruit taste, high citrate to malate ratios are preferred.

Tomato cultivars such as 'DoubleRich' with double the normal vitamin C content (ca. 50 mg/100g) were developed via an interspecific cross between the cultivated tomato and the ascorbic acid-rich wild species *S. peruvianum*. An association between high vitamin C and poor yield, due primarily to small fruit size, has stymied commercial use of these cultivars (Stevens and Rick 1986). Fruit of *high pigment* (*hp*) genotypes also contain increased levels of vitamin C. Unfortunately, undesirable effects of *hp* on yield have likewise limited its usefulness in increasing vitamin C. Andrews et al. (2004) proposed that increased levels of ascorbic acid in developing fruit may increase the tolerance of *hp-1* fruit to photooxidative injury. Modern tomato hybrids have come under scrutiny for alleged lower vitamin C content and inferior nutritional quality. Contrary to this notion, a controlled study demonstrated that newer cultivars contained approximately 25% more vitamin C than those developed twenty years earlier (Burge et al. 1975, Matthews et al. 1973). This suggests that further incremental gains in genetic improvement of tomato vitamin C content may yet be realized without compromising yield.

Despite the wealth of information documenting the importance of vitamin C to human health, relatively little information has been available about the pathway(s) leading to its biosynthesis in plants. This contrasts with a well-understood pathway in animals (Burns 1967). Wheeler et al. (1998) provided evidence that mannose and galactose are efficient precursors for ascorbic acid in plants and demonstrated the existence of a galactose dehydrogenase which is active in converting galactose to ascorbate precursors, thus filling a major gap in plant carbohydrate metabolism. Agius et al. (2003) recently isolated the gene *GalUR* which encodes an NADPH-dependent D-galacturonate reductase and demonstrated that biosynthesis of ascorbic acid in strawberry fruit occurs through galacturonic acid, a component of cell wall pectins. Overexpression of *GalUR* in *Arabidopsis thaliana* enhanced vitamin C levels two- to three-fold, thus demonstrating the potential to engineer increased vitamin C levels in tomato. In light of the negative effects on yield often associated with high vitamin C lines produced through conventional hybridization, characterization of transgene effects on fruit yield and other quality attributes must be evaluated.

Folate

Folate in the diet comes mainly from plant sources. Whereas green leafy vegetables and legume seeds are folate-rich, folate concentrations in other vegetables are considerably lower (Scott et al. 2000, Konings et al. 2001). Dietary fortification and supplementation with folic acid is practiced to offset birth defects, anemia, and increased risk of vascular disease and certain cancers (Lucock 2000, Krishnaswamy and Nair 2001, Molloy and Scott 2001). Nonetheless, folate deficiency is a serious concern in poor countries and is common in wealthy countries as well (de Bree et al. 1997, Konings et al. 2001, Krishnaswamy and Nair 2001). The biosynthetic pathway of folate is well characterized (Hanson and Gregory 2002, Goyer et al. 2004), and thus provides opportunities to engineer enhanced folate content in dietary foods (Zhang et al. 2003, Hossain et al. 2004).

Folates are tripartite molecules composed of pteridine and *p*-aminobenzoate (PABA) plus one or more glutamate moieties. Overexpression of GTP cyclohydrolase I in fruit of tomato transformants resulted in a 3- to 140-fold increase in levels of the folate precursor, pteridine, and an average 2-fold increase in folate content (de la Garza et al. 2004). Transformants with intermediate levels of pteridine exhibited the highest levels of folate. Transformants that accumulated increased folate also exhibited depleted levels of PABA. Exogenous supply of PABA resulted in additional 10-fold increases in folate content, suggesting that additional modifications in the folate biosynthetic pathway to increase levels of this limiting compound may further boost fruit folate content.

Glycoalkaloids

Tomato belongs to the poisonous nightshade family and was long regarded as toxic. Although widely consumed in Europe by the mid-eighteenth century, tomatoes did not gain acceptance in the United States until the late 1700's due to the presumption that the fruits were poisonous. Glycoalkaloids and their toxic effects are commonly associated with Solanaceous species. Tomato accumulates the glycoalkaloids α-tomatine and dehydrotomatine in a 10 : 1 ratio (Madhavi and Salunkhe 1998). These compounds act as toxins or feeding deterrents to insect pests of tomato and are toxic to several pathogenic microorganisms that affect the crop. In striking contrast with potato glycoalkaloids, the tomatine compounds in tomato appear to be less toxic for human consumption, presumably because they are eliminated from the body as an insoluble tomatine-cholesterol complex formed in the digestive tract (Kozukue and Friedman 2003). Friedman and Levin (1995) note that pickled and fried green tomato fruit have relatively high levels of tomatine in comparison to red ripe fruit and are widely consumed in many countries. Tomatine has a high affinity for cholesterol *in vitro*, and when administered orally to hamsters, significantly reduced plasma cholesterol (Friedman et al. 2000a, b). High tomatine containing green tomatoes had a significantly greater effect on reducing LDL cholesterol and triglyceride levels than feeding low tomatine-containing red tomatoes. Friedman (2002) notes the need for additional studies to further substantiate studies suggesting a beneficial role for dietary tomatine in enhancing the immune system, and in cancer chemotherapy.

Levels of tomatine compounds are influenced by the stage of tomato fruit maturity. Immature fruit contain up to 500 mg of β-tomatine per kg fresh weight of tomato fruit. Coincident with reduction in chlorophyll content of ripening tomato fruit, tomatine concentration declines 100-fold in breaker stage or ripe fruit to approximately 5 mg per kg fresh fruit weight (Kozukue and Friedman 2003). Courtney and Lambeth (1977) identified elevated levels of tomatine in fruit of the green-fruited species noting that the highest levels were found in an *S. chmielewskii*, *S. habrochaites*, *S. cheesmaniae*, *S. pimpinellifolium*, and *S. peruvianum*, noting that the highest levels were found in an *S. chmielewskii* accession. According to Rick et al. (1994), an unusual *S. lycopersicum* mutant that produces bitter fruit retains high tomatine levels in ripe fruit (500-5000 mg/kg dry weight). Retention of tomatine in these fruits is controlled by a single recessive gene and presumably encodes a defective tomatine-degrading enzyme that is normally active in ripening fruit. Consumption of these fruits by natives of Peru, apparently without acute toxic effects, further supports the notion that tomatine may be a lesser dietary toxin than once believed.

Considerable attention has been devoted to study of tomatine in vegetative tomato tissue. Juvik et al. (1982) identified considerable variation for foliar tomatine concentration, noting that levels were highest in an accession of *S. peruvianum*. Tomatine concentrations of *S. lycopersicum* and *S. pimpinellifolium* accessions were moderately higher than levels in *S. lycopersicum*. Inheritance studies demonstrated that foliar tomatine concentration was controlled by two codominant alleles at a single locus (Juvik and Stevens 1982). Tomatine levels in *S. lycopersicoides* were up to 3.5% of tissue dry weight (Oleszek et al. 1986).

Minerals

The roles of minerals in plant foods that have a positive effect on human health or nutrition are well established. There is a wealth of literature that provides evidence of their importance from research in animal model systems as well as from clinical and epidemiological studies (Lachance 1998).

Tomato mineral composition is greatly influenced by plant nutrition, and as a result, has been well characterized in the context of mineral deficiency and the effect of these conditions on plant health. There is significant genotypic variation for mineral content in tomato fruit. Potassium, together with nitrate and phosphorous, constitutes approximately 93% of the total inorganic fruit constituents (Hobson and Davies 1971). Phosphorous and potassium are the two most well studied constituents. The concentration of other minerals in the fruit is low. Hobson and Davies (1971) summarize prior reviews of tomato mineral composition. Although characterizations of mineral content, and in some cases, inheritance studies have been conducted, little effort has been made to improve fruit mineral composition.

Phosphorous

Phosphorous is the major inorganic anion in tomato fruit. Phosphorous levels in 25 divergent tomato accessions ranged from 3.1 to 6.7 mM (Paulson and Stevens 1974). Uptake of phosphorous is highly dependent on root growth. An inheritance study of tomato fruit phosphorous concentration found a strong genotype-environment interaction that was not attributed to variation in available soil phosphorous (Stevens and Paulson 1973). The study suggested that few genes are involved in genetic control of fruit phosphorous levels and that additive and dominance effects, in addition to epistatic interactions, contribute to observed phenotypes.

Potassium

The major inorganic cation in tomato fruit is potassium. Stevens (1972) reported potassium concentrations that ranged from 45.2 to 86.7 meq/liter among 55 divergent tomato lines. Potassium deficiency may contribute to poor fruit color and reduced acid content (Bradley 1946, Carangal et al. 1954). Significant positive correlations between potassium content and the total and titratable acidities have been reported (Hobson and Davies 1971). Factors which increase potassium content of the fruit result in a corresponding increase in organic acids in an effort to maintain fruit pH.

Calcium

Over 30 physiological plant disorders have been associated with calcium deficiency (Maynard 1979). In tomato, calcium deficiency contributes to fruit blossom end rot. Good cultural practices have been the primary means for producers to control blossom end rot. Heritable differences for calcium utilization efficiency in tomato have been reported (Giordano et al. 1982). The potential to manipulate mineral composition in tomato using transgene approaches is well-demonstrated by a recent report wherein calcium content was increased up to 50% via expression of an *Arabidopsis* H^+/Ca^{2+} transporter in carrot (Park et al. 2004).

Amino Acids

Reports on the amino acid composition of tomato fruit vary considerably. Variation in amino acid composition is likely due to both genotypic differences and varying plant nutrition (Freeman and Woodbridge 1960, Davies 1966a). Glutamic acid is the predominant amino acid of ripe fruit. Glutamic acid, α-aminobutyric acid, glutamine, and aspartic acid account for approximately 80% of the total free amino acids (Freeman and Woodbridge 1960). Glutamic acid content rises sharply, and aspartic acid increases to a lesser extent, during fruit ripening. Interest in the amino acid content of tomato centers on the influence of these nutrients on fruit flavor.

Attempts to modify fruit amino acid composition have been limited to studies evaluating the effect of different fertilizer regimes. Applications of nitrogen generally increased glutamic and aspartic acid content (Carangal et al. 1954, Davies 1964). Apart from serine and threonine, which reach peak levels before fruit fully ripen, other amino acids decrease during ripening, presumably for protein synthesis. Free amino acids comprise a significant portion of the total amino acids of immature and mature tomato fruit (Friedman 2002). Free lysine and methionine constitute 20-25% and 12-18% of total lysine and methionine, respectively. Lysine:arginine ratios are more than double that reported for cereal proteins and are comparable to that reported for legume proteins. Methionine:glycine ratios are greater than those reported for legume proteins and are similar to that for animal proteins. The essential amino acid content of tomato is considered to be of good quality, being similar to soy protein (Friedman and Brandon 2001). Considerable diversity in amino acid content was identified among tomato and its wild relatives (Schauer et al. 2004). Levels of amino acids in fruit of wild species were generally lower in wild species in comparison to the cultivated form and amino acid metabolite ratios varied across species.

Phenolic Compounds

More than 4,000 phenolic phytochemicals have been identified (King and Young 1999). Flavonoids, phenolic acids, and polyphenols are the main classes of dietary phenolics. Flavonoids, which include anthocyanins, are the largest group of plant phenols and have been the subject of considerable research since they impart color to many horticultural commodities. Various health-promoting effects have been ascribed to plant phenolic constituents. Vinson et al. (1998) determined that vegetables high in phenolic compounds have antioxidant quality comparable to that of pure phenols and superior to that of the antioxidant vitamins A, C and E. In a survey of *S. lycopersicum* and *S. pimpinellifolium* accessions, total phenolics content was most closely associated with measures of antioxidant activity (Hanson et al. 2004). Consumption of foods rich in dietary phenolics are believed to contribute to reduced radical-mediated pathogeneses such as carcinogenesis and atherosclerosis (Ames et al. 1993, Sawa et al. 1999). Phenolic compounds extracted from plant tissue have also been demonstrated to have significant hypolipidemic action (Sudheesh et al. 1997). The average per capita consumption of vegetable phenols in the U.S. is estimated at approximately 218 mg per day of catechin equivalents, which equates to three times the recommended intake of vitamins C, E and β-carotene antioxidants (Vinson et al. 1998).

Flavonoids

Flavonoids comprise a large group of secondary plant metabolites and include anthocyanins, flavonols, flavones, catechins, and flavonones (Harborne 1994, Harborne and Williams 2000). Many are present in plants as sugar conjugates. As food constituents, flavonoids are believed to have health promoting properties because of their antioxidant and free radical scavenging activity (Shahidi and Wanasundara 1992, Pietta 2000). A multitude of epidemiological and *in vitro* studies provide increasing evidence that flavonoids impart dietary health benefits. Epidemiological studies suggest that their consumption is associated with a reduced risk of cardiovascular disease (Hertog et al. 1997), cancer (Knekt et al. 1997, Wattenburg 1990), and dementia (Commenges et al. 2000). *In vitro* studies demonstrate that flavonols can induce human protective enzyme systems (Nijveldt et al. 2001), have effects on antiplatelet aggregation (Rice-Evans et al. 1997), and reduce blood viscosity. Flavonols are also associated with reduced inflammatory responses and allergic reactions (Cook and Samman 1996), and antiviral, antituberculosis, and antimalarial activities (Harborne and Williams 2000). Daily flavone and flavonol intake in western countries, expressed as aglycones, is estimated at approximately 25 mg per day (Hertog et al. 1993).

Nearly 98% of flavonols detected in tomato fruit were found to occur in the peel (Stewart et al. 2000). Tomatoes contain low levels of quercetin and kaempferol conjugates. The principal quercetin conjugate is rutin. Recent studies demonstrate that naringenin chalcone, a precursor to quercetin and kaempferol conjugates, is the main tomato fruit flavonoid and that a metabolic block at this step in the biosynthetic pathway results in low quercetin and kaempferol content (Muir et al. 2001). Total flavonol content in fruit of fresh market cultivars ranged from 1.3 to 203 microgram per gram of fresh weight (Stewart et al. 2000, Crozier et al. 1997). Flavonol concentrations in fruit of large-fruited genotypes were low, whereas the highest concentration of flavonols occurred in small cherry tomato fruit produced in warm sunny climates. Tomato juice and tomato puree contained 14 to 16 microgram per ml and 70 microgram per gram fresh weight, respectively. Whereas fresh tomatoes contained primarily flavonol conjugates, processed tomato products contained significant amounts of free flavonols. Cooking reduced levels of conjugated quercetin 35% to 82%, depending on the cooking method employed (Crozier et al. 1997). This may be a critical factor for bioavailability of different flavonols. Quercetin-β-glucoside, for example, is more easily absorbed than its aglycone, quercetin (Aziz et al. 1998).

In addition to colorless flavonols, novel heirloom cultivars such as 'Cherokee Purple' may exhibit red/purple flesh attributable to fruit anthocyanin accumulation (Table 6.2). Whereas anthocyanins normally occur in vegetative tissues of tomato, anthocyanin pigmentation in fruit is atypical. Fruit of tomato genotypes with the dominant *anthocyanin fruit* (*Aft*; formerly *Af*) gene accumulate elevated levels of anthocyanin in the fruit skin and outer pericarp tissues as well, resulting in purple fruit pigmentation (Giorgiev 1972, Jones et al. 2003). Anthocyanins in the *Aft* mutant are predominantly petunidin, and lesser amounts of malvidin and delphinidin (Jones et al. 2003). The recessive *atroviolacium* (*atv*; Rick 1964) and dominant *Aubergine* (*Abg*; Rick et al. 1994) loci also result in varying degrees of anthocyanin accumulation and purple pigmentation in fruit epidermal tissues.

Numerous studies have been directed toward manipulation of the flavonoid biosynthetic pathway. The majority of these have focused on alteration of anthocyanin pigmentation for flower and foliage color in ornamental crops (Mol et al. 1998, Dixon and Steele 1999, Forkmann and Martens 2001). Naturally occurring genetic diversity for these compounds in tomato, and a well-characterized biosynthetic pathway, make tomato an attractive crop in which to develop cultivars with enhanced levels of flavonoids and expanded range of health benefiting properties.

Heterologous overexpression in tomato of a *Petunia chi-a* gene encoding chalcone isomerase resulted in up to 78-fold increased levels of flavonols in fruit peel (Muir et al. 2001). Increased flavonol content was due to significant increases in peel concentration of quercetin glycosides and smaller increases in kaempferol glycosides. The altered phenotype exhibited stable inheritance and produced no demonstrable negative effects on plant growth and development. Flavonol levels in fruit of transgenic tomato lines overexpressing chalcone isomerase contained flavonol levels similar to those found in onions, a crop with naturally high flavonol content. Further studies revealed that concomitant expression of the sequences encoding chalcone synthase and flavonol synthase from *Petunia* were sufficient to achieve kaempferol accumulation in tomato flesh (Colliver et al. 2002). Introduction of the maize transcription factor genes *LC* and *C1*similarly resulted in upregulation of the flavonoid pathway in tomato fruit flesh (Bovy et al. 2002). These fruits accumulated high levels of kaempferol and lesser amounts of naringenin in their flesh. Absence of anthocyanins in *LC/C1* fruit was believed to be attributable to insufficient expression of the gene encoding flavonone-3'5'-hydroxylase, although other explanations including altered enzyme substrate specificity are plausible (Bovy et al. 2002).

Conventional approaches to develop flavonoid-rich tomato genotypes have been reliant on phenotypic evaluation of fruit color and/or biochemical analysis of flavonoid constituents. Utilizing information from molecular genetic studies of anthocyanin biosynthesis in tomato, Willits et al. (2005) devised an alternative strategy based upon gene expression assays of candidate loci to identify exotic tomato genotypes that may be useful in a traditional breeding program for improving fruit flavonoid content. Previously discussed transgene studies demonstrated that chalcone isomerase (*CHI*) is the primary block to synthesis of quercetin in tomato peel and that additional blocks may occur at chalcone synthase (*CHS*), flavanone hydroxylases, and flavonol synthase (*FS*) in fruit flesh (Muir et al. 2001, Colliver et al. 2002). Screening of a diverse set of *Solanum* section *Lycopersicon* accessions for flavonol gene expression identified two *S. chilense* accessions and a single *S. pennellii* accession that expressed *CHI* in the fruit peel, and also expressed *CHS*, *CHI*, and *FLS* in the fruit flesh (Willits et al. 2005). Introgression of the *S. pennellii* accession into tomato produced progeny that accumulated high levels of quercetin in both the fruit flesh and the fruit peel. The study is novel in its approach to identifying valuable accessions for enhancing fruit nutritive value and further highlights the untapped genetic diversity available in wild tomato germplasm.

The rich genetic diversity present in tomato and its wild relatives, together with the potential to manipulate biosynthetic pathways via introduction of foreign genes, provides potentially limitless opportunities to alter or introduce tomato fruit constituents that may contribute to fruit nutritive value. Resveratrol, an antioxidant with non-vitamin activity, is abundant in grape and several other species. It has recently been induced in tomato via expression of a grape stilbene synthase cDNA (Giovinazzo et al. 2005). Similar to a number of antioxidants, consumption of resveratrol-rich foods is associated with reduced atherosclerosis and carcinogenesis (Pace et al. 1995, Jang et al. 1997). Tomato plants do not normally produce resveratrol and do not have the stilbene synthase gene. However, the precursors for resveratrol synthesis are present and are also substrates for the flavonoid enzyme chalcone synthase. Resveratrol production in transformed fruit had no significant effect on levels of the associated flavonoid naringenin or on chlorogenic acid production (Giovinazzo et al. 2005).

Phenolic Acids

Phenolic acids form a diverse group that includes the widely distributed hydroxybenzoic and hydroxycinnamic acids. The hydroxycinnamic acids are phenolic acids included in the large class of secondary metabolites known as phenylpropanoids. Hydroxycinnamic acids in fruit tissues are typically esterfied to other polyhydroxylated compounds such as quinic acid, tartaric acid and glucose. Esters of caffeic acid predominate in Solanaceous species such as tomato, eggplant and tomato. Chlorogenic acid (5-O-caffeoylquinic acid) is typically the most abundant in these plants (Molgaard and Ravn 1988). This abundant phenolic acid is highly bioavailable yet somewhat overlooked dietary bioactive compound. Chlorogenic and related caffeoyl esters are among the most potent free radical scavengers found in plant tissues (Sawa et al. 1999, Nakatani et al. 2000). Chlorogenic acid has been shown to act as an antioxidant in human erythrocytes and for low-density lipoproteins *in vitro* (Nardini et al. 1995, Lekse et al. 2001). This class of hydroxycinnamic acid esters has also been reported to have antiviral activity (Cheminat et al. 1988).

Recent studies have investigated the biosynthesis of chlorogenic acid in plants. Gene silencing studies demonstrated that hydroxycinnamoyl transferase (HQT) is the primary enzyme required for chlorogenic acid accumulation in Solanaceous species (Niggeweg et al. 2004). Overexpression of HQT in tomato foliage caused plants to accumulate higher levels of chlorogenic acid, with no demonstrable side effects on the levels of other soluble phenolics or lignin. This research offers promise for development of tomato fruit with enhanced benefits for human health.

MACRONUTRIENTS

Plants contribute to the availability of dietary macronutrients including protein, fats, and carbohydrates. A large database exists for these macronutrients in agronomic crops wherein these well-characterized constituents are significant crop yield components. Modification of plant carbohydrate composition is a research area receiving closer scrutiny by plant researchers and the health community in light of the positive effects that fiber and complex carbohydrates can have on aspects of human health such as lipid metabolism and diabetes (Anderson 1990). Dietary plant fibers are derived from plant structural components and include both water-soluble and water-insoluble nonstarch polysaccharides and lignin. Genetic variability for the type and amount of soluble carbohydrates, pectin, gum, lignin, and other fiber components exists within tomato. These constituents have been studied principally for their effects on tomato fruit quality. Little attention has been given to their contribution to fruit nutritive value.

Carbohydrates

Considerable research effort has been dedicated to improving tomato fruit solids content because of the influence of solids on fruit quality (Stevens

1986). Sugars comprise 55% to 65% of the total soluble solids fraction and approximately 50% of the total solids in tomatoes. Considerable variability for soluble solids concentration is present within the cultivated tomato and its wild relatives. Soluble solids concentration of commercial hybrid cultivars generally ranges from 4.5% to 6.0% and can approach 15% in fruit of wild tomato species (Hewitt and Garvey 1987). Transient starch accumulation, which occurs prior to fruit maturation, contributes to sink strength and solids accumulation in developing fruit by maintaining a concentration gradient for sucrose between the leaves and the fruits (Dinar and Stevens 1981). Schaffer et al. (2000) demonstrated a relationship between increased levels of starch in immature *S. habrochaites* fruit and ADPglucose pyrophosphorylase activity. Increased starch levels in fruits inherited from *S. lycopersicum* x *S. habrochaites* plants were attributed to a *S. habrochaites* derived introgression coding for the large subunit of ADPglucose pyrophosphorylase (*AGPaseL1*).

Fruit of the cultivated tomato and those of red-fruited wild tomato species accumulate the reducing sugars glucose and fructose as the principal storage sugars during fruit development. Little or no sucrose is detectable in the mature fruit. In contrast, fruit of the green-fruited wild tomato species accumulate significant quantities of sucrose in addition to reducing sugars (Davies 1966b). Biochemical factors associated with sucrose accumulation in *S. chmielewskii* (Yelle et al. 1991), *S. habrochaites* (Miron and Schaffer 1991), and *S. peruvianum* (Stommel 1992b) have been described. Inheritance studies (Table 6.2) demonstrated that sucrose accumulation is controlled by a single recessive gene, *sucr*, in the green-fruited species *S. habrochaites* (Stommel and Haynes 1993) and *S. chmielewskii* (Chetelat et al. 1993). This locus maps to the pericentromeric region of chromosome 3. Transgenic tomato plants expressing a constitutive antisense invertase transgene exhibited increased sucrose and decreased hexose storage concentrations and reduced levels of acid invertase in ripe fruit (Klann et al. 1996). Accumulated evidence indicates that *sucr* encodes an inactive invertase allele.

In typical ripe fruit of *S. lycopersicum*, slightly higher amounts of fructose than glucose result in glucose:fructose (G:F) ratios of 0.8 to 1.0 (Davies 1966b). In hexose-accumulating fruit derived from interspecific crosses with *S. habrochaites*, glucose concentrations are commonly low relative to those of fructose and result in much lower G:F ratios than those typically noted in *S. lycopersicum*. Observed segregation in *S. lycopersicum* x *S. habrochaites* populations indicated that G:F ratios were controlled by at least two genes (Stommel and Haynes 1993). More recent investigations demonstrated that a major locus (*FGR*) located on chromosome 4 influences G:F ratios in tomato fruit and that an additional genetic factor may be involved in determining the ratio of hexose sugars (Schaffer et al. 1999, Levin et al. 2000). *FGR* increases levels of fructose, relative to glucose, and exhibits an allelic dosage effect. An additional locus (*FK2*) located on chromosome 6 is epistatic to *FGR* and may decrease G:F (Levin et al. 2000). *FK2* is subject to marked genotype x environment interaction.

Numerous studies have identified QTL introgressed into tomato from wild species that influence fruit-soluble solids content. Many of these QTL have a positive impact on solids content, but negatively influence fruit yield (Eshed and Zamir 1994, Tanksley et al. 1996, Chen et al. 1999, Yates et al. 2004). Additional studies identified chromosomal segments from *S. chmielewskii* and *S. cheesmaniae* that had a positive influence on fruit soluble solids while maintaining acceptable fruit size, pH, and yield (Triano and St. Clair 1995, Yousef and Juvik 2001). Collectively, these and other QTL studies underscore the contribution that multiple, non-allelic loci have on controlling soluble solids in tomato. Recent efforts to characterize a QTL that increases the hexose sugar component of soluble solids revealed a likely regulatory role for an apoplastic invertase gene (*LIN5*) and highlight the importance of intragenic recombination in genetic variability in soluble solids and other quantitatively inherited traits (Fridman et al. 2000).

Insoluble Solids

Tomato fruit insoluble solids contribute to fruit viscosity and firmness. Insoluble solids comprises water insoluble solids (WIS) and alcohol insoluble solids (AIS), the former being slightly larger. A strong relationship exists between AIS and viscosity, with fruit pericarp accounting for high correlation coefficients (Janoria and Rhodes 1974). The inheritance of AIS in a cross of high and low AIS cultivars demonstrated high heritability (0.68 and 0.75), additive genetic variation, and that less than three genes influence AIS levels (Stevens 1976). Fractionation of AIS into polyuronides and polysaccharide fractions demonstrated that water soluble polyuronides and water insoluble polyuronides accounted for approximately 90% of the difference in viscosity between a high and low viscosity cultivar (Stevens 1976). Water-soluble polyuronides comprise short and intermediate length chains as found in the fruit serum fraction. Water-insoluble polyuronides represent the protopectin fraction. An increase in water-insoluble polysaccharides had the greatest potential for increasing fruit viscosity.

Genetic variation for texture and factors that contribute to AIS in tomato results from the interaction of numerous QTLs. QTL associated with AIS and its constituents have been described (Fulton et al. 2000, Causse et al. 2002, Frary et al. 2003, Yates et al. 2004). These studies have shown that a few regions on chromosomes 2 and 4 have a large influence on fruit biochemical composition and organoleptic quality as determined by both physical and sensory measures (Causse et al. 2002, Yates et al. 2004).

Early work on the molecular genetics of fruit ripening and corresponding changes in fruit softening focused on polygalacturonase and its effects on ripening fruit. Antisense suppression of polygalacturonase accumulation demonstrated that the enzyme has only a minor effect on fruit softening, but has substantial effects on increasing viscosity of processed products and the integrity of stored fruit (Schuch et al. 1991, Kramer et al. 1992, Langley et al. 1994). Related efforts directed towards suppression of pectin methylesterase activity likewise had little influence on fruit firmness, but it increased soluble solids of juice and serum viscosity, paste viscosity, and serum separation of processed juice (Tieman et al. 1992, Thakur et al. 1996).

At least seven tomato β-galactosidase genes are expressed during tomato fruit development, six of which are expressed during ripening and may influence fruit textural properties (Smith and Gross 2000). Antisense suppression of the tomato β-galactosidase 3 gene did not improve fruit firmness but resulted in fruit that processed into pastes with an increased proportion of insoluble solids and slightly increased viscosity (de Silva and Verhoeyen 1998). Similar studies examining the tomato β-galactosidase 4 gene produced fruit from antisense lines that were 40% firmer than controls. Ongoing studies of a number of tomato ripening mutants (*rin, nor, Nr, Cnr*) that exhibit altered fruit textural properties offer promise for further elucidation of fruit AIS constituents (Seymour 2002). These mutants are discussed in more detail elsewhere in this volume.

Lipids

Phytosterols are important structural components of plant membranes and stabilize phospholipid bilayers in plant cell membranes. More than 200 different types of phytosterols have been reported in plant tissues. Phytosterols have received increased attention in the last ten years because of their cholesterol-lowering properties. Moreau et al. (2002) and Piironen et al. (2000) have recently reviewed phytosterols in foods and their health promoting properties. The primary interest in tomato phytosterols has focused on ripening related changes in content, composition, and conjugation that coincide with phospholipid catabolism mediated by phospholipase D (Whitaker 1988, Whitaker et al. 2001, Pinhero et al. 2002). Phospholipids decrease during tomato fruit ripening coincident with an increase in phospholipase D gene expression (Whitaker et al. 2001, Pinhero et al. 2002). The degradation of membrane lipids is an important feature of ethylene signal transduction pathways that take place in response to hormones, environmental stress, and senescence (Paliyath and Droillard 1992). Transformation of tomato with an antisense phospholipase D cDNA construct reduced phospholipase D activity 30 to 40% and resulted in firmer fruit with enhanced lycopene content, vitamin C, and flavor (Oke et al. 2003). In addition, juice and sauce prepared from the transgenic fruit exhibited improved viscosity and increased levels of major flavor volatiles. Increased membrane stability due to decreased phospholipase D activity may account for the observed enhancement in fruit quality attributes.

Related studies highlight the potential to modify the composition of fatty acids in tomato fruit and, more importantly, effect changes in fatty acid composition that alter levels of flavor volatile compounds. Wang et al. (2001) demonstrated that expression of the yeast Δ 9-desaturase gene in tomato resulted in changes in leaf fatty acid profiles that were accompanied by changes in volatiles derived from fatty acids. Using a similar strategy, Cook et al. (2002) expressed a Δ 6-desaturase transgene in tomato that also modified tomato fatty acid metabolism. Transformants produced γ-linolenic acid and octadecatetraenoic acid in both leaf and fruit tissue. Fruit tissue also contained reduced levels of linoleic acid and an increased percentage of α-linolenic acid. Additional studies suggest that a chloroplast-targeted lipoxygenase, TomloxC, can utilize both linoleic and linolenic acids as substrates to generate volatile flavor compounds (Chen et al. 2004). Gene silencing or antisense inhibition of TomloxC led to a marked reduction in levels of known flavor volatiles, including hexanal, hexenal, and hexenol. Volatile compounds in plants are derived from three different biosynthetic pathways (Croteau and Karp 1991). Enhanced levels of the aroma and flavor compound S-linalool were generated in transgenic plants by diverting metabolic flow normally committed to the biosynthesis of carotenoids without adverse effects on accumulation of lycopene or tocopherols (Lewinsohn et al. 2001).

PHYTOANTINUTRIENTS AND ALLERGENS

Genetic improvement of food crop nutritive value also encompasses variability present for food allergens and antinutrients. Although relatively uncommon, allergic reactions to fruit, vegetable, and grain crops do occur. Numerous studies have been conducted to characterize potential allergens in tomato fruit (e.g. Kondo et al. 2001). Whereas pollen-mediated allergies are relatively common, fruit consumption induced allergies are infrequent.

Tomatoes are a rich source of carotenoids that are valued for their beneficial effects on human health. Nonetheless, they may also be considered phytoantinutrients in special cases. Excessive dietary intake of carotenoid-containing foods may cause various afflictions characterized by aberrant discoloration of the skin. Carotenemia is commonly associated with ingestion of excessive amounts of β-carotene from carrots, but may occur with other yellow and green vegetables (Lascari 1981). Likewise, lycopenaemia is a rare cutaneous disease resulting from excessive consumption of lycopene-containing fruits and vegetables (La Placa et al. 2000).

Well publicized health studies evaluating the effects of β-carotene supplementation demonstrated that β-carotene does not prevent lung cancer in older men who smoke (Albanes et al. 1996) or in asbestos workers (Omenn et al. 1996). Supplementation at pharmacologic levels modestly increased lung cancer incidence in smokers, and this effect may have been associated with heavier smoking and higher alcohol intake. Whereas media reports focused upon the negatives of β-carotene supplementation; base-line concentrations of plasma β-carotene were inversely correlated with subsequent incidence of lung cancer. This suggests that β-carotene that was derived presumably from foods may be protective or that β-carotene is a marker for some unidentified protective dietary or lifestyle factor (Clevidence et al. 2000). The results of human health studies and genetic variability in food crops for potential antinutrients highlight the opportunities to breed for optimal levels of phytochemical constituents.

BIOPHARMACEUTICALS

Transgenic plants provide a novel opportunity for the development of biopharmaceuticals. Edible vaccines including those for rabies, hepatitis B, and enterotoxigenic E. coli have been expressed in tobacco, tomato, or potato and have been the focus of significant research and development (Yusibov et al. 1997, Haq et al. 1995, Mason et al. 1992). These plant-derived vaccines offer the opportunity to deliver inexpensive, orally administered vaccines and thus are promising for use in developing countries where cost and efficacious administration of available vaccines are typically of serious concern. The greatest challenge in developing edible vaccines via transgenic plants is to express sufficiently high levels of the foreign protein antigens in plant tissue. In tomato, expression of the gene for the respiratory syncytial virus (RSV) antigenic protein resulted in production of antigen that was active as an oral immunogen (Korban et al.

2002). Antigen accumulation was confined to seed tissue. Likewise, Mor et al. (2001) expressed recombinant acetylcholinesterase in tomato to provide therapeutic protein for protection against organophosphate poisoning.

Ruf et al. (2001) described a stable plastid transformation system for tomato. Transgenes of transformants exhibited maternal inheritance as expected for a plastid-encoded trait. Transgene expression in chromoplasts of tomato fruit was approximately 50% of the expression levels observed in leaf chloroplasts. In light of the very high foreign protein accumulation observed in transgenic tomato plastids (greater than 40% of soluble proteins), this system offers new possibilities for more efficient production of edible vaccines, pharmaceuticals, and antibodies in tomato.

PERSPECTIVES

Plant breeding has been the mainstay for genetic enhancement of horticultural and agronomic crop nutritive value. Breeding strategies for improved nutritional composition are similar to those for other traits in a germplasm development program. Based on knowledge gained through investigations on the characters heritability, the mode of inheritance, and existing genetic variability, appropriate breeding strategies based on phenotypic selection of individuals or families have been implemented to realize improvements in crop nutritional quality. Development of gene-specific probes and identification of markers tightly linked to phytonutrient constituent loci enables implementation of marker-assisted selection strategies for genotypic selection of well-studied traits. Detailed knowledge of the biochemical pathways for a number of phytonutrients has assisted in identification of structural and regulatory genes responsible for metabolite accumulation. The first restriction fragment length polymorphism (RFLP) map in tomato was constructed in 1986 with only 57 loci (Bernatzky and Tanksley 1986). Many additional RFLP, and PCR-based amplified fragment length polymorphism (AFLP), random amplified polymorphic DNA (RAPD), microsatellites, and expressed sequence tags (EST) have since been added to tomato linkage maps(Van der Hoeven et al. 2002). These dominant or codominant markers can be used for DNA fingerprinting, gene tagging, high-density genome mapping, and positional gene cloning. The availability of *S. pennellii* and *S. habrochaites* introgression lines (Eshed and Zamir 1995, Monforte and Tanksley 2000) further facilitates rapid screening and identification of major and minor QTL present in wild tomato species that influence fruit nutritive quality.

In tomato, there is a wealth of genetic variability within modern and heirloom cultivars, land races, and wild species for improvement of fruit nutritive value. Progress in breeding for improved tomato nutritional value is largely influenced by availability of sufficient genetic diversity and knowledge of gene action. The feasibility of a program focused on enhancement of phytonutrient content is also reliant upon methods for measuring phytonutrient constituents and evaluation of the relative importance of genotype and environment in the expression of phytonutrient content. Recent studies demonstrate renewed interest in characterizing *Solanum* section *Lycopersicon* germplasm collections for fruit phytonutrients and devising selection indices to maximize breeding efficiency (Hanson et al. 2004, Schauer et al. 2004, Willits et al. 2005). Where existing genetic variability for phytonutrient content is insufficient or difficult to introduce into adapted materials due to crossing barriers, appropriate gene constructs relevant to the biosynthesis of a specific phytonutrient or class of phytonutrients may be developed and genetically modified (GM) plants produced. The potential to introduce genetic material from unrelated organisms into tomato via these technologies provides an important source of new genetic variation. Unfortunately, GM plants are associated with modern agriculture, which is also viewed negatively. Barring long-term negative public opinion of GM crops, the wide range of products being developed and tested in commercial and academic laboratories suggests that the rate at which GM plants are introduced will increase. The introduction of GM plants with enhanced nutritive properties may foster increased acceptance of transgene technology in new value-added cultivars (Dunwell 2002, Lindsay 2002, Mehta et al. 2002). Ironically, the tools of modern molecular genetics that have been used to develop transgenic plants are now enabling traditional breeding to transcend limits imposed by conventional phenotypic selection. Although GM plant development has generally been considered a more expedient route to develop superior cultivars, Zamir (2001) argues that approximately ten years are required to create a transgenic cultivar for testing and that this timeframe is similar to that needed for development and testing of new lines developed in traditional breeding programs using exotic germplasm.

The evidence of accumulated studies indicates a positive link between fruit and vegetable consumption and improved health. Dietary studies in turn have stimulated interest in trying to define the constituents in fruits and vegetables that are responsible for their positive health effects. Considerable research still needs to be done to characterize these compounds. Cataloging the demonstrated/presumed benefits of these compounds will be an important step in sustaining research to improve phytonutrient content. Investigation of potential phytonutrient properties for well characterized fruit quality constituents such as organic acids, simple sugars, and specific carotenoids that have not been previously recognized for their phytonutrient properties will add important information to this database. Such recognition will enhance the intrinsic value of the characterized nutrient constituents. In addition to phytonutrient content, knowledge of their digestibility, absorption, and utilization is important. Whereas nutrition studies often focus on a single phytonutrient, current research clearly indicates that there are many bioactive compounds in tomato products and that it may be the combination of compounds in tomatoes that confer the beneficial health effects described (Laquatra et al. 2005). In addition to enhancing tomato fruit nutritive value, development of novel nutritionally enhanced germplasm provides new opportunities for dietary studies to examine phytonutrient constituents in a common food matrix for characterization of phytonutrient interactions.

In the mid-1970s, and again in the late 1980s, commentaries on genetic improvement of horticultural crops concluded that nutritional improvement may be impractical because of the lack of demonstrated dietary need or availability of alternate food choices (Kelly and Rhodes 1975, Tigchelaar 1987). In developing countries, overriding demands for research that enhances crop production and thus, food availability may supersede development of locally adapted cultivars with enhanced nutritional quality. Nonetheless, research to develop locally adapted, nutrient-dense cultivars is an important tactic in developing countries that complements related efforts to improve dietary nutrition (Hanson et al. 2004). Phytonutrient levels must be altered at the same time that numerous traits including yield, quality, disease resistance and many others are being selected for in a cultivar development program. In addition, negative relationships may exist between phytonutrient levels and other economically important traits such as fruit size and yield. Negative correlations between soluble solids, ascorbic acid, and total antioxidant content, for example, and fruit size or yield limit genetic enhancement of nutritive value. Successful breaking of these negative linkages may be critical to the successful commercialization of a new phytonutrient-enriched tomato cultivar. Although phytonutrients are typically viewed as positive constituents, it will be important to determine if there are any liabilities associated with the consumption of phytonutrient enhanced crops, particularly the possibility of negative health consequences that result from consuming plant parts containing phytonutrients at levels far above "normal". In collaboration with nutritionists and other health professionals, these issues should be raised early in the genetic improvement process since they too will limit successful utilization of phytonutrient-enhanced cultivars.

Consumer demand for horticultural products is a function of product price, alternative products, income, population, socioeconomic and demographic factors, as well as consumer tastes and preferences (Tomek and Robinson 1972). A market trial conducted in 1985 demonstrated that a segment of U.S. consumers developed loyalty to a premium quality fresh tomato and purchased them frequently at premium prices (Goldman 1988). A similar occurrence is evident in today's market wherein premium quality cluster tomatoes sold on the vine, vine-ripened greenhouse-grown fruit, and specialty cherry tomato cultivars enjoy growing sales and command premium prices. Although often viewed as burdensome, labeling can be a positive aspect for phytonutrient-enriched crops to inform consumers of the products nutrient enhanced status. Consumer education is critical to acceptance of food products with enhanced phytonutrient value. Similarly, food producers and distributors will not be apt to produce and market a new variety without evidence that it will be saleable and profitable. Clearly, the scientific community must not work in isolation of the marketplace. As consumers gain increased knowledge of nutrients in fruits and vegetables and recognize value-added products, markets for enhanced phytonutrient enriched crops will expand.

SUMMARY

During the twentieth century, plant breeding and genetics have improved the nutritive value of horticultural and agronomic crops. Tomatoes are a major dietary source of vitamins A and C, and lycopene. In addition to these well-known vitamins and antioxidants, other compounds in tomato fruit with antioxidant properties include chlorogenic acid, rutin, plastoquinones, tocopherols, and xanthophylls. Tomatoes also contribute carbohydrates, fiber, flavor compounds, minerals, proteins, and glycoalkaloids to the diet. Considerable genetic variation exists in tomato for micronutrients with antioxidant activity. A number of these micronutrients, particularly carotenoids, have been the subject of considerable research due to their contribution to the quality of fresh and processed tomato products. Increased recognition of their health promoting properties has stimulated new research and provided added justification for investigation in this area. Plants also contribute to the availability of dietary macronutrients including protein, fats, and carbohydrates. In tomato, these constituents have been studied principally for their effects on fruit quality. Little attention has been given to their contribution to fruit nutritive value. There is a wealth of genetic variability within modern and heirloom tomato cultivars, land races, and wild species for improvement of fruit nutritive quality. Where existing genetic variability for phytonutrient content is insufficient or difficult to introduce into adapted materials due to crossing barriers, appropriate gene constructs relevant to the biosynthesis of a specific phytonutrient or class of phytonutrients may be developed and genetically modified plants produced. Development of gene-specific probes and identification of markers tightly linked to phytonutrient constituent loci enables implementation of marker-assisted selection strategies for genotypic selection of well-studied traits. Detailed knowledge of the biochemical pathways for a number of phytonutrients has assisted in identification of structural and regulatory genes responsible for metabolite accumulation. As consumers gain increased knowledge of the nutrients of fruits and vegetables, and recognize value-added products, markets for phytonutrient enriched crops will expand.

REFERENCES

Albanes, D., O.P. Heinoonen, P.R. Taylor, J. Virtamo, B.K. Edwards, M. Rautalahti, A.M. Hartman, J. Palmgren, L.S. Freedman, J. Haapakoski, M.J. Barrett, P. Pietinen, N. Malila, E. Tala, K. Liippo, E.R. Salomaa, J.A. Tangrea, L. Teppo, F.B. Askin, E. Taskinen, Y. Erozan, P. Greenwald, and J.K. Huttunen. 1996. Alpha-Tocopherol and beta-carotene supplements and lung cancer incidence in the Alpha-Tocopherol, Beta-Carotene Cancer Prevention Study: Effects of base-line characteristics and study compliance. J Nat Canc Inst 88:1560-1570.

Agius, F., R. Gonzalez-Lamothe, J.L. Caballero, J. Munoz-Blanco, M.A. Botella, and V. Valpuesta. 2003. Engineering increased vitamin C levels in plants by overexpression of a D-galacturonic acid reductase. Nature Biotech 21:177-181.

Ames, B.N., M.K. Shigenaga, and T.M. Hagen. 1993. Oxidants, antioxidants, and the degenerative diseases of aging. Proc Natl Acad Sci USA 90:7915-7922.

Anderson, J.W. 1990. Dietary fiber and human health. HortScience 25:1488-1495.

Andrews, P.K., D.A. Fahy, and C.H. Foyer. 2004. Relationships between fruit exocarp antioxidants in the tomato (*Lycopersicon esculentum*) high pigment-1 mutant during development. Physiol Plant 120:519-528.

Aziz, A.A., C.A. Edwards, M.E.J. Lean, and A. Crozier. 1998. Absorption and excretion of conjugated flavonols, including quercitin-4-O-β-glucoside and isorhamnetin –4-O-β-glucoside by human volunteers after the consumption of onions. Free Rad Res 29:257-269.

Baker, L.R. and M.L. Tomes. 1964. Carotenoids and chlorophylls in two tomato mutants and their hybrids. Proc Amer Soc Hort Soc 85:507-513.

Barry, C. S., R.P. McQuinn, , A. J. Thompson, G.B. Seymour, D. Grierson, and J. Giovannoni. 2005. Ethylene insensitivity conferred by the *Green-ripe* and *Never-ripe Z* ripening mutants of tomato. Plant physiol 138: 267-275.

Bartley, G.E. and P.A. Scolnik.1993. cDNA cloning, expression during development, and genome mapping of PSY2, a second tomato gene encoding phytoene synthase. J Biol Chem 268:25718-25721.

Bartley, G.E. and P.A. Scolnik.1994. Molecular biology of carotenoid biosynthesis in plants. Ann Rev Plant Physiol Plant Mol Biol 45:287-301.

Beecher, G.R. Nutrient content of tomatoes. Proc Soc Exp Biol Med 218:98-100.

Bernacchi, D., T. Beck-Bunn, Y. Eshed, J. Lopez, V. Petiard, J. Uhlig, D. Zamir, and S. Tanksley. 1998. Advanced backcross QTL analysis in tomato. I. Identification of QTLs for traits of agronomic importance from *Lycopersicon hirsutum*. Theor Appl Genet 97:381-397.

Bernatzky, R. and S. Tanksley. 1986. Toward a saturated linkage map of tomato based on isozymes and random cDNA sequences. Genetics 112:887-898.

Berry, S.Z. and M.R. Uddin. 1991. Breeding tomato for quality and processing attributes. In: G. Kalloo (ed.), Genetic Improvement of Tomato. Springer-Verlag, Berlin Heidelberg, pp. 197-206.

Block, G., B. Patterson, and A. Subar. 1992. Fruits, vegetable, and cancer prevention: A review of the epidemiological evidence. Nutr Cancer 18:1-29.

Boileau, A.C., N.R. Merchen, K. Wasson, C.A. Atkinson, and J.W. Erdman. 1999. Cis-lycopene is more bioavailable than trans-lycopene *in vitro* and *in vivo* in lymph-cannulated ferrets. J Nutr 129:1176-1181.

Bovy, A., R. de Vos, M. Kemper, E. Schijlen, M.A. Pertejo, S. Muir, G. Collins, S. Robinson, M. Verhoeyen, S. Hughes, and C. Santos-Buelga. 2002. High-flavonol tomatoes resulting from the heterologous expression of the maize transcription factor genes *LC* and *C1*. Plant Cell 14:2509-2526.

Bradley, D.B. 1946. Varietal and location influence on acid composition of tomato fruit. J Agric Food Chem 12:213-216.

Bramley, P. 1992. Carotenoid biosynthesis. *In:* P.J. Lea [ed.], Methods in Plant Biochemistry. Academic Press, London.

Burge, J., O. Mickelsen, C. Nicklow, and G.L. Marsh. 1975. Vitamin C in tomatoes: Comparison of tomatoes developed for mechanical or hand harvesting. Ecol Food Nutr 4:27-31.

Burns, J.J. 1967. Ascorbic acid. In: D.M. Greenberg (ed.), Metabolic pathways 3rd ed. Vol. 1. Academic Press, New York, pp. 394-411.

Carangal, A.R., E.K. Alban, J.E. Varner, and R.C. Burrell. 1954. The influence of mineral nutrition on the organic acids of the tomato, *Lycopersicon esculentum*. Plant Physiol 29:355-360.

Causse, M., V. Saliba-Colombani, L. Lecomte, P. Duffe, P. Rousselle, and M. Buret. 2002. QTL analysis of fruit quality in fresh market tomato: a few chromosome regions control the variation of sensory and instrumental traits. J Expt Bot 53:2089-2098.

Chalukova M. 1988. Carotenoid composition of the fruits of hybrids between *Lycopersicon esculentum* and some wild species of the genus *Lycopersicon*, IV. Progenies of lycopene and β-carotene BC_1P_1 hybrids of *L. chmielewskii*. Genet Breed 21:49-57.

Cheminat, A., R. Zawatzsky, H. Becker, and R. Brouillard. 1988. Caffeoyl conjugates from *Echinacea* species: Structures and biological activity. Phytochemistry 27:2787-2794.

Chen, F.Q., M.R. Foolad, J. Hyman, D.A. St. Clair, and R.B. Beelaman. 1999. Mapping of QTLs for lycopene and other fruit traits in a *Lycopersicon esculentum* x *L. pimpinellifolium* cross and comparison of QTLs across tomato species. Mol Breeding 5:283-299.

Chen, G., R. Hackett, D. Walker, A. Taylor, Z. Lin, and D. Grierson. 2004. Identification of a specific isoform of tomato lipoxygenase (Tomlox C) involved in the generation of fatty acid-derived flavor compounds. Plant Physiol 136:2641-2651.

Chetelat, R.T., E. Klann, J.W. DeVerna, S. Yelle, and A.B. Bennett. 1993. Inheritance and genetic mapping of fruit sucrose accumulation in *Lycopersicon chmielewskii*. Plant J 4:643-650.

Clayberg, C.D., L. Butler, C.M. Rick, and P.A. Young. 1960. Second list of known genes in the tomato. J Hered 51:167-174.

Clevidence, B., I. Paetau, and J.C. Smith. 2000. Bioavailability of carotenoids from vegetables. HortScience 35:585-588.

Clinton, S.K. 1998. Lycopene: Chemistry, biology, and implications for human health and disease. Nutr Rev 56:35-51.

Colliver, S., A. Bovy, G. Collins, S. Muir, S. Robinson, C.H.R. de Vos, and M.E. Verhoeyen. 2002. Improving the nutritional content of tomatoes through reprogramming their biosynthetic pathway. Phytochem Rev 1:113-123.

Commenges, D., V. Scotet, S. Renaud, H. Jacqmin-Gadda, P. Barberger-Gateau, and J.F. Dartigues. 2000. Intake of flavonoids and risk of dementia. Eur J Epidemiol 16:357-363.

Cook, D., D. Grierson, C. Jones, A. Wallace, G. West, and G. Tucker. 2002. Modification of fatty acid composition in tomato (*Lycopersicon esculentum*) by expression of a borage Δ6-desaturase. Mol Biotech 21:123-128.

Cook, N.C. and S. Samman. 1996. Flavonoids – Chemistry, metabolism, cardioprotective effects, and dietary sources. Nutr Biochem 7:66-76.

Cookson, P.J., J.W. Kiano, C.A. Shipton, P.D. Fraser, S. Romer, W. Schuch, P.M. Bramley, and K.A. Pyke. 2003. Increases in cell elongation, plastid compartment size and phytoene synthase activity underlie the phenotype of the *high pigment-1* mutant of tomato. Planta 217:896-903.

Courtney, W.H.I. and V.N. Lambeth. 1977. Glycoalkaloid content of mature green fruit *Lycopersicon* species. HortScience 12:550-551.

Croteau, R. and F. Karp. 1991. Origin of natural odorants. *In:* P.M. Muller and D. Lamparsky (eds.), Perfumes: Art, Science and Technology. Elsevier Applied Science, London, pp. 101-126.

Crozier, A., M.E.J. Lean, M.S. McDonald, and C. Black. 1997. Quantitative analysis of the flavonoid content of commercial tomatoes, onions, lettuce, and celery. J Agric Food Chem 45:590-595.

Cunningham, F.X. Jr. and E. Gantt. 1998. Genes and enzymes of carotenoid biosynthesis in plants. Annu Rev Plant Physiol Plant Mol Biol 49:557-583.

Darby, L.A. 1978. Isogenic lines of tomato fruit colour mutants. Hort Res 18:73-84.

Davies, J.N. 1964. Effect of nitrogen, phosphorous and potassium fertilizers on the non-volatile organic acids of tomato fruit. J Sci Food Agric 15:665-673.

Davies, J.N. 1965. The effect of variety on the malic and citric acid content of tomato fruit. *In:* Rept Glasshouse Crops Res Inst, pp. 139-141.

Davies, J.N. 1966a. Changes in the non-volatile organic acids of tomato fruit during ripening. J Sci Food Agric 17:396-400.

Davies, J.N. 1966b. Occurrence of sucrose in the fruit of some species of *Lycopersicon*. Nature 209:640-641.

Davies, J.N. and G.E. Hobson. 1981. Constituents of tomato fruit – the influence of environment, nutrition, and genotype. CRC Crit Rev Food Sci Nutr 15:205-280.

de Bree, A., M. van Dusseldorp, I.A. Brouwer, K.H. van het Hof, and R.P. Steegers-Theunissen. 1997. Folate intake in Europe: recommended, actual and desired intake. Eur J Clin Nutr 51:643-660.

de la Garza, R.D., E.P. Quinlivan, S.M.J. Klaus, G.J.C. Basset, J.F. Gregory, and A.D. Hanson. 2004. Folate biofortification in tomatoes by engineering the pteridine branch of folate synthesis. Proc Natl Acad Sci 101:13720-13725.

Dharmapuri, S., C. Rosati, P. Pallara, R. Aquilani, F. Bouvier, B. Camara, and G. Giuliano.

2002. Metabolic engineering of xanthophyll content in tomato fruits. FEBS Letters 519:30-34.

Dinar, M. and M.A. Stevens. 1981. The relationship between starch accumulation and soluble solids content of tomato fruits. J Amer Soc Hort Sci 106:415-418.

Dixon, R.A. and C.L. Steele. 1999. Flavonoids and isoflavonoids – A gold mine for metabolic engineering. Trends Plant Sci 4:394-400.

Doll, R. 1990. Symposium on diet and cancer. An overview of the epidemiological evidence linking diet and cancer. Proc Nutr Soc 49:119-131.

Dumas, Y., M. Dadomo, G. Di Lucca, and P. Grolier. 2003. Effects of environmental factors and agricultural techniques on antioxidant content of tomatoes. J Sci Food Agric 83:369-382.

Dunwell, J.M. 2002. Future prospects for transgenic crops. Phytochemistry Rev 1:1-12.

Eshed, Y. and D. Zamir. 1994. Introgressions from *Lycopersicon pennellii* can improve the soluble-solids yield of tomato hybrids. Theor Appl Genet 88:891-897.

Eshed, Y. and D. Zamir. 1995. An introgression line population of *Lycopersicon pennellii* in the cultivated tomato enables the identification and fine mapping of yield-associated QTL. Genetics 141:1147-1162.

Food and Agricultural Organizations of the United Nations. 2004. Agricultural data. FAOSTAT. 6 May 2004. http://faostat.fao.org/faostat/collections?version=ext& has bulk=0& subset=agriculture

Forkmann, G. and S. Martens. 2001. Metabolic engineering and applications of flavonoids. Curr Opin Biotechnol 12:155-160.

Fraser, P.D., S. Romer, C.A. Shipton, P.B. Mills, J.W. Kiano, N. Misawa, R.G. Drake, W. Schuch, and P.M. Bramley. 2002. Evaluation of transgenic tomato plants expressing an additional phytoene synthase in a fruit-specific manner. Proc Natl Acad Sci 99:1092-1097.

Fraser, P.D., M.R. Truesdale, C.R. Bird, W. Schuch, and P.M. Bramley. 1994. Carotenoid biosynthesis during tomato fruit development. Evidence for tissue-specific gene expression. Plant Physiol 105:405-413.

Frary, A., S. Doganlar, A. Frampton, T. Fulton, J. Uhlig, H. Yates, and S. Tanksley. Fine mapping of quantitative trait loci for improved fruit characteristics from *Lycopersicon chmielewskii* chromosome 1. Genome 2003:46: 235-243.

Fray, R.G. and D. Grierson. 1993. Identification and genetic analysis of normal and mutant phytoene synthase genes of tomato by sequencing, complementation and co-suppression. Plant Mol Biol 22:589-602.

Fray, R.G., A. Wallace, P.D. Fraser, D. Valero, P. Hedden, P.M. Bramley, and D. Grierson. 1995. Constitutive expression of a fruit phytoene synthase gene in transgenic tomatoes causes dwarfism by redirecting metabolites from the gibberellin pathway. Plant J 8:693-701.

Freeman, J.A. and C.G. Woodbridge. 1960. Effect of maturation, ripening and truss position on the free amino acid content in tomato fruits. Proc Am Soc Hort Sci 76:515-523.

Fridman, E., T. Pleban, and D. Zamir. 2000. A recombination hotspot delimits a wild-species quantitative trait locus for tomato sugar content to 484 bp within an invertase gene. Proc Natl Acad Sci 97:4718-4723.

Friedman, M. 2002. Tomato glycoalkaloids: Role in the plant and in the diet. J Agric Food Chem 50:5751-5780.

Friedman, M. and D.L. Brandon. 2001. Nutritional and health benefits of soy proteins. J Agric Food Chem 49:1069-1086.

Friedman, M., T.E. Fitch, C.E. Levin and W.H. Yokoyama. 2000a. Feeding tomatoes to hamsters reduces their plasma low density lipoprotein cholesterol and triglycerides. J Food Sci 65:897-900.

Friedman, M., T.E. Fitch, and W.H. Yokoyama. 2000b. Lowering of plasma LDL cholesterol in hamsters by the tomato glycoalkaloids tomatine. Food Chem Toxicol 38:549-553.

Friedman, M. and C.E. Levin. 1995. α-Tomatine content in tomato and tomato products determined by HPLC with pulsed amperometric detection. J Agric Food Chem 43:1507-1511.

Fulton, T.M., T. Beck-Bunn, D.Emmatty, Y. Eshed, J. Lopez, V. Petriard, J. Uhlig, D. Zamir, and S. D. Tanksley. 1997. QTL analysis of an advanced backcross of *Lycopersicon peruvianum* to the cultivated tomato and comparisons with QTLs found in other wild species. Theor Appl Genet 95:881-894.

Fulton, T.M., S. Grandillo, T. Beck-Bunn, E. Fridman, A. Frampton, J. Lopez, V. Petriard, J. Uhlig, D. Zamir, and S. D. Tanksley. 2000. Advanced backcross QTL analysis of a *Lycopersicon esculentum* x *L. parviflorum* cross. Theor Appl Genet 100:1025-1042.

Gerster, H. 1997. The potential role of lycopene for human health. J Amer College Nutr 16:109-126.

Giordano, L.B., W.H. Gabelman, and G.C. Gerloff. 1982. Inheritance of differences in calcium utilization by tomatoes under low-calcium stress. J Amer Soc Hort Sci 107:664-669.

Giorgiev, C. 1972. Anthocyanin fruit tomato. Tomato Genet Coop Rpt. 22:10.

Giovinazzo, G., L. D'Amico, A. Paradiso, R. Bollin, F. Sparvoli, and L. DeGara. 2005. Antioxidant metabolite profiles in tomato fruit constitutively expressing the grapevine stilbene synthase gene. Plant Biotech J 3:57-69.

Giovannucci, E. 1999. Tomatoes, tomato-based products, lycopene, and cancer: Review of the epidemiologic literature. J Natl Canc Inst 91:317-331.

Giovannucci, E., A. Ascherio, E.B. Rimm, M.J. Stampfer, G.A. Colditz, and W.C. Willett. 1995. Intake of carotenoids and retinol in relation to risk of prostrate cancer. J Natl Cancer Inst 87:1767-1776.

Goldman, A. 1988. Consumer response to premium quality branded produce: the case of Israeli glasshouse tomatoes. Appl Agr Res 3:264-268.

Goodwin, T.W. 1971. Biosynthesis. In: O. Isler (ed.), Carotenoids. Birkhauser, Basel, Switzerland. pp. 577-636.

Gould, W.A. 1992. Tomato Production, Processing, and Technology, 3rd Ed. CTI Pub., Baltimore, MD, 536 pp.

Goyer, A., V. Illarionova, S. Roooje, M. Fischer, A. Bacher, and A.D. Hanson. 2004. Folate biosynthesis in higher plants. cDNA cloning, heterologous expression, and characterization of dihydroneopterin aldolases. Plant Physiol 135:103-111.

Gross, J. 1991. Pigments in Vegetables. Chlorophylls and Carotenoids. Van Nostrand Reinhold, New York, NY.

Gundersen, V., D. McCall, and I.E. Bechmann. 2001. Comparison of major and trace element concentrations in Danish greenhouse tomatoes (*Lycopersicon esculentum*) Cv. Aromata (F₁) cultivated on different substrates. J Agric Food Chem 49:3808-3815.

Hanson, A.D. and J.F. Gregory. 2002. Synthesis and turnover of folates in plants. Curr Opin Plant Biol 5:244-249.

Hanson, P.M., R. Yang, J. Wu, J. Chen, D. Ledesma, S.C.S. Tsou, and T. Lee. 2004. Variation for antioxidant activity and antioxidants in tomato. J Amer Soc Hort Sci 129:704-711.

Haq, T.A., H.S. Mason, J.D. Clements, and C.J. Arntzen. 1995. Oral immunization with a recombinant bacterial antigen produced in transgenic plants. Science 268:714-716.

Harborne, J.B. 1994. The Flavonoids. Advances in research since 1986, 1st ed., London: Chapman Hall.

Harborne, J.B. and C.A. Williams. 2000. Advances in flavonoid research since 1992. Phytochemistry 55:481-504.

Hassan, H.H. and J.P. McCollum. 1954. Factors affecting the content of ascorbic acid in tomatoes. University of Illinois Agric Expt Sta, Bull 573.

Hertog, M.G.L., E.J. Feskens, and D. Kromhout. 1997. Antioxidant flavonols and coronary heart disease risk. Lancet 349:699.

Hertog, M.G.L., E.J.M. Fesens, P.C.H. Hollman, M.B. Katan, and D. Kromhout. 1993. Dietary antioxidant flavonoids and the risk of coronary heart disease: The Zutphen Elderly Study. Lancet 342:1007.

Hewitt, J.D. and T.C. Garvey. 1987. Wild sources of high soluble solids in tomato. *In:* D.J. Nevins and R.A. Jones (eds.), Plant Biology, vol. 4. Tomato biotechnology. A.R. Liss, New York, pp. 45-54.

Hirschberg, J. 2001. Carotenoid biosynthesis in flowering plants. Curr Opin Plant Biol 4:210-218.

Hobson, G.E. and J.N. Davies. 1971. The tomato. *In:* A.C. Holme (ed.), The Biochemistry of Fruits and Their Products. Vol. 2. Academic Press, New York, pp. 437-482.

Hossain, T., I. Rosenberg, J. Selhub, G. Kishore, R. Beachy, and K. Schubert. 2004. Enhancement of folates in plants through metabolic engineering. Proc Nat Acad Sci 101:5158-5163.

Isaacson, T., G. Ronen, D. Zamir, and J. Hirschberg. 2002. Cloning of *tangerine* from tomato reveals a carotenoid isomerase essential for the production of β-carotene and xanthophylls in plants. Plant Cell 14:333-342.

Jang, M., L. Cai, G.O. Udeani, K.V. Slowing, C.F. Thomas, C.W.W. Beecher, H.H.S. Fong, N.R. Farnsworth, A.D. Kinghorn, R.G. Metha, R.C. Moon, and J.M. Pezzuto. 1997. Cancer chemopreventive activity of resveratrol, a natural product derived from grapes. Science 275:218-220.

Janoria, M.P. and A.M. Rhodes. 1974. Juice viscosity as related to various juice constituents and fruit characters in tomatoes. Euphytica 23:533-562.

Jarret, R.L., H. Sayama, and E.C. Tigchelaar. 1984. Pleiotropic effects associated with the chlorophyll intensifier mutations high pigment and dark green in tomato. J Amer Soc Hort Sci 109:873-878.

Jenkins, J.A. and G. Mackinney. 1955. Carotenoids of the apricot tomato and its hybrids with yellow and tangerine. Genetics 40:715-720.

Jones, C.M., P. Mes, and J.R. Myers. 2003. Characterization and inheritance of the anthocyanin fruit (*Aft*) tomato. J Hered 94:449-456.

Jones, B.L. and J.W. Porter. 1986. Biosynthesis of carotenoids in higher plants. CRC Crit Rev Plant Sci 3:295-324.

Juvik, J.A. and M.A. Stevens. 1982. Inheritance of foliar α-tomatine content in tomatoes. J Amer Soc Hort Sci 107:1061-1065.

Juvik, J.A., M.A. Stevens, and C.M. Rick. 1982. Survey of the genus *Lycopersicon* for variability in α-tomatine content. HortScience 17:764-766.

Kabelka, E., W. Yang, and D.M. Francis. 2004. Improved tomato fruit color within an inbred backcross line derived from *Lycopersicon esculentum* and *L. hirsutum* involves the interaction of loci. J Amer Soc Hort Sci 129:250-257.

Kelly, J. and B.R. Rhodes. 1975. The potential for improving the nutrient composition of horticultural crops. Food Tech 29:134-140.

Kerr, E.A. 1958a. Linkage relations of *gf*. Tomato Genet Coop Rpt 8:21.

Kerr, E.A.. 1958b. Mutations for chlorophyll retention in ripe fruit. Tomato Genet Coop Rpt 8: 22.

Khudairi, A.K. 1972. The ripening of the tomato. Amer Sci 60:696-707.

King, A. and G. Young. 1999. Characteristics and occurrence of phenolic phytochemicals. J Amer Dietetic Assoc 99:213-218.

Kerckhoffs, L.H.J. and R.E. Kendrick. 1997. Photocontrol of anthocyanin biosynthesis in tomato. J Plant Res 110:141-149.

Klann, E.M., B. Hall, and A.B. Bennett. 1996. Antisense acid invertase (*TIV1*) gene alters soluble sugar composition and size in transgenic tomato fruit. Plant Physiol 112:1321-1330.

Knekt, P., R. Jarvinen, R. Seppaanen, M. Heliovaara, L. Teppi, E. Pukkala, and A. Aromaa. 1997. Dietary flavonoids and the risk of lung cancer and other malignant neoplasms. Am J Epidemiol 146:223-230.

Kohler G.W., R.E. Lincoln, J.W. Porter, F.P. Zscheile, R.M. Caldwell, R.H. Harper, and W. Silver. 1947. Selection and breeding for high beta-carotene content (provitamin A) in tomato. Bot Gaz 109:219-225.

Kondo, Y., A. Urisu, and R. Tokuda. 2001. Identification and characterization of the allergens in the tomato fruit by immunoblotting. International-Archives-of-Allergy-and-Immunology. 126: 294-299.

Konings, E.J., H.H. Roomans, E. Dorant, R.A. Goldbohm, W.H. Saris, and P.A. van den Brandt. 2001. Folate intake of the Dutch population according to newly established liquid chromatography data for foods. Am J Clin Nutr 73:765-776.

Konsler, T.R. 1973. Three mutants appearing in 'Manapal' tomato. HortScience 8:331-333.

Korban, S.S., S.F. Kransnyanski, and D.E. Buetow. 2002. Foods as production and delivery vehicles for human vaccines. J Amer Coll Nutr 21:212S-217S.

Kozukue, N. and M. Friedman. 2003. Tomatine, chlorophyll, β-carotene and lycopene content in tomatoes during growth and maturation. J Sci Food Agric 83:195-200.

Kramer, M., R. Sanders, H. Bolkan, C. Waters, R.E. Sheehy, and W.R. Hiatt. 1992. Postharvest evaluation of transgenic tomatoes with reduced levels of polygalacturonase: processing, firmness and disease resistance. Postharvest Biol Technol 1:241-255.

Krishnaswamy, K. and K. Madhavan Nair. 2001. Importance of folate in human nutrition. Br J Nutr 85:S115-S124.

Lachance, P.A. 1998. Overview of key nutrients: micronutrient aspects. Nutr Rev 56:S34-S39.

Lambeth, V.N., E.F. Straten, and M.L. Fields. 1966. Fruit quality attributes of 250 foreign and domestic tomato accessions. Univ of Missouri, Agr Expt Sta Res Bull 908.

Langley, K.R., A. Martin, R. Stenning, A.J. Murray, G.E. Hobson, W.W. Schuch, and C.R. Bird. 1994. Mechanical and optical assessment of the ripening of tomato fruit with reduced polygalacturonase activity. J Sci Food Agric 66:547-554.

La Placa, M., M. Pazzaglia, and A. Tosti. 2000. Lycopenaemia. J Eur Acad Dermatol Venereol 14:311-312.

Laquatra, I., D.L. Yeung, M. Storey, and R. Forshee. 2005. Health benefits of lycopene in tomatoes – conference summary. Nutr Today 40:29-36.

Lascari, A.D. 1981. Carotenemia. A review. Clin Pediatr (Phila.). 20:25-29.

Lee, C.Y. and R.W. Robinson. 1980. Influence of the crimson gene (*og^c*) on vitamin content in tomato. HortScience 15:260-261.

Lekse, J.M., L. Xia, J.D. Morrow, and J.M. May. 2001. Plant catechols prevent lipid peroxidation in human plasma erythrocytes. Mol Cell Biochem 226:89-95.

Leonardi, C., P. Ambrosino, F. Esposito, and V. Fogliano. 2000. Antioxidative activity and carotenoid and tomatine contents in different typologies of fresh consumption tomatoes. J Agric Food Chem. 48:4723-4727.

Levin, I., P. Frankel, N. Gilboa, S. Tanny, and A. Lalazar. 2003. The tomato *dark green* mutation is a novel allele of the tomato homolog of the *DEETIOLATED1* gene. Theor Appl Genet 106:454-460.

Levin, I., N. Gilboa, E. Teselson, S. Shen, and A.A. Schaffer. 2000. *Fgr*, a major locus that modifies fructose to glucose ratio in mature tomato fruits. Theor Appl Genet 100:256-262.

Levin, I., A. Lalazar, M. Bar, and A.A. Schaffer. 2004. Non GMO fruit factories. Strategies for modulating metabolic pathways in the tomato fruit. Indust Crops Prod 20:29-36.

Lewinsohn, E., F. Schalechet, J. Wilkinson, K. Matsui, Y. Tadmor, K. Nam, O. Amar, E. Lastochkin, O. Larkov, U. Ravid, W. Hiatt, and S. Gepstein. 2001. Enhanced levels of the aroma and flavor compound S-Linalool by metabolic engineering of the terpenoid pathway in tomato fruits. Plant Physiol 127:1256-1265.

Lincoln, R.E. and J.W. Porter. 1950. Inheritance of beta-carotene in tomatoes. Genetics 35:206-211.

Lincoln, R.E., F.P. Zscheile, J.W. Porter, G.W. Kohler, and R.M. Caldwell. 1943. Provitamin A and vitamin C in the genus *Lycopersicon*. Bot Gaz 105:113-115.

Lindsay, D.G. 2002. The challenges facing scientists in the development of foods in Europe using biotechnology. Phytochemistry Rev 1:101-111.

Liu, Y., S. Roof, Z. Ye, C. Barry, A. van Tuinen, J. Vrebalov, C. Bowler, and J. Giovannoni. 2004. Manipulation of light signal transduction as a means of modifying fruit nutritional quality in tomato. Proc Nat Acad Sci 101:9897-9902.

Lois, L.M., M. Rodriguez-Concepcion, F. Gallego, N. Campos, and A. Boronat. 2000. Carotenoid biosynthesis during tomato fruit development: regulatory role of 1-deoxy-D-xylulose 5-phosphate synthase. Plant J 22:503-513.

Lucock, M. 2000. Folic acid: Nutritional biochemistry, molecular biology, and role in disease processes. Mol Genet Metab 71:121-138.

MacArthur, J.W. 1934. Linkage groups in the tomato. J Genet 29:123-133.

Madhavi, D.L. and D.K. Salunkhe. 1998. Tomato. In: D.K. Salunkhe and S.S. Kadam (ed.), Handbook of Vegetable Science, Marcel Dekker, New York, pp. 171-201.

Malewski, W. and P. Markakis. 1971. Ascorbic acid content of the developing tomato. J Food Sci 36:537.

Manuelyan, H., M. Yordanov, Z. Yordanova, and Z. Ilieva. 1975. Studies on β-carotene and lycopene content in the fruits of *Lycopersicon esculentum* Mill.× *L. chilense* Dun. hybrids. Qual Plant Foods Hum Nutr 25:205-210.

Mason, H.S., D.M.K. Lam, and C.J. Arntzen. 1992. Expression of hepatitis B surface antigen in transgenic plants. Proc Natl Acad Sci USA 89:11745-11749.

Matthews, R.F., P. Crill, and D. Burgis. 1973. Ascorbic acid content of tomato varieties. Proc Florida State Hort Soc 86:242-250.

Maynard, D.N. 1979. Nutritional disorders of vegetable crops: A review. J Plant Nutr 1:1-23.

McClendon, J.H., C.W. Woodmansee, and G.F. Somers. 1959. On the occurrence of free galacturonic acid in apples and tomatoes. Plant Physiol 34:389.

Mehta, R., T. Cassol, N. Li, N. Ali, A.K. Handa, and A.K. Mattoo. 2002. Engineered polyamine accumulation in tomato enhances phytonutrient content, juice quality, and vine life. Nature Biotech 20:613-618.

Miron, D. and A.A. Schaffer. 1991. Sucrose phosphate synthase, sucrose synthase, and invertase activities in developing fruit of *Lycopersicon hirsutum* Humb. and Bonpl. Plant Physiol 95:623-627.

Mitchell, C.A. 1990. Crucial issues facing horticulture: Present and future. HortScience 25:598-602.

Mochizuki, T. and S. Kaminura. 1984. Inheritance of vitamin C content and its relation to other characters in crosses between *hp* and *og* varieties of tomatoes. Eucarpia Tomato Working Group, Synopsis IX, Meeting 22-24 May, Wageningen, The Netherlands, pp. 8-13.

Mol, J., E. Grotewold, and R. Koes. 1998. How genes paint flowers and seeds. Trends Plant Sci 3:212-217.

Molgaard, P. and H. Ravn. 1988. Evolutionary aspects of caffeoyl ester distribution in dicotyledons. Phytochemistry 27:2411-2421.

Molloy, A.M. and J.M. Scott. 2001. Folates and prevention of disease. Public Health Nutr 4:601-609.

Monforte, A.J. and S. D. Tanksley. 2000. Development of a set of near isogenic and backcross recombinant inbred lines containing most of the *Lycopersicon hirsutum* genome in a *L. esculentum* genetic background: a tool for gene mapping and gene discovery. Genome 43:803-813.

Mor, T.S., M. Sternfeld, H. Soreq, C.J. Arntzen, and H.S. Mason. 2001. Expression of recombinant human acetylcholinesterase in transgenic tomato plants. Biotech Bioeng 75:259-266.

Moreau, R.A., B.D. Whitaker, and K.B. Hicks. 2002. Phytosterols, phytostanols, and their conjugates in foods: structural diversity, quantitative analysis, and health-promoting uses. Prog Lipid Res 41:457-500.

Muir, S.R., G.J. Collins, S. Robinson, S. Hughes, A. Bovy, C.H.R. de Vos, A.J. van Tunen, and M.E. Verhoeyen. 2001. Nature Biotech 19:470-474.

Mustilli, A.C., F. Fenzi, R. Ciliento, F. Alfano, and C. Bowler. 1999. Phenotype of the tomato *high pigment-2* mutant is caused by a mutation in the tomato homolog of *DEETIOLATED1*. Plant Cell 11:145-157.

Nakatani, N., S. Kayano, H. Kikuzaki, K. Sumino, K. Katagiri, and T. Mitani. 2000. Identification, quantitative determination, and antioxidative activities of chlorogenic acid isomers in prune (*Prunus domestica* L.). J Agr Food Chem 48:5512-5516.

Nardini, M., M. D'Aquino, G. Tomassi, V. Gentili, M. Di Felice, C. Scaccini. 1995. Inhibition of human low-density lipoprotein oxidation by caffeic acid and other hydroxycinnamic acid derivatives. Free Radical Biol Med 19:541-552.

Niggeweg, R., A.J. Michael, and C. Martin. 2004. Engineering plants with increased levels of the antioxidant chlorogenic acid. Nature Biotech. 22:746-754.

Nijveldt, R.J., E. van Nood, D.E.C. van Hoorn, P.G. Boelens, K. van Norren, and P.A.M. van Leeuwen. 2001. Flavonoids: A review of probable mechanisms of action and potential applications. Am J Clin Nutr 74:418-425.

Oke, M., R.G. Pinhero, and G. Paliyath. 2003. The effects of genetic transformation of tomato with antisense phospholipase D cDNA on the quality characteristics of fruit and their processed products. Food Biotech 17:163-182.

Oleszek, W., S. Shannon, and R.W. Robinson. 1986. Steroidal alkaloids of *Solanum lycopersicoides*. Acta Soc Bot Pol 55:653-657.

Omenn, G.S., G. Goodman, M. Thornquist, J. Grizzle, L. Rosenstock, S. Barnhart, J. Balmes, M.G. Cherniack, M.R. Cullen, A. Glass, J. Keogh, F. Meyskens, B. Valanis, and J. Williams. 1994. The β-carotene and retinol efficacy trail (CARET) for chemoprevention of lung cancer in high risk populations: smokers and asbestos-exposed workers. Cancer Res (suppl) 54:2038s-2043s.

Omenn, G.S., G. Goodman, M. Thornquist, J. Balmes, M.R. Cullen, A. Glass, J. Keogh, F. Meyskens, B. Valanis, J. Williams, S. Barnhart, and S. Hammar. 1996. Effects of a combination of beta-carotene and vitamin A on lung cancer and cardiovascular disease. New Engl J Med 334:1150-1155.

Pace-Asciak, C.R., S. Hahn, E.P. Diamandis, G. Soleas, and D.M. Goldberg. 1995. The red wine phenolics trans-resveratrol and quercetin block human platelet aggregation and eicosanoid synthesis: implication for protection against coronary heart disease. Clin Chim Acta 235:207-219.

Paiva, S.A.R. and R.M. Russell. 1999. Beta-carotene and other carotenoids as antioxidants. J Am Coll Nutr 18:426-433.

Paliyath, G. and M.J. Droillard. 1992. The mechanisms of membrane deterioration and disassembly during senescence. Plant Physiol Biochem 30:789-812.

Palmieri, S., P. Martiniello, and G.P. Sorressi. 1978. Chlorophyll and carotene content in *high pigment* and *green flesh* fruits. Tomato Gen Coop Rpt 28:10.

Pateau, I., F. Khachik, E.D. Brown, G.R. Beecher, T.R. Kramer, J. Chittams, and B.A. Clevidence. 1998. Chronic ingestion of lycopene-rich tomato juice or lycopene supplements significantly increases plasma concentrations of lycopene and related tomato carotenoids in humans. Am J Clin Nutr 68:1187-1195.

Park, S., C.K. Kim, L.M. Pike, R.H. Smith, and K.D. Hirschi. 2004. Increased calcium in carrots by expression of an *Arabidopsis* H⁺/Ca²⁺ transporter. Mol Breed 14:275-282.

Paulson, K.N. and M.A. Stevens. 1974. Relationships among titratable acidity, pH and buffer composition of tomato fruits. J Food Sci 39:354-357.

Piechulla, B. K.R.C. Imlay, and W. Gruissem 1985. Plastid gene expression during fruit ripening in tomato. Plant Mol Biol 5:373-384.

Pietta, P.G. 2000. Flavonoids as antioxidants. J Nat Prod 63:1035-1042.

Piironen V., D.G. Lindsay, T.A. Miettinen, J. Toivo, and A.M. Lampi. 2000. Plant sterols: biosynthesis, biological function and their importance to human nutrition. J Sci Food Agric 80:939-966.

Pinhero, R.G., K.C. Almquist, Z. Novotna, and G. Paliyath. 2003. Developmental regulation of phospholipase D in tomato fruits. Plant Physiol Biochem 41:223-240.

Porter, J.W. and D.G. Anderson. 1962. The biosynthesis of carotenes. Arch Biochem Biophys 97: 520-528.

Porter, J.W. and D.G. Anderson. 1967. Biosynthesis of carotenes. Ann Rev Plant Physiol 18:197-228.

Porter, J.W. and R.E. Lincoln. 1950. I. *Lycopersicon* selections containing a high content of carotenes and colorless polyenes. II. The mechanisms of carotene biosynthesis. Arch Biochem Biophys 27: 390-403.

Rice-Evans, C.A., N.J. Miller, and G. Paaganga. 1997. Antioxidant properties of phenolic compounds. Trends Plant Sci 2:152-159.

Rick, C.M. 1956. Genetic and systematic studies on accessions of *Lycopersicon* from the Galapagos Islands. Am J Bot 43:687-696.

Rick, C.M. 1964. Biosystematic studies on Galapagos Island tomatoes. Occ Pap Calif Acad Sci 44:59-77.

Rick, C.M. 1967. Fruit and pedicel characters derived from Galapagos tomatoes. Eco Bot 21:171-184.

Rick, C.M. 1974. High soluble-solids content in large-fruited tomato lines derived from a wild green-fruited species. Hilgardia 42:493-510.

Rick, C.M. and L. Butler. 1956. Cytogenetics of the tomato. Adv Genet 8:267-382.

Rick, C.M., P. Cisneros, R.T. Chetelat, and J.W. DeVerna. 1994. *Abg* – A gene on chromosome 10 for purple fruit derived from *S. lycopersicoides*. Tomato. Genet Coop Rpt 44:29-30.

Rick, C.M., A.E. Thompson, and O. Brauer. 1959. Genetics and development of an unstable chlorophyll deficiency in *Lycopersicon esculentum*. Am J Bot 46:1-11.

Rick, C.M., J.W. Uylig, and A.D. Jones. 1994. High α-tomatine content in ripe fruit of Andean *Lycopersicon esculentum* var. *cerasiforme*: developmental and genetic aspects. Proc Natl Acad Sci 91:12877-12881.

Romer, S., P.D. Fraser, J.W. Kiano, C.A. Shipton, N. Misawa, W. Schuch, and P.M. Bramley. 2000. Elevation of the provitamin A content of transgenic tomato plants. Nature Biotech 18:666-669.

Ronen, G., L. Carmel-Goren, D. Zamir, and J. Hirschberg. 2000. An alternative pathway to β-carotene formation in plant chromoplasts discovered by map-based cloning of *Beta* and *old-gold* color mutations in tomato. Proc Natl Acad Sci 97:11102-11107.

Ronen, G.L., M. Cohen, D. Zamir, and J. Hirschberg. 1999. Regulation of carotenoid biosynthesis during tomato fruit development: expression of the gene for lycopene epsilon-cyclase is down-regulated during ripening and is elevated in the mutant *Delta*. Plant J 17:341-351.

Rosati, C., R. Aquilani, S. Dharmapuri, P. Pallara, C. Marusic, R. Tavazza, F. Bouvier, B. Camara, and G. Giuliano. 2000. Metabolic engineering of beta-carotene and lycopene content in tomato fruit. Plant J 24:413-419.

Ruf, S., M. Hermann, I.J. Berger, H. Carrer, and R. Bock. 2001. Stable genetic transformation of tomato plastids and expression of a foreign protein in fruit. Nature Biotech 19:870-875.

Saliba-Colombani, V., M. Causse, D. Langlois, J. Philouze, and M. Buret. 2001. Genetic analysis of organoleptic quality in fresh market tomato. 1. Mapping of QTLs for physical and chemical traits. Theor Appl Genet 102:259-272.

Sawa, T., M. Nakao, T. Akaike, K. Ono, and H. Maeda. 1999.Alkylperoxyl radical scavenging activity of various flavonoids and other phenolic compounds: Implications for the anti-tumor-promoter effect of vegetables. J Agr Food Chem 47:397-402.

Schaffer, A.A., I. Levin, I. Oguz, M. Petreikov, F. Cincarevsky, E. Yeselson, S. Shen, N. Gilboa, and M. Bar. 2000. ADPglucose pyrophosphorylase activity and starch accumulation in immature tomato fruit: the effect of a *Lycopersicon hirsutum*-derived introgression encoding the large subunit. Plant Sci 152:135-144.

Schaffer, A.A., M. Petreikov, D. Miron, M. Fogelman, M. Spiegelman, Z. Bnei-Moshe, S. Shen, D. Granot, R. Hadas, N. Dai, I. Levin, M. Bar, M. Friedman, M. Pilowsky, N. Gilboa, and L. Chen. 1999. Modification of carbohydrate content in developing tomato fruit. HortScience 34:1024-1027.

Schauer, N., D. Zamir, and A.R. Fernie. 2005. Metabolic profiling of leaves and fruit of wild species tomato: a survey of the *Solanum lycopersicum* complex. J Expt Bot 56:297-307.

Schuh, W., J. Kanczler, D. Robertson, G. Hobson, G. Tucker, D. Grierson, S. Bright, and C. Bird. 1991. Fruit quality characteristics of transgenic tomato fruit with altered polygalacturonase activity. HortScience 26:1517-1520.

Scolnik, P.A., P. Hinton, I.M. Greenblatt, G. Giuliano, M.R. Delanoy, D.L. Spector, and D. Pollock. 1987. Somatic instability of carotenoid biosynthesis in the tomato *ghost* mutant and its effect on plastid development. Planta 171:1-18.

Scott, J., R. Rebeille, and J. Fletcher. 2000. Folic acid and folates: the feasibility for nutritional enhancement in plant foods. J Sci Food Agric 80:795-824.

Senti, F.R. and R.L. Rizek.1975. Nutrient levels in horticultural crops. HortScience 10:243-246.

Seymour, G.B., K. Manning, E.M. Eriksson, A.H. Popovich, and G. J. King. 2002. Genetic identification and genomic organization of factors affecting fruit texture. J Expt Bot 53:2065-2071.

Shahhidi, F. and P.K. Wanasundara. 1992. Phenolic antioxidants. Crit Rev Food Sci Nutr 32:67-103.

Sommer, A. 1997. Vitamin A deficiency, child health, and survival. Nutrition 13:484-485.

Spooner, D. 2005. New species of wild tomatoes (*Solanum* section *Lycopersicon:* Solanaceae). Systematic Botany 30(2) : 424-434.

Stevens, M.A. 1972. Citrate and malate concentrations in tomato fruits: Genetic control and maturational effects. J Amer Soc Hort Sci 97:655-658.

Stevens, M.A. 1976. Inheritance of viscosity potential in tomato. J Amer Soc Hort Sci 101:152-155.

Stevens, M.A. 1986. Inheritance of tomato fruit quality components. In: J. Janick (ed.), Plant Breeding Reviews, Vol. 4, AVI Publishing, Westport, Conn, pp. 273-312.

Stevens, M.A. and K.N. Paulson. 1973. Phosphorous concentrations in tomato fruits: inheritance and maturity effects. J Amer Soc Hort Sci 98:607-610.

Stevens, M.A. and C.M. Rick. 1986. Genetics and Breeding. In: Atherton, J.G. and Rudich, J. (eds.) The Tomato Crop: A Scientific Basis for Improvement. Chapman and Hall, New York, NY, pp. 35-109.

Stewart, A.J., S. Bozonnet, W. Mullen, G.I. Jenkins, M.E. Lean, and A. Crozier. 2000. Occurrence of flavonols in tomatoes and tomato-based products. J Agric Food Chem 48:2663-2669.

Stommel, J.R. 1992a. Tomato nutritional quality: genetic improvement of carotenoid content. Agro Food Industry Hi-tech. 3:7-11.

Stommel, J.R. 1992b. Enzymatic components of sucrose accumulation in the wild tomato species *Lycopersicon peruvianum*. Plant Physiol 99:324-328.

Stommel, J.R. 2001. USDA 97L63, 97L66, and 97L97: Tomato breeding lines with high fruit beta-carotene content. HortScience 36: 387-388.

Stommel, J.R., J.A. Abbott, and R.A. Saftner. 2005. USDA 02L1058 and 02L1059: Cherry tomato breeding lines with high fruit beta-carotene content. HortScience 40 : 1569-1570.

Stommel, J.R., Abbott, J., Saftner, R.A., and M. Camp. 2005. Sensory and objective quality attributes of beta-carotene- and lycopene-rich tomato fruit. J Amer Soc Hort Sci 130:244-251.

Stommel, J.R. and K.G. Haynes. 1993. Genetic control of fruit sugar accumulation in a cross of *Lycopersicon esculentum* x *L. hirsutum*. J Amer Soc Hort Sci 118:859-863.

Stommel J.R., and K.G. Haynes. 1994. Inheritance of beta-carotene content in the wild tomato species *Lycopersicon cheesmanii*. J Hered 85:401-404.

Sudheesh, S., G. Presaannakumar, S. Vijayakumar, and N.R. Vijayalakshmi. 1997. Hypolipidemic effect of flavonoids from *Solanum melongena*. Plant Foods Human Nutr. 51:321-330.

Tanksley, S.D., S. Grandillo, T.M. Fulton, D. Zamir, Y. Eshed, V. Petiard, J. Lopez, and T. Beck-Bunn. 1996. Advanced backcross QTL analysis in a cross between an elite processing line of tomato and its wild relative *L. pimpinellifolium*. Theor Appl Genet 92:213-224.

Tanksley, S.D. and J.C. Nelson. 1996. Advanced backcross QTL analysis: A method for the simultaneous discovery and transfer of valuable QTLs from unadapted germplasm into elite breeding lines. Theor Appl Genet 92:191-203.

Tee, E.S. 1992. Carotenoids and retinoids in human nutrition. Crit Rev Food Sci Nutr 31:103-163.

Thakur, B.R., R.K. Singh, and A.K. Handa. 1996. Effect of an antisense pectin methylesterase gene on the chemistry of pectin in tomato (*Lycopersicon esculentum*) juice. J Agric Food Chem 44:628-630.

Thompson, A.E., M.L. Tomes, H.T. Erickson, E.V. Wann, and R.J. Armstrong. 1967. Inheritance of crimson fruit color in tomatoes. Proc Amer Soc Hort Sci 91:495-504.

Tieman, D.M., R.W. Harriman, G. Ramamohan, and A.K. Handa. 1992. An antisense pectin methylesterase gene alters pectin chemistry and soluble solids in tomato fruit. Plant Cell 4:667-679.

Tigchelaar, E.C. and M..L. Tomes. 1974. 'Caro-Rich' tomato. HortScience 9:82.

Tigchelaar, E.C. 1987. Genetic improvement of tomato nutritional quality. *In:* B. Quebedeaux and F. Bliss, [eds.], Proc. 1st Int. Symp. on Hort. and Human Health. Prentice Hall, Englewood Cliffs, NJ, pp.185-190.

Tomek, W.G. and K.L. Robinson. 1972. Agricultural product prices. Cornell Unv. Press, Ithaca, NY.

Tomes, M.L. 1963. Temperature inhibition of carotene synthesis in tomato. Bot Gaz 124:180-185.

Tomes, M.L., F.W. Quackenbush, and M. McQuistan. 1954. Modification and dominance of the gene governing formation of high concentrations of beta-carotene in the tomato. Genetics 39:810-817.

Tomes, M.L. and F.W. Quackenbush. 1958. Caro-Red, a new provitamin A rich tomato. Eco Bot 12:256-260.

Triano, S.R. and D.A. St. Clair. 1995. Processing tomato germplasm with improved fruit soluble solids content. HortScience 30:1477-1478.

United States Department of Agriculture. 2004. National Nutrient Database for Standard Reference, Release 17. 27 April, 2005. <http://www.nal.usda.gov/fnic/foodcomp/search/>

United States Department of Agriculture. 2005. Briefing room, tomatoes. 27 April, 2005. http://www.ers.usda.gov/briefing/tomatoes/

United States Department of Agriculture. 2002. Nutritive value of foods. Home and Garden Bulletin 72.

Unlu, Z., S.K. Clinton, and S.J. Schwartz. 2003. Absorption of lycopene isomers following single meals containing tomato sauces with varying isomer patterns. FASEB-Journal. 17: Abstract No. 456.9.

Van der Hoeven, R., C. Ronning, J. Giovannoni, G. Martin, and S, Tanksley. 2002. Deductions about the number, organization, and evolution of genes in the tomato genome based on analysis of a large expressed sequence tag collection and sellective genomic sequencing. Plant Cell 14:1441-1456.

Van Tuinen, A., M. Cordonnier-Pratt, L.H. Pratt, R. Verkerk, P. Zabel, and M. Koorneef. 1997. The mapping of phytochrome genes and photomorphogenic mutants of tomato. Theor Appl Genet 94:115-122.

Vinson, J.A., Y. Hao, X. Su, and L. Zubik. 1998. Phenol antioxidant quantity and quality in foods: Vegetables. J Agr Food Chem 46:3630-3634.

Wang, C.L., J.S. Xing, C.K. Chin, C.T. Ho, and C.E. Martin. 2001. Modification of fatty acids changes the flavor volatiles in tomato leaves. Phytochemistry 58:227-232.

Warnock, S.J. 1983. Ascorbic acid content of fresh and processed purees from twelve tomato cultivars. HortScience 18:728-730.

Wattenburg, L. 1990. Inhibition of carcinogenesis by minor anutrient constituents of the diet. Proc Nutr Soc 49:173-183.

Wheeler, G.L., M.A. Jones, and N. Smirnoff. 1998. The biosynthetic pathway of vitamin C in higher plants. Nature 393:365-369.

Whitaker, B. D. 1998. Changes in the steryl lipid content and composition of tomato fruit during ripening. Phytochemistry 27: 3411- 3416.

Whitaker, B.D., D.L. Smith, and K.C. Green. 2001. Cloning, characterization and functional expression of a phospholipase Dα cDNA from tomato fruit. Physiol Plant 112:87-94.

Willits, M.G., C.M. Kramer, R.T.N. Prata, V. De Luca, B.G. Potter, J.C. Steffens, and G. Graser. 2005. Utilization of the genetic resources of wild species to create a nontransgenic high flavonoid tomato. J Agric Food Chem 53:1231-1236.

World Health Organization. 2005. Vitamin A. 6 May, 2005. <http://www.who.int/vaccines/en/vitaminamain.shtml>

Yates, H.E., A. Frary, S. Doganlar, A. Frampton, N.T. Eaannetta, J. Uhlig, and S.D. Tanksley. 2004. Comparative fine mapping of fruit quality QTLs on chromosome 4 introgressions derived from two wild species. Euphytica 135:283-296.

Yelle, S., R.T. Chetelat, M. Dorais, J.W. DeVerna, and A.B. Bennett. 1991. Sink metabolism in tomato fruit. IV. Genetic and biochemical analysis of sucrose accumulation. Plant Physiol 95:1026-1035.

Yen, H.C., B.A. Shelton, L.R. Howard, S. Lee, J. Vrebalov, and J. Giovannoni. 1997. The tomato high-pigment (*hp*) locus maps to chromosome 2 and influences plastome copy number and fruit quality. Theor Appl Genet 95:1069-1079.

Young, P.A. 1956. *ry*, A modifier gene for red color in yellow tomato fruit. Rpt Tomato Genet Coop 6:33.

Yousesf, G.G. and J.A. Juvik. 2001. Evaluation of breeding utility of a chromosomal segment from *Lycopersicon chmielewskii* that enhances cultivated tomato soluble solids. Theor Appl Genet 103:1022-1027.

Yusibov, V., A. Modelska, K. Steplewski, M. Agadjanyan, D. Weiner, D.C. Hooper, and H. Koprowski. 1997. Antigens produced plants by infection with chimeric plant viruses immunized against rabies virus and HIV-1. Proc Natl Acad Sci 94:5784-5788.

Zamir, D. 2001. Improving plant breeding with exotic genetic libraries. Nat Genet Rev 2:983-989.

Zhang, G.F., K.E. Maudens, S. Storozhenko, K.A. Mortier, D. Van Det Straeten, and W.E. Lambert. 2003. Determination of total folate in plant material by chemical conversion into *para*-aminobenzoic acid followed by high performance liquid chromatography combined with on-line postcolumn derivatization and fluorescence detection. J Agric Food Chem 51:7872-7878.

Zhang, Y. and J.R. Stommel. 2000. RAPD and AFLP tagging and mapping of *Beta* (*B*) and *Beta* modifier (*Mo*$_B$), two genes which influence β-carotene accumulation in fruit of tomato (*Lycopersicon esculentum* Mill.). Theor Appl Genet 100:368-375.

Ziegler, R.G. 1989. A review of epidemiological evidence that carotenoids reduce the risk of cancer. J Nutr 119:116-122.

Zscheile, F.P. and J.W. Lesley. 1967. Pigment analysis of sherry. J Hered 58:193-194.

7

Molecular Markers in Selection of Tomato Germplasm

Mikel R. Stevens[1] and Matthew D. Robbins[2]

[1]Department of Plant and Animal Sciences, 287 Widstoe Building,
Brigham Young University, Provo, UT 84602, USA
email: mikel_stevens@byu.edu
[2]Department of Horticulture, University of Wisconsin at Madison
1575 Linden Drive, Madison, WI 53706, USA
email: mdrobbins@wisc.edu

AN OVERVIEW OF GENETIC MARKERS IN TOMATO

For millennia we have selected plants to be more adapted to our needs, environments, and markets. These efforts were very much an art form. Roughly a century ago, Mendel's rediscovered pea experiments gave us an understanding of basic genetic principles and moved plant breeding from a qualitative 'art' towards a more quantitative science. Through most of the twentieth century, however, plant breeders heavily depended on their qualitative art skills for selection of the desired phenotype. The plethora of genes and alleles in their myriad of possible combinations mixed with environmental interactions has required a high degree of art, experience, and science to develop the most current 'elite cultivar' for the changing, competitive markets of today. No doubt, even without the more recent molecular tools, improvements in cultivars would continue through traditional, science/art based plant breeding methods because of the ubiquitous number of genes to reshuffle. However, most contemporary plant breeders are, to some extent, using molecular markers as aids to produce their newest, elite cultivar.

The tomato (*Lycopersicon esculentum* Mill.) has been in the forefront of modern plant breeding. Hedrick and Booth (1907) reported results from tomato hybrid experiments to verify Mendel's laws in tomato. In 1917, Jones employed Hedrick and Booth's (1907) data to construct the first tomato genetic map. MacArthur (1934) required 10 years and more than 48,000 segregating plants to map 21 phenotypic markers on 10 of the 12 tomato (2n=2x=24) chromosomes. By 1986, Stevens and Rick estimated that 1200 monogenic markers had been identified in tomato. Only a handful of these markers could be considered to be 'molecular,' of which, the majority were isozymes. Tanksley and Mutschler (1990) released a comprehensive isozyme map which included 12 enzyme systems with 31 isozyme loci and 205 phenotypic markers scattered across the tomato genome. An additional 98 phenotypic markers were mapped to specific chromosomes, but their precise position was unidentified. The localization of phenotypic markers to a specific chromosomal location required crossing a mutant phenotype to a number of parents with well characterized and mapped phenotypic markers, then evaluating the progeny. This required enormous effort along with land and/or greenhouse space, as attested by MacArthur's (1934) monumental 10 year study.

In addition to isozyme and phenotypic marker maps, Tanksley and Mutschler (1990) published a map of 162 RFLP (restriction fragment length polymorphism) markers identified from a single interspecific cross (*L. esculentum* x *L. pennellii* (Corr.) D'Arcy). Although RFLP mapping is time consuming and costly, the resources required are trivial compared to mapping phenotypic markers. Tanksley and Mutschler (1990) required less than three years and a single segregating population of less than 100 plants to develop and map 162 RFLP markers, in contrast to well over 50 years of effort by many research groups and thousands of populations to map 205 phenotypic markers. Five years after the first tomato RFLP map, Tanksley et al. (1992) published a high-density molecular map with over 1000 RFLP markers spanning 1200 cM (centiMorgans), which covered the approximately 950 Mbp (mega base pairs) genome of tomato (Arumuganathan and Earle 1991).

Ever since Tanksley et al. (1992) published the first comprehensive molecular map of the tomato genome, a number of studies have reported the fine mapping of specific areas, and rearrangements of specific genomic regions. For example, Haanstra et al. (1999) significantly augmented the Tanksley et al. (1992) map with the addition of hundreds of AFLP (amplified fragment length polymorphism) markers. A careful search of the current literature would provide a listing of thousands of molecular markers that have been positioned in the tomato genome. A comprehensive compellation of these markers is beyond the scope of this report. The objective of this chapter, however, is to show that the use of these molecular markers has permanently altered our approach to tomato breeding today.

ISOZYMES AS TOMATO MOLECULAR MARKERS

Isozymes were the first forays into tomato molecular markers. Charles Rick initiated the understanding of *Lycopersicon* isozyme genetics in 1973 (Rick 1983). Using four isozymes, Rick et al. (1974) studied the genetic relationship of *L. pimpinellifolium* Mill. to *L. esculentum* var. *cerasiforme*. They found that the isozyme peroxidase (*Prx*) loci discriminated the accessions according to geographic distribution and they also demonstrated tight genetic linkage between phenotypic *Ge* (gamete-eliminator) and *Prx-1*. These findings prompted a search for isozyme diversity in tomato and its relatives, which revealed a paucity of polymorphisms within the cultivated tomato. Conversely, the wild relatives were replete with polymorphic isozyme loci (Rick and Fobes 1975).

Out of these isozyme studies, two things became clear, both of which still impact tomato marker research today. First, the dearth of allelic variation within the cultivated tomato is due to its genetic 'bottlenecking' during domestication (Tanksley and McCouch 1997). This general homogeneity within the cultivated tomato was "discouraging" (Rick 1983). However, the second finding of those studies, the cornucopia of polymorphic isozymes within *Lycopersicon*, laid the foundation for future molecular marker work. Interspecific crossing within *Lycopersicon* was well understood and utilized in tomato breeding by the early 1970's, albeit with difficulty between the *esculentum* and *peruvianum* complexes (Rick 1979). Examples of genes derived from *L. peruvianum* (L.) Mill. before 1970 include *Mi*, root-knot nematode (*Meloidogyne* spp.) resistance (Smith 1944); *Tm-2* and *Tm-2²*, tobacco mosaic virus/tomato mosaic virus (TMV/ToMV) resistance (Alexander 1963, Laterrot and Pecaut 1969, Yamakawa and Nagata 1975); and *Sw-5*, tomato spotted wilt virus (TSWV) resistance (Stevens 1964, Stevens et al. 1992). These, and a number of other exotic genes introduced into tomato, concurrently increased tomato isozyme heterogeneity due to linkage drag.

In 1980, Medina-Filho reported the isozyme allele *Aps-1* (acid phosphatase) was tightly linked to *Mi* in cultivated tomatoes and polymorphic to *Mi⁺*. This discovery was quickly adopted in many tomato breeding programs and is still utilized today due to its ease, speed, tight linkage, and low cost. However, as Tanksley (1983b) discussed, the odds of finding other isozymes tightly linked to economically important genes would be remote. This is because most genes are not introgressed from interspecific crosses and, therefore, do not carry polymorphic isozymes as linkage drag. Nonetheless, having a simple laboratory test for a number of genes was intriguing, to the point of creativity. Tanksley et al. (1984) deliberately combined the rare Prx-2^1 isozyme allele with the nuclear male sterile locus, ms-10, in a tight linkage of 1.5 cM on chromosome 2. This innovative approach allowed the identification of plants heterozygous at the Prx-2^1 locus, providing a high degree of confidence in maintaining ms-10 during backcrossing into an elite hybrid parent. The use of markers for more efficient introduction of male sterility into breeding material is valuable in allowing breeders the possibility of producing hybrid seed without the use of expensive, hand-labor emasculation.

Even with deliberate introgression and serendipitous discovery of linked isozymes from exotic germplasm, the ideal situation of having isozyme markers available for all characteristics of interest was impractical for four reasons. First, each isozyme required different staining procedures with a wide range of reproducibility and expense (Vallejos 1983). Second, only 15 isozyme protocols had been developed with a total of 41 alleles for tomato by 1985 (Tanksley 1985). None of the alleles were mapped to chromosome 11, and chromosomes 1 and 4 carried the maximum, each with five. Furthermore, isozyme markers were often clumped together on a chromosome, thus reducing their effectiveness in plant breeding. Third, many of the isozyme markers were specific to tissue or environmental conditions, further complicating their full implementation (Tanksley 1980, Tanksley 1983b, Vallejos 1983, Tanksley 1985). Finally, DNA markers were beginning to show they could mitigate, to some degree, all the problems of isozymes (Tanksley 1983b).

TOMATO DNA MARKERS: THE EARLY YEARS

By 1980, a wide range of restriction endonucleases, the core of DNA-based RFLP markers, were available. Several research groups had identified the usefulness of RFLPs in monitoring genetic traits in humans (Botstein et al. 1980, Wyman and White 1980, Ruddle 1981, Kao et al. 1982, Murray et al. 1982). Tomatoes, however, were among the first crops to be utilized for RFLP markers with a plant breeding objective (Helentjaris et al. 1985). This initial study demonstrated RFLP scarcity in the cultivated tomato and an abundance of these markers in its wild relatives, comparable to the results of the isozyme research. The first DNA based tomato maps were reported in 1986 (Bernatzky and Tanksley 1986d, Helentjaris et al. 1986). Helentjaris et al. (1986) developed 50 leaf tissue cDNA probes to genotype 50 F_2 inter-specific plants from a 'Manpal' x *L. hirsutum* Humb. & Bonpl. (PI1123317) cross, resulting in a map with 19 linkage groups. Bernatzky and Tanksley

(1986d) used an interspecific (*L. esculentum* [LA1500] x *L. pennellii* [LA716]) backcross and F_2 population to map 57 RFLP cDNA probes, 21 isozyme markers, and other previously mapped DNA markers, resulting in 112 mapped loci on the 12 tomato chromosomes. This single study produced more than twice the number of isozyme markers previously mapped, effectively demonstrating that, although both marker systems are codominant, RFLPs have three distinct advantages over isozymes. First, RFLP markers are much more abundant. Second, one protocol can be used instead of a number of different staining methods. Lastly, the sampling method for RFLPs is potentially less plant destructive compared to some of the isozyme methods (Vallejos 1983).

In the next few years tomato mapping, marker development, and genome understanding moved at an unprecedented pace using RFLPs. Bernatzky and Tanksley (1986a,c) identified and mapped 10 actin-related loci and quickly developed a method to detect single/low copy, noncoding DNA sequences. They also demonstrated that most random cDNA clones are single loci in the tomato genome (Bernatzky and Tanksley 1986b). Zamir and Tanksley (1988) used 50 RFLP probes derived from sheared tomato DNA to learn that the tomato genome is mostly composed of "fast-evolving, low copy-number sequences." Nienhuis et al. (1987) associated RFLP markers with quantitative trait loci (QTL) linked to insect resistance derived from *L. hirsutum*. Using tomato trisomic lines, Young et al. (1987) correlated the RFLP marker map to the well-understood tomato cytogenetic map, tying past research to the information discovered through RFLPs.

As RFLPs began to prove their worth, the late 1980's brought reports focusing on yet another application of markers in tomato, marker-assisted selection (MAS). Young et al. (1988) identified an RFLP marker tightly linked to *Tm-2a* based on the idea of linkage drag from *L. peruvianum*, similar to the *Mi* and *Aps-1* study (Medina-Filho 1980). Using *Tm-2a* near isogenic lines (NILs), they searched for probes present in one line and absent in the other. This innovative study not only provided RFLP markers linked to *Tm-2a*, but produced evidence that multiple backcrossing did not necessarily reduce linkage drag as quickly as traditionally thought. In theory, over 99% of the genome should be from the recurrent parent after backcrossing six generations and selecting only for the introgressed character of interest. Young and Tanksley (1989b) demonstrated that *Tm-2a* was introgressed into eight unique cultivars via backcrossing. 'Craigella' (*Tm-2a/Tm-2a*) still had 51 cM (the entire short arm of chromosome 9) of exotic DNA remaining after many backcross generations, while three other cultivars had as little as 4 cM of linkage drag. These results inspired the concept that RFLPs could be used to graphically illustrate genomic regions originating from specific parents (Young and Tanksley 1989a). Tanksley et al. (1989) hypothesized that by using molecular markers, it would be possible to eliminate excess linkage drag surrounding an introgressed exotic gene in no more than two backcross generations. Today, private breeding companies have, on multiple occasions, introgressed alleles from the wild species within a year or two of initiating the project by using graphical genotyping and selection for specific crossover events. Without markers, exotic gene introgression and cultivar development could take a decade or longer.

It was abundantly clear by 1990 that DNA markers had the potential to become one of the most powerful breeding tools since the rediscovery of Mendel's laws. Just six years after Bernatzky and Tanksley (1986d) and Helentjaris et al. (1986) introduced the first tomato DNA molecular maps, Tanksley et al. (1992) released a high-density tomato genetic map. This map has provided the foundation for thousands of studies and hundreds of scientific papers resulting in a dramatic evolution of the understanding of the tomato genome. One of the more remarkable changes has been in the area of using exotic germplasm in tomato breeding.

MOLECULAR MARKERS AND EXPANDING THE TOMATO GERMPLASM BASE

Vavilov (1940) championed the improvement of crop genetics by utilizing their wild relatives. When studying a specific tomato problem, especially disease, the approach often includes the possibility of exploiting exotic germplasm. For instance, resistance to TSWV was first reported in the wild species *L. pimpinellifolium* Mill. (Samuel et al. 1930). A single *L. peruvianum* gene (*Sw-5*) has provided the most effective resolution to this disease. It took over two decades to introgress this gene into tomato in one program (Stevens et al. 1992) and almost a decade in another, independent program (Watterson 1993). This example is rather simplistic in that most traits of horticultural importance are controlled by QTL, rather than a single gene. The introgression of several QTL from exotic germplasm seemed insurmountable when compared to the time it took to introgress a simple dominant gene of such importance as TSWV resistance into tomato.

Although it was generally impractical, it was well understood that QTL could be monitored with phenotypic markers. A pioneering study using phenotypes to monitor QTL was reported by Thoday in 1961. In 1983, Tanksley (1983a) proposed using isozyme markers to overcome time and linkage drag issues. However, a high-density molecular map, essential to effectively manipulate multiple loci and linkage drag, would not be ready for almost a decade (Tanksley et al. 1992). Even before the high-density molecular map was completed, Tanksley et al. (1982) and Weller et al. (1988) used isozymes and morphological markers in interspecific tomato studies demonstrating the feasibility of linking QTL to molecular markers. Lander and Botstein (1989) reported on statistical procedures to handle the data generated from the developing high-density molecular marker maps and QTLs. Paterson et al. (1988) utilized the expanding inventory of RFLP markers with these statistical methods to successfully determine the exotic, introgressed regions (from *L. chmielewskii* Rick et al.), responsible for altering soluble solids, fruit mass, yield, and pH in tomato.

By the early 1990's, molecular markers, statistical procedures, and computers were in place and to consider innovative breeding techniques to enhance the harvesting of tomato genes from exotic germplasm (Rick 1988, Hille et el. 1989, Tanksley et al. 1989, Tanksley and McCouch 1997). Tanksley and Nelson (1996) proposed the "advanced backcross QTL analysis" (AB-QTL) method. This strategy was to cross tomato with a wild relative, backcross the F_1 to the elite parent, randomly select 150-200 BC_1 plants, then carry out single seed decent through the succeeding BC_2 or BC_3 generations. The anticipated result was for the entire exotic genome to be represented throughout the single seed decent lines with each line mostly containing the genetic background of cultivated tomato. With one or two selfing generations, these lines could end up being QTL-NILs. Tanksley et al. (1996) provided the data of such a project using *L. pimpinellifolium* as the exotic parent. They discovered that genetics of improved fruit size and shape, among other characteristics, was in complete contrast to the exotic and inferior phenotype of the parent. Thus, a succession of AB-QTL allele mining studies involving several wild species of tomato (e.g. *L. hirsutum, L. parviflorum* Rick et al., *L. pennellii, L. peruvianum*) followed (Bernacchi et al. 1998a,b,c, Fulton et al. 1997a,b, 2000, Frary et al. 2004). These studies demonstrated that multiple alleles and loci, even within different accessions of the same species, were available from exotic germplasm in most of the economically important traits in tomato. Examples of unexpected characteristics found in exotic germplasm were improved red color alleles from green fruited species, improved rooting from inferior rooting species, and many others.

Besides the AB-QTL method, which focused on detecting "additive, dominant, partially dominant, or overdominant QTLs" from exotic species (Tanksley and Nelson 1996), other methodologies using wild species were utilized to identify genetic inheritance patterns. Using an interspecific (*L. esculentum* x *L. pennellii*) F_2 population, transgressive segregation of some QTLs was identified (deVicente and Tanksley 1993). Eshed and Zamir (1996) investigated epistatic gene action from QTL, exploiting a set of tomato NILs that contained small, overlapping introgressions from *L. pennellii* (prepared by Eshed and Zamir 1994). Although they determined that additive gene action predominated, clear evidence was provided for epistasis in fruit mass and total soluble solids.

Molecular markers have enabled breeding programs to not only dissect quantitatively inherited traits, but have additionally allowed the introgression of these characteristics into modern cultivars. Furthermore, careful mapping of QTL have opened the door for a better understanding of the genetics in tomato and scrutinizing the physiology of economically important phenotypes.

MOLECULAR MARKERS IN GENE CLONING

Prior to the development of high-density molecular maps, gene products (proteins) were usually required previous to cloning tomato genes responsible for a given phenotype. With efficient marker development methods and large insert genomic libraries of tomato such as yeast artificial chromosomes (YACs) (Martin et al. 1992) and bacterial artificial chromosomes (BACs) (Hamilton et al. 1999), gene cloning has become much more feasible. The first gene cloned from genetic mapping in tomato was *Pto*, which imparts resistance to *Pseudomonas syringae* pv. *tomato* (Martin et al. 1993b). Using more than one technique, Martin et al. (1993a) identified molecular markers so tightly linked to *Pto* that they "landed" on the specific YAC clone in which *Pto* resided. The cloning of this gene was much more efficient compared to the map-based cloning, or "chromosome walking," efforts used in human and other organisms. As a result, the technique "chromosome landing" was suggested as a paradigm for future gene cloning efforts (Tanksley et al. 1995). In addition to chromosome landing/walking methods, Jones et al. (1994) developed a clever scheme to clone tomato genes by using molecular markers in combination with tomato lines transgenic for the transposable elements *Ac* and *Ds* from maize (*Zea mays* L.). This method was used to effectively clone *Cf-9*, the gene for *Cladosporium fulvum* resistance.

Although some genes have remained recalcitrant to cloning efforts short of sequencing the genome, a number of unique and economically important genes have been cloned through the aid of molecular markers via permutations of map-based cloning. Examples include the jointless (*j*) gene (Zhang et al. 1994), the TSWV resistance gene *Sw-5* (Folkertsma et al. 1999, Brommonschenkel et al. 2000), the fruit size QTL *fw2.2* (Frary et al. 2000), the self-pruning (*sp*) gene (Carmel-Goren et al. 2003) and the alleles involved with increased lycopene content known as high pigment (*hp-2*) and dark green (*dg*) (Mustilli et al. 1999, Levin et al. 2003). The impacts of cloning these and other genes on tomato breeding will be felt in transgenics for increased human nutrition, better understanding of yield, and many others. Most likely the greatest influences are yet to come.

MOLECULAR MARKERS WITHIN SOLANACEAE FOR TOMATO IMPROVEMENT

Tomatoes have led the way in developing genetic maps in Solanaceae. As could be expected, a steady 'spillover' effect has benefited the related genera. Early on, tomato DNA markers demonstrated their utility in Solanaceous crops such as peppers (*Capsicum* spp.) (Tanksley et al. 1988) and potatoes (*Solanum* spp.) (Bonierbale et al. 1988, Gebhardt et al. 1991). These studies demonstrated general conservation of gene order with broad chromosomal rearrangements between the genera of Solanaceae. Livingstone et al. (1999) provided clear evidence that tomato, pepper, and potato are essentially the same genetically, but differ in a number of chromosomal rearrangements with interesting evolutionary implications. Utilizing this understanding, Grube et al. (2000) searched for homologs of the cloned resistance genes *Sw-5*, *Pto*, *N* (TMV resistance from *Nicotiana tabacum* L.), *Prf* (required for *Pto*) and *I2* (resistant to *Fusarium oxysporum*) among tomato, pepper and potato. They found homologs in syntenic positions in every species for all five genes tested. Moreover, these genes were mapped, in several instances, to genomic regions where previously mapped resistance genes resided within a specific species. The implications of these results will be studied for sometime to come.

One of the benefits of molecular markers and their applicability across Solanaceae has been in tomato intergeneric crosses. For example, attempts to introgress potato genes into tomato were monitored via molecular markers following both protoplast fusion (Jacobsen et al. 1994, Shikanai et al. 1998) and gamma-irradiation experiments (Schoenmakers et al. 1994). Other tomato intergeneric work has included eggplant (*S. melongena* L.) (Samoylov and Sink 1996, Samoylov et al. 1996) and the weedy species *S. sitiens* Johnson (Syn. *S. rickii* Corr.) (DeVerna et al. 1990). However, tomato x *S. lycopersicoides* Dun. has been the most successful intergeneric cross. Using molecular markers, Chetelat and Meglic (2000) and Ji and Chetelat (2003) examined the location and amount of *S. lycopersicoides* genome incorporated into specific lines of tomato. Although difficult, similar progress has been made in introgressing sizeable chromosomal sections of *S. sitiens* into the tomato genome (Pertuz et al. 2003). The impact on tomato of these diverse genetic elements can be expected to be significant, especially in light of the contributions from the more closely related *Lycopersicon* species (Rick and Chetelat 1995).

Intra-Solanaceae molecular studies have spurred interest in reexamining the systematics within the family. A sampling of some of those studies include discussions on the merits of the present designation of species within *Lycopersicon* (Miller and Tanksley 1990, Bretó et al. 1993, see chapter 1) and whether additional species from *Solanum* should be included in *Lycopersicon* (Marshall et al. 2001). Due to the wealth of molecular genetic information developed within this family, the idea of using "Solanaceae as a model for linking genomics with biodiversity" is being considered (Knapp et al. 2004).

MOLECULAR MARKER DRAWBACKS IN TOMATOES

With all the progress that has been made to improve the incorporation of molecular markers into tomato breeding, there are two fundamental areas that need to be addressed. First, we need to better employ molecular markers in managing elite tomato germplasm. Tomato breeding has utilized more exotic germplasm than any other cultivated crop and molecular markers have greatly facilitated that work. However, little is understood about the diversity of the elite alleles within *L. esculentum*. The second need is to establish high throughput molecular marker systems that are economical. At present, the cost of molecular marker analysis is usually calculated in dollars per data point.

Due to these two limiting factors, most tomato breeding programs are using MAS for few alleles, mostly exotic disease resistant genes. The majority of the tomato breeding programs are using PCR (polymerase chain reaction) markers in proxy for disease screenings in elite germplasm breeding. Examples of available disease resistance genes linked to PCR markers include *Ol-1* (resistant to powdery mildew, *Oidium lycopersicum*) (Huang et al. 2000), *py-1* (resistant to corky root rot, *Pyrenochaeta lycopersici*) (Doganlar et al. 1998), *Mi* (Williamson et al. 1994), *Sw-5* (Stevens et al. 1995, Stevens et al. 1996), *Tm-1, Tm-2, Tm-2²* (resistance to TMV and/or TMoV) (Ohmori et al. 1996, Sobir et al. 2000, and Dax et al. 1998 respectively) and *Frl* (resistant to fusarium crown and root rot, *Fusarium oxysporum* f.sp. *radicis-lycopersici*) (Fazio et al. 1999), to name a few. Unfortunately, the effectiveness and convenience of each of these PCR markers vary from reliable to problematic.

In addressing the first drawback of molecular markers, we simply need to remember a fundamental of plant breeding. Breeding for release of the most recent cultivar probably began with the most elite and adapted tomato lines as parental material. The reason for this is that elite lines represent millennia of selecting and assembling thousands of the best alleles. All experienced breeders know that combining two, well adapted cultivars can and do provide the underpinnings for the next 'more elite' cultivar. This suggests, with foundation, that improved, elite allele combinations are yet to be discovered solely within the narrow tomato germplasm base, resulting in tomato improvement. Therein lies the challenge. How do we use molecular techniques to uncover and streamline the discovery of the untapped allele combinations to improve elite material (Yang et al. 2004)? Both Rick (1983) and Helentjaris et al. (1985) reported the dearth of polymorphisms in tomato and we have only made limited progress since those early molecular discoveries.

Over the last two decades, several DNA marker systems have revealed polymorphisms among the modern cultivars. Paran et al. (1995), Noli et al. (1999), and others demonstrated that cultivars can be distinguished using randomly amplified polymorphic DNA (RAPD) markers. Broun and Tanksley (1996) reported that microsatellites, or simple sequence repeats (SSRs), can also be used to distinguish between tomato cultivars. Areshchenkova and Ganal (2002) established that microsatellite markers could be created from expressed sequence tag (EST) data available on public databases. Nevertheless, the majority of essential horticultural characteristics in tomato are quantitatively inherited and lack molecular markers. A few examples include yield, flavor, shipping quality, cracking, and a number of ripening disorders. Unfortunately, at present, our marker technology is inadequate to address these questions.

We have made modest progress in the second deficiency of molecular markers, the ability to develop cost-effective, high-throughput techniques. The most outstanding improvement in marker throughput was the development of the PCR technique (Saiki et al. 1985). PCR has reduced the cost and increased the speed at which all DNA markers are identified and developed. Nevertheless, genotyping a given plant sample is limited, in most instances, to a maximum of three or four markers linked to specific characteristics in each PCR reaction (multiplexed). Time, reagents, equipment and supplies alone can overwhelm a breeding program trying to keep up with new markers linked to new traits, especially when only two to three markers can be examined at a time. The pressure is steadily increasing to utilize more markers and improve the utility of each linked marker.

FUTURE OF MOLECULAR MARKERS AND TOMATOES

Progress is currently being made on the two most limiting factors in molecular markers from: first, the rapid accumulation of tomato DNA sequences, second, the increasing range of single nucleotide polymorphism (SNP) detection methodologies, and third, from the dramatic and steady improvements of microarray techniques. Combinations of these advancements with computers are preparing the way for economical high throughput of tomato molecular markers.

Accumulation of tomato sequences, especially EST's (over 30,0000; Fei et al. 2004) has allowed various SNPs techniques (Rafalski 2002) to distinguish differences between elite cultivars. Yang et al. (2004) searched the tomato EST databases and found one SNP for every 8,500 bases. In a selection of 44 genes, 101 candidate SNPs were identified between two cultivated lines. A subset of these *"in silico"* SNPs were verified in the laboratory and then genetically mapped. Labate and Baldo (2005) further verified the value of SNPs in a comparison between 15 cultivated tomato lines. Validation of the functionality of SNPs in elite x elite breeding was accomplished when these and other molecular markers were used to successfully identify the QTLs conferring field resistance to bacterial spot of tomato (*Xanthomonas campestris* pv. *vesicatoria* race T1; Yang et al. 2005). Furthermore, to aid the tomato community in the discovery of additional SNPs a lower cost method of sequencing ten's of thousands genomic sequences within an afternoon was recently reported (Margulies et al. 2005). At present, only about 100 bases are sequenced with accuracy, however, it is anticipated this number will increase by several fold and the cost per reaction will go down! Once understanding how this pyrosequencing sequencing process is accomplished it is quickly possible to imagine studies where SNP discovery between elite tomato cultivars will be much more simplistic. These studies demonstrate that thousands of SNPs can be discovered between cultivars and provide a solution to one of the difficulties of implementing MAS in elite tomato germplasm selection.

With a burgeoning set of EST and potential sequences data, the next obstacle is to economically handle SNP discovery and implementation to manipulate QTL in elite tomato breeding. There are already at least two techniques that approach this challenge. The first technique, known as SNP Wavetm (van Eijk et. al. 2004), has already shown 'proof of concept' in tomatoes and is commercially available. This technique is based on PCR and sequencing technology and has demonstrated the ability to monitor 100 separate *Arabidopsis thaliana* (L.) Heynh. loci in a single reaction and 40 loci of the more complex tomato genome. The second method is to use microarray technology, which is currently hampered by cost, but may become more available in the future.

Microarray technology presents the possibility of examining tens-of-thousands of genomic fragments in one experiment. Array based SNP studies have already been conducted for human genetic mapping (Cutler et al. 2001, Huber et al. 2002, Shi 2001). In order to conduct similar studies, EST tomato arrays must be developed. Recently, a company (NimbleGen) has developed a photolithography based microarray chip consisting of 15,925 ESTs derived from Fei et al. (2004). Several techniques promise to be adaptable to EST based MAS in plant breeding of crops where vast amounts of EST sequence data are available (Huber et al. 2002, Stears et al. 2003, Wong et al. 2004). The deterrent to the widespread use of this technology at present is that an individual tomato photolithography chip can cost well over a $1000. However, each chip can provide well over 15,000 data points and the study conducted by Yang et al. (2004) suggests that well over 5000 SNPs may exist between two tomato cultivars. If these hypotheses prove valid, QTL studies between elite tomato lines is possible.

In addition to the use of SNPs, other techniques are being explored for microarray technology in relationship to plant breeding. The microarray technique was first described and utilized for monitoring the up and down regulation of genes using ESTs in a complex genome (Shalon et al. 1996, Winzeler et al. 1998). However, Jaccoud et al. (2001) demonstrated that microarray technology could be utilized to identify polymorphisms from uncharacterized genomic DNA fragments. The polymorphisms are largely derived from diversity in restriction enzyme recognition sites, and the technique was dubbed "diversity array technology" (DArT). Wenzl et al. (2004) demonstrated the utility of DArT for MAS in barley (*Hordeum vulgare* L.) breeding. In a set of experiments from an array of Fla7613 x *L. pennellii* (LA716 or LA2963) crosses, well over 400 markers were identified and mapped in tomato utilizing DArT. As expected, however, fewer polymorphic markers were identified differentiating 20 NILs of 'Moneymaker,' each with differing resistance genes (Laterrot 1996, Stevens, unpublished data). The power of DArT is the ability to quickly explore many loci in the genomes of little understood crops and species. As demonstrated in this tomato study, the throughput was phenomenal; hundreds of polymorphic markers were discovered and mapped in a wide cross in a single experiment. Furthermore, it took a matter of approximately three months to develop these markers compared to the first 50 years for approximately 200 phenotypic mapped markers (Tanksley and Mutschler 1990) and seven years for the first 1000 RFLP markers (Tanksley et al. 1992). The colossal amount of data from this technology presents a new challenge. However, the computing power and algorithms to manipulate the overwhelming quantities of data are rapidly being developed. Once the technical aspects of DArT are established for a specific genome, the cost is expected to be in the neighborhood of cents per data point. Nonetheless, the initial setup investment of both technical expertise and equipment is large.

Before we dismiss these advanced high-throughput techniques, a perspective needs to be taken. Tomato marker technologies have evolved from phenotypic to isozyme to the current DNA markers, and what was once unthinkable is now the norm. In a few years, the current challenges of molecular markers will most likely be overcome. Now that a tomato EST 'chip' is available, research has been proposed to utilize them for the discovery of polymorphic ESTs between specific elite tomato parents, enabling the mass discovery of SNPs. With increased demand, the cost of these technologies will inevitably drop. New technological adjustments, increased understanding of the power of SNPs, and the tomato genome sequencing project underway, will eventually lead to application of QTL MAS within elite x elite crosses to create new cultivars. Once these skills are perfected, tomato breeders will be able to make quantum leaps towards developing their latest 'elite' cultivar. Furthermore, they will be working towards resolving more recalcitrant problems such as flavor, environmental tolerances, and others utilizing combination of both exotic and elite alleles in the most precise expeditious manner.

SUMMARY

Tomatoes have a rich history of interfacing genetics and markers in germplasm development. The first tomato marker studies were reported in 1907 and the first genetic maps were developed by 1917. Over 200 phenotypic markers were mapped throughout 12 chromosomes of tomato by the early 1980's. Isozymes, the earliest molecular markers, demonstrated a paucity of polymorphic loci within cultivated tomato, which impeded their full implementation in breeding. However, these early studies demonstrated a profusion of polymorphic isozymes among exotic relatives of tomato. Isozyme markers were replaced with DNA markers beginning with RFLPs. Utilizing cultivated by exotic tomato crosses, over 1000 molecular markers were mapped by the early 1990's. Because cultivated by exotic crosses were employed, the majority of tomato DNA based markers, even today, are located in regions associated with linkage drag from exotic introgressions. Historically, exotic germplasm was typically utilized for disease resistance gene mining. However, with the advent of DNA based markers, mining for QTL from exotic germplasm is altering the future of tomato breeding. Mining for QTL is facilitated by molecular markers as well as innovative statistical procedures, computers, and breeding methodologies which are tools for dissecting a quantitative trait into individual loci. These tools have aided in the identification of untapped exotic alleles for yield, flavor, color, fruit shape, and many other unexpected discoveries. Furthermore, molecular markers have facilitated the cloning of both cultivated and exotic genes which would have otherwise remained elusive to biochemical and physiological research. Although DNA based molecular markers have dramatically impacted tomato breeding, two challenges impede their broader use: first, the lack of polymorphisms within elite germplasm and second, the need for economical high-throughput marker systems. SSRs and SNPs are providing some possibilities for working with elite x elite germplasm. However, the identification of new markers as part of these two DNA marker systems requires sequencing vast amounts of the tomato genome. Recently, considerable quantities of tomato sequence data have become available to the public. The use of sequence data, along with microarray based technologies, may resolve high throughput issues. Although microarray research is expensive and still in its infancy, the cost of this technology will inevitably subside. Microarray based technology offers the possibilities of examining thousands of markers in a single test, outperforming sequence based techniques and potentially reducing costs to a few cents per data point. SNP based marker assisted selection within tomato breeding offers unprecedented power for identifying and recombining QTL. Thus, we are possibly nearing the threshold of unparalleled advances in tomato breeding.

Acknowledgements

We wish to thank JoLynn J. Stevens for her patience during the writing of this document and her willingness to help in reviewing the manuscript.

REFERENCES

Alexander, L.J. 1963. Transfer of a dominant type of resistance to the four known Ohio pathogenic strains of tobacco mosaic virus (TMV), from *Lycopersicon peruvianum* to *L. esculentum*. Phytopathology 53:869.

Areshchenkova, T. and M.W. Ganal. 2002. Comparative analysis of polymorphism and chromosomal location of tomato microsatellite markers isolated from different sources. Theor Appl Genet 104:229-235.

Arumuganathan, K. and E.D. Earle. 1991. Nuclear DNA content of some important plant species. Plant Mol Biol Rept 9:208-218.

Bernacchi, D., T. Beck-Bunn, D. Emmatty, Y. Eshed, S. Inai, J. Lopez, V. Petiard, H. Sayama, J. Uhlig, D. Zamir, and S. Tanksley. 1998a. Advanced backcross QTL analysis of tomato. II. Evaluation of near-isogenic lines carrying single-donor introgressions for desirable wild QTL-alleles derived from *Lycopersicon hirsutum* and *L. pimpinellifolium*. Theor Appl Genet 97:170-180.

Bernacchi, D., T. Beck-Bunn, D. Emmatty, Y. Eshed, S. Inai, J. Lopez, V. Petiard, H. Sayama, J. Uhlig, D. Zamir, and S. Tanksley. 1998b. Advanced backcross QTL analysis of tomato. II. Evaluation of near-isogenic lines carrying single-donor introgressions for desirable wild QTL-alleles derived from *Lycopersicon hirsutum* and *L. pimpinellifolium*. Theor Appl Genet 97:1191-1196.

Bernacchi, D., T. Beck-Bunn, Y. Eshed, J. Lopez, V. Petiard, J. Uhlig, D. Zamir, and S. Tanksley. 1998c. Advanced backcross QTL analysis in tomato. I. Identification of QTLs for traits of agronomic importance form *Lycopersicon hirsutum*. Theor Appl Genet 97:381-397.

Bernatzky, R. and S.D. Tanksley. 1986a. Genetics of actin-related sequences in tomato. Theor Appl Genet 72:314-321.

Bernatzky, R. and S.D. Tanksley. 1986b. Majority of random cDNA clones correspond to single loci in the tomato genome. Mol Gen Genet 203:8-14.

Bernatzky, R. and S.D. Tanksley. 1986c. Methods for detection of single or low copy sequences in tomato on Southern blots. Plant Mol Biol Rep 4:37-41.

Bernatzky, R. and S.D. Tanksley. 1986d. Toward a saturated linkage map in tomato based on isozymes and random cDNA sequences. Genetics 112:887-898.

Bonierbale, M.W., R.L. Plaisted, and S.D. Tanksley. 1988. RFLP Maps based on a common set of clones reveal modes of chromosomal evolution in potato and tomato. Genetics 120:1095-1103.

Botstein, D., R.L. White, M. Skolnick, and R.W. Davis. 1980. Construction of a genetic linkage map in man using restriction fragment length polymorphisms. Amer J Hum Genet 32:314-331.

Bretó, M.P., M.J. Asins, and E.A. Carbonell. 1993. Genetic variability in *Lycopersicon* species and their genetic relationships. Theor Appl Genet 86:113-120.

Brommonschenkel, S.H., A. Frary, and S.D. Tanksley. 2000. The broad-spectrum tospovirus resistance gene *Sw-5* of tomato is a homolog of the root-knot nematode resistance gene *Mi*. Mol Plant-Microbe Interact 13:1130-1138.

Broun, P. and S.D. Tanksley. 1996. Characterization and genetic mapping of simple repeat sequences in the tomato genome. Mol Gen Genet 250:39-49.

Carmel-Goren, L., Y.S. Liu, E. Lifschitz, and D. Zamir. 2003. The *self-pruning* gene family in tomato. Plant Mol Biol 52:1215-1222.

Chetelat, R.T. and V. Meglic. 2000. Molecular mapping of chromosome segments introgressed from *Solanum lycopersicoides* into cultivated tomato (*Lycopersicon esculentum*). Theor Appl Genet 100:232-241.

Cutler, D.J., M.E. Zwick, M.M. Carrasquillo, C.T. Yohn, K.P. Tobin, C. Kashuk, D.J. Mathews, N.A. Shah, E.E. Eichler, J.A. Warrington, and A. Chakravarti. 2001. High-throughput variation detection and genotyping using microarrays. Genome Res 11:1913-1925.

Dax, E., O. Livneh, E. Aliskevicius, O. Edelbaum, N. Kedar, N. Gavish, J. Milo, F. Geffen, A. Blumenthal, H.D. Rabinowich, and I. Sela. 1998. A SCAR marker linked to the ToMV resistance gene, $Tm2^2$, in tomato. Euphytica 101:73-77.

DeVerna, J.W., C.M. Rick, R.T. Chetelat, B.J. Lanini, and K.B. Alpert. 1990. Sexual hybridization of *Lycopersicon esculentum* and *Solanum rickii* by means of a sesquidiploid bridging hybrid. Proc Natl Acad Sci USA 87:9486-9490.

deVicente, M.C. and S.D. Tanksley. 1993. QTL analysis of transgressive segregation in an interspecific tomato cross. Genetics 134:585-596.

Doganlar, S., J. Dodson, B. Gabor, T. Beck-Bunn, C. Crossman, and S.D. Tanksley. 1998. Molecular mapping of the *py-1* gene for resistance to corky root rot (*Pyrenochaeta lycopersici*) in tomato. Theor Appl Genet 97:784-788.

Eshed, Y. and D. Zamir. 1994. A genomic library of *Lycopersicon pennellii* in *L. esculentum*: A tool for fine mapping of genes. Euphytica 79:175-179.

Eshed, Y. and D. Zamir. 1996. Less-than-additive epistatic interactions of quantitative trait loci in tomato. Genetics 143:1807-1817.

Fazio, G., M.R. Stevens, and J.W. Scott. 1999. Identification of RAPD markers linked to fusarium crown and root rot resistance (*Frl*) in tomato. Euphytica 105:205-210.

Fei, Z., X. Tang, R.M. Alba, J.A. White, C.M. Ronning, G.B. Martin, S.D. Tanksley, and J.J. Giovannoni. 2004. Comprehensive EST analysis of tomato and comparative genomics of fruit ripening. Plant J 40:47-59.

Folkertsma, R.T., M.I. Spassova, M. Prins, M.R. Stevens, J. Hille, and R.W. Goldbach. 1999. Construction of a bacterial artificial chromosome (BAC) library of *Lycopersicon esculentum* cv. Stevens and its application to physically map the *Sw-5* locus. Mol Breed 5:197-207.

Frary, A., T.C. Nesbitt, A. Frary, S. Grandillo, E. van der Knaap, B. Cong, J. Liu, J. Meller, R. Elber, K.B. Alpert, and S.D. Tanksley. 2000. *fw2.2*: A quantitative trait locus key to the evolution of tomato fruit size. Science 289:85-88.

Frary, A., T.M. Fulton, D. Zamir, and S. D. Tanksley. 2004. Advanced backcross QTL analysis of a *Lycopersicon esculentum* x *L. pennellii* cross and identification of possible orthologs in the Solanaceae. Theor Appl Genet 108:485-496.

Fulton, T.M., J.C. Nelson, and S.D. Tanksley. 1997a. Introgression and DNA marker analysis of *Lycopersicon peruvianum*, a wild relative of the cultivated tomato, into *Lycopersicon esculentum*, followed through three successive backcross generations. Theor Appl Genet 95:895-902.

Fulton, T.M., S. Grandillo, T. Beck-Bunn, E. Fridman, A. Frampton, J. Lopez, V. Petiard, J. Uhlig, D. Zamir, and S.D. Tanksley. 2000. Advanced backcross QTL analysis of a *Lycopersicon esculentum* x *Lycopersicon parviflorum* cross. Theor Appl Genet 100:1025-1042.

Fulton, T.M., T. Beck-Bunn, D. Emmatty, Y. Eshed, J. Lopez, V. Petiard, J. Uhlig, D. Zamir, and S.D. Tanksley. 1997b. QTL analysis of an advanced backcross of *Lycopersicon peruvianum* to the cultivated tomato and comparisons with QTLs found in other wild species. Theor Appl Genet 95:881-894.

Gebhardt, C., E. Ritter, A. Barone, T. Debener, B. Walkemeier, U. Schachtschabel, H. Kaufmann, R.D. Thompson, M.W. Bonierbale, M.W. Ganal, S.D. Tanksley, and F. Salamini. 1991. RFLP maps of potato and their alignment with the homoeologous tomato genome. Theor Appl Genet 83:49-57.

Grube, R.C., E.R. Rabwanski, and M. Jahn. 2000. Comparative genetics of disease resistance within the Solanaceae. Genetics 155:873-887.

Haanstra, J.P.W., C. Wye, H. Verbakel, F. Meijer-Dekens, P. van den Berg, P. Odinot, A.W. van Heusden, S. Tanksley, P. Lindhout, and J. Peleman. 1999. An integrated high-density RFLP-AFLP map of tomato based on two *Lycopersicon esculentum* x *L. pennellii* F_2 populations. Theor Appl Genet 99:254-271.

Hamilton, C.M., A. Frary, Y. Xu, S.D. Tanksley, and H-B. Zhang. 1999. Construction of tomato genomic DNA libraries in a binary BAC (BIBAC) vector. Plant J 18:223-229.

Helentjaris, T., G. King, M. Slocum, C. Siedenstrang, and S. Wegman. 1985. Restriction fragment polymorphisms as probes for plant diversity and their development as tools for applied plant breeding. Plant Mol Biol 5:109-118.

Helentjaris, T., M. Slocum, S. Wright, A. Schaefer, and J. Nienhuis. 1986. Construction of genetic linkage maps in maize and tomato using restriction fragment length polymorphisms. Theor Appl Genet 72:761-769.

Hedrick, U.P., and N.O. Booth. 1907. Mendelian characters in tomatoes. Proc Amer Soc Hort Sci 25:19-24.

Hille, J., M. Koornneef, M.S. Ramanna, and P. Zabel. 1989. Tomato: a crop species amenable to improvement by cellular and molecular methods. Euphytica 42:1-23.

Huang, C.C., Y.Y. Cui, C.R. Weng, P. Zabel, and P. Lindhout. 2000. Development of diagnostic PCR markers closely linked to the tomato powdery mildew resistance gene *Ol-1* on chromosome 6 of tomato. Theor Appl Genet 101:918-924.

Huber, M., A. Mundlein, E. Dornstauder, C. Schneeberger, C.B. Tempfer, M.W. Mueller, and M.W. Schmidt. 2002. Accessing single nucleotide polymorphisms in genomic DNA by direct multiplex polymerase chain reaction amplification on oligonucleotide microarrays. Anal Biochem 303:25-33.

Jaccoud, D., K. Peng, D. Feinstein, and A. Kilian. 2001. Diversity arrays: a solid state technology for sequence information independent genotyping. Nucl Acids Res 29:No. 4 e25.

Jacobsen, E., M.K. Daniel, J.E.M. Bergervoet-van Deelen, D.J. Huigen, and M.S. Ramanna. 1994. The first and second backcross progeny of the intergeneric fusion hybrids of potato and tomato after crossing with potato. Theor Appl Genet 88:181-186.

Ji, Y. and R.T. Chetelat. 2003. Homoeologous pairing and recombination in *Solanum lycopersicoides* monosomic addition and substitution lines of tomato. Theor Appl Genet 106:979-989.

Jones, D.A., C.M. Thomas, K.E. Hammond-Kosack, P.J. Balint-Kurti, and J.D.G. Jones. 1994. Isolation of the tomato *Cf-9* gene for resistance to *Cladosporium fulvum* by transposon tagging. Science 266:789-793.

Jones, D.F. 1917. Linkage in *Lycopersicon*. Amer Natural 51:608-621.

Kao, F., J.A. Hartz, M.L. Law, and J.N. Davidson. 1982. Isolation and chromosomal localization of unique DNA sequences form a human genomic library. Proc Natl Acad Sci USA 79:865-869.

Knapp, S., L. Bohs, M. Nee, and D.M. Spooner. 2004. Solanaceae – a model for linking genomics with biodiversity. Comp Funct Genom 5:285-291.

Labate, J.A., and A.M. Baldo. 2005. Tomato SNP discovery by EST mining and resequencing. Mol Breed 16:393-349.

Lander, E.S. and D. Botstein. 1989. Mapping Mendelian factors underlying quantitative traits using RFLP linkage maps. Genetics 121:185-199.

Laterrot, H. 1996. Stock list. Rep Tom Genet Coop 46:34.

Laterrot, H. and P. Pecaut. 1969. Gene *Tm-2*: new source. Rep Tom Genet Coop 19:13-14.

Levin, I., P. Frankel, N. Gilboa, S. Tanny, and A. Lalazar. 2003. The tomato *dark green* mutation is a novel allele of the tomato homolog of the *DEETIOLATED1* gene. Theor Appl Genet 106:454-460.

Livingstone, K.D., V.K. Lackney, J.R. Blauth, R. van Wijk, and M.K. Jahn. 1999. Genome mapping in *Capsicum* and the evolution of genome structure in the Solanaceae. Genetics 152:1183-1202.

MacArthur, J.W. 1934. Linkage groups in the tomato. J Genet 29:123-133.

Margulies, M., M. Egholm, W.E. Altman, S. Attiya, J.S. Bader, L.A. Bemben, J. Berka, M.
S. Braverman, Y.J. Chen, Z.T. Chen, S.B. Dewell, L. Du, J.M. Fierro, X.V. Gomes,
B.C. Godwin, W. He, S. Helgesen, C.H. Ho, G.P. Irzyk, S.C. Jando, M.L.I. Alenquer,
T.P. Jarvie, K.B. Jirage, J.B. Kim, J.R. Knight, J.R. Lanza, J.H. Leamon, S.M. Lefkowitz,
M. Lie, J. Li, K.L. Lohman, H. Lu, V.B. Makhijani, K.E. McDade, M.P. McKenna, E.
W. Myers, E. Nickerson, J.R. Nobbile, R. Plant, B.P. Puc, M.T. Ronan, G.T. Roth, G.
J. Sarkis, J.F. Simons, J.W. Simpson, M. Srinivasan, K.R. Tartaro, A. Tomasz, K.A.
Vogt, G.A. Volkmer, S.H. Wang, Y. Wang, M.P., Weiner, P.G. Yu, R.F. Begley, and
J.M. Rothverg. 2005. Genome sequencing in microfabricated high-density picolitre
reactors. Nature 437:376-380.
Marshall, J.A., S. Knapp, M.R. Davey, J.B. Power, E.C. Cocking, M.D. Bennett, and A.V.
Cox. 2001. Molecular systematics of *Solanum* section *Lycopersicum* (*Lycopersicon*)
using the nuclear ITS rDNA region. Theor Appl Genet 103: 1216-1222.
Martin, G.B., M.C. de Vicente, and S.D. Tanksley. 1993a. High-resolution linkage analysis
and physical characterization of the *Pto* bacterial resistance locus in tomato. Mol
Plant-Microbe Interact 6:26-34.
Martin, G.B., M.W. Ganal, and S.D. Tanksley. 1992. Construction of a yeast artificial
chromosome library of tomato and identification of cloned segments linked to two
disease resistance loci. Mol Gen Genet 233:25-32.
Martin, G.B., S.H. Brommonschenkel, J. Chunwongse, A. Frary, M.W. Ganal, R. Spivey, T.
Wu, E.D. Earle, and S.D. Tanksley. 1993b. Map-based cloning of a protein kinase gene
conferring disease resistance in tomato. Science 262:1432-1436.
Medina-Filho, H.P. 1980. Linkage of *Aps-1*, *Mi* and other markers on chromosome 6. Rep
Tom Genet Coop 30:26-28.
Miller, J.C. and S.D. Tanksley. 1990. RFLP analysis of phylogenetic relationships and genetic
variation in the genus *Lycopersicon*. Theor Appl Genet 80:437-448.
Murray, J.M., K.E. Davis, P.S. Harper, L. Meredith, C.R. Mueller, and R. Williamson. 1982.
Linkage relationship of a cloned DNA sequence on the short arm of the X chromosome
to Duchenne muscular dystrophy. Nature 300:69-71.
Mustilli, A.C., F. Francesca, R. Ciliento, F. Alfano, and C. Bowler. 1999. Phenotype of the
tomato *high pigment-2* mutant is caused by a mutation in the tomato homolog of
DEETIOLATED1. Plant Cell 11:145-157.
Nienhuis, J., T. Helentjaris, M. Slocum, B. Ruggero, and A. Schaefer. 1987. Restriction
fragment length polymorphism analysis of loci associated with insect resistance in
tomato. Crop Sci 27:797-803.
Noli, E., S. Conti, M. Maccaferri, and M.C. Sanguineti. 1999. Molecular characterization of
tomato cultivars. Seed Sci Technol 27:1-10.
Ohmori, T., M. Murata, and F. Motoyoshi. 1996. Molecular characterization of RAPD and
SCAR markers linked to the *Tm-1* locus in tomato. Theor Appl Genet 92:151-156.
Paran, I., M. Horowitz, D. Zamir, and S. Wolf. 1995. Random amplified polymorphic DNA
markers are useful for purity determination of tomato hybrids. HortScience 30:377.
Paterson, A.H., E.S. Lander, J.D. Hewitt, S. Peterson, S.E. Lincoln, and S.D. Tanksley. 1988.
Resolution of quantitative traits into Mendelian factors by using a complete linkage
map of restriction fragment length polymorphisms. Nature 335:721-726.
Pertuz, R.A., Y. Ji, and R.T. Chetelat. 2003. Transmission and recombination of homeologous
Solanum sitiens chromosomes in tomato. Theor Appl Genet 107:1391-1401.
Rafalski, A. 2002. Applications of single nucleotide polymorphisms in crop genetics. Curr
Opin Plant Biol 5:94-100.
Rick, C.M. 1979. Biosystematic studies in *Lycopersicon* and closely related species of *Solanum*.
In: J.G. Hawkes, R.N. Lester, and A.D. Skelding. [eds.], The Biology and Taxonomy
of the Solanaceae. Academic Press, New York, USA, pp. 667-678.

Rick, C.M. 1983. Tomato (*Lycopersicon*). In: S.D. Tanksley and T.J. Orton. [eds.], Isozymes in Plant Genetics and Breeding. Part B. Elsevier Science Publishers B.V., Amsterdam, The Netherlands, pp. 147-165.

Rick, C.M. 1988. Molecular markers as aids for germplasm management and use in *Lycopersicon*. HortScience 23:55-57.

Rick, C.M. and J.F. Fobes. 1975. Allozyme variation in the cultivated tomato and closely related species. Bull Torr Bot Club 102:376-384.

Rick, C.M. and R.T. Chetelat. 1995. Utilization of related wild species for tomato improvement. Acta Hort 412:21-38.

Rick, C.M., R.W. Zobel, and J.F. Fobes. 1974. Four peroxidase loci in red-fruited tomato species: Genetics and geographic distribution. Proc Natl Acad Sci USA 71:835-839.

Ruddle, F.F. 1981. A new era in mammalian gene mapping: Somatic cell genetics and recombinant DNA methodologies. Nature 294:115-120.

Saiki, R.K., S. Scharf, F. Faloona, K.B. Mullis, G.T. Horn, H.A. Erlich, and N. Arnheim. 1985. Enzymatic amplification of ß-globin genomic sequences and restriction site analysis for diagnosis of sickle-cell anemia. Science 230:1350-1354.

Samoylov, V.M. and K.C. Sink. 1996. The role of irradiation dose and DNA content of somatic hybrid calli in producing asymmetric plants between an interspecific tomato hybrid and eggplant. Theor Appl Genet 92:850-857.

Samoylov, V.M., S. Izhar, and K.C. Sink. 1996. Donor chromosome elimination and organelle composition of asymmetric somatic hybrid plants between an interspecific tomato hybrid and eggplant. Theor Appl Genet 93:268-274.

Samuel, G., J.G. Bald, and H.A. Pittman. 1930. Investigations on spotted wilt of tomatoes. Aust Council Sci Ind Res Bull 44.

Schoenmakers, H.C.H., A.-M.A.Wolters, A. de Haan, A.K. Saiedi, and M. Koornneef. 1994. Asymmetric somatic hybridization between tomato (*Lycopersicon esculentum* Mill) and gamma-irradiated potato (*Solanum tuberosum* L.): a quantitative analysis. Theor Appl Genet 87:713-720.

Shalon, D., S.J. Smith, and P.O. Brown. 1996. A DNA microarray system for analyzing complex DNA samples using two-color fluorescent probe hybridization. Genome Res 6:639-645.

Shi, M.M. 2001. Enabling large-scale pharmacogenetic studies by high-throughput mutation detection and genotyping technologies. Clin Chem 47:164-172.

Shikanai, T., H. Kaneko, S. Nakata, K. Harada, and K. Watanabe. 1998. Mitochondrial genome structure of a cytoplasmic hybrid between tomato and wild potato. Plant Cell Rep 17:823-836.

Smith, P. G. 1944. Embryo culture of a tomato species hybrid. Proc Amer Soc Hort Sci 44: 413-416.

Sobir, T., M. Ohmori, F. Murata, and F. Motoyoshi. 2000. Molecular characterization of the SCAR markers tightly linked to the *Tm-2* locus of the genus *Lycopersicon*. Theor Appl Genet 101:64-69.

Stears, R.L., T. Martinsky, and M. Schena. 2003. Trends in microarray analysis. Nat Med 9:140-145.

Stevens, J.M. 1964. Tomato breeding. Project Report W-Vv1, Dept. Agricultural Technical Services. Republic of South Africa.

Stevens, M.A. and C.M. Rick. 1986. Genetics and breeding. In: J.G. Atherton and J. Rudich [eds.], The Tomato Crop, a Scientific Basis for Improvement. Chapman and Hall, New York, pp. 35-100.

Stevens, M.R., D.K. Heiny, D.D. Rhoads, P.D. Griffiths, and J.W. Scott. 1996. A linkage map of the tomato spotted wilt virus resistance gene *Sw-5* using isogenic lines and an interspecific cross. Acta Hort 431:385-392.

Stevens, M.R., E.M. Lamb, and D.D. Rhoads. 1995. Mapping the *Sw-5* locus for tomato spotted wilt virus resistance in tomatoes using RAPD and RFLP analyses. Theor Appl Genet 90:451-456.

Stevens, M.R., S.J. Scott, and R.C. Gergerich. 1992. Inheritance of a gene for resistance to tomato spotted wilt virus (TSWV) from *Lycopersicon peruvianum* Mill. Euphytica 59:9-17.

Tanksley, S.D. 1980. *Pgi-1*, a single gene in tomato responsible for a variable number of isozymes. Can J Gent Cytol 22:271-278.

Tanksley, S.D. 1983a. Introgression of genes from the wild species. In: S.D. Tanksley and T.J. Orton. [eds.], Isozymes in Plant Genetics and Breeding. Vol. Part A. Elsevier Science Publishers B.V., Amsterdam, The Netherlands, pp. 331-337.

Tanksley, S.D. 1983b. Molecular markers in plant breeding. Plant Mol Biol Rep 1:3-8.

Tanksley, S.D. 1985. Enzyme-coding genes in tomato (*Lycopersicon esculentum*). Isozyme Bull 18:43-45.

Tanksley, S.D. and J.C. Nelson. 1996. Advanced backcross QTL analysis: A method for the simultaneous discovery and transfer of valuable QTLs from unadapted germplasm into elite breeding lines. Theor Appl Genet 92:191-203.

Tanksley, S.D. and M.A. Mutschler. 1990. Linkage map of the tomato (*Lycopersicon esculentum*) (2 N = 24). In: S.J. O'Brien. [ed.], Genetic maps: Locus Maps of Complex Genomes 5th Edition. Cold Spring Harbor Laboratory Press, Cold Spring Harbor, New York, USA, pp 6.3-6.15.

Tanksley, S.D. and S.R. McCouch. 1997. Seed banks and molecular maps: Unlocking genetic potential from the wild. Science 277:1063-1066.

Tanksley, S.D., C.M. Rick, and C.E. Vallejos. 1984. Tight linkage between a nuclear male-sterile locus and an enzyme marker in tomato. Theor Appl Genet 68:109-113.

Tanksley, S.D., H. Medina-Filho, and C.M. Rick. 1982. Use of naturally-occurring enzyme variation to detect and map genes controlling quantitative traits in an interspecific backcross of tomato. Heredity 49:11-25.

Tanksley, S.D., M.W. Ganal, and G.B. Martin. 1995. Chromosome landing: a paradigm for map-based gene cloning in plants with large genomes. Trends Genet 11:63-68.

Tanksley, S.D., M.W. Ganal, J.P. Prince, M.C. de Vicente, M.W. Bonierbale, P. Broun, T.M. Fulton, J.J. Giovannoni, S. Grandillo, G.B. Martin, R. Messeguer, J.C. Miller, L. Miller, A.H. Paterson, O. Pineda, M.S. Röder, R.A. Wing, W. Wu, and N.D. Young. 1992. High density molecular linkage maps of the tomato and potato genomes. Genetics 132:1141-1160.

Tanksley, S.D., N.D. Young, A.H. Paterson, and M.W. Bonierbale. 1989. RFLP mapping in plant breeding: New tools for an old science. BioTechnol 7:257-264.

Tanksley, S.D., R. Bernatzky, N.L. Lapitan, and J.P. Prince. 1988. Conservation of gene repertoire but not gene order in pepper and tomato. Proc Natl Acad Sci USA 85:6419-6423.

Tanksley, S.D., S. Grandillo, T.M. Fulton, D. Zamir, Y. Eshed, V. Petiard, J. Lopez, and T. Beck-Bunn. 1996. Advanced backcross QTL analysis in a cross between an elite processing line of tomato and its wild relative *L. pimpinellifolium*. Theor Appl Genet 92:213-224.

Thoday, J.M. 1961. Location of polygenes. Nature 191:291-296.

Vallejos, C.E. 1983. Enzyme activity staining. In: S.D. Tanksley and T.J. Orton [eds.], Isozymes in Plant Genetics and Breeding. Part B. Elsevier Science Publishers B.V., Amsterdam, The Netherlands, pp. 469-516.

van Eijk, M.J.T., J.L.N. Broekhof, H.J.A. van der Poel, R.C.J. Hogers, H. Schneiders, J. Kamerbeek, E. Verstege, J.W. van Aart, H. Geerlings, J.B. Buntjer, A.J. van Oeveren, and P. Vos. 2004. SNPWave™: A flexible multiplexed SNP genotyping technology. Nucl Acids Res 32:No. 4 e47.

Vavilov, N.I. 1940. The new systematics of cultivated plants. In: J. Huxley [ed.], The New Systematics. Oxford University Press, London, Great Britain, pp. 549-566.

Watterson, J.C. 1993. Development and breeding of resistance to pepper and tomato viruses. In: M.M. Kyle [ed.], Resistance to Viral Diseases of Vegetables. Timber Press, Portland, USA, pp. 80-101.

Weller, J.I., M. Soller, and T. Brody. 1988. Linkage analysis of quantitative traits in an interspecific cross of tomato (*Lycopersicon esculentum* x *Lycopersicon pimpinellifolium*) by means of genetic markers. Genetics 118:329-339.

Wenzl, P., J. Carling, D. Kudrna, D. Jaccoud, E. Huttner, A. Kleinhofs, and A. Kilian. 2004. Diversity arrays technology (DArT) for whole-genome profiling of barley. Proc Natl Acad Sci USA 101: 9915-9920.

Williamson, V.M., J.-Y. Ho, F.F. Wu, N. Miller, and I. Kaloshian. 1994. A PCR-based marker tightly linked to the nematode resistance gene, *Mi*, in tomato. Theor Appl Genet 87:757-763.

Winzeler, E., D.R. Richards, A.R. Conway, A.L. Goldstein, S. Kalman, M.J. McCullough, J.H. McCusker, D.A. Stevens, L. Wodicka, D.J. Lockhart, and R.W. Davis. 1998. Direct allelic variation scanning of the yeast genome. Science 281:1194-1197.

Wong, K-K., Y.T.M. Tsang, J. Shen, R.S. Cheng, Y-M. Chang, T-K. Man, and C.C. Lau. 2004. Allelic imbalance analysis by high-density single nucleotide polymorphic allele (SNP) array with whole genome amplified DNA Nucl Acids Res 32:No. 9 e69.

Wyman, A.R. and R. White. 1980. A highly polymorphic locus in human DNA. Proc Natl Acad Sci USA 77:6754-6758.

Yamakawa, K. and N. Nagata. 1975. Three tomato lines obtained by the use of chronic gamma radiation with combined resistance to TMV and fusarium race J-3. Tech News No. 16 Inst Radiat Breeding Jpn October.

Yang, W.C., and D.M. Francis. 2005. Marker-assisted selection for combining resistance to bacterial spot and bacterial speck in tomato. J Am Soc Sci 130:716-721.

Yang, W.C., X.D. Bai, E. Kabelka, C. Eaton, S. Kamoun, E. van der Knaap, and D. Francis. 2004. Discovery of single nucleotide polymorphisms in *Lycopersicon esculentum* by computer aided analysis of expressed sequence tags. Mol Breed 14:21-34.

Young, N.D. and S.D. Tanksley. 1989a. Restriction fragment length polymorphism maps and the concept of graphical genotypes. Theor Appl Genet 77:95-101.

Young, N.D. and S.D. Tanksley. 1989b. RFLP analysis of the size of chromosomal segments retained around the *Tm-2* locus of tomato during backcross breeding. Theor Appl Genet 77:353-359.

Young, N.D., D. Zamir, M.W. Ganal, and S.D. Tanksley. 1988. Use of isogenic lines and simultaneous probing to identify DNA markers tightly linked to the *Tm-2a* gene in tomato. Genetics 120:579-585.

Young, N.D., J.C. Miller, and S.D. Tanksley. 1987. Rapid chromosomal assignment of multiple genomic clones in tomato using primary trisomics. Nucl Acids Res 15:9339-9348.

Zamir, D. and S.D. Tanksley. 1988. Tomato genome is comprised largely of fast-evolving, low copy-number sequences. Mol Gen Genet 213:254-261.

Zhang, H.-B., G.B. Martin, S.D. Tanksley, and R.A. Wing. 1994. Map-based cloning in crop plants: tomato as a model system II. Isolation and characterization of a set of overlapping yeast artificial chromosomes encompassing the *jointless* locus. Mol Gen Genet 244:613-621.

8

Molecular Genetics of Drought Resistance Response in Tomato and Related Species

MARY A. O'CONNELL, ANDREA L. MEDINA, PEDRO SÁNCHEZ PEÑA,
AND MARCELA B. TREVIÑO
Department of Plant and Environmental Sciences
New Mexico State University, P.O. Box 3003, MSC 3Q, Las Cruces, NM, 88003, USA
e-mail: moconnel@nmsu.edu

DROUGHT AND DROUGHT RESISTANCE PHYSIOLOGY

Drought or water deficit stress is the most common and most severe limitation to crop plant productivity (Boyer 1982, Kramer and Boyer 1995). Just under half of the earth's land surface is considered dryland, and these regions are found on all the continents (Middleton and Thomas 1997). Increasing drought resistance in crop plants grown in semi-arid regions, as well as developing drought resistance in mesophytic crop plants, are common goals in agricultural research programs around the world. Unfortunately, drought is not a simple homogeneous environmental condition. Some semi-arid regions have seasonal summer rains, while in other semi-arid regions the rains come in winter months. Relative humidity and temperature can also be markedly different in regions with similar precipitation rates. All of these climatic factors influence the severity of the drought stress on the plant. Thus, different physiological, anatomical and biochemical mechanisms of drought resistance are likely to be more successful in these different water stressed climates.

Plants that grow in areas with limited water usually use one or more of the following three general mechanisms to resist drought stress (Kramer and Boyer 1995). Some plants complete their life cycle within a seasonal

---

Corresponding author: Mary O'Connell

rain; they essentially *avoid* drought. Other plants develop deep roots or reduce transpiration area or increase wax layers on leaves; they essentially *postpone* drought. Finally there are plants that continue to grow with lower tissue water content; these plants *tolerate* dehydration. These classifications are not mutually exclusive; plants will demonstrate multiple components in their response to drought stress.

Genetic manipulation of all three types of mechanisms can be expected to improve or maintain crop yields in areas with limited water availability. The first two classes of mechanisms involve organ-specific gene expression, e.g. timing of reproduction, or changes in root development. The third class is often accomplished by osmotic adjustment and would require changes in gene expression at a global cell-level, i.e. most, if not all, of the cells throughout the plant body would be involved. The current challenge is to identify the genes that confer drought-resistant phenotypes. Comparative assays to study drought responses are likely to identify genes in the dehydration tolerance class of plants. Drought resistance strategies that involve adapted or unique developmental patterns of gene expression or pre-existing (constitutive) expression are not likely to be identified by differential treatment screens. Genetic comparisons may be productive for finding developmentally based drought resistance mechanisms, presuming that there are well characterized sources for these intra or inter-species differences in drought resistance.

GENETICS AND DROUGHT RESISTANCE CHARACTERISTICS OF *Lycopersicon*

Cultivated tomato, *Lycopersicon esculentum*, is mesophytic and plants of the species are not significantly drought resistant. So there is variability for drought and/or osmotic stress resistance within *Lycopersicon esculentum* and this variability is limited (Foolad and Lin 1999, Srinivasa Rao et al. 2000, Fellner and Sawhney 2001, Foolad et al. 2003). The best genetic sources of drought resistance for cultivated tomato are from other species in the genus, namely *L. pennellii* and *L. chilense* (Fig. 8.1). These species are indigenous to arid and semi-arid environments in South America (Rick 1973). *L. pennellii* is adapted to the coastal cliffs of Peru. This location has very low precipitation, but lots of early morning dew. *L. chilense* is adapted to the Atacama desert in northern Chile (Maldonado et al. 2003). This is the most arid temperate desert on the planet. Plants of *L. chilense* are often found in dry arroyos with no other vegetation in the area (Rick 1973).

Fig. 8.1 Plants of *Lycopersicon esculentum*, cv. UC82B (**A**); *L. pennellii*, LA716 (**B**); and *L. chilense*, LA2880 (**C**).

The three species (*L. chilense, L. esculentum, L. pennellii*) are quite different in morphology. Unlike cultivated tomato *L. esculentum*, *L. pennellii* and *L. chilense* produce small green fruit, and they both have an indeterminate growth habit. *L. pennellii*, however, has small thick rounded leaves, light green in color, with very small root to shoot ratio; whereas *L. chilense* has thin finely divided leaves and a well developed root system.

These morphological differences are well matched to the environments in which they grow. *L. pennellii* appears adapted to arid environments by virtue of high water-use efficiency (Table 8.1, Martin and Thorstenson 1988) and an ability of its leaves to take up dew (Rick 1973), compensating for a strikingly low root to shoot biomass ratio. This species increases water-use efficiency under water deficit conditions, 25% vs 100% field capacity, while *L. esculentum* has the same values for water-use efficiency regardless of

Table 8.1 Physiological Characteristics of *Lycoperison spp.*

Species, accession	WUE[a] g/kg	leaf wilt (% fresh wt lost/h)	$\Psi water$ (MPa ± sd)	$\Psi osmotic$ (MPa ± sd)	$\Psi turgor$ (MPa ± sd)
L. esculentum, UC82b	2.43 ± 0.21 (25%FC[b]) 2.23 ± 0.16 (100%FC)	6.3	T_0[c] −0.55 ± 0.12 T_w[d] −1.97 ± 0.06	T_0 −0.98 ± 0.05 T_w −1.62 ± 0.13	T_0 0.43 ± 0.11 T_w −0.34 ± 0.11
L. pennellii, LA716	3.42 ± 0.91 (25% FC) 2.71 ± 0.34 (100% FC)	1.2	T_0 −0.17 ± 0.03 T_w −0.57 ± 0.24	T_0 −1.00 ± 0.00 T_w −1.20 ± 0.01	T_0 0.83 ± 0.03 T_w 0.63 ± 0.24
L. chilense, LA2880	nd	4.2	T_0 −0.73 ± 0.06 T_w −1.42 ± 0.26	T_0 −1.30 ± 0.08 T_w −2.37 ± 0.28	T_0 0.56 ± 0.02 T_w 0.95 ± 0.18

[a]WUE, water use efficiency, [b]FC, field capacity, [c]T_0, time zero, [d]T_w, at wilt.

field capacity (Table 8.1). Three gene loci have been linked to the water-use phenotype in *L. pennellii* (Martin et al. 1989). A second feature of the drought response of *L. pennellii* is stomatal response (Kebede et al. 1994). Stomata close rapidly upon water deficit stress; the detached leaf wilt rate is the lowest of all three species (Table 8.1). In contrast, *L. chilense* has a very well developed root system, presumably to explore deep soil layers in arroyos following seasonal rains. During drought treatments, leaf area and seedling growth rates are reduced while stressed roots are better developed (Chen and Tabaeizadeh 1992a). This enhanced root development is also observed under non-stress conditions (Table 8.2). *L. chilense* has a longer primary root, and twice the number of secondary roots like cultivated tomato. No differences were observed in distance of the root tip to the youngest secondary root or distance from the root base to oldest secondary root.

Table 8.2 *Root characteristics in 10-day old* Lycopersicon *seedlings.*

Species, accessions	Length 1° root	Root tip to 2° root	Root base to 2° root	# 2° roots	Total length of 2° roots
L. esculentum, UC82b	7.4**	2.32	0.26	7.67**	6.66**
L. chilense, LA2880	14.6**	3.28	1.71	11.41**	10.28**

n = 9, all units in cm, **LSD at 0.05

Measurements of plant water status during drought episodes also indicate that the two species *(L. pennellii* and *L. chilense)* are more drought resistant than cultivated tomato (Kahn et al. 1993, Sánchez Peña et al. 1995). Representative values for leaf water, osmotic and turgor potentials for cultivated tomato, *L. pennellii* and *L. chilense* are presented in Table 8.1. These data were collected from flowering plants grown under greenhouse conditions. *L. chilense* is the most drought resistant species, as evidenced by wilting time under controlled drought stress conditions (Fig. 8.2), and its ability to continue to function with leaf at low water potential (Table 8.1).

DROUGHT-INDUCED CHANGES IN GENE EXPRESSION

During drought stress synthesis of specific proteins gets altered. In general, drought stress results in the accumulation of several novel proteins along with the increased accumulation and/or reduction of several already existing proteins. It does not result, however, in a complete alteration in the pattern of protein synthesis, as is often seen in response to heat shock. Recent reviews of the changes in gene expression in different plants as a

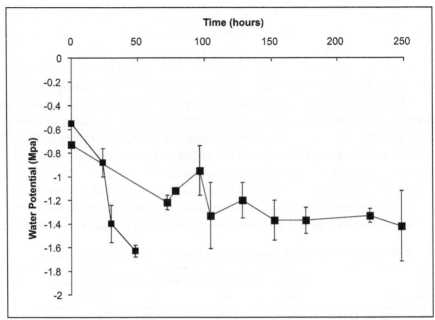

Fig. 8.2   Leaf water potential in *Lycopersicon* during a drought cycle. Leaf water potential in MPa was measured using a pressure bomb in *L. esculentum* cv. UC82B
(open squares) and *L. chilense*, LA2880, (closed squares) at the indicated times
in hours. Values are reported as the average plus or minus the standard
deviation (n = 3).

consequence of drought stress summarize these studies (O'Connell 1995,
Tabaeizadeh 1998, Cushman and Bohnert 2000, Seki et al. 2001, Bray 2002a,
2002b) and a scheme of the events common to all plant responses to water
deficit stress is presented in Fig. 8.3. The first event is the perception of the
water deficit state, followed by a signal transduction cascade that may be
initiated by altered levels or sensitivities to plant hormones like abscisic
acid (ABA) or through other non-ABA based path-ways. Based on these,
altered patterns of transcription are then observed, presumably in response
to increased levels of specific transcription factors. The new transcripts
then result in the accumulation of proteins/enzymes that synthesize
osmoticum, stabilize membranes, bind water, thicken wax layers, alter ion
flux, close stomata, etc.

  Transcription factors, or other types of chromatin proteins, may also
repress transcription and reduce rate of photosynthesis during drought
stress. This physiological change is mediated at a number of levels, but in
particular, transcription of the genes for light harvesting chlorophyll
binding protein (*cab*) and Rubisco small subunit (*RbcS*) is reduced during

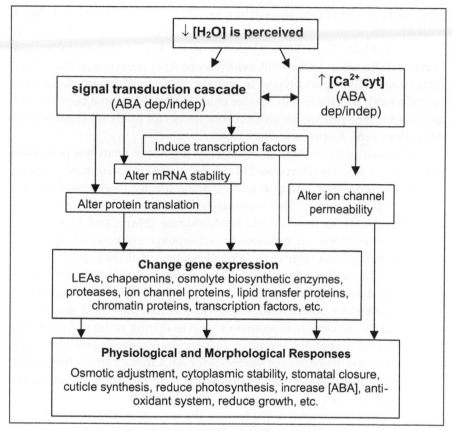

Fig. 8.3 Model of drought responses in plants.

drought stress (Bartholomew et al. 1991). Comparisons of transcriptional and translational profiles suggest that there is post-transcriptional regulation in response to drought stress and that mechanisms function to translate only specific transcripts within the pool of transcripts that accumulate in response to drought stress. Very little is known about how this regulation is achieved.

ABA REGULATES SOME OF THE DROUGHT-INDUCED CHANGES IN GENE EXPRESSION

Many of the transcripts that accumulate in vegetative tissues in response to drought stress are also developmentally regulated and accumulate during the later stages of embryogenesis (Skriver and Mundy 1990). Both these processes are regulated to some extent by ABA and some progress has been made in identifying *trans*-acting factors and *cis*-elements associated with ABA-dependent regulation of transcription (Chandler and Roberston 1994, Bray 2002a, Rock 2000). There are a number of well-characterized mutants deficient in either ABA synthesis or ABA perception. These mutants have "wilty" phenotypes. In tomato, these mutants are *flacca* (*flc*), *notabalis* (*not*), and *sitiens* (*sit*) (Taylor et al. 2000), which have been used to identify drought-responsive genes whose transcript levels are regulated by ABA (Bray 1988, Cohen and Bray 1990).

The biosynthesis of ABA is essentially regulated from the precursor zeaxanthin, a 40 carbon terpenoid (Taylor et al. 2000). The tomato mutants for ABA accumulation characterized have the specific biochemical lesions identified: *flc* deficient in MoCo-sulfurylase, an activity necessary to produce the cofactor for the ABA-aldehyde oxidase (Marin and Marion-Poll 1997, Sagi et al. 1999), *not* deficient in 9-*cis*-epoxycarotenoid dioxygenase (NCED) activity (Burbridge et al. 1999, Thompson et al. 2000, 2004), and *sit* deficient in the ABA-aldehyde oxidase (Marin and Marion-Poll 1997). ABA is considered to be a key regulator for two different processes associated with reduced water potential, (a) the developmental process of seed maturation, and (b) the closure of stomata in leaves during water deficit stress, and regulation of changes in numerous patterns of gene expression (Rock 2000). One of the major sites of synthesis for ABA in the plant is the root. It has been modeled for at least 20 years that the drying soil is perceived by the root, where ABA levels increase as a result, which is then transported via the xylem from the root to the leaves. The increased ABA concentration in leaves causes the stomatal closure. Recent results indicate that the pH of the different tissues in the plant play a significant role in partitioning of ABA (Hartung et al. 2002). Indeed, pH changes may be sufficient to mimic soil drying and/or increases in ABA concentration for the regulation of stomatal closure (Wilkinson and Davies 1997).

EFFECT OF DROUGHT STRESS ON GENE EXPRESSION IN *LYCOPERSICON*

Alterations in gene expression as a result of drought stress have been measured in *Lycopersicon* using *in vivo* and *in vitro* translation assays (Bray 1988, Ho and Mishkind 1991, Chen and Tabaeizadeh 1992a, Chen and Tabaeizadeh 1992b, Jin et al. 2000) and differential screening of cDNA libraries (Cohen and Bray 1990, Chen and Tabaeizadeh 1992a). As expected, leaves and roots of *L. esculentum* and *L. chilense* exhibit induction and repression in the expression of specific genes. Changes in protein synthesis have been detected after a 4 h drought stress using a detached leaf assay in tomato (Bray 1988) with approximately 20 spots on 2D gels identified as drought responsive. In a whole plant assay, drought stress imposed by withholding water resulted in fewer changes in leaves as only ten spots could be detected on 2D gels, (Chen and Tabaeizadeh 1992b). This whole plant assay, though, was able to detect repressed genes. Using *in vitro* labeling, alterations in gene expression in response to drought were observed as early as 20 min after imposition of the stress, in a detached leaf method of drought stress (Ho and Mishkind 1991). These studies altogether demonstrated that the steady state levels of specific transcripts either increased or decreased in response to drought stress.

Drought responsive genes have been cloned from *L. esculentum* (Lusem et al. 1993, Fray et al. 1994, Imai et al. 1995, Harrak et al. 2001) and from the drought-resistant species *L. pennellii* (Wei and O'Connell 1996, Treviño and O'Connell 1998) and *L. chilense* (Chen and Tabaeizadeh 1992a, Chen et al. 1993, Chen et al. 1994, Yu et al. 1996 Harrak et al. 1999, Frankel et al. 2003). All of these genes from *Lycopersicon spp* (GenBank late summer 2002) that were annotated with keywords related to water deficit stress are listed in Table 8.3; approximately 35 entries from three species were retrieved. The number of distinct drought responsive genes, however, is lower as heterologous genes were identified among the three species. For example, histone H1 was found in all three species. In addition, there were several biochemical functions with different gene family members identified as drought responsive, e.g. SAM synthetase and leucine aminopeptidase. This list includes genes whose function is predicted to participate in signal transduction events. The expression of many of these genes has been characterized by patterns of expression in different organs, during other abiotic stresses, and for ABA regulation (Cohen and Bray 1990, Kahn et al. 1993, Thompson and Corlett 1995, Cohen et al. 1999, Maskin et al. 2001, Doczi et al. 2005).

How many genes are expected to be drought responsive? The 2D gel analyses described earlier identified 20 spots uniquely accumulated in drought-stressed leaves. The list of genes in Table 8.3 suggests fewer than 30 unique genes. Are there ~ 25 drought-responsive genes in tomato? The DNA sequence of *Arabidopsis thaliana* has been completed, this plant could be used, as a benchmark for how many genes might be drought-responsive in an average plant. Like cultivated tomato, *Arabidopsis* is also mesophytic. Table 8.4 lists the numbers of genes in GenBank that were retrieved using different queries in December 2005. For example, there were 300,439 genes or sequences annotated with the word "ABA", and there were 409,583

Table 8.3 *GenBank entries retrieved with the search terms:* Lycopersicon + *Drought/water-deficit/ABA/salt stress.*

Species	GenBank ID	Annotation
L. chilense	AF253416	histone H1
	L19342	endochitinase
	U19098	proline-rich protein (PRP13)
	M97211	dehydrin (pLC3015)
L. esculentum	AF172856	cysteine protease TDI-65 (tdi-65)
	AF261140	unknown mRNA
	AW036653	EST278696 tomato fruit mature green, cLEF37M17
	AW030922	EST274229 tomato callus, cLEC5M17
	AW033878	EST277449 tomato callus, cLEC27O13
	U77719	ethylene-responsive late embryogenesis-like protein (ER5)
	U81996	non specific lipid transfer protein (le16)
	X56040	TSW12
	Z11842	Histone H1-like protein
	M76552	accumulating in developing seeds and drought-stressed leaves LE25
	AF162854	leucine aminopeptidase LapA1
	AF162855	leucine aminopeptidase LapA2
	AF218774	water channel protein (Aqp2)
	X63093	GAST1
	AF419320	SNF4 gene
	AI637360	Giant cell specific cDNA similar to water-stress aquoporin
	U86130	Asr1 gene
	L20756	Asr2 gene
	X74908	Asr3 gene
	X73848	tomato ripening membrane protein, clone pNY507
	X73847	tomato ripening membrane protein, clone pTOM75
	AF263917	plasma membrane proton ATPase (LHA8)
	U26423	dehydrin TAS14
	X51904	TAS14, inducible by abscisic acid and environmental stress
	Z24741	S-adenosyl-L-methionine synthetase
	Z24742	S-adenosyl-L-methionine synthetase
	Z24743	S-adenosyl-L-methionine synthetase
L. pennellii	U01890	histone H1
	U66465	lipid transfer protein 1 (LpLTP1)
	U66466	lipid transfer protein 2 (LpLTP2)

Table 8.4 GenBank Nucleotide database entries annotated as drought responsive (December 2005)

Plant	ABA (300,439)*	Drought (171,324)	Water-deficit (23,731)	Water stress (11,625)	Salt stress (51,531)
Lycopersicon (409,583)	16	14	112	6	15
Arabidopsis (1,015,299)	2,819	179	4	24	782

*(total # of entries in each category)

genes or sequences for the organisms in the genus *Lycopersicon*. The intersection of those two sets, ABA annotated genes in the organism *Lycopersion*, was 16 genes. There are at least 179 genes identified as drought responsive in *Arabidopsis thaliana* compared to only 14 in *Lycopersicon*. This suggests that fewer than 10% of the drought responsive genes in tomato have been identified. ABA responsive genes are even less well studied in tomato. While the number of genes entered into GenBank will be much higher by the time this chapter is published, the number of genes annotated as drought responsive will not have changed proportionately.

The genomics community has begun to functionally classify genes into one of thirteen general categories, eg. metabolism, transport, translation, energy, oxidative stress, unclassified, etc. Using microarray screens, ~130 drought responsive genes have been identified in *Arabidopsis* (Reymond et al. 2000, Seki et al. 2001). The drought responsive genes have been placed into functional classes using the predicted amino acid sequence of the gene to identify possible gene functions (Bray 2002b). Drought responsive genes are found in eleven of the thirteen possible functional classes, which means that essentially all of a plant's physiological and metabolic processes are sensitive to drought stress.

Selected Examples of Drought-induced Genes and their Possible Roles in Drought Resistance

In the following sections a detailed description of two drought-responsive genes will be presented. These two genes are drought-induced in many plants including tomato, and at least some level of biochemical function for the corresponding gene products is understood. The drought-induced H1 histone represents a gene product modeled to function in regulation of changes in gene expression in response to drought stress. Lipid transfer proteins have been modeled to function in deposition of thicker wax layers in response to drought stress.

H1 Histone

Gene transcription for this protein takes place in an environment in which the genome is packaged into hierarchial structures of chromatin composed of nucleosome units. The nucleosomal core consists of two copies each of histones H2A, H2B, H3 and H4. These cores, with DNA wrapped around them, are condensed by linker or H1 histones to generate 30 nm chromatin fibers (Wolffe and Kurumizaka 1998). While the H1 histones are the most divergent of the histones in amino acid sequence, they share a tripartite structure with a highly conserved central globular region and a lysine-rich

C-terminal domain (Gantt and Lenvik 1991, Wells and Brown 1991). The association of H1 histones with nucleosomal arrays is believed to be primarily responsible for the formation of higher orders of chromatin compaction, thus inactivating the transcription of genes in that chromatin region. The H1 histones are responsible for the positioning of the nucelosome in a sequence specific manner (Wolffe and Pruss 1996). Further, there are developmentally controlled H1 histone variants, suggesting that these different H1 subtypes play regulatory roles in plant and animal development by selectively repressing transcription of genes via an alteration of the chromatin organization in the region that contains them (Cole 1984).

The DNA sequences of less than 20 plant H1 histone genes have been published. This family of proteins can be divided into two groups based on size, amino acid sequence and expression characteristics (Ascenzi and Gantt 1997). The larger group, equivalent to the somatic linker histones in animals, is comprised of the H1 histones associated with the bulk of the chromatin. These genes are expressed in a cell cycle dependent manner. The second group is populated with H1 variant or subtypes that have a slightly smaller carboxy terminal domain, and more importantly, the members of this group are drought responsive (Wei and O'Connell 1996, Ascenzi and Gantt 1997, Bray et al. 1999). These drought-responsive H1 histones form a separate clade in a dendrogram of H1 histones (Fig. 8.4). These drought responsive H1 histones accumulate in the nucleus of drought-stressed plants (Ascenzi and Gantt 1997, Scippa et al. 2000). The abundance of the protein however is not sufficient to account for a general replacement of the H1 histones in the chromatin (Ascenzi and Gantt 1999b). Rather the abundance and pattern of expression of these variant H1 histones are likely to remodel the chromatin of selected regions of the genome, presumably around genes whose transcription is altered.

As mentioned earlier, repression of transcription of specific genes occurs during drought stress, notably the transcription of *cab* and *rbcS* (Bartholomew et al. 1991). Transcripts for the drought responsive H1 histone accumulated prior to the repression of *rbcS* transcription in tomato leaves during a drought cycle (Wei and O'Connell 1996). These results are consistent with the hypothesis that the function of the drought induced H1 histone is to repress transcription of genes during drought stress. However there were no changes in transcription of selected drought responsive genes in transgenic *Arabidopsis* over-expressing the drought-induced H1 histone (Ascenzi and Gantt 1999a). These plants did not demonstrate any alterations in water content or any other phenotypic differences during drought stress. Similar results were obtained when anti-sense constructs

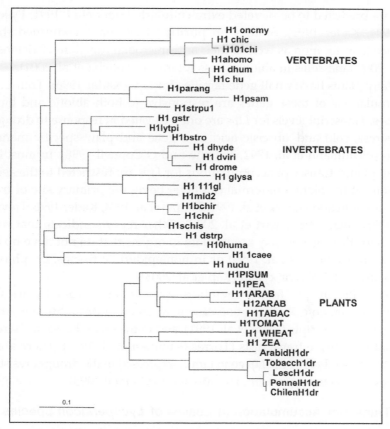

Fig. 8.4 Dendrogram of H1 histones. Amino acids sequences of selected H1 histones from plants and animals were used to generate the neighbor-joining tree. The distance representing 0.1 substitutions per site is shown. All of the drought-responsive H1 histones are indicated with the suffix dr.

were used to reduce the expression of drought-induced H1 histone. The only phenotype was a change in leaf anatomy and stomatal physiology (Scippa et al. 2004). If the expression of drought-induced H1 histones is associated with specific alterations in transcription, then the target genes are yet to be identified.

Lipid Transfer Proteins

Lipid transfer proteins (LTPs) in plants have the ability to transfer lipids between membrane vesicles and exhibit a broad range of substrate specificity (Bourgis and Kader 1997). While lipid transfer activity is observed *in vitro*, the *in vivo* function(s) of these proteins in plants is debated (Kader 1997). LTPs are small basic proteins; they usually have a signal sequence and are predicted to be secreted extra-cellularly (Sterk et al. 1991, Pyee et al. 1994, Kader 1996). Studies on purified LTPs have confirmed their extracellular location as well as their anti-microbial functions (Kristensen et al. 2000) and roles in abiotic stress tolerance (Hincha et al. 2001).

Many plants have small gene families for *Ltps* (Kader 1996). Transcript accumulations of these genes are responsive to both abiotic and biotic stresses. Transcript levels for *Ltps* are often elevated in response to drought, salt stress, cold and abscisic acid in a gene and plant-specific manner (Torres-Schumann et al. 1992, Treviño and O'Connell 1998). In most but not all plants, transcript accumulations for *Ltps* are restricted to the aerial portions of the plant, epidermal cell layer being the primary site of transcript accumulation (Sterk et al. 1991, Thoma et al. 1994, Kader 1996, Treviño and O'Connell 1998, Smart et al. 2000). Other reports indicate that transcripts for these genes may accumulate in organs that do not develop cutin, e.g. roots (Sohal et al. 1999) or in non-epidermal tissues, i.e. phloem (Horvath et al. 2002) or xylem (Rep et al. 2003).

Two different *Ltp* family members were isolated from libraries of cultivated tomato cDNAs, in one case in a screen of drought- and ABA-responsive transcripts (Plant et al. 1991), and in the second case in a screen for salt responsive transcripts (Torres-Schuman et al. 1992). There are at least three members of the *Ltp* gene family expressed in the drought resistant species *Lycopersicon pennellii* (Treviño and O'Connell 1998).

Ltp Transcript Accumulation in Leaves of *Lycopersicon* Species

Gene specific probes for three members of the *Ltp* family were used to demonstrate that transcripts for specific gene family members accumulated in response to developmental and drought stress signals (Fig. 8.5). In *L. pennellii,* as well as in *L. esculentum*, *Ltp2* was responsible for most of the developmental expression while *Ltp3* was responsible for most of the drought-induced expression. The *Ltp* family members are drought and developmentally regulated in other species and accessions of *Lycopersicon*. Table 8.5 represents the quantification of northern blot data for experimental results generated as described in Fig. 8.5. A phosphorimager was used to quantify the abundance of transcripts and this signal was normalized to RNA load. *L. chilense* also has drought responsive transcription of *Ltp* family members, with *Ltp3* again responsible for the majority of the drought-induced gene expression. Developmental regulation of the *Ltp* family is also demonstrated, but in *L. chilense*, *Ltp2* is not the only gene family member that is developmentally responsive.

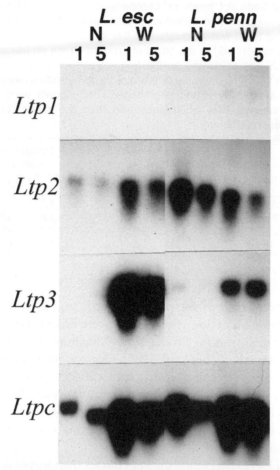

Fig. 8.5   Transcript accumulation of specific Ltp gene family members in response to drought in leaves of different developmental stages. To compare the influence of drought stress and development on transcript accumulation, leaf samples were collected from well-watered plants (N) or from drought-stressed plants (W). These leaves were also sorted by developmental stage, so that RNA was prepared from either fully expanded leaves (5) or from leaf primordia (1). Northern blots containing all of these samples were prepared and probed with each of the gene specific probes.

Inheritance of *Ltp* alleles in *L. esculentum* x *L. pennellii* Plants and Leaf Cuticle Thickness

RFLPs for each of the three *Ltps* were identified in *Hind*III digests of genomic DNA using the gene specific probes (Treviño and O'Connell 1998). The inheritance of either the *L. pennellii* allele or the *L. esculentum* allele for

Table 8.5 Ltp transcript levels in *Lycopersicon spp.* The effect of leaf blade development on transcript accumulation of *Ltp1, Ltp2, Ltp3* and all *Ltps* (cod) was calculated as the fold increases in relative transcript levels (N1/N5) or the effect of drought stress on transcript accumulation was calculated as the fold increase in relative transcript levels (W1/N5). Northern blots prepared and hybridized as in Fig. 8.5 were quantified by a phosphorimager. The signals were normalized by ribosomal RNA abundance and then the ratios N1/N5 or W1/N5 were calculated.

	Developmental influence on transcript levels (n1/n5)				Drought stress influence on transcript levels (w5/n5)			
Genotype	*Ltp1*	*Ltp2*	*Ltp3*	*cod*	*Ltp1*	*Ltp2*	*Ltp3*	*cod*
L. chilense, LA2880	7.5	1.3	9.0	1.3	1.7	0.3	6.0	0.8
L. chilense, LA2748	1.1	5.3	1.5	3.3	1.6	3.0	7.0	8.0
L. pennellii, LA1376	3.6	4.6	3.0	10.0	2.1	0.9	8.0	1.5
L. pennellii, LA716	2.7	2.3	1.3	2.7	1.0	1.0	13.0	3.3
F1 (UC82 x LA716)	2.0	25.0	1.4	2.0	1.0	33.0	8.4	5.2
L. esculentum, cv. UC82	0.7	4.3	0.4	1.0	1.7	6.0	14.0	6.1

each of the three *Ltps* was determined in the F_1 and in F_2 36 individuals. It was demonstrated that these three genes were linked (Treviño 1997).

The leaves of *L. pennellii* have 20 times the amount of epicuticular lipid as cultivated tomato (Fobes et al. 1985). Electron microscopy was used to investigate the existence of ultrastructural differences in leaf cuticles of *L. esculentum, L. pennellii*, the F_1 and selected F_2 individuals (Fig. 8.6A). The surface of all leaves appeared to have the same morphology, i.e., a smooth surface, (Fig. 8.6B). Little to no fibrillar wax crystals were apparent even at higher magnifications.

L. pennellii leaves had much thicker cuticles than leaves of *L. esculentum*, 5.5 µm vs. 1.5 µm. The F_1 was intermediate but closer to the *L. esculentum* parent, i.e. 2.0 µm. Cuticle thickness was determined in F_2 18 individuals. This group included five individuals scored as E, two individuals that scored as P, and 11 individuals scored as hybrid at the *Ltp* locus. No F_2 individual in this population had a cuticle thicker than the *L. pennellii* parent, but several of them had cuticles thinner than the *L. esculentum* parent. Most of the F_2 individuals, therefore, had intermediate cuticle dimensions with a range of cuticle thickness observed in the population. These results are consistent with multiple genes controlling the character of leaf cuticle thickness.

To test whether there was any correlation between *Ltp* genotype and cuticle thickness, the F_2s were grouped into genotype clusters and the average cuticle thickness of the group was determined. From this analysis it

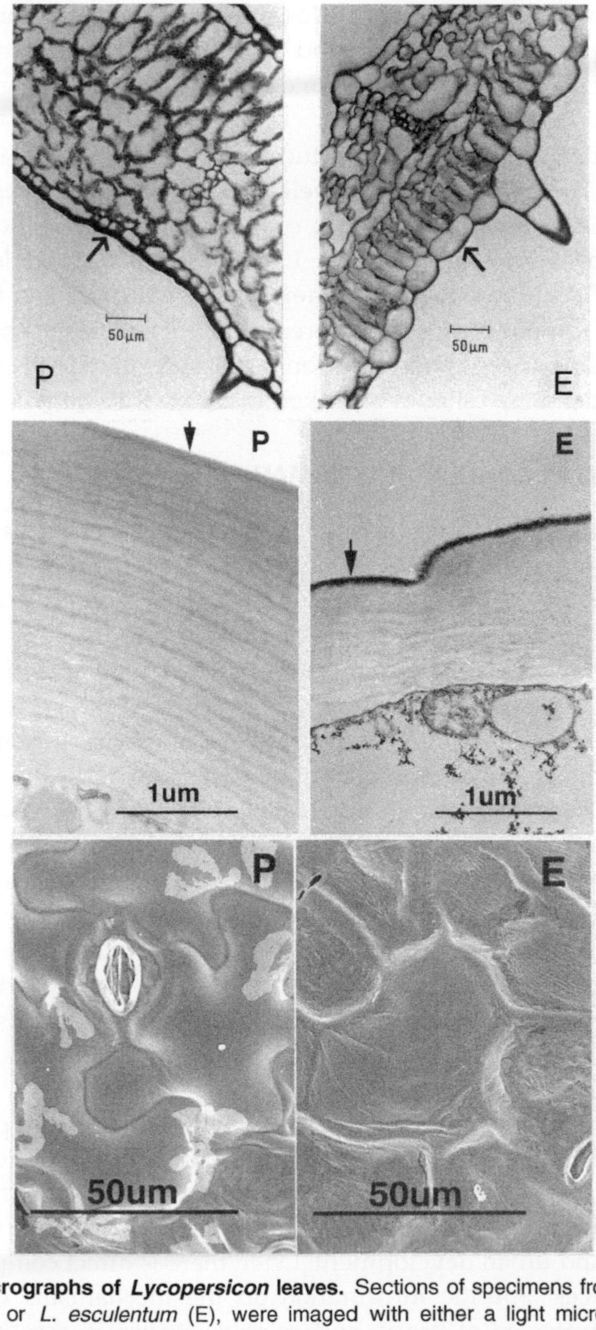

Fig. 8.6 **Micrographs of *Lycopersicon* leaves.** Sections of specimens from *L. pennellii* (P), or *L. esculentum* (E), were imaged with either a light microscope (upper row), with TEM (middle row) or SEM (lower row). Arrows indicate the external surface of the cuticle.

was clear that there was no relationship between inheritance of the esculentum parent alleles for *Ltp*s and thinner cuticles. Infact, there was a complete overlap in range of cuticle thickness regardless of the genotype of the *Ltp* locus (Treviño 1997).

The function of the drought-induced expression of *Ltp* is still not understood. There is much more developmentally regulated expression of these genes in *L. pennellii* than in *L. esculentum*. However, there was no correlation between inheritance of *L. pennellii* alleles and leaf cuticle thickness. It is still possible that differences in abundance of these gene products are responsible for the differences in leaf cuticle thickness. In that case, allelic differences in transcription factor(s) regulating the expression of *Ltp1*, *Ltp2*, and *Ltp3* should be linked to cuticle thickness.

SUMMARY AND CONCLUDING REMARKS

Drought stress is the most common significant limitation to plant growth. To best utilize a molecular approach, a full understanding of the physiological basis for the drought resistant phenotype is critical. The best genetic sources of drought resistance for cultivated tomato are from other species in the genus. There are two species with very different mechanisms for drought resistance; *Lycopersicon chilense* invests in deep root growth, while *Lycopersicon pennellii* regulates stomatal aperture efficiently during drought stress. Drought-induced alterations in gene expression, most commonly assayed at the transcriptional level, have been observed in a number of systems. Identification of genes associated with drought responses in plants, tomato in particular has been achieved through a variety of differential screens and comparisons. Inter-specific comparisons with drought-resistant species have been especially informative. The diversity of well characterized phenotypes in *Lycopersicon* coupled with current molecular genetic approaches are likely to result in a full understanding of the variety of ways that plants continue to grow in the face of limited water.

Understanding at the molecular level the various strategies plants have evolved to tolerate and/or flourish in environments with reduced water availability is valuable now and will become crucial in the future. Worldwide, communities face conflicting demands on available water, food production and urban development. Often there is direct competition for these resources. Specific information on the mechanisms of drought resistance strategies of crops and related species is essential as the agricultural industry continues to improve the element of sustainability in the design of elite germplasm and crop management practices. Tomato is mostly produced under irrigation. Any improvement in drought resistance for this crop will have significant effects on the cost of its production. Development of crops with decreased water requirements further decreases application of other inputs, e.g. pesticides and herbicides. Drought-resistance crops should be more competitive in their environment, less likely to be attacked by insects, and because of reduced water availablity diseases associated with humid environments will be diminished.

Molecular analysis of drought responses in tomato and related species has resulted in identification of a number of very interesting genes. The full potential of genomic analysis of drought response has yet to be accomplished. The comparative analyses between cultivated tomato and *Lycopersicon* species that have evolved to survive in arid and semi-arid environments should continue to be a productive research tool. These analyses if conducted using microarray technology are likely to give a close to complete picture of how wild tomatoes perform under drought stress conditions.

Acknowledgements

This work was supported in part by the New Mexico Agricultural Experiment Station, and grants from the NIH GM S06 GM08136, USDA CSREES 98-34387-5897. We thank Jerry Rodriguez, University of Capetown, for his discussions on histones and the phylogenetic analysis of drought-responsive H1 histones.

REFERENCES

Ascenzi, R. and J. S. Gantt. 1997. A drought-stress-inducible histone gene in *Arabidopsis thaliana* is a member of a distinct class of plant linker histone variants. Plant Mol Biol 34:629-641.
Ascenzi, R. and J. S. Gantt. 1999a. Molecular genetic analysis of the drought-inducible linker histone variant in *Arabidopsis thaliana*. Plant Mol Bio 141:159-169.
Ascenzi, R. and J. S. Gantt. 1999b. Subnuclear distribution of the entire complement of linker histone variants in *Arabidopsis thaliana*. Chromosoma 108:345-355.
Bartholomew, D. M., G. E. Bartley, and P.A. Scolnik. 1991. Abscisic acid control of *rbcS* and *cab* transcription in tomato leaves. Plant Physiol 96:291-296.
Bourgis, F. and J.-C. Kader. 1997. Lipid-transfer proteins: Tools for manipulating membrane lipids. Physiol Plant 100:78-84.
Boyer, J. S. 1982. Plant productivity and environment. Science 218:443-448.
Bray, E. A. 1988. Drought- and ABA-induced changes in polypeptide and mRNA accumulation in tomato leaves. Plant Physiol 88: 1210-1214.
Bray, E.A. 2002a. Abscisic acid regulation of gene expression during water-deficit stress in the era of the *Arabidopsis* genome. Plant Cell Environ 25:153-161.

Bray, E.A. 2002b. Classification of genes differentially expressed during water-deficit stress in *Arabidopsis thaliana*: an analysis using microarray and differential expression data. Ann Bot 89:803-811.

Bray, E.A., T.-Y. Shih, M.S. Moses, A. Cohen, R. Imai, and A.L. Plant. 1999. Water-deficit induction of a tomato H1 histone requires abscisic acid. Plant Growth Reg 29: 35-46.

Burbridge, A., T. Grieve, A. Jackson, A. Thompson, D. McCarty, and I. Taylor. 1999. Characterization of the ABA-deficient tomato mutant *notabilis* and its relationship with maize *Vp14*. Plant J 17:427-431.

Chandler, P. and M. Robertson. 1994. Gene expression regulated by abscisic acid and its relation to stress tolerance. Annu Rev Plant Physiol Plant Mol Biol 45:113-141.

Chen, R.-D. and Z. Tabaeizadeh. 1992a. Expression and molecular cloning of drought-induced genes in the wild tomato *Lycopersicon chilense*. Biochem Cell Biol 70:199-206.

Chen, R.-D. and Z. Tabaeizadeh. 1992b. Alteration of gene expression in tomato plants (*Lycopersicon esculentum*) by drought and salt stress. Genome 35:385-391.

Chen, R.-D., L.-X. Yu, A.F. Greer, H. Cheriti, and Z. Tabaeizadeh. 1994. Isolation of an osmotic stress- and abscisic acid-induced gene encoding an acidic endochitinase from *Lycopersicon chilense*. Mol Gen Genet 245:195-202.

Chen, R.-D., N. Campeau, A.F. Greer, G. Bellemare, and Z. Tabaeizadeh. 1993. Sequence of a novel abscisic acid- and drought-induced cDNA from wild tomato (*Lycopersicon chilense*). Plant Physiol 103:301.

Cohen, A. and E. A. Bray. 1990. Characterization of three mRNAs that accumulate in wilted tomato leaves in response to elevated levels of endogenous abscisic acid. Planta 182:27-33.

Cohen, A., M.S. Moses, A.L. Plant, and E.A. Bray. 1999. Multiple mechanisms control the expression of abscisic acid (ABA)-requiring genes in tomato plants exposed to soil water deficit. Plant Cell Environ 22:989-998.

Cole, R. D. 1984. A minireview of microheterogeneity in H1 histone and its possible significance. Anal Biochem 136:24-30.

Cushman, J. and H. Bohnert. 2000. Genomic approaches to plant stress tolerance. Curr Opin Plant Biol. 3:117-124.

Doczi, R., M. Kondrak, G. Kovacs, F. Beczner, and Z. Banfalvi. 2005. Conservation of the drought-inducible, DS2 genes and divergences from their ASR paralogues in solanaceous species. Plant Physiol Biochem 43:269-276.

Fellner, M. and V.K. Sawhney. 2001. Seed germination in a tomato male-sterile mutant is resistant to osmotic, salt and low-temperature stresses. Theor Appl Genet 102:215-221.

Fobes, J.F., J.B. Mudd, and M.P.F. Marsden. 1985. Epicuticular lipid accumulation on the leaves of *Lycopersicon pennellii* (Corr.) D'Arcy and *Lycopersicon esculentum* Mill. Plant Physiol 77:567-570.

Foolad, M.R. and G.Y. Lin. 1999. Relationships between cold- and salt-tolerance during seed germination in tomato: germplasm evaluation. Plant Breed 118:45-48.

Foolad, M.R., L.P. Zhang, and P. Subbiah. 2003. Genetics of drought tolerance during seed germination in tomato: inheritance and QTL mapping. Genome 46:536-545.

Frankel, N., E. Hasson, N.D. Iusem, and M.S. Rossi. 2003. Adaptive evolution of the water stress-induced gene *Asr2* in *Lycopersicon* species dwelling in arid habitats. Mol Biol Evol 20:1955-1962.

Fray, R. G., A. Wallace, D. Grierson, and G.W. Lycett. 1994. Nucleotide sequence and expression of a ripening and water stress-related cDNA from tomato with homology to the MIP class of membrane channel proteins. Plant Mol Biol 24:539-543.

Gantt, J.S. and T.R. Lenvik. 1991. *Arabidopsis thaliana* H1 histones: analysis of two members of a small gene family. Eur J Biochem 202:1029-1039.

Harrak, H., H. Chamberland, M. Plante, G. Bellemare, J.G. Lafontaine, and Z. Tabaeizadeh. 1999. A proline-, threonine-, and glycine-rich protein down-regulated by drought is localized in the cell wall of xylem elements. Plant Physiol 121:557-564.

Harrak, H., S. Azelmat, E.N. Baker, and Z. Tabaeizadeh. 2001. Isolation and characterization of a gene encoding a drought-induced cysteine protease in tomato (*Lycopersicon esculentum*). Genome 44:368-374.

Hartung, W., A. Sauter, and E. Hose. 2002. Abscisic acid in the xylem: where does it come from, where does it go to? J Exp Bot 53:27-32.

Hincha, D.K., B. Neukamm, H.A.M. Sror, F. Sieg, W. Weckwarth, M. Ruckels, V. Lullien-Pellerin, W. Schroder, and J.M. Schmitt. 2001. Cabbage cryoprotectin is a member of the nonspecific plant lipid transfer protein gene family. Plant Physiol 125:835-846.

Ho, T.-Y. and M. L. Mishkind. 1991. The influence of water deficits on mRNA levels in tomato. Plant Cell Environ 14:67-75.

Horvath, B.M., C.W.B. Bachem, L.M. Trindade, M.E.P. Oortwijn, and R.G.F. Visser. 2002. Expression analysis of a family of nsLTP genes tissue specifically expressed throughout the plant and during potato tuber life cycle. Plant Physiol 129:1494-1506.

Imai, R., M.S. Moses, and E.A. Bray. 1995. Expression of an ABA-induced gene of tomato in transgenic tobacco during periods of water deficit. J Exp Bot 46:1077-1084.

Iusem, N.D., D.M. Bartholomew, W.D. Hitz, and P.A. Scolnik. 1993. Tomato (*Lycopersicon esculentum*) transcript induced by water deficit and ripening. Plant Physiol 102:1353-1354.

Jin, S., C.C.S. Chen, and A.L. Plant. 2000. Regulation by ABA of osmotic-stress-induced changes in protein synthesis in tomato roots. Plant Cell Environ 23:51-60.

Kader, J.-C. 1996. Lipid-transfer proteins in plants. Annu Rev Plant Physiol Plant Mol Biol 47:627-654.

Kader, J.-C. 1997. Lipid-transfer proteins: a puzzling family of plant proteins. Trends Plant Sci 2:66-70.

Kahn, T. L., S. E. Fender, E.A. Bray and M.A. O'Connell. 1993. Characterization of expression of drought- and ABA-regulated tomato genes in the drought-resistant species *Lycopersicon pennellii*. Plant Physiol 103:597-605.

Kebede, H., B. Martin, J. Nienhuis, and G. King. 1994. Leaf anatomy of two *Lycopersicon* species with contrasting gas exchange properties. Crop Sci 34:108-113.

Kramer, P. and J. Boyer. 1995. Water Relations of Plants and Soils. San Diego, Academic Press.

Kristensen, A.K., J. Brunstedt, K.K. Nielsen, P. Roepstorff, and J.D. Millelsen. 2000. Characterization of a new antifungal non-specific lipid transfer protein (nsLTP) from sugar beet leaves. Plant Sci 155:31-40.

Maldonado, C., F.A. Squeo, and E. Ibacache. 2003. Phenotypic response of *Lycopersicon chilense* to water deficit. Revista Chilena Historia Natural 76:129-137.

Marin, E. and A. Marion-Poll. 1997. Tomato *flacca* mutant is impaired in ABA aldehyde oxidase and xanthine dehydrogenase activities. Plant Physiol Biochem 35:369-372.

Martin, B. and Y. Thorstenson. 1988. Stable carbon isotope composition ($?^{13}C$), water use efficiency and biomass productivity of *Lycopersicon esculentum*, *Lycopersicon pennellii* and the F_1 hybrid. Plant Physiol 88:213-217.

Martin, B., J. Nienhuis, G. King, and A. Schaefer. 1989. Restriction fragment length polymorphisms associated with water use efficiency in tomato. Science 243:1725-1728.

Maskin, L., G.E. Gudesblat, J.E. Moreno, F.O. Carrari, N. Frankel, A. Sambade, M. Rossi, and N.D. Iusem. 2001. Differential expression of the members of the *Asr* gene family in tomato (*Lycopersicon esculentum*). Plant Sci 161:739-746.

Middleton, N. and D. Thomas. 1997. World atlas of desertification. London, John Wiley & Sons, Inc.

O'Connell, M. 1995. The role of drought-responsive genes in drought resistance. AgBiotech News Info 7:143N-147N.

Plant, A.L., A. Cohen, M.S. Moses, and E.A. Bray. 1991. Nucleotide sequence and spatial expression pattern of a drought- and ABA-induced gene of tomato. Plant Physiol 97:900-906

Pyee, J., H. Yu, and P.E. Kolattukudy. 1994. Identification of a lipid transfer protein as the major protein in the surface wax of broccoli (*Brassica oleracea*) leaves. Arch Biochem Biophys 311:460-468.

Ramanjulu, S. and D. Bartels. 2002. Drought and desiccation-induced modulation of gene expression in plants. Plant Cell Environ. 25:141-151.

Rep, M., H.L. Dekker, J.H. Vossen, A.D. de Boer, P.M. Houterman, C.G. de Koster, and B.J.C. Cornelissen. 2003. A tomato xylem sap protein represents a new family of small cysteine-rich proteins with structural similarity to lipid transfer proteins. FEBS Lett 534:82-86.

Reymond, P., H. Weber, M. Damond, and E.E. Farmer. 2000. Differential gene expression in response to mechanical wounding and insect feeding in *Arabidopsis*. Plant Cell 12:707-720.

Rick, C. M. 1973. Potential genetic resources in tomato species: clues from observation in native habitats. In: A. M. Srb (ed.), Genes, Enzymes and Populations. Plenum Press, New York, pp. 255-269.

Rock, C. 2000. Pathways to abscisic acid-regulated gene expression. New Phytol 148:357-396.

Sagi, M., R. Fluhr, and S.H. Lips. 1999. Aldehyde oxidase and xanthine dehydrogenase in a *flacca* tomato mutant with deficient abscisic acid and wilty phenotype. Plant Physiol 120:571-577.

Sánchez Peña, P., S. Fender, and M.A. O'Connell. 1995. Leaf water relations of *Lycopersicon chilense* during a drought cycle. Tomato Genet Rep 45:40-41.

Scippa, G.S., A. Griffiths, D. Chiantante, and E.A. Bray. 2000. The H1 histone variant of tomato, H1-S, is targeted to the nucleus and accumulates in chromatin in response to water deficit stress. Planta 211:173-181.

Scippa, G.S., M. Di Michele, E. Onelli, G. Patrignani, D. Chiatante, and E.A. Bray. 2004. The histone-like protein H1-S and the response of tomato leaves to water deficit. J Exp Bot 55:99-109.

Seki, M., M. Narusaka, H. Abe, M. Kasuga, K. Yamaguchi-Shinozaki, P. Carninci, Y. Hayashizaki, and K. Shinozaki. 2001. Monitoring the expression pattern of 1300 *Arabidopsis* genes under drought and cold stresses by using a full-length cDNA microarray. Plant Cell 13:61-72.

Skriver, K. and J. Mundy. 1990. Gene expression in response to abscisic acid and osmotic stress. Plant Cell 2:503-512.

Smart, L.B., K.D. Cameron, and A.B. Bennett. 2000. Isolation of genes predominantly expressed in guard cells and epidermal cells of *Nicotiana glauca*. Plant Mol Biol 42:857-869.

Sohal, A., J. Pallas, and G.I. Jenkins. 1999. The promoter of a *Brassica napus* lipid transfer protein gene is active in a range of tissues and stimulated by light and viral infection in transgenic *Arabidopsis*. Plant Mol Biol 41:75-87.

Srinivasa Rao, N.K., R.M. Bhatt, and A.T. Sadashiva. 2000. Tolerance to water stress in tomato cultivars. Photosyn 38: 465-467.

Sterk, P., H. Booij, G.A. Schellenkens, A. Van Kammen, and S.C. de Vries. 1991. Cell-specific expression of the carrot EP2 lipid transfer protein gene. Plant Cell 3:907-921.

Tabaeizadeh, Z. 1998. Drought-induced responses in plant cells. Intl Rev Cytol 182:193-247.

Taylor, I.B., A. Burbridge, and A.J. Thompson. 2000. Control of abscisic acid synthesis. J Exp Bot 51:1563-1574.

Thoma, S., U. Hecht, A. Kippers, J. Botella, S.C. de Vries, and C. Somerville. 1994. Tissue-specific expression of a gene encoding a cell wall-localized lipid transfer protein from *Arabidopsis*. Plant Physiol 105:35-45.

Thompson, A.J. and J. E. Corlett. 1995. mRNA levels of four tomato (*Lycopersicon esculentum* Mill. L.) genes are related to fluctuating plant and soil water status. Plant Cell Environ 18:773-780.

Thompson, A.J., A.C. Jackson, R.A. Parker. D.R. Morpeth, A. Burbridge, and I.B. Taylor. 2000. Abscisic acid biosynthesis in tomato: regulation of zeaxanthin epoxidase and 9-*cis*-epoxycarotenoid dioxygenase mRNAs by light/dark cycles, water stress and abscisic acid. Plant Mol Biol 42:833-845.

Thompson, A.J., E.T. Thorne, A. Burbidge, A.C. Jackson, R.E. Sharp, and I.B. Taylor. 2004. Complementation of *notabilis*, an abscisic acid-deficient mutant of tomato: Importance of sequence context and utility of partial complementation. Plant Cell Environ 27:459-471.

Torres-Schumann, S., J. A. Godoy, and J.A. Pintor-Toro. 1992. A probable lipid transfer protein gene is induced by NaCl in stems of tomato plants. Plant Mol Biol 18:749-757.

Treviño, M.B. 1997. Drought-induced gene expression in plants: analysis of the non-specific lipid transfer protein gene family in the drought-tolerant tomato species *Lycopersicon pennellii*. PhD Dissertation, New Mexico State University.

Treviño, M.B. and M.A. O'Connell. 1998. Three drought-responsive members of the nonspecific lipid-transfer protein gene family in *Lycopersicon pennellii* show different developmental patterns of expression. Plant Physiol 116:1461-1468.

Wei, T. and M.A. O'Connell. 1996. Structure and characterization of a putative drought-inducible H1 histone gene. Plant Mol Biol 30:255-268.

Wells, D. and D. Brown. 1991. Histone and histone gene compilation and alignment update. Nucl Acids Res 19:2173-2188.

Wilkinson, S. and W.J. Davies. 1997. Xylem sap pH increase: a drought signal received at the apoplastic face of the guard cell that involves the suppression of saturable abscisic acid uptake by the epidermal symplast. Plant Physiol 113:559-573.

Wolffe, A. P. and D. Pruss. 1996. Deviant nucleosomes: the functional specialization of chromatin. Trends in Genet 12:58-62.

Wolffe, A.P. and H. Kurumizaka. 1998. The nucleosome: a powerful regulator of transcription. Prog Nucl Acid Res 61:379-422.

Xiong, L. and J-K. Zhu. 2002. Molecular and genetic aspects of plant responses to osmotic stress. Plant Cell Environ 25:131-139.

Yu, L.-X., H. Chamberland, J.G. Lafontaine, and Z. Tabaeizadeh. 1996. Negative regulation of gene expression of a novel proline-, threonine-, and glycine-rich protein by water stress in *Lycopersicon chilense*. Genome 39:1185-1193.

9

Applications of Genetic Engineering in Tomato

**M.V. Rajam[1]*, P. Madhulatha[1], R. Pandey[1],
P.J. Hazarika[1] and M.K. Razdan[2]****

[1]*Department of Genetics, University of Delhi – South Campus, Benito Juarez Road, New Delhi 110021, India;* [2]*Department of Botany, Ramjas College, University of Delhi (Main Campus), Delhi 110007, India*

INTRODUCTION

Tomato (*Lycopersicon esculentum* Miller, recently also called *Solanum lycopersicum*) is one of the most important vegetable crop all over the world and constitutes an important source of human diet. It is grown in practically every country in outdoor fields, green-houses, and net-houses, for fresh market and processing purposes. It ranks second (after potato) among all the vegetables and its worldwide production is approximately 97 million tons (FAO production year book 2001). Tomato belongs to Solanaceae family and its origin is from South America and Mexico. The wild species of tomato are: *L. pimpinellifolium, L. cheesmanii, L. peruvianum, L. chilense, L. hirsutum, L. parviflorum, L. chmielewskii* and *L. pennellii*. They are all diploids with 24 chromosomes (2n=24).

Tomatoes are a good source of many nutrients. They are rich in minerals, fibers, vitamins (A and C), lycopene pigment (a powerful antioxidant) and health acids. Although a variety of tomato varieties are cultivated worldwide under warm seasons, countries such as USA, several European countries, Japan and China are the most important tomato growing countries (Kalloo 1991).

Apart from being a major vegetable crop, tomato also serves as a model plant for studying fruit development and this is because of the availability of valuable germplasm, monogenic mutants (genomic resources of tomato such as inbred lines and mutagenized populations), enzymatic (isozyme)/

---

*Corresponding author: M.V. Rajam; Email: mv_rajam@hotmail.com
**Present Address : Principal, Shyam Lal College (University of Delhi), Delhi - 110032, India

molecular markers, and expressed sequence tags (EST) that are associated
with many useful traits, including disease resistance, plant architecture,
male sterility, fruit development, colour and ripening, and nutritional quality
(Kalloo 1991). Moreover, tomato has modest-sized diploid genome making
it amenable for tissue culture and genetic transformation studies (Bhatia et
al. 2004). Further, development of resources such as a full-length cDNA
clones will expand the potential usefulness of tomato for use in genetic
and functional genomic approaches (Taneaki et al. 2005).

Tomato is highly susceptible to various pathogens (fungi, bacteria and
viruses) and pests (Table 9.1) as well as environmental stresses (e.g. chilling,
frost, heat, excessive moisture, drought and salinity), and these stresses
cause a colossal loss of tomato yield and quality (Kalloo 1991, Bhatia et al.
2004). Therefore, there is need to reduce such losses. This objective can be
achieved by various ways: i) increasing the present area under tomato
cultivation, ii) adopting better agricultural techniques, iii) developing the
improved varieties by breeding methods, and iv) producing transgenic
tomatoes with improved characteristics by deploying native and
heterologous genes. The first two approaches already being exploited
strategies, yet there is much scope of tomato improvement using these
strategies. Although plant breeding has contributed for improvement of
crop plants, including tomato, it has certain limitations regarding
improvement of this fruit crop because of the availability of limited gene

Table 9.1   Some of the important diseases and pests of tomato.

Disease	Pathogen/pest
Pathogens	
Fusarium wilt	*Fusarium oxysporum* f.sp. *lycopersici*
Verticillium wilt	*Verticillium alboatrum*
Damping-off	*Pythium aphanidermatum*
Early blight	*Alternaria solani*
Late blight	*Phytophthora infestans*
Leaf mould	*Cladosporium fulvum*
Anthracnose	*Colletotrichum coccodes*
Fruit rot	*Phytophthora* spp.
Bacterial wilt	*Pseudomonas solanacearum*
Bacterial stem and fruit canker	*Corynebacterium michiganense*
Bacterial spot	*Xanthomonas campestris*
Mosaic	Tobacco mosaic virus (TMV)
Leaf curl	Tomato leaf curl virus
Bid-bud	*Mycoplasma*
Pests	
Gram caterpillar	*Heliothis armigera*
Tobacco caterpillar	*Spodoptera litura*
Leaf eating beetles	*Epilachna* spp.
Mites	*Tetranychus cucurbitae*
Root-knot nematode	*Meloidogyne* spp.

pool (i.e. transfer of traits only from closely related tomato species). Furthermore, gene transfer is through cluster of genes, which includes the gene of interest (i.e. gene transfer is not precise), and takes more time (usually 10-15 years) to develop a new variety. Thus, genetic engineering has gained relevance for crop improvement as this strategy can overcome the underlined problems.

Biotechnological approaches have in recent added a new impetus to tomato improvement programs, and good progress has been made in this direction. Efficient plant regeneration protocols now available for tomato have helped in developing transgenic tomatoes with new traits. This chapter reviews aspects of biotechnology and its applications in tomato.

PLANT REGENERATION

Efficient *in vitro* plant regeneration protocol is necessary for raising transgenic crops with useful traits. As far as tomato is concerned, a good deal of tissue culture work has been done to regenerate plants from *in vitro* culture systems. Tomato is quite amenable and responsive to *in vitro* regeneration (Fari et al. 1992, Izadpanah and Khosh Khui 1992), generation of somaclonal variation, development of haploids (Gresshoff and Doy 1972, Zagorska et al. 1982, 1998, Chlyah and Taarji 1984, Shtereva et al. 1998), selection for biotic and abiotic stresses (Toyoda et al. 1984, 1985, 1989, Rahman and Kaul 1989), distant hybridization (Sink et al. 1986, Wijbrandi et al. 1988) and an efficient genetic transformation. Various cultivars and lines for resistance to various pathogens and improved quality have been released using *in vitro* anther culture techniques (Foroughi-Wehr and Friedt 1984, Friedt et al. 1986).

Researchers have used various types of explant sources viz. cotyledon (Schutze and Wieczorrek 1987), hypocotyl (Plastira and Perdikaris 1997, Gunay and Rao 1980), pedicel/peduncle (Compton and Veilleux 1991), leaf (Duzyaman et al. 1994), stem sections and inflorescence (Applewhite et al. 1994). Most tissues of tomato seem to have high totipotency, however, the choice of the right explant may vary with the genotype.

Plant regeneration and micropropagation of tomato has been attempted through the use of various techniques, such as shoot tip culture (Novak and Maskova 1979, Padliskikh and Yarmishin 1990, Fari et al. 1992, Izadpanah and Khosh Khui 1992), somatic embryogenesis (Chen and Adachi 1994, Gill et al. 1995, Kaparakis and Alderson 2002), direct organogenesis from intact explants (El-Farash et al. 1993, Davis et al. 1994, Duzyaman et al. 1994, Ichimura and Oda 1998) or protoplast culture (Muhlbach 1980, Sink et al. 1986). Shoot-tip culture (meristem with leaf primordia) has been used as a model system to study tomato shoot development and it was noted that many factors including gibberellic acid, coconut water and kinetin (Hussey 1971). Shoot regeneration from tomato cotyledon explants were also obtained on several kinds of supporting material from polyester, ceramics, wood pulp and cotton fibres (Kazua Ichimura et al. 1995). In tomato, adventitious shoot regeneration can be achieved either directly (Dwivedi et al. 1990) or indirectly through an intermediate callus phase (Behki and Lesley 1980, Geetha et al. 1998). Indeed, both callus and shoots may be produced together (Bhatia 2004). A simple and efficient organogenetic mechanism of shoot regeneration via seedling decapitation method (SDM) has been reported for tomato (Fari et al. 1991). Tomato root culture dates back to 1934, when White successfully cultured them *in vitro*.

The success in tomato regeneration response has been found to depend largely on the genotype, explant carbon source, and plant growth regulators (PGRs) used in the culture medium (El-Farash et al. 1993). Compared to cultivated tomato, its wild counterparts such as *L. pimpinellifolium*, *L. peruvianum* and *L. glandulosum* show better regeneration capabilities (Lech et al. 1996). Successful attempts have been made to transfer the superior regeneration capacity of *L. peruvianum* into cultivated tomato through backcrossing (Koorneef et al. 1993). Furthermore, *L. pimpinellifolium* has been successfully used as a donor parent to introgress *in vitro* regeneration capacity in recalcitrant tomato cultivars (Faria et al. 2002). Other unconventional breeding procedures such as protoplast fusion have also been attempted to transfer regeneration capability from *L. peruvianum* to *L. esculentum* (Wijbrandi et al. 1988, 1990). Amongst the *Lycopersicon* species, *L. peruvianum* is considered to be highly organogenic species as it regenerates shoots from the roots, whereas shoot induction from roots is comparatively low in *L. esculentum* (Peres et al. 2001).

Most researchers avoid using multifarious nutrient media for tomato tissue culture. Preferably either MS or modified MS medium is used (Kartha et al. 1976, Compton and Veilleux 1988, 1991, Chandel and Katiyar 2000, Park et al. 2001). Sucrose is almost universally used for regeneration and micropropagation purposes as carbon source, as it is readily utilisable by cells. However, others have also tried glucose, maltose, ribose, palatinose, and furanose (Locy 1995, El-Bakry 2002, Bhatia et al. 2004). It is notable that maltose followed by glucose has been found to be a better source than sucrose (El-Bakry 2002). For regeneration, a wide variety of PGRs have been used. The concentration and combination of growth regulators employed is dependent on the cultivar being cultured and the particular cytokinin or auxin being employed. Four major cytokinins, viz. zeatin, N-isopentenylamino purine (2-iP), benzylaminopurine (BAP), and kinetin, can be used either separately or in combination with auxins for organogenesis. Vnuchkova (1977 a, b) examined 150 different media and concluded that combinations of kinetin and indole-3-acetic acid (IAA) are the most suitable for meristem formation in tomato explants. In later studies, a combination of IAA-BAP was found to be superior to IAA-kinetin for shoot regeneration (Gunay and Rao 1980). BAP or zeatin alone induced shoot formation from leaf callus and these were found to be superior to kinetin (Kartha et al. 1976). Pulse treatment with cytokinin was also tried and found to be not beneficial for shoot production (Villiers et al. 1993). For root induction, tomato does not seem to require any exogenous PGR because of a high endogenous auxins level in tomato explants (De Langhe and De Bruijne 1976, Mensuali-Sodi et al. 1995).

In order to develop a very efficient and reliable procedure for tomato regeneration, Madhulatha et al. (2006a) found that MS medium amended with 2.5 mg/l BAP and 0.5 mg/l IAA, along with 3% maltose (in place of 3% sucrose) and 0.5 mM diamine putrescine (or 0.1 mM triamine spermidine) greatly enhanced shoot regeneration in cotyledonary explants of tomato.

GENETIC ENGINEERING

The most widely used method for transferring genes into tomato plants is *Agrobacterium*-mediated transformation, except a recent report on chloroplast transformation using particle bombardment (Ruf et al. 2001). The later report demonstrated efficient transgene expression in both chloroplasts of green leaves as well as in chromoplasts of tomato fruits, demonstrating the potential of this system for biotechnological applications. In tomato transformation, reporter and marker genes have been used and these are summarized in Table 9.2.

Since 1986, a number of reports have been published describing the use of *Agrobacterium tumefaciens*-mediated transformation and regeneration of different tomato cultivars (McCormick et al. 1986, Fillati et al. 1987, Chyi et al. 1987). In most cases, neomycin phosphotransferase (*NPTII*) has been used as plant selection marker and ß-glucuronidase (*GUS*) as a reporter gene. Transformation of tomato shows widely variable rates of success, depending on different factors like plant variety (genotype) (Koorneef et al. 1986, Ultzen et al. 1995, Ling et al. 1998, Ellul et al. 2003) explant type (Fillati et al. 1987, Ohki et al. 1978, McCormick et al. 1986, Bird et al. 1988), explant size (Shanin et al. 1986, Beck and Camper 1991, Davis et al. 1991, Van Roekel et al. 1993, Frary and Earle 1996), explant orientation (Frary and Earle 1996) and the age of explant (Patil et al. 2002, Hamza and Chupeau 1993), plant growth regulators (PGRs) (Yakuwa et al. 1973, Ohki et al. 1978, McCormick et al. 1986, Pfitzner 1998), carbon source (Madhulatha et al. 2006b), bacterial concentration (Shanin et al. 1986) and *Agrobacterium vir* gene inducers (Garfinkel and Nester (1980), Bolton et al. 1986, Stachel et al. 1986a,b).

Table 9.2 Genetic transformation of tomato using reporter and marker genes

Cultivar	Reporter gene	Marker gene	Selection agent (Concentration)	Gene transfer method	References
Marglobe, Rutgers Red cherry, UC-82, Improved Pearson, Heinz 2152 & ONT 7710	–	NPT II	Kanamycin 100mg/l	Agrobacterium tumefaciens (GV3111, A208)	McCormick et al. (1986)
	–	NPT II	Kanamycin 50 mg/l	Agrobacterium rhizogenes (LBA4404)	Shanin et al. (1986)
UC82B	GUS	NPT II	Kanamycin 100 mg/l	Agrobacterium tumefaciens (–)	Hamza and Chupeau (1993)
Moneymaker	GUS	NPT II	Kanamycin 30 mg/l	Agrobacterium tumefaciens (MOG101, MOG301, EHA105)	Van Roekel et al. (1993)
Ailsa Craig	–	NPT II	Kanamycin 50 mg/l	Agrobacterium tumefaciens (C58C1Rif: pGSFR1161)	Lipp Jao and Brown (1993)
Moneymaker WC1, H2274 & SC2121	GUS	NPT II	Kanamycin 75 mg/l	Agrobacterium tumefaciens (LBA4404)	Frary and Earle (1996)
Moneymaker	GUS	NPT II	Kanamycin 50 mg/l	Agrobacterium tumefaciens (EHA105)	Oktem (1998)
	–	NPT II	Kanamycin 30 mg/l	Agrobacterium tumefaciens (LBA4404)	Ling et al. (1998)
IAC-Santa Clara	–	Aad A	Spectinomycin 300-500 mg/l	Particle bombardment	Ruf et al. (2001)
Pusa Ruby	–	NPT II	Kanamycin 50-100 mg/l	Agrobacterium tumefaciens (–)	Patil et al. (2002)
	–	NPT II	Kanamycin 30 mg/l		Madhulatha et al. (2006a)
UC82B	–	NPT II	Kanamycin 100 mg/l	Agrobacterium tumefaciens (LBA4404)	Cortina and Culianez-Macia (2004)
Pixie	GUS	NPT II	Kanamycin 100 mg/l	Agrobacterium tumefaciens (EHA105)	Moon and Callahan (2004)

Tomato Cultivars that have been extensively used in transformation studies via *Agrobacterium* are "UC 82b" (well known for its regenerating capacity: McCormick et al. 1986, Fillati et al. 1987, Hamza and Chupeau 1993), "Moneymaker" (Van Roekel et al. 1993, Frary and Earle 1996, Ling et al. 1998) and the cv "Aisla Craig" (Bird et al. 1988, Joao and Brown 1993). All tomato varieties, known for their transformation competence, however, lack tolerance to TYLCV (Tomato Yellow Leaf Curl Virus). Since TYLCV attack on transgenic plants has become a severe problem for the transgenic work done in Israel, several TYLCV-tolerant tomato lines were tested for their transformation competence. In this regard, MP-1, which is named as "Ady", has been found to be highly amenable to transformation compared with the commonly utilized tomato cultivars (Barg et al. 1997).

The transformation frequency significantly varies with type of explants. Different explant tissues used for transformation studies include cotyledons (McCormick et al. 1986, Fillati et al. 1987, Hamza and Chupeau 1993, Pfitzner 1998, Oktem et al. 1999), hypocotyls (Ohki et al. 1978), stem (Chyi and Phillips 1987, Bird et al. 1988) and leaf segments (McCormick et al. 1986, Davis and Miller 1991).

Regarding the choice of PGRs, IAA is generally used for tomato regeneration, but cytokinin choice is still unresolved (Ohki et al. 1978, McCormick et al. 1986, Fillati et al. 1987, Hamza and Chupeau 1993). However, a study on shoot differentiation medium with the combination of IAA and cytokinins - BAP, 2-iP and kinetin, showed that 2-iP gave best results on shoot formation from transformed hypocotyls and leaf segments (Ohki et al. 1978). Shoot formation on hypocotyl segments of tomato was observed in the presence of naphthaleneacetic acid (NAA) combined with BAP (Yakuwa et al. 1973). Improved results on transformed explant source have also been reported using zeatin (McCormick et al. 1986, Fillati et al. 1987, Hamza and Chupeau 1993) or zeatin riboside (Pfitzner 1998).

In vitro morphogenetic response of transformed plant tissues are also affected by the different components of the culture media and it is important to evaluate their effects on plant regeneration. MS medium (Murashige and Skoog 1962) is the commonly used medium in plant regeneration. However, a high percentage of tomato cotyledon explants have been found to develop damaging necrosis in it. Increasing the vitamin (thiamine) concentration in the MS medium decreased the expansion of necrotic lesions and promoted cell growth (Cortina and Culianez-Macia 2004). As previously mentioned (Madhulatha et al. 2006a) conditions have been optimized for an efficient *Agrobacterium*-mediated transformation for cotyledonary explants of tomato and were able to minimize the necrotic problem as well as recover more number of transformants on the improved shoot regeneration medium amended with 3% maltose and 0.1 or 0.5 mM polyamines (putrescine and spermidine); this has led to the increased transformation frequency (Madhulatha et al. 2006b).

The most critical factors affecting the regeneration of the number of transformed explants recovered following co-cultivation were, the pre-incubation period, bacterial density and co-cultivation time. The decrease in the transformation rate at high bacterial density and longer than optimal co-cultivation period has been reported which probably may be due to bacterial-induced stress and not due to a decrease in the virulence of *Agrobacterium tumefaciens*. This stress threshold may be related to the growth rate of bacteria, the initial inoculum of the bacteria per explant, and the sensitivity of the tissue used. This reflects that co-cultivation conditions should be optimized for each bacterial strain according to the plant species (Fillati et al. 1987).

Acetosyringone, a naturally occurring *vir*-inducing phenolic compound, improves the transformation efficiency in plants, which naturally produce insufficient amounts of *vir*-inducing compounds (Sheikholeslam and Weeks 1987, Joao and Brown 1993). The enhancement of the transformation frequency has been observed when acetosyringone is added after the pre-incubation treatment. This may be due to the accumulation of substances that induce the *vir* genes or due to an increase in the rate of explant cell division (Fillati et al. 1987). Alternatively, some authors have also used feeder cells from species like tobacco, as phenolic source to increase transformation rate (McCormick et al. 1986, Fillati et al. 1987, Hamza and Chupeau 1993).

For successful transformation using *Agrobacterium*, effective elimination of bacteria from the culture is necessary as soon as their presence is no longer required. Carbenicillin, cefotaxime and augmentin are extensively used antibiotics for this purpose. An ideal antibiotic for inhibiting *Agrobacterium* species should be highly effective, inexpensive, without a negative effect on plant growth and regeneration (Cheng et al. 1998). Among the antibiotics, timentin, which has recently been made available, shows a marked efficiency for eliminating *Agrobacterium* and also has stimulatory effects on organogenesis. Timentin has been used for eliminating *Agrobacterium* after explant transformation (Frary and Earle 1986, Schroeder et al. 1993, Frary 1995, Zimmerman 1995, Ling et al. 1998), but its possible effects on regeneration have not been described.

A liquid culture system for *Agrobacterium*-mediated transformation of tomato has been developed and found to be better for transformation, shoot regeneration and selection of transformed tomato plants as compared to solid media, because of the better distribution of the selective agent in the liquid cultures (Velcheva et al. 2005).

Tomato transformation with *Agrobacterium rhizogenes* has also been undertaken. This system could be used to produce transgenic tomato plants expressing the *rol* genes as well as the genes present in *A. rhizogenes*-based binary vectors. Hairy roots have the potential to produce useful materials such as enzymes (Kato et al. 1991, Uozumi et al. 1991) and secondary metabolites (Flores et al. 1987). To date plant regeneration from hairy roots has been reported for several plant species (Tepfer 1984, Ooms et al. 1985, Lambert and Tepfer 1991, Uozumi et al. 1996). In addition, the use of *A. rhizogenes* offers the opportunity to introduce foreign genes into plant genomes when the hairy root is induced, enabling the alteration of a given plant's properties by genetic manipulation (Hamill et al. 1987).

Genetic manipulation via *Agrobacterium* has helped in increasing the plant productivity and also the nutritional content of food, which was not an easy task with traditional breeding approaches. This approach has utilized strategies applied in genetic engineering including antisense RNA technology (Stone et al. 1994), over-expression of genes (Mehta et al. 2002) and gene pyramiding (Mandaokar et al. 1999) to increase the shelf life, improve the fruit quality and nutrients, and also to develop value-added traits in tomato (Table 9.3).

Antisense RNA technology has been used successfully to manipulate the expression of several tomato ripening and softening associated genes – polygalacturonase (PG) (Sheehy et al. 1988, Smith et al. 1988, 1990; Watson et al. 1994), pectin methylesterase (PME) (Tieman et al. 1992, Hall et al. 1993), 1-aminocyclopropane-1-carboxylic acid (ACC) synthase (Oeller et al. 1991), ACC oxidase (Hamilton et al. 1990), sucrose synthase (SuSy) (Marc-Andre 'D' Aoust et al. 1999), non-ripening gene in Nr mutant (Hackett et al. 2000), ß-galactosidase (TBG4) (Smith et al. 2002), Rab11GTPase (Lu et al. 2001), phospholipase D (PLD) (Pinhero et al. 2003) and phytoene synthase (Bird et al. 1991). Over-expression of various genes has also been attempted for improvement of fruit quality, expansin (Le Exp1) to increase shelf life (Brummell et al. 2002), linalool synthase (LIS) (Lewinsohn et al. 2001) and carotenoid cleavage dioxygenase (LeCCD1) (Simkin et al. 2004) to enhance flavour and aroma. To enhance the carotenoid content in tomato fruit, transgenic lines were developed by introducing bacterial carotenoid gene (*crt1*) encoding the enzyme phytoene desaturase, which converts phytoene to lycopene (Romer et al. 2000) and *crt B* gene (Fraser et al. 2002).

Fruit taste is also an important component of fruit quality. Several genes have been introduced into tomato to improve the fruit characteristics of tomato. Petunia chalcone isomerase (*Petunia chi-a*) gene was introduced into tomato and the resulting transgenic tomato lines produced increased amounts of flavonols (up to 78 fold) in fruit peel, mainly due to an

Table 9.3 Transgenic tomatoes engineered with agronomically important genes

Cultivar	Gene	Gene Source	Marker Gene	Trait introduced	Reference
UC 82b	aro A	Salmonella typhimurium	NPTII	Glyphosate-Herbicide tolerance	Fillatti et al. (1987)
Rutgers	PME	L. esculentum	NPTII	Increased soluble solid content	Tieman et al. (1992)
Ailsa Craig	PG (antisense)	L. esculentum	NPTII	Flavr Savr (increased lycopene content & shelf-life)	Watson et al. (1994)
Vollendung	Vst1 & vst2	Vitis vinifera	NPTII	Phytophthora infestans Resistance	Thomzik et al. (1997)
UC82B	HAL2	Saccharomyces cereviciae	NPTII and uid A	Salt tolerance	Arillaga et al. (1998)
CM, L.276	iaaM	Pseudomonas syringae	NPTII	Seedless fruits (parthenocarpic)	Ficcadenti et al. (1999)
Summerset	SuSy (antisense)	L. esculentum	NPTII	Reduced fruit setting and development	Marc-Andre'D'Aoust et al. (1999)
P73	HAL 1	Saccharomyces cereviciae	NPTII	Salt tolerance	Gisbert et al. (2000)
Ailsa Craig	NR (antisense)	L. esculentum	NPTII	Restored ripening pattern in NR mutant	Hackett et al. (2000)
Pusa Ruby	Cry1Ac	Bacillus thuringiensis	NPTII	Resistant to fruit borer	Mandaokar et al. (2000)
Ailsa Craig	Crt 1	Erwinia uredovora	NPTII	Increased levels of provitamin A	Romer et al. (2000)
Heinz 902, Heinz 1439	ACC deaminase	Enterobacter cloacae UW4	NPTII	Flood tolerance	Grichko and Glick (2001)
UC 82b, CB3	LIS	Clarkia breweri	NPTII	Enhanced levels of aroma and flavour	Lewinsohn et al. (2001)
Ailsa Craig and mutants Alcobacca and Never- ripe	Rab11GTPase (antisense)	L. esculentum	NPTII	Reduced fruit softening	Lu et al. (2001)

(Contd.)

(Contd.)

		PCR amplified product	NPTII		
FM62003 (unilever commercial variety)	chi		NPTII	Increased levels of flavonol	Muir et al. (2001)
Ailsa Craig	Le Exp 1	L. esculentum	NPTII	Increased shelf-life	Brummell et al. (2002)
Ailsa Craig	crt B	E. uredovora	NPTII	Increased carotenoid content	Fraser et al. (2002)
CL5915-93 D₄-1-0-3	CBF1	Arabidopsis thaliana	NPTII	Tolerance to water deficit stress	Hsieh et al. (2002)
Bailichun Asc/Asc,	BADH	Atriplex hortensis	NPTII	Salt tolerance	Jia et al. (2002)
VFNT cherry	p35	Baculovirus	NPTII	Broad-spectrum disease resistance	Lincoln et al. (2002)
L. esculentum Mill	Samdc	S. cerevisiae	NPTII	Increased phytonutrient, carotene in fruit	Mehta et al. (2002)
Rutgers	TBG4 (antisense)	L. esculentum	NPTII	Decreased fruit softening	Smith et al. (2002)
—	Thaumatin	Thaumatococcus. daniellie Benth	NPTII	Improved fruit taste	Bartoszewski et al. (2003)
Minitomato	gdh A	Aspergillus nidulans	NPTII	Improved fruit taste (increased glutamate levels)	Kisaka and Kida (2003)
LEPA	CrylAb	B. thuringiensis	NPTII	Resistant to fruit borer	Kumar and Kumar (2004)
L. 276-76, L.149-88, INB777	Nucleoprotein gene	Tomato Spotted Wilt Virus	NPTII	Resistant to TSWV	Nervo et al. (2003)
Microtom, Celebrity	PLD (antisense)	A. thaliana	NPTII	Increased fruit firmness and red colour	Pinhero et al. (2003)
Ventura	TBI-HBS	Hepatitis B Virus	NPTII	Edible vaccine against HBV & HIV	Shchelkunov et al. (2004)
M82	LeCCD1	L. esculentum	NPTII	Improved flavour and aroma	Simkin et al. (2004)
	CtrB	Vibrio cholerae	NPTII	Edible Vaccines	Tyagi et al. (2002)
M82	LeCCD1	L. esculentum	NPTII	Improved flavour and aroma	Simkin et al. (2004)
Moneymaker	PI-II and PCI	S. tuberosum	NPTII	Multiple insect resistance	Abdeen et al. (2005)

(Contd.)

(Contd.)

	Gene	Source	Marker	Trait	Reference
—	*Ep5c* (antisense)	*Pseudomonas syringae*	—	Resistant to P.s tomato	Alberto et al. (2005)
Hezuo 906	*ACCO* (double stranded RNA)	PCR amplified product	*modified NPTII*	Prolonged shelf life	Xiong et al. (2005)
Pusa Ruby	*CP*	Tomato leaf Curl Virus (TLCV)	*NPTII*	Resistance against TLCV	Raj et al. (2005)
Moneymaker	*spike*	SARS-corona virus	*NPTII*	Recombinant vaccine	Pogrebnyak et al. (2005)
UC82B	*TPS1*	*S. cereviciae*	*NPTII*	Increased tolerance of abiotic stress	Cortina et al. (2005)
L. esculentum	*APX*	*Pisum sativum*	*NPTII*	Tolerance to chilling an d salt stress	Wang et al. (2005)
L. esculentum	*TPS1*	*S. cereviciae*	*NPTII*	Increased tolerance to abiotic stress	Cortina et al. (2005)
L. esculentum	*CAX*	*A. thaliana*	—	Increased shelf life and Ca^{+2} levels	Park et al. (2005)

accumulation of rutin. No gross phenotypical differences were observed between high-flavonol transgenic and control lines. The phenotype segregated with the transgene and demonstrated a stable inheritance pattern over four subsequent generations. Whole-fruit flavonol levels in the best of these lines are similar to those found in onions, a crop with naturally high levels of flavonol compounds. Processing of high-flavonol tomatoes demonstrated that 65% of flavonols present in the fresh fruit were retained in the processed paste, supporting their potential as raw materials (Shelagh et al. 2001). A double antisense construct containing ACC oxidase and ACC synthase fusion gene with *NPTII* as a marker gene was introduced by means of *Agrobacterium*-mediated transformation for longer shelf life of fruits. The transgenic lines showed remarkable delay in fruit ripening and increased shelf life (Xiong et al. 2005).

Fruits from tomato plants expressing Arabidopsis H^+/cation exchangers (*CAX*) demonstrate modest increase in Ca $^{2+}$ levels and prolonged shelf life but no deterious effects on plant growth. These findings suggest that *CAX* expression may fortify plants with Ca^{2+} and may serve as an alternative to the application of $CaCl_2$ used to extend the shelf life of numerous agriculturally important commodities (Park et al. 2005).

A gene for thaumatin (sweet tasting, flavour enhancing protein) has been introduced into tomato to improve the sweetness of fruit (Bartoszewski et al. 2003). Free amino acids are essential nonvolatile compounds involved in the overall taste of many foods and glutamate, in particular, contributes to the taste of tomatoes. (Fuke and Konosu 1991). The enzymes involved in the synthesis of amino acids during ripening of tomato fruits have been investigated. The over-expression of the *gdh A* gene from *Aspergillus nidulans*, which encodes NADP-GDH (Hawkins et al. 1989, Kisaka and Kida 2003), can modulate nitrogen metabolism in tomato, with resultant increase in levels of some free amino acids, in particular, glutamate in fruits.

RNA interference (RNAi), a new technology has also been applied for the improvement of tomato fruit characteristics. PCR amplified ACC oxidase double stranded RNA with modified *NPTII*, which contained the catalase intron in the coding sequence, as a marker gene was used to transform tomato. The resultant transgenic lines showed remarkable delay in fruit ripening and increased shelf life (Xiong et al. 2005). A recent study on fruit-specific RNAi-mediated suppression of *DET1* (DE-ETIOLATED1), a regulatory gene, which represses several signaling pathways controlled by light, enhanced carotenoid and flavonoid content in tomato (Ganga Rao et al. 2005).

Some of the genes in tomato have been cloned successfully using targeted transposan tagging (Jones et al. 1994, Bishop et al. 1996, Keddie et al. 1996, Van der Biezen et al. 1996, Takken et al. 1998). A reverse genetics approach was also developed using Ds element insertions creating a collection of 2932 families of a miniature tomato Micro-Tom (Meissner et al. 2000). An efficient reverse genetic approach to assess gene function is however still lacking.

Virus-induced gene silencing (VIGS) offers an attractive and alternative way to knockout the expression of genes without the need of genetically transforming the plants. In this method, recombinant virus carrying a partial sequence of a host gene is used to infect the plant. When the virus spreads systematically, the endogenous gene transcripts, which are homologous to the insert gene in the viral vector, are degraded by post-transcriptional gene silencing (PTGS) (Baulcombe 1999). Using this system, several genes like *PDS* (phytoene desaturase), *tCTR1* and *tCTR2* (constitutive triple response 1and 2) have been suppressed in tomato. Several reporter genes related to ethylene responses and fruit ripening, including LeCTR1 and LeEILs genes, were also successfully silenced by this method during fruit development. In addition, silencing of LeEIN2 gene resulted in the suppression of tomato fruit ripening (Fu et al. 2005). This system will facilitate large-scale functional anlysis of tomato ESTs (expressed sequence tags) (Liu et al. 2002).

Production of ethylene has been shown to be involved in the initiation, modulation, and co-ordination of expression of many genes required for the ripening process. Functionally and metabolically ethylene and polyamines (putrescine, spermidine and spermine) seem inter-related as they share a common precursor, S-adenosylmethionine (SAM). Polyamines, ubiquitous, aliphatic cations, have all been implicated in a myriad of physiological and developmental processes, including fruit ripening (Rajam 1997). It has been reported that the exogenous application of polyamines can retard fruit ripening and prolong shelf life of fruits (Kumar and Rajam 2004). Transgenic approach has been undertaken with yeast *samdc* gene, which has led to increased lycopene content, prolonged vine life and enhanced fruit quality (Mehta et al. 2002). Over-expression of polyamine biosynthesis genes – *adc, odc, samdc* and *spd syn* under the control of fruit specific promoter (2A11) was achieved and the developed independent tomato transgenic lines revealed delayed fruit ripening and parthenocarpic fruits (Rajam et al. unpublished results).

Tomato fruits are consumed either fresh or processed. Processing tomatoes account for most of the tomato production. So, it is very important to improve the fruit quality and reduce processing costs. Parthenocarpy (seedless fruits) is a valuable trait because seeds are usually difficult to digest, and often their presence is undesirable. Parthenocarpic fruits are reported to have a higher percentage of soluble solids, improved yield and flavour. Phytohormonal sprays cause parthenocarpic fruit development but excess of exogenous phytohormones causes malformations of the tomato fruit (Santangelo and Soressi 1990). Interestingly, the introduction of *iaaM* gene (which codes for tryptophan monooxygenase and is involved in auxin synthesis) into tomato plants has allowed for the combination of parthenocarpy (upon emasculation), high yield, and high fruit quality in tomato fruits for the fresh market (Ficcadenti et al. 1999).

A wide range of insect pests and pathogens are known to attack tomato. Hence several studies have been undertaken to confer significant resistance against such biotic stresses to remedy significant yield losses in commercial tomato via *Agrobacterium*-mediated transformation. Tomato is severely damaged by lepidopteran insect pest *Helicoverpa armigera* Hubner, also called tomato fruit borer (Atwal 1986). The introduction of the synthetic *cry1Ac* gene (Mandaokar et al. 2000), *cry1Ab* gene (Kumar and Kumar 2004) into tomato conferred high levels of protection against *H. armigera* infestations (Mandaokar et al. 1999). Combined expression of defense genes such as potato protease inhibitors (*PI-II*) and carboxypeptidase inhibitors (*PCI*) was reported in tomato against multiple insect resistance (Abdeen et al. 2005). In addition to the development of a refugium strategy, pyramiding of genes encoding insecticidal proteins that differ in their mode of action might effectively curtail resistance development (Gould 1998). Genes encoding inhibitors of insect proteases and vegetative insecticidal proteins (VIP) were considered for introduction into transgenic tomatoes in conjunction with *cry1Ac* gene (Mandaokar et al. 1999). Broad-spectrum disease resistance in tomato has been achieved by expression of antiapoptotic baculovirus p35 gene (Lincoln et al. 2002). Likewise, the introduction of viral nucleoprotein gene into tomato conferred resistance against tomato spotted wilt virus (Nervo et al. 2003) and coat protein (CP) gene of tomato leaf curl virus (TLCV) against TLCV (Raj et al. 2005). There are very few reports of transgenic tomato for fungal resistance. For instance, stilbene synthase (Vst1 & 2) gene (Thomzik et al. 1997) and Thi 2.1 gene (Chan et al. 2005) have been used to create resistance against *Phytophthora infestans* and phytopathogens, respectively. Glyphosate (herbicide) tolerance using *aro A* gene has also been attempted in tomato (Fillatti et al. 1987). Bacterial speck caused by the pathogen *Pseudomonas syringae* pv tomato (P.s tomato) is a devastating disease of tomato plant. The inhibition of *Ep5c* represents a novel form of disease resistance based on a loss-of-gene function in the plants. *Ep5c* expression is rapidly induced by H_2O_2, a reactive oxygen intermediate normally generated during plant-pathogen interaction (Alberto et al. 2005).

Most of the tomato cultivars are moderately sensitive to salts (Cuartero and Munoz 1999, Foolad 1999). There are very few reports published on the engineering of salt tolerance in tomatoes. Since tomato has no glycinebetaine synthesis pathway (Weretilnyk et al. 1989) to synthesize glycinebetaine (a osmoprotectant), betaine aldehyde dehydogenase (BADH) gene from *Atriplex hortensis* (Xiao et al. 1995, Jia et al. 2002) has been introduced into tomato that allowed the biosynthesis of glycinebetaine to maintain an osmotic balance with the environment and also to withstand the salinity stress (Zhang and Blumwald 2001, Robinson and Jones 1986). Other examples for salt-tolerance in tomato include introduction of *HAL1* (Gisbert et al. 2000) and *HAL2* (Arillaga et al. 1998) to maintain high internal K^+ concentration and decreased intracellular Na^+ concentration during salt-stress. Waterlogging is another stress for which ACC deaminase gene has been introduced to confer flood tolerance (Grichko and Glick 2001). Transgenic tomato that could withstand water-deficit condition has also been developed by introducing *C*-repeat/ dehydration responsive element binding factor gene (CBF1) by Hsieh et al. (2002). Tomato transgenic lines have been developed to combat abiotic stress by engineering the trehalose biosynthetic pathway. The introduction of the yeast trehalose-6-phosphate synthase gene (*TPS1*) resulted in pleiotropic changes such as thick shoots, rigid dark-green leaves, erected branches and an aberrant root development in transgenic plants. These plants showed improved tolerance under drought, salt and oxidative stress conditions as compared to wild type plants (Cortina and Carolina 2005).

Edible vaccine is becoming a reality as scientists have found a way to incorporate the protein gene-HIV antigen in tomatoes. This discovery has become popular when it was found that the protein needed for the vaccine could be derived from both tomato leaves and the fruit. Moreover, tomatoes are edible and immune to any thermal process, which helps retain its healing capabilities. Tomatoes producing edible vaccines were found to grow at a high rate of success in Russia, compared to bananas, which are also used to produce vaccines. A candidate edible vaccine against hepatitis B and HIV has been successfully developed in tomato by introducing TBI-HBS gene (Shchelkunov et al. 2004). Tomato-based edible vaccines against cholera have also been developed by transforming tomato plants with the gene encoding cholera toxin B subunit (*ctx*B) along with an endoplasmic reticulum retention signal (SEKDEI) under the control of the CaMV 35S promoter (Jani et al. 2002). The recent spread of severe acute respiratory syndrome (SARS) has heightened demand for SARS vaccine. SARS-coronavirus (CoV) spike protein (S protein) and its truncated fragments are considered to be the best candidates for generation of recombinant vaccine in tomato and low nicotine tobacco plants (Pogrebnyak et al. 2005).

Due to decreasing public acceptance of antibiotic and herbicide resistance genes in food crops, alternative selectable markers and even complete removal of marker genes (Hohn et al. 2001, Penna et al. 2002) have been attempted. One such alternative approach utilizes mannose as a selective agent as many plants cannot utilize mannose as a carbohydrate source. Such cells will not grow when cultured on mannose containing media. However, when cells are transformed with the phosphomannose isomerase (PMI) gene, they can then survive by utilizing mannose as carbon source. Unlike standard antibiotic or herbicide selection, an efficient mannose selection protocol for tomato that has no adverse effect on ploidy levels of transgenic plants has been successfully developed. This selection does not result in a direct, acute toxic effect, but rather provides a physiological advantage to transformed cells over non-transformed cells as "positive selection" (Sigareva et al. 2004).

The major technical challenge facing the practical application of plant transformation is the development of a method that produces a high proportion of transgenic plants without collateral genetic variations. In case of the transformed cotyledon explants of cv ATV847, only 60% of the transgenic tomato retained the diploid level (Ultzen et al. 1995). The high frequency of polyploid transgenic plants (40%) could be due to the mixoploid nature of the cotyledon tissue (Smulders et al. 1994, 1995). The confirmation of ploidy level before performing an evaluation of transgenic material is particularly important when a polysomatic tissue is to be used as an explant source (Ellul et al. 2003).

CONCLUSIONS AND FUTURE DIRECTIONS

Tomato is a major vegetable crop and serves an important source of nutrients for human consumption. As the crop is highly susceptible to various pests and diseases as well as environmental extremes, its improvement through the use of traditional breeding methods coupled with biotechnological approaches is extremely important. Besides, the improvement of nutrient quality and quantity is also very important in tomato.

Tissue culture and genetic engineering of tomato has come a long way since the 1930s. A large number of publications have appeared in tomato biotechnology. Tomato is quite amenable for both tissue culture regeneration and genetic transformations. In general, the cotyledons are most preferred explants for plant regeneration on MS medium with BAP and IAA hormonal combination. Tomato transformation is usually done via *A. tumefaciens* containing binary vectors with specific selection markers.

A good number of transgenic tomatoes have been produced for introducing the traits like slow ripening, fruit softness, improvement of fruit taste, aroma, flavour and nutrient content, seedless fruits, resistance against insect pests and pathogens, herbicide tolerance, salt and drought tolerance, and antibody production. Full potential is yet to be exploited for improvement of tomato yield and quality by genetic engineering.

Further, the antisense RNA technology may be replaced with the latest and potent RNAi (RNA interference) technology for effective suppression of ethylene biosynthesis genes to achieve efficient delay of fruit ripening. The areas that need to be strengthened in tomato genetic engineering are tolerance to various abiotic stresses (e.g. salinity, drought and cold), production of a variety of edible vaccines/antibodies against human and animal diseases, and stacking of transgenes for introduction of multiple and complex traits.

Acknowledgements

The research work on tomato biotechnology has been generously supported by the Department of Biotechnology (Grant no. BT/PR/2990/ Agr/16/232/2002), New Delhi to MVR. Research fellowships from the Council of Scientific and Industrial Reseach, New Delhi to PM, RP and PH is gratefully acknowledged.

REFERENCES

Abdeen, A., A. Virgos, E. Olivella, X. Villanueva, Aviles R. Gabarra, and S. Prat. 2005. Multiple insect resistance in transgenic tomato plants over-expressing two families of plant proteinase inhibitors. Plant Mol Biol 57: 189-202.

Alberto, C., R. Vicente, E. Phillipe, M. Esther, and V. Pablo. 2005. The H_2O_2 -regulated *Ep5C* gene encodes a peroxidase required for bacterial speck susceptibility in tomato. Plant J 42: 283.

Applewhite, P.B., R. Kaur-Sawhney, and A.W. Galston. 1994. Isatin as an auxin source favoring floral and vegetative shoot regeneration from calli produced by thin layer explants of tomato pedicel. Plant Growth Regul 15: 17–21

Arillaga, I., R. Gil-Mascarell, C. Gisbert, E. Sales, C. Montesinos, R. Serrano, and V. Moreno. 1998. Expression of the yeast *HAL2* gene in tomato increases the *in vitro* salt tolerance of transgenic progenies. Plant Sci 136: 219-226.

Atwal, A.S. 1986. Agricultural pests of India and South East Asia. Kalyani Publishers, New Delhi.

Barg, R., M. Pilowsky, S. Shabtai, N. Carmi, A.D. Szetchman, B. Dedicova, and Y. Salts. 1997. The TYLCV-tolerant tomato line MP-1 is characterized by superior transformation competence. J Exp Bot 48: 1919–1923.

Bartoszewski, G., A. Niedziela, M. Scwacka, and K. Niemirowicz-Szczytt. 2003. Modification of tomato taste in transgenic plants carrying a thaumatin gene from *Thaumatococcus daniellii* Benth. Plant Breed 122: 347-351.

Baulcombe, D.C. 1999. Fast forward genetics based on virus-induced gene silencing. Curr Opinion Plant Biol 2: 109-113.

Beck, M.J. and N.D. Camper. 1991. Shoot regeneration from petunia leaf discs as a function of explants size, configuration and benzyladenine exposure. Plant Cell Tiss Org Cult 26: 101-106

Behki, R.M. and S.M. Lesley. 1980. Shoot regeneration from leaf callus of *Lycopersicon esculentum*. Z. Pflanzenphysiol 98: 83–87

Bhatia P., N. Ashwath, T. Senaratn, and D. Midmore. 2004. Tissue culture studies of tomato (*Lycopersicon esculentum*). Plant Cell Tiss Org Cult 78: 1-21.

Bird, C.R., C.J.S. Smith, J.A. Ray, P. Moureau, M.J. Bevan, A.S. Birds, S. Hughes, P.C. Morris,

D. Grierson, and W. Schuch. 1988. The tomato polygalacturonase gene and ripening specific expression in transgenic plants. Plant Mol Biol 11: 651–662.

Bird, C.R., J.A. Ray, J.D. Fletcher, J.M. Boniwell, A.S. Bird, C. Teulieres, I. Blain, P.M. Bramley, and W. Schuch. 1991. Using antisense RNA to study gene function: Inhibition of carotenoid biosynthesis in transgenic tomatoes. Bio/Technol 9: 635-639.

Bishop, G.J., K. Harrison, and J.D.G. Jones. 1996. The tomato *dwarf* gene isolated by heterologous transposan tagging encodes the first member of a new cytochrome P_{450} family. Plant Cell 8: 959-969.

Bolton, G.W., E.W. Nester, and M.P. Gordon. 1986. Plant phenolic compounds induce the expression of the *Agrobacterium tumefaciens* loci needed for virulence. Science 232: 983-985.

Brummell, D.A., W.J. Howie, C. Ma, and P. Dunsmuir. 2002. Post-harvest fruit quality of transgenic tomatoes suppressed in expression of a ripening-related expansin. Post-harvest Biol Technol 25: 209-220.

Chan, Y.L., V. Prasad, Sanjaya, K.H. Chen, P.C. Liu, M.T. Chan, and C.P. Cheng. 2005. Transgenic tomato plants expressing an *Arabidopsis thionin* (Thi2.1) driven by fruit-inactive promoter battle against phytopathogenic attack. Planta 19: 1432-2048.

Chandel, G. and S.K. Katiyar. 2000. Organogenesis and somatic embryogenesis in tomato (*Lycopersicon esculantum* Mill.). Adv Plant Sci 13: 11–17.

Chen, L.Z. and T. Adachi. 1994. Plant regeneration via somatic embryogenesis from cotyledon protoplasts of tomato (*Lycopersicon esculentum* Mill). Breed Sci 44: 337–338.

Cheng, Z.M., J.Á. Schnurr, and J.Á. Kapaun. 1998. Timentin as an alternative antibiotic for suppression of *Agrobacterium tumefaciens* in genetic transformation. Plant Cell Rep 17:646–649.

Chlyah, A. and H. Taarji 1984. Androgenesis in tomato. Plant tissue and cell culture application to crop improvement. Czechoslovak Ac: 241–242.

Chyi, Y.S. and G.C. Phillips. 1987. High efficiency *Agrobacterium*-mediated transformation of *Lycopersicon* based on conditions favourable for regeneration. Plant Cell Rep. 6: 105–108.

Compton, M.E. and R.E. Veilleux. 1988. Morphogenesis in tomato thin cell layers. HortScience 23: 754.

Compton, M.E. and R.E. Veilleux. 1991. Shoot, root and flower morphogenesis on tomato inflorescence explants. Plant Cell Tiss Org Cult 24: 223–231.

Cortina, C and F.A. Culianez-Macia. 2004. Tomato transformation and transgenic plant production. Plant Cell Tiss Org Cult 76: 269-275.

Cortina C., F.A. Culianez-Macia. 2005. Tomato abiotic stress enhanced tolerance by trehalose biosynthesis. Plant Sci 169: 75-82.

Cuartero J. and R.F. Munoz. 1999. Tomato and salinity. Sci Horti 78: 83-125.

Davis, M.R. and L.D. Miller. 1991. Temporal competence for transformation of *Lycopersicon esculentum* L. Mill cotyledons by *Agrobacterium tumefaciens*: relation to wound healing and soluble plant factors. J Exp Bot 42: 359–364.

Davis, D.G., K.A. Breiland, D.S. Frear, and G.A. Secor. 1994. Callus initiation and regeneration of tomato (*Lycopersicon esculentum*) cultivars with different sensitivities to metribuzin. Plant Growth Regul Soc Am Quart 22: 65–73.

DeLanghe, F. and E. DeBruijne. 1976. Continuous propagation of tomato plants by means of callus culture. Sci Hort 4: 221–227.

Duzyaman, E., A. Tanrisever and Gunver G. 1994. Comparative studies on regeneration of different tissues of tomato *in vitro*. Acta Hort 235–242.

Dwivedi, K., P. Srivastava, H.N. Verma, and H.C. Chaturvedi. 1990. Direct regeneration of shoots from leaf segments of tomato (*Lycopersicon esculentum*) cultured *in vitro* and production of plants. Indian J Exp Biol 28: 32–35.

El- Bakry, A.A. 2002. Effect of genotype, growth regulators, carbon source, and pH on shoot induction and plant regeneration in tomato. *In vitro* Cell Dev Biol 38: 501-507.

El-Farash, E.M., H.I. Abdalla, A.S. Taghian, and M.H. Ahmad. 1993. Genotype, explant age and explant type as effecting callus and shoot regeneration in tomato. Assiut J Agri Sci 24: 3-14.

Ellul, P., B. Garcia-Sogo, B. Pineda, G. Rios, L.A. Riog, and V. Moreno 2003. The ploidy level of transgenic plants in *Agrobacterium*-mediated transformation of tomato cotyledons (*Lycopersicon esculentum* L. Mill.) is genotype and procedure dependent. Theor Appl Genet 106 (2): 231-238.

Fari, M., A. Banki-Peredi, and M. Toth-Csanyi. 1991. Highly efficient *in vitro* shoot regeneration system in tomato and egg plant via seedling decapitation method (SDM). ISHS Acta Hort: International symposium on plant biotechnology and its Contribution to Plant Development, Multiplication and Improvement pp. 238.

Fari, M., A. Szasz, J. Mityko, I. Nagy, M. Csanyi, and A. Andrasfalvy. 1992. Induced organogenesis via the seedling decapitation method (SDM) in three solanaceous vegetable species. *Capsicum* Newsl pp. 243–248.

Faria, RTd, D. Destro, J.C. Bespalhok Filho, and R.D. Illg. 2002. Introgression of *in vitro* regeneration capability of *Lycopersicon pimpinellifolium* Mill. into recalcitrant tomato cultivars. Euphytica 124: 59–63.

Ficcadenti, N., S. Sestili, T. Pandolfini, C. Cirillo, L.G. Rotino, and A. Spena. 1999. Genetic engineering of parthenocarpic fruit development in tomato. Mol. Breed 5: 463-470.

Fillati, J.J., J. Kiser, R. Ronald, and L. Comai. 1987. Efficient transfer of glyphosate tolerance gene into tomato using a binary *Agrobacterium tumefaciens* vector. Biotechnology 5: 726–730.

Flores H.E., M.W. Hoy, and J.J. Pickard. 1987. Secondary metabolites from root cultures. Trends Biotechnol 5: 64-69.

Foolad, M.R. 1999. Genetics of salt and cold tolerance in tomato: quantitative analysis and QTL mapping. Plant Biotechnol. 16: 55-64.

Foroughi-Wehr, B. and W. Friedt. 1984. Rapid production of recombinant barley yellow mosaic virus resistant *Hordium vulgare* by anther culture. Theor Appl Genet 67: 377-382.

Frary, A. 1995. The use of *Agrobacterim tumefaciens*-mediated transformation in the map-based cloning of tomato genes and an analysis of factor affecting transformation efficiency. PhD Thesis, Cornell University, Ithaca, NY.

Frary, A. and E.D. Earle. 1996. An examination of factors affecting the efficiency of *Agrobacterium*-mediated transformation of tomato. Plant Cell Rep 16: 235–240.

Fraser, P.D., R. Susanne, A.S. Cathie, B.M. Philippa, J.W. Kiano, N. Misawa, R.G. Drake, Schuch, and P.M. Bramley. 2002. Evaluation of transgenic tomato plants expressing an additional phytoene synthase in a fruit-specific manner. Proc Natl Acad Sci, USA 99: 1092-1097.

Friedt, W., B. Foroughi-Wehr, and J.W. Snape. 1986. The significance of biotechnology for the evolution of barley breeding methods. In: H. Caul (ed.), Proc. 5th Int. Barley Genet Symp Garching Barley Genet, Okayama, Japan, pp. 367-373.

Fu, D.Q., B.Z. Zhu, H.L. Zhu, W.B. Jiang, and Y.B. Luo. 2005. Virus-induced gene silencing in tomato fruit. Plant J 43: 299-308.

Fuke, S. and S. Konosu. 1991. Taste active components in some foods: a review of Japanese research. Physiol Behav 49: 863-868.

Ganga Rao, D., A. van Tuinen, P.D. Fraser, A. Manfredonia, R. Newman, D. Burgess, D.A. Brummell, S.R. King, J. Palys, J. Uhlig, P.M. Bramley, H.M.J. Pennings, and C. Bowler 2005. Fruit-specific RNAi-mediated suppression of *DET1* enhances carotenoid and flavonoid content in tomatoes. Nature Biotechnol 23: 890-895.

Garfinkel, D.J. and E.W. Nester. 1980. *Agrobacterium tumefaciens* mutants affected in crown gall tumorigenesis and octopine catabolism. J Bacteriol 144: 732-743.

Geetha, N., P. Venkatachalam, P.S. Reddy, and G. Rajaseger 1998. *In vitro* plant regeneration from leaf callus cultures of tomato (*Lycopersicon esculentum* Mill.). Adv Plant Sci 11: 253-257.

Gill, R., K.A. Malik, M.H.M. Sanago, and P.K. Saxena. 1995. Somatic embryogenesis and plant regeneration from seedling cultures of tomato (*Lycopersicon esculentum* Mill.). J Plant Physiol 147: 273–276.

Gisbert, C., A.M. Rus, C. Bolarin, J.M. Lopez-Coronado, I. Arrillaga, C. Montesinos, M. Caro, R. Serrano, and V. Moreno. 2000. The yeast *HAL1* gene improves salt tolerance of transgenic tomato. 123: 393-402.

Gould, F. 1998. Sustainability of transgenic insecticidal cultivars: integrating pest genetics and ecology. Annu Rev Entomol 43: 701-726.

Gresshoff, P.M. and C.H. Doy. 1972. Development and differentiation of haploid *Lycopersicon esculentum* (tomato). Planta 107: 161–170.

Grichko, V.P. and B.R. Glick. 2001. Flooding tolerance of transgenic tomato plants expressing the bacterial enzyme ACC deaminase controlled by the 35S, rolD or PRB-1b promoter. Plant Physiol. Biochem 39: 19-25.

Gunay, A.L. and P.S. Rao. (1980). *In vitro* propagation of hybrid tomato plants (*Lycopersicon esculentum* L.) using hypocotyl and cotyledon explants. Ann Bot 45: 205–207.

Hackett, R.M., C.W. Ho, Z. Lin, H.C.C. Foote, R.G. Fray, and D. Grierson. 2000. Antisense inhibition of the Nr gene restores normal ripening to the tomato Never-ripe mutant, consistent with the ethylene receptor-inhibition model. Plant Physiol 124: 1079-1085.

Hall, L.N., G.A.Tucker, C.J.S. Smith, C.F. Watson, G.B. Seymour, Y. Bundick, J.M. Boniwell, J.D. Fletcher, J.A. Ray, W. Schuch, C.R. Bird, and Greirson D. 1993. Antisense inhibition of pectinesterase gene expression in transgenic tomatoes. Plant J 3: 121-129.

Hamill, J.D., A . Perscott, and C. Martin. 1987. Assessment of the efficiency of co-transformation of the T-DNA of disarmed binary vectors derived from *Agrobacterium tumefaciens* and the T-DNA of *A. rhizogenes*. Plant Mol Biol 9: 573-584.

Hamilton, A.J., G.W. Lycett, and D. Grierson. 1990. Antisense gene that inhibits synthesis of the plant hormone ethylene in transgenic plants. Nature 346: 284-287.

Hamza, S. and Y. Chupeau. 1993. Re-evaluation of conditions for plant regeneration and *Agrobacterium*-mediated transformation from tomato (*Lycopersicon esculentum*). J Exp Bot 269: 1837–1845.

Hawkins, A., S.J. Gurr, P. Montague, and J.R. Kinghorn. 1989. Nucleotide sequence and regulation of expression of the *Aspergillus nidulans* gdhA gene encoding NADP-dependent glutamate dehydrogenase. Mol Gen Genet 218: 105-111.

Hohn, B., A.A. Levy, and H. Puchta. 2001. Elimination of selection markers from transgenic plants. Curr Opin Biotechnol 12:139-143.

Hsieh, T.H., J.T. Lee, Y.Y. Charng, and M.T. Chan. 2002. Tomato plants ectopically expressing *Arabidopsis* CBF1 show enhanced resistance to water deficit stress. Plant Physiol. 130: 618-626.

Hussey, G. 1971. *In vitro* growth of vegetative tomato shoot apices. J Exp Bot 22: 688–700.

Ichimura, K. and M. Oda. 1998. Stimulation of phenotypically normal shoot regeneration of tomato (*Lycopersicon esculentum* Mill.) by commercial filter paper extract. J Jap Soc Hort Sci 67: 378–380.

Izadpanah, M. and M. Khosh-Khui. 1992. Comparisons of *in vitro* propagation of tomato cultivars. Iran Agric Res 8: 37–47.

Jacinto, T., K.V.S. Fernandes, O.L.T. Machado, and C.L. Siqueira-Junior. 1998. Leaves of transgenic tomato plants overexpressing prosystemin accumulate high levels of cystatin. Plant Sci 138: 35-42.

Jani, D., L.S. Meena, Q.M. Rizman-ul-Haq, Y. Singh, A.K. Sharma, and A.K. Tyagi 2002. Expression of cholera toxin B subunit in transgenic tomato plants. Transg Res 11: 447-454.

Jia, G.X., Z.Q. Zhu, F.Q. Chang, and Y.X. Li. 2002. Transformation of tomato with the *BADH* gene from *Atriplex* improves salt tolerance. Plant Cell Rep 21: 141-156.

Joao, K.H.L. and T.A. Brown. 1993. Enhanced transformation of tomato co-cultivated with *Agrobacterium tumefaciens* C58 C1 Rif^r::pGSFR1161 in the presence of acetosyringone. Plant Cell Rep 12: 422-425.

Jones, D.A., C.M. Thomas, K.E. Hammond-Kosack, P.J. Balint-Kurti, and J.D.G. Jones. 1994. Isolation of the tomato *cf-9* gene for resistance to *Cladosporium fulvum* by transposon tagging. Science 266: 789-793.

Kalloo, G. 1991. Introduction. In: G. Kalloo (ed), Monographs on Theoretical and Applied Genetics 14, Genetic Improvement of Tomato. Springer-Verlag, Berlin. pp. 1–9.

Kaparakis, G and P.G. Alderson. 2002. Influence of high concentrations of cytokinins on the production of somatic embryos by germinating seeds of tomato, aubergine and pepper. J Hort Sci Biotechnol 77: 186–190.

Kartha, K.K., O.L. Gamborg, J.P. Shyluk, and F. Constabel 1976. Morphogenetic investigations on *in vitro* leaf culture of tomato (*Lycopersicon esculentum* Mill. cv. Starfire) and high frequency plant regeneration. Z. Pflanzenphysiol. 77: 292–301.

Kato, Y., N. Uozumi, T. Kimura, H. Honda, and T. Kobayashi. 1991. Enhancement of peroxidase production and excertion from horseradish hairy roots by light, NaCl, and peroxidase-adsorption *in situ*. Plant Cell Tiss Org Cult Lett 8: 158-165.

Kazua, I., U. Toshiko, T. Kenkou, O. Masayuki, and K.E.A. Masaaki-Nagaoka. 1982. *Never ripe-2 (Nr-2)* a slow ripening mutant resembling *Nr* and *Gr*. Rep Tomato Genet Coop 32: 33.

Keddie, J.S., B. Carroll, J.D.G. Jones, and W. Gruissem. 1996. The *DCL* gene of tomato is required for chloroplast development and palisade cell morphogenesis in leaves. Embo J 15: 4208-4217.

Kisaka H. and T. Kida. 2003. Transgenic tomato plant carrying a gene for NADP-dependent glutamate dehydrogenase (*gdh A*) from *Aspergillus nidulans*. Plant Sci 164: 35-42.

Koorneef, M., C. Hanhart, M. Jongsma, T.I.R. Weide, P. Zabel, and J. Hille. 1986. Breeding of a tomato genotype readily accessible to genetic manipulation. Plant Sci 45: 201–208.

Koornneef, M., J. Bade., C. Hanhart, K. Horsman, J. Schel, W. Soppe, R. Verkerk, and P. Zabel. 1993. Characterization and mapping of a gene controlling shoot regeneration in tomato. Plant J 3: 131–141.

Kumar, H. and V. Kumar . 2004. Tomato expressing Cry1A (b) insecticidal protein from *Bacillus thurigiensis* protected against tomato fruit borer, *Helicoverpa armigera* (Hubner) (Lepidoptera: Noctuidae) damage in the laboratory, green house and field. Crop Protection 23:135-139.

Kumar, S.V. and M.V. Rajam. 2004. Polyamine-ethylene nexus: a potential target for post-harvest biotechnology. Indian J Biotech 3: 299-304.

Lambert, C. and D. Tepfer. 1991. Use of *Agrobacterium rhizogenes* to create chimeric apple trees through genetic grafting. Biotechnol 9: 80-83.

Lech, M., K. Miczynski, and A. Pindel. 1996. Comparison of regeneration potentials in tissue cultures of primitive and cultivated tomato species (*Lycopersicon* sp.). Acta Soc Bot Poloniae 65: 53–56.

Lewinsohn, E., F. Schalechet, J. Wilkinson, K. Matsui, Y. Tadmor, K.H. Nam, O. Amar, E. Lastochkin, O. Larkov, U. Ravid, W. Hiatt, S. Gepstein, and E. Pichersky. 2001. Enhanced levels of the aroma and flavor compound S-Linalool by metabolic engineering of the terpenoid pathway in tomato fruits. Plant Physiol. 127: 1256-1265.

Lincoln, J.E., C. Richael, B. Overduin, K. Smith, R. Bostock, and D.G. Gilchrist. 2002. Expression of the antiapoptotic baculovirus p35 gene in tomato blocks programmed cell death and provides broad-spectrum resistance to disease. Proc Natl Acad Sci USA 99: 15217-15221.

Ling, H.Q., D. Kriseleit, and M.W. Ganal. 1998. Effect of ticarcillin/potassium clavulanate on callus growth and shoot regeneration in *Agrobacterium*-mediated transformation of tomato (*Lycopersicon esculentum* Mill.). Plant Cell Rep. 17: 843–847.

Liu, Y., M. Schiff, and S.P. Dinesh-Kumar. 2002. Virus-induced gene silencing in tomato. Plant J. 31(6): 777-786.

Locy, R.D. 1995, Selection of tomato tissue cultures able to grow on ribose as the sole carbon source. Plant Cell Rep. 14: 777–780.

Lu, C., Z. Zainal, G.A. Tucker, and G.W. Lycett. 2001. Developmental abnormalities and reduced fruit softening in tomato plants expressing an antisense Rab11 GTPase gene. Plant Cell 13: 1819-1833.

Madhulatha, P., R. Pandey, P. Hazarika, and M.V. Rajam. 2006a. Polyamines and maltose significantly enhance shoot regeneration in tomato. Physio Mol Biol Plants (Communicated).

Madhulatha, P., R. Pandey, P. Hazarika, and M.V. Rajam. 2006b. High transformation frequency in *Agrobacterium*-mediated genetic transformation of tomato by using polyamines and maltose in shoot regeneration medium. Plant Sci (communicated).

Mandaokar, A.D., P.A. Kumar, R.P. Sharma, and V.S. Malik. 1999. Bt-transgenic crop plants-Progress and Prospectus. In: V.L. Chopra, V.S. Malik, and S.R. Bhat (Eds.), Applied Plant Biotechnology. Oxford & IBH publishing Co, New Delhi, pp. 285-300.

Mandaokar, A.D., R.K. Goyal, A. Shukla, Bisaria, R. Bhalla, V.S. Reddy, A. Chaurasia, R.P. Sharma, I. Altosaar, and P. Ananda Kumar. 2000. Transgenic tomato plants resistant to fruit borer (*Helicoverpa armigera* Hubner). Crop Protection 19: 307-312.

Marc-Andre', D' Aoust, S. Yelle, and A. Nguyen-Quoc. 1999. Antisense inhibition of tomato fruit sucrose synthase decreases fruit setting and sucrose unloading capacity of young fruit. Plant Cell 11: 2407-2418.

McCormick, S., J. Niedermeyer, B. Fry, A. Barnason, R. Horch, and R. Farley. 1986. Leaf disk transformation of cultivated tomato (*L. esculentum*) using *Agrobacterium tumefaciens*. Plant Cell Rep 5: 81-84.

Mehta, R.A., T. Cassol, N. Li, N. Ali, A.K. Handa, and A.K. Mattoo. 2002. Engineered polyamine accumulation in tomato enhances phytonutrient content, juice quality, and vine life. Nature Biotechnol 20: 613-618.

Meissner, R., V. Chague, Q. Zhu, E. Emmanuel, Y. Elkind, and A.A. Levy. 2000. A high throughput system for transposon tagging and promoter trapping in tomato. Plant J 22: 265-274.

Mensuali-Sodi, A., M. Panizza, and F. Tognoni. 1995. Endogenous ethylene requirement for adventitious root induction and growth in tomato cotyledons and lavandin microcuttings *in vitro*. Plant Growth Reg 17: 205–212.

Moon, H. and A.M. Callahan. 2004. Developmental regulation of peach ACC oxidase promoter-GUS fusions in transgenic tomato fruits. J Exp Bot 55: 1519-1528.

Muhlbach, H.P. 1980. Different regeneration potentials of mesophyll protoplasts from cultivated and a wild species of tomato. Planta 148: 89–96.

Murashige, T. and F. Skoog. 1962. A revised medium for rapid growth and bioassays with tobacco tissue cultures. Physiol Plant 15: 473-497.

Nervo, G., C. Cirillo, G.P. Accotto, and A.M. Vaira. 2003. Characterisation of two tomato lines highly resistant to tomato spotted wilt virus following transformation with the viral nucleoprotein gene. J Plant Pathol 85: 139-144.

Novak, F.J. and I. Maskova. 1979. Apical shoot tip culture of tomato. Sci Hort 10: 337–344.

Oeller, P.W., L.M. Wong, L.P. Taylr, D.A. Pike, and A. Theologis. 1991. Reversible inhibition of tomato fruit senescence by antisense RNA. Science 254: 437-439.

Ohki, S., C. Bigot, and J. Mousseau. 1978. Analysis of shoot-forming capacity *in vitro* in two lines of tomato (*Lycopersicon esculentum* Mill) and their hybrids. Plant Cell Physiol 19: 27–42.

Oktem, H.A., Y. Bulbul, E. Oktem, and M. Yucel. 1999. Regeneration and *Agrobacterium*-mediated transformation studies in tomato (*Lycopersicon esculentum* Miller). Tr. J. Bot. 23: 345-348.

Ooms, G., A. Karp, M.M. Burrell, D. Twell, and J. Roberts. 1985. Genetic modification of potato development using Ri T-DNA. Theor Appl Genet 70: 440-446.

Padliskikh, V.L. and A.P. Yarmishin. 1990. Features of microclonal propagation in tomato. Vestsi Akademii Navuk BSS, Reryya Biyalagichnykh Navuk 6: 52–54.

Park, J., B. Yi, and C. Lee 2001. Effects of plant growth regulators, bud length, donor plant age, low temperature treatment and glucose concentration on callus induction and plant regeneration in anther culture of cherry tomato 'Mini-carol'. J Korea Soc Hort 42: 32-37.

Park, S., N. Cheng, J.K. Pittman, K.S. Yoo, J. Park, R.H. Smith, and K.D. Hirschi. 2005. Increased calcium levels and prolonged shelf life in tomatoes expressing *Arabidopsis* H^+/Ca^{2+} Transporters. Plant Physiol. 139: 1194-1206.

Patil, R.S., M.R. Davey, J.B. Power, and E.C. Cocking. 2002. Effective Protocol for *Agrobacterium*-mediated Transformation in Tomato (*Lycopersicon esculentum* Mill.). Indian J Biot 1: 339-343.

Penna, S., L. Sagi, and R. Swennen. 2002. Positive selectable marker genes for routine plant transformation. *In vitro* Cell Dev Biol Plant 38:125–128.

Peres, L.E.P., P.G. Morgante, C. Vecchi, J.E. Kraus, and MAv Sluys. 2001. Shoot regeneration capacity from roots and transgenic hairy roots of tomato cultivars and wild related species. Plant Cell Tiss. Org Cult 65: 37–44.

Pfitzner, A.J.P. 1998. Transformation of tomato. Meth Mol Biol 81:359–363.

Pinhero, R.G., K.C. Almquist, Z. Novotna, and G. Paliyath. 2003. Developmental regulation of phospholipase D in tomato fruits. Plant Physiol Biochem 41: 223-240.

Plastira, V.A. and A.K. Perdikaris. 1997. Effect of genotype and explant type in regeneration frequency of tomato *in vitro*. Acta Hort 231–234.

Pogrebnyak, N., M. Golovkin, V. Andrianov, S. Spitsin, Y. Smirnov, R. Egolf, and H. Koprowski. 2005. Severe acute respiratory syndrome (SARS) S protein production in plants: development of recombinant vaccine. Proc Natl Acad Sci USA 102: 9062-9667.

Rahman, M.M. and K. Kaul. 1989. Differentiation of sodium chloride tolerant cell lines of tomato (*Lycopersicon esculentum* Mill.) cv. Jet Star. J Plant Physiol 133: 710–712.

Raj, S.K., S. Rachana, S.K. Pandey and B.P. Singh. 2005. *Agrobacterium*-mediated tomato transformation and regeneration of transgenic plants expressing tomato leaf curl virus coat protein gene for resistance against TLCV infection. Curr Sci 33: 1674-1679.

Rajam, M.V. 1997. Polyamines. In: M.N.V. Prasad (ed.), Plant Ecophysiology. John Wiley & Sons, New York, pp. 343-374.

Robinson, S.P. and G.P. Jones. 1986. Accumulation of glycinebetaine in chloroplasts provides osmotic adjustment during salt stress. Aust J Plant Physiol 13: 659–668.

Romer, S. 2000. Elevation of the provitamin A content of transgenic tomato plant. Nature Biotechnol 18:666–669.

Ruf, S., M. Hermann, I.J. Berger, H. Carrer, and R. Bock. 2001. Stable genetic transformation of tomato plastids and expression of a foreign protein in fruit. Nature Biotechnol 19: 870-875.

Santangelo, E. and G.P. Soressi. 1990, La partenocarpia nel pomodoro. *Colture Protette* 3: 29-33.

Schroeder, H.E., A.H. Schotz, T. Wardley-Ricardson, D. Spencer, and T.J.V. Higgins. 1993 Transformation and regeneration of two cultivars of pea (*Pisum sativum* L.). Plant Physiol 101:751–757.

Schutze, R and G. Wieczorrek. 1987. Investigations into tomato tissue cultures. I. Shoot regeneration in primary explants of tomato. Arch Zuchtungsforschung 17: 3–15.

Shanin, E.A., K. Sukhapinda, and R.B. Simpson. 1986. Transformation of cultivated tomato by a binary vector in *Agrobacterium rhizogenes*: transgenic plants with normal phenotypes harbour binary vector T-DNA, but no Ri -plasmid T-DNA. Theor Appl Genet 72: 770–777.

Sheehy, R.E., M. Kramer, and W.R. Hiatt. 1988. Reduction of polygalacturonase activity in tomato fruit by antisense RNA. Proc Natl Acad Sci 85: 8805-8809.

Shchelkunov, S.N., R.K. Saliaev, T.S. Ryzhova, S.G. Pozdniakov, A.E. Nesterov, N.I. Rekoslavskaia, V.M. Sumtsova, N.V. Pakova, U.O. Mishutina, T.V. Kopytina, and R.V. Hammond. 2004. Designing of a candidate edible vaccine against hepatitis B and HIV on the basis of a transgenic tomato. Vestn Ross Akad Med Nauk 11: 50-55.

Sheikholeslam, S.N. and D.P. Weeks. 1987. Acetosyringone promotes high efficiency transformation of *Arabidopsis thaliana* explants by *Agrobacterium tumefaciens*. Plant Mol Biol 8: 291-298.

Shelagh, R.M., J.C. Geoff, R. Susan, H. Stephen, C.H. Ric De Vos, A.J. van Tunen, and M.E. Verhoeyen 2001. Over-expression of petunia chalcone isomerase in tomato results in fruit containing increased levels of flavonols. Nature Biotechnol 19: 470-474.

Shtereva, L.A., N.A. Zagorska, B.D. Dimitrov, M.M. Kruleva, and H.K. Oanh. 1998. Induced androgenesis in tomato (*Lycopersicon esculentum* Mill.) II. Factors affecting induction of androgenesis. Plant Cell Rep 18: 312–317.

Sigareva, M., R. Spivey, M.G. Willits, C.M. Kramer, and Y. Chang. 2004. An efficient mannose selection protocol for tomato that has no adverse effect on the ploidy level of transgenic plants. Plant Cell Rep 23: 236–245.

Simkin, A.J., S.H. Schwartz, M. Auldridge, M.G. Taylor, and H.J. Klee. 2004. The tomato carotenoid cleavage dioxygenase 1 genes contribute to the formation of the flavor volatiles b-ionone, pseudoionone, and geranylacetone. Plant J 40: 882-892.

Sink, K.C., L.W. Handley, R.P. Niedz, and P.P. Moore. 1986. Protoplast culture and use of regeneration attributes to select tomato plants. Genet Manip 405–413.

Smith, C.J.S., C.F. Watson, J. Ray, C.R. Bird, P.C. Morris, W. Schuch, and D. Grierson. 1988. Antisense RNA inhibition of polygalacturonase gene expression in transgenic expression in transgenic tomatoes. Nature 334: 724-726.

Smith, C.J.S., C.F. Watson, P.C. Morris, C.R. Bird, G.B. Seymour, J.E. Gray, C. Arnold, G.A. Tucker, W. Schuch, S. Harding, and D. Grierson. 1990. Inheritance and effects on ripening of antisense polygalacturonase genes in transgenic tomatoes. Plant Mol Biol 14: 369-379.

Smith, D.L., A.A. Judith, and K.C. Gross. 2002. Down regulation of tomato b- galactosidase 4 results in decreased fruit softening. Plant Physiol 129: 1755-1762.

Smulders, M.J.M., W. Rus-Kortekaas, and L.J.W. Gilissen. 1994. Development of polysomaty during differentiation in diploid and tetraploid tomato (*Lycopersicon esculentum*) plants. Plant Sci 97: 53-60.

Smulders, M.J.M., W. Rus-Kortekaas, and L.J.W. Gilissen. 1995. Natural variation in patterns of polysomaty among individual tomato plants and their regenerated progeny. Plant Sci 106: 129-139.

Stachel, S.E., E.W. Nester, and P.C. Zambryski. 1986a. A plant cell factor induces *Agrobacterium tumefaciens vir* gene expression. Proc Nat Acad Sci 83: 379-383.

Stachel, S.E., E.W. Nester, and P.C. Zambryski. 1986b. Generation of single stranded T-DNA molecules during the initial stages of T-DNA transfer from *Agrobacterium tumefaciens* to plant cells. Nature 322: 706-712.

Stone, B. 1994. The Flavr Savr arrives. http:// www.Accessexcellence. Org/AB/BA/Flavr_Savr_Arrives.html, Date of access 25 June 2003.

Takken, F., D. Schipper, H. Nijkamp, and J. Hile. 1998. Identification and Ds-tagged isolation of a new gene at the cf-4 locus of tomato involved in disease resistance to *Cladosporium fulvum* race 5. Plant J. 14: 401-411.

Tepfer, D. 1984. Genetic transformation of special species of higher plants by *Agrobacterium rhizogenes*: phenotypic consequences and sexual transmission of transformed genotype and phenotype. Cell 37: 959-967.

Taneaki, T., W. Manabu, Y. Kentaro, S. Nozomu, S. Hideyuki, and S. Daisuke. 2005. Expressed sequence tags of full-length cDNA clones from the miniature tomato (*Lycopersicon esculentum*) cultivar Micro-Tom. Plant Biotechnol 22: 161-165.

Thomzik, J.E., K. Stenzel, R. Stocker, P.H. Schreier, R. Hain, and D.J. Stahl. 1997. Synthesis of a grapevine phytoalexin in transgenic tomatoes (*Lycopersicon esculentum* Mill.) conditions resistance against *Phytophthora infestans*. Physiol Mol Plant Pathol 51: 265-278.

Tieman, D.M., R.W. Harriman, G. Ramamohan, and A.K. Handa. 1992. An antisense pectin methylesterase gene alters pectin chemistry and soluble solids in tomato fruit. Plant Cell 4: 667-679.

Toyoda, H., N. Tanaka, and T. Hirai. 1984. Effects of the culture filtrate of *Fusarium oxysporum* f. sp. *lycopersicon* tomato callus growth and the selection of resistant callus cells to the filtrate. Ann Phytol Soc Jap 50: 53–62.

Toyoda, H., Y. Matsuda, and T. Hirai. 1985. Resistance mechanism of cultured plant cells to tobacco mosaic virus (III). Efficient microinjection of tobacco mosaic virus into tomato callus cells. Ann Phytol Soc Jap 51: 32–38

Toyoda, H., K. Shimizu, K. Chatani, N. Kita, Y. Matsuda, and S. Ouchi. 1989. Selection of bacterial wilt-resistant tomato through tissue culture. Plant Cell Rep 8: 317–320.

Ultzen, T., J. Gielen, F. Venema, A. Westerbroek, P. De Haan, M. Tan, A. Schram, M. Van Grinsven, and R. Golbach. 1995. Resistance to tomato spotted wilt virus in transgenic tomato hybrids. Euphytica 85: 159–168.

Uozumi, N., K. Kohketsu, O. Kondo, H. Honda, and T. Kobayashi. 1991. Fed-bach culture of hairy root using fructose as a carbon source. J Ferment Bioeng 72: 457-460.

Uozumi, N., Y. Ohatake, Y. Nakashimada, Y. Morikawa, N. Tanaka, and T. Kobayashi. 1996. Efficient regeneration from GUS-transformed Ajuga hairy root. J Ferment Bioeng 5: 374-378.

Van Roekel, J.S.C., B. Damm, L.S. Melchers, and A. Hoekema. 1993. Factors influencing transformation frequency of tomato (*Lycopersicon esculentum*). Plant Cell Rep 12: 644–647.

Van der Biezen, E., B. Brandwagt, W. van Leeuwen, and J.H. Nijkamp. 1996. Identification and isolation of the *FEEBLY* gene from tomato by transposon tagging. Mol Gen Genet 12: 267-280.

Velcheva, M., Z. Faltin, M. Flaishman, Y. Eshdat, and A. Perl. 2005. A liquid culture system for *Agrobacterium*-mediated transformation of tomato (*Lycopersicon esculentum* L. Mill.). Plant Sci 168: 121-130.

Villiers, RPd, R.J.v. Vuuren, D.I. Ferreira, and J.v. Staden. 1993. Regeneration of adventitious buds from leaf discs of *Lycopersicon esculentum* cv. Rodade: optimization of culture medium and growth conditions. J South African Soc Hort Sci 3: 24–27.

Vnuchkova, V.A. 1977a. Development of a method for obtaining regenerated tomato plants under tissue culture conditions. Fiziol Rast 24: 1094–1100.

Vnuchkova, V.A. 1977b. Elaboration of methods for obtaining tomato plants by tissue culture. Fiziol Rast 24: 1095–1100.

Wang, Y., W. Michael, M. Richard, C. Richard, M. Cui, and W. Robert. 2005. Overexpression of cystolic ascorbate peroxidase in tomato confers tolerance to chilling and salt stress. J Amer Soc Hort Sci 130: 167-168.

Watson, C.F., L. Zheng, and D. Della Penna. 1994. Reduction of tomato polygalacturonase b subunit expression affects pectin solubilization and degradation during fruit ripening. Plant Cell 6: 1623-1634.

Weretilnyk, E.A., S. Bednarek, K.F. McCue, and D. Rhodes. 1989. Comparative biochemical

development (Table 10.1). The growth and maturation of tomato fruit follow a single sigmoidal growth curve (Fig. 10.1), with the fruit development being broadly divided into four phases (Tanksley 2004). *Phase I* represents floral development (including ovary), fertilization and fruit set. It is during this phase that a decision is made to either abort or proceed with cell division leading to the fruit set. *Phase II* involves cell division that lasts for 7-14 days after pollination (Mapelli et al. 1978). Although during this phase most of fruit cells are established, fruit growth is slow, reaching only about 10% of the final fruit weight (Fig. 10.1). *Phase III* primarily comprises of cell expansion which, depending upon the genotype, continues for 3-5 weeks and is responsible for attainment of the maximum fruit size (Ho and Hewitt 1986). This phase is followed by slow growth and intense metabolic changes resulting in fruit ripening (*Phase IV*). On the basis of gel formation and lycopene accumulation, the ripening phase of the tomato fruit has been subdivided into several stages including MG1 (mature green; firm locular tissue), MG2 (small amount of gel), MG3 (gel formation complete), MG4 or breaker (fruit pigment detectable), turning (10-30% red fruit), pink (30-60 % red), light-red (60-90 % red) and red (over 90% red) (Su et al. 1984, Lincoln et al. 1987). The various structures and tissues of the fruit can be traced back to that of the ovary. Tomato fruit is composed of the flesh (pericarp walls and skin) and pulp (placenta and locular tissues including seeds), the ovary wall comprising the flesh and placental tissue expanding into locules and seeds comprising the pulp (Ho and Hewitt 1986). Life cycle of tomato fruit spans 49-70 days from fertilization to the "red ripe" stage. Many factors such as the cultivar, position on the cluster, climatic conditions and cultural practices influence the overall fruit development. The genetic, developmental and molecular bases of fruit size and shape variation in tomato have been reviewed elsewhere (Tanksley 2004).

HORMONAL FLUX AND INTERACTIONS DURING FRUIT DEVELOPMENT

Several investigators have quantified levels of plant hormones in developing tomato fruits (Abdel-Rahman 1977, Mapelli et al. 1978, Sjut and Bangerth, 1982, 1983, Buta and Spaulding 1994). However, there are discrepancies between the reported patterns. For example, some investigators have reported that AUXs levels peak during the cell expansion phase (Abdel-Rahman 1978, Sjut and Bangerth 1982, 1983) whereas others reported it to peak during the early phase (5-10 days after anthesis) of tomato fruit development (Mapelli et al. 1978, Buta and Spaulding 1994).

Table 10.1 Hormonal and developmental mutants of tomato

Mutant	Description	Phenotype	Hormonal deviations	Reference
pat	Parthenocarpy	Short anthers, aberrant ovules, autonomous ovary growth, aberrant pollen-tube-ovary interactions. High PA content during flower development	GA metabolism pathway may be affected. Accumulation of PAs, especially spermine and spermidine, during pre-anthesis floral stages.	Mazzucato et al. (1999, 2000), Antognoni et al. (2002)
pat-2	Parthenocarpy (Russian cv. Severianin)	Seedless fruit set with complete locular development and normal fruit weight and size.	High GA_{20} levels in the ovaries	Lin et al. (1983), Fos et al. (2003)
pat-3/4	Parthenocarpy (German line RP75/59)	-Same as above-	Enhanced early 13-hydroxylation pathway of GA biosynthesis	Fos et al. (2001)
gib-1	GA deficient	Extreme dwarfism, reduced germination, and abnormal flower development.	GA biosynthetic pathway affected. (Lesion in *ent-copalyl diphosphate synthase*, gene)	Bensen and Zeevart (1990), Koorneef et al. (1990)
gib-3	GA deficient	Same as above	GA biosynthetic pathway affected. (Lesion in *ent-kaurene synthase*, gene)	Bensen and Zeevart, (1990), Koorneef et al. (1990)
hp/ dg (allelic)	High pigment/ dark green mutation	Exaggerated light responsiveness, high fruit and foliar pigmentation.	Reduced GA levels	Wann, (1995)
R-mutant	Carotenoid – deficient ripening mutant	Yellow fruit flesh and skin color, pale yellow corolla; absence of ripening associated carotenoid increase.	Reduced ABA levels, elevated GA levels	Fraser et al. (1995), Fray and Grierson (1993)
sitiens (sit^w)	ABA deficient	Precocious seed germination	ABA deficient	Groot and Karssen (1992), Liu et al. (1996, 1997)
rin	Ripening inhibitor (*LeMADS-RIN*)	Non-climacteric, severely impaired ripening.	Low ethylene production, impaired putrescine decline during ripening.	Tigchelaar et al. (1978), Hong and Lee, (1996), Vrebalov, (2002)
nor	Non-ripening	Non-climacteric, impaired ripening.	Low ethylene and ABA production, delayed ABA accumulation.	Tigchelaar et al. (1978), Hong and Lee, (1996)

(Contd.)

(Contd.)

Nr	never ripe (ETR1)	Failure to ripen	Insensitive to ethylene.	Tigchelaar et al. (1978). Lanahan et al. (1994)
alc	alcobaca	Delayed fruit ripening.	Reduced ethylene, high PA contents.	Dibble et al. (1986)
cnr	colorless, non-ripening	Ripening inhibited, fruit white at maturity, turns yellow later; remains firm.	Reduced ethylene.	Thompson et al. (1999)
poc	polycotyledonous	Multiple cotyledons.	Enhanced polar auxin transport.	Al-Hammadi et al. (2003)
ls*	Lateral suppressor	Absence of side shoots, lower number of flowers per inflorescence, absence of petals, reduced male and female fertility, limited seed yield and low rate of seed germination.	High GA and auxin levels, lower cytokinin levels.	Schumacher (1999)
pro*	procera	Giant mutant, with altered leaf morphology, less vigorous root system.	Lower GA levels.	Jones, (1987), van Tuinen et al. (1999)
flc*	flacca	Wilty phenotype.	Reduced ABA levels.	Sagi et al. (1999)
not*	notabilis	Wilty phenotype.	Reduced ABA levels.	Burbidge et al. (1999)
d*	extreme dwarf	Dwarf plants with dark green rugose leaves.	Brassinosteroid deficient. (biosynthesis mutant)	Bishop et al. (1999)
dpy*	dumpy	Short stature, reduced axillary branching, and altered leaf morphology.	Brassinosteroid insensitive (signaling mutant)	Koka et al. (2002)
cu3*	curl3	Altered leaf morphology, de-etiolation, and reduced fertility.	Brassinosteroid insensitive. (signaling mutant)	Koka et al. (2003)
dgt*	diageotropica	Stunted, flowering time, fruit set, weight locule and seed number altered.	Blocked rapid auxin reaction of tomato hypocotyls.	Balbi and Lomax (2003) Coenen et al. (2003)
7B-1*	photoperiod sensitive	Photoperiod-dependent male sterility, reduced de-etiolation of hypocotyls in long days and increased seed size and weight.	High ABA levels.	Fellner et al. (2001)

* Effects on fruit growth and development not yet described.

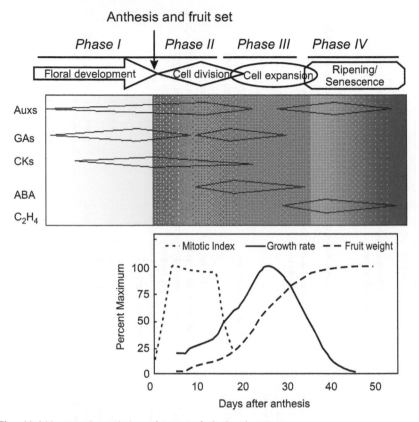

Fig. 10.1 Hormonal regulation of tomato fruit development;

Upper panel: Developmental phases of tomato fruit.

Middle panel: Changes in hormonal levels observed during fruit development (redrawn with approval from Gillaspy et al. (1993), copyright American Society for Plant Biologists).

Lower panel: Cell division as mitotic index is redrawn with approval from Cong et al. (2002). Copyright (2002) National Academy of Sciences, U.S.A.; fruit weight gain and growth rate (weight gained/day) are redrawn and calculated from Abdel-Rahman (1977) with approval. All data are normalized as percent maximum observed during fruit development.

These discrepancies need to be resolved before the correlation between endogenous levels of phytohormones and fruit development can be established. In spite of these discrepancies, some generalization can be drawn (Fig. 10.1) about the patterns of appearance and changes in the levels of the five classic plant hormones namely, AUXs, CKs, GAs, ABA and ethylene during tomato fruit development. *Phase I* that comprises ovary development, anthesis, fertilization and fruit set events, witnesses increase

flux modulating IAA oxidase activity (Haard 1977, Thomas and Jen 1980). IAA is considered a natural fruit ripening retardant. Increase in IAA oxidase activity prior to respiration climacteric has been associated essentially with a decrease in IAA levels that is necessary for induction of ripening (Frenkel 1972, Thomas et al. 1981). IAA oxidation by peroxidases in tomato could also be enhanced by phosphate, oxalate, pyrophosphate, malate and citrate, and decreased by the presence of Mn^{2+} (Pressey 1990). Collectively, these studies indicate that CK and AUX interact at various levels and thus affect each other's responses in the developing fruit.

Emerging evidences indicate that GAs and AUXs affect each other's biosynthesis (O'Neill and Ross 2002, Swarup et al. 2002). Auxin maintains bio-reactive GA levels in elongating pea hypocotyls by promoting synthesis of GA_1 from GA_{20} by enhancing a GA_3-oxidase activity and simultaneously inhibiting deactivation of GA_1 (Ross et al. 2003). AUX-specific regulation of gibberellin 3β-hydroxylase has been reported during pollination and development of pea fruit and seeds; however, these mechanisms have not yet been demonstrated in tomato fruit. GA and ABA have been shown to have antagonistic effects during fruit development and seed maturation, whereas ABA has been shown to have a stimulatory effect on ethylene production. This stimulatory effect is thought to be the result of an enhancement in the synthesis of 1-aminocyclopropane-1-carboxylic acid (ACC), a key enzyme involved in ethylene biosynthesis, rather than due to the induction of senescence (Riov et al. 1990). Ethylene synthesis is also affected by AUX levels and during the post-pollination phase. Investigations with polyamines (Mehta et al. 2002, Cassol and Mattoo 2003), brassinosteroids (Schlagnhaufer and Arteca 1985a, b; 1991) and jasmonic acids (Miyamoto et al. 1997) have also shown their possible role in regulation of hormonal activity.

Hormonal Regulation of Early Fruit Development

The signals for fruit set are generated during pollination and the decision to set fruit is dependent upon successful fertilization. The manifestations of pollination involve two diametrically opposite events: (1) Wilting and abscission of the corolla and calyx; and (2) preparation of other organs for fertilization, embryogenesis and fruit development (O'Neill 1997). Although pollination-regulated development is initiated at the stigma surface, it is hypothesized that secondary inter-organ signals amplify and transmit the primary pollination signals to bring about a plethora of changes witnessed after pollination. Increase in ethylene production in response to pollination has been observed in many flowers (O'Neill 1997) and has been implicated in coordinating many of the pollination-associated events such as ovary growth and senescence of the perianth (Larsen et al. 1993, Woltering et al. 1994). In contrast to observations in many ornamental flowers where ethylene levels increased simultaneously or before pollen germination (see O'Neill 1997), increase in ethylene levels in tomato pistil occurs after pollen germination and penetration of the pollen tubes into the stylar tissue (Lloup-Tous et al. 2000). Using *dialytic* (*dl*) and *Neverripe* (*Nr*) tomato mutants, Llop-Tous et al. (2000) reported that pollination-induced enhanced expression of *LeACS1A* gene, a member of the ACC synthase (ACS) gene family, is independent of ethylene, whereas the expression of *LeACO1, 2* and *3*, members of ACC oxidase (*ACO*) gene family, is ethylene dependent. In contrast, the expression of *LeACO4* is ethylene independent. *LeACS1A* seems to be the sole *ACS* gene responsible for elevated ethylene levels in tomato pistils as well as the ethylene-mediated expression of *LeACO1* and *LeACO3*, which seems essential for the timely ethylene production in response to pollination. Since ethylene perception is necessary for the decrease in ethylene production after reaching the maximum, it has been suggested that ethylene acts as a regulator of its own biosynthesis following pollination. Ethylene is known to stimulate (auto-activation) or inhibit (auto-inhibition) its own biosynthesis (see Mattoo and White 1991).

TOMATO MUTANTS: A WINDOW TO HORMONAL REGULATION OF FRUIT DEVELOPMENT

The effect of pollination can be mimicked by hormones, in particular AUXs and GAs, resulting in fruits without seeds (Goodwin 1978, Schwabe and Mills 1981, George et al. 1984, Gillaspy et al. 1993). The phenomenon of seed-less fruit development is referred to as parthenocarpy and may occur due to lack of pollination, pollination not leading to fertilization, or embryo abortion. Parthenocarpy has provided much information about the possible roles of hormones during early fruit development (George et al. 1984). Several lines of evidences indicate that AUXs and GAs play roles in setting and development of parthenocarpic fruits. Parthenocarpic lines show higher endogenous levels of AUXs and GAs in the ovaries as compared to the normal tomato lines (Gustafson 1939, Nitsch et al. 1960, Mapelli et al. 1979, Mapelli and Lombardi 1982). Also, exogenous application of AUXs, GAs, or their inhibitors, to the flower/ovary before fertilization can cause parthenocarpy (Robinson et al. 1971). The accumulation of sufficient levels of AUXs within the ovary effects fruit set and activation of cell division in the absence of fertilization indicating that incorrect temporal and/or spatial regulation of AUX synthesis may be responsible for parthenocarpy (Gillaspy et al. 1993). GAs produced by pollens have been implied in the signal transduction pathway leading to increased AUX synthesis during pollination. Application of GAs to unpollinated tomato flowers causes increased levels of AUXs in the ovary (Sastry and Muir 1963). The developmental control of seeded tomato fruit, at least in part, by GAs is supported by studies involving GA application to unpollinated ovaries (Bunger-Kibler and Bangerth 1982, Sjut and Bangerth 1982, Sawheny 1984, Alabadi et al. 1996, Fos et al. 2000, 2001), identification and quantification of GAs in parthenocarpic and seeded fruits (Bohner et al. 1988, Koshioka et al. 1994) and the detection of transcripts of genes encoding for enzymes in GA biosynthesis in developing fruits (Rebers et al. 1999).

Genetic evidence for the role of GAs in development of fertile tomato flowers comes from studies with the dwarf, GA-deficient mutant, *gib-1* (Groot et al. 1987). The floral development in this mutant is not completed in the absence of GAs but a single application of GA_3 or GA_{4+7} restores its fertility resulting in seed set. The isolation and characterization of many GA-biosynthetic genes (Hedden et al. 2002) has begun to provide a new insight into the role of GAs in fruit set (Nester and Zeevart 1988, Rebers et al. 1999). Nester and Zeevart (1988) have reported that the GA-deficient tomato mutant *gib-2* shows improper ovary development and degeneration of tapetal cells. Based on expression patterns and localization in anther tapetum cells and placental tissue of a *GA 20 oxidase* gene, namely, *Le20ox-2*, during flower bud and early fruit development of tomato, Rebers et al. (1999) have suggested that *Le20ox-2* may be responsible for the required GA biosynthesis in these tissues. The transcript levels of *Le20ox-1* were higher during anthesis, whereas the expression of copalyl diphosphate synthase (*LeCPS*) was considerably increased during flower senescence and early fruit development. Based on increased GA_{20} oxidase activity in the unpollinated ovaries of tomato lines carrying *pat-2* mutation, it has been suggested that the *pat-2* gene induces parthenocarpy by enhancing synthesis of GA_{20}, the precursor of an active GA (Fos et al. 2000). Two flower specific cDNAs (*tgas100* and *tgas105*) up-regulated by GAs have been isolated from GA deficient *gib-1* mutant of tomato (van den Heuvel et al. 2002). The deduced TGAS105 polypeptide shows homology to extensin-like proteins, whereas TGAS100 polypeptide is similar to a stamen specific gene from *Antirrhinum*.

Expressions of *GAD3*, a short chain alcohol dehydrogenase like-gene (Jacobsen and Olszewski 1996) and H1 and H2B tomato histone (van den Heuvel et al. 1999) genes have been reported to be GA responsive in shoot and leaf tissues of tomato, respectively. Differential display amongst parthenocarpic mutants and their near isogenic wild type lines has resulted in isolation of three genes namely *LeH2A-2*, *GAD3* and *LeSPH1* (*L. esculentum S protein homologue*). Among these genes *LeH2A-2* and *GAD3* showed GA responsiveness in wild types but not in the *pat* lines (Testa et al. 2002). *LeH2A-2*, which belongs to a small group of plant histone H2As, showed increased transcript levels in tomato ovaries and is speculated to play important role in cell division phase of fruit growth with developmental regulation and tissue-specific expression patterns (Dong et al. 1998, Joubès et al. 1999, van den Heuvel et al. 1999). *GAD3* expression is strongly up-regulated in ovary at the time of pollination and is maintained throughout the phase of active ovary growth. Accumulation of gene transcripts during active cell division phase and their GA responsiveness suggest their involvement in the early fruit development (Testa et al. 2002).

Ovary growth in *pat* mutants initiates before anthesis (Mazzucato et al. 1998). The complete machinery for fruit set and development was found to be switched on before, and independently from pollen shedding, pollination and fertilization in these mutants. Organ identity and development was found to be affected in *pat* mutants with malformed male and female organs. The androecium is typically short, irregular and the anthers are not fused which leave the stigma exerted; dehiscence is preferentially external and although the pollen is fertile, the number of grains was found to be reduced (Mazzucato et al. 1998). Since the ovule development is found to be aberrant in *pat* mutants, it is suggested that parthenocarpy is a secondary effect of a mutated gene whose primary function is to regulate floral organ development. It is further speculated by these authors that the anther aberrancy, especially the occurrence of adaxial carpel-like structures bearing external ovules, is indicative of *pat* mutation affecting a homeotic gene. Defective cell elongation of *pat* organs and reversion of *pat* anther phenotype by application of GA_3 led Mazzucato et al. (1998) to propose that this mutation may have interactive roles with the GA metabolic pathway.

MADS-box genes have been recognized to play a central role in plant development, especially during flower development (Lohmann and Weigel 2002). At least seven MADS-box genes expressed through the first stages of fruit and seed development in tomato have been identified (Busi et al. 2003). These include three previously identified flower identity genes (*TAG1*, *TDR4* and *TDR6*; Pnueli et al. 1991, 1994a, b). The other four are novel MADS-box genes (*TAGL1*, *TAGL2*, *TAG11* and *TAG12*). In view of the hypothesis that successful pollination and fertilization generates signals essential for normal fruit development, it has been suggested that induction of certain MADS box genes by these signals initiates early developmental program in the fruit (Busi et al. 2003). The proposed MADS box candidates for this process include *TDR4, TAGL2, TAGL11* and *TAGL1* as these are induced immediately after anthesis in the ovary wall. The antisense down-regulation of TM29, a tomato SEPALLATA homolog, causes parthenocarpic fruit development and floral reversion (Ampomah-Dwamena et al. 2002). *LeMADS-RIN*, a MADS box gene has been shown to play a decisive role in fruit ripening (Vrebalov et al. 2002). Further characterization of hormonal regulation of MADS box genes should help understand the roles of these genes in fruit development.

HORMONAL REGULATION IN CELL DIVISION, SEED FORMATION AND EARLY EMBRYO DEVELOPMENT

The final fruit size is dependent on the number of cells within the ovary before fertilization, the number of seeds, the number of cell divisions that occur in the developing fruit after fertilization, and the extent of cell expansion (Gillaspy et al. 1993). Several hormones and growth regulators including AUXs, CKs, GAs, BRs, ABA, PAs and sugar play significant roles in these processes (Dewitte and Murray 2003). Following fertilization, tomato ovary undergoes a period of active cell division (*Phase II*) which continues for approximately 7-14 days (Mapelli et al. 1978, Bohner and Bangerth 1988). In the very early stages of this phase, the mitotic activity is higher in outer pericarp and the placental tissue while cell division in the developing seeds is confined to the peripheral integument layers rather than the embryos (Suzuka et al. 1989, Daidoji et al. 1992). The highest CKs levels in the developing seed during *Phase II* correlate with cell division activity in the surrounding tissue (Abdel-Rahman 1977, Bohner and Bangerth 1988). Fruits from wild-type normal and *rin* mutant show little change in the endogenous CK levels during ripening (Hong and Lee 1993), indicating that although CKs play crucial role during the early cell division phase, they have a limited role in fruit ripening. CKs primarily accumulate in developing seed, with low levels in the rest of the fruit. The molecular basis of CKs accumulation in seed is not understood. However, based on low accumulation of CKs in parthenocarpic fruits, CKs have been hypothesized to be transported from the site of synthesis with developing seeds signaling this transport (Mapelli 1981, Bohner and Bangerth 1988). The rapid seed growth during cell division phase is attributed to increased CKs levels (Bohner and Bangerth 1988). Frequently, a positive correlation between seed number in the fruit and fruit size is observed, with the distribution of the seed throughout the fruit affecting fruit shape (Ho 1992).

Auxins produced in the seeds are also implied in the cell division or cell enlargement of the tissue surrounding the seeds (Ho 1992) and are thought to be important for the seed growth (Bohner and Bangerth 1988). The molecular analysis of plant cell cycle progression has revealed cyclin-dependent kinases (CDKs) as one of the key regulators of this process (Dewitte and Murray 2003). CDKs represent a highly conserved superfamily of serine/threonine protein kinases whose activity requires binding to a cyclin. During cell cycle, the G1–S and G2–M phase transitions in plants are under the control of several hormones including AUX, CKs, ABA, GAs, BRs and sugar. These growth regulators control the expression of members of CDKs and cyclins families. In plants, five distinct classes of CDKs (CDKA through CDKE) have been defined on the basis of phylogenetic, structural and functional similarities with animal and yeast CDKs (Joubès et al. 2000). In developing tomato fruit, the differential expression of CDKA and cyclins is correlated with the temporal and spatially regulated mitotic activity. The accumulation of Lyces;CDKA1 and Lyces; CDKA2 transcripts and proteins were observed between anthesis and 5 d post anthesis (Joubes et al. 1999). The overall expression patterns of three additional CDKs, (CDKC;1, CDKB1;1, CDKB2;1) were shown to be similar to the expression of CDKA, i.e. the expression was higher in dividing tissues (Joubes et al. 2001). However, in contrast to that of CDKA and CDKBs, the expression of CDKC;1 in tomato cell suspension cultures and excised roots was found to be independent of sugar or hormonal supply. Taken together these results indicate a hormonal regulation of CDKs and cell division during early fruit development.

The AUX-resistant diageotropica (*dgt*) mutation dramatically alters early fruit development including increases in the time to flowering and the time from anthesis to the onset of fruit, fruit weight, fruit set, and the numbers of locules and seeds, but fruit ripening remains normal (Balbi and Lomax 2003). Some of the long-term effects of CKs application to wild-type seedlings, such as stunting of root and shoot growth, reduced elongation of internodes, reduced apical dominance, and reduced leaf size and complexity, are similar to that caused by *dgt* mutation (Coenen et al. 2003). Additionally, CK treatment inhibits AUX-stimulated elongation, H^+ secretion, and ethylene synthesis in wild-type hypocotyl segments, and thus mimics the impaired AUX responsiveness found in *dgt* hypocotyls. However, CK treatment inhibits the AUX-induced expression of only one of the two ACS genes that exhibit impaired AUX inducibility in *dgt* hypocotyls. Also, CK treatment fails to inhibit AUX induced *LeSAUR* gene expression, an AUX response that is exhibited by the *dgt* mutant. These results suggest that CKs inhibits only a subset of the AUX responses impaired in *dgt* hypocotyls and has led the authors to propose that CK effects on AUX responses are mediated through interactions between specific AUX- and CK-signaling pathways rather than through global effect of CKs on active AUX levels or responses. Based on the effects of *dgt* mutation on fruit morphology and differential expression of subsets of *LeIAA* and *LeACS* gene family members it has been suggested that AUX- and ethylene-mediated gene expression play significant roles during the early stages of fruit development. (Balbi and Lomax 2003).

REGULATION OF CELL EXPANSION AND EMBRYO DURING FRUIT MATURATION

The period of active cell division is followed by *Phase-III* in which the growth of the fruit is mostly due to cell expansion. In tomato, there is a 10- to 20-fold increase in the volumes of cells in placenta, locular tissue, and mesocarp tissue but cells comprising the exo- and the endocarp, which continue to divide, expand less (Gillaspy et al. 1993). There is a general consensus that AUXs play a pivotal role during cell expansion (see Hager 2003). AUXs have been implicated in the elongation of stem and coleoptile cells by promoting wall loosening via cleavage of load-bearing bonds in the cell wall (Rayle and Cleland 1992). The mode of action of the symplast located AUXs is thought to be through the release of protons from AUXs exposed cells into the apoplast (acid growth theory). The resulting decrease in the pH has been hypothesized to activate pH-sensitive enzymes and proteins within the wall causing cell-wall loosening and extension growth (see Hager 2003). Some of the early AUX-response genes involved in AUX signaling have been identified and grouped into four major classes viz., *Aux/IAAs*, *SAURs*, *GH3s*, and *OsARF1* (Abel and Theologis 1996, Hagen and Guilfoyle 2002, Waller et al. 2002). Some of the cell wall loosening proteins/enzymes activated at acidic pH include expansins (Cosgrove et al. 2002), xyloglucan hydrolase (XGH) and xyloglucan endotransglycosylase (XET) (Fry et al. 1992) and yieldins (Okamoto-Nakazato 2002). Expression of an expansin gene (*LeExp2*), xyloglucan endotransglycolase (*LeEXT1*) and endo-1, 4-β- glucanase (*Cel7*) are correlated with the increases in AUX levels during tomato fruit growth with negligible expression in ripening fruit (Catala et al. 2000). *LeExp2* was isolated from AUX-treated etiolated tomato hypocotyls whereas *LeEXT1* and *Cel7* have been reported to be AUX-regulated in etiolated hypocotyls (Catala et al. 1997). The transcript levels of both *LeEXP2* and *LeEXT1* peaked during the stages of higher rates of fruit growth, while *Cel7* expression increased and remained high during later stages of fruit expansion. The down regulation of these genes coincides with the cessation of the cell expansion phase and entry of the fruit into the ripening phase. The patterns of *LeEXP2*, *LeEXT1* and *Cel7* expression during the period of rapid growth and increased AUX levels have been interpreted to indicate their roles in developmental and/or hormonal signal transduction network controlling cell expansion in fruit. On the other hand, expression of ripening-related endo-1, 4-β- glucanases, XET, and expansin genes have been shown to be ethylene up-regulated in tomato fruit (Lashbrook et al. 1994, Arrowsmith and de Silva 1995, Rose et al. 1997). These data suggest that different hormones regulate expression of the cell wall modifying proteins/enzymes and thereby coordinate the growth and ripening processes.

The exogenous application of AUXs causes delayed tomato fruit ripening (Vendrell 1985, Cohen 1996) indicating a role for AUXs in regulating the capacity of fruit to ripen. AUXs have been shown to modulate plant growth and development through transcriptional regulation of specific genes (Ulmasov et al. 1999). The functional analyses of the promoter regions of AUX-regulated genes have identified a conserved AUX-responsive *cis*-element (AuxRe) such as TGTCTC (Ulmasov et al. 1995) and TGTCCCAT (Oeller et al. 1993). AUX response factors (ARFs) that bind to these *cis*-elements have been characterized and shown to regulate transcription in an AUX-dependent manner and to interact with short-lived nuclear proteins belonging to the *Aux/IAA* family of AUX-responsive transcription factors (Abel et al. 1994, Kim et al. 1997, Ulmasov et al. 1997, 1999). Differential screening of gene expression during fruit development has resulted in isolation of four developmentally regulated cDNAs (DR1, DR3, DR4 and DR8), which show homologies to *Aux/IAA* family and one cDNA (DR12) belonging to ARF transcription factor family (Jones et al. 2002). Ethylene regulates accumulation of the *Aux/IAA* like transcripts in tomato fruits but not in leaves, suggesting that these ARFs participate in the ethylene-dependent, developmentally regulated gene expression. DR12 shows nuclear localization and its transcripts are most abundant in ripening fruit. Antisense and sense co-suppressed DR12-inhibited transgenic tomato lines display pleiotropic phenotypes that include dark-green immature fruits, unusual cell division in pericarp, blotchy ripening and enhanced fruit firmness, upward curling of leaves and increased hypocotyl and cotyledon growth (Jones et al. 2002).These authors have suggested that the observed phenotypes of DR12 transgenic lines could be due to increased responsiveness of tissues to AUXs or alternatively, the inhibition of DR12 affects response to the endogenous CKs and ethylene.

CKs are involved in regulation of cell division and biological processes such as active growth, metabolism and plant development, all of which show enhanced carbohydrate demand. Because of this association, CKs have been linked (Balibrea et al. 2004) to regulation of assimilate partitioning (Brenner and Cheikh 1995), sink strength (Kuiper 1993), and source-sink relations (Roitsch and Ehness 2000). The extracellular invertases have been implicated in supplying carbohydrate to various sink tissues (Koch 2004). GAs, AUXs and ethylene are also known to regulate extracellular invertases (Roitsch et al. 2003). Amongst the four tomato invertase genes identified (*Lin5, Lin6, Lin7* and *Lin8*), *Lin6* transcripts accumulate in actively growing sink tissues such as seedlings, roots, flower buds and tumors (Godt and Roitsch 1997). Since the expression of *Lin6* is induced by zeatin, it has been suggested that CKs regulate the expression of extracellular invertases in sink tissues. Extracellular invertases also play an essential role in CK-mediated delay of senescence. Tobacco transgenic plants harboring an invertase inhibitor gene under the control of CK-inducible *Lin6* promoter failed to show a delay in leaf senescence in the presence of CK (Balibrea-Lara et al. 2004). *Lin6* has also been shown to be induced by BRs (Goetz et al. 2000). *Lin7* expression is localized in tapetum and pollen and *Lin5* expression is higher in non-mature green tomato fruit than mature or red tomato fruits (Proels et al. 2003). The *Lin5* promoter sequence contains GA-, AUX- and ABA-response motifs that confer hormone inductivity to a truncated *nos* promoter-*GUS* fusion. *Lin5* induction by AUX has been suggested to mediate a higher carbohydrate supply to developing flower and fruit tissue, especially in the early developmental stages. The inductivity of *Lin5* promoter with GAs further suggests that GAs affect the extracellular invertase activity during flower development.

During *Phase III*, the seed growth does not parallel cellular expansion of fruit. As the embryo develops from globular to bilateral structure it shows well-developed cotyledons and an established root-shoot axis (Smith et al. 1935). AUX concentration that is higher in the seeds than in the surrounding fruit cells has been interpreted to play a role in the cell wall extensibility and establishment of fruit tissue as the sink. Polar auxin transport (PAT) has been implicated in determination of pattern specificity and embryological studies in *Arabidopsis* mutants. These results suggest roles for elements involved in AUX perception and distribution during transition from globular to heart shape and the ultimate appearance of the cotyledons, marking the establishment of the bilateral symmetry (see Al-Hammadi et al. 2003). The polycotyledon (*poc*) mutant of tomato, which has multiple cotyledons and displays several abnormalities during vegetative development, shows enhanced PAT (Al-Hammadi et al. 2003). The *poc* is located on the 9th chromosome of tomato and has been implicated in the separation of the two cotyledons (Hadfi et al. 1998)

Abscisic acid (ABA) is known to play important roles in seed maturation and dormancy. Developing seeds in contact with the sheath and the locular tissues do not germinate inside the fruit (Gillaspy et al. 1993). However, the removal of tomato seeds even at an early stage of fruit development does not prevent their germination in water, suggesting that seed germination is suppressed within the fruit. Osmotic potential, along with ABA, which induces dormancy to the seed, has been suggested to prevent precocious seed germination (Berry and Bewley 1991). Characterization of gib-1 and ABA-deficient *sit*[w] tomato mutants indicates that the precocious seed germination is prevented by the action of the fruit's osmotic environment and ABA on the seed tissues surrounding the embryo and not the embryo itself (Liu et al. 1996). It has been proposed that the combined action of these two components prevents early germination by preventing endosperm weakening which is an essential feature in germination as it facilitates water uptake by the embryo.

ABA and sugars show similar or antagonistic effects on diverse physiological processes during seed development, germination, and seedling growth. A number of *Arabidopsis* mutants that exhibit germination ability and seedling development on inhibitory concentrations of sugars have been identified in *Arabidopsis*, and interestingly, are found to be allelic to known mutations in ABA synthesis (Bradford et al. 2003). These and other observations indicate interactions between sugar and hormonal signaling (Finkelstein and Gibson 2002). Sucrose non-fermenting 1 (*SNF1*)-related kinase (SnRK1) complex has been implicated in both sugar and ABA sensing (Himmelbach et al. 2003, Lunn and MacRae 2003). The central component of the sugar sensing and response mechanism is thought to be the *SNF1*-related kinase (SnRK1) complex which was first identified in yeast (Halford and Hardie 1998, Halford et al. 2000, 2003). cDNAs corresponding to the kinase (*LeSNF1*), regulatory (*LeSNF4*), and localization (*LeSIP1* and *LeGAL83*) subunits of the SnRK1 complex were identified and their expression characterized during seed development in tomato (Bradford et al. 2003). These studies revealed that *LeSNF4* expression is influenced by ABA and GA levels, along with other factors influencing seed germination. These authors have suggested that during the seed maturation binding of

LeSNF4 to LeSNF1/LeGAL83 (or other SIP proteins) alters the kinase activity of the complex, thereby promoting metabolic pathways involved in the accumulation or maintenance of storage reserves and blocking those involved in the mobilization or utilization of stored reserves. After imbibition, expression of *LeSNF4* was found to be reduced in seeds that were not dormant or stimulated by GA, which potentially altered LeSNF1 kinase activity to de-repress genes encoding enzymes required for reserve mobilization and metabolism (Bradford et al. 2003). This study provides an insight into the possible mechanism of sugar and hormonal regulation of seed maturation and germination.

ROLES OF OTHER PLANT GROWTH REGULATORS

Brassinosteroids (BRs), a group of plant steroid hormones, play diverse roles in plant growth and development (see Bishop 2003). Exogenous BRs application to tomato pericarp discs leads to elevated levels of lycopene, sugars (reduced and total) and ethylene, and decreased ABA levels (Vardhini and Rao 2002). BR regulated genes including cell wall modifying xyloglycan endotransglycosylases (Zurek and Clouse 1994, Oh et al. 1998) are implied in BR-induced cell elongation. BR induced growth responses have been correlated with increased carbohydrate supply brought about by the increased levels of the extracellular invertase, *Lin6* (Goetz et al. 2000). Characterization of tomato *extreme dwarf* (d^x) and *dumpy* (*dpy*) mutants have revealed genes that are involved in the biosynthetic pathway of BRs (Bishop et al. 1999, Koka et al. 2000). Tomato homologue of BRI1, an essential component of the BR receptor complex that was isolated from the signaling mutants *curl3* (*cu3*) and *abs* (*altered brassinolide sensitivity 1*), showed over 99% identity with the SR160 systemin receptor (see Szekeres 2003). Systemin is a polypeptide hormone known to regulate wound inducible genes in tomato (Pearce et al. 1991) and activates a lipid-based signaling cascade, which causes release of linolenic acid from the membrane which is then converted to oxylipins and jasmonic acids (Bergey et al. 1996). Both BRI1 and SR160 are implicated in BR and systemin signaling (see Szekeres 2003).

Jasmonic acid (JA) and its methyl ester (methyl jasmonate, MeJA) are derivatives of linolenic acid and are known to modulate aspects of fruit ripening, pollen viability, root growth and resistance to insect and pathogen attack (Creelman and Mullet 1997). JA-induced defense responses in tomato have profound effect on their reproductive fitness. Treatment of plants with higher levels of JA produce fewer but larger fruits with fewer seed per unit of the fresh weight (Redman et al. 2001). Leucine aminopeptidase (LapA) expression increases under the influence of systemins, MeJA, ABA, ethylene and stress conditions such as water deficit and salinity in tomato. *LapA1-promoter:GUS* transgenic tomato plants revealed that this promoter is active during floral and fruit development (Chao et al. 1999). However, at present little is known about the role of JA in fruit growth and maturation.

Polyamines (PAs) are ubiquitous organic cations that affect a large number of developmental and physiological responses in a number of organisms, including plants (see Walters 2003). They have also been implicated to influence early fruit development and ripening. In tomato, both ornithine decarboxylase (ODC) and arginine decarboxylase (ADC) pathways for PA biosynthesis are active (Alabadi and Carbonell 1998). Application of PAs to wild type unpollinated ovaries results in partial parthenocarpy. The higher PA levels in unpollinated *pat-2* ovaries are correlated with the activation of the ODC pathway, which in turn is influenced by elevated GA levels found in these ovaries (Fos et al. 2003). Decrease in arginase levels and ODC activity after tomato fruit set is hypothesized to indicate that the ADC pathway is involved in cell expansion and the ODC pathway in cell division during early fruit growth of tomato (Cohen et al. 1982, Alabadi et al. 1996). Developmentally regulated increases in spermidine and spermine in transgenic tomato overexpresing a yeast SAM-decarboxylase enhanced lycopene and ethylene levels and increased fruit juice viscosity in tomato fruit (Mehta et al. 2002). PAs, along with salicylic acid, which is an inhibitor of wound-responsive genes in tomato, have been suggested to regulate ethylene biosynthesis at the level of ACC synthase transcript accumulation (Li et al. 1992).

FUTURE PERSPECTIVE

Fruit development has emerged as an important area of research in plant development. Much progress has been made in identifying the growth regulators involved in fruit development and ripening and the gene and protein receptor players. Most of these studies have implicated cross talks and signaling among the classic hormones. However, there seem to be other dimensions to these processes, involving other plant growth regulators such as polyamines, brassinosteroids and methyl jasmonates. Only recently, molecular work with other plant growth regulators such as polyamines and its role in these processes has begun (Mehta et al. 2002). The hormonal regulation of plant development is a complex process, especially because of the interactive nature of plant hormones. Likely, many more signaling compounds play role in fruit set, development and ripening. The primary question regarding fruit growth and development is not only which genes regulate these processes but how the genetic and molecular circuitries determine the differential expression of the genes underlying the development of fruit phenotype. In addition to genes that are induced during fruit ripening, several groups have cloned and characterized genes that are differentially expressed during fruit cell division, expansion and ripening phases (Tieman and Handa 1996, Lemaire-Chamley et al. 2000, Alexander and Grierson 2002, Seymour et al. 2002, White, 2002, Testa et al. 2002, Giovannoni 2001, 2004). The DNA microarray analyses, proteomics and functional genomics should, in not-too-distant-future, reveal genes whose expression is intimately associated with fruit development. We will also understand how the temporal and spatial expression of these genes is regulated by various developmental cues, including hormones, other growth regulators, and by environmental extremes. Such an information base will begin to provide specific information that will ultimately translate into the development of designer fruits with enhanced quality and longer shelf-life.

Acknowledgment

We thank Sangita Handa for helpful comments on the manuscript. This work was supported by grants from the U.S. Department of Agriculture IFAFS program (Award No 741740) and the United States-Israel Binational Agriculture Research and Development Fund (Grant No. US-3132-99).

REFERENCES

Abdel Rahman, M. 1977. Patterns of hormones, respiration and ripening enzymes during development, maturation and ripening of cherry tomato fruits. Physiol Plant 39: 115-118.

Abel, S. and A. Theologis. 1996. Early genes and auxin action. Plant Physiol 111: 9–17.

Abel, S., Oeller, P. W., and A. Theologis. 1994. Early auxin-induced genes encode short-lived nuclear proteins. Proc Natl Acad Sci USA 89: 326-330.

Abeles, F. B., P. W. Morgan, and M. E. Jr. Saltveit. 1992. Ethylene in Plant Biology, 2nd ed. Academic Press, San Diego.

Ahrens, M. J. and D. J. Huber. 1990. Physiology and firmness determination of ripening tomato fruit. Physiol Plant 78: 8-14.

Alabadi, D., M. S. Agüero, M. A. Pérez-Amador, and J. Carbonell. 1996. Arginase, arginine decarboxylase, ornithine decarboxylase, and polyamines in tomato ovaries: changes in unpollinated ovaries and parthenocarpic fruits induced by auxin and gibberellin. Plant Physiol 112: 1237-1244.

Alabadi, D. and J. Carbonell. 1998. Expression of ornithine decarboxylase is transiently increased by pollination, 2, 4-D, and GA_3 in tomato ovaries. Plant Physiol 118: 323-328.

Al-Hammadi, A. S. A., Y. Sreelakshmi, S. Negi, I. Siddiqi, and R. Sharma. 2003. The polycotyledon mutant of tomato shows enhanced polar auxin transport. Plant Physiol 133: 113-125.

Alexander, L. and D. Grierson. 2002. Ethylene biosynthesis and action in tomato: a model for climacteric fruit ripening. J Exp Bot 53: 2039-2055.

Ampomah-Dwamena C., B. A. Morris, P. Sutherland, B. Veit, and J. L. Yao. 2002. Down-regulation of *TM29*, a tomato *SEPALLATA* homolog, causes parthenocarpic fruit development and floral reversion. Plant Physiol 130: 605-617.

Antognoni, F., F. Ghetti, A. Mazzucato, M. Franceschetti, and N. Bagn. 2002. Polyamine pattern during flower development in the parthenocarpic fruit (*pat*) mutant of tomato. Physiol Plant 116: 539-547.

Arrowsmith, D. A. and J. de Silva. 1995. Characterization of two tomato fruit-expressed cDNAs encoding xyloglucan endo-transglycosylases. Plant Mol Biol 28: 391-403.

Balibrea Lara, M. E., M. C. Garcia, T. Fatima, R. EhneR, T. K. Lee, R. Proels, W. Tanner, and T. Roitsch. 2004. Extracellular invertase is an essential component of cytokinin-mediated delay of senescence. Plant Cell 16: 1276–1287.

Balbi, V. and T. L. Lomax. 2003. Regulation of early tomato fruit development by the Diageotropica gene. Plant Physiol 131: 186-197.

Bensen, R. J. and J. A. D. Zeevaart. 1990. Comparison of *ent-kaurene synthetase A* and *B* activities in cell-free extracts from young tomato fruits of wild-type and *gib-1*, *gib-2*, and *gib-3* tomato plants. J Plant Growth Reg 9: 237-242.

Bergey, D. R., G. A. Howe, and C. A. Ryan. 1996. Polypeptide signaling for plant defensive genes exhibits analogies to defense signaling in animals. Proc Natl Acad Sci USA 93: 12053–12058.

Berry, T. and J. D. Bewley. 1991. Seeds of tomato (*Lycopersicon esculentum* Mill.) which develop in a fully hydrated environment in the fruit switch from a developmental to a germinative mode without a requirement for desiccation. Planta 186: 27-34.

Bishop, G. J., T. Nomura, T. Yokota, K. Harrison, T. Noguchi, S. Fujioka, S. Takatsuto, J. D.G. Jones, and Y. Kamiya. 1999. The tomato DWARF enzyme catalyses C-6 oxidation in brassinosteroid biosynthesis. Proc Natl Acad Sci USA 96: 1761-1766.

Bishop, G. J. 2003. Brassinosteroid mutants of crops. J Plant Growth Reg 22, 325-335.

Bohner, J., P. Hedden, E. Bora-Haber, and F. Bangerth. 1988. Identification and quantitation of gibberellins in fruits of *Lycopersicon esculentum*, and their relationship to fruit size in *L. esculentum* and *L.pimpinellifolium*. Physiol Plant 73: 348-353.

Bohner, J. and F. Bangerth. 1988. Cell number, cell size and hormone levels in semi-isogenic mutants of *Lycopersicon pimpinefollium* differing in fruit size. Physiol Plant 72: 316-320.

Bradford, K. J., A. B. Downie, O. H. Gee, V. Alvarado, H. Yang, and P. Dahal. 2003. Abscisic acid and gibberellin differentially regulate expression of genes of the *snf1*-related kinase complex in tomato seeds. Plant Physiol 132: 1560-1576.

Brenner, M. L. and N. Cheikh. 1995. The role of hormones in photosynthate partitioning and seed filling. *In*: P.J. Davies (ed), Plant Hormones, Physiology, Biochemistry and Molecular Biology. Kluwer Academic Publishers, Dordrecht, The Netherlands, pp. 649-670.

Brown, K.M. 1997. Ethylene and abscission. Physiol Plant 100: 567–576.

Bunger-Kibler, S. and F. Bangerth. 1983. Relationship between cell number, cell size and fruit size of seeded fruits of tomato (*Lycopersicon esculentum* Mill.) and those induced parthenocarpically by application of growth regulators. Plant Growth Reg 1: 143-154.

Burbidge A, T. M. Grieve, A. Jackson, A. Thompson, D. R. McCarty, and I. B. Taylor. 1999. Characterization of the ABA-deficient tomato mutant *notabilis* and its relationship with maize Vp14. Plant J 17: 427-431.

Busi, M.V., C. D' Bustamante, C. Angelo, M. Hidalgo-Cuevas, S. B. Boggio, M. Valle Estela, and E. Zabaleta. 2003. MADS-box genes expressed during tomato seed and fruit development. Plant Mol Biol 52: 801-815.

Cassol, T. and A. K. Mattoo. 2003. Do polyamines and ethylene interact to regulate plant growth, development and senescence? *In*: P., Nath, A. K. Mattoo, S. A. Ranade and J. H. Weil (eds), Molecular Insight in Plant Biology. Science Publishers, Inc., Enfield, pp. 121-132.

Catala, C., J. K. C. Rose, and A. B. Bennett. 1997. Auxin regulation and spatial localization of an endo-1, 4-b-d-glucanase and a xyloglucan endotransglycosylase in expanding tomato hypocotyls. Plant J 12: 417–426.

Catala, C., J. K.C. Rose, and A. B. Bennett. 2000. Auxin-regulated genes encoding cell wall-modifying proteins are expressed during early tomato fruit growth. Plant Physiol 122: 527–534.

Chao, W. S., Y. Q. Gu, V. Pautot, E. A. Bray, and L. L. Walling. 1999. Leucine aminopeptidase RNAs, proteins, and activities increase in response to water deficit, salinity, and the wound signals systemin, methyl jasmonate, and abscisic acid. Plant Physiol 120: 979–992.

Coenen, C., M. Christian, H. Lüthen, and T. L. Lomax. 2003. Cytokinin inhibits a subset of *diageotropica*-dependent primary auxin responses in tomato. Plant Physiol 131: 1692-1704.

Cohen, E., S. M. Arad, Y. M. Heimer, and Y. Mizrahi. 1982. Participation of ornithine decarboxylase in early stages of tomato fruit development. Plant Physiol 70: 540–543.

Cohen, J. D. 1996. *In vitro* tomato fruit cultures demonstrate a role for indole-3-acetic acid in regulating fruit ripening. J Am Soc Hort Sci 121: 520-524.

Cong, B., J. Liu, and S. D. Tanksley. 2002. Natural alleles at a tomato fruit size quantitative trait locus differ by heterochronic regulatory mutations. Proc Natl Acad Sci USA 99: 13606-13611.

Coombe, B. G. 1976. The development of fleshy fruits. Ann Rev Plant Physiol 27: 507-528.

Cosgrove, D. J., L. C. Li, H. T. Cho, S. H. Benning, R. C. Moore, and D. Blecker. 2002. The growing world of expansins. Plant Cell Physiol 43: 1436-1444.

Creelman, R. A. and J. E. Mullet. 1997. Biosynthesis and action of jasmonates in plants. Ann Rev Plant Physiol Plant Mol Biol 48: 355–381.

Daidoji, H., Y. Takasaki, and P. Nakane. 1992. Proliferating-cell-nuclear antigen PCNA/ cyclin in proliferating cells: immunohistochemical and quantitative analysis using antibody and murine monoclonal antibodies to PCNA. Cell Biochem Func 10: 123-132.

Dewitte, W. and J. A. H. Murray. 2003. The plant cell cycle. Ann Rev Plant Biol 54: 235-264.

Dibble, A. R. G., P. Davies, and M. A. Mutschler. 1988. Polyamine content of long-keeping *alcobaca* tomato fruit. Plant Physiol 86: 338–340.

Dong, Y. H., A. Kvarnheden, J. L. Yao, P. W. Sutherland, R. G. Atkinson, B. A. Morris, and R. C. Gardner. 1998. Identification of pollination-induced genes from the ovary of apple (*Malus domestica*). Sex Plant Reprod 11: 277-283.

Fellner, M., R. Zhang, R. P. Pharis, and V. K. Sawhney. 2001. Reduced de-etiolation of hypocotyl growth in a tomato mutant is associated with hypersensitivity to, and high endogenous levels of, abscisic acid. J Exp Bot 52: 725-738.

Finkelstein, R. and S. I. Gibson. 2002. ABA and sugar interactions regulating development: "Cross-talk" or "voices in a crowd"? Curr Opin Plant Biol 5: 26–32.

Fluhr, R. and A. K. Mattoo. 1996. Ethylene-biosynthesis and perception. Crit Rev Plant Sci 15: 479-523.

Fos, M., F. Nuez, and J. L. García-Martínez. 2000. The *pat-2* gene, which induces natural parthenocarpy, alters the gibberellin content in unpollinated tomato ovaries. Plant Physiol 122: 471-479.

Fos, M., K. Proaño, D. Alabadí, F. Nuez, J. Carbonell, and J. L. García-Martínez. 2003. Polyamine metabolism is altered in unpollinated parthenocarpic *pat-2* tomato ovaries. Plant Physiol 131: 359-366.

Fos, M., K. Proaño, F. Nuez, and J. L. García-Martínez. 2001. Role of gibberellins in parthenocarpic fruit development induced by the genetic system *pat-3/pat-4* in tomato. Physiol Plant 111: 545-550.

Fraser, P. D., P. Hedden, D. T. Cooke, C. R. Bird, W. Schuch, and P. Bramley. 1995. The effect of reduced activity of phytoene synthase on isoprenoid levels in tomato pericarp during fruit development and ripening. Planta 196: 321-326.

Fray R. G. and D. Grierson 1993. Identification and genetic analysis of normal and mutant phytoene synthase genes of tomato by sequencing, complementation and co-suppression. Plant Mol Biol 22: 589-602.

Frenkel, C. 1972. Involvement of peroxidase and indole-3-acetic acid oxidase isozyme from pear, tomato, and blueberry fruit in ripening. Plant Physiol 49: 757.

Fry, S. C., R. C. Smith, K. F. Renwick, D. J. Martin, S. K. Hodge, and K. J. Matthews. 1992. Xyloglucan endotransglucosylase, a new wall-loosening enzyme activity from plants. Biochem J 282: 821-828.

Fulton, T. M., T. Beck-Bunn, D. Emmatty, Y. Eshed, J. Lopez, V. Petiard, J. Uhlig, D. Zamir, and S. D. Tanksley. 1997. QTL analysis of an advanced backcross of *Lycopersicon peruvianum* to the cultivated tomato and comparisons with QTLs found in other wild species. Theor Appl Genet 95: 881-894.

Galuszka, P., I. Frebort, M. Sebala, and P. Pavel. 2000. Degradation of cytokinins by cytokinin oxidase in plants. Plant growth Reg 32: 315-327.

Garcia-Martinez, J. L., I. Lopez-Diaz, M. J. Sanchez-Beltran, A. L. Phillips, D. A. Ward, P. Gaskin, and P. Hedden. 1997. Isolation and transcript analysis of gibberellin 20-oxidase genes in pea and bean in relation to fruit development. Plant Mol Biol 33: 1073-1084.

George, W. L., J. W. Scott, and W. E. Splittstoesser. 1984. Parthenocarpy in tomato. Hort Rev 6: 65-84.

Gillaspy, G., H. Ben-David, and W. Gruissem. 1993. Fruits: A developmental perspective. Plant Cell 5: 1439-1451.

Giovannoni, J. 2001. Molecular biology of fruit maturation and ripening. Ann Rev Plant Physiol Plant Mol Biol 52: 729-752.

Giovannoni, J. 2004. Genetic regulation of fruit development and ripening. Plant Cell 16: S170-S180.

Godt, D. E. and T. Roitsch. 1997. Regulation and tissue-specific distribution of mRNAs for three extracellular invertase isoenzymes of tomato suggests an important function in establishing and maintaining sink metabolism. Plant Physiol 115: 273–282.

Goetz, M., D. E. Godt, and T. Roitsch. 2000. Tissue-specific induction of the mRNA for an extracellular invertase isoenzyme of tomato by brassinosteroids suggests a role for steroid hormones in assimilate partitioning. Plant J 22: 515–522.

Goodwin, P. B. 1978. Phytohormones and fruit growth. *In*: D.S. Letham, P.B. Goodwin and T.J.V. Higgins (eds.), Phytohormones and Related Compounds-a Comprehensive Treatise, Vol. 2. Elsevier North Holland Biomedical Press, Amsterdam, pp 175–214.

Grandillo, S., H. M. Ku, and S. D. Tanksley. 1996. Characterization of *fs8.1* a major QTL influencing fruit shape in tomato. Mol Breed 2: 251–260.

Groot, S. P. C., J. Bruinsma, and C. M. Karssen. 1987. The role of endogenous gibberellin in seed and fruit development of tomato: studies with a gibberellin-deficient mutant. Physiol Plant 71: 184-190.

Groot, S. P. C. and C. M. Karssen. 1992. Dormancy and germination of abscisic acid-deficient tomato seeds. Plant Physiol 99: 952-958

Gustafson, F. G. 1937. Parthenocarpy induced by pollen extracts. Am J Bot 24: 102-107.

Gustafson, F. G. 1936. Inducement of fruit development by growth promoting chemicals. Proc Natl Acad Sci USA 22: 628-636.

Gustafson, F. G. 1939. The cause of natural parthenocarpy. Am J Bot 26: 135-138.

Haard, N. F. 1977. Physiological roles of peroxidase in posharvest fruits and vegetables. *In*: R. L. Ory and St. Angelo A. J. (eds), Enzymes in Food and Beverages Processing, Advances in Chemistry Series No. 47, American Chemical Society, Washington, DC. 132-143.

Hackett, R. M., C. W. Ho, Z. Lin, H. C. C. Foote, R. G. Fray, and D. Grierson. 2000. Antisense inhibition of the *Nr* Gene restores normal ripening to the tomato never-ripe mutant, consistent with the ethylene receptor inhibition model. Plant Physiol 124: 1079-1085.

Hadfi, K., V. Speth, and G. Neuhaus. 1998. Auxin-induced developmental patterns in *Brassica juncea* embryos. Development 125: 879-887.

Hagen, G. and T. Guilfoyle. 2002. Auxin-responsive gene expression: genes, promoters and regulatory factors. Plant Mol Biol 49: 373-385.

Hager, A. 2003. Role of the plasma membrane H+-ATPase in auxin-induced elongation growth: historical and new aspects. J Plant Res 116: 483-505.

Halford, N. G., J. P. Bouly, and M. Thomas. 2000. SNF1-related protein kinases (SnRKs): regulators at the heart of the control of carbon metabolism and partitioning. Adv Bot Res 32: 405-434.

Halford, N. G. and D. G. Hardie. 1998. SNF1-related protein kinases: global regulators of carbon catabolism in plants? Plant Mol Biol 37: 735-748.

Halford, N. G., S. Hey, D. Jhurreea, S. Laurie, R. S. McKibbin, M. Paul, and Y. Zhang. 2003. Metabolic signaling and carbon partitioning: role for Snf1-related (SnRK1) protein kinase. J Exp Bot 54: 467-475.

Hamilton, A. J., G. W. Lycett, and D. Grierson. 1990. Antisense gene that inhibits synthesis of the hormone ethylene in transgenic plants. Nature 346: 284-287.

Hamilton, C. M. 1997. A binary-BAC system for plant transformation with high-molecular weight DNA. Gene 200: 107-116.

Hamilton, C. M., A. Frary, C. Lewis, and S. D. Tanksley. 1996. Stable transfer of intact high molecular weight DNA into plant chromosomes. Proc Natl Acad Sci USA 93: 9975-9979.

Hedden, P., A. L. Phillips, M. C. Rojas, E. Carrera, and B. Tudzynski. 2002. Gibberellin biosynthesis in plants and fungi: A case of convergent evolution? J Plant Growth Reg 20: 319-331.

Himmelbach, A., Y. Y. Yang, and E. Grill. 2003. Relay and control of abscisic acid signaling. Curr Opin Plant Biol 6: 470-479.

Ho, L. C. 1992. Fruit growth and sink strength. *In*: C. Marshall and J. Grace (eds), Fruit and Seed Production. Aspects of Development, Environmental Physiology and Ecology. Society for Experimental Biology, Seminar series: 47, Cambridge University Press, pp. 206.

Ho, L. C. and J. D. Hewitt. 1986. Fruit development. *In*: J. G. Atherton and J. Rudich (eds), The Tomato Crop. A scientific Basis for Improvement. Cambridge University Press, Chapman and Hall Ltd., pp. 201-240.

Hobson, G. E. and J. E. Harman. 1986. Tomato fruit development and the control of ripening. Acta Hort 190: 167-174.

Hong, S. J. and S. K. Lee. 1996. Changes in endogenous putrescine and the relationship to the ripening of tomato fruits. J Korean Soc Hortic Sci 37: 369-373.

Jacobsen S. E. and N. E. Olszewski. 1996. Gibberellins regulate the abundance of RNAs with sequence similarity to proteinase inhibitors, dioxygenases and dehydrogenases. Planta 198: 78-86.

Jones, R. J. and B. M. N. Schreiber. 1997. Role and function of cytokinin oxidase in plants. Plant Growth Reg 23: 123–134.

Jones, B., P. Frasse, E. Olmos, H. Zegzouti, Z. G. Li, A. Latché, J. C. Pech, and M. Bouzayen. 2002. Down-regulation of DR12, an auxin-response-factor homolog, in the tomato results in a pleiotropic phenotype including dark green and blotchy ripening fruit. Plant J 32: 603-613.

Jones, M. G. 1987. Gibberellins and the procera mutant of tomato. Planta 172: 280–284.

Joubès, J., C. Chevalier, D. Dudits, E. Heberle-Bors, D. Inze, M. Umeda, and J. P. Renaudin. 2000. CDK-related protein kinases in plants. Plant Mol Biol 43: 607-620.

Joubes, J., M. Lemaire-Chamley, F. Delmas, J. Walter, M. Hernould, A. Mouras, P. Raymond, and C. Chevalier. 2001. A new C-type cyclin-dependent kinase from tomato expressed in dividing tissues does not interact with mitotic and G1 cyclins. Plant Physiol 26: 1403–1415.

Joubès, J., T.H. Phan, D. Just, C. Rothan, C. Bergounioux, P. Raymond, and C. Chevalier. 1999. Molecular and biochemical characterization of the involvement of cyclin-dependent kinase A during the early development of tomato fruit. Plant Physiol 121: 857-869.

Kaminek, M., V. Motyka, and R. Vankova. 1997. Regulation of cytokinin content in plant cells. Physiol Plant 101: 689-700.

Kim, J., K. Harter, and A. Theologis. 1997. Protein-protein interactions among the Aux/IAA proteins. Proc Natl Acad Sci USA 94: 11786-11791.

Koch, K. 2004. Sucrose metabolism: regulatory mechanisms and pivotal roles in sugar sensing and plant development. Curr Opin Plant Biol 7: 235-246.

Koka, C. V., R. E. Cerny, R. G. Gardner, T. Noguchi, S. Fujioka, S. Takatsuto, S. Yoshida, and S. D. Clouse. 2000. A putative role for the tomato genes DUMPY and CURL-3 in brassinosteroid biosynthesis and response. Plant Physiol 122: 85-98.

Koornneef, M., T. D. G. Bosma, C. J. Hanhart, J. A. D. Zeevaart, and J. H. Van-der Veen. 1990. The isolation and characterization of gibberellin-deficient mutants in tomato. Theor Appl Genet 80: 852-857.

Koshioka, M., T. Nishijima, H. Yamazaki, M. Nonaka, and L. N. Mander. 1994. Analysis of gibberellins in growing fruits of Lycopersicon esculentum after pollination or treatment with 4-chlorophenoxyacetic acid. J Hortic Sci 69: 171-179.

Kuiper, D. 1993. Sink strength: established and regulated by plant growth regulators. Plant Cell Environ 16: 1025–1026.

Lanahan, M. B., H. C. Yen, J. J. Giovannoni, and H. J. Klee. 1994. The never ripe mutation blocks ethylene perception in tomato. Plant Cell 6: 521-530.

Larsen, P. B., E. N. Ashworth, M. L. Jones, and W. R. Woodson. 1995. Pollination-induced ethylene in carnation: role of pollen tube growth and sexual compatibility. Plant Physiol 108: 1405-412.

Larsen, P. B., E. J. Woltering, and W. R. Woodson. 1993. Ethylene and interorgan signaling in flowers following pollination. In: I. Raskin and J. Schultz, (eds), Plants Signals in Interactions with Other Organisms. American Society of Plant Physiologists, Rockville, MD, pp. 112-122.

Lashbrook, C. C., C. Gonzalez-Bosch, and A. B. Bennett. 1994. Two divergent endo-β-1, 4-glucanase genes exhibit overlapping expression in ripening fruit and abscising flowers. Plant Cell 6: 1485-1493.

Li, N., B. L. Parsons, D. R. Liu, and A. K. Mattoo. 1992. Accumulation of wound-inducible ACC synthase transcript in tomato fruit is inhibited by salicylic acid and polyamines. Plant Mol Biol 18: 477-87.

Lin, S., W. E. Splittstoesser, and W. L. George. 1983. A comparison of normal seeds and pseudoembryos produced in parthenocarpic fruit of Severianin tomato. HortScience 18: 75-76.

Lincoln, J. E., S. Cordes, E. Read, and R. L. Fischer. 1987. Regulation of gene expression by ethylene during *Lycopersicon esculentum* (tomato) fruit development. Proc Natl Acad Sci USA 84: 2793-2797.

Liu, Y. Q., Z. M. Luo, H. W. M. Hilhorst, and C. M. Karssen. 1996. Effects of endogenous GA and ABA on water relations in tomato fruit and seeds during development. Acta Phytophysiol Sinica 22: 19-26.

Liu, Y., H.W.M. Hilhorst, S.P.C. Groot, and R. J. Bino. 1997. Amounts of nuclear DNA and internal morphology of gibberellin and abscisic acid-deficient tomato (*Lycopersicon esculentum* Mill.) seeds during maturation, imbibition and germination. Ann Bot 79:161-168.

Llop-Tous, I., S. B. Cornelius, and D. Grierson. 2000. Regulation of ethylene biosynthesis in response to pollination in tomato flowers. Plant Physiol 123: 971–978.

Lohmann, J. U. and D. Weigel. 2002. Building beauty: the genetic control of floral patterning. Dev Cell 2: 135-142.

Lunn, J. E. and E. MacRae. 2003. New complexities in the synthesis of sucrose. Curr Opin Plant Biol 6: 208-214.

Lu, C., H. L. Xu, and L. X. Zhou. 1995. Effect of PG activity, ACC and ethylene production in fruit ripening of tomato. Acta Hort Sinica 22: 57-60.

Mapelli, S. 1981. Changes in cytokinin in the fruits of parthenocarpic and normal tomatoes. Plant Sci Lett 22: 227-233.

Mapelli, S. and L. Lombardi. 1982. A comparative auxin and cytokinin study in normal and to-2 mutant tomato plants. Plant Cell Physiol 23: 751-757.

Mapelli, S., C. Frova, G. Tort, and G. Soressi. 1978. Relationship between set, development and activities of growth regulators in tomato fruit. Plant Cell Physiol 19: 1281-1288.

Mapelli, S., G. Torti, M. Badino, and G. P. Soressi. 1979. Effects of GA_3 on flowering and fruit-set in a mutant of tomato. HortScience 14: 736-737.

Mattoo, A. K. and J. C. Suttle. 1991. The plant hormone ethylene. CRC Press, Boca Raton, Florida.

Mattoo, A. K. and W. B. White. 1991. Regulation of ethylene biosynthesis. *In*: A. K. Mattoo and J. C. Suttle (eds), The Plant Hormone Ethylene. CRC Press, Boca Raton, FL, pp. 21-42

Mazzucato, A., A. R. Taddei, and G. P. Soressi. 1998. The parthenocarpic fruit (*pat*) mutant of tomato (*Lycopersicon esculentum* Mill.) sets seedless fruits and has aberrant anther and ovule development. Development 125: 107-114.

McGlasson, W. B. and M.J. Franklin. 1979. Influence of the *Nr, rin*, and *nor* genes in changes in abscisic acid, phaseic acid, and gibberellin activity during growth and senescence of tomato fruits. J Am Soc Hort Sci 104: 455-459.

Miyamoto, K., M. Oka, and J. Ueda. 1997. Update on the possible mode of action of the jasmonates: Focus on the metabolism of cell wall polysaccharides in relation to growth and development. Physiol Plant 100: 631–638.

Mutschler, M. A., R.W. Doerge, S. C. Liu, J. P. Kuai, B. E. Liedl, and J. A. Shapiro. 1996. QTL analysis of pest resistance in the wild tomato *Lycopersicon pennellii*: QTLs controlling acylsugar level and composition. Theor Appl Genet 92: 709–718.

Nester, J. E. and J. A. D. Zeevaart. 1988. Flower development in normal tomato and a gibberellin-deficient (*ga-2*) mutant. Am J Bot 75: 45-55.

Nitsch, J.1970. Hormonal factors in growth and development. *In*: A. C. Hulme, (ed.), The Biochemistry of Fruits and Their Products. Academic Press, London. pp 427-472.

Oeller, P. W., J. A. Keller, J. E. Parks, J. E. Silbert, and A. Theologis. 1993. Structural characterization of the early indoleacetic acid-inducible genes, PS-IAA4/5, of pea (*Pisum sativum*). J Mol Biol 233: 789-798.

Oeller, P. W., L. Min-Wong, L. Taylor, D. Pike, and A. Theologis. 1991. Reversible inhibition of tomato fruit senescence by antisense RNA. Science 254: 437-439.

Oh, M. H., W. G. Romanow, R. C. Smith, E. Zamski, J. Sasse, and S. D. Clouse. 1998. Soybean BRU1 encodes a functional xyloglycan endotransglycosylases that is highly expressed in inner epicotyl tissues during brassinosteroid-promoted elongation. Plant and Cell Physiol 39: 124-130.

Okamoto-Nakazato, A. 2002. A brief note on the study of yield in, a wall-bound protein that regulates the yield threshold of the cell wall. J Plant Res 115: 309-313.

O'Neill, S. D. 1997. Pollination regulation of flower development. Annu Rev Plant Physiol Plant Mol Biol 48: 547-574.

Ozga J. A. and D. M. Reinecke. 2003. Hormonal interactions in fruit development. J Plant Growth Reg 22: 73-81.

Ozga, J. A., M. L. Brenner and D. M. Reinecke. 1992. Seed effects on gibberellin metabolism in pea pericarp. Plant Physiol 100: 88-94.

Pearce, G., D. Strydom, S. Johnson, and C. A. Ryan. 1991. A polypeptide from tomato leaves induces wound-inducible proteinase protein. Science 253: 895–898.

Pnueli, L., M. Abu-Abeid, D. Zamir, W. Nacken, Z. Schwarz-Sommer, and E. Lifschitz. 1991. The MADS box gene family in tomato: temporal expression during floral development, conserved secondary structures and homology with homeotic genes from *Antirrhinum* and *Arabidopsis*. Plant J 1: 255-266.

Pnueli, L., D. Hareven, S. D. Rounsley, M. F. Yanofsky, and E. Lifschitz. 1994a Isolation of the tomato *AGAMOUS* gene *TAG1* and analysis of its homeotic role in transgenic plants. Plant Cell 6:163-173.

Pnueli, L., D. Hareven, L. Broday, C. Hurwitz, and E. Lifschitz. 1994b. The *TM5* MADS box gene mediates organ differentiation in the three inner whorls of tomato flowers. Plant Cell 6: 175-186.

Proels, R. K., B. Hause, S. Berger, and T. Roitsch. 2003. Novel mode of hormone induction of tandem tomato invertase genes in floral tissues. Plant Mol Biol 52: 191-201.

Rayle, D. L. and R. E. Cleland. 1992. The acid growth theory of auxin-induced cell elongation is alive and well. Plant Physiol 99: 1271-1274.

Rebers, M., T. Kaneta, H. Kawaide, S. Yamaguchi, Y.Y. Yang, R. Imai, H. Sekimoto, and Y. Kamiya. 1999. Regulation of gibberellin biosynthesis genes during flower and early fruit development of tomato. Plant J 17: 241-250.

Redman, A. M., D. F. Jr. Cipollini, and J. C. Schultz. 2001. Fitness costs of jasmonic acid-induced defense in tomato, *Lycopersicon esculentum*. Oecologia. 126: 380–385.

Riov, J., E. Dagan, R. Goren, S. F. Yang. 1990. Characterization of abscisic acid-induced ethylene production in citrus leaf and tomato fruit tissues. Plant Physiol 92: 48-53.

Robinson, R., D. Cantliffe, and S. Shannon. 1971. Morphactin induced parthenocarpy in cucumber. Science 171: 1251-1252.

Roitsch, T. and R. Ehness. 2000. Regulation of source/sink relations by cytokinins. Plant Growth Reg 32: 359-367.

Roitsch, T., M. E. Balibrea, M. Hofmann, R. Proels, and A. K. Sinha. 2003. Extracellular invertase: key metabolic enzyme and PR protein. J Exp Bot 54: 513-524.

Rose, J. K. C., H. H. Lee, and A. B. Bennett. 1997. Expression of a divergent expansin gene is fruit-specific and ripening-regulated. Proc Natl Acad Sci USA 94: 5955-5960.

Ross, J. J., D. P. O'Neill, and D. A. Rathbone. 2003. Auxin-gibberellin interactions in pea: Integrating the old with the new. J Plant Growth Reg 22: 99-108.

Sagi, M., R. Fluhr, and S. H. Lips. 1999. Aldehyde oxidase and xanthine dehydrogenase in a *flacca* tomato mutant with deficient abscisic acid and wilty phenotype. Plant Physiol 120: 571-578.

Sastry K. K. S. and R. M. Muir. 1963. Gibberellin: effect on diffusible auxin in fruit development. Science 140: 494-495.

Sawheny, V. K. 1984. Gibberellins and fruit formation in tomato: A review. Sci Hort 22: 1-8.

Schlagnhaufer, C. D. and R. N. Arteca. 1985a. Brassinosteroid-induced epinasty in tomato [*Lycopersicon esculentum* cultivar Heinz 1350] plants. Plant Physiol 78: 300-303.

Schlagnhaufer, C. D. and R. N. Arteca. 1991. The uptake and metabolism of brassinosteroid by tomato (*Lycopersicon esculentum*) plants. J Plant Physiol 138: 191-194.

Schlagnhaufer, C. D. and R. N. Arteca. 1985b. Inhibition of brassinosteroid-induced epinasty in tomato [*Lycopersicon esculentum* cultivar Heinz 1350] plants by aminooxyacetic acid and cobalt. Physiol Plant 65: 151-155.

Schumacher, K., T. Schmitt, M. Rossberg, G. Schmitz and K.Theres. 1999. The Lateral suppressor (Ls) gene of tomato encodes a new member of the VHIID protein family. Proc Natl Acad Sci USA. 5: 290-29.

Schwabe, W. W. and J.J. Mills. 1981. Hormones and parthenocarpic fruit set. Hort Abstracts. 51: 161-198.

Seymour, G. B., K. Manning, E. M. Eriksson, A. H. Popovich, and G. J. King. 2002. Genetic identification and genomic organization of factors affecting fruit texture. J Exp Bot 53: 2065-2071.

Sjut, V. and F. Bangerth. 1982. Induced parthenocarpy: a way of changing the levels of endogenous hormones in tomato fruits (*Lycopersicon esculentum* Mill.): 1. Extractable hormones. Plant Growth Reg 1: 243-251.

Smith, O. 1935. Pollination and life history studies of the tomato. Cornell Univ Agricul Exp Sta Mem 184: 13-16.

Su, L., T. McKeon, D. Grierson, M. Cantwell, and S. F. Yang. 1984. Development of 1-aminocyclopropane-1-carboxylic acid synthase and polygalacturonase activities during the maturation and ripening of tomato fruit. HortScience 19: 576-578.

Suzuka, I., H. Daidoji, M. Matsuoka, K. Kadowski, Y. Takasaki, P. Nakane, and T. Moriuchi. 1989. Gene for proliferating-cell-nuclear antigen (DNA polymerase delta auxiliary protein) is present in both mammalian and higher plant genomes. Proc Natl Acad Sci USA. 86: 3189-3193.

Swarup, R., G. Parry, N. Graham, T. Allen, and M. Bennett. 2002. Auxin cross-talk: integration of signaling pathways to control plant development. Plant Mol Biol 49: 411-426.

Szekeres, M. 2003. Brassinosteroid and systemin: two hormones perceived by the same receptor. Trends Plant Sci 8: 102-104.

Tang, X. and W.R. Woodson. 1996. Temporal and spatial expression of 1-aminocyclopropane-1-carboxylate oxidase mRNA following pollination of immature and mature petunia flowers. Plant Physiol 112: 503-511.

Tanksley, S. D. 2004. The genetic, developmental, and molecular bases of fruit size and shape variation in tomato. Plant cell 16: S181-S189.

Testa, G., R. Caccia, F. Tilesi, G. Soressi, and A. Mazzucato. 2002. Sequencing and characterization of tomato genes putatively involved in fruit set and early development. Sex Plant Rep 14: 269-277.

Thomas, R. L. and J. J. Jen. 1980. Preparation of a homogenous tomato peroxidase. Prep Biochem 10: 581.

Thomas, R. L., J. J. Joseph, and C. V. Morr. 1981. Changes in soluble and bound peroxidase-IAA oxidase during tomato fruit development. J Food Sci 47: 158-161.

Thompson, A. J., M. Tor, C. S. Barry, J. Vrebalov, C. Orfila, M. C. Jarvis, J. J. Giovannoni, D. Grierson, and G. B. Seymour. 1999. Molecular and genetic characterization of a novel pleiotropic tomato-ripening mutant. Plant Physiol 120: 383-389.

Ulmasov, T., G. Hagen, and T. J. Guilfoyle. 1997. ARF1, a transcription factor that binds to auxin response elements. Science 276: 1865-1868.

Ulmasov, T., G. Hagen, and T. J. Guilfoyle. 1999. Dimerization and DNA binding of auxin response factors. Plant J 19: 309-319.

Ulmasov, T., Z. B. Liu, G. Hagen, and T. J. Guilfoyle. 1995. Composite structure of auxin response elements. Plant Cell 7: 1611-1623.

Van den Heuvel K. J. P. T., R. J. Esch, van, G. W. M. Barendse, and G. J. Wullems. 1999. Isolation and molecular characterization of gibberellin-regulated H1 and H2B histone cDNAs in the leaf of a gibberellin-deficient tomato. Plant Mol Biol 39: 883-890.

Van den Heuvel, K. J. P. T., R. H. Van Lipzig, G. W. Barendse and G. J. Wullems. 2002. Regulation of expression of two novel flower-specific genes from tomato (*Solanum lycopersicum*) by gibberellin. J Exp Bot 53: 51-59.

Van Tuinen A., A. H. L. J. Peters, R. E. Kendrick, J. A. D. Zeevaart, and M. Koornneef. 1999. Characterisation of the *procera* mutant of tomato and the interaction of gibberellins with end-of-day far-red light treatments. Physiol Plant 106: 121–128.

Vardhini, V. B. and S. S. Rao. 2002. Acceleration of ripening of tomato pericarp discs by brassinosteroids. Phytochem 61: 843-847.

Vendrell, M. 1985. Dual effect of 2, 4-D on ethylene production and ripening of tomato fruit tissue. Physiol Plant 64: 559-563.

Vrebalov, J., D. Ruezinsky, V. Padmanabhan, R. White, D. Medrano, R. Drake, W. Schuch, and J. Giovannoni. 2002. A MADS-box gene necessary for fruit ripening at the tomato ripening-inhibitor (*rin*) locus. Science 296: 342-346.

Waller, F., M. Furuya, and P. Nick. 2002. OsARF1, an auxin response factor from rice, is auxin-regulated and classifies as a primary auxin responsive gene. Plant Mol Biol 50: 415–425.

Wann, E. Van. 1995. Reduced plant growth in tomato mutants high pigment and dark green partially overcome by exogenous gibberellin. HortScience 30: 379.

Weyers J. D. B. and N. W. Paterson. 2001. Plant hormones and the control of physiological processes. New Phytologist 152: 375–407.

White, P. J. 2002. Recent advances in fruit development and ripening: an overview. J Exp Bot 53: 1995-2000.

Woltering, E. J., A. J. De Jong, and E.T. Yakimova. 1999. Apoptotic cell death in plants: the role of ethylene. *In*: A. K. Kanellis, (ed), Biology and Biotechnology of the Plant Hormone Ethylene II. Kluwer, Dordrecht, pp 209–216.

Woltering, E. J., D. Somhorst, and Van, P. Der Veer. 1995. The role of ethylene in interorgan signaling during flower senescence. Plant Physiol 109: 1219-1225.

Zhang, X. S. and S. D. O'Neill. 1993. Ovary and gametophyte development are coordinately regulated by auxin and ethylene following pollination. Plant Cell 5: 403-418.

Zurek, D. M. and S. D. Clouse. 1994. Molecular cloning and characterization of a brassinosteroid regulated gene from elongating soybean (*Glycine max* L.) epicotyls. Plant Physiol 104: 161–170.

11

Genetic Control of Fruit Ripening

ELIZABETH FOX[1] **AND JIM GIOVANNONI**[2,3*]

[1] *Department of Plant Biology, Cornell University, Ithaca, NY 14853, USA*
[2] *USDA-ARS Plant, Soil and Nutrition Lab, Tower Rd., Ithaca, NY 14853, USA*
[3] *Boyce Thompson Institute for Plant Research, Tower Rd., Ithaca, NY 14853, USA*
e-mail: jjg33@cornell.edu

INTRODUCTION

Fruit ripening is an especially significant aspect of plant biology because of the importance of fruit in the human diet. Plant species devote vast amounts of energy and resources to the maturation of fruit organs (Gray et al. 1992). The ripening of fruit can be generally defined as the summation of changes in texture, flavor, aroma, and color marking the complete maturation of the organ in terms of attraction for seed dispersing organisms (Brady 1992, Grierson et al. 1992, Giovannoni 2001, Adams et al. 2004, Giovannoni 2004). Lipid metabolism, fiber content, and vitamin levels are also affected over the course of ripening, ultimately impacting the nutritional quality of mature fruit tissues. It has also been shown that levels of antioxidant compounds, capable of modifying enzyme activities and detoxifying potentially damaging free radical compounds, can be altered substantially during ripening, with effects on both fruit color and nutrient quality (Ronen et al. 1999, Verhoeyen et al. 2002).

Ripening has also been defined as a "functionally modified, protracted form of senescence" as it is the final stage of fruit development and often terminates in tissue decay and death (Huber et al. 1987, Grierson 1987) thereby leading some to surmise that ripening represents in its essence a degradative process. In our view, ripening can more accurately be compared to other active developmental processes involving *de novo* modification of gene expression (Brady 1992, Grierson and Schuch 1993).

From a structural perspective, fruits are typically either fleshy (non-dehiscent) or dry (dehiscent) with tomato receiving the most scientific

---

*Corresponding author: Jim J. Giovannoni

attention as a model for fleshy fruit development, while *Arabidopsis* (siliques) are the best studied dehiscent fruit at the molecular level (reviewed in White 2002).

Arabidopsis has been used as a model genetic system for plants due in part to its small stature, compact genome, short life cycle, simple genetics, and efficiency of transformation. Tomato has been used as a model system for fleshy fruit development and ripening for many of the same reasons mentioned for *Arabidopsis*, especially as compared to other species yielding fleshy fruit. In addition, numerous tomato mutations have been identified in which the normal ripening program is altered. These mutations include *Never-ripe* (*Nr*), *ripening-inhibitor* (*rin*), *non-ripening* (*nor*) (Tigchellar et al. 1973) and *Colourless non-ripening* (*Cnr*) (Thompson et al. 1999) which have universal effects on ripening. *Nr*, *rin*, *nor*, and *Cnr* display the most complete inhibition on ripening although numerous additional mutants altering subsets of ripening phenotypes, most notably pigment accumulation, have been described (see Gray et al. 1994, Giovannoni 2001, Giovannoni 2004, Barry et al. 2005).

Human manipulation of fruit ripening dates back thousands of years as documented in Biblical references on piercing figs to induce ripening. Greater molecular understanding of the ripening process will allow for improvement in fruits with respect to increase in yield, quality, nutrition and storage attributes. A number of attempts have been made to modify fruit quality and ripening characteristics via biotechnological approaches at the level of "proof-of-concept", and in some instances, in pursuit of commercial objectives.

CLIMACTERIC VS NON-CLIMACTERIC RIPENING

Ripening fleshy fruits are classically divided into two physiological categories, climacteric and non-climacteric, based on respiration patterns associated with ripening (Biale et al. 1981). Climacteric fruit displays a spike of respiration and ripen on or off the plant if mature when picked. Climacteric fruits are also typified by increased ethylene production at the onset of ripening (Abeles et al. 1992). Ethylene is required for the ripening of climacteric fruit and the addition of exogenous ethylene is a window just prior to the climacteric typically induced ripening. The necessity of ethylene in climacteric ripening has been shown through the action of a number of ethylene biosynthesis and action inhibitors (Yang 1985). Silver thiosulfate (STS or AgSTS) retarded ripening of these sectors of tomato fruit to which it had been infiltrated (Hobson et al. 1984).

Aminoethoxyvinylglycine (AVG) is an inhibitor of 1-aminocyclopropane-1-carboxylic acid synthase (ACS), a limiting step in ethylene synthesis (Boller et al. 1979, Yu and Yang 1979). 1-methycyclopropene (1-MCP) inhibits ethylene action by competitively binding ethylene receptors (Sisler and Serek 1997) and has been shown to retard ripening in many agriculturally important fruit species including apple (Watkins et al. 2000) and banana (Jiang et al. 2002). Examples of important climacteric fruits include tomato, apple, pear, avocado, melon, most stone fruits, and many tropical fruits including papaya, mango and banana.

Non-climacteric fruits include strawberry, grape, cucumber, pineapple, cherry, and citrus. Fruits that are non-climacteric do not display the burst of respiration or the increase in autocatalytic ethylene characteristic of their climacteric counterparts (McMurchie et al. 1972), and typically do not ripen as fully as climacteric fruit after harvest. While these fruits generally do not require ethylene for ripening, some non-climacteric fruits do respond to exposure to exogenous ethylene. Non-climacteric fruit, with the exception of cherries, display a rise in respiration upon addition of exogenous ethylene; however this increase reverts upon removal of the hormone (Lelievre et al. 1997). In strawberry fruit, addition of exogenous ethylene causes softening and enhances color development (Tian et al. 2000). In non-climacteric pepper, ethylene has been shown to induce biosynthesis of carotenoids (Ferrarese et al. 1995, Harpster et al. 1997). Exogenous ethylene also causes increased sensitivity to chilling injury, enhanced senescence, and a higher susceptibility to pathogen infection in non-climacteric fruit (Kader 1985). It is important to note that the ripening classification of a number of species is not completely clear. While some varieties of melon and pepper are clearly climacteric, others behave physiologically as non-climacteric fruits. It remains unclear as to whether the non-climacteric varieties display a truly different ripening program or simply produce and/or perceive less ethylene than their climacteric counterparts.

MODEL RIPENING SYSTEMS

Strawberry

Strawberry has become one of the most important model systems for non-climacteric fruit ripening. The true fruit (ovaries) of the strawberry are the achenes lining the outside of a fleshy receptacle (Seymour et al. 1993). Strawberry fruits have a rapid growth cycle reaching full size approximately 30 days post anthesis, but time to ripening can vary considerably depending on temperature (Seymour et al. 1993). The process of ripening for a strawberry consists of accumulation of anthocyanins, sucrose, hexoses, volatiles, and the concomitant loosening of cell wall integrity leading to tissue softening (Manning 1998). Auxin released from the achenes has been shown to regulate maturation and development of the receptacle and its loss in later fruit development is associated with ripening (Given et al. 1988, Civello et al. 1999). Much like its climacteric counterparts, strawberry ripening involves a complex group of processes, which rely on numerous changes in the expression of ripening related genes (Manning 1998).

Wilkinson et al. (1995) reported a cDNA (RJ4), discovered through differential display of unripe and ripe strawberry fruit, with homology to the annexin super-family. Annexins are described as phospholipid-binding proteins which are calcium dependent and have been associated with voltage-gated ion channels, signaling molecules, and regulation of enzyme activities (e.g. callose synthase in cotton; Andrawis et al. 1993, Wilkinson et al. 1995, Verma and Hong 2001). Annexins may contribute to changes in cell wall structure and membrane properties, both of which are affected by calcium (Ferguson 1984, Brett and Waldron 1996). Wilkinson et al. (1995) also identified a cDNA (RJ5) with homology to chalcone synthase (CHS). CHS catalyzes conversion of 3 molecules of 4-coumaroyl-CoA and 1 molecule of p-coumaroyl to naringenin chalcone, representing the first committed step in flavanoid biosythesis, leading to anthocyanins responsible for ripe strawberry color (Clegg and Durbin 2000, Hadacek 2002).

Manning (1998) used differential screening of two cDNA libraries constructed from either white (unripe) or red (ripe) strawberry to isolate additional ripening related genes. Of the gene families recovered, six were related to expression of enzymes responsible for phenylpropanoid synthesis of the red pigmentation of ripe strawberry fruit, present in low amounts in the white stage while elevated in immature and ripening fruit (Manning 1998). Manning (1998) also isolated an enzyme involved in cell wall metabolism (EGase). EGases (endo-Beta-(1,4)-glucanase) are enzymes which hydrolyze (1,4)-Beta linkages, flanking un-substituted glucose residues, and thought to act on xyloglucan (a component of plant cell walls; Brummell and Harpster 2001) which may contribute to softening of strawberry and other fleshy fruit.

A small strawberry micro-array was constructed containing 1701 randomly selected strawberry cDNAs, plus 480 cDNAs from corollas of petunia (Aharoni et al.2000). This array was probed with RNA from multiple stages consisting of green, turning, white with red, and red strawberry fruit. A number of cDNAs were shown to display differential expression when compared to the different stages with the largest differences occurring between red versus green stages. From these clones one cDNA was identified as the *SAAT* (strawberry alcohol aclytransferase) gene impacting flavor biogenesis in ripe strawberry fruit (Aharoni et al. 2000). The expression pattern of SAAT showed a 16-fold increase in red versus green fruit and was limited to receptacle tissue (Aharoni et al. 2000). These results correlate well with the belief that AATs are involved in ester volatile formation by catalyzing their production from acyl CoA and alcohol (Aharoni et al. 2000).

Aharoni et al. (2002) performed a second microarray analysis to gain knowledge regarding large scale gene expression of maturing and ripening strawberry fruit, specifically in response to oxidative stress and auxin treatment. In response to oxidative stress, 20 ripening-related cDNAs were recovered including chalcone synthase (Aharoni et al. 2002). Auxin treatment resulted in observation of ripening-related genes involved in flavonoid metabolism which are repressed. In addition, expression of auxin-independent genes, such as annexin, was also observed thereby suggesting a complex regulatory arrangement involved in non-climacteric ripening and an entry point for further elucidation of this process.

A key question to be addressed in continuing molecular analysis is "how and at what level are climacteric and non-climacteric ripening regulatory mechanisms related?" The continued discovery of genes involved in ripening of both non-climacteric and climacteric fruit will allow for cross-comparison among species of different ripening types. The recent cloning of a MADS-box transcription factor necessary for tomato ripening at the *rin* locus, indicates a well conserved regulator of ripening acting upstream of ethylene with an apparent homologue expressed in fruit from the strawberry genome (Vrebalov et al. 2002). Efforts are underway to repress the strawberry RIN homologue in transgenic strawberry to test for function in ripening (K. Manning, G. Seymour, J. Giovannoni, unpublished). Conservation of RIN homologues across species representing diverse ripening types suggests a more universal (ethylene-independent) mechanism of ripening control. The characterization of RIN homologues from other fruit species will provide possible tools for controlling fruit ripening in a variety of agriculturally important plant species in a manner that does not focus on ethylene synthesis or responses (Vrebalov et al. 2002).

Charentais melons (a climacteric variety) were transformed with antisense 1-aminocyclopropane-1-carboxylic acid oxidase (ACO) which resulted in melon fruit displaying reduction of ethylene production to <1% of wild-type (Ayub et al. 1996). These antisense fruit also showed inhibition of ripening regardless of whether they were on or off the vine. When crossed to lines of agronomic importance, the aroma-related volatile ester content of the resulting fruit was shown to be reduced (Bauchot et al. 1998). These results indicate ethylene has an important regulatory role in many aspects of climacteric ripening, and while one desirable trait may be achieved via manipulation of ethylene, another equally important trait may suffer. Molecular tools, such as *RIN* gene, for comprehensive regulation of ripening at a step upstream of ethylene will be both more universal in application to both climacteric and non-climacteric species resulting in climacteric fruit with better quality than those whose ripening is regulated only at the level of ethylene.

Tomato

Tomato belongs to the Solanaceae, or nightshade family, and is native to South America and Mexico (Kalloo and Bergh 1993). The importance of tomato as a crop plant can be readily seen in the vast amount of land under cultivation for commercial tomato production amounting to over 2,000,000 hectares in recent years (Kalloo and Bergh 1993). Declining consumer satisfaction with the flavor/appearance/spoiling of many fruits and vegetables, including tomato, has prompted considerable interest in the field of ripening and additional emphasis on this model system.

Tomato possesses a diploid genome comprised of 12 chromosomes and is enhanced by a number of tools that contribute to its molecular and genetic characterization. Mutants involved in various aspects of tomato growth and development, including ripening (as mentioned above), have been isolated and characterized (Giovannoni 2001) along with numerous Quantitative Trait Loci (QTL) identified in tomato, including QTLs responsible for variation in time of maturation to ripening (Doganlar et al. 2000). Over one thousand molecular markers have been mapped in the tomato genome spacing at an average less than 2cM (Tansksley et al. 1992). These markers have contributed to chromosome walks to a number of important loci including fw2.2 (Frary et al. 2000), *ovate* (Liu et al. 2002), *jointless-1* (Mao et al. 2000), Pto (Martin et al. 1993), and *rin* (Vrebalov et al. 2002).

Eshed and Zamir (1995) produced a series of 50 ordered introgression lines (IL) from a cross between *Lycopersicon esculentum* and the wild tomato species *Lycopersicon pennellii* which represents a library of the complete *L. pennellii* genome through ordered and overlapping introgressions. Each introgression line contains a small portion (avg. 10 -50 cM) of the *L. pennellii* genome in the background of the M82 *L. esculentum* parent, resulting in abundant phenotypic variation including fruit ripening and color traits that are linked to variation at the corresponding introgressed *L. pennellii* chromosomal segment. Recently, additional IL lines have been isolated to define smaller introgressions for a total of 76 IL lines (Liu and Zamir 1999). These ILs have been used to map quantative trait loci (QTLs) for important fruit quality traits, such as fruit brix, or total soluble solids like fructose and sucrose (Fridman et al. 2002).

Tomato is also benefiting from a NSF-funded expressed sequence tag (EST) development and utilization project (Van der Hoeven et al. 2002, Alba et al. 2004, Fei et al. 2004). Numerous cDNA libraries were constructed and sequenced from a comprehensive selection of tissues and physiological treatments to yield an extensive EST database (Moore et al. 2002 and Van der Hoeven et al. 2002). The EST collection has been annotated into a unigene set based on combining multiple copy ESTs into intersecting contigs totaling 27,274 unigenes (Van der Hoeven et al. 2002). This EST collection along with a set 6 of sequenced bacterial artificial chromosome (BAC) clones has allowed the prediction of the number of genes within the tomato genome (Van der Hoeven et al. 2002). The Arabidopsis Genome Initiative (2000) has shown the estimate of Arabidopsis genes based on an EST unigene set was 35% overestimated. Therefore when this percentage was applied to tomato, it was determined that the unigene set would stand for 17,500 genes. This number represents half of the number of predicted genes in the BAC clones resulting in an estimate of 35,000 genes in the tomato genome (Van der Hoeven et al. 2002).

Finally, microarrays are being employed to explore the expression profile of tomato developmental processes such as fruit ripening (Moore et al. 2002). A microarray has been constructed using the above mentioned EST libraries enriched with genes implicated in ethylene biosynthesis and fruit ripening including representative of a wide range of pathways and processes (Moore et al. 2002). Tomato microarrays are available to the research public at http://bti.cornell.edu/CGEP/CGEP.html. A comprehensive expression profile for ripening fruit based on a 10-stage time course from 7 days post anthesis to 15 days post breaker is currently under development (R. Alba, S. Moore, P. Payton and J. Giovannoni, unpublished; Moore et al. 2002, Alba et al. 2005).

BIOSYNTHESIS AND SIGNALING OF ETHYLENE

Ethylene is synthesized via the Yang cycle (Fig. 11.1) employing methionine as the initial substrate (reviewed in Kende 1993, Yang and Baur 1969). The pathway proceeds by conversion of methionine to S-adenosine-L-methionine (SAM). The enzyme 1-aminocyclopropane-1-carboxylic acid synthase (ACC synthase), or ACS, catalyzes the conversion of SAM to ACC. ACS is

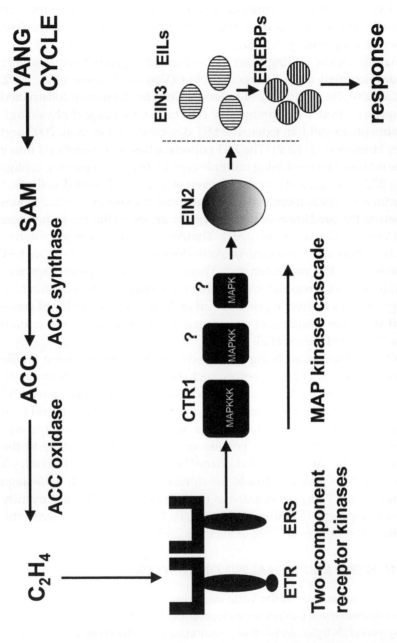

Fig. 11.1　Ethylene Biosynthesis and Signal Transduction Pathway (Modified from Giovannoni 2001).

considered to be a rate-limiting step and is encoded by multigene families varying in size among species and whose members demonstrate differential expression and include ripening-related ACS genes (Yip et al. 1992). The final step in the synthesis of ethylene is generation of ethylene from ACC via the enzyme ACC oxidase (ACO, reviewed in Fluhr and Mattoo 1996). A multigene family consisting of four tomato genes, ACO1 (Smith et al. 1986) to ACO4 (Nakatsuka et al. 1998) display variation in expression both temporally and spatially (Barry et al. 1996, Nakatsuka et al. 1998). ACO has also been found to represent a second rate-limiting step in ethylene biosynthesis (Barry et al. 1996). Repression of ACS or ACO results in decreased ethylene synthesis and inhibition of ripening in tomato (Theologis et al. 1993, Fray and Grierson 1993). Genes encoding both ACS and ACO have been cloned from a number of additional plant species such as rice (ACS: Zhou et al. 2002), apple (ACS: Lay-Yee and Knighton 1995), apple (ACO: Ross et al. 1992, Atkinson et al. 1998), banana (ACO: Lopez-Gomez et al. 1997), and melon (ACO: Balague et al. 1993).

Arabidopsis has proven to be of exceptional importance in elucidating mechanisms of ethylene signal transduction (Fig. 11.1). This is due in large part to the ease of screening for seedling-triple response phenotypes in mutagenized seeds which revealed key steps within the signaling pathway (reviewed in Ecker 1995, Kieber 1997, Chang and Shockey 1999). A number of ethylene signaling genes identified in *Arabidopsis* have been shown to have homologues in crop species including tomato (reviewed in Watkins 2002). These include homologs to the *CTR1* kinase, the *EIN3* family of transcription factors, as well as members of the ethylene receptor family (Leclercq et al., 2002, Tieman et al. 2001, Yen et al. 1995). While *Arabidopsis* has been used to study many aspects of general ethylene signaling, it clearly can not be used to study the role of ethylene signaling in fleshy climacteric fruit development.

The *Nr* mutation was originally characterized as a dominant mutation displaying delayed and incomplete ripening and reduced softening (Rick and Butler 1956, Hobson 1967). Lanahan et al. (1994) found the *Nr* mutant fruit is unable to ripen in response to either endogenous or exogenously applied ethylene. *Nr* seedlings exposed to exogenous ethylene failed to show the "triple response" (Fig. 11.2) suggesting a greater perception of effects of ethylene (Lanahan et al. 1994, Yen et al. 1995). In support of this hypothesis, the *Nr* mutant also demonstrated delayed floral abscission and impaired senescence of leaves and petals, in addition to reduced expression of ethylene regulated genes (Lanahan et al. 1994, Yen et al. 1995). The *NR* gene was isolated and is homologous to ethylene receptors found

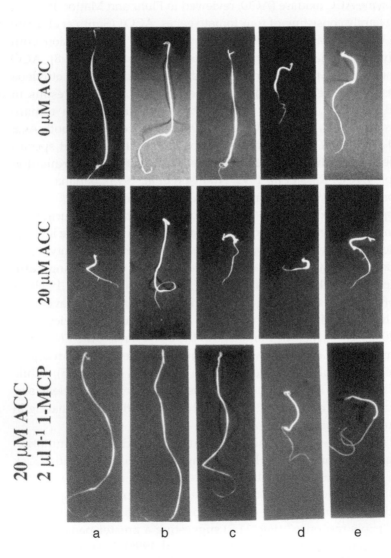

Fig. 11.2 Tomato seedling ethylene-responses phenotypes. Seeds were germinated on water-agar in baby food jars for 12 days in the dark with or without the ethylene precursor ACC (20 ppm) which results in de facto ethylene exposure. Genotypes are: a) *nr/nr*, b) *Nr/Nr*, c) *Epi/Epi*, d) *epi/epi*, and e) *epi/epi;Nr/Nr* (Modified from Barry et al. 2001).

in *Arabidopsis*, including *ETR1*, though with most similarity to *ERS* (Wilkinson et al. 1995, Hua et al. 1995, reviewed in Lashbrook et al. 1998a). The *ETR1* gene encodes a protein homologous to bacterial two-component regulators (Chang et al. 1993), which is sufficient to confer dominant ethylene insensitivity when mutated and transferred to the *Arabidopsis* or other plant genomes (Wilkinson et al. 1997). Unlike bacterial two-component systems, ETR1 contains a sensor and a response-regulator domain in one protein. The *ERS1* and *ERS2* genes are distinguished by structural variation compared to *ETR1* in that they do not contain the receiver domain (reviewed in Stepanova and Ecker 2000). It is thought receptors missing this domain may utilize the receiver domain of other receptors such as ETR1 (Stepanova and Ecker 2000). Homologues for the ethylene receptors have also been cloned from non-climacteric fruit including an *ERS* homologue from *Citrus sinensis* (Li et al. 1998). The presence of ethylene receptor homologues in non-climacteric fruit indicate either use of ethylene in processes other than ripening or ethylene presence influencing processes that are exerted with lower ethylene concentrations.

Along with *Nr* (also known as *LeETR3*), additional genes (*LeETR1, 2, 4, 5*) belonging to the ethylene receptor family have been isolated from tomato (Lashbrook et al. 1998a, Tieman et al. 1999, reviewed in Klee and Tieman 2002). *NR* and *LeETR4* are distinct among others in this gene family from their sharp induction during fruit ripening (Tieman et al. 2000). *NR* ripening induction is due in part to its own responsiveness to ethylene (Wilkinson et al. 1995), suggesting multiple levels of ripening influence by this hormone. Transgenic lines showing repression of *LeETR4* exhibited a wide range of phenotypes including extreme epinasty of leaves, increased senescence of flowers, and a reduction in ripening time (Tieman et al. 2000), all of which are consistent with increased sensitivity to ethylene. Conversely, lines having reduced expression of *NR* showed normal sensitivity to ethylene. While counter-intuitive in isolation, this result accounted for a concomitant increase in accumulation of *LeETR4* mRNA (Tieman et al. 2000). The converse proved true in transgenic lines engineered to be deficient in *LeETR4* expression, suggesting that *NR* and *LeETR4* are redundant in function (Tieman et al. 2000). This result is distinct from observations in *Arabidopsis*, where single gene receptor knock outs result in no discernible phenotypes and only multiple knockouts of at least three receptor family members display a phenotype consistent with constitutive response to ethylene (Hua and Meyerowitz 1998). Comparative functional and expression analyses of the tomato and *Arabidopsis* ethylene receptor families suggest similar gene/predicted peptide structures under regulatory constraints that likely reflect differential selective pressures on the Solanaceae versus the Brassicaceae.

In addition to *Nr*, a spontaneous tomato mutation called *epinastic (epi)* is believed to be involved in the ethylene signaling pathway. This single gene mutation was originally described as a semi-dominant mutation (Ursin 1987), but is now thought to behave as recessive (Barry et al. 2001). This mutation leads to profound epinasty of leaves, thickening of petioles and stems, a shorter yet highly branched root system, overproduction of ethylene in vegetative tissues, and normal ripening (Ursin 1987, Fujino et al. 1988). Constitutive seedling triple response in the absence of ethylene is also a characteristic of the *epi* mutant (Fig. 11.2; Lee 1999) and is comparable to the *constitutive triple response (ctr1)* mutant of *Arabidopsis* (Kieber et al. 1993). The CTR1 gene encodes a MAP kinase kinase kinase (MAP KKK) through which all measured ethylene responses must flow in *Arabidopsis* (Kieber et al., 1993). Although homologs to *Arabidopsis CTR1* have been found in tomato (Leclercq et al., 2002; Adams and Giovannoni unpublished), *epi* does not appear to map to any SlCTR loci (Fig. 11.3; Barry, Adams, and Giovannoni unpublished).

To further characterize *epi,* and to help elucidate the mechanism of its effects, it was crossed with the ethylene insensitive receptor mutant *Nr* to create a *Nr/Nr;epi/epi* line (Barry et al. 2001). Wild-type seedlings displayed a typical triple response with the addition of ACC synthase while *Nr* showed characteristic insensitivity to ethylene. *Nr/Nr;epi/epi* exhibited a typical *epi* response regardless of treatment (Barry et al. 2001). In the presence of the ethylene action inhibitor 1-MCP the triple response of wild-type seedlings was inhibited while *Nr* seedlings remained insensitive. *epi* and *Nr/Nr;epi/epi* were not reverted to wild-type (Barry et al. 2001). *Nr/Nr;epi/epi* plants displayed constitutive ethylene response phenotypes following 1-MCP treatment in leaves as well (Barry et al 2001). Conversely, petal senescence and ripening of *Nr/Nr;epi/epi* was ethylene insensitive similar to the *Nr* parent (Barry et al. 2001) suggesting *epi*, unlike *ctr1*, is involved in only a subset of ethylene responses (Barry et al. 2001). Alternatively, *epi* may represent a step in a separate pathway involved in cross-talk to the ethylene signal transduction pathway (Barry et al. 2001). Microarray analysis of *epi* versus its NIL normal control could provide insights into the role of *epi* in tomato ethylene signal transduction.

RIPENING-RELATED GENES AND MUTANTS

A variety of tomato genes related to ripening have been identified and studied in recent years. For example, the pTOM series of clones were isolated

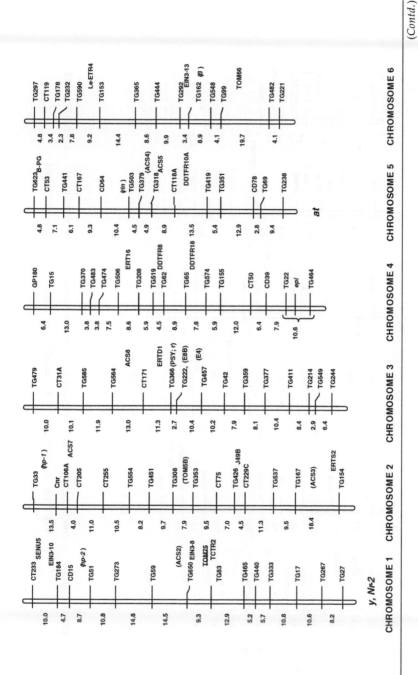

(Contd.)

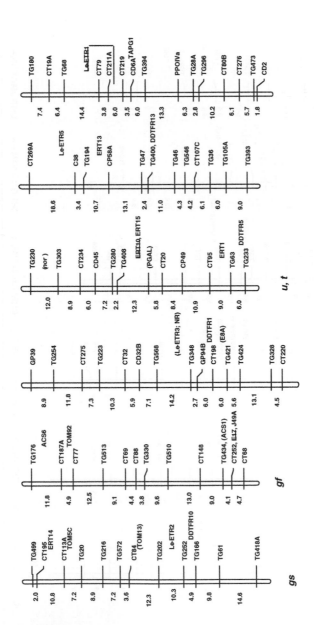

Fig. 11.3 Genetic map of tomato showing location of ripening-related loci.

by differential screening of mature green versus ripening fruits (Slater et al. 1985) and number well over 100 independent ripening-related tomato genes (reviewed in Gray et al. 1992, Grierson and Schuch 1993, Gray et al. 1994; Table 11.1). The pTOM clones were either present or absent in unripe fruit, but displayed a marked increase as fruit ripened. A number of the pTOM series proved to be important in key aspects of tomato fruit ripening.

pTOM6 corresponded to polygalacturonase (PG) which catalyzes hydrolytic cleavage of alpha-1,4 linkages of polygalacturonic acid (pectin) whose solubilization has been associated with ripening-related changes in fruit texture (Grierson and Schuch 1993). Antisense expression of PG, under CaMV 35s, did not result in significant delay effects in softening of ripening fruit but did limit supply and action of the enzyme (Smith et al. 1988, Sheehy et al. 1988, Carington et al. 1993), suggesting PG is not the only significant player in cell wall breakdown during ripening (Smith et al. 1988, reviewed in Theologis et al. 1992).

pTOM13 was shown as ACC oxidase in antisense technology used to functionally define the activity of a plant gene (Hamilton et al.1990, 1991). pTOM13 was also shown to be regulated at the mRNA level by ethylene in both fruit and leaves (Maunders et al. 1987, Smith et al. 1988). Several additional clones from the pTOM series of note include TOM5 phytoene desaturase (Bird et al. 1991), TOM66 low M.Wt. heat-shock protein (Fray et al.1990), and TOM75 membrane-spanning solute channel (Fray et al. 1994).

Lincoln et al. 1987 also identified a series of clones through differential screens based on ethylene induction including E4, E8, and E17. E4 was shown to have homology to a methionine sulphoxide reductase, though the role of this gene in ethylene response remains mysterious (Montgomery et al. 1993). E8 encodes a member of a small multi gene family that shares homology with dioxygenase (Kniessel and Deikman 1996). E8 is itself transcriptionally activated by ethylene (Lincoln et al. 1988), and on analysis of antisense lines has shown to play a significant regulatory role in the synthesis of ethylene during fruit ripening (Penarrubia et al. 1992). Although E8 is activated by ethylene, it has been shown that along with PG and ACOE8 is expressed in transgenic plants characterized for supressed biosynthesis of ethylene, thus, further supporting the existence of non-ethylene mediated regulatory factors controlling aspects of climacteric ripening (Oeller et al. 1991 and Theologis et al. 1993).

There are a substantial number of mutations that impact ripening of tomato fruit (Table 11.1). Included are the previously mentioned *Nr* (Rick and Butler 1956), *rin* (Robinson and Tomes 1968), *nor* (Tigchelaar et al.

Table 11.1 Tomato mutants and ripening-related genes (subset)

Mutant	Chromosome	Identification of Gene and/or Mutant	Fruit Phenotype
Ripening			
1. ripening inhibitor (rin)	5	Mads-box (Vrebalov et al. 2002)	yellow and firm
2. non ripening (nor)	10	Tigchelaar et al. (1973)	severely delayed ripening
3. Never-ripe (Nr)	9	Ethylene receptor (Wilkinson et al. 1994)	orange
4. Never-ripe-2 (Nr-2)	1	Rick and Butler (1956)	yellow-green
5. alcobaca (alc)	10	Kopeliovich et al. (1981)	delayed ripening
6. Colorless non-ripening (Cnr)	2	Thompson et al. (1999)	yellow skin/non-pigmented pericarp
7. Green-ripe (gr)	Barry and Giovannoni unpublished	Kerr (1958)	green
8. high-pigment 1 (hp 1)	2	Peters et al. (1998), Liu et al. (2004)	deep red
Carotenoid			
1. apricot (at)	5	Jenkins and Mckinney (1955)	yellow-pink hue
2. green-flesh (gf)	8	Clayberg et al. (1967)	green
3. tangerine (t)	10	Carotenoid isomerase (Issacson el. al. 2002)	orange
4. old gold crimson (ogc)	6	Clayberg et al. (1966)	dark red
5. uniform (u)	10	Rick and Butler (1956)	darker green shoulder
Ripening-Related Genes	**Gene ID**	**Reference**	
1. pTOM5	phytoene synthase	Bird et al. (1991)	
2. pTOM6	polygalacturonase	Grierson et at. (1986)	
3. pTOM13	ACC oxidase	Hamilton et al. (1990)	
4. pTOM66	heat-shock protein	Fray et al. (1990)	
5. pTOM75	solute channel	Fray et al. (1994)	
6. E8	dioxygenase homology fruit specific	Kniessel and Deikman (1996)	
7. E4	met.sulphoxide reduct.	Montgomery et al. (1993)	
8. LeExp1	expansin (in ripening fruit)	Rose et al. (1999)	
9. ACS	ACC synthase	Yip et al. (1992)	

1973) and *Cnr* (Thompson et al. 1999) mutants, in addition to *Green-ripe (Gr)* (Kerr 1958), *Never-ripe2 (Nr2)* (Kerr 1982), and *alcobaca (alc)* (Kopeliovich et al. 1981). Many of these mutations have been placed on the tomato genetic map (Fig. 11.3) (Giovannoni et al. 1999). The mutation *Nr* is a lesion in one of the genes in the family of ethylene receptors which results in the inability of the receptor to bind ethylene (Klee 2002). Some of the remaining mutations appear to impact non-ethylene signaling aspects of ripening control. For example, the phenotype of the spontaneous recessive and non-ripening *rin* mutation includes the inability of fruit of this mutation to ripen when supplied with exogenous ethylene, yet ethylene responsive genes are induced in such ethylene treated fruits (Giovannoni et al. 1989). Furthermore, *rin* seedlings display the triple response to exogenous ethylene, indicating that non-fruit ethylene responses also remain intact (Yen et al. 1995). These phenotypes of *rin* have been interpreted to indicate a lesion in a regulatory component that impacts both ripening-related ethylene production and non-ethylene mediated components of climacteric fruit ripening (Vrebalov et al. 2002). As stated above, *RIN* encodes a MADS-box transcription factor which, in plants, have been primarily associated with floral development (reviewed in Soltis et al. 2002). A floral development function is consistent with the predicted developmental ripening regulation implied by the *rin* phenotype (Vrebalov et al. 2002).

The single *rin* mutant allele, in addition to ripening impaired fruit, also results in enlarged sepals (macrocalyx) and both traits map in perfect linkage on tomato chromosome 5 (Fig. 11.3; Giovannoni et al. 1995). A separate recessive mutation bearing large sepals, termed *macrocalyx (mc)*, was also mapped to a similar location on chromosome 5 (Fig. 11.3; Rick and Butler 1956), suggesting the possibility that the enlarged sepal phenotype displayed by the *rin* mutation was a result of a mutation in an adjacent *(MC)* gene (Robinson and Tomes 1968). *MC*, like *RIN*, was also cloned and shown to be a MADS-box gene (Vrebalov et al. 2002).

Another mutation closely resembling *rin* in phenotype is the recessive *non-ripening (nor)* mutation (Tigchelaar et al. 1973). The latter mutation displays ethylene synthesis/response phenotypes of the whole fruit and the level of gene expression is similar to that described for *rin* (Giovannoni et al. 1995). This suggests that both gene products may participate in the same or similar regulatory circuits. *NOR* gene isolation is now complete and functional characterization is in progress (J. Vrebalov and J. Giovannoni, unpublished).

LIGHT SIGNAL TRANSDUCTION AND
RIPENING-RELATED PIGMENTATION

All photosynthetic organisms depend on light as their source of energy. Plants are also extremely dependent on their respective light environment for cues impacting optimal growth and development. At the molecular level, plants discern various wavelengths of light using photoreceptor molecules which impact plant development as a result of quality, quantity, direction, and period of light (Alba et al. 2000). Photoreceptors that absorb red or far-red light are phytochromes, while blue-light receptors are termed cryptochromes. Phytochromes in particular are known to regulate the accumulation of a number of pigments in various plant tissues (Kendrick and Kronenberg 1994), and phytochromes localized to ripening tomato fruit influence the accumulation of lycopene (Alba et al. 2000).

The change in fruit pigmentation from green (chlorophyll) to red (lycopene and beta-carotene) is a hallmark of tomato fruit ripening and is influenced by light. Genes representing steps involved in biosynthesis of certain pigments have been cloned from tomato and in some cases through screens for ripening-related genes. Indeed this was precisely the case for phytoene synthase-pTOM5 (Fray and Grierson 1993).

A number of fruit color mutants in tomato have been shown to reflect lesions in various steps of the carotenoid biosynthesis pathway. Many of these mutants have been mapped (Fig. 11.3) and most of the corresponding genes recently been cloned (Table 11.1). Practical use of a number of carotneoid mutants in tomato breeding confirms the potential for manipulating fruit carotenoid levels for fruit quality and nutritional impact (reviewed in Hirschberg 2001). Carotenoids are antoixidant pro-vitamin A compounds capable of quenching oxygen radicals produced as a consequence of photosynthesis. As such, carotenoid synthesis, including that associated with fruit ripening, can be strongly influenced by light quality and intensity (Alba et al. 2000). Manipulation of light perception and response may serve as an additional avenue through which carotenoid accumulation may be enhanced in fruit color and nutrient quality in crop species. High levels of dietary lycopene have been associated with lower prostate cancer cases in men (reviewed in Giovannucci 2002).

The tomato *high-pigment (hp1)* mutant is reported to represent lesion in a repressor of light signal transduction resulting in light hypersensitivity and elevated carotenoid accumulation in *hp1/hp1* fruit (Peters et al. 1989). The mutation is a single recessive gene mutation (Reynard 1956) displaying an increase in the amount of hypocotyl anthocyanin, reduced hypocotyl elongation, and increased leaf chlorophyll content, in addition to elevated ripe fruit carotenoids (Kerr 1965, Mochizuki and Kamimura 1985, Peters et al. 1989, Jarret et al. 1984, Thompson 1962). The seedling phenotypes of *hp1* have been shown to be red light enhanced, and far-red reversible, serving as the basis of the hypothesis of a lesion resulting in hyper-activity of phytochrome responses (Kerckhoffs and Kendrick 1997). *hp1* also specifically alters expression of genes encoding proteins involved in photosynthesis (CAB chlorophyll *a/b*-binding protein and RBCS Rubisco small subunit) as might be predicted for a hypersensitive light response mutant (Peters et al. 1998). In further support of a role in phytochrome hyper-signaling, over-expression of oat phytochrome (phyA) in tomato resulted in phenotypes similar to those observed in *hp1* (Boylan and Quail 1989), indicating responses mediated by phytochromes are expressly affected by the *hp* mutations (Mustilli et al. 1999). Although to date there have been no reported mutants in *Arabidopsis* corresponding directly to *hp1* at the phenotypic level, *Arabidopsis cop, det*, and *fus* light signal transduction mutants share some similar seeding characteristics (reviewed in Hardtke and Deng 2000, Schwechheimer and Deng 2000). The gene for *HP1* has been mapped to chromosome 2 in tomato (Yen et al. 1997) and is currently a target for a chromosome walk (Y. Liu and J. Giovannoni, unpublished).

A second mutation phenotypically similar to *hp1* and termed *hp2*-mutant has also been isolated (Soressi 1975). *HP-2*-wild type gene was cloned and determined to be the tomato homolog of the *Arabidopsis* DET1 gene (Mustilli et al. 1999). DET1 encodes a nuclear localized protein implicated in regulation of gene expression and development in response to light (Pepper et al. 1994). This was despite the fact *hp-2* mutants (Mustilli et al. 1999) did not display many of the phenotypes which characterize the mutant *det1* (e.g. no discernable seedling phenotypes in the dark). The differences between *hp2* and *det1* mutants are postulated to represent the possibility that tomato and *Arabidopsis* contain modifications in regulatory pathways which control photomorphogenesis (Mustilli et al. 1999).

Changes in fruit color and carotenoid levels may also be achieved by influencing levels of polyamines (PAs) (Mehta et al. 2002). Polyamines are organic cations of low molecular weight derived from amino acids via decarboxylation and associated with numerous physiological processes across diverse taxa (Kakkar and Sawhney 2002). Two PAs in particular, spermidine and spermine, are synthesized by adding aminopropyl groups removed from decarboxylated *S*-adenosyl methionine (SAM) to the precursor putrescine (Tiburcio et al. 1997, Kakkar and Sawhney 2002). In plants, polyamines are said to be involved in cell division, formation of tubers, embryogenesis, flower development, root initiation, and ripening though the precise relationship to carotenoid accumulation in fruit remains unclear (Mehta et al. 2002, reviewed in Kakkar and Sawhney 2002).

In Mehta et al. (2002), a yeast S-adenosylmethionine decarboxylase (ySAMdc) gene was introduced into tomato under the fruit-specific E8 promoter (Deikman et al. 1992). The resulting transgenic plants displayed increased levels of spermine and spermidine throughout fruit ripening, with corresponding reduction in levels of putrescine (Mehta et al. 2002). Ripening fruit on transgenic lines also appear to have extended "vine life" (Mehta et al. 2002). An increase of 50% in precipitate weight ratio (PPT), a gauge of the quality of processing tomatoes (Takada and Nelson 1983), was seen in juice and enhanced juice quality (Mehta et al. 2002). Lycopene was also increased by 200-300% yielding deeper red transgenic fruit (Mehta et al. 2002). Lastly, the transgenic fruit displayed elevated levels of ethylene but with no effect on time of ripening, suggesting the increase in polyamines may countermand the effects of amplified ethylene levels (Mehta et al. 2002). These results highlight new avenues of possible manipulation of fruit pigmentation and associated nutritional properties in crop plants and downstream products.

CELL WALL MODIFICATIONS AND RIPENING

Fruit ripening and cell wall metabolism are clearly interconnected processes. Cell walls in tomato consist of cellulose microfibrils, a hemicellulose matrix (primarily xyloglucan in dicots), and a pectin network (reviewed in Carpita and Gibeaut 1993, Cosgrove et al. 2000b). Cell wall-related changes occuring in ripening fruit include: reduction of turgor, decrease in cell-cell adhesion, decline in apoplast pH, depolymerization/solubilization of pectins and hemicelluloses, and increase in cell wall porosity (Brett and Waldron 1995, Brummell and Harpster 2001). Decline of cell wall internal integrity, and loss of attachments to adjacent cells, governs the structural integrity and texture of many ripening fruit (Redgwell and Fischer 2002). During ripening, cell wall compounds including polyuronides are solubilized resulting in cell wall destabilization and loss of side chain sugars-galactan and arabinan (Gross 1984, Seymour et al. 1990). This loss in cell wall integrity results in swelling of the wall space and loss of cell to cell adhesion leading to softening of the pericarp and locule liquefication upon ripening. A key enzyme long implicated in cell wall degradation and associated textural changes during tomato ripening is polygalacturonase (PG). There is a family of tomato genes encoding PG enzymes involved in various developmental processes including fruit ripening (Sitrit and Bennett, 1998). The PG involved in ripening is known to be expressed only in fruit with transcription limited to the ripening process (Bird et al. 1988, Della-Penna et al. 1989, Montgomery et al. 1993). It had been thought previously that PG alone was responsible for the majority of cell wall degradation and associated softening in ripening fruit (Bennett and Della-Penna 1987). However, repression of PG resulted in degredation of pectin, but did not affect softening (Sheehy et al 1988). Furthermore, PG coding sequence was also fused to a fruit specific promoter E8 (Deikman et al. 1992) in the sense orientation and transformed into a cultivar homozygous for the *rin* allele to determine whether PG activity was sufficient to restore softening in the mutant (Giovannoni et al. 1989). The resulting transgenics displayed accumulation up to 60% of PG activity, but did not restore softening to any appreciable degree (Giovannoni et al. 1989, reviewed in Brummel and Harpster 2001). This result, in concert with data from PG antisense experiments, provided the first evidence that cell wall modifications leading to tomato softening are more complicated than activation of a single cell wall hydrolytic enzyme.

Additional cell wall metabolizing enzymes have been cloned and characterized in ripening tomato fruit, revealing a more complex view of cell wall ultrastructural changes associated with ripening than simple pectin degradation (reviewed in Brummell and Harpster 2001). Pectinmethylesterase (PME) has been found in a variety of plant species as well as bacteria and fungi that are pathogenic to plants (Rexova-Benkova and Markovic 1976, Huber 1983, Collmer and Keen 1986). PME is involved in the removal of methyl groups from pectin (Grierson and Schuch 1993) allowing pectins to form Ca^{2+} crosslinked gel structures and providing additional susceptibility to PG activity (reviewed in Brummell and Harpster 2001). Antisense PME tomato lines with <10% normal PME activity yielded the anticipated higher MW pectins but had no other notable impact on softening or other aspects of ripening (Tieman et al. 1992). These results indicate that like PG, PME is not sufficient for softening of ripening tomato fruit (Carington et al. 1993).

Expansins are a group of integral cell wall proteins originally identified as contributing to cell wall extension of cucumber hypocotyls (McQueen-Mason et al. 1992). Expansins have been shown to be capable of loosening and stretching of isolated cell walls (reviewed in Cosgrove 1999) and two distinct families (α-expansins and β-expansins) have been identified

(Cosgrove 2000a). α-expansins are tightly associated with cell elongation and wall loosening (Cosgrove 2000a), while β-expansins are among the grass pollen allergens (Cosgrove 1999).

The tomato α-expansin family has been extensively characterized (Reinhardt et al. 1998, Brummell et al. 1999a, reviewed in Brummell and Harpster 2001). *LeEXP1* is the primary member expressed in ripening fruit (Rose et al. 1997), and homology to *LeEXP1* has been employed to isolate homologous genes from melon and strawberry (Rose et al. 1997). *LeEXP1* expression is regulated by ethylene and found at 1-2% of wild-type levels in *rin* and *nor* (Rose et al. 1997). Also, *LeEXP1* was not induced upon treatment of mature green (MG) *rin* and *nor* fruit given exogenous ethylene (Rose et al. 1997). These results correspond well with the decrease in fruit softening characteristic of both mutations and indicate an influence on *LeEXP1* by the pathway involving *rin* and *nor* (Rose et al. 1997). In the receptor mutant *Nr*, expression of *LeEXP1* was observed at comparable levels to wild-type, but did not increase upon addition of exogenous ethylene (Rose et al. 1997). These results may be due to leakiness of the *Nr* allele in the Alisa Craig background or may represent compensation by other members of the tomato ethylene receptor family. Rose et al. (2000) defined a subclass of α-expansins based on a dendrogram constructed from amino acid sequence consisting of *LeEXP1*, *LeEXP4* present in growing flowers and fruit (Brummell et al. 1999a), and *LeEXP18* expressed primarily in meristamatic tissues (Reinhardt et al. 1998). These alignments were used to isolate antibodies for *LeEXP1* to detect expansin activity in actively growing and ripening fruit as well as to indicate characteristic expansin activity along with the hydrolytic capabilities of expansin proteins (Rose et al. 2000). The function of *LeEXP1* in ripening fruit has been studied via over-expression under a 35sCaMV promoter. The results displayed increased softening at all stages of ripening including typically firm MG fruit. Antisense suppression of the gene displayed inhibition of breakdown of polyuronide and an increase in fruit firmness during the course of ripening, but was insufficient in impeding breakdown of hemicelluloses (Brummell et al. 1999a).

The endo-1,4-Beta-Glucanases (EGases or Celluases-CEL), mentioned above are implicated in a number of physiological processes such as abscission of organs as well as ripening in both climacteric and non-climacteric ripening fruit. EGases have been isolated from strawberry (Harpster et al. 1998, Manning 1998, Llop-Tous et al. 1999, Trainotti et al. 1999, Spolaore et al. 2003), pepper (Ferrarese et al. 1997, Harpster et al. 1997), and tomato (Brummell et al. 1997, Lashbrook et al. 1994; reviewed in Rose and Bennett 1999). Ethylene has also been found to either be required or induce accumulation of EGases (Brummell et al. 1999b). Two EGases, Cel1 (Lashbrook et al. 1994) and Cel2 (Gonzalez-Bosch et al. 1996), increase in accumulation upon ripening pointing to a possible role in cell wall disassembly during ripening.

In Lashbrook et al. (1998b), Cel1 was placed under CaMV 35s promoter in antisense orientation to determine the effects of suppression in tomato. Transgenic lines displayed a decrease in Cel1 expression to 0.1% of wild-type levels in fruit resulting in no measurable decrease in softening during ripening, but a decrease of 5-6% in abscission zones caused reduction in abscission of transgenic flowers (Lashbrook et al. 1998b). A second suppression study using Cel2 also under 35s resulted in transgenic tomatoes whose expression was reduced by greater than 95% in fruit and ~80% in abscission zones (Brummell et al. 1999b). Transgenics exhibited no discernable changes in softening of ripening fruit, but did display an increase in break-strength (e.g. the amount of pressure required to remove abscission zone 4 days post breaker) (Brummell et al. 1999b). Taken together, these studies indicate that during ripening Cel1 and Cel2 may compensate for absence of the other. Harpster et al. (2002a) showed suppression of a ripening-related EGase (CaCel1) in pepper did not cause a decrease in the softening of pepper fruit upon ripening and did not affect depolymerization of matrix glycans. CaCel1 was also placed under CaMV 35s promoter and transformed into tomato (Harpster et al. 2002b). Transgenics did not show any increases in depolymerization of xyloglucans or an increase in softening of ripening fruit. These results once again indicate that suppression/over-expression of only one EGase is not sufficient for significant changes in cell wall softening, suggesting either compensation by other family members and/or involvement of other enzymes.

A third enzyme believed to contribute to cell wall breakdown in ripening fruit is xyloglucan endotransglycosylase (XET), or XTH (Rose et al. 2002). This enzyme was purified originally from nasturtium seed due to its capability to hydrolyze xyloglucan in the absence of supplemental xyloglucan sugars (Edwards et al. 1986, Faik et al. 1998). XHTs catalyze cleavage of xyloglucan backbone internal linkages in the (1à4) Beta-D orientation and shift the xyloglucan molecule to another xyloglucan with a non-reducing end (Steele et al. 1999). Several genes encoding XHTs have been isolated from tomato (tXET-B1, tXET-B2, and LeExt2, down-regulated by auxin in hypocotyls; Arrowsmith and de Silva 1995, Catala et al. 2001), kiwi (AdXET1-6, linked with ripe kiwi fruit, Schroder et al. 1998) and grape (VXET1, coupled with grape softening; Ishimaru and Kobayshi 2002).

The results suggest a role of these genes in ripening of both climacteric and non-climacteric fruit. *LeEXTB1* (expressed in ripening fruit) and *LeEXGT1* (expressed in green fruit during cell expansion) have both been expressed in transgenic tomato (de Silva et al. 1994). In the over-expression of *LeEXGT1*, abundance of the mRNA transcript increased the size of fruit along with higher levels of sugars, indicating an important role for *LeEXGT1* in young expanding fruit for affecting the final size of the fruit (Asada et al. 1999). de Silva et al. (1994) showed suppression of *LeXETB1* under a fruit specific *PG* promoter, but found no parallels between suppression and softening. These results would indicate possible redundancy of other family members while reiterating the belief that multiple enzymes must be involved in the breakdown of cell walls during ripening.

SUMMARY AND FUTURE PROSPECTS

The growth and maturation of fruits is a unique aspect of plant development which serves as a significant component of human diets with associated impacts on nutrition and health. Progress has been made in recent years in both biochemical and molecular genetics of fruit ripening leading to enhanced understanding of the molecular basis of fruit development and ripening control. Various model systems have greatly contributed to the understanding of fruit ripening of both climacteric and non-climacteric fruits and recent efforts have demonstrated molecular connections between these physiologically distinct two ripening types. Extrapolation of rapid molecular advances in ethylene and light signaling in *Arabidopsis* has led to important insights pertaining to how these signaling pathways impact maturing fruit. Advancing molecular genetic, genomic and biotechnological approaches have yielded additional insights into universal control of fruit ripening besides specific regulatory and biosynthetic pathways impacting distinct quality characteristics. This chapter summarizes recent developments pertaining to the molecular and genetic characterization of maturation and ripening of fleshy fruits.

Fruit ripening is a complex process which includes changes in physiology, biochemistry, color, flavor, and nutritional content. Many fruit species are of great economic importance, and thereby are targets for extensive study and manipulation at both the molecular and biochemical levels. While the division between fruits that ripen climacterically and those that are non-climacteric is clear at the physiological level, current work on genes present in both types of fruit is beginning to piece together possible universal regulation mechanisms for non-ethylene mediated ripening. Recent technical advances such as microarrays and EST databases are allowing further comparisons between ripening profiles of model systems in both climacteric tomato and non-climacteric strawberry. Numerous homologues of genes involved in ethylene biosynthesis have been cloned in several species allowing a broader understanding of ethylene biosynthesis. Comprehensive knowledge of ethylene signal transduction has greatly benefited from studies on both *Arabidopsis* and tomato. Cloning of genes encoding enzymes thought to be involved in ripening has contributed to better understanding of cell wall breakdown during this process. These enzymes have become useful in the quest to manipulate softening of many fruits. Understanding of the function of light during fruit ripening has been advanced through considerable innovations in the field of light signaling and color development. Several genes involved in light signaling and fruit pigment accumulation have provided targets for manipulation of ripening color as well as nutritional content of fruit. Further elucidation of the biosynthesis of ripening pigments, light signaling, and pathways discussed in this chapter will increase the capacity for nutritional value and attractiveness for the consumer.

Acknowledgements

Research described herein was funded by the various agencies to whom the author is exceptionally grateful: USDA-NRI (92- 37300-7653, 95-37300-1575), NSF (IBN-9604115, DBI-9872617), Zeneca Agrochemicals (Syngenta), Lipton Foods, and USDA-ARS.

REFERENCES

Abeles, F.B., P.W. Morgan, and M.E. Saltveit. 1992. Ethylene in plant biology (2nd ed). Academic Press Inc., CA, USA.

Adams, D.O. and S.F. Yang. 1979. Ethylene biosythesis: identification of 1-aminocyclopropane-1-carboxylic acid is an intermediate in the conversion of methionine to ethylene. Proc Natl Acad Sci USA 76: 170-174.

Adams-Phillips, L., C. Barry, and J. Giovannoni. 2004. Single transduction systems regulating fruit ripening. *Trends Plant* Sci 9: 331-338.

Aharoni, A., C.P. Leopold, H.C. Keizer, H.J. Bouwmeester, Z. Sun, M. Alvarez-Huerta, H.A. Verhoeven, J. Blaas, A.M.M.L. van Houwelingen, R.C.H. De Vos, H. van der Voet, R.C. Jansen, M. Guis, J. Mol, R.W. Davis, M. Schena, van A.J. Tunen, and A.P. O'Connell. 2000. Identification of the SAAT gene involved in strawberry biogenesis by use of DNA microarrays. Plant Cell 12: 647-661.

Aharoni, A., C.P. Leopold, H.C. Keizer, R. B.-P. Van Den Broeck, J. Munoz-Blanco, G. Bois, P. Smit, R.C.H. De Vos, and A.P. O'Connell. 2002. Novel insight into vascular, stress, and auxin-dependent and –independent gene expression programs in strawberry, a non-climacteric fruit. Plant Physiol 129: 1019-1031.

Alba, R., M.M. Cordonnier-Pratt, and L.H. Pratt. 2000. Fruit-localized phytochromes regulate lycopene accumulation independently of ethylene production in tomato. Plant Physiol 123: 363-370.

Alba, R., Z. Fei, P. Payton, Y. Liu, S. Moore, P. Debbie, J., Gordon, J., Rose, G., Martin, S., Tanksley, M., Bouzayen, M. Jahn, and J. Giovannoni. 2004. ESTs, cDNA microarrays, and gene expression profiling: tools for dissecting plant physiology and development. *Plant* 39: 697-714.

Alba, R., P. Payton, Z. Fei, R. Mc Quinn, P. Dbbie, G. Martin, S. Tanksley, and J. Giovannoni. 2005. Transcriptome and selected fruit metabolite analysis reveal multiple points of ethylene regulatory control during tomato fruit development. *Plant Cell* 17: 2954-2965.

Alexander, L. and D. Grierson. 2002. Ethylene biosynthesis and action in tomato: a model for climacteric ripening. J Exp Bot 53: 2039-2055.

Andrawis, A., M. Solomon, and D.P. Delmer. 1993. Cotton fiber annexins: a potential role in regulation of callose synthase. Plant J 3:763-772.

Arabidopsis Initiative. 2000. Analysis of the genome sequence of the flowering plant *Arabidopsis thaliana*. Nature 406: 796-815.

Arrowsmith, D.A. and J. de Silva. 1995. Characterization of two tomato fruit-expressed cDNAs encoding xyloglucan endotransglycosylaes. Plant Mol Biol 28: 391-403.

Atkinson, R.G., K.M. Bolitho, M.A. Wright, B.T. Iturriagagoitia, S.J. Reid, and G.S. Ross. 1999. Apple ACC-oxidase and polygalacturonase: Ripening-specific gene expression and promoter analysis in transgenic tomato. Plant Mol Biol 38: 449-460.

Ayub, R., M. Guis, Ben M. Amor, L. Gillot, J.P. Roustan, A. Latche, M. Bouzayen, and J.C. Pech. 1996. Expression of ACC oxidase antisense gene inhibits ripening of cantaloupe melon fruit. Nature Biotech 14: 862-866.

Barry, C., R. Mc Quinn, A. Thompson, G. Seymour, D. Grierson, and J. Giovannoni. 2005. Ethylene insensitivity conferred by the *Green-ripe* (*Gr*) and *Never-ripe* 2 (*Nr-2*) ripening mutants of tomato. *Plant Physiol* 138: 267-275.

Balague, C., C.F. Watson, A.J. Turner, P. Rouge, S. Picton, J.C. Pech, and D. Grierson. 1993. Isolation of a ripening and wound-induced cDNA from Cucumis melo L. encoding a protein with homology to the ethylene-forming enzyme. Eur J Biochem 212: 27-34.

Barry, C.S., B. Blume, M. Bouzayen, W. Cooper, A.J. Hamilton, and D. Grierson. 1996. Differential expression of the 1-aminocyclopropane-1-carboxylate oxidase gene family of tomato. Plant J 9: 525-535.

Barry, C.B., E.A. Fox, H.C. Yen, S. Lee, T.J. Ying, D. Grierson, and J.J. Giovannoni. 2001. Analysis of the ethylene response in the epinastic (epi) mutant of tomato. Plant Physiol 127: 58-66.

Bauchot, A.D., D.S. Mottram, A.T. Dodson, and P. John. 1998. Effect of Aminocyclopropane-1-carboxylic acid oxidase antisense gene on the formation of volatile esters in cantaloupe Charentais melon (Cv. Vedranais). J Agric Food Chem 46: 4787-4792.

Beecher, G.R. 1998. Nutrient content of tomatoes and tomato products. Proc Soc Exp Biol Med 218: 98-100.

Bennett, A.B. and D. Della-Penna. 1987. Polygalacturonase gene expression in ripening tomato fruit, In: D. Nevins and R. Jones [eds.], Tomato Biotechnology. Alan R. Liss, NY, USA, pp.299-308.

Biale, J.B., and R.E. Young. 1981. Respiration and ripening in fruits-retrospect and prospect, In: J. Friend, and M.J.C. Rhodes [eds.], Recent Advances in the Biochemistry of Fruits and Vegetables. Academic Press, London, UK, pp1-39.

Bird, C.R., C.J.S. Smith, J.A. Ray, P. Moureau, M.W. Bevan, A.S. Bird, S. Hughes, P.C. Morris, D. Grierson, and W. Schuch. 1988. The tomato polygalacturonase gene and ripening-specific expression in transgenic plants. Plant Mol Biol 11: 651-662.

Bird, C.R., J.A. Ray, J.D. Fletcher, J.M. Boniwell, A.S. Bird, C. Teulieres, I. Blain, P.M. Bramley, and W. Schuch. 1991. Using antisense RNA to study gene function: Inhibition of carotenoid biosynthesis in transgenic tomatoes. Biol Tech 9: 635-639.

Boller, T., R.C. Herner, and H. Kende. 1979. Assay for an enzymatic formation of ethylene precursor 1-aminocyclopropane-1-carboxylic acid. Planta 145: 293-304.

Boylan, M.T. and P.H. Quail. 1989. Oat phytochrome is biologically active in transgenic tomatoes. Plant Cell 1: 765-773.

Brady, C.J. 1992. Molecular approaches to understanding fruit ripening. HortScience 20: 107-117.

Brett, C.T. and K.W. Waldron. 1996. Physiology and biochemistry of plant cell walls. Cambridge University Press, Cambridge, United Kingdom.

Brummell, D.A., M.H. Harpster, P.M. Civello, J.M. Palys, and A.B. Bennett. 1999a. Modification of expansin protein abundance in tomato fruit alters softening and cell wall polymer metabolism during ripening. Plant Cell 11: 2203-2216.

Brummell, D.A., B.D. Hall, and A.B. Bennett. 1999b. Antisense suppression of tomato endo-1,4-β-glucanase Cel2 mRNA accumulation increases the force required to break fruit abscission zones but does not affect fruit ripening. Plant Mol Biol 40: 615-622.

Brummell, D.A. and M.H. Harpster. 2001. Cell wall metabolism in fruit softening and quality and its manipulation in transgenic plants. Plant Mol Biol 47: 311-340.

Buchanan, B., W. Gruissem, and R.L. Jones. 2000. Biochemistry and Molecular Biology of Plants. American Society of Plant Physiology, MD, USA.

Catala, C., J.K.C. Rose, and A.B. Bennett. 2001. Auxin-regulated genes encoding cell-wall modifying proteins are expressed during early tomato fruit growth. Plant Physiol 122: 527-534.

Carpita, N.C. and D.M. Gibeaut. 1993. Structural models of primary cell walls in flowering plants consistency of molecular structure with the physical properties of the walls during growth. Plant J 3: 1-30.

Carrington, C.M.S., L.C. Greve, and J.M. Labavitch. 1993. Cell wall metabolism in ripening fruit. VI. Effects of the antisense polygalacturonase gene on cell wall changes accompanying ripening in transgenic tomatoes. Plant Physiol 103: 429-434.

Chang, C., S.F. Kwok, A.B. Bleecker, and E.M. Meyerowitz. 1993. *Arabidopsis* ethylene-response gene ETR1: Similarity of product to two-component regulators. Science 262: 539-545.

Chang, C. and J.A. Shockey. 1999. The ethylene-response pathway:signal perception to gene regulation. Curr Opin In Plant Biol 2: 352-358.

Chory, J. 1997. Light modulation of vegetative development. Plant Cell 9:1225-1234.

Civello, P.M., A.L.T. Powell, A. Sabehat, and A.B. Bennett. 1999. An expansin gene expressed in ripening strawberry fruit. Plant Physiol 121: 1273-1279.

Clayberg, C.D., L. Butler, E.A. Kerr, C.M. Rick, and R.W. Robinson. 1966. Third list of known genes in tomato. J Hered 57: 189-196.

Clayberg, C.D., L. Butler, E.A. Kerr, C.M. Rick, and R.W. Robinson. 1967. Supplementary list of tomato genes as of January, 1967. Rep Tomato Genet Coop 17: 2-11.

Clegg, M.T. and M.L. Durbin. 2000. Flower color variation: a model for the experimental study of evolution. Proc Natl Acad Sci USA. 97: 7016-7023.

Collmer, A. and N.T. Keen. 1986. The role of pectic enzymes in plant pathogenesis. Ann Rev Phytopath 24: 383-409.

Cosgrove, D.J. 1999. Enzymes and other agents that enhance cell wall extensibility, *In*: R.L. Jones [ed.], Ann Rev Plant Physiol Plant Mol Biol Ann Rev Inc., pp. 391-417

Cosgrove, D.J. 2000a. New genes and new biological roles for expansins. Curr Opin, In Plant Biol 3: 73-78.

Cosgrove, D.J. 2000b. Expansive growth of plant cell walls. Plant Physiol Biochem 38: 109-124.

Deikman, J., R. Kline, and R.L. Fischer. 1992. Organization of ripening and ethylene regulatory regions in a fruit-specific promoter from tomato. Plant Physiol 100: 2013-2017.

Della-Penna, D., J.E. Lincoln, R.L. Fischer, and A.B. Bennett. 1989. Transcriptional analysis of polygalacturonase and other ripening associated genes in rutgers, rin, nor, and Nr tomato fruit. Plant Physiol 90: 1372-1277.

de Silva, J., D. Arrowsmith, A. Hellyer, S. Whiteman, and S. Robinson. 1994. Xyloglucan endotransglycosylase and plant growth. J Exp Bot 45: 1693-1701.

Doganlar, S., S.D. Tanksley, and M.A. Mutschler. 2000. Identification and molecular mapping of loci controlling fruit ripening time in tomato. Theor Appl Genet 100: 249-255.

Dogbo, O., A. Laferriere, A. D'harlingue, and B. Camara. 1988. Carotenoid biosynthesis: Isolation and characterization of a bifunctional enzyme catalyzing the synthesis of phytoene. Proc Natl Acad Sci USA 85: 7054-7058.

Ecker, J.R. 1995. The ethylene signal transduction pathway in plants. Science 268: 667-675.

Edwards, M., I.C.M. Dea., P.V. Bulpin, and J.S.G. Reid. 1986. Purification and properties of a novel xyloglucan-specific endo-1-4-â-D-glucanase from germinated nasturtium seeds Tropaeolum majus J Biol Chem 261: 9489-9494.

Fei, Z., X. Tang, R. Alba, J. White, C. Ronning, G. Martin, S. Tanksley, and J. Giovannoni. 2004. Comprehensive EST analysis of tomato and comparative genomics of fruit ripening. Plant J 40: 47-59.

Ferrarese, L., L. Trainotti, P. Moretto, P. Polverino de Laureto, N. Rascio and G. Casadoro. 1995. Differential ethylene-inducible expression of cellulase in pepper plants. Plant Mol Bio 29: 735-747.

Faik, A., D. Desveaux, and G. MacClachlan. 1998. Enzymatic activities responsible for xyloglucan depolymerization in extracts of developing tomato fruit. Phytochemistry 49: 365-376.

Flurh, R. and A. K. Mattoo. 1996. Ethylene-biosynthesis and perception. Crit Rev Plant Sci 15: 479-523.

Frary, A., C. Nesbitt, A. Frary, S. Grandillo, E. van der Knaap, B. Cong, J. Liu, J. Meller, R. Elber, K.B. Alpert, and S.D. Tanksley. 2000. fw2.2: A quantative trait locus key to the evolution of tomato fruit size. Science 289: 85-88.

Fray, R.G., G.W. Lycett, and D. Grierson. 1990. Nucleotide sequence of a heat-shock and ripening-related cDNA from tomato. Nucl Acids Res 18: 7148.

Fray, R.G. and D. Grierson. 1993. Identification and genetic analysis of normal and mutant phytoene synthase genes of tomato by sequencing, complementation, and co-supression. Plant Mol Biol 22: 589-602.

Fray, R.G., A. Wallace, D. Grierson, and G.W. Lycett. 1994. Nucleotide sequence and expression of a ripening and water stress-related cDNA from tomato with homology to the MIP class of membrane channel proteins. Plant Mol Biol 24: 539-543.

Fridman, E., Y.-S. Liu, L. Carmel-Goren, A. Gur., M. Shoresh, T. Pleban, Y. Eshed. and D. Zamir. 2002. Two tightly linked QTLs modify tomato sugar content via different physiological pathways. Mol Gene Genom 266: 821-826.

Fujino, D.W., S.J. Nissen, D.A. Jones, D.W. Burger, and K.J. Bradford. 1988. Quantification of indole-3-acetic acid in dark-grown seedlings of the Diageotropica and Epinastic mutants of tomato (Lycopersicon esculentum Mill.). Plant Physiol 88: 780-784.

Fujino, D.W., D.W. Burger, and K.J. Bradford. 1989. Ineffectiveness of ethylene biosynthetic and action inhibitors in phenotypically reverting the epinastic mutant of tomato (Lycopersicon esculentum Mill.). J Plant Growth Reg 8: 53-61.

Giovannoni, J.J. 1993. Molecular and Genetic Analysis of tomato fruit development and ripening. *In*: J. Byrant [ed.], Methods in Plant Biochemstry. Academic Press, London, UK, pp. 253-287.

Giovannoni, J.J. 2001. Molecular biology of fruit maturation and ripening. Ann Rev Plant Physiol Plant Mol Biol 52: 725-749.

Giovannoni J. 2004. Genetic regulation of fruit development and ripening. The *Plant Cell* 16: S170-180.

Giovannoni, J.J., D. DellaPenna, A.B. Bennett, and R.L. Fischer. 1989. Expression of a chimeric polygalacturonase in transgenic rin (ripening inhibitor) tomato fruit results in polyuronide degradation but not fruit softening. Plant Cell 1: 53-64.

Giovannoni, J.J., E.N. Noensie, D.M. Ruezinsky, X. Lu, S.L. Tracy, M.W. Ganal, G.B. Martin, K. Pillen, K. Alpert, and S.D. Tanskley. 1995. Molecular genetic analysis of the ripening inhibitor and non-ripening loci of tomato: a first step in the genetic map-based cloning of fruit ripening genes. Mol Gen Genet 248: 195-206.

Giovannoni, J.J., H. Yen, B. Shelton, S. Miller, J. Vrebalov, P. Kannan, D. Tieman, R. Hackett, D. Grierson, and H. Klee. 1999. Genetic Mapping of ripening and ethylene-related loci in tomato. Theor App Genet 98: 1005-1013.

Giovannucci, E. 2002. A review of epidemiologic studies of tomatoes, lycopene, and prostate cancer. Experimental Biol and Med 227: 852-859.

Given, N.K., M.A. Venis, and D. Grierson. 1988. Hormonal regulation of ripening in the strawberry, a non-climacteric fruit. Planta 174: 402-406.

Gonzalez-Bosch, C., D.A. Brummell, and A.B. Bennett. 1996. Differential expression of two endo-1,4-â-glucanase genes in pericarp and locules of wild-type and mutant tomato fruit. Plant Physiol 111: 1313-1319.

Gray, J.E., S. Picton, J. Shabbeer, W. Schuch, and D. Grierson. 1992. Molecular biology of fruit ripening and its manipulation with antisense genes. Plant Mol Biol 19: 69-87.

Gray, J.E., S. Picton, J.J. Giovannoni, and D. Grierson. 1994. The use of transgenic and naturally occurring mutants to understand and manipulate tomato fruit ripening. Plant Cell Env 17: 557-571.

Grierson, D. 1984. The appearance of polygalacturonase messenger RNA in tomatoes Lycopersicon 1 of a series of changes in gene expression during development and ripening. Planta 163: 263-271.

Grierson, D., G.A. Tucker, J. Keen, J. Ray, C.R. Bird, W. Schuch, M.J. Holdsworth, and J.E. Knapp. 1986. Sequencing and identification of a cDNA clone for tomato polygalacturonase. Nucl Acids Res 14: 8595-8603.

Grierson, D. 1987. Senescence in Fruits. HortScience 22: 859-862.

Grierson, D., A.J. Hamilton, M. Bouzayen, M. Kock, G.W. Lycett, and S. Barton. 1992. Regulation of gene expression, ethylene synthesis and ripening in transgenic tomatoes. *In*: J.L. Wray [ed.], Inducible Plant Proteins. Cambridge University Press, UK, pp. 155-174.

Grierson, D. and W. Schuch. 1993. Control of ripening. Phil Trans R Soc Lond 342: 241-250.

Gross, K.C. 1984. Fractionation and partial characterization of cell walls from normal and non-ripening mutant tomato fruit. Physiol Plant 62: 25-32.

Hadacek, F. 2002. Secondary metabolites as plant traits: current assessment and future perspectives. Crit Rev in Plant Sci 21: 273-322.

Hall, L.N., G.A. Tucker, C.J.S. Smith, C.F. Watson, G.B. Seymour, Y. Bundick, J.M. Boniwell, J.D. Fletcher, J.A. Ray, W. Schuch, C. Bird, and D. Grierson. 1993. Antisense inhibition of pectin esterase gene expression in transgenic tomatoes. Plant J 3: 121-129.

Hamilton, A.J., G.W. Lycett, and D. Grierson. 1990. Antisense gene that inhibits synthesis of ethylene in transgenic, plants. Nature 346: 284-287.

Hamilton, A.J., M. Bouzayen, and D. Grierson. 1991. Identification of a tomato gene for the ethylene-forming enzyme by expression in yeast. Proc Natl Acad Sci USA 88: 7434-7437.

Hardtke, C.S. and X.-W. Deng. 2000. The cell biology of the COP/DET/FUS proteins. Regulating proteolysis in photomorphogenesis and beyond. Plant Physiol 124: 1548-1557.

Harpster, M.H., K.L. Lee, and P. Dunsmuir. 1997. Isolation and characterization of a gene encoding endo-b-1,4-glucanase from pepper (*Capiscum annuum* L.). Plant Mol Biol 33: 47-59.

Harpster, M.H., D.A. Brummell, and P. Dunsmuir. 2002a. Suppression of a ripening-related endo-1,4-β-glucanase in transgenic pepper fruit does not prevent depolymeriztion of cell wall polysaccharides during ripening. Plant Mol Biol 50: 345-355.

Harpster, M.H., D.M. Dawson, D.J. Nevins, P. Dunsmuir, and D.A. Brummell. 2002b. Constitutive overexpression of a ripening-related pepper endo-1,4-β-glucanase in transgenic tomato fruit does not increase xyloglucan depolymerization or fruit softening. Plant Mol Biol 50: 357-369.

Hauser, B.A., M-M. Cordonnier-Pratt, F. Daniel-Vedele, and L.H. Pratt. 1995. The phytochrome gene family in tomato includes a novel subfamily. Plant Mol Biol 29: 1143-1155.

Hirschberg, J. 2001. Carotenoid biosynthesis in flowering plants. Curr opin Plant Biol 4: 210-218.

Hobson, G. 1967. The effects of alleles at the "Never-ripe" locus on the ripening of tomato fruit. Phytochemistry 6: 1337-1341.

Hobson, G.E., R. Nichols, J.N. Davies, and P. Atkey. 1984. The inhibition of tomato fruit ripening by silver. J Plant Physiol 116: 21-29.

Hua, J., C. Chang, Q. Sun, and E.M. Meyerowitz. 1995. Ethylene insensitivity conferred by *Arabdiopsis* ERS gene. Science 269:1712-1714.

Hua, J. and E.M. Meyerowitz. 1998. Ethylene responses are negatively regulated by a receptor gene family in *Arabidopsis*. Cell 94: 261-271.

Huber, D.J. 1983. The role of cell wall hydrolases in fruit softening. Hortic Rev 5: 169-219.

Isaacson, T., G. Ronen, D. Zamir, and J. Hirschberg. 2002. Cloning of tangerine from tomato reveals a carotenoid isomerase essential for the production of b-carotene and xanthophylls in plants. Plant Cell 14: 333-342.

Ishimaru, M. and S. Kobayashi. 2002. Expression of a xyloglucan endo-transglycosylase gene is closely related to grape berry softening. Plant Sci 161: 621-628.

Jarrett, R.L., H. Sayama, and E.C. Tigchelaar. 1984. Pleiotropic effects associated with the chlorophyll intensifier mutations high-pigment and dark-green in tomato. J Am Soc Hort Sci 109: 873-878.

Jeffrey, D., C. Smith, P. Goodenough, I. Prosser, and D. Grierson. 1984. Ethylene-independent and ethylene-dependent biochemical changes in ripening tomatoes. Plant Physiol 74: 32-38.

Jenkins, J. A. and G. MacKinney. 1995. Carotenoids of the apricot tomato and its hybrids with yellow and tangerine. Genetics 40: 715-720.

Jian, Y., D.C. Joyce, and A.J. Macnish. 2002. Softening response of banana fruit treated with 1-methylcyclopropene to high temperature exposure. Plant Growth Reg 36: 7-11.

Kader, A. A. 1985. Ethylene-induced senescence and physiological disorders in harvested horticultural crops. HortScience 20: 54-57.

Kakkar, R.K. and V.K. Sawhney. 2002. Polyamine research in plants—a changing perspective. Physiol Plant 116: 281-292.

Kalloo, G. and B.O. Bergh. 1993. Genetic improvement of vegetable crops. Pergamon Press, NY, USA.

Kendrick, R.E. and G.H.M. Kronenberg. 1994. Photomorphogenesis in plants, Ed.2. Kluwer Academic Publishers, Dordrecht, The Netherlands.

Kerckhoffs, L.H.J. and R.E. Kendrick. 1997. Photocontrol of anthocyanin biosynthesis in tomato. J Plant Res 110: 141-149.

Kerr, E.A. 1958. Mutations of chlorophyll retention in ripe fruit. Rept Tomato Genet Co-Op 8: 22.

Kerr, E.A. 1965. Identification of high-pigment, hp, tomatoes in the seedling stage. Can J Plant Sci 45: 104-105.

Kerr, E.A. 1982. Never ripe-2 (Nr-2) a slow ripening mutant resembling Nr and Gr. Rept Tomato Genet Co-Op 32: 33.

Kieber, J.J., M. Rothenberg, G. Roman, K.A. Feldman, and J.R. Ecker. 1993. CTR1, a negative regulator of the ethylene response pathway in Arabidopsis, encodes a member of the raf family of protein kinases. Cell 72: 411-427.

Kieber, J.J. 1997. The ethylene response pathway in Arabidopsis. In: R.L. Jones [ed.], Ann Rev of Plant Physiol and Plant Mol Biol Ann Rev, Inc., CA, USA, vol. 48 pp. 277-296.

Kinet, J.M. and M.M. Peet. 1997. Tomato. In: H.C. Wien [ed.], The Physiology of Vegetable Crops. CAB International, England, UK, pp. 207-258.

Klee, H. and D. Tieman. 2002. The tomato ethylene receptor gene family: form and function. Phyiol Plant 115: 336-341.

Klee, H. 2002. Control of ethylene-mediated processes in tomato at the level of receptors. J Exp Bot 53: 2057-2063.

Knapp, J., P. Moureau, W. Schuch, and D.Grierson. 1989. Organisation and expression of polygalacturonase and other ripening genes in Alisa Craig 'Never-ripe' and 'ripening-inhibitor' tomato mutants. Plant Mol Biol 12: 105-116.

Kneissel, M.L. and J. Deikman. 1996. The tomato E8 gene influences ethylene biosynthesis in fruit but not in flowers. Plant Physiol 112: 537-547.

Kopeliovich, E., H.D. Rabinowitch, Y. Mizrahi, and K. Kedar. 1981. Mode of inheritance of alcobaca, a tomato ripening mutant. Euphytica 30: 223-225.

Kumar, A., M.A. Taylor, S.A.M. Arif, and H.V. Davies. 1996. Potato plants expressing antisense and sense S-adenosylmethionine decarboxylase (SAMDC) transgenes show altered levels of polyamines and ethylene: antisense plants display abnormal phenotypes. Plant J 9: 147-158.

Lanahan, M.B., H.C. Yen, J.J. Giovannoni, and H.J. Klee. 1994. The Never-ripe mutation blocks ethylene preception in tomato. Plant Cell 6: 521-530.

Lashbrook, C.C., D.M. Tieman, and H.J. Klee. 1998a. Differential regulation of the tomato ETR gene family throughout plant development. Plant J 15: 243-252.

Lay-Yee, M. and M.L. Knighton. 1995. A full-length cDNA encoding 1-aminocyclopropane-1-carboxylate synthase from apple. Plant Physiol 107: 1017-1918.

Lee, S. 1999. Genetic and physiological analysis of ethylene signal transduction in tomato and identification of DNA markers for the tomato high-pigment (hp) mutant. PhD. Dissertation. Texas A&M University.

Lelievre, J.M., A. Latche, B. Jones, M. Bouzayen, and J.C. Pech. 1997. Ethylene and fruit ripening. Physiol Plant 101: 727-739.

Leclercq, J., L.C. Adams-Phillips, H. Zegzouti, B. Jones, A. Latche, J.J. Giovannoni, J-C. Pech, and M. Bouzayen. 2002. LeCTR1, a tomato CTR1-like gene, demonstrates ethylene signaling ability in Arabidopsis and novel expression patterns in tomato. Plant Physiol 130: 1132-1142.

Li, C.Y., D. Jacob-Wilk, G.Y. Zhong, R. Goron, and D. Holland. 1998. A full-length cDNA encoding an ethylene receptor ERS homologue from Citrus (Ascession No. AF092088). Plant Physiol 118: 1534.

Lincoln, J.E., S. Cordes, E. Read, and R.L. Fischer. 1987. Regulation of gene expression ethylene during *Lycoperisicon esculentum* (tomato) fruit development. Proc Natl Acad Sci USA. 84: 2793-2797.

Liu, J., J. Van Eck, B. Cong, and S.D. Tanksley. 2002. A new class of regulatory genes underlying the cause of pear-shaped tomato fruit. Proc Natl Acad Sci 99:13302-13306.

Liu, Y., S. Roof, Z. Ye, C. Barry, A. van Tuinen, J. Vrebalov, C. Bowler, and J. Giovannoni. 2004. Manipulation of light signal transduction as a means of modifying fruit nutritional quality in tomato. Proc. Natl Acad Sci USA. 26: 9897-9902

Liu, Y.-S. and D. Zamir. 1999. Second generation *L.pennellii* introgression lines and the concept of bin mapping. Tomato Genet Coop 49: 26-30.

Llop-Tous, I., C.S. Barry, and D.Grierson. 2000. Regulation of ethylene biosynthesis in response to pollination in tomato flowers. Plant Physiol 123: 971-978.

Lopez-Gomez, R., A. Cambell, J.G. Dong, S.F. Yang, and M.A. Gomez-Lim. 1997. Ethylene biosynthesis in banana fruit: isolation of a genomic clone to ACC oxidase and expression studies. Plant Sci 123: 123-131.

Lyons, J.M. and H.K. Pratt. 1964. Effect of stage of maturity and ethylene treatment on respiration and ripening of tomato fruits. Proc Am Soc Hort Sci 84: 491-500.

Mann, V., I. Pecker, and J. Hirschberg. 1994. Cloning and characterization of the gene for phytoene desaturase (Pds) from tomato (*Lycopersicon esculentum*). Plant Mol Biol 24: 429-434.

Manning, K. 1998. Identification of a set of ripening-related genes from strawberry: their identification and possible relationship to fruit quality traits. Planta 205: 622-631.

Mansson, P.E., D. Hsu, and D. Stalker. 1985. Characterization of fruit specific cDNAs from tomato. Mol Gen Genet 200: 356-361.

Mao, L., D. Begum, H.W. Chuang, M.A. Budiman, E.J. Szymkowlak, E.E. Irish and R.A. Wing. 2000. JOINTLESS is a MADS-box gene controlling tomato flower abscission zone development. Nature 406: 910-913.

Marin-Rodriguez, M.C., J. Orchard, and G.B. Seymour. 2002. Pectate lyases, cell wall degredation and fruit softening. J Exp Bot 53: 2115-2119.

Martin, G.B., S.H. Brommonschenkel, J. Chunwongse, A. Frary, M.W. Ganal, R. Spivey, R. Wu, E.D. Earle, and S.D. Tanskley. 1993. Map-based cloning of a protein kinase gene conferring disease resistance in tomato. Science 262: 1432-1436.

Maunders, M.J., M.J. Holdsworth, A. Slater, J.E. Knapp, C.R. Bird, W. Schuch, and D. Grierson. 1987. Ethylene stimulates the accumulation of ripening-related mRNAs in tomatoes. Plant Cell and Envir 10: 177-184.

McChurchie, E.J., W.B. McGlasson, and I.L. Eaks. 1972. Treatment of fruit with propylene gives information about the biogenesis of ethylene. Nature 237: 235-236.

McGlasson, W.B. 1985. Ethylene and Fruit ripening. HortScience 20: 51-54.

McQueen-Mason, S.J., D.M. Durachko, and D.J. Cosgrove. 1992. Two endogenous proteins that induce cell wall extension in plants. Plant Physiol 107: 87-100.

Medina-Escobar, N., J. Cardenas, E. Moyano, J.L. Caballero, and J. Munoz-Blanco. 1997. Cloning, molecular characterization and expression pattern of a strawberry ripening-specific cDNA with sequence homology to pectate lyase from higher plants Plant Mol Biol 34: 867-877.

Mehta, R.A., T. Cassol, N. Li, N. Ali, A.K. Handa, and A.K. Mattoo. 2002. Engineered polyamine accumulation in tomato enhances phytonutrient content, juice quality, and vine life. Nature Biotech 20: 613-618.

Mochizuki, T. and S. Kamimura. 1985. Photoselective method for selection of hp at the cotyledon stage. Rep Tomato Genet Co-Op 35: 12-13.

Montgomery, J., S. Goldman, J. Deikman, L. Margossian, and R.L. Fischer. 1993. Identification of an ethylene-responsive region in the promoter of a fruit ripening gene. Proc Natl Acad Sci USA. 90: 5939-5943.

Moore, S., J. Vrebalov, P. Payton, and J. Giovannoni. 2002. Use of genomic tools to isolate key ripening genes and analyse fruit maturation in tomato. J Exp Bot 53: 2023-2030.

Mustilli, A.C., F. Fenzi, R. Ciliento, F. Alfano, and C. Bowler. 1999. Phenotype of the tomato high-pigment-2 mutant is caused by a mutation in the tomato homolog of DETIOLATE1. Plant Cell 11: 145-157.

Nakatsuka, A., S. Murachi, H. Okunishi, S. Shiomi, R. Nakano, Y. Kubo, and A. Inuba. 1998. Differential expression and internal feedback regulation of 1-aminocyclopropane-1-carboxylate synthase, 1-aminocyclopropane-1-carboxylate oxidase, and ethylene receptor genes in tomato fruit ripening during development and ripening. Plant Physiol 118: 1295-1305.

Oeller, P.W., L.M. Wong, L.P. Taylor, D.A. Pike, and A. Theologis. 1991. Reversible inhibition of tomato fruit ripening. Science 254: 437-439.

Pepper, A., T. Delaney, T. Washburn, D. Poole, and J. Chory. 1994. DET1, a negative regulator of light-mediated development and gene expression in arabidopsis, encodes a novel nuclear-localized protein. Cell 78: 109-116.

Peters, J.L., A. van Tuinen, P. Adamse, R.E. Kendrick, and M. Koorneef. 1989. High-pigment mutants of tomato exhibit high sensitivity for phytochrome action. J Plant Physiol 134: 661-666.

Peters, J.L., M.E.L. Schreuder, S.J.W. Verduin, and R.E. Kendrick. 1992. Physiological characterization of a high-pigment mutant of tomato. Photochem Photobiol 56: 75-82.

Peters, J.L., M. Szell, and R.E. Kendrick. 1998. The expression of light-regulated genes in the High-Pigment-1 mutant in tomato. Plant Physiol 117: 797-807.

Penarrubia, L., M. Aguilar, L. Margossian, and R.L. Fischer. 1992. An antisense gene stimulated ethylene hormone production during tomato fruit ripening. Plant Cell 4: 681-687.

Picton, S., S.L. Barton, M. Bouzayen, A.J. Hamilton, and D.Grierson. 1993. Altered fruit ripening and leaf senescence in tomatoes expressing an antisense ethylene-forming enzyme transgene. Plant J 3: 469-481.

Redgwell. R.J. and M. Fischer. Fruit texture, cell wall metabolism and consumer perceptions. In: Brett, C.T. and K.W. Waldron[eds.], Physiology and Biochemistry of Plant Cell Walls. Chapman & Hall, London, UK, pp 47-88.

Reinhardt, D., F. Wittwer, T. Mandel, and C. Kuhlemeier. 1998. Localized upregulation of a new expansin gene predicts the site of leaf formation in the tomato meristem. Plant Cell 10: 1427-1437.

Rexova-Benkova, L. and O. Markovic. 1976. Pectic enzymes, In: R.S. Tipton and D. Horton [eds.], Advances in Carbohydrate Chemistry and Biochemistry, Vol. 33, New York Academic Press, NY, USA. pp. 323-385.

Rick, C.M. and L. Butler. 1956. Phytogenetics of the tomato. Adv Genet 8: 267-382.

Rick, C.M. 1978. The tomato. Sci Am 239: 67-76.

Robinson, R.W. and M.L. Tomes. 1968. Ripening inhibitor a gene with multiple effects on ripening. Rept Tomato Genet Co-Op 18: 36-37.

Ronen, G., M. Cohen, D. Zamir, and J. Hirschberg. 1999. Regulation of carotenoid biosynthesis during tomato fruit development:expression of the gene for lycopene epsilon-cyclase is down-regulated during ripening and is elevated in the mutant delta. Plant J 17: 341-351.

Rose, J.K.C., H.H. Lee, and A.B. Bennett. 1997. Expression of a divergent expansin gene is fruit-specific and ripening-regulated. Proc Natl Acad Sci USA 94: 5955-5960.

Rose, J.K.C and A.B. Bennett. 1999. Cooperative disassembly of the cellulose-xyloglucan network of plant cell walls: parallels between cell expansion and fruit ripening. Trends Plant Sci 4: 176-183.

Rose, J.K.C., J. Braam, S.C. Fry, and K. Nishitani. 2002. The XTH family of enzymes involved in xyloglucan endtransglycosylation and endohydrolysis: current perspectives and a new unifying nomenclature. Plant Cell Physiol 43: 1421-1435.

Ross, G.S., M.L. Knighton, and M. Lay-Yee. 1992. An ethylene regulated cDNA from ripening apples. Plant Mol Biol 19: 231-238.

Schaller, G.E. and A.B. Bleeker. 1995. Ethylene-binding sites generated in yeast expressing the *Arabidopsis* ETR1 gene. Science 270: 1086-1089.

Schwechheimer, C. and X.-W. Deng. 2000. The COP/DET/FUS proteins-regulators of eukaryotic growth and development. Sem In Cell and Dev Biol 11: 495-503.

Schroder, R., R.G. Atkinson, G. Langenkamper, and R.J. Redgwell. 1998. Biochemical and molecular characterisation of xyloglucan endotransglycosylase from ripe kiwifruit. Planta 204: 242-251.

Seymour, G.B., J.E. Taylor, and G.A. Tucker. 1993. Biochemistry of fruit ripening. Chapman & Hall, London, UK.

Sheehy, R.E., M. Kramer, and W.R. Hiatt. 1988. Reduction of polygalacturonase activity in tomato fruit by antisense RNA. Proc Natl Acad Sci USA 85: 8805-8809.

Sisler, E.C., M. Serek, and E. Dupille. 1995. Comparison of cyclopropene, 1-methylcyclopropene, and 3,3-dimethylcyclopropene as ethylene antagonists in plants. Plant Growth Reg. 17: 1-6.

Sisler, E.C. and M. Serek. 1997. Inhibitors of ethylene responses in plants at the receptor level: recent developments. Physiol Plant 100: 577-582.

Sitrit. Y. and A.B. Bennett. 1998. Regulation of tomato fruit polygalacturonase mRNA accumulation by ethylene: a re-examination. Plant Physiol 116: 1145-1150.

Slater, A., M.J. Maunders, K. Edwards, W. Schuch, and D. Grierson. 1985. Isolation and characterization of complementary DNA clones for tomato. Plant Mol Biol 5: 137-148.

Smith, C.J.S., A., Slater, and D. Grierson. 1986. Rapid appearance of a mRNA correlated with ethylene synthesis encoding a protein of MW 35,000. Planta 168: 94-100.

Smith, C.J.S., C.F. Watson, J. Ray, C.R. Bird, P.C. Morris, W. Schuch, and D.Grierson. 1988. Antisense RNA inhibition of polygalacturonase gene expression in transgenic tomatoes. Nature 334: 724-726.

Smith, C.J.S., C.F. Watson, P.C. Morris, C.R. Bird, G.B. Seymour, J.E. Gray, C. Arnold, G.A. Tucker, W. Schuch, S. Harding, and D. Grierson. 1990. Inheritance and effect on ripening of antisense polygalacturonase genes in transgenic tomatoes. Plant Mol Biol 14: 369-379.

Soltis, D.E., P.S. Soltis, V.A. Albert, D.G. Oppenheimer, C.W. dePamphilis, M. Hong, M.W. Frohlich, and G. Theissen. 2002. Missing links: the genetic architecture of flower and floral diversification. Trends Plant Sci 7: 22-31.

Soressi, G.P. 1975. New spontaneous or chemically-induced fruit ripening tomato mutants. Rep Tomato Genet Co-Op 25: 21-22.

Spolaore, S., L., Trainotti, A., Pavanello, and G. Casadoro. 2003. Isolation and promoter analysis of two genes encoding different endo-α-1,4-glucanases in the non-climacteric strawberry. J Exp Bot 54: 271-277.

Steele, N.M. and S.C. Fry. 1999. Purification of xyloglucan endotransglycosylases (XETs): a generally applicable and simple method based on reversible formation of an enzyme-substrate complex. Biochem J 340: 207-211.

Stepanova, A.N. and J.R. Ecker. 2000. Ethylene signaling: from mutants to molecules. Curr Opin Plant Biol 3: 353-360.

Takada, N. and P. Nelson. 1983. New consistency method for tomato products-the precipitate weight ratio. J Food Sci 48: 1460-1462.

Tanksley, S.D., M.W. Ganal, J.P. Prince, de M.C. Vincente, and M.W. Bonierbale. 1992. High density molecular linkage maps of the tomato and potato genomes. Genetics 132: 1141-1160.

Theologis, A., T.I. Zarembinski, P.W. Oeller, X. Liang, and S. Abel. 1992. Modification of fruit ripening by supressing gene expression. Plant Physiol 100: 549-551.

Theologis, A., P.W. Oeller, L.M. Wong, W.H. Rottmann, and D.M. Gantz. 1993. Use of a tomato mutant constructed with reverse genetics to study fruit ripening, a complex developmental process. Dev Gen 14: 282-295.

Thompson, A.E. 1962. A comparison of fruit quality constituents of normal and high pigment tomatoes. Proc Am Soc Hort Sci 78: 464-473.

Thompson, A.J., M. Tor, C.S. Barry, J. Vrebalov, C. Orfila, M.C. Jarvis, J.J. Giovannoni, D. Grierson, and G.B. Seymour. 1999. Molecular and genetic characterization of a novel pleiotropic tomato-ripening mutant. Plant Physiol 120: 383-389.

Tian, M.S., S. Prakash, H.J. Elgar, H. Young, D.M. Burmeister, and G.S. Ross. 2000. Responses of strawberry fruit to 1-Methylcyclopropene (1-MCP) and ethylene. Plant Growth Reg 32: 83-90.

Tiburcio A.F., T. Altabella, A. Borrel, and C. Masgrau. 1997. Polyamine metabolism and its regulation. Physiol Plant 100: 664-674.

Tieman, D.M., R.W. Harriman, G. Ramamohan, and A.K. Handa. 1992. An antisense pectin methylesterase gene alters pectin chemistry and soluble solids in tomato fruit. Plant Cell 4: 667-679.

Tieman, D.M. and H.J. Klee. 1999. Differential expression of two novel members of the tomato ethylene-receptor family. Plant Physiol 120: 165-172.

Tieman, D.M., M.G. Taylor, J.A. Ciardi, and H.J. Klee. 2000. The tomato ethylene receptors NR and LeETR4 are negative regulators of ethylene response and exhibit functional compensation within a multigene family. Proc Natl Acad Sci USA 97: 5663-5668.

Tieman, D.M., J.A. Ciardi, M.G. Taylor, and H.J. Klee. 2001. Members of the tomato LeEIL (EIN3-like) gene family are functionally redundant and regulate ethylene responses throughout plant development. Plant J 26: 47-58.

Tigchelaar, E.C., M.L. Tomes, E.A. Kerr, and R.J. Barman. 1973. A new fruit ripening mutant, non-ripening (nor). Rep Tomato Genet Co-Op. 23: 33.

Tigchelaar, E.C., W.B. McGlasson, and R.W. Buescher. 1978. Genetic Regulation of tomato fruit ripening. HortScience 13: 508-513.

Trainotti, L., S. Spolaore, A. Pavanello, B. Baldan, and G. Casadoro. 1999. A novel E-type endo â-1,4-glucanase with putative cellulose binding domain is highly expressed in ripening strawberry fruits. Plant Mol Biol 40: 323-332.

Tucker, G.A. and D. Grierson. 1982. Synthesis of polygalacturonase during tomato fruit ripening. Planta 155: 64-67.

Ursin, V.M. 1987. Morphogenetic and physiological analyses of two developmental mutants of tomato (Lycopersicon esculentum MILL.), Epinastic and diageotropica. PhD dissertation, University of California, Davis.

Van der Hoeven, R., C. Ronning, J. Giovannoni, G. Martin, and S. Tanksley. 2002. Deductions about the number, organization, and evolution of genes in the tomato genome based on analysis of a large expressed sequence tag collection and selective genomic sequencing. Plant Cell 14: 1441-1456.

Van der Knaap, E. and S.D. Tanksley. 2001. Identification and characterization of a novel locus controlling early fruit development in tomato. Theor Appl Genet 103: 353-358.

van Tuinen, A., M-M. Cordonnier-Pratt, L.H. Pratt, R. Verkerk, P. Zabel, and M. Koorneef. 1997. The mapping of phytochrome genes and photomorphogenic mutants in tomato. Theor Appl Genet 94: 115-122.

Verhoeyen, M.E., A. Bovy, G. Collins, S. Muir, S. Robinson, C.H.R. de Vos, and S. Colliver. 2002. Increasing antioxidant levels in tomatoes through modification of the flavanoid biosynthesis pathway. J Exp Bot 53: 2099-2106.

Verma, D.P.S. and Z. Hong. 2001. Plant callose synthase complexes. Plant Mol Biol 47: 693-701.

Vrebalov, J., D. Ruezinsky, V. Padmanabhan, R. White, D. Medrano, R. Drake, W. Schuch, and J.J. Giovannoni. 2002. A MADS-box gene necessary for fruit ripening at the tomato Ripening-inhibitor (Rin) Locus. Science 296: 343-346.

Watkins, C.B., J.F. Nock, and B.D. Whitaker. 2000. Responses of early, mid and late season apple cultivars to postharvest application of 1-methylcyclopropene(1-MCP) under air and controlled atmosphere storage conditions. Postharvest Biol Technol 19: 17-32.

Watkins, C.B. 2002. Ethylene synthesis, mode of action, consequences and control, In: M.Knee [ed.], Fruit Quality and its Biological Basis. CRC Press. Boca Raton, FL, pp. 180-224.

Wann, E.V., E.L. Jourdain, R. Pressey, and B.G. Lyon. 1985. Effect of mutant genotypes hp og^c and dg og^c on tomato fruit quality. J Am Soc Hort Sci 110: 212-215.

White, P.J. 2002. Recent advances in fruit development and ripening: an overview. J Exp Bot 53: 1995-2000.

Wilkinson, J.Q., M.B. Lanahan, T.W. Conner, and H.J. Klee. 1995. Identification of mRNAs with enhanced expression in ripening strawberry fruit using polymerase chain reaction differential display. Plant Mol Biol 27: 1097-1108.

Wilkinson, J.Q., M.B. Lanahan, D.G. Clark, A.B. Bleecker, and C. Chang. 1997. A dominant mutant receptor from *Arabidopsis* confers ethylene insensitivity in heterologous plants. Nature Biotechnol 15: 444-447.

Yang, S.F. and A.H. Baur. 1969. Pathways of ethylene biosynthesis. Qualitas Plantarum et Materiae Vegetabiles. 19: 201-220.

Yang, S.F. 1985. Biosynthesis and action of ethylene. HortScience 20: 41-45.

Yang, S.F. 1987. The role of ethylene and ethylene synthesis in fruit ripening, . In: W.W. Thompson, E.A. Nothnagel, and R.C. Huffaker [eds.], Plant Senescence: Its Biochemistry and Physiology. Am Soc Plant Physiol, pp. 156-166:

Yen, H.C., S. Lee, S. Tanksley, M.B. Lanahan, H.J. Klee, and J.J. Giovannoni. 1995. The tomato Never-ripe locus regulates ethylene-inducible gene expression and is linked to a homolog of the *Arabidopsis* ETR1 gene. Plant Physiol 107: 1343-1353.

Yen, H.C., A. Shelton, L. Howard, J. Vrebalov, and J.J. Giovannoni. 1997. The tomato high-pigment (hp) locus maps to chromosome 2 and influences plastome copy number and fruit quality. Theor and App Genet 95: 1069-1079.

Yip, W.-K., T. Moore, and S.F. Yang. 1992. Differential accumulation of transcripts for four tomato 1-aminocyclopropane-1-carboxylate synthase homologs under various conditions. Proc Natl Acad Sci USA 89: 2475-2479.

Yu, Y.B. and S.F. Yang. 1979. Auxin-induced ethylene production and its inhibition by aminoethoxyvinylglycine and cobalt ion. Plant Physiol 64: 1074-1077.

Zhang, H.B., M.A. Budiman, and R.A. Wing. 2000. Genetic mapping of jointless-2 to tomato chromosome 12 using RFLP and RAPD markers. Theor Appl Genet 100: 1183-1189.

Zhou, Z., J. de Almeida Engler, D. Rouan, F. Michiels, M. D. Van Montagu, and Van Der Straeten. Tissue localization of submergence-induced 1-Aminocyclopropane-1-Carboxylic Acid Synthase in rice. Plant Physiol 129: 72-84.

12

Genetics and Breeding for Resistance to Bacterial Diseases in Tomato: Prospects for Marker-assisted Selection

WENCAI YANG[1] AND DAVID M. FRANCIS[2]

[1]*Department of Vegetable Science, College of Agronomy and Biotechnology, China Agricultural University, No.2 Yuanmingyuan Xi Lu, Haiden District, Beijing 100094, The People's Republic of China*
E-mail: yangwencai@cau.edu.cn
[2]*Dept. of Horticulture and Crop Science, The Ohio State University, OARDC. 1680 Madison Ave, Wooster, OH 44691 U.S.A.*
E-mail: francis.77@osu.edu

INTRODUCTION

Bacterial pathogens of tomato (*Lycopersicon esculentum* Mill. syn. *Solanum lycopersion*) cause serious and destructive diseases affecting both field and greenhouse grown crops. In this review we focus on bacterial canker caused by *Clavibacter michiganensis* subsp. *michiganensis*, bacterial speck caused by *Pseudomonas syringae* pv. *tomato*, bacterial spot caused by *Xanthomonas* species, and bacterial wilt caused by *Ralstonia solanacearum*. These pathogens account for the most serious bacterial diseases in tomato (Jones et al. 1991).

Control of bacterial diseases has not been effective once epidemics start. The frequency and severity of epidemics is compounded by multiple sources of inoculum, multiple species and races causing disease, marginal efficacy of commonly applied chemicals, development of resistance to these chemicals in bacterial populations, and a lack of available disease resistance traits in commercial cultivars. The source of inoculum is often varied. Bacterial diseases can be transmitted through seed, populations

---
*Corresponding author: D.M. Francis

may persist as epiphytes on asymptomatic seedlings and mature plants, and over-wintering may occur on debris in the field and greenhouse. Ethylene-bis-dithiocarbamate (EBDC) fungicides, which are commonly used to increase the efficacy of copper-containing sprays and applied for reduction of bacterial populations on tomatoes, have raised public health questions and have been voluntarily banned by tomato processors. Plasmid borne resistance to copper is well documented (Bender and Cooksey 1986, Bender et al. 1990, Kearney and Staskawicz 1990) and a high incidence of copper insensitivity is documented for pathogenic populations (Sahin and Miller 1996). Effective treatment of disease must therefore be based on prevention. With few exceptions, resistance has not yet been an effectively deployed tool for the prevention and management of bacterial diseases.

A number of reviews covering the biology and epidemiology, host-parasite interactions and genetic basis of resistance in tomato to bacterial canker, bacterial speck, bacterial spot and bacterial wilt have appeared over the years. The most recent (Scott, 1997) provides a thorough background of these diseases, the genetic basis of resistance and breeding progress up to 1997. Since numerous studies relevant to these four diseases have been reported since 1997, recent insight into identification of new pathogen races, new sources of resistance, and molecular markers linked to bacterial disease resistance will be presented in this chapter. We have chosen to emphasize the integration of marker-assisted selection (MAS) with traditional breeding strategies as a unifying concept in breeding for resistance to bacterial pathogens.

The practical tomato breeder seeking to initiate a resistance-breeding project will be faced with abundant literature from academic studies and the challenge of sorting out reliable information. Furthermore, descriptions of resistant germplasm often lack the most basic of genetic information, such as the heritability and degree of dominance. Genetic studies, when described, often conflict with previous reports. Although conflicting information may be due to interpretation of results, they are more often due to more subtle experimental differences involving methods of evaluation and experimental technique. In this chapter we attempt to call the readers' attention on the effects of multiple inoculation techniques, difficulty in interpreting different environmental effects on the expression of resistance, use of reference varieties or populations as controls, and the frequent lack of defined information on the bacterial strains used. All these experimental issues affect the expression of resistance and the interpretation of results.

Given the diversity of methodology and controls, it is not uncommon to see reports in the scientific literature that are of little value for the practical breeder. Wherever possible, we have filtered the literature through our own experience to highlight germplasm and experimental methods that show promise for developing varieties with improved resistance to bacterial disease.

Breeding for resistance depends on the knowledge of the pathogen population structure, especially as it relates to variation in virulence. Of the four bacterial diseases, breeding for resistance to bacterial speck has been successful because the resistance is simply inherited and pathogen variation is limited. However, breeding for resistance to bacterial canker, bacterial spot and bacterial wilt is difficult using traditional approaches due to the extensive variation in the aggressiveness of pathogen isolates, the existence of species complexes, and/or multiple pathogen races. With variable pathogen populations combined with a complicated inheritance of resistance, breeding progress gets slowed. Combining indirect selection using DNA-based markers with breeding population structures that allow replicated testing against diverse populations may provide a method for selecting resistant individuals. However, significant barriers to progress remain to be overcome.

A limitation in applying marker-assisted selection (MAS) to breeding tomato varieties is their low level of polymorphism among *L. esculentum* varieties (Miller and Tanksley 1990, Williams and St. Clair 1993; see also chapter 10). Although the initial work on identification of molecular markers linked to the loci conferring resistance to bacterial diseases started in early 1990s, marker-assisted selection has not been widely applied in breeding for resistance to bacterial canker, bacterial spot and bacterial wilt. Despite the cloning of the major gene for resistance to bacterial speck, *Pto* (Martin et al. 1993a), indirect selection has continued to focus on the reaction to Fenthion, an organophosphate insecticide, which is conferred by a tightly linked homolog to *Pto* (Laterrot and Moretti 1989, Carland and Staskawicz 1993, Martin et al. 1994). Recent success in developing molecular markers that can be used within *L. esculentum* populations (Suliman-Pollatschek et al. 2002, Yang et al. 2004) have invigorated efforts to both characterize bacterial resistance and apply MAS to breeding for resistance to bacterial diseases. Marker-assisted selection alternated with field selection has been successfully used to develop desirable lines with partial resistance to race T1 of bacterial spot (Francis and Miller 2004, Yang et al. 2005) and resistance to bacterial speck and bacterial spot (Yang and Francis 2005).

Although new markers provide tomato breeders with a promising approach to using MAS for disease resistance breeding, there remains a need to balance the strengths of DNA-based selection with classical approaches. MAS will not offer increased efficiency, defined as gain under selection, over replicated field evaluation (Francis et al. 2003). Even in the absence of MAS, population structures that favor replicated field evaluation against variable pathogen populations offer the most promise for improving resistance to bacterial disease.

BACTERIAL CANKER

Bacterial canker is one of the most difficult diseases of tomato to control. It is challenging to detect infected seed and plants prior to establishing fields. The disease can be highly infectious and there are multiple sources of inoculum. The organism is seed-borne and can survive for short periods on plant debris in soil, greenhouse structures, and equipment. Although seed transmission is reported to be inefficient (Grogan and Kendrick 1953), transmission in as few as 5 per 10,000 seed may contribute to epidemics (Chang et al. 1991). The absence of effective chemicals for treatment once epidemics are underway makes prevention a key element of control. Treatment of seed with hot water or acid (Thyr et al. 1973, Fatmi et al. 1991) and the preventative application of copper sprays to seedling transplants in the greenhouse (Hausbeck et al. 2000, Werner et al. 2002) are effective approaches to reduce bacterial populations as well as the incidence and severity of epidemics. Breeding for resistance to bacterial canker has been underway in several countries for decades and lines/populations with improved resistance or tolerance have been introduced (Table 12.1). However, few cultivars combine significant tolerance to canker with desirable fruit quality and yield necessary for a commercial variety. Progress in breeding for resistance to bacterial canker has been hampered by variations in pathogen aggressiveness, variable inoculation techniques, and a poor understanding of the genetic basis of resistance.

Bacterial canker of tomato was first reported in Michigan, USA, in 1909 (Smith 1910), and is recognized as a serious disease that occurs in all tomato-growing regions of the world. The incidence of bacterial canker is increasing where greenhouse-grown transplants are used to establish field stands. Amplification of bacterial populations on asymptomatic transplants in the greenhouse appears to be a major contributor to epidemics (Hausbeck et al. 2000, Werner et al. 2002). Estimates of yield loss in field-grown tomatoes range from 10% annual regional losses (Hibberd et al. 1992, Strider

Table 12.1 Sources of resistance to bacterial canker.

Genotype	Source	Plant Habit	Reference
Wild species [a]			
LA407	L. hirsutum	indeter.	Francis et al. (2001)
LA2157	L. peruvianum	indeter.	Sandbrink et al. (1995)
Improved sources of resistance			
Bulgaria 12[b]	R from L. pimpinellifolium	indeter.	Elenkov, (1965; cited by Laterrot 1992)
IRAT L3	R from L. pimpinellifolium	indeter.	Daly, (1976; cited by Laterrot 1984), Henderson and Jenkins,
NC 72	R from L. esculentum var. cerasiforme	indeter.	(1972; cited by Laterrot 1992)
Cm 190	R from L. chilense	indeter.	Vulkova and Sotirova,
Cocabul	R from Bulgaria 12, NC 72	indeter.	(1992), Laterrot, (1992)
Breeding lines and varieties			
Heinz 2990 [c]	R from Bulgaria 12	sp	Emmatty and John, (1973)
Heinz 9144 (F₁)	R from Bulgaria 12	sp	Ricker et al. (1996)

[a] When possible, sources of resistance are indicated by *Lycopersicon* accession (LA) number at the C. M. Rick Tomato Genetics Resource Center (TGRC), U. C. Davis, or by Plant Introduction (PI) or National Seed Lab (NSL) number from the United States Department of Agriculture (USDA), Agricultural Research Service (ARS), National Plant Germplasm System. Accessions with LA, PI, or NSL numbers are available through the respective stock centers.

[b] Bulgaria 12 is also known as Plovdiv 8/12 and is available through the NPGS as PI 330727.

[c] Heinz 2990 was deposited with the National Center for Genetic Resources Preservation (NCGRP), USDA, ARS, Fort Collins, Colorado as accession NSL 92634.

1969) to 80% for individual growers (Gleason et al. 1993, Strider 1969). Losses in greenhouse-grown tomatoes of 10% to 25% have been reported (Huang and Tu 2001). Yield losses attributable to bacterial canker in controlled studies range from 11% to 99% (Emmatty and John 1973, Chang et al. 1992, Ricker and Riedel 1993). Detailed reviews of the symptom expression and epidemiology may be found in Jones et al. (1991) and Gleason et al. (1993).

Bacterial canker is caused by the gram-positive pathogen *Clavibacter michiganensis* subsp. *michiganensis* (*Cmm*) (Davis et al. 1984), which was previously classified as *Corynebacterium michiganense* (Smith) Jensen. It is a disease with a wide array of symptoms, such as loss of photosynthetic area, unilateral leaf wilt, stunting, open stem cankers, plant defoliation and premature death. The most devastating symptoms occur when infections become systemic with invasion of the vascular tissue and movement in the xylem. Subsequent destruction of the xylem, phloem, pith, and cortex leads to characteristic stem cankers. Invasion of the xylem results in translocation along the length of the stem into the fruit, where seed infection is likely to occur. The production of unmarketable fruit due to "bird's-eye" spots also contribute to the economic impact of disease (Sherf and MacNab 1986, Jones et al. 1991, Gleason et al. 1993, Medina-Mora et al. 2001). These superficial symptoms are caused when high populations are present at the time of flowering and initial fruit set, and are not due to systemic infection (Medina-Mora et al. 2001).

Interpretation of genetic studies requires a standardization of methodology that is only now beginning to occur. Different manifestations of symptoms combined with strain-to-strain variation in aggressiveness, differences in the expression of disease with plant maturity, and environmental conditions that affect symptom development have slowed both breeding progress and the characterization of the genetic basis of resistance. In the field, bacterial canker symptoms are manifested as systemic wilt (primary canker) that may kill plants and/or as a foliar phase (secondary canker) with characteristic "firing" symptoms due to marginal necrosis of leaflets. Anecdotal evidence suggests that some germplasm, e.g. Hawaii 7998, may be tolerant of foliar disease (secondary canker) but susceptible to the systemic phase (primary canker). Methods of screening germplasm and progeny may favor development of primary or secondary symptoms. There are three common methods of applying inoculum. A reliable technique for inducing primary canker is the removal of a petiole followed by the addition of inoculum to the wound (Laterrot et al. 1978, Berry et al. 1989). Spray inoculation of plants with 6-8 true leaves is an effective means of inducing secondary canker symptoms. In general, the expression of symptoms is reduced in older plants and symptoms can shift from the primary to the secondary phase. Field epidemics that result from both primary and secondary canker can be induced by spray inoculation of seedlings at the 2-4 true leaf stage of development in the greenhouse prior to transplanting to the field (Hausbeck et al. 2000, Kabelka et al. 2002). Correlations between greenhouse screens using the petiole clip technique and field screens based on spray inoculation are generally significant and positive, but correlation coefficients are generally low (0.35 to 0.46), suggesting that inoculation techniques strongly influence whether evaluations emphasize primary or secondary canker (Francis et al. 2001). Other inoculation techniques reported to induce symptoms include addition of bacteria to the root zone (Huang and Tu 2001) and inoculation of flowers at the time of anthesis to induce superficial fruit symptoms (Medina-Mora et al. 2001). Effective inoculation requires application of 10^7-10^8 bacteria per ml, while treatment with lower concentrations produce inconsistent results. Evaluations of disease are generally based on subjective assessment of symptom development (e.g. Sandbrink et al. 1995, Francis et al. 2001), colonization percentages and estimation of bacterial populations in infected plants (Dreier et al. 1997, Coaker et al. 2004).

Variation in the aggressiveness and pathogenicity of *Cmm* strains remains an area of active investigation. Strain-to-strain comparisons result in significant and positive linear correlations, but correlation coefficients range from 0.27 to 0.36 using the same inoculation technique (Francis et al. 2001). Thus it is prudent to consider evaluation against multiple strains. Knowledge of pathogen population structure is limited to studies based on DNA-fingerprint patterns. There are no race designations for isolates causing bacterial canker on tomato owing to a lack of well-defined major gene resistance to the disease. In the absence of race designations based on virulence, classification of *Cmm* isolates has been possible using the REP-PCR technique for DNA fingerprint analysis (Louws et al. 1994, 1995). At least four distinct genetic types of *Cmm* occur in North America (Louws et al. 1998). A high proportion of *Cmm* isolates found on processing tomatoes in the states surrounding the North American Great Lakes are type A isolates. Type C isolates represent approximately 5%. In addition, type B and D isolates have been identified, primarily from fresh market tomatoes. There is no evidence suggesting that this distribution reflects adaptation to specific genotypes; it is more likely that the division reflects differences in seed production and greenhouse transplant production between fresh market and processing tomatoes.

Differences in aggressiveness exist within each DNA fingerprint class. For example, isolates used by Berry et al. (1989) were described as differing in aggressiveness, yet all were type C strains when genotyped with the REP-PCR technique. A second, more extreme example of within class variation occurs within the type A strains. Pathogenicity factors in the type A *Cmm* strains may be plasmid borne and subject to loss. Non-pathogenic strains isolated from the field lack the plasmid, and when the plasmid is cured from pathogenic type A strains, they lose the ability to infect plant hosts. Plasmid borne pathogenicity factors have been previously described for *Cmm* strains (Meletzus et al. 1993, Dreier et al. 1997, Gartemann et al. 2003). Despite variation in aggressiveness within DNA fingerprint classes, the REP-PCR technique remains the most relevant classification system for epidemiology and strain selection (Louws et al. 1994, 1995, 1998).

Resistance to bacterial canker has been reported in several wild relatives of tomato (Table 12.1), including *L. hirsutum* (Elenkov 1965, Hassan et al. 1968, Thyr 1969, Francis et al. 2001), *L. peruvianum* (Lindhout et al. 1987, Sandbrink et al. 1995), *L. pimpinellifolium* (Ark 1944, Elenkov 1965, Thyr 1968, 1969, Berry et al. 1989), and *L. chilense* (Yordanov and Stamova 1977). Despite many years of work, lines listed as "improved sources of resistance" (Table 12.1) are of marginal improvement status relative to commercial varieties. Of the diverse sources of resistance, only resistance from *L. peruvianum* accession LA 2157 and *L. hirsutum* accession LA407 has been accompanied by detailed genetic analysis. Of the few commercial varieties and breeding lines claiming resistance or tolerance, none have been genetically characterized for resistance that facilitates effective manipulation in subsequent crosses or through MAS.

An example of the difficulty in interpreting classical genetic studies involving resistance is provided by results involving Bulgaria 12 (PI 330727). This source of resistance has been developed the furthest, resulting in varieties and breeding lines with tolerance or partial resistance (Table 12.1). Initially, Elenkov (1965) reported that resistance in Bulgaria 12 was controlled by one dominant character. Laterrot (1974) confirmed that hybrids of susceptible cultivars with Bulgaria 12 were as resistant as Bulgaria 12. However, subsequent studies failed to confirm single locus inheritance and indicated that the resistance in Bulgaria 12 was controlled by quantitative factors or by a number of partially dominant major genes. According to De Jong and Honma (1976), crosses between Bulgaria 12 and a susceptible line MSU 72-279 result in a susceptible F_1 and F_2 segregation consistent with a four-gene model when challenged with a mild strain, "H". Crosses between Bulgaria 12 and Earliana resulted in a partially resistant F_1, whereas F_2 segregation was again consistent with a four-gene model (De Jong and Honma 1976). In these studies, a minimum of four genes were postulated to explain resistance in *L. pimpinellifolium* accessions PI344102 and PI344103, and one-gene to three-gene models were proposed to explain the resistance in *L. hirsutum* (PI251305) (De Jong and Honma 1976). Thyr (1976) reported that the resistance to strain *Cm*15 in Bulgaria 12 was controlled by one or two incompletely dominant genes that are influenced by modifying genes. Resistance in *L. pimpinellifolium* accessions PI344102 and PI 344103 appeared to involve more loci, with a minimum estimate of four (Thyr 1976). Taken together, these studies suggest that genetic background (e.g. MSU 72-279 vs Earliana), strain differences, and methodology can affect the expression of resistance and the interpretation of results.

Genetic analyses that are more easily interpreted and applied due to the incorporation of DNA-based molecular markers have been conducted for the *L. peruvianum* source of resistance, LA 2157. Initial studies reported results obtained using an intraspecific BC_1 population of resistant *L. peruvianum* LA2157 × susceptible *L. peruvianum* LA2127. These studies identified five possible quantitative trait loci (QTL) associated with resistance to strain *Cm*542 (Sandbrink et al. 1995). The bacterium has not been characterized relative to the REP-PCR classification and the germplasm resulting from these studies are 100% *L. peruvianum*. These QTL are summarized in Fig. 12.1. A second segregating F_2 population derived from the interspecific cross between *L. peruvianum* LA2157 and a susceptible *L. esculentum* variety "Solentos" revealed only three QTL linked with resistance (van Heusden et al. 1999). A locus on chromosome 7 may overlap with those discovered in the intraspecific population, suggesting that although genetic background may affect the detection of QTL, some of the loci detected may be robust across genetic backgrounds (Figure 12.1).

Resistance in LA407 has been characterized using *Cmm* strains from the A and C DNA-fingerprint groups, and the magnitude of resistance appears to be similar to that of *L. peruvianum* LA 2157 (Francis et al. 2001). Kabelka et al. (2002) identified two major loci (Rcm2.0 and Rcm 5.1), on chromosome 2 and 5, originating from *L. hirsutum* LA407 (Figure 12.1). These loci confer resistance to both strains tested, but the response of strains in the B and D fingerprint classes remains untested. Subsequent crosses and progeny testing of recombinant plants narrowed the QTL location of Rcm2.0 to a 4.4 cM interval between TG537-TG091 and to a 2.2 cM interval between CT202-TG358 for Rcm5.1 (Coaker and Francis 2004). These two loci explained 25.7-34.0% and 25.8-27.9% of total phenotypic variation, respectively. When both loci are present and epistatic interactions are accounted for, Rcm 2.0 and Rcm 5.1 explain as much as 68% of the variation in symptom expression (Coaker and Francis 2004). It is possible that Rcm 5.0 from LA2157 and Rcm 5.1 from LA407 are allelic loci, but this hypothesis has not been formally tested.

MAS provides a potential for increasing selection efficiency by allowing for earlier selection and reducing plant population size used during selection. Restriction fragment length polymorphisms (RFLP) linked to QTLs from LA2157 and PCR-based DNA markers linked to QTLs from LA407 are summarized in Tables 12.2 and 12.3. For both LA2157 and LA407, breeding for resistance into elite genetic backgrounds remains problematic and will require minimizing linkage to undesirable genes and

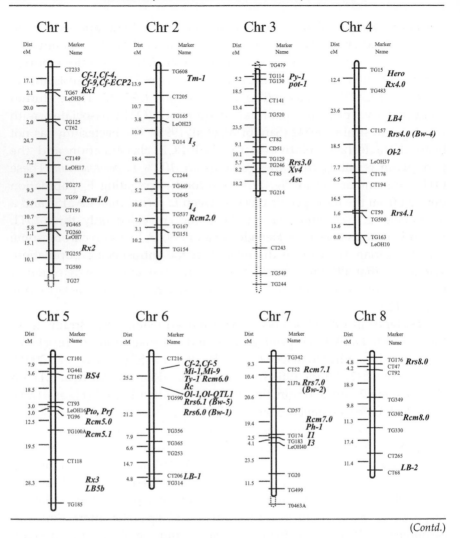

Fig. 12.1 Genetic map indicating the approximate location of disease-resistance genes (R genes) and quantitative resistance loci (QTLs) obtained from published research. The framework map was adapted from Yang et al. (2004) and extended according to marker positions from SGN (http://www.sgn.cornell.edu). Descriptions of the R genes and QTLs are as follows: *Asc*, resistance to stem canker (*Alternaria alternata*) (van der Biezen et al. 1995); *Cf* (*Cf-1, Cf-2, Cf-4, Cf-5, Cf-9, Cf-ECP2*), resistance to leaf mold (*Cladosporium fulvum*) (Haanstra et al. 1999, Thomas et al. 1998); *Cmr*, resistance to cucumber mosaic virus (Stamova and Chetelat 2000); *Fr1*, resistance to Fusarium crown rot (*Fusarium oxysporum* f.sp. *radicis-lycopersici*, Laterrot and Moretti 1995); *Hero*, resistance to potato cyst nematode (*Globodera rostochiensis*, Ganal et al. 1995); *I* (*I, I2, I3*), resis-

(Contd.)

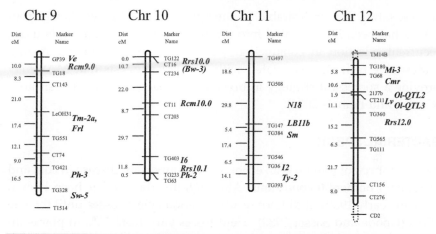

(Contd.)

tance to races of Fusarium wilt (*Fusarium oxysporum* f.sp. *lycopersici*)
(Bournival et al. 1989, Sarfatti et al. 1989, 1991, Hemming et al. 2004; *LB* (*LB-1*, *LB-2*, *LB4*, *LB5b*, *LB11b*), QTLs for resistance to tomato late blight
(*Phytophthora infestans*; Frary et al. 1998, Brouwer and St. Clair 2004); *Lv*,
resistance to powdery mildew (*Leveuillula taurica*) (Chunwongse et al. 1994);
Mi (*Mi-1*, *Mi-3*, *Mi-9*, resistance to root knot nematodes (*Meloidogyne* spp.
Yaghoobi et al. 1995, Veremis et al. 1999, Ammiraju et al. 2003); *N18*, resis-
tance to tobacco mosaic virus (Whitham et al. 1994), *Ol* (*Ol-1*, *Ol-2*, *Ol-QTL1*,
QI-QTl2, *QI-QTL3*), resistance to powdery mildew (*Oidium lycopersicum*; van
der Beek et al. 1994, Bai et al. 2003, De Giovanni et al. 2004); *Ph* (*Ph-1*, *Ph-2*, *Ph-3*), resistance to late blight (*Phytophthora infestans*) in tomato
(Chunwongse et al. 1998, Moreau et al. 1998); *Pot-1*, resistance to potyviruses
(Parrella et al. 2002); *Pto* and *Prf*, resistance to bacterial speck (*Pseudomonas syringae* pv. *tomato*) (Martin et al. 1993b; Salmeron et al. 1996); *Py-1*,
resistance to corky root rot (*Pyrenochaeta lycopersici*; Doganlar et al. 1998);
Rcm (*Rcm1.0*, *Rcm2.0*, *Rcm5.0*, *Rcm5.1*, *Rcm6.0*, *Rcm7.0*, *Rcm7.1*, *Rcm8.0*,
Rcm9.0, *Rcm10.0*), QTLs for resistance to bacterial canker (*Clavibacter
michiganensis subsp. michiganensis*; Sandbrink et al. 1995, van Heusden et
al. 1999, Coaker and Francis 2004); *Rrs* (*Rrs3.0*, *Rrs4.0/Bw-4*, *Rrs4.1*, *Rrs6.0/
Bw-1*, *Rrs6.1/Bw-5*, *Rrs7.0/Bw-2*, *Rrs8.0*, *Rrs10.0/Bw-3*, *Rrs10.1*, *Rrs12.0*),
QTLs for resistance to bacterial wilt (*Ralstonia solanacearum*; Danesh et al.
1994, Thoquet et al. 1996a,b; Mangin et al. 1999, Wang et al. 2000); *Rx* (*Rx1*,
Rx2, *Rx3*, *BS4*, *Rx4.0*, *Xv4*,)), resistance to bacterial spot (four species of
Xanthomonas; *X. euvesicatoria*, *X. vesicatoria*, *X. perforans*, *X. gardneri*) (Yu
et al. 1995, Astua-Monge et al. 2000, Ballvora et al. 2001, Yang et al. 2005);
Sm, resistance to gray leaf spot (*Stemphyllum solani*, *S. lycopersici*; Behare et
al. 1991); *Sw-5*, resistance to tomato spotted wilt virus (Stevens et al. 1995);
Tm-1 and *Tm-2a*, resistance to tobacco mosaic virus (Young and Tanksley
1989; Levesque et al. 1990); *Ty-1* and *Ty-2*, *Rc*, resistance to tomato yellow
leaf curl virus (Zamir et al. 1994, Chague et al. 1997, Hanson et al. 2000); *Ve*,
resistance to Verticillium wilt (*Verticillium dahliae*; Diwan et al. 1999).

compensating for undesirable plieotropic effects. The resistance from LA407 has been incorporated into improved genetic backgrounds using recurrent selection in conjunction with MAS. Progress in minimizing undesirable fruit size associated with Rcm 5.1 suggests that MAS may provide a successful strategy in breeding for bacterial canker resistance (Coaker and Francis 2004).

BACTERIAL SPECK

Bacterial speck of tomato is an economically important disease in cool and moist environmental conditions. Although this disease has been known since the early 1930s, it did not result in serious yield losses until the late 1970s (Goode and Sasser 1980). Yield losses vary from 75% in plants infected at an early stage of growth to 5% in plants infected later in the season (Yunis et al. 1980).

Control of bacterial speck using copper-based bactericides is only marginally effective because the pathogen can quickly develop plasmid-mediated copper-tolerance (Alexander et al. 1999, Bender and Cooksey 1986, Cooksey 1987, Silva and Lopes 1995). Combining copper-based bactericides with EBDC fungicides (maneb or mancoseb) provided improved levels of control even with copper-tolerant populations (Conlin and McCarter 1983), but the effectiveness is moderate. The use of resistant varieties is only an effective approach to control this disease.

Bacterial speck is caused by *Pseudomonas syringae* pv. *tomato* Young, Dye and Wilkie (Okabe 1933, Dye et al. 1980). The optimum conditions for disease development and spread include temperatures between 13 and 28°C, high relative humidity, and free water on leaves (Yunis et al. 1980, Smitley and McCarter 1982). The symptoms of speck occur on foliage and fruit and consist of small black lesions, often with a discrete yellow halo (Pitblado and MacNeill 1983). It is difficult to distinguish foliar symptoms of bacterial speck from bacterial spot; the small dark fruit lesions are therefore diagnostic of bacterial speck as they contrast with the larger, brown, scabby fruit symptoms associated with spot. The pathogen can be seed-borne and can survive in soil for a limited period, on plant debris, weeds, nonhost plants, or as an epiphyte on symptomless tomato transplants for longer periods of time (Chambers and Merriman 1975, schneider and Grogan 1977, Bashen et al. 1978, McCarter et al. 1983, Bonn et al. 1985).

Of the four diseases treated in this chapter, the variation of *Pseudomonas syringae* pv. *tomato* populations is relatively limited, but the pathogen population structure may be changing gradually. Two races, 0 and 1, have been reported to date (Scott, 1997). In the absence of a differential series of cultivars, it may be more appropriate to consider race 1 strains as non-race 0 strains. New sources of resistance are likely to further divide race 1. A survey conducted in Ontario, Canada, indicated that race 0 was dominant but the occurrence of non-race 0 strains in the population increased from 8% to 25.2% between 1982 and 1985 (Lawton and MacNeill, 1986). The recent occurrence of non-race 0 strains has been reported in several countries (Buonaurio et al. 1996, Donner and Barker, 1996, Arredondo and Davis 2000). The wide introduction of cultivars containing the *Pto* gene for resistance is possibly driving a change in the pathogen population structure.

Techniques for screening for resistance to bacterial speck include infiltration to induce hypersensitive reaction (HR), saturating the foliage with a suspension of bacteria, and drenching of bacteria in a solution containing surfactant. Field inoculations generally involve spraying plants to run-off with 10^7 to 10^8 CFU/ml of bacteria. Infiltration through the back of a fully expanded leaflet generally involves 10^7 to 10^8 CFU/ml using a 3 ml syringe without needle. Preparation of the plants involves misting with water one hour before inoculation and maintaining plants at 20-25°C in a humid environment in the light for at least four hours following infiltration. The HR can normally be recorded 24 hours after inoculation for bacterial speck. We have found that concentrations of 4×10^7 CFU/ml are appropriate for strain DC3000. Evaluation of colonization and bacterial populations in infected leaves are performed following immersion in a suspension of bacteria containing the surfactant Silwet L-77 (Union Carbide) (Salmeron et al. 1994). These inoculation techniques appear to produce similar results with respect to determination of resistance and susceptibility, though formal analysis of correlations are lacking. Conclusions regarding gene action may differ depending on the technique used. For example, inoculation of 15-25 cm plants with a spray of 10^7 CFU/ml followed by a subjective evaluation of symptoms lead to the conclusion of dominant gene action (Pitblado and MacNeill 1983), while dipping plants in a suspension of 2×10^8 bacteria containing Silwet L-77 and measuring bacterial populations suggests that gene action is additive (Carland and Staskawicz 1993).

Resistant sources have been identified in many wild species though the *L. pimpinellifolium* gene, *Pto*, has been the only source of resistance widely bred into cultivated tomato. There are no symptoms in many accessions of *Lycopersicon hirsutum, L. peruvianum, L. glandulosum,* and *L. pimpinellifolium*

(Lawson and Summer 1984, Pilowsky and Zutra 1986, Stockinger and Walling 1994, Tanksley et al. 1996, Krause et al. 2001). Resistance to race 0 of bacterial speck in cultivated tomato is controlled by *Pto* (*Pto-1*), a single gene with additive to dominant action. The source of the *Pto* gene in most modern varieties is Ontario 7710 (Pitblado and MacNeill 1983), with resistance probably derived from an accession of *L. pimpinellifolium* (PI 370093) through the cultivar Farthest North (Pitblado and Kerr 1980). The gene *Pto* was mapped to chromosome 5 (Pitblado et al. 1984, Figure 12.1) and was subsequently characterized through map-based cloning (Martin et al. 1993a). Other sources of resistance from wild species are less well characterized. A single dominant gene allelic to *Pto* conditions the resistance in PI112215 of *L. pimpinellifolium* and PI129157 of *L. hirsutum f. glabratum* (Lawson and Summer 1984). The resistance to race 0 in PI126430 is conditioned by one single dominant gene designated as *Pto-2*, which is non-allelic to *Pto* (Pilowsky and Zutra 1986). *Pto-3* is an incompletely dominant gene conferring resistance to race 0 and *Pto-4* is a dominant gene conferring resistance to race 1 in PI134417 (Stockinger and Walling 1994). However, the allelism between *Pto-3* and *Pto-2* and the genetic information about *Pto-4* are not very clear. The resistance to race 0 in PI134418 is controlled by *Pto*[h], an allele of *Pto*, that is not linked to the gene conferring sensitivity to insecticide fenthion (Tanksley et al. 1996), whereas *Pto* is tightly linked to the gene that confers sensitivity to fenthion (Laterrot and Moretti 1989). The resistance in Rehovot-13 is also conditioned by one single dominant gene, but further information regarding map-position and allelic relationships is not available (Fallik et al. 1983).

Breeding for resistance to bacterial speck has been successful because the resistance conferred by *Pto* is simply inherited and efficient and accurate evaluation techniques are available (Emmatty et al. 1982). Using MAS has the potential of accelerating breeding progress when the goal is to pyramid resistance to multiple pathogens. Molecular markers linked to *Pto* have been available for over ten years (e.g., Martin et al. 1993b). The RFLP marker TG538 on chromosome 5 co-segregates with *Pto* and provides breeders a tool for MAS (Table 12.2). Recently, PCR-based DNA markers have been developed based on the cloned gene, with polymorphisms detected as cleaved amplified polymorphisms using restriction enzyme *Rsa* I (Coaker and Francis 2004, see Table 12.3). Whether MAS for *Pto* is more efficient than direct selection or other indirect selection techniques remains to be tested. As noted above, resistance conferred by *Pto* is tightly linked to the gene that confer sensitivity to the organophosphate insecticide, Fenthion (Laterrot and Moretti 1989; Carland and Staskawicz 1993). Selection

Table 12.2   RFLP markers linked to loci for resistance to bacterial diseases in tomato

Diease	Race/ strain	Source	Chrom.	Locus[1]	Marker	Reference
Canker	Cm542	LA2157	1	Rcm1.0	TG059	Sandbrink et al. (1995)
	Cm542	LA2157	5	Rcm5.0	TG363	van Heusden et al. (1999)
	Cm542	LA2157	6	Rcm6.0	TG178	Sandbrink et al. (1995)
	Cm542	LA2157	7	Rcm7.0	TG061, TG174, TG210A	Sandbrink et al. (1995)
	Cm542	LA2157	7	Rcm7.1	TG061, TG342	van Heusden et al. (1999)
	Cm542	LA2157	8	Rcm8.0	TG261	Sandbrink et al. (1995)
	Cm542	LA2157	9	Rcm9.0	TG254	van Heusden et al. (1999)
	Cm542	LA2157	10	Rcm10.0	TG103	Sandbrink et al. (1995)
Speck	0	Rio Grande-PtoR	5	Pto	TG538	Martin et al. (1993b)
Spot	T1	Hawaii 7998	1	Rx1	TG236	Yu et al. (1995)
	T1	Hawaii 7998	1	Rx2	TG157	Yu et al. (1995)
	T1	Hawaii 7998	5	Rx3	TG351	Yu et al. (1995)
	T1	Moneymaker	5	BS4	TG432	Ballvora et al. (2001)
	T3	LA716	3	Xv4	TG134	Astua-Monge et al. (2000)
Wilt	GMI8217	Hawaii 7996	3	Rrs3.0	GP226	Thoquet et al. (1996b)
	GMI8217	Hawaii 7996	4	Rrs4.0/Bw-4	TG268	Thoquet et al. (1996a)
	GMI8217	Hawaii 7996	4	Rrs4.1	GP165	Thoquet et al. (1996a)
	UW364	L285	6	Rrs6.0/Bw-1	CT184	Danesh et al. (1994)
	GMI8217	Hawaii 7996	6	Rrs6.0/Bw-1	TG118, CP18	Thoquet et al. (1996a)
	GMI8217	Hawaii 7996	6	Rrs6.0/Bw-1	TG118, TG240	Mangin et al. (1999)
	GMI8217	Hawaii 7996	6	Rrs6.1/Bw-5	Cf-2, TG153	Mangin et al. (1999)
	Pss4	Hawaii 7996	6	Rrs6.0/Bw-1	TG73	Wang et al. (2000)
	UW364	L285	7	Rrs7.0/Bw-2	TG135	Danesh et al. (1994)
	GMI8217	Hawaii 7996	8	Rrs8.0	CP112	Thoquet et al. (1996b)
	UW364	L285	10	Rrs10.0/Bw-3	CT225b, TG230	Danesh et al. (1994)
	GMI8217	Hawaii 7996	10	Rrs10.1	CP105	Thoquet et al. (1996a)
	Pss4	Hawaii 7996	12	Rrs12.0	TG564	Wang et al. (2000)

[1]Locus names are either adapted from the reference or assigned in this review using R (resistance) followed by the initials corresponding to the pathogen (or disease name), chromosome, and a decimal number indicating the order of the locus identified on the chromosome.

based on this sensitivity is a widely practised indirect technique for identifying resistant plants. It is now known that a gene with 80% nucleotide identity to *Pto*, *Fen*, is responsible for sensitivity to Fenthion, Fensulfothion and Fenitrothion. As *Fen* is separated from *Pto* by less than 50 Kb (Martin et al. 1994), it is unlikely that recombination will separate Fenthion sensitivity and resistance in most breeding applications. The cloning of *Pto* (Martin et al. 1993a) opens up the possibility of directly transferring resistance into susceptible varieties using transformation. Given the efficiency of direct and indirect selection, it is unlikely that this approach

Table 12.3 PCR-based markers linked to loci for resistance to bacterial disease in tomato

Disease	Race /strain	Resistance source	Chrom.	Locus[1]	Marker	Marker Type[2]	Primer sequence	Annel. Temp.	Rest. enzyme	Reference
Canker	A & C	LA407	2	Rcm2.0	TG537	CAP	f: TACCCGAGGCTCAGAAACAC r: CATCAACAGGAGATCGGTTTT	57	Hinf I	Coaker and Francis (2004)
					TG492	CAP	f: TGGAGAAGGTTCAAAGGGAACG r: GGGCCAAGGAATATTTCTCAAGG	57	Mnl I	Coaker and Francis (2004)
					TG091	CAP	f: TGCAGAGCTGTAATATTTAGAC r: CGGTCTCAGTTGCAACTCAA	60	Dra I	Coaker and Francis (2004)
	A & C	LA407	5	Rcm5.1	CT202	CAP	f: TAATCCGAGAAGGTGATCCG r: GGCTTATAACCCATGCCAAAAG	60	Tsp45 I	Coaker and Francis (2004)
					TG318		f: CAAGCCATAGAAATTGCCGTA r: TGCTCTCTGTGATGGAAGC	57	-	Coaker and Francis (2004)
					TG358	CAP	f: CAACTTTCCAGGTTCATTTTCTC r: ACACCTACATGCTACTAAGGGGTC	53	Hha I	
Speck	0 -PtoR	Rio Grande	5	Pto r:	PTO GTGCATACTCCAGTTTCCAC	CAP	f: ATCTACCCACAATGAGCATGAGCTG	62	Rsa I	Coaker and Francis (2004)
Spot	T1	Hawaii 7998	5	Rx3	Rx3-L1	SNP	f: CTCCGAGCGAAGAGTCTAGAGTC r: GAAGGCAAAAGGAAAAGGAGAAGGATGG	60	BsrB I	Yang et al. (2005)
			5	Rx3	Cos OH73	SNP	f: CTTCCCGACAAGCACAAAAA r: CGAATGCTCTGTACCATTTCC	56	Alu I	Yang et al. (2005)
			5	Rx3	TOM49	SSR	f: AAGAAACTTTTTGAATGTTGC r: ATTACAATTTAGAGAGTCAAGG	45	-	Yang et al. (2005)
			4	Rx4.0	SSR43	SSR	f: CTCCAAATTGGGCAATAACA r: TTAGGAAGTTGCATTAGGCCA	45	-	Yang et al. (2005)
	T4	LA716	3	Xv4	TG599	CAP	f: TGTTGATCCTTGCTTGCTGT r: TTGTATGGTGCAACTTCCC	50	Hind III -EcoR I	Astua-Monge et al. (2000)

[1]Locus names are either adapted from the reference or assigned in this review using R (resistance) followed by the initials corresponding to the pathogen (or disease) name, chromosome, and a decimal number indicating the order of the locus identified on the chromosome.
[2]CAP, cleaved amplification polymorphism; SNP, single nucleotide polymorphism; SSR, simple sequence repeat.

will provide a viable alternative. Finally, breeding for resistance to bacterial speck should be viewed as a long-term disease-management strategy. The use of less well developed germplasm resources will depend on population dynamics and the emergence of new races. In general, the widespread use of a few resistance genes might accelerate the selection of new pathogenic races and consequently result in destabilization of crop production (Browning and Frey 1969, National Academy of Sciences 1972). The apparent trend toward a change in the *Pseudomonas syringae* pv. *tomato* population structure may be an indication of over deploying varieties with the *Pto* gene. This suggests that developing varieties containing genes for resistance to non-race 0 strains should be part of the near-term goal of bacterial speck resistance breeding.

Although highly resistant *L. esculentum* sources of resistance have been reported (Pitblado and MacNeill 1983, Gitaitis et al. 1985), little is known about the genetic basis of this resistance. The variety Campbell 28, C28 (LA 3317), has been noted as a source of quantitative resistance to bacterial speck (Gitaitis et al. 1985). The resistance in this variety is independent of *Pto* and appears to be easily transferred since varieties with C28 in the pedigree have also been noted for their resistance. For example, Ohio 7663 (Berry and Gould 1979) has been noted as a highly resistant variety (Gitaitis et al. 1985). Campbell 28 also contributes significantly to Ohio 7870 (Berry and Gould 1982) and Ohio 8245 (Berry et al. 1991), which we have observed as partially resistant to both speck and spot. Partial resistance to bacterial spot (e.g. Somodi et al. 1994) and bacterial canker (Dick and MacNeill 1981) have been reported previously for C28, suggesting that this variety may be an interesting source for future genetic studies.

BACTERIAL SPOT

Bacterial spot was first reported in South Africa (Doidge 1921) and Indiana, USA (Gardner and Kendrick 1921). It occurs throughout the world wherever tomatoes are grown and environmental conditions are favorable for disease development (Stall 1995). Maximum infection occurs under warm, moist conditions with the incidence especially high when night temperatures are also warm. Bacterial spot enters healthy plants through stomata or wounds and can attack leaves, stems and fruits, generating economic losses due to defoliation and the development of brown scabby lesions on fruit (Scott 1997). The pathogen can survive in debris and in association with certain weeds such as black nightshade. Although no data on yield loss are available, surveys of the processing tomato industry in the Midwest of USA have ranked bacterial spot among the top five diseases (Foster et al. 1993, Francis et al. 1995).

Since first reported, it has been apparent that considerable phenotypic variation exists within the bacteria that cause bacterial spot on tomatoes and peppers (see Jones et al. 2000). Although causal organisms of bacterial spot had been classified as single species, *Xanthomonas campestris* pv *vesicatoria*, since 1978, taxonomic studies have suggested that more than one species in the genus *Xanthomonas* is responsible for epidemics. Four phenotypic groups have been identified that are pathogenic on tomato (Jones et al. 2000). Classification is based on carbon substrate utilization, enzymatic activity, a panel of monoclonal antibodies (serology), RFLP profiles separated by Pulse Field Gel electrophoresis, and sequence comparisons of rRNA genes and intragenic spacers (Jones et al. 2000). The body of evidence is consistent with four distinct groups. Group A includes strains that are the same as described by Doidge (1921). Group B includes strains that are the same as described by Gardner and Kendrick (1921). The A and B groups were classified as *X. axonopodis* pv. *vesicatoria* and *X. vesicatoria*, respectively (Jones et al. 2000), though this classification has been subsequently revised (Jones et al. 2004b, see below). Group C strains were first isolated in Florida, USA, in the early 1990s, and were associated with the emergence of a new race, T3, which displaced race T1 strains in Florida (Jones et al. 1995). The Group C strains have distinct serological profiles and DNA restriction patterns (Jones et al. 2000), yet share 70% similarity based on DNA-DNA hybridization with Group A strains. Group D strains, first isolated in Yugoslavia (Sutic 1957), are weakly amylolytic and pectolytic. They have a distinct pattern of reaction to a panel of monoclonal antibodies (serological profile) and a distinct RFLP banding pattern (Bouzar et al. 1994). Strains in Group D are also found in Costa Rica and Brazil and are classified as *X. gardneri*. DNA-DNA hybridization data indicate that Group B and D strains share less than 10% homology with each other and with Group A and C strains. More recently, Jones et al. (2004b) proposed four species of *Xanthomonas* for the four phenotypic groups causing bacterial spot. *X. euvesicatoria* has been proposed for Group A, *X. vesicatoria* for Group B, *X. perforans* for Group C, and *X. gardneri* for Group D. Details about systematic analysis and reclassification of the xanthomonads causing bacterial spot may be found in Jones et al. (2000, 2004b).

In resistance breeding, the pathogenicity of the species causing bacterial spot is an important consideration. Five races designated T1 through T5 have been defined by their virulence on tomato (Jones et al. 2000, Astua-Monge et al. 2000, Jones et al. 2004a). Races T1, T2, and T3 were identified based on their reaction on three tomato varieties: Hawaii 7998, Hawaii 7981, and Bonny Best. Race T1 induces a hypersensitive reaction (HR) on Hawaii 7998 (Whalen et al. 1993), T3 induces HR on Hawaii 7981 (Minsavage et al. 1996), and T2 does not induce HR on any of these cultivars (Table 12.4). Races T4 and T5 were identified as strains that are able to overcome resistance from Hawaii 7981, and T4 strains elicit an HR on the *L. pennellii* accession LA 716 (Astua-Monge et al. 2000, Jones et al. 2004a). The phenotypic groups appear to coincide with races. T1 strains belong to Group A, or the proposed *X. euvesicatoria*. Most T2 strains belong to the proposed *X. vesicatoria*, while T3, T4, and T5 belong to Group C, the proposed *X. perforans*. Group D strains are *X. gardneri*, with these also being classified as race T2 based on their inability to elicit an HR on Hawaii 7998 or Hawaii 7981. It is clear that breeding for resistance to bacterial spot will require combining resistance to multiple species with diverse pathogenicity and virulence on tomato.

Table 12.4   Sources of resistance to bacterial spot

Resistance Genotype Sources	Species Race	X. euvesicatoria T1	X. vesicatoria T2	X. perforans T3	T4
Greenhouse Infiltration					
Hawaii 7998	*L. esculentum*	HR	S	S	S
Hawaii 7981	*L. esculentum*	S	S	HR	S
LA716	*L. pennellii*	S	S	HR	HR
PI128216	*L. pimpinellifolium*	S	S	HR	S
PI126932	*L. pimpinellifolium*	S	S	HR	S
Field Evaluation of Resistance					
Hawaii 7998	*L. esculentum*	R	S	S	S
Hawaii 7981	*L. esculentum*	S	S	R	S
PI128216	*L. pimpinellifolium*	S	PR	PR	PR
PI126932	*L. pimpinellifolium*	S	S	PR	R
PI114490	*L. esculentum var. cerasiformae*	R	R	PR	R
Advanced lines Field Evaluation of Resistance					
Ohio 9834	Hawaii 7998	PR (Rx3)	S	S	S
Ohio 9816	Hawaii 7998	PR (Rx3)	S	S	S
FL7770	Hawaii 7998	R/PR	S	S	S
FL8000	Hawaii 7981	S	S	R	S

As is the case with several of the bacterial diseases described in this review, inoculation techniques and assessment of symptoms influence the outcome of disease evaluations. For bacterial spot, inoculation strategies parallel those used for bacterial speck. Infiltration of bacteria is used to evaluate the development of HR; immersion of young plants in a suspension of bacteria containing surfactant is used to evaluate colonization and bacterial growth, and spray inoculation is used to establish field epidemics. Inoculum is generally produced by growing the bacteria on YDC agar medium (Lelliot and Stead 1987) at 28 °C for 48-72 h. Bacterial cells are suspended in water supplemented with 10 mM $MgSO_4 \cdot 7H_2O$ and suspensions are standardized to A_{600}=0.15, at a concentration of approximately 3×10^8 CFU/ml. For evaluating colonization of bacterial populations in the greenhouse, plants are immersed in the bacterial suspension in the presence of Silwet L-77 as described by Salmeron et al. (1994). For establishing disease in the field, inoculum may be applied to plants with a backpack sprayer in the late evening approximately two weeks post-transplant of seedlings. As an alternative, seedling transplants can be spray-inoculated in the greenhouse one to two weeks prior to transplanting in the field. Both approaches are effective at establishing populations, but greenhouse inoculation requires less labor.

Although evaluation of symptoms generally follows standard procedures, there are nuances that deserve mention. For example, we have found that the evaluation of HR to T1 strains is unambiguous when approximately 2×10^8 CFU/ml are applied and ratings of HR on Hawaii 7998-derived resistance are taken 48-72 hours following infiltration. This time point is somewhat longer than common for HR ratings. In contrast, the QTL studies reported by Wang et al. (1994a,b) and Yu et al. (1995), rely on a scale that incorporates the temporal progression of HR. In the field, plants are rated for disease severity using the scale of Horsfall and Barratt (1945), which translates percentage of diseased tissue to numbers as follows: 1 = 0%, 2 = 0%-3%, 3 = 3%-6%, 4 = 6%-12%, 5 = 12%-25%, 6 = 25%-50%, 7 = 50%-75%, 8 = 75%-87%, 9 = 87%-94%, 10 = 94%-97%, 11 = 97%-100%, and 12 = 100%. Reported correlation coefficients for regression between HR and field symptoms are generally low. For example the correlation (r) between HR and field symptoms was only 0.37 in progeny of HA 7998 x LA 716 (Wang 1992). In crosses between two *L. esculentum* lines, HA 7998 and the cultivar Walter, the correlation between HR and field symptoms was r = 0.31 (Somodi et al. 1996). Thus it is likely that loci other than those reported for HR play a role in resistance to bacterial spot in the field and practical breeding for resistance to bacterial spot must rely on field evaluation.

Sources of resistance from wild species and cultivated tomatoes, *Lycopersicon esculentum, L. pimpinellifolium,* and *L. pennelli* have been identified and confirmed for races T1-T3 (Scott and Jones 1986, Sotirova and Bogatzevska 1993, Somodi et al. 1994, Scott et al. 1995, 1997; Astua-Monge et al. 2000, Ballvora et al. 2001, Scott et al. 2003). The *L. esculentum* breeding line, Hawaii 7998, is considered the most reliable source of resistance to T1 (Jones and Scott 1985), while Hawaii 7981 provides a robust source of HR-based resistance to T3 strains (Scott et al. 1995). The indeterminate yellow cherry tomato PI 114490 (*L. esculentum var. cerisforme*) is resistant to races T1, T2, and T3 (Scott et al. 2003). Resistance to race T4 in PI 114490 was also observed in field trails (Scott, personal communication). Resistance based on HR to T3 and T4 strains has been reported in *L. pennellii* LA 716 (Astua-Monge et al. 2000).

Bacterial strains that cause bacterial spot on pepper but not on tomato are distinguished by the presence of a bacterial avirulence gene, *AvrBs4*, that activates HR in most *L. esculentum* germplasm. After screening germplasm in multiple accessions of *L. esculentum, L. cheesmanii, L. peruvianum, L. pimpinellifolium,* and *L. pennellii,* susceptibility to strains carrying *AvrBs4* was detected in two *Lycopersicon* accessions, *L. pennelli* LA2963 and LA1282. Recognition of *AvrBs4* is controlled by a single dominant gene designated *Bs4* (Ballvora et al. 2001). This study points out the formal possibility that susceptibility could be inadvertently bred into resistant *L. esculentum* germplasm, though the probability of such an occurrence is low.

Genetic characterization of the interaction between resistant germplasm and diverse races suggests that resistance is often quantitative. Scott and Jones (1989) studied the inheritance of T1 resistance by crossing Hawaii 7998 with susceptible cultivar Walter and found intermediate resistance in the F_1 and that genetic models involving multiple loci with both additive and dominance were necessary to explain resistance in segregating progenies. According to Whalen et al. (1993), the inheritance of resistance in Hawaii 7998 was governed by more than one locus based on F_2 populations derived from crosses between Hawaii 7998 and Florida 7060 and between Hawaii 7998 and Alisa Craig. Using an F_2 population obtained from a cross between Hawaii 7998 and Florida 7060, Wang et al. (1994a) found that the HR in Hawaii 7998 was controlled by multiple genetic factors. Application of molecular markers to studying the genetic basis of resistance from Hawaii 7998 was slowed by the low levels of polymorphism between the *L. esculentum* source of resistance and elite breeding material. By developing populations based on Hawaii 7998 and the susceptible wild species accession, LA 716, three loci, Rx1 and Rx2 on each arm of chromosome 1 and Rx3 on chromosome 5, were described based on linkage to RFLP markers (Wang et al. 1994b, Yu et al. 1995, Fig. 12.1; Tables 12.2 and 12.3). These non-dominant genes act independently. They also detected three regions on chromosomes 3, 9 and 11 of the genome carrying *L. pennellii* (susceptible to race T1) alleles that appear to modify the HR. Unfortunately the polymorphic markers identified in the HA 7998 × *L. pennellii* cross are inherited from the wild parent, linked to susceptibility, and are of limited use for breeding purposes. Furthermore, the use of a rate measurement for HR response for these genetic studies places some doubt on the practical value of the QTL. In follow up studies, we demonstrated that *Rx3* contributes to resistance based on analysis of bacterial populations and field resistance in an advanced backcross population derived from a cross of Ohio 88119 × Hawaii 7998 (Yang et al. 2005). The locus Rx3 explained 41% of total phenotypic variation, but Rx1 did not contribute significantly to resistance. The role of Rx2 in field resistance remains untested. A locus on chromosome 4, linked to marker SSR43, was associated with susceptibility in the resistant parent, Hawaii 7998. The discovery of this QTL that explains 11% of total phenotypic variation serves to emphasize the importance of genetic background effects and suggests that Ohio 88119 contains genes that can contribute to resistance.

Classical genetic analysis of T2 resistance in F_2 and inbred backcross (IBC) populations derived from PI114490 suggested that genetic control is conferred by a minimum of two loci (Scott et al. 2003). The poor correlation between T2 and T3 resistance in the IBC population suggested that resistance to these two races are not controlled by the same genes in PI 114490. Molecular characterization of resistance in an IBC population developed from PI 114490, Ohio 9242, and Florida 7600 demonstrated that a QTL on chromosome 11 confers resistance to races T2, T3, and T4 (Yang et al. unpublished data). The QTL appears in an allelic series with FL 7600, providing more resistance than PI 114490 relative to susceptible alleles from Ohio 9242. Several QTLs from PI114490 that confer resistance to races T2, T3, or T4 remain to be confirmed as do the apparent epistatic interactions between QTL. Alleles from Ohio 9242 that contribute to resistance have also been identified. Again, these preliminary findings emphasize the importance of genetic background effects and epistasis in resistance to bacterial spot.

The HR to race T3 found in Hawaii 7981 (*L. esculentum*) and PI 128216 (*L. pimpinellifolium*) is controlled by the interaction between *avrXv3* and a single gene *Xv3* (Minsavage et al. 1996, Scott et al. 1996). Field resistance to race T3 in Hawaii 7981 is conferred quantitatively by *Xv3* and other loci (Scott et al. 2001). Resistance to T3 has not been the subject of mapping studies and the location of *Xv3* and modifying QTL remains unknown. Races T4 and T5 are taxonomically related to T3 and involve the loss of *avrXv3* through distinct mutations (Jones et al. 2004a). Resistance based on HR to races T4 and T5 was reported in *L. pennellii* accession LA716 (Astua-Monge et al. 2000, Jones et al. 2004a). Although a map position on chromosome 3 between TG599 and TG134 has been reported for *Xv4*, the map position was not verified in subsequent studies. The genetic characterization of resistance to T4 and T5 strains from LA716 remains to be clarified.

Breeding for resistance to bacterial spot in tomato using field and greenhouse selection has met with limited success due to the existence of multiple species and races of the bacterial spot pathogen and the quantitative inheritance of resistance. Identification of molecular markers linked to the resistance QTL and the application of MAS offer an opportunity to pyramid resistance in a breeding line or hybrid. However the potential of MAS has been limited for bacterial spot resistance because of the predominance of *L. esculentum* sources of resistance. The three lines Hawaii 7998, Hawaii 7981, and PI114490 are either *L. esculentum* or *L. esculentum* var *cerasiformae*. The majority of markers available for wide cross mapping are not polymorphic in crosses between elite breeding material and *L. esculentum* sources of resistance. Efforts have been made to develop molecular markers that can be used within *L. esculentum* varieties and molecular markers linked to resistance have been identified (Tables 12.2 and 12.3). By using the PCR-based DNA markers, we were able to select desirable individuals carrying the Rx3 locus on an accelerated time-scale (Yang et al. 2005). In addition, we have applied MAS to combine *Rx3* with *Pto* (Yang and Francis 2005). The experience gained in these studies will help refine strategies for developing cultivars resistant to multiple bacterial diseases. Due to the importance of phenotypic verification, genetic background effects and epistasis, it is clear that MAS strategies will be most successful when combined with population structures that facilitate replicated testing of resistance in the field.

BACTERIAL WILT

Bacterial wilt of tomato, described by Smith (1896) over 100 years ago, ranks among the most serious diseases of tomato in tropical, subtropical, and some warm temperature regions of the world (Hayward 1991). Reports of cold-tolerant strains of the bacterium capable of infecting potato and tomato in Northern Europe and North America (Janse 1996, Elphinstone 1994, Williamson et al. 2003, Kim et al. 2003, Carmeille et al. 2004) further raised concern about the spread of the disease into temperate regions. Bacterial wilt is now the major constraint on production of tomato in many parts of the world.

Bacterial wilt is caused by the soilborne bacterium *Ralstonia solanacearum* (previously *Pseudomonas solanacearum* E. F. Smith (Yabuuchi et al. 1995). *R. solanacearum* causes a systemic vascular disease on tomato. The pathogen invades tomato plant where secondary roots emerge, and infects root tips (Schmit 1978, Vasse et al. 1995). It colonizes the xylem vessels, and spreads rapidly and systemically inside the plant. Infection disrupts water transport, which causes dropping of the lower leaves and wilting of the entire plant. The destructiveness of the pathogen is attributed to the broad host range, widespread distribution, and vast genetic variation of the pathogen (Hayward 1991). The pathogen has an exceptional ability to survive in soil and in the roots of non-host plants. However, there is no evidence that the bacterium survives as an epiphyte on leaf and other plant surfaces (Kelman et al. 1994).

Ralstonia solanacearum is a complex species with a large variability, illustrated by the subdivision of the species into races, biovars and phylotypes. The pathogen varies considerably according to climate, cropping practices, soil type, and geographic location (Hayward 1991). Thousands of strains have been reported to date. They can be classified into five races, based on the host range (Buddenhagen et al. 1962), and five biovars according to their ability to utilize and/or oxidize several hexose alcohols and disaccharides (Hayward 1991). The genetic diversity of *R. solanacearum* has also been illustrated through molecular approach. Using Restriction Fragment Length Polymorphisms (RFLPs), as the basis of classification, Cook et al. (1989) separated 62 strains representing all races and biovars into two major divisions. There was a strong correlation between RFLP data and the geographical origin of strains. This classification was supported by using PCR-RFLP analysis of the *hrp* gene region, AFLP and 16S rRNA sequence (Poussier et al. 2000a), and by partial sequencing of the *hrp*B and endoglucanase genes (Poussier et al. 2000b). More recently, Prior and Fegan (2004) reported that *R. solanacearum* species complex could be subdivided into four phylotypes based on phylogenetic analysis of sequence data for 16S-23S internal transcribed spacer region. The four phylotypes correspond well to the geographical origin of strains.

Phylotype I is assigned to strains of Asian origin; phylotype II to strains from the Americas; phylotype III, to strains from Africa; and phylotype IV to strains from Indonesia. The strains that cause bacterial wilt of tomato generally belong to races 1, 2, and 3 (Allen 2004). Race 1 strains are highly diverse and these have been historically the most important causes of bacterial wilt (Scott 1997). Race 1 strains from biovar 1, 3 or 4 and from all four phylotypes have been associated with disease on tomato. More recently, cold tolerant strains classified as race 3 phylotype II have been associated with disease on tomato (Carmeille et al. 2004).

Control of the pathogen using cultural and chemical methods is not highly effective. This is in part due to the ability of *R. solanacearum* to survive in the soil for long periods of time. As the bacterial colonization in vascular tissues correlates to frequencies of wilt in the field (Grimault et al. 1995), controlling populations in the field is an important goal. However, application of soil fumigants is becoming environmentally unacceptable and the need to use resistance to control disease is becoming more pressing. Unfortunately, the development of resistant tomato varieties that are horticulturally acceptable has been problematic and there are few varieties that combine desirable fruit size and yield with resistance (Scott et al. 2004).

Techniques for screening for resistance to bacterial wilt include application of bacteria to the root zone in greenhouse screens and assessment in fields known to have a high population of bacteria and a high incidence of disease. For greenhouse evaluations, bacteria are grown on solid media, transferred to liquid media or water and adjusted to a concentration of approximately 10^8 bacteria per ml. Thirty ml of inoculum are used for 7.6 cm (approx. 3 inch) size pots. Various solid media have been used to initiate inoculum cultures, such as tetrazolium chloride (Kelman 1954, Hanson 1996), BGT (Boucher et al. 1985, Thoquet et al. 1996a), and 523 medium (Kado and Heskett 1970, Wang et al. 2000). Optical density readings of 0.3 at 600 nm correspond to 10^8 bacteria per ml for water suspensions (Wang et al. 2000) while OD_{660} readings of 0.1 correspond to 10^8 bacteria per ml for minimal media (Thoquet et al. 1996a). Studies relying on inoculum present in the open field have been reported (e.g. Mohamed et al. 1997) and variation due to unequal distribution should be accounted for through increased replication and appropriate blocking incorporated into the experimental design. One approach to field evaluation that may overcome uneven distribution is to inoculate seedlings two days prior to transplant to infected fields (Thoquet et al. 1996b).

Resistance to bacterial wilt was first reported in *Lycopersicon pimpinellifolium* PI 127805A (Acosta et al. 1964). Numerous resistant sources have been found since (Henderson and Jenkins 1972, Wang et al. 1996, Hanson et al. 1998, Sharma et al. 2002), and these are summarized by Scott (1997). It is generally believed that the resistance to bacterial wilt in cultivated tomato originated from *L. esculentum* var. *cerasiforme* or *L. pimpinellifolium* (Prior et al. 1994). Most resistance in improved *L. esculentum* lines and cultivars are derived from three major sources: PI127805A (*L. pimpinellifolium*), CRA66 (*L. esculentum* var. *cerasiforme*) and PI129080 (*L. pimpinellifolium*), and a few additional resistant sources (Hanson et al. 1998). An international collaboration to evaluate resistance determined that Hawaii 7996 contained the highest level of resistance (Wang et al. 1998). It is thought that PI 127805A is the source of resistance in Hawaii 7996 (Rouamaba et al. 1988, Mangin et al. 1999), though there remains some doubt about this origin (Scott et al. 2004).

The genetics of resistance to bacterial wilt has been studied for a number of resistant sources. In general most of these are polygenic in inheritance (Thurston 1976). Exceptions include the suggestion of a single dominant gene for bacterial wilt resistance from Hawaii 7998 (Scott et al. 1988) and Hawaii 7996 (Grimault et al. 1995). The genetics of resistance from Hawaii 7996 and related sources will be treated in more detail in this section. The dominance reported for resistance from Hawaii 7998 and Hawaii 7996 is not generalizable to all sources of resistance and all studies. The resistance in *L. pimpinellifolium* is partially dominant only at the seedling stage (Acosta 1972), while the resistance in LA 1421 (*L. esculentum*) is completely recessive (Mohamed et al. 1997).

Resistance in Hawaii 7996 has been the subject of numerous studies, using diverse strains, and quantitative models are generally invoked. Analyses using molecular markers indicate that a major QTL on chromosome 6 functions against Race 1 strains from Biovar 1 (Thoquet et al. 1996a, Thoquet et al. 1996b) and Biovar 3 (Wang et al. 2000). Analysis of resistance to strain GMI8217 (Race 1, Biovar 1, Phylotype II) in progeny derived from Hawaii 7996 x West Virginia 700 identified a QTL on chromosome 6, two QTL on chromosome 4, and a putative QTL on chromosome 10, all of which account for 30 to 56% of the phenotypic variation observed in Hawaii 7996 (Thoquet et al. 1996a, Fig. 12.1). Evaluation of the Hawaii 7996 x West Virginia 700 population using strain GMI8217 in the field led to the detection of six QTL conferring resistance to bacterial wilt (Thoquet et al. 1996b). In addition to the QTL identified in the greenhouse, they found new QTL on chromosomes 3 and 8 (Fig. 12.1).

Subsequent studies using this population-strain combination suggest that the QTL detected on chromosome 6 can be resolved into two linked QTL (Mangin et al. 1999). More recently, Wang et al. (2000) confirmed that QTL from Hawaii 7996 on chromosome 6 confer resistance to strain Pss4 (Race 1, Biovar 3, phylotype I). They also found a strain-specific locus on chromosome 12 (Wang et al. 2000, Fig. 12.1). It is interesting to note that a major QTL on chromosome 6 conferring resistance to a strain UW364 (Race 1, Biovar 3, Phylotype I) was found in a population derived from L285, a resistant wild tomato (*L. esculentum* var. *cerasiforme*; Danesh et al. 1994). QTL on chromosomes 7 and 10 were also identified in L285 (Fig. 12.1). Although the importance of a QTL or linked QTL on chromosme 6 emerges from these studies, the importance of other QTL and strain specific responses cannot be overlooked.

Breeding for resistance to bacterial wilt has been very difficult due to the extensive variation of pathogen, environmental effects on screening, and linkage of resistance QTL to small fruit size, among other factors. Cultivars and breeding lines that appear resistant in one location may be susceptible at other locations (Hanson et al. 1996, Prior et al. 1994, Wang et al. 1998) and recovery of desirable horticultural traits with high level of resistance has been elusive (Scott et al. 2004). One strategy may be to pyramid resistance by crossing *L. pimpinellifolium* (e.g. Hawaii 7996) and *L. esculentum* var. *cerasiforme* (e.g. L285) resistance sources. The Hawaii 7996 x L285 cross would serve to concentrate on non-chromosome 6 QTL, though it would not facilitate the improvement of fruit size and yield in resistant material. Pyramiding approaches to be successful will require the genetic characterization of other sources of resistance using molecular markers. Selecting the appropriate sources for further characterization also remains problematic. For example, BF Okitsu, which ranked highly in trials world-wide derives its resistance from North Carolina NC 19/53-64N (Wang et al. 1998), but it is possible that BF Okitsu derives its resistance from the Hawaii material (Scott et al. 2004). Thus it is not entirely clear which sources of resistance will maximize new information. Another strategy to develop cultivars with acceptable fruit size and yield would be to concentrate on introgression of Hawaii 7996 QTL while breaking up undesirable linkages. Markers summarized in Tables 12.2 and 12.3 may facilitate this approach, but linkage of QTL with a large number of other potential QTL is likely to slow the progress. In the end, it will be necessary to develop populations that favor replicated testing in order to confirm resistance to multiple strains in multiple environments.

CONCLUSION AND FUTURE PROSPECTS

The genetic analysis of resistance to diverse bacterial pathogens in tomato has revealed several common themes. Sources of resistance to the important diseases treated in this review have been identified, yet pathogen diversity, quantitative inheritance of resistance, and low levels of polymorphism between resistant germplasm and elite lines have slowed progress in deploying resistance in commercially elite varieties. For many of the diseases, the recovery of fruit size and yield in a resistant background remains a barrier. A further complication is the dilemma faced by those seeking to exploit MAS. Wide crosses maximize genetic polymorphism, but carry a much higher cost in terms of desirable fruit size, yield and quality. Although it is theoretically possible to reduce the negative effects of the donor through repeated marker-assisted introgression, the progress in practice has been slow. Crosses between resistant *L. esculentum* sources and elite lines do not produce progeny with acceptable horticultural traits, and lack the necessary polymorphism to make MAS an efficient strategy. We expect progress in developing appropriate markers for exploiting variation within *L. esculentum* since current genome sequencing projects have opened up new opportunities for marker discovery.

The potential role played by epistatic interactions and of genetic background effects in maximizing resistance also appears to be an important theme. It is difficult to determine whether methodology has contributed to different genetic interpretations, but our experience indicates that genetic background is an important consideration. Given these observations, it is disappointing that genetic analyses of bacterial resistance using molecular markers have not been conducted with more diverse germplasm. Despite numerous genetic studies, there is a need to characterize further sources of resistance.

Another theme that emerges from the review of these bacterial diseases is the tendency for resistant germplasm to be more effective against multiple diseases (Laterrot and Kaan 1978, Scott et al. 2004). Noteworthy varieties and breeding lines include Campbell 28 (LA 3317) with partial resistance to bacterial speck (Gitaitis et al. 1985), bacterial spot (e.g. Somodi et al. 1994) and bacterial canker (Dick and MacNeill 1981). The breeding line Hawaii 7998 has resistance to T1 strains of bacterial spot (Jones and Scott 1985), partial resistance to bacterial wilt (Scott et al. 1988), and partial resistance to the foliar phase of bacterial canker (Randy Gardner, N. Carolina State University, personal communication). Development of inbred backcross populations from varieties with multiple resistance would facilitate genetic analysis for multiple diseases in replicated trials. This population structure is less than ideal for detecting epistatic interactions, although alternatives that balance the needs of genetic studies with the needs of breeding programs should be considered.

Though the lack of genetic polymorphism within *L. esculentum* has prevented a detailed study of many economically important traits, including bacterial diseases in cultivated tomato, we still believe that MAS is a promising approach to breeding for disease resistance. Efforts have been made to discover PCR-based DNA markers that can be used within *L. esculentum*. We estimate that genomic approaches to marker discovery could yield thousands of makers that can be used within cultivated tomato germplasm (Yang et al. 2004). Although whole genome coverage may be difficult within individual breeding programs with regard to the fixation of loci, marker coverage sufficient to exploit the multiple *L. esculentum* sources of bacterial resistance should be a reality.

MAS as a tool is not a substitute for field evaluation; it can be best used for rapid generation turn-over when alternated with cycles of field selection. A strategy that combines MAS selection with populations that facilitate replicated field trials offers an opportunity to pyramid resistance in a single line or hybrid. Cycles of field selection alternated with cycles of MAS could facilitate the development of resistant germplasm on an accelerated time-scale. Using this strategy, we were able to select and develop breeding lines with reduced introgressions surrounding bacterial canker resistance QTL (Coaker and Francis 2004) and lines with resistance to race T1 of bacterial spot (Yang et al. 2005).

As part of a gene pyramiding strategy for bacterial resistance, it will be necessary to consider more than a single disease. To effectively pyramid resistance to different pathogens, it is necessary to select for desirable recombination events in addition to bringing unlinked genes together. MAS provides an advantage in this situation. With the markers recently developed, we were able to combine the gene *Pto* conferring resistance to race 0 of bacterial speck and a locus Rx3 conferring resistance to race T1 of bacterial spot (Yang et al. 2005).

The experiences gained in our studies show that MAS is promising in tomato bacterial disease resistance breeding. As discovery of molecular markers continues, we expect that MAS will be widely used to facilitate the development of tomato varieties with improved resistance.

SUMMARY

Four major bacterial diseases, bacterial canker caused by *Clavibacter michiganensis* subsp. *michiganensis*, bacterial speck caused by *Pseudomonas syringae* pv. *tomato*, bacterial spot caused by as many as four *Xanthomonas* species, and bacterial wilt caused by *Ralstonia solanacearum*, have received breeding attention for decades due to their impact on tomato (*Lycopersicon esculentum* Mill.) production and quality. Despite various breeding efforts, control of bacterial diseases through resistance has been restricted to the wide use of a single gene, *Pto*, which confers resistance to bacterial speck race 0. Breeding for resistance to the other three diseases has been slowed by the existence of multiple races and the quantitative inheritance of resistance. Marker-assisted selection offers an opportunity to circumvent some of the problems associated with phenotypic selection for resistance to multiple bacterial pathogens and races. Previous reviews provide a thorough background of these diseases, their genetic basis of resistance and advances in breeding programs up to 1997. During the intervening years, progress has been made to identify molecular markers linked to genes conferring resistance to bacterial canker, bacterial spot and bacterial wilt. The combination of new tools with renewed efforts to use the power of traditional breeding is beginning to demonstrate progress. This chapter reviews the progress made and also discusses the promise marker-assisted selection holds for breeding for resistance to bacterial diseases.

REFERENCES

Acosta, J.C. 1972. Genetic analysis for bacterial wilt resistance in a tomato cross, *Lycopersicon esculentum* Mill. ´ *L. pimpinellifolium* Mill. Proc. Third Annu Meeting Crop Sci Soc Philippines, pp. 183-190.

Acosta, J.C., J.C. Gilbert, and V.L. Quinon. 1964. Heritability of bacterial wilt resistance in tomato. Proc Am Soc Hort Sci 84: 455-462.

Alexander, S.A., S.H. Kim, and C.M. Waldenmaier. 1999. First report of copper-tolerant *Pseudomonas syringae* pv. *tomato* in Virginia. Plant Dis 83: 964.

Allen, C. 2004. Bacterial wilt: worldwide distribution, importance, and review. 1st International Symposium on Tomato Diseases and 19th Annual Tomato Disease Workshop. Orlando, Florida, USA June 20-24, 2004, p.22.

Ammiraju, J.S.S., J.C. Veremis, X. Huang, P.A. Roberts, and I. Kaloshian. 2003. The heat-stable root-knot nematode resistance gene *Mi-9* from *Lycopersicon peruvianum* is localized on the short arm of chromosome 6. Theor Appl Genet 106 (3): 478-484.

Anand, N., A.T. Sadashiva, S.K. Tikoo, Ramkishun, and K. M. Reddy. 1993. Resistance to bacterial wilt in tomato: gene dosage effects. In: G. L. Hartman and A. C. Haywood (eds), Bacterial wilt. Proceedings of an international conference held at Kaohsiung, Taiwan, 28-31 Oct. 1992. ACIAR Proc No 45: 142-148.

Ark, P.A. 1944. Studies on bacterial canker of tomato. Phytopathology 34: 394-400.

Arredondo, C.R. and R.M. Davis. 2000. First report of *Pseudomonas syringae* pv. *tomato* race 1 on tomato in California. Plant Dis 84: 371.

Astua-Monge, G., G.V. Minsavage, R.E. Stall, C.E. Vallejos, M.J. Davis, and J.B. Jones. 2000. *Xv4-vrxv4*: A new gene-for-gene interaction identified between *Xanthomonas campestris* pv. *vesicatoria* race T3 and the wild tomato relative *Lycopersicon pennellii*. Mol Plant-Microbe Interact 13: 1346-1355.

Bai, Y.L., C.C. Huang, R. van der Hulst, F. Meijer-Dekens, G. Bonnema, and P. Lindhout. 2003. QTLs for tomato powdery mildew resistance (*Oidium lycopersici*) in *Lycopersicon parviflorum* G1.1601 co-localize with two qualitative powdery mildew resistance genes. Mol Plant-Microbe Interact 16 (2): 169-176.

Ballvora, A., M. Pierre, G. van den Ackerveken, S. Schornack, O. Rossier, M. Ganal, T. Lahaye, and U. Bonas. 2001. Genetic mapping and functional analysis of the tomato *Bs4* locus governing recognition of the *Xanthomonas campestris* pv. *vesicatoria* AvrBs4 protein. Mol Plant-Microbe Interact 14: 629-638.

Bashen, Y.Y., Y. Okon, and Y. Henis. 1978. Infection studies of *Pseudomonas tomato*, causal agent of bacterial speck of tomato. Phytoparasitica 6: 135-243.

Behare, J., H. Laterrot, M. Sarfatti, and D. Zamir. 1991. Restriction-fragment-length-polymorphism mapping of the *stemphylium* resistance gene in tomato. Mol Plant-Microbe Interact 4 (5): 489-492.

Bender, C.L. and D.A. Cooksey. 1986. Indigenous plasmids in *Pseudomonas syringae* pv. *tomato*: Conjugative transfer and role in copper resistance. J Bacteriol 165: 534-541.

Bender, C.L., D.K. Malvick, K.E. Conway, S. George, and P. Pratt. 1990. Characterization of pXV10A, a copper resistance plasmid in *Xanthomonas campestris* pv. *vesicatoria*. Appl Environ Microbiol. 56: 170-175.

Berry, S.Z. and W. A. Gould. 1979. Ohio 7663, New tomato variety for machine harvest. Ohio Agricultural Research and Development Center Report 64: 22-23.

Berry, S.Z. and W. A. Gould. 1982. 'Ohio 7870' Tomato. HortScience 17: 266-267.

Berry, S.Z., G.G. Madumadu, M.R. Uddin, and D.L. Coplin. 1989. Virulence studies and resistance to *Clavibacter michiganensis* ssp. *michiganensis* in tomato germplasm. HortScience 24: 362-365.

Berry, S.Z., W. A. Gould, and K. L. Wiese. 1991. 'Ohio 8245' Processing Tomtato. HortScience 26: 1093.

Bonn, W.G., R.D. Gitaitis, and B.H. MacNeill. 1985. Epiphytic survival of *Pseudomonas syringae* pv. *tomato* on tomato transplants shipped from Georgia. Plant Dis 69: 58-60.

Boucher, C. A., P. A. Barberis, A. P. Trigalet, D. A. Demery. 1985. Transposon mutagenisis of *Pseudomonas solanacearum*: isolation of Tn5-induced avirulent mutants. J Gen Microbiol 131:2449-2457.

Bournival, B.L., C.E. Vallejos, and J.W. Scott. 1989. Genetic analysis of resistance to races 1 and 2 of *Fusarium oxysporum* f.sp. *lycopersici* from the wild tomato *Lycopersicon pennellii*. Theor Appl Genet 79: 641-645.

Bouzar, H., J.B. Jones, G.V. Minsavage, R.E. Stall, and J.W. Scott. 1994. Proteins unique to phenotypically distinct groups of *Xanthomonas campestris* pv. *vesicatoria* revealed by silver staining. Phytopathology 84: 39-44.

Brouwer, D.J. and D.A. St Clair. 2004. Fine mapping of three quantitative trait loci for late blight resistance in tomato using near isogenic lines (NILs) and sub-NILs..Theor Appl Genet 108 (4): 628-638.

Browning, J.A., and K.J. Frey. 1969. Multiline cultivars as a means of disease control. Ann Rev Phytopathol 7: 355-382.

Buddenhagen, I.W., L. Sequeria, and A. Kelman. 1962. Designation of races of *Pseudomonas solanacearum*. Phytopathology 2: 203-230.

Buonaurio, R., V.M. Stravato, and C. Cappelli. 1996. Occurrence of *Pseudomonas syringae* pv. *tomato* race 1 in Italy on *Pto* gene-bearing tomato plants. J Phytopathol 144: 437-440.

Carland, F.M., and B.J. Staskawicz. 1993. Genetic-characterization of the Pto locus of tomato - semi-dominance and cosegregation of resistance to pseudomonas-syringae pathovar tomato and sensitivity to the insecticide fenthion. Mol Gen Genet 239: 17-27.

Carmeille, A., F. Chiroleu, J. Dintinger, P. Prior, C. Caranta, P. Besse, H. Kodja, and J. Luisetti. 2004. Evaluation of resistance to a race 3 strain of *Ralstonia solanacearum* in tomato germplasm. 1st International Symposium on Tomato Diseases and 19th Annual Tomato Disease Workshop. Orlando, Florida, USA. June 20-24, 2004. p.23.

Chague, V., J.C. Mercier, M. Guenard, A. deCourcel, and F. Vedel. 1997. Identification of RAPD markers linked to a locus involved in quantitative resistance to TYLCV in tomato by Bulked Segregant Analysis. Theor Appl Genet (4): 671-677.

Chambers, S.C., and P.R. Merriman. 1975. Perennation and control of *Pseudomonas tomato* in Victoria. Aust J Agric Res 26: 657-663.

Chang, R.J., S.M. Ries, and J.K. Pataky. 1991. Dissemination of *Clavibacter michiganensis* subsp. *michiganensis* by practices used to produce tomato transplants. Phytopathology 81: 1276-1281.

Chang, R.J., S.M. Ries, and J.K. Pataky. 1992. Reductions in yield of processing tomato and incidence of bacterial canker. Plant Dis 76: 805-809.

Chunwongse, J., C. Chunwongse, L. Black, and P. Hanson. 2002. Molecular mapping of the *Ph-3* gene for late blight resistance in tomato. J Hort Sci Biotechnol 77 (3): 281-286.

Chunwongse, J., T.B. Bunn, C. Crossman, J. Jiang, and S.D. Tanksley. 1994. Chromosomal localization and molecular marker tagging of the powdery mildew resistance gene (*lv*) in tomato. Theor Appl Genet 89 (1): 76-79.

Coaker G., B. Willard, M. Kinter, E.J. Stockinger, and D. Francis. 2004. Proteomic analysis of resistance to bacterial canker of tomato. Mol Plant Microbe Interact 17: 1019-1028.

Coaker, G.L. and D.M. Francis. 2004. Mapping, genetic effects, and epistatic interaction of two bacterial canker resistance QTLs from *Lycopersicon hirsutum*. Theor Appl Genet 108: 1047-1055.

Conlin, K.C., and S.M. McCarter. 1983. Effectiveness of selected chemicals in inhibiting *Pseudomonas syringae* pv. *tomato* in vitro and in controlling bacterial speck. Plant Dis 67: 639-644.

Cook, D., E. Barlow, and L. Sequeira. 1989. Genetic diversity of *Pseudomonas solanacearum*: detection of restriction fragment length polymorphisms that specify virulence and the hypersensitive response. Mol Plant-Microbe Interact 2: 113-121.

Cooksey, D. A. 1987. Characterization of a copper resistance plasmid conserved in copper-resistant strains of *Pseudomonas syringae* pv. *tomato*. Appl Environ Microbiol 53: 454-456.

Danesh, D., S. Aarons, G.E. McGill, and N.D. Young. 1994. Genetic dissection of oligogenic resistance to bacterial wilt in tomato. Mol Plant-Microbe Interact 7: 464-471.

Davis, M.J., A.G. Gillaspie, A.K. Vidaner Jr., and R.W. Harris. 1984. *Clavibacter*: A new genus containing some phytopathogenic coryneform bacteria, including *Clavibacter xyli* subsp. *xyli* sp. nov., subsp. nov. and *Clavibacter xyli* subsp. *cynodontis* subsp. nov., pathogens that cause ratoon stunting disease of sugarcane and bermudagrass stunting disease. Int J Syst Bacteriol 34: 107-117.

De Giovanni, C., P. Dell' Orco, A. Bruno, F. Ciccarese, C. Lotti, and L. Ricciardi. 2004. Identification of PCR-based markers (RAPD, AFLP) linked to a novel powdery mildew resistance gene (*ol-2*) in tomato. Plant Sci 166 (1): 41-48.

De Jong, J., and S. Honma. 1976. Inheritance of resistance to *Corynebacterium michiganense* in the tomato. J Hered 67: 79-84.

Dick, J. and B. H. MacNiell. 1981. A differential response to *Corynebacterium michiganense* in tomato seedlings. Phytopathology (Abstract) 71: 214.

Diwan, N., R. Fluhr, Y. Eshed, D. Zamir, and S.D. Tanksley. 1999. Mapping of Ve in tomato: a gene conferring resistance to the broad-spectrum pathogen *Verticillium dahliae* race 1. Theor Appl Genet 98 (2): 315-319.

Doganlar, S., J. Dodson, B. Gabor, T. Beck-Bunn, C. Crossman, S.D. Tanksley. 1998. Molecular mapping of the *py-1* gene for resistance to corky root rot (Pyrenochaeta lycopersici) in tomato. Theor Appl Genet 97 (5-6): 784-788.

Doidge, E.M. 1921. A tomato canker. Ann Appl Biol 7: 407-430.

Donner, S.C. and S.J. Barker. 1996. *Pto* resistance will not be effective in Australia. (Abstr.) 8th Int Congr Mol Plant-Microbe Interact Knoxville, TN.

Dreier, J., D. Meletzus, and R. Eichenlaub. 1997. Characterization of the plasmid endoded virulence region pat-1 of phytopathogenic *Clavibacter michiganensis* subsp. *michiganensis*. Mol Plant-Microbe Interact 10: 195-206.

Dye, D.W., J.F. Bradbury, M. Goto, A.C. Hayward, R.A. Lelliott, and M.N. Scroth. 1980. International standards for naming pathovars of phytopathogenic bacteria and a list of pathovar names and pathotype strains. Rev Plant Pathol 59: 153-168.

Elenkov, E. 1965. Die Selektion von tomaten auf resistenz gegen die bakterienwelke. Int Z Landwirtsch 594-597.

Elphinstone, J. G. 1994. Virulence of isolates of *Pseudomonas solanaceearum* from worldwide sources on resistant and susceptible tomato cultivars. Institute National de la Recherche Agronomique (ed.), Proc 8th Intl Conf Plant Pathogenic Bacteria. Institut National de Recherche Agronomique, Paris, pp 599-604.

Emmatty, D.A., and C.A. John. 1973. Comparison of yield loss to bacterial canker of tomato in a resistant and a susceptible variety. Plant Dis Rep 57: 787-788.

Emmatty, D.A., M.D. Scott, and B.F. George. 1982. Inoculation technique to screen for bacterial speck resistance of tomatoes. Plant Dis 66: 993-994.

Fallik, E., Y. Bashan, Y. Okon, A. Cahaner, and N. Kedar. 1983. Inheritance and sources of resistance to bacterial speck of tomato caused by *Pseudomonas syringae* pv. *tomato*. Ann Appl Biol 102: 365-371.

Fatmi, M., N. W. Schaad, H. A. Bolkan. 1991. Seed treatment for eradicating *Clavibacter michiganensis* subsp. *michiganensis* from naturally infected tomato seeds. Plant Dis 75:383-385.

Foster, R.E., R. X. Latin and S.C. Weller. 1993. Pesticide use on processing tomatoes grown in Indiana. Purdue Univ Coop Ext Bull ID-193 (1160).

Francis, D. M. and S. A. Miller. 2004. Ohio 9834 and Ohio 9816: processing tomato breeding lines with partial resistance to race T1 of bacterial spot. Rep Tomato Genet Coop. 54: 49.

Francis, D. M., A. R. Miller, Z. Chen, A. M. Bongue Bartelsman, C. A. Barringer. 2003. State of the art of genetics and breeding of processing tomato: a comparison of selection based on molecular markers, biochemical pathway, and phenotype for the improvement of fruit color and juice viscosity. Acta Hort 613: 273-282.

Francis, D.M., E. Kabelka, J. Bell, B. Franchino, and D. St. Clair. 2001. Resistance to bacterial canker in tomato (*Lycopersicon hirsutum* LA407) and its progeny derived from crosses to *L. esculentum*. Plant. Dis 85:1171-1176.

Francis, DM, TS Aldrich, SZ Berry, KL Scaife, BS Schult, and WD Bash. 1995. Evaluation of processing tomato breeding lines and cultivars for mechanical harvesting and quality in 1995. Horticulture and Crop Science Dept, Series No. 648.

Frary, A., E. Graham, J. Jacobs, R.T. Chetelat, and S.D. Tanksley. 1998. Identification of QTL for late blight resistance from *L. pimpinellifolium* L3708. Rep Tomato Genetics Coop 48: 19-21.

Ganal, M.W., R. Simon, S. Brommonschenkel, S.D. Tanksley, and A. Kumar. 1995. Genetic mapping of a wide spectrum nematode resistance gene (*Hero*) against *Globodera rostochiensis* in tomato. Mol Plant-Microbe Interact 8: 886-891.

Gardner, M.W. and J.B. Kendrick. 1921. Bacterial spot of tomato. J Agr Res 21: 123-156.

Gartemann, K.H., O. Kirchner, J. Engemann, I. Grafen, R. Eichenlaub, and A. Burger. 2003. *Clavibacter michiganensis* subsp. Michiganensis: first steps in the understanding of virulence of a Gram-positive phytopathogenic bacterium. J Biotech 106: 179-191.

Gitaitis, R.D., J.B. Jones, C.A. Jaworski, and S.C. Phatak. 1985. Incidence and development of *Pseudomonas syringae* pv. *syringae* on tomato transplants in Georgia. Plant Dis 69: 32-35.

Gleason, L.G., R.D. Gitaitis, and M.D. Ricker. 1993. Recent progress in understanding and controlling bacterial canker in tomato in eastern North America. Plant Dis 77: 1069-1076.

Goode, M.J. and M. Sasser. 1980. Prevention—the key to controlling bacterial spot and bacterial speck of tomato. Plant Dis 64: 831-834.

Grimault, V., P. Prior, and G. Anaïs. 1995. A monogenic dominant resistance of tomato to bacterial wilt in Hawaii 7996 is associated with plant colonization by *Pseudomonas solanacearum*. J Phytopathol 143: 349-352.

Grogan, R. C. and J. B. Kendrick. 1953. Seed transmission, mode of overwintering and spread of bacterial canker of tomato caused by *Corynebacterium michiganenese*. Phytopathology (Abstract) 43: 473.

Haanstra, J.P.W., R.Lauge, F. Meijer-Dekens, G. Bonnema, P.J.G.M. de Wit, and P. Lindhout. 1999. The *Cf-ECP2* gene is linked to, but not part of, the *Cf-4/Cf-9* cluster on the short arm of chromosome 1 in tomato. Mol Gen Genet 262 (4-5): 839-845.

Hanson, P.M., D. Bernacchi, S. Green, S.D. Tanksley, V. Muniyappa, S. Padmaja, H.M. Chen, G. Kuo, D. Fang, and J.T. Chen. 2000. Mapping a wild tomato introgression associated with tomato yellow leaf curl virus resistance in a cultivated tomato line. J Am Soc Hort Sci 125 (1): 15-20.

Hanson, P.M., J.F. Wang, O. Licardo, Hanudin, S.Y. Mah, G.L. Hartman, Y.C. Lin, and J.T. Chen. 1996. Variable reaction of tomato lines to bacterial wilt evaluated at several locations in Southeast Asia. HortScience 31: 143-146.

Hanson, P.M., O. Licardo, Hanudin, J.F. Wang, and J.T. Chen. 1998. Diallel analysis of bacterial wilt resistance in tomato derived from different sources. Plant Dis 82: 74-78.

Hassan, A.A., D.L. Strider, and T.R. Konsler. 1968. Application of cotylendonary symptoms in screening for resistance to tomato bacterial canker and in host range studies. Phytopathology 58: 233-239.

Hausbeck, M.K., J. Bell, C. Medina-Mora, R. Podolsky, and D.W. Fulbright. 2000. Effect of bactericides on population sizes and spread of *Clavibacter michiganensis* subsp. *michiganensis* on tomatoes in the greenhouse and on disease development and crop yield in the field. Phytopathology 90: 38-44.

Hayward, A.C. 1991. Biology and epidemiology of bacterial wilt caused by *Pseudomonas solanacearum*. Ann Rev Phytopathol 29: 65-87.

Hemming, M.N., S. Basuki, D.J. McGrath, B.J. Carroll, D.A. Jones. 2004. Fine mapping of the tomato *I-3* gene for fusarium wilt resistance and elimination of a co-segregating resistance gene analogue as a candidate for *I-3*. Theor Appl Genet 109 (2): 409-418.

Henderson, W.R., and S.F. Jenkins, Jr. 1972. 'Venus' and 'Saturn' tomato varieties resistant to southern bacterial wilt. HortScience 7: 346.

Hibberd, A.M., J.B. Heaton, G.P. Finlay, and S.R. Dullahide. 1992. A greenhouse method for selecting tomato seedlings resistance to bacterial canker. Plant Dis 76: 1004-1007.

Horsfall, J. G. and Barratt, R. W. 1945. An improved grading system for measuring plant diseases. Phytopathology 35:655.

Huang, R. and J. C. Tu. 2001. Effects of nutrient solution pH on the survival and transmission of *Clavibacter michiganensis* ssp. *michiganensis* in hydroponically grown tomatoes. Plant Pathology 50: 503-508

Janse, J. 1996. Extermination of brown rot is a good option. Potato Leaves 2: 2-3.

Jones, J. B., R.E. Stall, G.C. Somodi, H. Bouzar, N. C. Hodge. 1995. A third tomato race of *Xanthomonas campestris* pv. *vesicatoria*. Plant Dis 79: 395-398.

Jones, J.B., and J.W. Scott. 1985. The hypersensitive response in tomato to *Xanthomonas campestris* pv. *vesicatoria*. Phytopathology 75: 1281.

Jones, J.B., G.H. Lacy, H. Bouzar, G.V. Minsavage, R.E. Stall, and N.W. Schaad. 2004a. Bacterial spot – World wide distribution, importance and review. 1st International Symposium on Tomato Diseases and 19th Annual Tomato Disease Workshop. Orlando, Florida, USA. June 20-24, 2004. p.21.

Jones, J.B., G.H. Lacy, H. Bouzar, R.E. Stall, and N.W. Schaad. 2004b. Reclassification of the xanthomonads associated with bacterial spot disease of tomato and pepper. Syst and Appl Microbiol 27:755-762.

Jones, J.B., H. Bouzar, R.E. Stall, E.C. Almira, P.D. Roberts, B.W. Bowen, J. Sudberry, P.M. Strickler, and J. Chun. 2000. Systematic analysis of xanthomonads (*Xanthomonas* spp.) associated with pepper and tomato lesions. Intl. J Syst Evol Microbiol 50:1211-1219.

Jones, J.B., J.P. Jones, R.E. Stall, and T.A. Zitter (eds). 1991. Compendium of tomato diseases. APS Press, Minnesota.

Kabelka, E., B. Franchino, and D.M. Francis. 2002. Two loci from *Lycopersicon hirsutum* LA407 confer resistance to strains of *Clavibacter michiganensis* subsp. *michiganensis*. Phytopathology 92: 504-510.

Kado, C. I. and M. G. Heskett. 1970. Selective media for isolation of *Agrobacterium*, *Corynebacterium, Erwinia, Pseudomonas*, and *Xanthomonas*. Phytopathology 60:696-676.

Kearney, B. and B. J. Staskawicz. 1990. Characterization of IS476 and its role in bacterial spot disease of tomato and pepper. J Bacteriol. 172:143-148. Erratum *In:* J Bacteriol. 172:2199.

Kelman, A. 1954. The relationship of pathogenicity in *Pseudomonas solanacearum* to colony appearance on a tetrazolium medium. Phytopathology 44: 693-695.

Kelman, A., G.L. Hartman, and A.C. Hayward. 1994. Introduction. Pages: 1-7 in: Bacterial Wilt: The Disease and Its Causative Agent *Pseudomonas solanacearum*. A. C. Hayward and G. L. Hartman, eds. CAB International, Wallingford, England.

Kim, S. H., T. N. Olson, W. Schaad, and G. W. Moorman. 2003. *Ralstonia solanacearum* Race 3, Biovar 2, the Causal Agent of Brown Rot of Potato, Identified in Geraniums in Pennsylvania, Delaware, and Connecticut. Plant Dis. 87:450

Krause, R., C. Kurozawa, J.W. Scott, and A. Cataneo. 2001. Evaluation of the response of tomato genotypes to bacterial speck. Summa Phytopathologica 27: 60-62.

Kruger, J., C.M. Thomas, C. Golstein, M.S. Dixon, M. Smoker, S.K. Tang, L. Mulder, and J.D.G. Jones. 2002. A tomato cysteine protease required for Cf-2-dependent disease resistance and suppression of autonecrosis. Science 296 (5568): 744-747.

Laterrot, H. 1974. Value of the resistance of the 8/12 Bulgarian tomato to *Corynebacterium michiganense* (E. F. Smith) jesen. Eucarpia Tomato Conf. Bari, Italy

Laterrot, H. 1984. Specific resistance of *Verticillium dahlias* race 2 in tomato. Rept Tomato Genet Coop 34:10.

Laterrot, H. and A. Moretti. 1989. Linkage between *Pto* and susceptibility to fenthion. Rpt Tomato Genet Coop 39: 21-22.

Laterrot, H. and F. Kaan. 1978. Resistance to *Corynebacterium michiganense* of lines bred fro resistance to *Pseudomonas solanacearum*. Rpt Tomato Genet Coop 28: 7-8.

Laterrot, H. and Moretti, 1995. A Confirmation of the linkage Tm-2^2, ah, Frl. Rept Tomato Genet Coop 45: 29.

Laterrot, H. and Moretti, A. 1992. *Pseudomonas* tomato resistance in *Lycopersicon hirutum* without Fenthion necrosis. Rep Tomato Genetics Cooperative 42: 26.

Laterrot., H., R. Brand, and M.-C. Danway. 1978. Etude bibliographique: La resistance a *Corynebacterium michiganense* chez la tomate. Ann Amelior Plantes 28: 579-591.

Laterrot, H. 1992. Creation of the population COCABUL with *Corynebacterium* resistance. Rep Tomato Genet Coop 42: 25.

Lauge, R., P.H. Goodwin, P.J.G.M. de Wit, and M.H.A.J. Joosten. 2000. Specific HR-associated recognition of secreted proteins from *Cladosporium fulvum* occurs in both host and non-host plants. Plant J 23 (6): 735-745.

Lawson, V.F. and W.L. Summer. 1984. Resistance to *Pseudomonas syringae* pv. *tomato* in wild *Lycopersicon* species. Plant Dis 68: 139-141.

Lawton, M.B. and B.H. MacNeill. 1986. Occurrence of race 1 of *Pseudomonas syringae* pv. *tomato* on field tomato in southwestern Ontario. Can J Plant Pathol 8: 85-88.

Lelliot, R. A. and Stead, D. E. 1987. Methods for the diagnosis of bacterial diseases of plants. Blackwell Scientific Publication Ltd, Oxford.

Levesque, H., F. Vedel, C. Mathieu, and A.G.L. Decourcel. 1990. Identification of a short rdna spacer sequence highly specific of a tomato line containing *Tm-1* gene introgressed from *Lycopersicon-hirsutum*. Theor Appl Genet 80 (5): 602-608.

Lindhout, P., C. Purimahua, and A. van der Giessen. 1987. New resistance to bacterial canker in the wild tomato species *Lycopersicon peruvianum*. Prophyta 41(5):100-102.

Louws F.J., D.W. Fulbright, C.T. Stephens, and F.J. Debruijn. 1994. Specific genomic fingerprints of phytopathogenic *Xanthomonas* and *Pseudomonas* pathovars and strains generated with repetitive sequences and PCR. Appl Environ Microbiol 60: 2286-2295.

Louws F.J., D.W. Fulbright, C.T. Stephens, and F.J. Debruijn. 1995. Differentiation of genomic structure by rep-PCR fingerprinting to rapidly classify *Xanthomonas-campestris* pv *vesicatoria*. Phytopathology 85: 528-536.

Louws, F.J., J. Bell, C.M. Medina-Mora, C.D. Smart, D. Opgenorth, C.A. Ishimaru, M.K. Hausbeck, F.J. de Bruijn, and D.W. Fulbright. 1998. rep-PCR-medicated genomic fingerprinting: A rapid and effective method to identify *Clavibacter michiganensis*. Phytopathology 88: 862-868.

Mangin, B., P. Thoquet, J. Olivier, and N.H. Grimsley. 1999. Temporal and multiple quantitative trait loci analyses of resistance to bacterial wilt in tomato permit the resolution of linked loci. Genetics 152: 1165-1172.

Martin G. B., A. Frary, T. Wu, S. Brommonschenkel, J. Chunwongse, E. D. Earle, and S. D. Tanksley. 1994 A member of the tomato Pto gene family confers sensitivity to fenthion resulting in rapid cell death. Plant Cell 6:1543-1552.

Martin, G.B., M. Carmen de Vicente, and S.D. Tanksley. 1993b. High-resolution linkage analysis and physical characterization of the *Pto* bacterial resistance locus in tomato. Mol Plant-Microbe Interact 6: 26-34.

Martin, G.B., S.H. Brommonschenkel, J. Chunwongse, A. Frary, M.W. Ganal, R. Spivey, T. Wu, E.D. Earle, and S.D. Tanksley. 1993a. Map-based cloning of a protein kinase gene conferring disease resistance in tomato. Science 262: 1432-1436.

McCarter, S.M., J.B. Jones, R.D. Gitaitis, and D.R. Smitley. 1983. Survival of *Pseudomonas syringae* pv. *tomato* in association with tomato seed, soil, host tissue and epiphytic weed hosts in Georgia. Phytopathology 73: 1393-1398.

Medina-Mora, C.M., M.K. Hausbeck, and D.W. Fulbright. 2001. Bird's eye lesions of tomato fruit produced by aerosol and direct application of *Clavibacter michiganensis* subsp. *michiganensis*. Plant Dis 85: 88-91.

Meletzus, D., A. Bermpohl, J. Dreier, and R. Eichenlaub. 1993. Evidence for plasmid encoded virulence factors in the phytopathogenic bacterium *Clavibacter michiganensis* subsp. *michiganensis* NCPPB382. J Bacteriol 175: 2131-2136.

Miller, J.C. and S.D. Tanksley. 1990. RFLP analysis of phylogenetic relationships and genetic variation in the genus *Lycopersicon*. Theor Appl Genet 80:437-448.

Minsavage, G.V., J.B. Jones, and R.E. Stall. 1996. Cloning and sequencing of an avirulence gene (*avrRxv3*) isolated from *Xanthomonas campestris* pv. *vesicatoria* tomato race 3. Phytopathology (Abstract) 86: S15.

Mohamed, M.El.S., P. Umaharan, and R.H. Phelps. 1997. Genetic nature of bacterial wilt resistance in tomato (*Lycopersicon esculentum* Mill.) accession LA 1421. Euphytica 96: 323-326.

Moreau, P., P. Thoquet, J. Olivier, H. Laterrot, and N. Grimsley. 1998. Genetic mapping of *Ph-2*, a single locus controlling partial resistance to *Phytophthora infestans* in tomato. Mol Plant-Microbe Interact 11 (4): 259-269.

National Academy of Sciences. 1972. Genetic vulnerability of major crops. National Academy of Sciences USA, Washington, D. C.

Parrella, G., S. Ruffel, A. Moretti, C. Morel, A. Palloix, and C. Caranta. 2002. Recessive resistance genes against potyviruses are localized in colinear genomic regions of the tomato (*Lycopersicon* spp.) and pepper (*Capsicum* spp.) genomes. Theor Appl Genet 105 (6-7): 855-861.

Pilowsky, M., and D. Zutra. 1986. Reaction of different tomato genotypes to the bacterial speck pathogen (*Pseudomonas syringae* pv. *tomato*). Phytoparasitica 14: 39-42.

Pitblado, R.E. and E.A. Kerr. 1980. Resistance to bacterial speck (*Pseudomonas* tomato) in tomato. Acta Hort 100: 379-382.

Pitblado, R.E., and B.H. MacNeill. 1983. Genetic basis of resistance to *Pseudomonas syringae* pv. tomato in field tomatoes. Can J Plant Pathol 5: 251-255.

Pitblado, R.E., B.H. MacNeill, and E.A. Kerr. 1984. Chromosomal identity and linkage relationship of *Pto*, a gene for resistance to *Pseudomonas syringae* pv. *tomato*. Can J Plant Pathol 6: 48-53.

Poussier, S., D. Trigalet-Demery, P. Vandewalle, B. Goffinet, J. Luisetti, and A. Trigalet. 2000a. Genetic diversity of *Ralstonia solanacearum* as assessed by PCR-RFLP of the *hrp* gene region, AFLP and 16S rRNA sequence analysis, and identification of an African subdivision. Microbiology-SGM 146: 1679-1692.

Poussier, S., P. Prior, J. Luisetti, C. Hayward, and M. Fegan. 2000b. Partial sequencing of the *hrp*B and endoglucanase genes confirms and expands the known diversity within the *Ralstonia solanacearum* species complex. Syst Appl Microbiol 23: 479-486.

Prior, P. and M. Fegan. 2004. Recent development in the phylogeny and classification of *Ralstonia solanacearum*. 1st International Symposium on Tomato Diseases and 19th Annual Tomato Disease Workshop. Orlando, Florida, USA. June 20-24, 2004. p.57.

Prior, P., V. Grimault, and J. Schmit. 1994. Resistance to bacterial wilt (*Pseudomonas solanacearum*) in tomato: Present status and prospects. In: A. C. Hayward and G. L. Hartman, (eds.), Bacterial Wilt: The Disease and Its Causative Agent *Pseudomonas solanacearum*. CAB International, Wallingford, England, pp. 209-223.

Ricker, M. D., D. A. Emmaty, M. D. Schott, and R. H. Ozminkowski. 1996. Evaluation of three hybrids for resistance to *Clavibacter michiganensis* subsp. *michiganensis*. Phytopathology (Abstract) 86:S61.

Ricker, M.D., and R.M. Riedel. 1993. Effect of secondary spread of *Clavibacter michiganensis* subsp. *michiganensis* on yield of Northern processing tomatoes. Plant Dis 77: 364-366.

Rouamaba, A., H. Laterrot, and A. Moretti. 1988. A case of relation between resistances to *Pseudomonas solanacearum* and *Verticillium* pathotype 2. Rep Tomato Genet Coop 38: 43-44.

Sahin, F. and S.A. Miller. 1996. Characterization of Ohio strains of *Xanthomonas campestris* pv *vesicatoria*, causal agent of bacterial spot of pepper. Plant Dis 80: 773-778.

Salmeron, J. M., Barker, S. J., Carland, F. M., Mehta, A. Y., and Staskawicz, B. J. 1994. Tomato mutants altered in bacterial disease resistance provide evidence for a new locus controlling pathogen recognition. Plant Cell 6:511-520.

Salmeron, J.M., G.E.D. Oldroyd, C.M.T. Rommens, S.R. Scofield, H.S. Kim, D.T. Lavelle, D. Dahlbeck, and B.J. Staskawicz. 1996. Tomato *Prf* is a member of the leucine-rich repeat class of plant disease resistance genes and lies embedded within the *Pto* kinase gene cluster. Cell 86 (1): 123-133.

Sandbrink, J.M., J.W. van Ooijen, C.C. Purimahua, M. Vrielink, R. Verkerk, P. Zabel, and P. Lindout. 1995. Localization of genes for bacterial canker resistance in *Lycopersicon peruvianum* using RFLPs. Theor. Appl. Genet. 90: 444-450.

Sarfatti, M., J. Katan, R. Fluhr, and D. Zamir. 1989. An RFLP Marker in tomato linked to the *Fusarium-Oxysporum* resistance Gene *I2*. Theor Appl Genet 78 (5): 755-759.

Sarfatti, M., M. Abu-Abied, J. katan, and D. Zamir. 1991. RFLP mapping of *I1*, a new locus in tomato conferring resistance against *Fusarium oxysporum* f.sp. *lycopersici* race 1. Theor Appl Genet 80: 22-26.

Schmit, J. 1978. Microscopic study of early stages of infection by *Pseudomonas solanacearum* E.F.S. on "in vitro" grown tomato seedlings. *In:* 4th Proc Int Conf Plant Pathol Bacteriol, pp. 841-856.

Schneider, R.W. and R.G. Grogan. 1977. Bacterial speck of tomato: sources of inoculum and establishment of a resident population. Phytopathology 67: 388-394.

Scott, J. W., J. B. Jones, G. C. Somodi, and R. E. Stall. 1995. Screening tomato accessions for resistance to *Xanthomonas campestris* pv. *vesicatoria* race T3. HortScience 30:579-581.

Scott, J. W., J.-F. Wang, and P. M. Hanson. 2004. Breeding tomatoes for resistance to bacterial wilt, a global view. 1st International Symposium on Tomato Diseases and 19th Annual Tomato Disease Workshop. Orlando, Florida, USA. June 20-24, 2004.

Scott, J.W. 1997. Tomato improvement for bacterial disease resistance for the tropics: a contemporaty basis and future prospects. Proceedings of the 1st International Symposium on Tropical Tomato Diseases. Recife, Brazil, 1997. ASHS Press, Alexandria, VA.

Scott, J.W. and J.B. Jones. 1989. Inheritance of resistance to foliar bacterial spot of tomato incited by *Xanthomonas campestris* pv. *vesicatoria*. J Am Soc Hort Sci 114: 111-114.

Scott, J.W., and J.B. Jones. 1986. Sources of resistance to bacterial spot in tomato. HortScience 21: 304-306.

Scott, J.W., G.C. Somodi, and J.B. Jones. 1988. Bacterial spot resistance is not associated with bacterial wilt resistance in tomato. Proc Fla State Hort Soc 101: 390-392.

Scott, J.W., D.M. Francis, S.A. Miller, G.C. Somodi, and J.B. Jones. 2003. Tomato bacterial spot resistance derived from PI 114490; Inheritance of resistance to Race T2 and relationship across three pathogen races. J Am Soc Hort Sci 128: 698-703.

Scott, J.W., J.B. Jones, and G.C. Somodi. 2001. Inheritance of resistance in tomato to race T3 of the bacterial spot pathogen. J Am Soc Hort Sci 126: 436-441.

Scott, J.W., J.B. Jones, G.C. Somodi, and R.E. Stall. 1995. Screening tomato accession for resistance to *Xanthomonas campestris* pv. *vesicatoria*, race T3. HortScience 30: 579-581.

Scott, J.W., R.E. Stall, J.B. Jones, and G.C. Somodi. 1996. A single gene controls the hypersensitive response of Hawaii 7981 to race 3 (T3) of the bacterial spot pathogen. Rep. Tomato Genet Coop 46: 23.

Scott, J.W., S.A. Miller, R.E. Stall, J.B. Jones, G.C. Somodi, V. Barbosa, D.L. Francis, and F. Sahin. 1997. Resistance to race T2 of the bacterial spot pathogen in tomato. HortScience 32: 724-727.

Sharma, K.C., V.K. Rathee, and S. Verma. 2002. Identification of resistance sources for bacterial wilt (*Ralstonia solanacearum*) in tomato (*Lycopersicon esculentum*). Indian J Agric Sci 73: 186-187.

Sherf, A.F. and A.A. MacNab. 1986. Vegetable diseases and their control. Wiley, New York, pp. 599-614.

Silva, V.L., da, and C.A. Lopes. 1995. *Pseudomonas syringae* pv. *tomato* resistant to copper in copper-sprayed tomato fields. Fitopatol Bras 20: 85-89.

Smith, E.F. 1896. A bacterial disease of tomato, pepper, eggplant and Irish potato (*Bacillus solanacearum* nov. sp.). United States Department of Agriculture, Division of Vegetable Physiology and Pathology Bulletin 12: 1-28.

Smith, E.F. 1910. A new tomato disease of economic importance. (Abstr.) Science (New Series) 31: 794-796.

Smitley, D.R., and S.M. McCarter. 1982. Spread of *Pseudomonas syringae* pv. *tomato* and role of epiphytic populations and environmental conditions in disease development. Plant Dis 66: 713-717.

Somodi, G.C., J.B. Jones, J.W. Scott, and J.P. Jones. 1994. Screening tomato seedlings for resistance to bacterial spot. HortScience 29: 680-682.

Somodi, G.C., J.B. Jones, J.W. Scott, J.F. Wang, and R.E. Stall. 1996. Relationship between the hypersensitive reaction and field resistance to tomato race 1 of *Xanthomonas campestris* pv. *vesicatoria*. Plant Dis 80: 1151-1154.

Sotirova, V., and N. Bogatzevska. 1993. New sources of resistance to *Xanthomonas campestris* pv. *vesicatoria* (Doidge) Dye in tomato. Phytopath Medit 32: 145-148.

Stall, R.E. 1995. *Xanthomonas campestris* pv. *vesicatoria*. in: U. S. Singh, R. P. Singh, and K. Kphmoto, (eds.), Pathogenesis and Host Specificity in Plant Diseases: Histopathological, Biochemical, Genetic and Molecular Bases. Elsevier Science, New York, pp. 167-181.

Stamova, B.S., and R.T. Chetelat. 2000. Inheritance and genetic mapping of cucumber mosaic virus resistance introgressed from *Lycopersicon chilense* into tomato. Theor Appl Genet 101: 527-537.

Stevens, M.R., E.M. Lamb, and D.D. Rhoads. 1995. Mapping the *sw-5* locus for tomato spotted wilt virus-resistance in tomatoes using RAPD and RFLP analyses. Theor Appl Genet 90 (3-4): 451-456.

Stockinger, E.J. and L.L. Walling. 1994. *Pto-3* and *Pto-4*: novel genes from *Lycopersicon hirsutum* var. *glabratum* that confer resistance to *Pseudomonas syringae* pv. *tomato*. Theor Appl Genet 89: 879-884.

Strider, D.L. 1969. Bacterial canker of tomato, a literature review and bibliography. N. C. Agric Exp Stn Tech Bull 193.

Suliman-Pollatschek S., K. Kashkush, H. Shats, J. Hillel, and U. Lavi. 2002. Generation and mapping of AFLP, SSRs and SNPs in *Lycopersicon esculentum*. Cell Mol Biol Letters 7: 583-597.

Šutic, D. 1957. Bakterioze crvenog patlidzana (Tomato bacteriosis). Posebna Izd. Inst. Zasht. Bilja Beograd (Spec. Edit. Inst. Plant Prot. Beograd) 6:1-65. English summary Rev Appl Mycol 36: 734-735.

Tanksley, S.D., S. Brommonschenkel, and G. Martin. 1996. *Pto*[h], an allele of *Pto* conferring resistance to *Pseudomonas syringae* pv. *tomato* (race 0) that is not associated with fenthion sensitivity. Rep. Tomato Gene Coop 46: 28-29.

Thomas, C.M., M.S. Dixon, M. Parniske, C. Golstein, and J.D.G. Jones. 1998. Genetic and molecular analysis of tomato *Cf* genes for resistance to *Cladosporium fulvum*. Philosophical Transactions of The Royal Society of London Series B-Biological Sciences 353 (1374): 1413-1424.

Thoquet, P, J. Olivier, C. Sperisen, P. Rogowsky, H. Laterrot, and N. Grimsley. 1996a. Quantitative trait loci determining resistance to bacterial wilt in tomato cultivar Hawaii7996. MolPlant-Microbe Interact 9: 826-836.

Thoquet, P, J. Olivier, C. Sperisen, P. Rogowsky, P. Prior, G. Anaïs, B. Mangin, B. Bazin, R. Nazer, and N. Grimsley. 1996b. Poloygenic resistance of tomato plants to bacterial wilt in the French West Indies. Mol Plant-Microbe Interact 9: 837-842.

Thurston, H. D. 1976. Resistance to bacterial wilt (*Pseudomonas solanacearum*). Proceedings of the First International Planning Conference and Workshop on the Ecology and Control of Bacterial Wilt caused by *Pseudomonas solanacearum*. North Carolina State University,Raleigh, USA pp 58-62.

Thyr, B. D., R. E. Webb, C. A. Jaworski, and T. J. Ratcliff. 1973. Tomato bacterial canker: control by seed treatment. Plant Dis Reporter 57: 974-977.

Thyr, B.D. 1968. Resistance to *Corynebacterium michiganense* measured in six *Lycopersicon* accessions. Phytopathology 61: 972-974.

Thyr, B.D. 1969. Additional sources of resistance to bacterial canker of tomato (*Corynebacterium michiganense*). Plant Dis Reporter 53:234-237.

Thyr, B.D. 1976. Inheritance of resistance to *Corynebacterium michiganense* in tomato. Phytopathology 66: 1116-1119.

Vakalounakis, D.J., H. Laterrot, A. Moretti, E.K. Ligoxigakis, and K. Smardas. 1997. Linkage between *Frl* (*Fusarium oxysporum* f sp *radicis-lycopersici* resistance) and *Tm-2* (tobacco mosaic virus resistance-2) loci in tomato (*Lycopersicon esculentum*). Ann Appl Biol 130 (2): 319-323.

van der Beek, J.G., G. Pet, and P. Lindhout. 1994. Resistance to powder mildew (*Oidium lycopersicum*) in *Lycopersicon hirsutum* is controlled by incompletely-dominant gene OI-1 on chromosome 6. Theor Appl Genet 89: 467-473.

van der Biezen, E.A., T. Glagotskaya, B. Overduin, H.J.J. Nijkamp, and J. Hille. 1995. Inheritance and genetic mapping of resistance to *Alternaria alternata* f. sp. *lycopersici* in *Lycopersicon pennellii*. Mol Gen Genet 247: 453-461.

Van Heusden, A.W., M. Koornneef, R.E. Voorrips, W. Brüggemann, G. Pet, R. Virelink-van Ginkel, X. Chen, and P. Lindout. 1999. Three QTLs from *Lycopersicon peruvianum* confer a high level of resistance to *Clavibacter michiganensis* ssp. *michiganensis*. Theor Appl Genet 99: 1068-1074.

Vasse, J., P. Frey, and A. Trigalet. 1995. Microscopic studies of inter-cellular infection and protoxylem invasion of tomato rests by *Pseudomonas solanacearum*. Mol Plant-Microbe Interact 8: 241-251.

Veremis, J.C., A.W. van Heusden, and P.A. Roberts. 1999. Mapping a novel heat-stable resistance to *Meloidogyne* in *Lycopersicon peruvianum*. Theor Appl Genet 98 (2): 274-280.

Vulkova, Z. and Sotirova,V. 1992. Genetic study of line Cm 190, resistant to Clavibacter michiganensis subsp. michiganensis (Smith) Davis et al. Rep Tomato Genet Coop 42: 45.

Wang, J. F. 1992. Resistance to *Xanthomonas campestris* pv. *vesicatoria* in tomato. Ph.D. diss. University of Florida, Gainesville, Florida.

Wang, J. F., Hanson, P. M., and Barnes, J. A. 1998. Worldwide evaluation of international set of resistance sources to bacterial wilt in tomato. in: P. Prior, C. Allen, and J. Elphinstone, (eds.), Bacterial Wilt Disease: Molecular and Ecological Aspects. Springer, Berlin, pp. 269-275.

Wang, J. F., Stall, R. E., and Vallejos, C. E. 1994b. Genetic analysis of a complex hypersensitive reaction to bacterial spot in tomato. Phytopathology 84:126-132.

Wang, J.F., J. Olivier, P. Thoquet, B. Maigin, L. Sauviac, and N.H. Grimsley. 2000. Resistance of tomato line Hawaii7996 to *Ralstonia solanacearum* Pss4 in Taiwan is controlled mainly by a major strain-specific locus. Mol Plant-Microbe Interact 13: 6-13.

Wang, J.F., J.B. Jones, J.W. Scott, and R.E. Stall. 1994a. Several genes in *Lycopersicon esculentum* control hypersensitivity to *Xanthomonas campestris* pv. *vesicatoria*. Phytopathology 84: 702-706.

Wang, J.F., P.M. Hanson, and J.A. Barnes. 1996. Preliminary results of worldwide evaluation of international set of resistance sources to bacterial wilt in tomatoes. Bact. Wilt Newsl. 13: 8-10.

Werner, N.A., D.W. Fulbright, R. Podolsky, J. Bell, and M.K. Hausbeck. 2002. Limiting populations and spread of *Clavibacter michiganensis* subsp. *michiganensis* on seedling tomatoes in the greenhouse. Plant Dis 86: 535-542.

Whalen, M.C., J.F. Wang, F.M. Carland, M.E. Heiskell, D. Dahlbeck, G.V. Minsavage, J.B. Jones, J.W. Scott, R.E. Stall, and B.J. Staskawicz. 1993. Avirulence gene *avr Rxv* from *Xanthomonas campestris* pv. *vesicatoria* specifies resistance on tomato line Hawaii 7998. Mol Plant-Microbe Interact 6: 616-627.

Whitham, S., S.P. Dineshkumar, D. Choi, R. Hehl, C. Corr, and B. Baker. 1994. The product of the tobacco mosaic-virus resistance gene-N - similarity to toll and the interleukin-1 receptor. Cell 78 (6): 1101-1115.

Williams, C.E. and D.A. St. Clair. 1993. Phenetic relationships and levels of variability detected by restriction fragment length polymorphism and random amplified polymorphic DNA analysis of cultivated and wild accessions of *Lycopersicon esculentum*. Genome 36:619-630.

Williamson, L., K. Nakaho, B. Hudelson, and C. Allen. 2003. *Ralstonia solanacearum* Race 3, Biovar 2 Strains Isolated from Geranium Are Pathogenic on Potato., Plant Dis 86:987-991.

Yabuuchi, E., Y. Kosako, I. Yano, H. Hotta, and Y. Nishiuki. 1995. Transfer of two *Burkholderia* and an *Alcaligenes* species to *Ralstonia* gen. nov.: Proposal of *Ralstonia pickettii* (Ralston, Palleroni and Duodoroff 1973) comb. nov. and *Ralstonia eutropa* (Davis 1969) comb. nov. Microbiol Immunol 39: 897-904.

Yaghoobi, J., I. Kaloshian, Y. Wen, and V.M. Williamson. 1995. Mapping a new nematode resistance locus in *Lycopersicon-peruvianum*. Theor Appl Genet 91 (3): 457-464.

Yang, W. and D. M. Francis. 2005. Marker assisted selection for combining resistance to bacterial spot and bacterial speck in tomato. J Amer Soc Hort Sci 130:716-721.

Yang, W. and D. M. Francis. 2005. Marker-assisted selection in pyramiding resistance to bacterial speck and bacterial spot in tomato. phytopathology 95: 519-527.

Yang, W., E. Sacks, M.L.L. Ivey, S.A. Miller, and D.M. Francis. 2005. Resistance to race T1 strains of bacterial spot of tomato in *L esculentum* × *L esculentum* corsses. Phytopathology (Accepted Dec. 14, 2004)

Yang, W., X. Bai, E. Kabelka, C. Eaton, S. Kamoun, E. van der Knaap, and D. Francis. 2004. Discovery of single nucleotide polymorphisms in Lycopersicon esculentum by computer aided analysis of expressed sequence tags. Mol Breed 14: 21-34.

Yordanov, M. and L. Stamova. 1977. A new source of resistance to *Corynebacterium michiganense* (E. F. Sm) Jensen. Rep. Tomato Genet Coop 27: 26.

Young, N.D. and S.D. Tanksley. 1989. RFLP analysis of the size of chromosomal segments retained around the *Tm-2* locus of tomato during backcross breeding. Theor Appl Genet 77 (3): 353-359.

Yu, Z.H., J.F. Wang, R.E. Stall, and C.E. Vallejos. 1995. Genomic localization of tomato genes that control a hypersensitive reaction to *Xanthomonas campestris* pv. *vesicatoria* (Doidge) Dye. Genetics 141: 675-682.

Yuan, Y.N., J. Haanstra, P. Lindhout, and G. Bonnema. 2002. The *Cladosporium fulvum* resistance gene *Cf-ECP3* is part of the Orion cluster on the short arm of tomato Chromosome 1. Mol Breed 10 (1-2): 45-50.

Yunis, H., Y. Bashan, Y. Okon, and Y. Henis. 1980. Weather dependence, yield losses, and control of bacterial speck of tomato caused by *Pseudomonas tomato*. Plant Dis 64: 937-939.

Zamir, D., I. Eksteinmichelson, Y. Zakay, N. Navot, M. Zeidan, M. Sarfatti, Y. Eshed, E. Harel, T. Pleban, H. Vanoss, N. Kedar, H.D. Rabinowitch, and H. Czosnek. 1994. Mapping and introgression of a tomato yellow leaf curl virus tolerance gene, *Ty-1*. Theor Appl Genet 88 (2): 141-146.

13

Breeding for Resistance to Fungal Pathogens

J.W. Scott¹ AND R.G. Gardner²

¹ *Gulf Coast Research and Education Center, University of Florida, Institute of Food and Agricultural Sciences, 14625 CR 672, Wimauma, FL 33598 USA*
email jwsc@mcsu.ufl.edu

² *Mountain Horticultural Crops Research and Extension Center, North Carolina State University 455 Research Drive, Fletcher, NC 28732 USA*
email randy-gardner@mcsu.edu

INTRODUCTION

Numerous fungal pathogens attack tomato (*Lycopersicon esculentum* Mill.) causing crop loss directly by reducing yields, or indirectly by reducing marketable yield as a result of defoliation and exposure of fruit to the elements. Profits for growers are also reduced by the costs of chemical sprays or fumigants used to prevent the diseases. These diseases, as well as bacterial and viral tomato diseases, are discussed by Watterson (1986). The reader is also referred to the chapter of Lukyanenko (1991), where resistances to fungal and other diseases are discussed. This chapter will focus only on diseases that have been subjected to conventional plant breeding for resistance. Molecular markers will be covered, but not genetic transformation. To date all disease resistance traits in commercial tomato varieties have been derived from conventional plant breeding. Multiple disease resistant varieties generally derive their resistance from single major genes that usually have dominant inheritance, and of these fungal resistances are the most common (Laterrot 1997, Scott 2005). The dominant inheritance of these disease resistant genes is one of the major reasons that hybrid varieties are so prevalent, other reasons for the prevalence of hybrids are discussed by Scott and Angell (1998). A summary of fungal diseases and resistance genes used to control them is presented in Table 13.1. A map showing approximate locations of resistance genes is depicted in

Corresponding author: J.W. Scott

Table 13.1 Summary of disease resistance genes and QTLs mapped on tomato chromosomes.
(from Foolad and Sharma 2005)

Disease	Gene/QTL	Pathogen	Origin	Chromosome Location	References
Fungal					
Alternaria Stem Canker	Asc	Alternaria alternata f. sp. lycopersici	L. pennellii	3	(Mesbah et al. 1999, van der Biezen et al. 1995).
Anthracnose ripe rot	Anthracnose (Q)[y]	Colletotrichum coccodes	L. esculentum	Different chrs.	(Stommel and Zhang 1998)
Blackmold	Blackmold (Q)	Alternaria alternate	L. cheesmanii	2, 3, 9, 12	(Robert et al. 2001)
Corky Root	py-1	Pyrenochaeta lycopersici	L. peruvianum	3	(Doganlar et al. 1998)
Early Blight	EBR (Q)	Alternaria solani	L. esculentum, L. hirsutum, L. pimpinellifolium	Different chrs.	(Foolad et al. 2002, Zhang et al. 2003), Foolad et al. (unpubl.), (Vakalounakis et al. 1997).
Fusarium Crown and Root Rot	Frl	Fusarium oxysporum f. sp. radicis-lycopersici	L. peruvianum	9	
Fusarium Wilt	I, I-1, I-2, I-2C, I-3	Fusarium oxysporum f. sp. lycopersici	L. pimpinellifolium, L. pennellii	7, 11	(Bournival et al. 1990, Ori et al. 1997, Sarfatti et al. 1991, Simons et al. 1998, Tanksley and Costello 1991)
Gray Leaf Spot	Sm	Stemphylium spp.	L. pimpinellifolium	11	(Behare et al. 1991)
Late Blight	Ph-1, Ph-2, Ph-3	Phytophthora infestans	L. pimpinellifolium	7, 9, 10	(Chunwongse et al. 2002, Moreau et al. 1998, Pierce 1971).
Late Blight	lb1-lb12 (Q)	Phytophthora infestans	L. hirsutum, L. pimpinellifolium	All 12 chrs.	(Brouwer et al. 2004, Brouwer and St. Clair 2004, Frary et al. 1998, Lough 2003)
Leaf Mould	Cf-1, Cf-2, Cf-4, Cf-5, Cf-9	Cladosporium fulvum	L. hirsutum, L. pimpinellifolium	1, 6	(Balint-Kurti et al. 1994, Jones et al. 1993, Lauge et al. 1998).
Powdery Mildew	Lv	Leveillula taurica	L. chilense	12	(Chunwongse et al. 1994).
Powdery Mildew	Ol-1, Ol-2	Oidium lycopersici	L. esculentum, L. hirsutum	4, 6	(De Giovanni et al. 2004, Huang

(Contd.)

(Contd.)

Disease	Gene	Pathogen	Source	Chromosome	Reference
Powdery Mildew	Ol (Q)-1, Ol (Q)-2, Ol (Q)-3	Oidium lycopersici	L. parviflorum	6, 12	et al. 2000, van der Beek et al. 1994). (Bai et al. 2003)
Verticillium Wilt	Ve	Verticillium dahliae	L. esculentum	9	(Diwan et al. 1999, Kawchuk et al. 1998).
Bacterial					
Bacterial Canker	Cm 1.1-10.1 (Q)	Clavibacter michiganensis subsp. michiganensis	L. peruvianum	1, 6, 7, 8, 9, 10	(Sandbrink et al. 1995).
Bacterial Canker	Rcm2.0 (Q), Rcm5.1(Q)	Calvibacter michiganensis subsp. michiganensis	L. esculentum, L. hirsutum	2, 5	(Coaker and Francis 2004, Kabelka et al. 2002).
Bacterial Canker	Three (Q)	Calvibacter michiganensis subsp. michiganensis	L. peruvianum	5, 7, 9	(van Heusden et al. 1999).
Bacterial Speck	Pto	Pseudomonas syringae pv. tomato	L. pimpinellifolium	5	(Martin et al. 1993).
	Prf	Required for Pto/Fen	L. pimpinellifolium	5	(Salmeron et al. 1996).
Bacterial Spot	Bs4	Xanthomonas compestris pv. vesicatoria	L. pennellii	5	(Ballvora et al. 2001).
Bacterial Wilt	Bw-1, Bw-3, Bw-4, Bw-5 (Q)	Ralstonia solanacearum	L. pimpinellifolium	4, 6, 10	(Danesh et al. 1994, Thoquet et al. 1996).
Bacterial Wilt	Two (Q)	Ralstonia solanacearum	L. peruvianum	6	(Mangin et al. 1999).
Viral					
Cucumber Mosaic Virus	Cmr	CMV	L. chilense	12	(Stamova and Chetelat 2000).
Potyviruses	Pot-1	Potyviruses	L. hirsutum	3	(Parrella et al. 2002).
Tobacco/Tomato Mosaic Virus	Tm-1, Tm-2[a] (allelic to Tm-2)	TMV/ToMV	L. hirsutum, L. peruvianum	2, 9	(Levesque et al. 1990, Vakalounakis et al. 1997, Young and Tanksley 1988).
Tomato Mottle Virus	Two genes	ToMoV	L. chilense	6	(Griffiths and Scott 2001).
Tomato Spotted	Sw-5	TSWV	L. peruvianum	9	(Brommenschenkel et al. 2000,

(Contd.)

424

Genetic Improvement of Solanaceous Crops: Tomato

(Contd.)

Wilt Virus					Brommenschenkel and Tanksley, 1997, Stevens et al. 1995)
Tomato Yellow Leaf Curl Virus	Ty-1, Ty-2, two (Q)	TYLCV	L. esculentum, L. chilense, L. pimpinellifolium, L. hirsutum	6, 11	(Chagué et al. 1997, Hanson et al. 2000, Zamir et al. 1994)
Nematode					
Potato Cyst Nematode	Hero	Globodera restochiensis	L. pimpinellifolium	4	(Ganal et al. 1995)
Root-Knot Nematode	Mi, Mi-1, Mi-3, Mi-9	Meloidogyne spp.	L. peruvianum	6, 12	(Ammiraju et al. 2003, Veremis et al. 1999, Williamson et al. 1994, Yaghoobi et al. 1995)

zReferences not found in Literature Cited sections of chapters (13-14) can be found in Foolad and Sharma, (2005) and chapter 16.
yQ in parentheses indicates quantitative trait loci (QTLs).

Fig. 12.1 of the bacterial resistance chapter and another is in the paper of Foolad and Sharma (2005) and chapter 16 of this book.

SOIL-BORNE DISEASES

Alternaria stem canker. This disease caused by *Alternata alternaria* (Fr.) Keissler f.sp. *lycopersici* was first described in California (Grogan et al. 1975). Infected plants have dark brown to black cankers with concentric zonation on the stems near the soil line or above ground. Tissue beneath the cankers has a brown dry rot in the pith area and discontinuous brown streaks may emanate in the xylem near the cankers. The lesions can girdle the plants destroying them by harvest time. Foliar symptoms include spotting and eventual necrosis of leaflets on one or both sides of the midrib. Susceptible varieties had 'Pearson' in their pedigree while varieties bred in Florida such as 'Florida MH-1', 'Walter', 'Homestead' and 'Manalucie' were resistant. Grogan et al. (1975) postulated that resistance was conferred by a single dominant gene. This was later confirmed by Vakalounakis (1988a), who also tested 105 varieties for resistance and found that 103 of them were resistant. Clouse and Gilchrist (1987) showed that resistance in isogenic lines was controlled by a single dominant gene at the *Asc* locus and that *Asc* was incompletely dominant to a host specific pathotoxin. There was complete co-segregation of simultaneously tested plants to both toxin and the pathogen. Witsenboer et al. (1989) found *Asc* was linked 10-17.8 cM from the solanifolium (*sf*) gene on chromosome 3. Later it was reported that *Asc* was tightly linked to RFLP markers TG 134 and TG 442 on chromosome 3 (Van der Biezen et al. 1995, Mesbah et al. 1999).

While inoculations can be done by several methods, the method of Grogan et al. (1975) will be described. The pathogen is cultured on PDA or CMA (cornmeal agar) where it maintained virulence for at least six weeks. Spores and mycelia were scraped from the plates and water was added to obtain a suspension concentration of 2.5 x 10^5 spores/ml. The suspension was sprayed on plants at the 2 leaf stage or older, since younger plants at the cotyledon stage were not infected. The inoculated plants were kept in a mist chamber for 36-48 h at 21-24°C and then removed to a greenhouse. Susceptible plants develop stem lesions and are girdled in 10-15 days while resistant plants remain healthy. A toxin can also be used in a detached leaf bioassay which is described by Clouse and Gilchrist (1987).

Corky root rot. This disease is caused by *Pyrenocaeyta lycopersici* Schn. & Ger. and has caused problems in European greenhouse production and is referred to as "old land disease" in Florida. Infected plants form root lesions that eventually become furrowed and develop a corky texture causing a progressive deterioration of the root system. Resistance has been reported from the wild species *L. hirsutum* and *L. peruvianum* (Hogenboom 1970). A single recessive gene *py-1* was introgressed from *L. peruvianum* (Laterrot 1983). However, greenhouse inoculation is unreliable and field tests are required. Thus, the transfer of resistance has been tedious and slow. Doganlar et al. (1998) grew segregating populations in California, USA and Italy and used RAPD and RFLP marker analysis to map the *py-1* gene in a 8.8 cM region between TG 40 and CT 31 on the short arm of chromosome 3. They were unable to obtain recombinants in this region from 100 F_2 plants and concluded recombination was suppressed in the region. CAPS markers were derived from three RFLP markers and the one derived from TG 324 R/F should facilitate future breeding efforts. Stamova (2004) reported on a tomato line called Pirelly 38 with resistance derived from the subgenus *Eriopersicon* that appears to have resistance from a single dominant gene when tested with infested soil from California farms. The *py-1* gene was not effective for the isolate(s) tested (Stamova, personnel communication). More inheritance work needs to be done with Pirelly 38 and its effectiveness against other isolates should be tested.

Recalcitrant sporulation of the pathogen has made it difficult to work with. McGrath and Campbell (1983) reported on methods to improve sporulation. V8 agar was superior to 18 other media they tested and sporulation and pathogenicity could be maintained for at least two years. Sporulation was enhanced: by 3-4 day incubation of isolates on water agar before transfer to V8 agar, by constant light from cool white fluorescent lamps over 12 or 8 hr photoperiods, by media at 5.5 pH, and with incubation temperatures of 20-24°C. Doganlar et al. (1998) inoculated plants by growing the pathogen on V8 agar, then 20 10 mm^2 pieces were added to 100g of sterilized millet seed in 120 ml water. After 30 days the millet seed was mixed in soil (50% sand and 50% peat moss) at a 1:20 dilution. The soil was placed in seedling trays and the tomato seeds directly planted in this soil. At the sixth true leaf stage the plants were transplanted to a field infested with the pathogen. When plants had 15% red fruit they were dug and the root systems rated for disease symptoms on a 1-5 scale where 1= no disease symptoms and 5 = extensive corking of the root system.

Foot and stem rot. This disease is a problem in western European greenhouses. The pathogen, *Didymella lycopersici* Kleb., causes dark brown sunken lesions on the stem that can girdle the plant. Resistance from *L. hirsutum* f. *glabratum* accession 61292 was reported to be dominant and controlled by more than one gene (Boukema 1982). Heritability was high suggesting selection efficiency, but no resistant cultivars are available and there does not appear to be much breeding work with this disease. Inoculation procedures were reported by Van Steekelenburg (1981). Pathogen isolates, which maintain virulence for at least five years of sub-culturing, are grown on cherry decoction agar at 18-20°C and exposed to near UV-light (Philips TL 20W F20 T12 bulb) to induce sporulation. Fourteen-day-old cultures (9 cm) are either homogenized with water in a blender or the conidia are washed off and the suspension adjusted to 10^6 conidia per ml. Four-week-old plants are inoculated either by dipping the roots for 5 sec in the suspension or 4 ml of the suspension is applied at the base of the plants with a syringe. Final scoring is done three weeks later. The inoculation was better with soil temperatures of 15°C than at higher soil temperatures. Wounding roots by transplanting was beneficial in inducing disease. Two-week-old plants showed as much infection as four-week-old plants, but the younger plants were prone to *Pythium* sp. infection, so the latter stage was recommended.

Fusarium wilt. The causal organism is *Fusarium oxysporum* f.sp. *lycopersici* (Sacc.) Snyder & Hansen. The disease has a world-wide distribution and can cause major crop losses. There are presently three races of the pathogen commonly referred to as races 1, 2, and 3 which are analogous to races 0, 1, and 2, respectively using the nomenclature of Gabe (1975). The first vertical gene for resistance (*I*) was reported by Bohn and Tucker (1939) in *L. pimpinellifolium* accession PI 79532. This gene was later assigned to chromosome 11 (Paddock 1950). Recently *I* was found to be linked to RFLP TG523 (P. Lindhout pers. comm.). A second race was discovered before 1940 (Alexander and Tucker 1945) but was not reported again until 1961 when the pathogen caused serious crop losses in Florida (Stall and Walter 1965). A gene for resistance to the second race (*I-2*) was discovered in a *L. pimpinellifolium-L. esculentum* F_1 accession, PI 126915 (Stall and Walter 1965). The *I-2* gene was mapped to chromosome 11 by morphological markers (Laterrot 1976) and by RFLP marker TG105 (Sarfatti et al., 1989). The *I-2* gene has been cloned and found to be a complex locus (Ori et al. 1997, Simons et al. 1998). Stall and Walter (1965) originally reported PI 126915 resistant to race 1, but it was not clear if this was due to *I-2* or to a linked gene, either *I* or another gene. Cirulli and Alexander (1966) reported that resistances to race 1 and race 2 from PI 126915 were controlled by separate genes. Later, Laterrot and Philouze (1984) reported recombination between the genes when they obtained a line that was resistant to race 2, but not race 1, although it was less susceptible to race 1 than the susceptible control line. It was concluded that the *I-2* gene gave a reduced susceptibility to race 1.

Meanwhile, a third race of fusarium wilt was discovered in Australia (Grattidge and O'Brien 1982) and has now spread to the southeastern and western USA, Mexico and Japan. Resistance to race 3 (*I-3* gene) was discovered in *L. pennellii* accessions PI 414773 (McGrath et al. 1987) and LA 716 (Scott and Jones 1989). In fact, all accessions of *L. pennellii* carry resistance to all three races of the Fusarium wilt organism (Scott and Jones 1990). The gene conferring race 3 resistance from LA 716 was determined by Bournival et al. (1989) to be linked to an allozyme of *Got-2* on chromosome 7. PI 414773 also had the same *Got-2* allozyme (Scott, unpublished data). This marker could also be used to select for race 1 and 2 resistance derived from LA 716 (Bournival et al. 1990). From this work, it was not clear whether the resistances to races 1 and 2 were conferred by *I-3* or genes linked to *I-3*. Tanksley et al. (1992) reported *I-3* was linked to RFLP markers TG128, TG217, and TG170. Sarfatti et al. (1991) reported gene *I1*, derived from LA 716, conferred race 1 resistance and was linked in a 22 cM region between RFLP markers TG20 and TG128 on chromosome 7. They reported this gene not to be an allele of *I-3* presumably due to differences in linkage estimates from *Got-2*. Scott et al. (2004) obtained four lines that were resistant to races 2 and 3 but susceptible to race 1. Two other lines resistant to race 3 had partial resistance to race 2 and were susceptible to race 1. All six had *L. esculentum* alleles distal to TG 639 which reduces the possible *I1* region by 13 cM from that of Sarfatti et al. (1991). However, Scott et al. (2004) speculated that *I1* was located near CT113 6.5 cM beyond TG20 which bordered the region reported by Sarfetti et al. (1991). It is evident that race 2 resistance is more tightly linked to *I-3* than race 1 resistance (Scott et al. 2004). The likely location of a putative race 2 locus is distal to the *I-3* locus, but further research is needed to verify the existence of such a locus.

The *I* and *I-2* genes have been two of the most widely deployed genes in tomato cultivars. It is difficult to find any modern cultivars without these genes, especially the *I* gene. Scott and Jones (1995) released two race 3 resistant breeding lines, and cultivars such as 'Floralina', 'Sunguard', 'Solar Fire' and 'Escudero' are now available that have the race 3 resistance conferred by the *I-3* gene with numerous other cultivars soon to be released. Delta Contender is a race 3 resistant cultivar that has been grown in Australia for a number of years. Numerous other accessions have resistance to various races of fusarium wilt (Scott and Jones 1986, Bournival and Vallejos 1991, Huang and Lindhout 1997). Some of these might also be resistant to new races of the pathogen that may emerge in the future.

Screening for fusarium wilt is done with seedlings and is quite routine. The pathogen is grown on potato dextrose agar at 28°C for 1 week and communited in a blender with dH_2O and adjusted to 6 x 10^7 spores/ml. Seedlings at the cotyledon stage are root dipped in the slurry and transplanted to flats under temperatures of 28°C or more with a soil media at 6.5 pH. High (>7) pH soils suppress fusarium wilt and should be avoided. Symptoms can be read 3 weeks later. Susceptible plants may die or express stunting, wilting, yellowing of foliage, epinasty, and/or have enlarged stems. When plant stems are cut longitudinally, susceptible plants have vascular browning whereas resistant plants do not. The penetrance of the *I* gene is sometimes reduced with this test (Alon et al. 1974) but varieties with the *I* gene are healthy under field conditions. The race 2 test is the most straightforward as the penetrance of *I-2* is close to 100% and there are few susceptible escapes. The *I-3* gene has penetrance similar to *I-2*, but the susceptible plants show symptoms a few days later than with race 2 and overall the symptoms are not as severe. Although dead plants are evident, there are more susceptible plants with lesser symptoms for race 3 than with race 2. Thus, more plants require cutting to verify if they are susceptible with vascular browning.

Fusarium crown and root rot. The causal organism of this pathogen is *Fusarium oxysporum* Schlecht. f. sp. *radicus-lycopersici*, Jarvis and Shoemaker. This disease has been found in both greenhouses and field production. Plants wilt similarly to those infected with fusarium wilt, but the vascular browning is confined to the crown of the plant whereas browning in plants infected with fusarium wilt can also be found in the apices. Fusarium crown rot thrives at cooler temperatures whereas fusarium wilt thrives under high temperature conditions. Resistance is conferred by the single dominant gene *Frl* (Berry and Oakes 1987a, Vakalounakis 1988b) that likely was introgressed from *L. peruvianum* (Yamakawa and Nagata 1975). The *Frl* gene has been mapped to chromosome 9 at 5.1cM from the *Tm-2* locus (Vakalounakis et al. 1997). Several researchers have found that germplasm selected for the *Tm2²* gene also carry resistance to fusarium crown rot (Laterrot and Moretti 1992 1996, Laterrot and Couteadier 1989). The *Tm2²* gene was introgressed from *L. peruvianum* accession PI 128650 (Alexander 1963). Resistance has also been identified in several accessions of *L. pennellii* (Scott and Jones 1990). Development of breeding lines with resistance from LA 1277 has been pursued but it is not clear how it is inherited despite two attempted genetic studies (Scott, unpublished data). Stamova (1996) reported resistance was found in lines derived from *L. pennellii, L. chilense,* and *L. pimpinellifolium.* The first greenhouse tomato cultivar with resistance was Ohio CR6 (Scott and Farley 1983). Since then numerous resistant varieties have been developed for greenhouse or field production.

A disease screening method has been described based on parameters defined by Jones et al. (1990). The pathogen is grown on PDA as with fusarium wilt. Plants at the cotyledon stage or with 2 true leaves are inoculated by dipping into a suspension with 2×10^7 spores/ml of the pathogen and transplanted to a soil amended to a pH of 5.0 in a growth chamber at 20 °C with 12 hr days. Results can be evaluated in a week when some plants have died. Living plants are dug and roots are rinsed. Susceptible plants are often stunted and have some brown roots while the roots of resistant plants remain white.

Phytophthora root and crown rot. This disease, caused primarily by *Phytophthora parasitica* Dast. and secondarily by *P. capsici* Leonian, has been a problem in processing tomatoes in California, especially in wet soils. High salts also exacerbate symptoms of this disease (Swiecki and MacDonald 1991). Seedlings can be killed, but the most severe damage occurs late in the season when fruit are ripening. The entire canopy can collapse and die leaving fruit exposed to sunburn (Blaker and Hewitt 1987). Moderate resistance was reported in the cultivar CX8303 and *L. esculentum* var. *cerasiforme* accessions LA 1312, 11-2A, 13-1A, and 27-1A (Blaker and Hewitt 1987a). Blaker and Hewitt (1987b) concluded that at least two types of resistance exist in tomato when they compared LA 1312, 27-1A, and CX8303 to susceptible genotypes. Both mechanisms resulted in less colonization of roots. The former two *Lycosperium esculentum* var. *cerasiforme* genotypes inhibited lesion extension and hyphal growth from the point of infection. CX8303 had fewer propagules per gram of root tissue than susceptible genotypes suggesting less extensive colonization with possibly only the outer cortical layers being infected, although this was not tested. The ability to regenerate roots after infection is another possible secondary resistance mechanism as 27-1A and CX803 regenerated more roots than Peto 343, the susceptible control. Inheritance of resistance from LA 1312 was determined to be quantitative (Kozik et al. 1991). The F_1 was intermediate to the parents and skewed slightly toward resistance. Additive effects were 96% but dominance and epistasis were also detected. Broad sense heritability was 36% and narrow sense heritability was 22%. The effective factor estimate suggested five genes control resistance. Since heritability was low, family selection rather than individual plant selections should be utilized. This resistance will have to be incorporated into both parents of a hybrid and this coupled with the low heritability has been a hindrance to developing commercial cultivars.

Bolkan (1985) described an inoculation technique that correlated well with field results (r=0.77). Other studies use close adaptations of this method and the interested reader may wish to check the references for the exact methods employed. The method of Kozik et al. (1991) will be described here. Cultures are maintained on V-8 agar and renewed monthly. Mycelia are incubated at 24°C for three days to form zoosporangia. Cultures are flooded with distilled water and transferred to 8°C for 30 min to induce zoospores. Discharge of zoosporangia is facilitated within 45 min by transfer to room temperature (20±3°C). The mycelial suspension is filtered through two layers of cheesecloth and zoospores quantified with a haemocytometer to a concentration of 10^3 zoospores /ml. Seedlings at the 2 to 3-leaf-stage are dipped into the inoculum suspension for 1 min and then transferred to 50 ml beakers with 20 ml distilled water and held at 20°C under fluorescent lights and 10 h days. Disease symptoms were read 11 days after inoculation on a 0-4 scale: 0= no symptoms; 1= superficial browning of tap root and/or lateral; 2= moderate browning of roots with lesions > 2 cm and discoloration not penetrating pith; 3= all root surface with confluent brown lesions, pith discolored with rotted roots; 4= stem and canopy collapsed, seedling dead.

Southern blight. The causal organism for this tropical disease is *Sclerotium rolfsii* Sacc. This pathogen thrives in warm (>29° C) acidic soils (Sherf and MacNab 1986). Symptoms include dark brown or black lesions near the soil surface, lesions are covered with white mycelium when soil moisture is high. Later 1-2 mm sclerotia are formed. Plants are killed or often stunted and wilted. Fruit can also be infected. Resistance was reported in *L. pimpinellifolium* accessions PI 126932 (Mohr and Watkins 1959) and PI 126432 (Phatak and Bell 1983). Resistance was attributed to the precocious development of secondary periderm on the basal stem (Mohr and Watkins 1959). However, it seems that this resistance has not been deployed in tomato. Six breeding lines were released from Paul Leeper's breeding program at Texas A & M, but the source of resistance was not specifically known (Leeper et al. 1992). All six had 100% resistance as did PI 126432 (Leeper et al. 1992). They were all derived from a cross made in the 1950's of STEP 54 from the University of Hawaii and 'Southland' from the USDA-ARS at Charleston, SC. One or both of these lines may be the source of resistance. 'Southland' has *L. pimpinellifolium* in its background (Scott 1983). No inheritance studies have been conducted.

Phatak and Bell (1983) described a field inoculation procedure. Isolates were grown for 85 days on whole crimped oat grains in Fernbach flasks containing 300 g oats and 300 ml water. The oats and water had been previously autoclaved at 121°C for one hour on two successive days before the isolates were added. The oat grains with mycelia and sclerotia were passed through a 7- mm diameter sieve to break up clumps. Ten percent (w/w) fresh media was mixed with the sieved inoculum and stored at 3°C for 16 h before use. Ten to 15 g of inoculum per 30 cm of row was mixed into the top 5 cm of soil with a rake immediately after transplanting.

Verticillium wilt. This disease, which has a broad host range, is caused by the soil fungi *Verticillium dahliae* (Kleb) and *V. albo-atrum*. Resistance to race 1 is conferred by the single dominant gene *Ve* that was discovered in 1932 in the accession Peru Wild (Schaible et al. 1951). The first cultivars with resistance were released in 1952 (Cannon and Waddoups 1952) and most modern cultivars have this resistance. There has been considerable confusion in mapping the *Ve* gene as various Tomato Genetics Cooperative reports have linked it to chromosomes 4, 7, 9 or 12 (see Diwan et al. 1999). Diwan et al. (1999) presented strong evidence based on recombinant inbred, F_2, and introgression line populations that *Ve* is linked to RFLP GP39 on the short arm of chromosome 9. A co-dominant SCAR marker linked less than 0.67 cM from *Ve* has been developed that should be useful for tomato improvement (Kawchuk et al. 1998).

Strains of the fungus that are virulent on cultivars with *Ve*, collectively called race 2, have been reported in California and North Carolina, USA; Canada, Australia, Brazil, Morocco, and Crete (see Baergen et al. 1993). Race 2 was also found in greenhouses in Ohio, USA (Berry and Oakes 1987b) and Africa according to Stamova (2005). However, there may in fact be several different strains of the fungus that are virulent on race 1 resistant cultivars. Baergen et al. (1993) tested six lines with reported resistance to race 2 against four isolates; two from North Carolina, one from Spain, and one from Brazil. They found that resistance sources provided partial resistance, primarily to the isolate for which they were tested originally, but not to most other isolates. The resistance was not as great as that provided by the *Ve* gene for race 1. Thus, race 2 does not fit a gene for gene system. Instead, several strains that overcome *Ve* are different from each other and are controlled to some extent by different genotypes. Recently, Stamova (2005) reported a high level of resistance to verticillium wilt race 2 against 11 California isolates. The resistance, in a line called VEDA, derived from the subgenus *Eriopersicon* appeared to have a dominant inheritance when tested against 2 of the isolates in a genetic study. The VEDA resistance needs to be tested against isolates from other regions.

Inoculum is grown on PDA for a week as with *Fusarium* cultures. Screening is done with a seedling dip technique where plants at the cotyledon stage are inoculated with a slurry of the pathogen at 10^7 or more spores/ml and disease symptoms are rated about three weeks later before field transplanting. Symptoms include stunting, yellowing of foliage, tan discoloration of vascular bundles, wilting, and death. Symptoms are more severe when day length is reduced to 4 hours (Jones et al. 1975). Although verticillium wilt is considered to be a cool weather pathogen, Jones et al. (1977) found that symptoms increased on the *Ve* resistant cultivar 'Tropic' as temperature was increased from 20 to 32°C under short day (4 hour) conditions.

FOLIAR DISEASES

Cladosporium leaf mold. This foliar disease most often causes problems in greenhouse production under high humidity conditions. The pathogen is *Cladosporium fulvum* (*Fulvia fulvum*) Cke. Much basic research has been done on this host-pathogen interaction (Joosten and DeWit 1999). Resistance was discovered in the 1930's when the first dominant gene, *Cf-1*, was identified (Langford 1937). After Langford, much resistance research was done by Ernie Kerr and coworkers in Canada, who discovered 24 resistance genes (Kanwar et al. 1980a,b, Table 13.2). However, Haanstra (2000) found five of the genes were identical in their specificites using a potato virus X (PVX) expression system. The genes were renamed according to the elicitor or to the avirulence protein which they recognize (Table 13.2). Haanstra (2000) found two new genes and thus concluded there are 21 *Cf* genes (Table 13.2). When cultivars with single resistance genes are deployed, a virulent race generally appears rather quickly. *Cladosporium* resistance genes are at the opposite end of the durability spectrum from the *Stemphyllium* resistance gene discussed below. Thus, it has been suggested that at least two genes that have not yet been overcome by current isolates be incorporated into varieties (Lindhout et al. 1989). The race situation can

Table 13.2   Sources of *Cladosporium* leaf mold resistance genes, the chromosome they are on and other reported information (based on Kanwar et al. 1980a, b unless noted).

Gene	Source	Species	Chromosome	Comments
Cf-1	PI 270247	L. esculentum	1	Horizontal resistance
Cf-2	PI 270254	L. pimpinellifolium[z]	6	Allele of Cf-5.
Cf-3	PI 270252	L. pimpinellifolium	11	
Cf-4	PI 370084	L. hirsutum[z,y]	1	
Cf-5	PI 187002	L. pimpinellifolium	6	
Cf-6	PI 211839	L. pimpinellifolium	12	
Cf-7	PI 124161	L. pimpinellifolium	9	
Cf-8	PI 124161	L. pimpinellifolium	9	Same as Cf-4 (Gerlagh et al. 1989).
Cf-9	PI 126933	L. pimpinellifolium	1	
Cf-10	PI 124161	L. pimpinellifolium	8	
Cf-11	Ont. 7716 Mass No. 2	L. peruvianum	12?	
Cf-12	PI 124161	L. pimpinellifolium	8	
Cf-13	PI 211839	L. pimpinellifolium	11	
Cf-14	PI 211839	L. pimpinellifolium	3	
Cf-15	PI 211840	L. pimpinellifolium	3	
Cf-16	PI 211840	L. pimpinellifolium	11	
Cf-17	PI 126947	L. pimpinellifolium	11	
Cf-18	PI 126947	L. pimpinellifolium	2	=Cf-ECP2 (Haanstra 2000).
Cf-19	PI 126947	L. pimpinellifolium	2	
Cf-20	PI 126947	L. pimpinellifolium	2	=Cf-ECP2 (Haanstra 2000).
Cf-21	PI 126947	L. pimpinellifolium	4	
Cf-22	PI 126947	L. pimpinellifolium	1	
Cf-23	PI 126947	L. pimpinellifolium	5	=Cf-ECP2 (Haanstra 2000).
Cf-24	PI 126947	L. pimpinellifolium	7	=Cf-ECP2 (Haanstra 2000).
Cf-ECP3	G1.1153	L. esculentum	1	Possible Cf-4 allele.
	LA 3271	L. esculentum	?	
Cf-ECP5	G1.1161	L. esculentum	1	Derived from L. pimpinellifolium CGN 15529.

[z]Also found in *L. pennellii* (Stamova 1985).
[y]Also found in *L. peruvianium* (Lukyanenko 1991).

be rather confusing and it does not appear that anyone has done a thorough characterization of the resistance genes against all the pathogen races. Kerr (1977) presented results for genes *Cf1* through *Cf-11* for 5 common Canadian races, and this summary has been presented elsewhere (Lukyanenko 1991, Stevens and Rick 1986). Lindhout et al. (1989) tested 11 resistant lines for 12 races and found *Cf-6* was resistant to all of them.

Theoretically there could be 2^n races of the pathogen where n is the number of resistance genes (Day 1956). The highest number of reported races has been 18 (Lukyanenko 1991). There are Canadian and European nomenclature systems for races that are cross referenced in Table 13.3. The Canadian system is based on the order of appearance while the European system is based on the gene-for-gene relationships between the host and the pathogen. Thus for the latter, a race called 2.4 would be virulent on hosts with resistance genes *Cf-2* and/or *Cf-4*.

Breeders use genes *Cf-4* and *Cf-5* according to Lukyanenko (1991), but there are races that are virulent on this gene combination (Table 13.3). Some seed companies use *Cf-2* and an unknown gene (van den Bosch personnal comm.). Given the data of Lindhout et al. (1989), the unknown gene might be *Cf-6*. According to Beek et al. (1991) *Cf-9* was still effective in the Netherlands and was being used in cultivar development. Jones et al. (1993) showed that *Cf-2* and *Cf-5* were allelic or tightly linked on chromosome 5, while *Cf-4* and *Cf-9* were allelic or tightly linked on chromosome 1. Balint-Kurti et al. (1994) reported *Cf-4* and *Cf-9* were allelic and located on chromosome 1 2.6 cM distal to CP46 and 3.7 cM proximal to TG236. Haanstra et al. (1999) reported *Cf-ECP2* was 11.3 cM proximal to *Cf-4* and

Table 13.3 Reported races of *Cladosporium fulvum* and comparisons between the European and Canadian nomenclature (adapted from Lindhout et al. 1989 and Lukyanenko 1991).

Race designation	
European	*Canadian*
0	1
1	2
2	5
1.2.3	6
2.3	7
1.2	8
3	9
4	10
2.3.4	12
2.4	-
2.4.11	-
2.4.5	-
2.4.5.11	-
2.4.5.9.11	-
2.5.9	-
4.11	-
2.4.11	-
2.4.9.11	-

had no recombination with CT116. Haanstra (2000) mapped *Cf-ECP5* 3.5 cM proximal to *Cf-4* and 3.9 cM distal to TG236.

Inoculation (Lindhout et al. 1989) is done on plants at the 1- or 2-leaf stage using a mixture of common pathogen races, usually five in number. A suspension of 10^6 spores/ml is sprayed on the plants and they are kept in a chamber at 100% RH and 20°C for 48 h. The plants are then kept in a greenhouse at 20°C for 2 weeks with the RH at >70% with a humidifier. Plants with no or arrested symptoms are considered resistant while those with extensive disease development are susceptible. Kerr (1977) used the following scale: 1= lesions 10 mm diameter, sporulating freely, surface whitish; 2= delayed sporulation, extensive chlorosis; 3= chlorotic spots 5-8 mm diameter, sporulation under low light and high humidity; 4=chlorotic flecks 2-3 mm diameter, no sporulation; 4.5= necrotic flecks 1 mm diameter; 4.8= pale flecks 1 mm diameter; and 5= immune, no visible infection.

Early Blight. Early blight of tomato, caused by the fungus *Alternaria solani* (Ellis and Martin) Jones and Grout, occurs worldwide in areas with frequent rainfall and/or heavy dew. Symptoms of early blight include stem lesions or cankers, referred to as collar rot in tomato seedlings, leaf spotting and blighting, and sunken lesions at the stem end of the fruit. Lesions are dark brown with concentric rings, producing a chlorotic halo as they expand on leaves to the blighting phase. Early blight is more severe on plants that are stressed nutritionally or bear a heavy fruit load (Barratt and Richards 1944). It is particularly severe on early maturing varieties with a small, determinate plant type. Varieties with the curled foliage trait are more susceptible to early blight, and many other foliar fungal and bacterial diseases, than varieties with non-curled foliage. Varieties with potato leaf foliage are reportedly less affected by early blight than are varieties with normal, serrated leaves. Later maturing varieties, particularly of the indeterminate type, are generally less affected by early blight than the determinate varieties, probably because of less stress resulting from lower fruit to foliage ratio. Highly vigorous sterile plants with little or no fruit generally develop little early blight in field plantings where plants with normal fruit load are severely affected by the disease (Gardner, personal observations).

Early breeding of resistant varieties was focused on incorporation of collar rot resistance because of the severity of this phase of the disease in field grown transplants before the advent of highly effective fungicides (Andrus et al. 1942a). The sources of resistance used for collar rot, primarily 'Campbell 1943', also confer a moderate level of leaf spot resistance and resistance to the fruit lesion phase of the disease (Andrus et al. 1942b, Barksdale and Stoner 1977, Gardner 1990). The breeding lines NC EBR-2, -3, and -4 and the hybrid 'Mountain Supreme' have resistance derived from 'Campbell 1943' (Gardner 1988, Gardner and Shoemaker 1999). In addition to stem lesion resistance, more recent breeding has included foliar resistance. The early blight resistant processing tomato, 71B2, developed by Barksdale and Stoner (1973, 1977), has moderate foliar resistance but is susceptible to the stem and fruit lesion phases of the disease. Recently developed fresh market plum tomatoes derived from the 71B2 resistance source include NC EBR-5, -6, -7, and -8 and the hybrids 'Plum Dandy' and 'Plum Crimson' (Gardner 2000).

Numerous small-fruited accessions of *L. hirsutum, L. pimpinellifolium* and other wild species with indeterminate growth habit and a high foliage to fruit ratio have been identified as highly resistant to early blight (Foolad et al. 2000). The high level of early blight resistance from these sources may be lowered as the resistance genes are advanced into an acceptable background. A good example of this is the resistance in NC EBR-1, derived from the *L. hirsutum* selection PI 126445 (Gardner 1988, Nash and Gardner 1988a, b). The original PI 126445 plants were free of early blight symptoms in the field and in greenhouse inoculations of seedlings as was the F_1 hybrid of *L. esculentum* x PI 126445. With additional backcrosses to *L. esculentum* and selection for determinate growth habit with a heavy set of large fruit, the resistance level became lower. Screening of early backcrosses to *L. esculentum* was possible in the greenhouse, but in more advanced backcrosses as the resistance level became lower, field screening had to be used because of the failure to clearly distinguish resistance in greenhouse tests. The lower level of resistance to early blight in NC EBR-1 compared to the original *L. hirsutum* selection is likely to have resulted from the additional stress imposed on the plant by advancing to a determinate background and earlier maturity with a heavy set of large fruit as well as the possible loss of resistance genes linked to undesirable traits. This situation is typical of the challenge repeatedly facing breeders in transferring quantitative traits from a highly vigorous wild background to commercially acceptable varieties.

Screening for stem lesion resistance can be done efficiently in a simple plastic-covered growth chamber in the greenhouse (Gardner 1990). The method developed by Barksdale (1969), or modifications thereof, are used to produce early blight spores. This method consists of growing the fungus in petri dishes of lima bean agar, scraping the mycelium to flatten it, and exposing the inverted, uncovered dishes to an alternating light and dark period to induce sporulation. Spores are washed from the dishes into distilled water and the inoculum concentration adjusted to 10,000 spores/ ml. Spores are atomized in late afternoon onto the stems of 6-7-week-old plants using a plastic household spray bottle. For the development of disease, plants need to be kept wet during the night. This is easily accomplished by misting the plants in late evening prior to closing the chamber and running a humidifier to introduce moisture into the chamber during the night. During the day the chamber is opened and the foliage allowed to dry. Disease ratings can be made approximately one week following inoculation. Resistance from sources such as 'Campbell 1943', which is highly resistant to stem lesions, is expressed as minute necrotic flecks on the stems compared to large expanding lesions on the stems of susceptible plants (Gardner 1990). Attempts to screen for moderate levels of foliar resistance in the greenhouse have been less successful than the stem lesion resistance because leaf spot development on resistant plants is not consistently limited enough to clearly distinguish susceptible from resistant plants. Screening for foliar resistance in the greenhouse, however, is useful for determining a high level of early blight resistance in wild tomato selections (Foolad et al. 2000).

Some believe that early blight resistance is associated with low fruit yield and late maturity. Yield trials with 'Mountain Supreme', an early blight resistant hybrid, showed that its yields over several seasons were not less than other commercial varieties (Gardner and Shoemaker 1999). However, attempts to produce early maturing tomato cultivars with an acceptable level of early blight resistance have not been successful, presumably because of the stress resulting from a heavy fruit load early in development of the plant.

Sources of early blight resistance incorporated into acceptable cultivars have been moderate, but appear to be durable. This resistance is useful in reducing the frequency of fungicide applications needed for control of early blight, but generally does not eliminate the need for chemical control (Shoemaker and Gardner 1986, Gardner and Shoemaker 1999). Early blight resistance in tomato follows the disease development pattern seen with other moderate, horizontal plant disease resistances. Under conditions of low inoculum pressure, the resistance is highly effective but can be overcome under conditions of high inoculum pressure and environmental factors favorable for the disease. Resistance is most effective when resistant cultivars are not grown in close proximity to susceptible cultivars and in fields where there is not a heavy carryover of inoculum from a previous tomato crop. Because of the additive nature of genes for early blight resistance

(Maiero et al. 1989, Nash and Gardner 1988a) both parents need to carry the resistance to achieve a useful resistance level in hybrids. Most disease resistances in commercial tomato hybrids are conferred by single dominant genes in heterozygous condition. This simplifies the breeding, allows for using genes that are associated with deleterious effects when they are homozygous but not heterozygous, and makes it easier to combine multiple disease resistance genes in hybrids. From a seed company standpoint, a further limitation of varieties with moderate horizontal resistance is that the resistance is subject to being overcome under severe conditions. As a result the company has to be more cautious about claiming disease resistance because of possible liability issues.

QTLs associated with early blight resistance in wild tomato species have been identified (Table 13.1; Zhang et al. 2003). Accordingly, Graham et al. (2005) did not find much resistance in any of the introgression lines developed from *L. hirsutum* assession LA/777 (Monforte and Tanksley 2000). If useful markers can be developed for these QTLs, the transfer of early blight resistance into cultivated backgrounds should be facilitated. Because single dominant genes conferring a high level of resistance to early blight have not been identified, the breeder is challenged with transferring and accumulating quantitative genes into a useful horticultural background. Based on the undesirable linkages and pleiotropic effects generally encountered in such breeding approaches, continued development of acceptable early blight resistant cultivars, particularly early maturing determinate types with high fruit to foliage ratio, will remain a challenge.

Gray leafspot. This warm weather foliar disease can be caused by four species of *Stemphyllium; S. solani* Weber, *S. floridanum* Hannon and Weber, *S. botryosum* Wallr. and *S. vesicarum* (Waller.) Simmons (Bashi et al. 1973). The circular lesions on the foliage are shiny, smooth and gray on the underside of the leaf. Under dry conditions, the lesions become tan in color. When the disease is severe, plants can be defoliated as lesions coalesce. Andrus et al. (1942b) identified resistance in *L. pimpinellifolium* PI 79532, which was the source of fusarium wilt race 1 resistance (*I*). Resistance is conferred by a single incompletely dominant gene *Sm* (Hendrix and Frazier 1949). The *Sm* gene is linked to the *I* gene with a crossover value of 36% (Dennett 1950). Behare et al. (1991) mapped *Sm* on chromosome 11 between RFLP markers T10 and TG110 at 4.1 and 6.8 cM from the markers, respectively. The distance of *Sm* from TG105, the marker linked to *I-2*, was 30.2 cM. The approximate distance to TG523 which marks the *I* gene is 31.1cM (Tanksley et al. 1992). These linkages coupled with preferential fertilization for *I* (Kedar et al. 1967) and *Sm* (Behare et al. 1991) have contributed to the prevalence of this resistance in tomato varieties. The *Sm* gene has been widely deployed in cultivars since 1950 and there are no reports of resistant fungal races overcoming the resistance. It is a rather rare example of a non-viral, durable, single gene (Parlevliet 2002).

A screening method originally developed by Blanchard and Laterrot (1986) was described by Behare et al. (1991). The pathogen is grown on a solid medium consisting of 200 ml of V8 vegetable juice, 2 g of $CaCO_3$, 18 g of gelose, and 800 ml of dH_2O. Seedlings three weeks past germination are sprayed to runoff with a suspension of 10^4 conidia/ml. The plants are transferred to a moist chamber at 24°C, sprayed twice a day with water, covered with a plastic sheet the first four days after inoculation, and illuminated for 12 h with fluorescent light at 4,000 lx. After 10 days plants are assessed for disease severity on a 1-5 scale where 1= coalescence of lesions, 2= numerous lesions, 3= few lesions, 4=rare lesions, and 5= no lesions.

Gray mold. This disease, named for the color of its fluffy mycelial growth, is most common in greenhouse culture under cool conditions. The necrotrophic causal organism *Botrytis cineraria* Pers.:Fr. affects all above ground plant parts with fruit rots often causing the most serious losses. However, if stems are infected, often from pruning wounds, girdling can occur and whole plants can be lost. This is especially serious for the intensive production of greenhouse tomatoes. Despite considerable interest in greenhouse tomato breeding, no resistant varieties have been developed due to a lack of resistant sources. Lobo et al. (1986) reported *L. esculentum* x *Lycopersicon esculentum* var. *cerasiforme* hybrid PI 119214 had fruit resistance after puncture inoculation of fruit. Supression of botrytis fruit rot has also been reported in *rin* and *nor* genotypes (Lavy-Meir et al. 1989). Hybrids heterozygous for *nor* also suppressed symptoms and this appears to be one advantage of *nor* hybrids in the market. Chetelat et al. (1997) inoculated stems of an intergeneric *L. esculentum* x *Solanum lycopersicoides* LA2951 hybrid and found it had resistance. Later, the same hybrid was also found resistant when detached leaves and entire seedlings were inoculated (Guimaraes et al. 2004). In this work the lesion size and spore production were reduced and hyphae were killed in the resistant germplasm. Egashira et al. (2000) tested six tomato cultivars, 44 wild tomato accessions, and *S. lycopersicoides* accession LA2386 for resistance using leaf and stem assays. They found no correlation between the assays, but *L. peruvianum* LA2745,

L. hirsutum LA2314 and *L. pimpinellifolium* LA1246 showed high resistance in both leaflets and stems. Nicot et al (2002) also found no correlation between stem and leaf reactions. *L. chilense* accession LA1969 and *L. chmielewskii* accession 731089 were in the most resistant categories for both stem and leaf assays of the 19 wild species accessions tested. In this study, LA2745 had no lesions two weeks after inoculation. It was suggested that this was a promising source of resistance in breeding.

Although several possible sources of resistance have been reported, much of their breeding application remains to be accomplished. Little is known about how strain affects on resistance. Inoculation methods need to be evaluated against performance in the field and an optimal screening technique adopted. The introgression lines derived from *S. lycopersicoides* LA2951 (Chetelat and Meglic 2000) should be tested for resistance to move the breeding forward. The reported resistance sources should be compared and those with useful resistance need to be introgressed and possibly combined.

Late Blight. Late blight, caused by *Phytopthora infestans* (Mont.) de Bary, is a destructive disease of potato and tomato occurring world wide. Wet, cloudy periods with cool temperatures are most conducive to development of the disease, and under such conditions, the disease develops rapidly, destroying all above ground plant parts.

Single dominant resistant (R) genes conferring a high level of resistance to late blight in potato and tomato are relatively common in wild species. Late blight resistance in potato follows the typical gene for gene interaction between host plant and disease organism (Al-Kherb et al. 1995). A major shortcoming of the R genes is the ability of the late blight organism to quickly overcome the resistance mechanisms making them ineffective. In potato in particular this is a severe problem since strains capable of overcoming R gene resistances singly or in combinations can readily carry over in potato tubers to infect subsequent crops. The airborne nature of late blight over long distances makes single gene resistance of little value because of the potential for rapid and widespread dissemination of strains capable of overcoming a particular resistance gene. Occurrence of both the A-1 and A-2 mating types of the late blight fungus outside Mexico makes possible sexual recombination and overwintering of oospores independent of live plant tissue. Moderate, or horizontal, resistance to late blight is of questionable value because of the short disease cycle, heavy spore production of the late blight organism, and the ability of the disease to be quickly spread by airborne spores over long distances.

Named genes currently being used in tomato breeding include *Ph-1* (Bonde and Murphy 1952, Gallegly 1952), *Ph-2* (Turkensteen 1973, Laterrot 1975), and *Ph-3* (Black et al. 1996). Because of the high level of resistance conferred by these genes, it is easy to screen for them in breeding. So far, no linkage to undesirable traits or adverse pleiotropic effects have been reported with the *Ph* genes. The *Ph-1* gene has been of limited value because of the rapid occurrence of strains capable of overcoming it (Conover and Walter 1953). *Ph-2* is considered to represent incomplete resistance (Turkensteen 1973, Laterrot 1975), showing good resistance in early season but sometimes failing in late season. *Ph-3* provides a very high level of resistance but like *Ph-1* and *Ph-2* has been overcome (Brusca 2003). Varieties with *Ph-1* and *Ph-2* have been developed (Walter 1967, Laterrot 1997) but do not occupy an important role in tomato production. All three *Ph* genes have been mapped (Fig. 12.1 – bacterial resistance Chapter). *Ph-1* was mapped to the distal end of chromosome 7 (Pierce 1971). *Ph-2* is flanked by CP105 and TG 233 on chromosome 10 (Moreau et al. 1998). *Ph-3* was mapped to TG 591A on chromosome 9 (Chunwongse et al. 2002). A co-dominant PCR marker for *Ph-3* has been developed (M. Mutschler, personal communication).

Numerous selections from the species *L. hirsutum* have been identified as highly resistant to late blight. Studies with the accession LA1033 using traditional breeding methods and quantitative trait analysis indicate that its resistance is based on two or more genes (Lough 2003) and that the resistance in an *L. esculentum* background is not as high as that conferred by the *Ph* genes (Gardner, unpublished). QTLs were identified for late blight resistance for the *L. hirsutum* accession LA2099 but were found to be linked to undesirable horticultural traits (Brouwer and St. Clair 2003, Brouwer et al. 2004), indicating a need to develop closely linked molecular markers for the QTLs to efficiently transfer resistance to an acceptable background.

Breeding in North Carolina is currently focused on combining late blight resistance with early blight resistance and on investigating the effects of combining various late blight resistance genes in an effort to increase durability of resistance. Hybrids with a combination of *Ph-2* and *Ph-3* appear to be useful in this breeding effort (Brusca 2003). However, under ideal environmental conditions for late blight, heavy inoculum pressure, and presence of other lines with the single genes in the same plantings; lines with resistance conferred by the combination of the two genes in both heterozygous and homozygous condition were overcome by late blight at two field locations in North Carolina in the summer of 2004 (Gardner, unpublished). In plantings in North Carolina, a hybrid combining heterozygous *Ph-2* and *Ph-3* when grown in the absence of lines carrying the single genes, resistance held up throughout the season.

Late blight resistance screening techniques using detached leaves or leaflets have been developed. Also, it is easy to select strains of late blight that differentially overcome the known *Ph* genes. In the North Carolina breeding program, strains specifically overcoming *Ph-2* and *Ph-3* are maintained by weekly transfers onto live tomato leaves of breeding lines that carry the specific gene overcome by the late blight strain. Whole detached leaves from 6 to 8-week-old plants are inoculated by atomizing leaves with a water suspension of 5000 sporangia/ml and incubating the leaves in clear sealed plastic containers in a growth chamber with 12 hours alternating light (21°C) and dark (16°C) . Disease ratings are made 5-7 days after inoculation. By inoculating an F_2 population derived from a hybrid heterozygous for *Ph-2* and *Ph-3* combined, plants with a very high level of resistance to combined isolates were identified in detached leaf inoculation tests. Progeny testing of F_3 lines in the field and F_4 lines detached leaf inoculations, verified homozygosity for *Ph-2* and *Ph-3* in some selections.

Because the genomes of potato and tomato are closely related and at least 15 R genes are present in potato (Umaerus et al. 1994, Van der Plank 1971), it is likely that there are numerous other R genes in tomato, particularly in *L. pimpinellifolium* from which the three currently identified *Ph* genes were derived. Although the use of R genes in potato breeding has met with limited success because of the lack of durability of resistance (Black and Gallegly 1957, Ross 1986), the potential usefulness of combinations of R genes in tomato should still be investigated. Breeding for late blight resistance in tomato presents a dilemma often faced in the development of useful resistance to plant diseases. The R genes confer a high level of resistance, are easy to screen and incorporate, and do not appear to be linked to genes with deleterious effects. Their usefulness, however, is limited by the lack of durability of resistance. Horizontal resistance to late blight is likely to be of limited value because the moderate level of resistance conferred is not sufficient to be of value with a disease as aggressive as late blight. In addition, the quantitative resistance sources are more difficult to breed because it is more difficult to develop simple, clearcut screening techniques and because of the increased chances for linkage to undesirable traits or deleterious pleiotropic effects associated with multiple genes. Pyramiding of resistance genes has been proposed for many years as a method of providing more durable disease resistance, and breeders have been criticized for using vertical genes, which can be easily overcome. Therefore, single dominant resistance genes continue to be used almost exclusively for major disease resistances in tomato because of the difficulty of developing commercially acceptable varieties with adequate levels of multiple gene resistance. Marker-assisted selection may be of value in developing improved late blight and other disease resistances based on multiple genes if markers tightly linked to QTLs can be identified that facilitate combining multiple resistance genes and reducing linkages to undesirable traits.

Powdery mildew. Historically, the causal organisms of the diseases referred to as powdery mildew were *Oidium lycopersici* and *Leveillula taurica*, the latter occurring in subtropical regions. Recently the nomenclature of the former species outside Australia has been changed to *O. neolycopersici* (Bai 2004). Morphologically the mycelium of *L. taurica* grows into the leaf mesophyll on the underside of the leaf, while *O. neolycopersici* grows mainly on the upper side and does not penetrate into the mesophyll (Lindhout et al. 1994). Powdery mildew has spread around the world often in greenhouse production but also in the field in some instances. Resistance to *Oidium* species has been found in several wild species (Lindhout et al. 1994, Ciccarese et al. 1998, Kozik 1999, Mieslerova et al. 2000). Incompletely dominant genes *Ol-1* and *Ol-3* were introgressed from *L. hirsutum* accessions G1560 and G1.1290, respectively, and they are probably allelic (Huang et al. 2000). They are flanked by RFLP markers TG153 and TG 164 on the long arm of chromosome 6 near the *Mi* gene. SCAR markers are available and are used for breeding purposes (Huang et al. 2000). The recessive *ol-2* gene was derived from *L. esculentum* var. *cerasiforme* and has been mapped to chromosome 4. CAPS and AFLP markers are available for marker-assisted selection (De Giovanni et al. 2004). Recently, Bai et al. (2003) found three QTLs linked to quantitative genes derived from *L. parviflorum* accession G1.1601 that could be added to other resistance genes to provide more durable resistance. One was on chromosome 6 in the same region as *Ol-1* and *Ol-3* and two were on chromosome 12 with one being in the region of *Lv* (see below). Bai et al. (2003) also mapped a gene (*Ol-4*) from *L. peruvianum* accession LA2172 to chromosome 6 in the region near markers *Aps-1*, TG153, and TG25 and close to *Ol-1*. Another gene (*Ol-5*) from *L. hirsutum* PI 247087 was mapped to chromosome 6 near marker TG 25 and between *Ol-4* and *Ol-1/Ol-3* (Bai 2004). Dominant *Ol* genes conferred isolate-dependent hypersensitive resistance that was whole the recessive *Ol-2* gene associated with papillae formation (Bai et al. 2005).

A resistance gene to *L. taurica* (*Lv*) was first introgressed from *L. chilense* accession LA 1969 by Yordanov et al. (1975). Later this gene was named

Lv by Stamova and Yordanov (1987). More recently it was fine mapped to chromosome 12 in a 0.84 cM interval between RFLP markers CT121 and CT129 with CT 121 being only 0.16 cM from the gene (Chunwongse et al. 1997). PCR markers are presently being used by tomato breeders to incorporate this resistance gene into tomato varieties (Barten, pers. comm.).

Disease screening is generally done by washing spores from infected leaves and adjusting the suspension to $2x10^4$ spores/ml. Plants to be tested are about one month old and have 3-4 true leaves. They are sprayed with the spore suspension and grown in greenhouses at about 20°C ± 3 with 30-70% RH. Fungal growth is rated at intervals from 10-20 days later on a scale where 0= no sporulation; 1 = slight sporulation with less than 5% of the foliage affected, 2= moderate sporulation with 5-30% of the area affected ; and 3 = abundant sporulation with over 30% of the area affected (Bai 2004).

Septoria leaf spot. This disease caused by *Septoria lycopersici* Speg. infects tomatoes in eastern Canada and the United States. Barksdale and Stoner (1978) reported resistance in *L. pimpinellifolium* accession PI 422397 that was controlled by a single dominant gene. Later, it was reported that this resistance was not incorporated into cultivars because the resistance was moderate and associated with small fruit size and lateness (Poysa and Tu 1993). To date we are not aware of any resistant cultivars.

Screening for resistance can be accomplished by a brushing method (Tu and Poysa 1990). Spore suspensions were prepared from 3-wk-old colonies on PDA; 5 ml of sterile distilled water was added to each plate and the surface of the culture was scraped to dislodge the spores. Conditions considered optimal for inoculation were an inoculum concentration of 10^6 spores/ml, temperature at 24°C, moisture period of 48 h, and a photoperiod of 14 h. The brushing of adaxial and abaxial leaf surfaces of plants at the three-leaf stage with a camel hair brush was superior to spraying or dipping methods tested. Disease scoring was done 8 to 15 days after inoculation on a 0-9 scale where 0= <10% leaf area with symptoms, 1 = 10-20%, 2= 20-30% to 9=90-100%. However, in a later report, the same authors used the method of Gardner (1990) to inoculate over 500 breeding lines and over 200 wild species to find resistance (Tu and Poysa 1993). Presumably this method was used because it is easier (see the early blight section for this procedure.). All tomato varieties and accessions tested were susceptible. Highest levels of resistance (<3) were found in *L. hirsutum* accessions LA 2100, LA 2650, LA 2204, and PE-36; in *L. peruvianum* accessions LA 1675, LA 1360, PI 251307, PE-33, PI 270435, PI 390655; and

L. pennellii accession PE-44. High resistance (3-3.9) was found in *L. hirsutum* accessions LA 1366, LA 2124, and LA 255; and in *L. peruvianum* accessions LA 1910, LA 1292, LA 1304, LA 1365, PI 128654, PI 390671, PI 365951, and PE-32. Poysa and Tu (1993) also found inter specific hybrids from resistant parents × sensitive parents were resistant, indicating dominant inheritance of this trait.

Target spot. This foliar disease is caused by *Corynespora cassiicola* (Berk & Curt.). It can be controlled by fungicides if detected early, but if not, it can cause serious problems with defoliation and fruit rots. Bliss et al. (1973) tested 242 accessions and reported high resistance from *L. esculentum* PI 120265 and *L. pimpinellifolium* PI 112215. Slight resistance was reported in 12 other accessions of the above two species. The high resistance lines have been effective in Florida (Scott, unpublished data). The resistance from PI 120265 and PI 112215 was conferred by the same incompletely dominant gene (Bliss et al. 1973). No highly resistant varieties have been developed, but Blazquez (1972) reported 'Homestead 24' was less susceptible than 'Bonny Best' indicating that some partial resistance may be present in some varieties.

Disease screening has been done in greenhouses by inoculating plants at the four-leaf stage with a suspension of 80,000 spores/ml mixed with one drop/100 ml Tween 80 as a wetting agent. The plants were sprayed until wet and placed in a mist chamber at 20-24°C for 24 h. The disease reaction could be scored 36 h later. Inoculum was prepared by washing canisters from 5-day old cultures on PDA plates exposed to continuous, white florescent light at 2,690 lux. Inoculum was then filtered through cheesecloth (Bliss et al. 1973).

FRUIT

Anthracnose. Anthracnose of tomato is caused by several species of the genus *Colletotrichum*, including *C. coccodes (Wallr.)*, *C. dematium* (Pers.) Grove, and *C. gloeosporoioides* (Penz.) Penz. & Sacc. *C. coccodes* is most commonly associated with fruit anthracnose symptoms which are sunken, dark, circular lesions. Symptoms express on ripe fruit and the disease is primarily a problem on processing tomatoes that are being ripened on the vine in humid regions such as the Northeast and Midwestern US. It is also found in Asia, Europe, Africa, and the East Indies. Screening for fruit rot is done by placing a suspension droplet of the pathogen at 5,000 spores/ml on red ripe fruit with a hypodermic needle and then pricking the epidermis under the droplet with the needle (Robbins and Angell 1970a). This can be done at room temperature and humidity. Lesion size is measured nine days after inoculation. Susceptible lines have larger lesions and resistant lines have smaller or no lesions. The method overcame the problem of other methods where lines selected for resistance turned out to be susceptible in the field. Resistant fruit using this technique apparently have an internal environment that is not conducive to growth of the pathogen.

Resistance to anthracnose caused by *C. coccodes* was first reported in *L. pimpinellifolium* PI 127833 and *L. esculentum* x *L. pimpinellifolium* PI 129027, in 1964, according to Robbins and Angell (1970b). Resistance reported in *L. esculentum* accession PI 272636 (Barksdale 1971) and PI 272636 was also found to be resistant to five other *Colletotrichum* spp. fungi that cause anthracnose (Barksdale 1972). Robbins and Angell (1970b) found resistance from PI 129027 and PI 127833 was partially dominant and controlled by six genes. All other genetic studies have been based on resistance derived from PI 272636. Miller et al. (1983) studied this resistance in a diallel using *C. dematium* as the pathogen and found general combining ability (GCA) was highly significant while specific combining ability (SCA) was smaller but also significant. There was partial dominance for susceptibility. Narrow sense heritability was 70% over the 2 locations in the experiment. A study comparing *C. coccodes* and *C. dematium* had similar but not identical results as SCA was not significant; narrow sense heritability was 78% for the former and 64% for the latter (Miller et al. 1984). Later Ng et al. (1990) did a generation means analysis using *C. dematium* as the pathogen and found inheritance was primarily additive but there was also dominance and homozygous x homozygous epistasis. Broad-sense heritability was 0.57 while narrow-sense heritability was 0.42. Stommel and Haynes (1998) did a genetic study using an advanced breeding line, 88B147, inoculated with *C. coccodes*. They found partial dominance for resistance. An additive-dominance model with an additive x additive (homozygous x homozygous) epistatic interaction best explained the data. Broad sense heritability was 0.42 and narrow sense heritability was only 0.004. One gene or linkage group was estimated to control resistance. However, Stommel and Zhang (1998, 2001) found RAPD's linked to resistance in an F_2 population derived from 88B147 and reported at least 3 regions were associated with resistance. Later, Stommel (2001) studied inheritance using three breeding lines at various stages of horticultural advancement, 88B147 being the most advanced line. Heritability and gene number estimates declined as the level of resistance declined in more horticulturally advanced lines.

Inheritance was primarily additive. This may help to explain some of the descrepencies between studies. It has been difficult to transfer resistance from the small fruited wild species into advanced breeding lines and varieties because of the polygenic nature of the resistance where each gene has a small effect (Stommel 2001). This is like an example of linkage drag problems that hamper the development of varieties with quantitative resistance as discussed by Scott (2005) for early blight.

Black mold and Phoma. As indicated for anthracnose, breeding for resistance to fruit rots is difficult, varieties with such resistances are lacking. There has been some work done with black mold incited by *Alternaria alternata* (Fr.) Keissler (Cassol and St. Clair 1994, Robert et al. 2001) and *Phoma andina* (Lobo et al. 1987, 1988). Readers interested in resistance to these diseases are referred to these references.

REFERENCES

Alexander, L. J. 1963. Transfer of a dominant type of resistance to the four known Ohio pathogenic strains of tobacco mosaic virus (TMV) from *L. peruvianum* to *L. esculentum*. Phytopathology 53:896.

Alexander, L. J. and C. M. Tucker. 1945. Physiological specialization in the tomato wilt fungus *Fusarium oxysporum* f. *lycopersici*. J Agr Res 70:303-313.

Al-Kherb, S. M., C. Fininsa, R. C. Shattock, and D. S. Shaw. 1995. The inheritance of virulence of *Phytophthora infestans* to potato. Plant Pathol 44:552.

Alon, H., J. Katan, and N. Kedar. 1974. Factors affecting penetrance of resistance to *Fusarium oxysporum* f.sp. *lycopersici* in tomatoes. Phytopathology 64:455-461.

Andrus, C. F., G. B. Reynard, H. Jorgensen, and J. Eades. 1942a. Collar rot resistance in tomatoes. J Agr Res 65:339-346.

Andrus, C. F., G. B. Reynard, and B. L. Wade. 1942b. Relative resistance of tomato varieties, selections and crosses to defoliation by *Alternaria* and *Stemphylium*. US Dep Agric Circ 652.

Baergen, K. D., J. D. Hewitt, and D. A. St. Clair. 1993. Resistance of tomato genotypes to four isolates of *Verticillium dahliae* Race 2. HortScience 28(8):833-836.

Bai, Y. 2004. The genetics and mechanisms of resistance to tomato powdery mildew (*Oidium neolycopersici*) in *Lycopersicon* species. Thesis. Wageningen University, The Netherlands.

Bai, Y., C. C. Huang, R. Van der Hulst, F. Meijer-Dekens, G. Bonnema, and P. Lindhout. 2003. QTLs for tomato powdery mildew resistance (*Oidium lycopersici*) in *Lycopersicon parviflorum* G1.1601 co-localize with two qualitative powdery mildew resistance genes. Mol. Plant Microbe Interact 16:169-176.

Bai, Y., R. Van der Hulst, C. C. Huang, L. Wei, P. Stam, and P. Lindhout. 2004. Mapping Ol-4, a gene conferring resistance to *Oidium neolycopersici* and originating from *Lycopersicon peruvianum* LA2172, requires multi-allelic single locus markers. Theor Appl Genet 109:1215-1223.

Balint-Kurti, P. J., M. S. Dixon, D. A. Jones, K. A. Norcott, and J. D. G. Jones.1994. RFLP linkage analysis of the Cf-4 and Cf-9 genes for resistance to *Cladosporium fulvum* in tomato. Theor Appl Gen 88:691-700.

Barksdale, T. H. 1969. Resistance of tomato seedlings to early blight. Phytopathology 59:443-446.

Barksdale, T. H. 1971. Inheritance of tomato anthracnose resistance. Plant Dis Reptr 55(3):253-256.

Barksdale, T. H. 1972. Resistance in tomato to six anthracnose fungi. Phytopathology 62:660-663.

Barksdale, T. H. and A. K. Stoner. 1973. Segregation for horizontal resistance to tomato early blight. Plant Dis Reptr 57:964-965.

Barksdale, T. H. and A. K. Stoner. 1977. A study of the inheritance of tomato early blight resistance. Plant Dis Reptr 61:63-65.

Barksdale, T. H. and A. K. Stoner. 1978. Resistance in tomato to *Septoria lycopersici*. Plant Dis Reptr 62:844-847.

Barratt, R. W. and M. C. Richards. 1944. Physiological maturity in relation to *Alternaria* blight in tomato. Phytopathology 34:997.

Bashi, E., M. Pilowski, and J. Rotem. 1973. Resistance in tomatoes to *Stemphylium floridanum* and *S. botryosum* f.sp. *lycopersici*. Phytopathology 63:1542-1544.

Beek, Hans van der, R. Verkerk, and P. Lindhout. 1991. Mapping gene Cf-9 (resistance to *Cladosporium fulvum*) on chromosome 1 of tomato. Rpt Tomato Genes Coop. 41:15.

Behare, J., H. Laterrot, M. Sarfatti, and D. Zamir. 1991. Restriction fragment length polymorphism mapping of the *Stemphylium* resistance gene in tomato. Mol Plant-Microbe Interact 4(5):489-492.

Berry, S. Z. and G. L. Oakes. 1987a. Inheritance of resistance to fusarium crown and root rot in tomato. HortScience 22(1):110-111.

Berry, S. Z. and G. L. Oakes. 1987b. Ohio 11 and 12, verticillium-wilt race-2-resistant greenhouse tomato breeding lines. HortScience 22(1):167.

Black, L. L., T. C. Wang, P. M. Hanson, and J. T. Chen. 1996. Late blight resistance in four wild tomato accessions: Effectiveness in diverse locations and inheritance of resistance. Phytopathology 86(11):S24 (abstr.).

Black, W. and M. E. Gallegly. 1957. Screening of *Solanum* species for resistance to physiological races of *Phytophthora infestans*. Amer Potato J 34:273-281.

Blaker, N. S. and J. D. Hewitt. 1987a. Comparison of seedling and mature plant resistance to *Phytophthora parasitica* in tomato. HortScience 22(1):103-105.

Blaker, N. S. and J. D. Hewitt. 1987b. A comparison of resistance to *Phytophthora parasitica* in tomato. Phytopathology 77:1113-1116.

Blanchard, D. and H. Laterrot. 1986. Les Stemphylia rencontres sur tomate. Phytopathol Medit 25:140-144.

Blazquez, C. H. 1972. Target spot of tomato. Plant Dis Reptr 56(3):243-245.

Bliss, F. A., P. T. Onesirosan, and D. C. Arny. 1973. Inheritance of resistance in tomato to target leaf spot. Phytopathology 63:837-840.

Bohn, G. W. and C. M. Tucker. 1939. Immunity to fusarium wilt in the tomato. Science 89:603-604.

Bolkan, H. A. 1985. A technique to evaluate tomatoes for resistance to Phytophthora root rot in the greenhouse. Plant Dis 69(8):708-709.

Bonde, R. and E. F. Murphy. 1952. Resistance of certain tomato varieties and crosses to late blight. Maine Agr Expt Sta Bul 497:5-15.

Boukema, I. W. 1982. Inheritance of resistance to *Didymella lycopersici* Klebb. in tomato (*Lycopersicon esculentum* Mill). Euphytica 31:981-989.

Bournival, B. L., J. W. Scott, and C. E. Vallejos. 1989. An isozyme marker for resistance to race 3 of *Fusarium oxysporum* f.sp. lycopersici in tomato. Theor Appl Genet 78:489-494.

Bournival, B. L., C. E. Vallejos, and J. W. Scott. 1990. Genetic analysis of resistances to races 1 and 2 of *Fusarium oxysporum* f.sp. lycopersici from the wild tomato *Lycopersicon pennellii*. Theor Appl Genet 79:641-645.

Bournival, B. L. and C. E. Vallejos. 1991. New source of genetic resistance to race 3 of *Fusarium* wilt of tomato. Plant Dis 75:281-284.

Brouwer, D. J., E. S. Jones, and D. A. St. Clair. 2004. QTL analysis of quantitative resistance to *Phytophthora infestans* (late blight) in tomato and comparisons with potato. Genome 47:475-492.

Brouwer, D. J. and D. A. St. Clair. 2003. Fine mapping of three quantitative trait loci for late blight resistance in tomato using near isogenic lines (NILs) and sub-NILS. Theor Appl Genet 108:628-638.

Brusca, J. P. 2003. Inheritance of tomato late blight resistance from 'Richter's Wild Tomato' and evaluation of late blight combinations in adapted fresh market tomato backgrounds. MS thesis, North Carolina State University, Raleigh.

Cannon, O. S. and V. Waddoups. 1952. Loran Blood and V.R. Moscow, two new Verticillium wilt resistance tomatoes for Utah. Utah Farm and Home Sci. 13(4):74.

Cassol, T. and D. A. St. Clair. 1994. Inheritance of resistance to blackmold (*Alternaria alternata* (Fr.) Keissler) in two interspecific crosses of tomato (*Lycopersicon esculentum* x *L. cheesmanii* f. *typicum*). Theor Appl Genet 88:581-588.

Chetelat, R. T., P. Cisneros, L. Stamova, and C. M. Rick. 1997. A male-fertile *Lycopersicon esculentum* x *Solanum lycopersicoides* hybrid enables direct backcrossing to tomato at the diploid level. Euphytica 95:99-108.

Chetelat, R. T. and V. Meglic. 2000. Molecular mapping of chromosome segments introgressed from *Solanum lycopersicoides* into cultivated tomato (*Lycopersicon esculentum*). Theor Appl Genet 100:232-241.

Chunwongse, J., C. Chunwongse, L. Black, and P. Hanson. 2002. Molecular mapping of Ph-3 gene for late blight resistance in tomato. J Hort Sci Biotechnol 77:281-286.

Chunwongse, J., S. Doganlar, C. Crossman, J. Jiang, and S. D. Tanksley. 1997. High-resolution genetic map of the Lv resistance locus in tomato. Theor Appl Genet 95:220-223.

Ciccarese, F., M. Amenduni, D. Schiavone, and M. Cirulli. 1998. Occurrence and inheritance of resistance to powdery mildew (*Oidium lycopersici*) in *Lycopersicon* species. Plant Pathol 47:417-419.

Cirulli, M. and L. J. Alexander. 1966. A comparison of pathogenic isolates of *Fusarium oxysporum* f. *lycopersici* and different sources of resistance in tomato. Phytopathology 56:1301-1304.

Clouse, S. D. and D. G. Gilchrist. 1987. Interaction of the asc locus in F_8 paired lines of tomato with *Alternaria alternata* f.sp. *lycopersici* and AAL-toxin. Phytopathology 77:80-82.

Conover, R. A. and J. M. Walter. 1953. The occurrence of a virulent race of *Phytophthora infestans* on late blight resistant tomato stocks. Phytopathology 43:344-345.

Day, P. 1956. Race names of *Cladosporium fulvum*. Rept Tomato Genet Coop. 6:13-14.

De Giovanni, C., P. Dell'Orco, A. Bruno, F. Ciccarese, C. Lotti, and L. Ricciardi. 2004. Identification of PCR-based markers (RAPD, AFLP) linked to a novel powdery mildew resistance gene (ol-2) in tomato. Plant Sci 166:41-48.

Dennett, R. K. 1950. The association of resistance to fusarium wilt and stemphylium leaf spot disease in tomato, *Lycopersicon esculentum*. Proc Amer Soc Hort Sci 56:353-357.

Diwan, N., R. Fluhr, Y. Eshed, D. Zamir, and S. D. Tanksley. 1999. Mapping of Ve in tomato: a gene conferring resistance to the broad-spectrum pathogen, *Verticillium dahliae* race 1. Theor Appl Genet 98:315-319.

Doganlar, S., J. Dodson, B. Gabor, T. Beck-Bunn, C. Crossman, and S. D. Tanksley. 1998. Molecular mapping of the py-1 gene for resistance to corky root rot (*Pyrenochaeta lycopersici*) in tomato. Theor Appl Genet 97:784-788.

Egashira, H., A. Kuwashima, and S. Imanishi. 2000. Screening of wild accessions resistant to gray mold (*Botrytis cinerea* Pers.) in *Lycopersicon*. Rpt Tomato Genet Coop 50:14-15.

Foolad, M. R. and A. Sharma. 2005. Molecular markers as selection tools in tomato breeding. Acta Hort 695(ISHS): 225-240.

Foolad, M. R., N. Ntahimpera, and B. J. Christ. 2000. Comparison of field, greenhouse and detached-leaflet evaluations of tomato germplasm for early blight resistance. Plant Disease 84:967-972.

Gabe, H. L. 1975. Standardization of nomenclature for pathogenic races of *Fusarium oxysporum* f. sp. *lycopersici*. Transaction of the British Mycological Society 64 (1):156-159.

Gallegly, M. E. 1952. Sources of resistance to two races of the tomato late blight fungus. Phytopathology 42:466.

Gardner, R. G. 1988. NC EBR-1 and NC EBR-2 early blight resistant tomato breeding lines. HortScience 23:779-781.

Gardner, R. G. 1990. Greenhouse disease screen facilitates breeding resistance to tomato early blight. HortScience 25:222-223.

Gardner, R. G. 2000. 'Plum Dandy', a hybrid tomato; and its parents, NC EBR-5 and NC EBR-6. HortScience 35:962-963.

Gardner, R. G. and P. B. Shoemaker. 1999. 'Mountain Supreme' early blight-resistant hybrid tomato and its parents, NC EBR-3 and NC EBR-4 tomato breeding lines. HortScience 34:745-746.

Gerlagh, M., W. H. Lindhout, and I. Vos. 1989. Allelic test proves genes Cf_4 and Cf_9 for resistance to *Cladosporium fulvum* (*Fulvia fulva*) on tomato to be undistinguishable. Neth J Plant Path 95:357-359.

Graham E., T. C. Wang, and P. Hanson. 2005. Preliminary evaluation of LA/777 introgression lines for early blight resistance. Rep Tomato Genet Coop 55: 15-18.

Grattidge, R. and R. G. O'Brien. 1982. Occurrence of a third race of fusarium wilt of tomatoes in Queensland. Plant Dis 66:165-166.

Grogan, R. G. K. A. Kimble, and I. Misaghi. 1975. A stem canker disease of tomato caused by *Alternaria alternata* f. sp. *lycopersici*. Phytopathology 65:880-886.

Guimaraes, R. L., R. T. Chetelat, and H. U. Stotz. 2004. Resistance to *B. cinerea* in *S. lycopersicoides* is dominant in hybrids with tomato, and involves induced hyphal death. Eur J Pl Path 110:13-23.

Haanstra, J. P. W., R. Lauge, F. Meijer-Dekens, G. Bonnema, P. J. G. M. de Wit, and P. Lindhout. 1999. The Cf-ECP2 gene is linked to, but not part of, the Cf-4/Cf-9 cluster on the short arm of chromosome 1 in tomato. Mol Gen Genet 262:839-845.

Haanstra, J. P. W., F. Meijer-Dekens, R. Lauge, D. C. Seetanah, M. H. A. J. Joosten, P. J. G. M. de Wit, and P. Lindhout. 2000. Mapping strategy for resistance genes against *Cladosporium fulvum* on the short arm of chromosome 1 of tomato: Cf-ECP5 near the Hcr9 Milky Way cluster. Theor Appl Genet 101:661-668.

Haanstra, J. P. W. 2000. Characterization of resistance genes to *Cladosporium fulvum* on the short arm of chromosome 1 of tomato. Thesis. Wageningen University, Wageningen, NL 119 pp.

Hendrix, J. . and W. A. Frazier. 1949. Studies of the inheritance of stemphylium resistance in tomatoes. Hawaii Agric Exp Stn Tech Bull 8.

Hogenboom, N. G. 1970. Inheritance of resistance to corky root in tomato. Euphytica 19:413-425.

Huang, C. C., Y.Y. Cui, C. R. Weng, P. Zabel, and P. Lindhout. 2000. Development of diagnostic PCR markers closely linked to the tomato powdery mildew resistance gene 0l-1 on chromosome 6 of tomato. Theor Appl Genet 101:918-924.

Huang, C. C. and P. Lindhout. 1997. Screening for resistance in wild *Lycopersicon* species to *Fusarium oxysporum* f. sp. *lycopersici* race 1 and race 2. Euphytica 93:145-153.

Jones, D. A., M. J. Dickinson, P. J. Balint-Kurti, M. S. Dixon, and J. D. G. Jones. 1993. Two complex resistance loci revealed in tomato by classical and RFLP mapping of Cf-2,

Cf-4, Cf-5, and Cf-9 genes for resistance to *Cladosporium fulvum*. Mol Plant-Microbe Interact 6:348-357.

Jones, J. P., P. Crill, and R. B. Volin. 1975. Effect of light duration on verticillium wilt of tomato. Phytopathology 65:647-648.

Jones, J. P., A. J. Overman, and P. Crill. 1977. Effect of temperature and short day on development of verticillium wilt of susceptible, tolerant, and resistant tomato cultivars. Proc Fla State Hort Soc 90:397-399.

Jones, John Paul, S. S. Woltz, and J. W. Scott. 1990. Factors affecting development of fusarium crown rot of tomato. Proc Fla State Hort Soc103:142-148.

Joosten, M. H. A. J. and P. J. G. M. De Wit. 1999. The tomato-*Cladosporium fulvum* interaction: a versatile experimental system to study plant-pathogen interactions. Ann Rev Phytopathol 37:335-367.

Kanwar, J. S., E. A. Kerr, and P. M. Harney. 1980a. Linkage of Cf-I to Cf-II genes for resistance to tomato leaf mold *Cladosporium fulvum* Cke. Rep Tomato Genet Coop 30:20-21.

Kanwar, J.S., E. A. Kerr, and P. M. Harney. 1980b. Linkage of Cf-12 to Cf-24 genes for resistance to tomato leaf mold *Cladosporium fulvum* Cke. Rep Tomato Genetics Coop 30:22-23.

Kawchuk, L. M., J. Hachey, and D. R. Lynch. 1998. Development of sequence characterized DNA markers linked to a dominant verticillium wilt resistance gene in tomato. Genome 41:91-95.

Kedar, N., N. Retig, and J. Katan. 1967. Non-random segregation of gene I for fusarium resistance in tomato. Euphytica 16:258-266.

Kerr, E. A. 1977. Problems and complications of developing new greenhouse varieties. In Proc. Can. Midwest Greenhouse Vegetable Grower. Ontario, Canada, pp. 45-47.

Kozik, E. U. 1999. Inheritance of resistance to powdery mildew (*Oidium lycopersicum* Cooke & Massee, emend. Noordeloos &Loerakker) in accessions of three wild species of *Lycopersicon*. J Appl Genet 40(3):175-183.

Kozik, E., M. R. Foolad, and R. A. Jones. 1991. Genetic analysis of resistance to phytophthora root rot in tomato (*Lycopersicon esculentum* Mill.). Plant Breed 106:27-32.

Langford, A. N. 1937. The parasitism of *Cladosporium fulvum* Cooke and the genetics of resistance to it. Can J Research 15:108-128.

Laterrot, H. 1975. Selection pour la résistance au mildiou, *Phytophthora infestans* Mont., de Bary chez la tomate. Ann Amél Plant 25(2):129-149.

Laterrot, H. 1976. Localisation chromosomique de 12 chez la tomate controlant la résistance au pathotype 2 de *Fusarium oxysporum* f. *lycopersici*. Ann Amel Plant 26:485-491.

Laterrot, H. 1983. La lutte genetique contre la maladie des racines liegueses de la tomate. Rev Hortic 238:143-150.

Laterrot, H. 1997. Breeding strategies for disease resistance in tomatoes with emphasis on the tropics: current status and research challenges. In: G. A. Maciel, G. M. B. Lopes, C. Hayward, R.R.L. Mariano, and E.A. de A Maranhao (eds.), First International Symposium on Tropical Tomato Diseases, ASHS Press, Alexandria, Va. USA, pp. 126-132.

Laterrot, H. and Conteadier. 1989. Linkage between TMV and FORL resistances. Rpt Tomato Genetics Coop 39:21.

Laterrot, H. and Moretti. 1992. FORL resistance in $Tm2^2$ -*ah* lines. Rpt Tomato Genet Coop 42:24-25.

Laterrot, H. and Moretti. 1996. FORL resistance in Tm2 and $Tm2^a$ lines from the TGRC stock. Rpt Tomato Genet Coop 46:17.

Laterrot, H. and J. Philouze. 1984. Recombination between pathotype 1 (I-2 allele) and susceptibility to pathotype 0 (I+allele) of *Fusarium oxysporum* f.sp. *lycopersici* in tomato (*Lycopersicon esculentum* Mill.). In Pays-Bas (ed.), Proc. Meeting of the

Eucarpia Tomato Working Group. May 1984. Wageningen, The Netherlands, pp. 70-74.

Lavy-Meir, G., R. Barkai-Golan, and E. Kopeliovitch. 1989. Resistance of tomato ripening mutants and their hybrids to *Botrytis cinerea*. Plant Dis 73:976-978.

Leeper, P. W., S. C. Phatak, D. K. Bell, B. F. George, E. L. Cox, G. E. Oerther, and B. T. Scully. 1992. Southern blight resistant breeding lines: 5635M, 5707M, 5719M, 5737M, 5876M and 5913M. HortScience 27(5):475-478.

Lindhout, Pim, Wanda Korta, Margonata Cislik, Ingrid Vos, and Thijs Gerlagh. 1989. Further identification of races of *Cladosporium fulvum* (*Fulvia fulva*) on tomato originating from the Netherlands, France and Poland. Neth J Plant Pathol 85:143-148.

Lindhout, P., G. Pet, and H. Van der Beek. 1994. Screening wild *Lycopersicon* species for resistance to powdery mildew (*Oidium lycopersicum*). Euphytica 72:43-49.

Lobo, M., H. Cortina, and R. Navarro. 1988. Inheritance of *Phoma andina* resistance derived from L. *hirsutum*. Rpt. Tomato Genetics Coop. 38:32-33.

Lobo, M. , R. Navarro, A. Bernal, and L. M. Llano. 1986. *Botrytis cinerea* resistance in tomato. Rpt Tomato Genet Coop 36:19.

Lobo, M., R. Navarro, M. M. Murilo, and A. C. Ruiz. 1987. *Phoma andina* resistance in *Lycopersicon hirsutum* accessions. Rpt Tomato Genet Coop 37:53-54.

Lough, R. C. 2003. Inheritance of tomato late blight resistance in *Lycopersicon hirsutum* LA 1033. Ph.D. Dissertation, North Carolina State University, Raleigh.

Lukyanenko, A. N. 1991. Disease resistance in tomato. *In:* G. Kalloo. (ed.), Genetic Improvement of Tomato. Monographs on Theoretical and Applied Genetics 14, Springer-Verlag, Berlin Heidelberg, Germany, pp. 99-119.

Maiero, M., T. J. Ng, and T. H. Barksdale. 1989. Combining ability estimates for early blight resistance in tomato. J Amer Soc Hort Sci 114:118-121.

McGrath, D. M. and R. N. Campbell. 1983. Improved methods for inducing sporulation of *Pyrenochaeta lycopersici*. Plant Dis 67:1245-1248.

McGrath, D. J., D. Gillespie, and L. Vawdrey. 1987. Inheritance of resistance to *Fusarium oxysporum* f. sp. *lycopersici* races 2 and 3 in *Lycopersicon pennelli*. Aust J Agric Res 38:729-733.

Mesbah, L. A., R. J. A. Kneppers, F. L. W. Takken, P. Laurent, J. Hille, and H. J. J. Nijkamp. 1999. Genetic and physical analysis of a YAC contig spanning the fungal disease resistance locus Asc of tomato (*Lycopersicon esculentum*). Mol Gen Genet 261:50-57.

Mieslerova, B., A. Lebeda, and R. T. Chetelat. 2000. Variation in response of wild *Lycopersicon* and *Solanum* spp. against tomato powdery mildew (*Oidium lycopersici*). J Phytopathology 148:303-311.

Miller, A. N., T. J. Ng, and T. H. Barksdale. 1983. Inheritance and heritability of resistance to tomato anthracnose caused by *Colletotrichum dematium*. J Amer Soc Hort Sci 108(6):1020-1023.

Miller, A. N., T. J. Ng, and T. H. Barksdale. 1984. Comparison of inheritance of resistance to tomato anthracnose caused by two *Colletotrichum* spp. Plant Dis 68(10):875-877.

Mohr, H. C. and G. M. Watkins. 1959. The nature of resistance to southern blight in tomato and the influence of nutrition on its expression. J Amer Soc Hort Sci 74:484-493.

Monforte, A. J. and S. O. Tanksley. 2000. Development of a set of isogenic and backcross recombinant inbred lines containing most of the *Lycopersicon hirsutum* genome in a L. *esculentum* genetic background: A tool for gene mapping and gene discovery. Gemome 43:803-813.

Moreau, P., P. Thoquet, J. Olivier, H. Laterrot, and N. Grimsley. 1998. Genetic mapping of Ph-2, a single locus controlling partial resistance to *Phytophthora infestans* in tomato. Mol Plant-Microbe Interact 11(4):259-269.

Nash, A. F. and R. G. Gardner. 1988a. Heritability of tomato early blight resistance derived from *Lycopersicon hirsutum* P. I. 126445. J Amer Soc Hort Sci 113:264-268.

Nash, A. F. and R.G. Gardner. 1988b. Tomato early blight resistance in a breeding line derived from *Lycopersicon hirsutum* P. I. 126445. Plant Dis 72:206-209.

Ng, T. J., A. N. Miller, and T. H. Barksdale. 1990. Inheritance of tomato resistance to *Colletotrichum dematium*. HortScience 25(5):562-564.

Nicot, P. C., A. Moretti, C. Romiti, M. Bardin, C. Caranta, and H. Ferriere. 2002. Differences in susceptibility of pruning wounds and leaves to infection by *Botrytis cinerea* among wild tomato accessions. Rpt Tomato Genet Coop 52:24-26.

Ori, N., Y. Eshed, I. Paran, G. Presting, D. Aviv, S. Tanksley, D. Zamir, and R. Fluhr. 1997. The I2C family from the wilt disease resistance locus I2 belongs to the nucleotide binding, leucine-rich repeat superfamily of plant resistance genes. Plant Cell 9:521-532.

Paddock, E. F. 1950. A tentative assignment of Fusarium-immunity locus to linkage group 5 in tomato. Genetics 35:683-684.

Parlevliet, J. E. 2002. Durability of resistance against fungal, bacterial and viral pathogens; present situation. Euphytica 124:147-156.

Phatak, S. C. and D. K. Bell. 1983. Screening for *Sclerotium rolfsii* resistance in the tomato. Proc. 4th Tomato Quality Workshop, Veg. Crops Res. Rpt., VEC-83-1, Dept. Veg. Crops, Inst. Food and Agr. Sci., Univ. of Florida, Gainesville, p. 107.

Pierce, L. C. 1971. Linkage tests with Ph conditioning resistance to race 0, *Phytophthora infestans*. Rpt Tomato Genet Coop 21:30.

Poysa, V. and J. C. Tu. 1993. Response of cultivars and breeding lines of *Lycopersicon* spp. to *Septoria lycopersici*. Can Plant Dis Sur 73:9-13.

Robbins, M. L. and F. F. Angell. 1970a. Tomato anthracnose: A hypodermic inoculation technique for determining genetic reaction. J Amer Soc Hort Sci 95(1):118-119.

Robbins, M. L. and F. F. Angell. 1970b. Tomato anthracnose: Inheritance of reaction to *Colletotrichum coccides* in *Lycopersicon* spp. J Amer Soc Hort Sci 95:469-471.

Robert, V. J. M., M. A. L. West, S. Inai, A. Caines, L. Arntzen, J. K. Smith, and D. A. St. Clair. 2001. Marker-assisted introgression of blackmold resistance QTL alleles from wild *Lycopersicon cheesmanii* to cultivated tomato (*L. esculentum*) and evaluation of QTL phenotypic effects. Molecular Breed 8:217-233.

Ross, H. 1986. Potato breeding – Problems and perspectives. Paul Parey Verlag, Hamburg, Germany.

Sarfatti, M., M. Abu-Abied, J. Katan, and D. Zamir. 1991. RFLP mapping of I1, a new locus in tomato conferring resistance against *Fusarium oxysporum* f.sp. *lycopersici* race 1. Theor Appl Genet 82:22-26.

Sarfatti, M., J. Katan, R. Fluhr, and D. Zamir. 1989. An RFLP marker in tomato linked to the *Fusarium oxysporum* resistance gene I2. Theor Appl Genet 78:755-759.

Schaible, L., O. S. Cannon, and V. Waddoups. 1951. Inheritance of resistance to verticillium wilt in a tomato cross. Phytopathology 41:986-990.

Scott, J. W. 1983. Genetic sources of tomato firmness. Proc. 4th Tomato Quality Workshop, Veg Crops Res Rpt, VEC-83-1, Dept Veg Crops Inst Food and Agr Sci, Univ of Florida, Gainesville, pp. 60-67.

Scott, J. W. 2005. Perspectives on tomato disease resistance breeding; past, present, and future. Acta Hort 695 (ISHS): 217-224.

Scott, J. W., H. A. Agrama, and J. P. Jones. 2004. RFLP-based analysis of recombination among resistance genes to fusarium wilt races 1, 2, and 3 in tomato. J Amer Soc Hort Sci 129(3):394-400.

Scott, J. W. and F. F. Angell. 1998. Tomato. In: S. S. Banga and S. K. Banga. (eds.), Hybrid Cultivar Development. Narosa Publishing House, New Delhi, India, pp. 451-475.

Scott, J. W. and J. D. Farley. 1983. 'Ohio CR-6' tomato. HortScience 18:114-115.

Scott, J. W. and J. P. Jones. 1986. Sources of resistance to fusarium wilt race 3. Rep Tomato Genet Coop 36:28-30.

Scott, J. W. and J. P. Jones. 1989. Monogenic resistance in tomato to *Fusarium oxysporum* f.sp. *lycopersici* race 3. Euphytica 40:49-53.

Scott, J. W. and J. P. Jones. 1990. Soil-borne fungal resistance in *Lycopersicon pennellii* accessions. HortScience 25:1068 (abstr.).

Scott, J. W. and J. P. Jones. 1995. Fla. 7547 and Fla. 7481 tomato breeding lines resistant to *Fusarium oxysporum* f . sp. *lycopersici* races 1, 2, and 3. HortScience 30:645-646.

Sherf, A. F. and A. A. McNab. 1986. Vegetable diseases and their control. Wiley, New York.

Shoemaker, P. B. and R. G. Gardner. 1986. Resistance and fungicide application interval for tomato foliar diseases, 1985. Biol Cult Tests 1:24.

Simons G., J. Groenendijk, J. Wijbrandi, M. Reijans, J. Groenen, P. Diergaarde, T. Van der Lee, M. Bleeker, J. Onstenk, M. de Both, M. Haring, J. Mes, B. Cornelissen, M. Zabeau, and P. Vos. 1998. Dissection of the fusarium I2 gene cluster in tomato reveals six homologs and one active gene copy. Plant Cell 10:1055-1068.

Stall, R. E. and J. M. Walter. 1965. Selection and inheritance of resistance in tomato to isolates of races 1 and 2 of the fusarium wilt organism. Phytopathology 55:1213-1215.

Stamova, L. 1996. Additional sources of resistance to brown and root rot disease. Rpt Tomato Genet Coop 46:24.

Stamova, L. 2004. Indication of dominant resistance to corky root. Proc Tomato Breeders Roundtable, Annapolis, MD, USA (Abstr.).

Stamova, L. 2005. Resistance to *Verticillium dahliae* Race 2 and its introgression into processing tomato cultivars. Acta Hort 695 (ISHS): 257-262.

Stamova, L. and M. Yordanov. 1987. Resistance to *Leveillula taurica* (Le.). Rpt Tomato Genet Coop 37:73.

Stamova, L., M. Yordanov, and S. Stamova. 1985. Genetic study of the genes in *S. pennellii* controlling resistance to *Cladosporium fulvum* Cke race mixture 2.4+5. Rpt Tomato Genet Coop 35:18.

Stevens, M. A. and C. M. Rick. 1986. Genetics and Breeding. In: J. G. Atherton and J. Rudich, (eds.), The Tomato crop: a Scientific Basis for Improvement. Chapman and Hall, London, pp. 35-109.

Stommel, J. R. 2001. Selection influences heritability estimates and variance components for anthracnose resistance in populations derived from an intraspecific cross of tomato. J Amer Soc Hort Sci 126:468-473.

Stommel, J. R. and K. G. Haynes. 1998. Inheritance of resistance to anthracnose caused by *Colletotrichum coccodes* in tomato. J Amer Soc Hort Sci 123:832-836.

Stommel, J. R. and Y. Zhang. 1998. Molecular markers linked to quantitative trait loci for anthracnose resistance in tomato. HortScience 33(3):514. (Abstr.)

Stommel, J. R. and Y. Zhang. 2001. Inheritance and QTL analysis of anthracnose resistance in the cultivated tomato (*Lycopersicon esculentum*). Acta Hort 542:303-310.

Swiecki, T.J. and J.D. MacDonald. 1991. Soil salinity enhances *Phytophthora* root rot of tomato but hinders asexual reproduction by *Phytophthora parasitica*. J Amer Soc Hort Sci 116(3):471-477.

Tanksley, S. D., M. W. Ganal, J. P. Prince, M. C. de Vincente, M. W. Bonierbale, P. Broun, T. M. Fulton, J. J. Giovannoni, S. Grandillo, G. B. Martin, R. Messeguer, J. C. Miller, L. Miller, A. H. Paterson, O. Pineda, M. S. Roder, R. A. Wing, W. Wu, and N. D. Young. 1992. High-density molecular linkage maps of the tomato and potato genomes. Genetics 132:1141-1160.

Tu, J. C. and V. Poysa. 1990. A brushing method of inoculation for screening tomato seedlings for resistance to *Septoria lycopersici*. Plant Dis 74(4):294-297.

Turkensteen, L. J. 1973. Partial resistance of tomato against *Phytophthora infestans*, the late blight fungus. Centre for Agricultural Publishing and Documentation, Wageningen.

Umaerus, V., M. Umaerus, J. E. Bradshaw, and G. R. Mackay. 1994. Inheritance of resistance to late blight. In: Potato Genetics, CAB International, Wallingford, U.K.

Vakalounakis, D. J. 1988a. Cultivar reactions and the genetic basis of resistance to alternaria stem canker (*Alternaria alternata* f. sp. *lycopersici*) in tomato. Plant Pathology 37:373-376.

Vakalounakis, D. J. 1988b. The genetic analysis of resistance to fusarium crown and root rot of tomato. Plant Pathology 37:71-73.

Vakalounakis, D. J., H. Laterrot, A. Moretti, E. K. Ligoxigakis, and K. Smardas. 1997. Linkage between Fr1 (*Fusarium oxysporum* f. sp. radicis-lycopersici resistance) and Tm-2 (tobacco mosaic virus resistance-2) loci in tomato (*Lycopersicon esculentum*). Ann Appl Biol 130:319-323.

Van der Biezen, E. A., T. Glagotskaya, B. Overduin, H. J. J. Nijkamp, and J. Hille. 1995. Inheritance and genetic mapping of resistance to *Alternaria alternata* f. sp. *lycopersici* from *Lycopersicon pennellii*. Mol Gen Genet 247:453-461.

Van der Hulst, R. Ist, G, Bonnema, T.C. Mercal, F. Meijer-Dekans, R. E. Niks and Bai, Y., and A. P. Livdhont 2005. Tomato defense to *Oidium neolycopersici* dominant *Ol* genes confer isolate-dependent resistance vin a different mechanism than recessive *Ol-2*, Mol PI MTc Int 18:354-362.

Van der Plank, J. E. 1971. Stability of resistance to *Phytophthora infestans* in cultivars without R genes. Potato Res 14:263-270.

Van Steekelenburg, N. A. M. 1981. Inoculation of tomato with *Didymella lycopersici*. In: J. Philouze (ed.), Genetics and Breeding of Tomato. EUCARPIA Tomato Working Group, Avignon, Fr, May 1981, pp. 277-284.

Walter, J. M. 1967. Hereditary resistance to disease in tomato. Ann Rev Phytopath 5:131-162.

Watterson, J. C. 1986. Diseases. In: J. G. Atherton and J. Rudich (eds.), The Tomato Crop: a scientific basis for improvement. Chapman & Hall, London, pp. 35-109.

Witsenboer, H. M. A., E. G. van de Griend, J. B. Tiersma, H. J. J. Nijkamp, and J. Hille. 1989. Tomato resistance to alternaria stem canker: localization in host genotypes and functional expression compared to non-host resistance. Theor Appl Genet 78:457-462.

Yamakawa, K. and N. Nagata. 1975. Three tomato lines obtained by the use of chronic gamma radiation with combined resistance to TMV and fusarium race J-3. Technical News, Institute of Radiation Breeding No. 16, 2 pp.

Yordanov, M., L. Stamova, and Z. Stoyanova. 1975. *Leveillula taurica* resistance in the tomato. Rpt Tomato Genet Coop 25:24.

Zhang, L., G. Y. Lin, D. O. Nino-Liu, and M. R. Foolad. 2003. Mapping QTLs conferring early blight (*Alternaria solani*) resistance in a *Lycopersicon esculentum* x *L. hirsutum* cross by selective genotyping. Mol Breed 12:3-19.

14

Breeding for Resistance to Viral Pathogens

J.W. Scott

Gulf Coast Research and Education Center, University of Florida
Institute of Food and Agricultural Sciences, 14625 CR 672,
Wimauma, FL 33598, USA
email: jws@mcsu.edu

INTRODUCTION

As with other tomato disease resistances, breeders have developed cultivars with virus resistances that are conferred by single dominant genes. In the previous chapter, Table 13.1 provides a summary of the virus resistance genes that have been utilized in breeding. Cultivars with tomato mosaic virus resistance are common, especially in greenhouse tomatoes. Some varieties with spotted wilt virus resistance have been developed although virulent races have overcome many of the genes deployed in the past. More recently, cultivars with resistance to tomato yellow leaf curl virus (TYLCV) have been developed and the best of these utilize major vertical genes as opposed to horizontal genes. More details are available in the sections for each of these diseases. There has been considerable work done on tomato virus resistance using genetic transformation and the potential for commercial impact from these approaches may be nearer than is the case for other types of pathogens. But cultivars have yet to be deployed with genetically engineered resistances. Some references involving genetic transformation will be mentioned but an adequate treatment of this topic is beyond the scope of this writing. Several book chapters have information on tomato virus and other resistance(s) (Dixon 1981, Tigchelaar 1986, Watterson 1986, Lukyanenko 1991). Herein, the primary goal will be to relay new information and provide reasonably thorough coverage of virus diseases where genetic information on resistance has been reported. Information on sources of resistance, inheritance, virus strains, inoculation techniques, and the status of breeding efforts will be given for each disease that are presented alphabetically.

Alfalfa mosaic virus (AMV). This virus is grouped in the genus *Alfamovirus* and through natural processes infects over 600 plant species of 250 genera belonging to 70 families (Parrella et al. 1997a). It is not considered a major tomato pathogen although infections have been reported in several countries and usually connected to growing tomatoes in the presence of nearby alfalfa crops (Zitter 1991). Recently, necrotic strains have emerged that have caused more serious disease outbreaks (Parrella et al. 2000). The disease is spread by 22 species of aphids. Symptoms range from bright yellow mosaic on leaves with necrotic patches and deformation of fruit for non-necrotic strains to severe necrosis of the leaves, and fruits in the case of necrotic strains. Early infection with necrotic strains can result in plant death. Strains have been grouped into two phylogenetic groups, subgroup I and subgroup II, based on coat protein sequences which reflect mainly a geographic structure (Parrella et al. 2000). Resistance was identified in three accessions of *L. hirsutum* PI 134417, LA 1777, and Bruinsma (Parrella et al. 1997a). A single dominant resistance gene, *Am*, from PI 134417 was described against necrotic strain LYH-1 (Parrella et al. 1998). Recently *Am* was mapped to the short arm of chromosome 6 near the centromere at 1.1 cM from RFLP markers CT21 and TG 232 (Parrella et al. 2004). The location of *Am* is in a major disease resistance hot spot (Fig. 12.1, bacterial resistance chapter). Preliminary indications in this work showed *Am* was resistant to 16 diverse strains of AMV, but did not provide protection to unrelated viruses CMV, TSWV, TMV, or PVY when co-inoculated.

The mechanical inoculation procedure was described by Parrella et al. (1997a). Virus strains are propagated in *Nicotiana tabacum* 'Xanthi n.c.' for 15 to 20 days before inoculation to obtain a high titer. Inoculum is prepared by grinding 1g of young tobacco leaves in 4 ml of a solution containing 0.03M Na_2HPO_4 with 0.2% sodium diethyldithiocarbamate. Before inoculation, 75 mg/ml of both carborundum and activated charcoal was added to avoid virus inactivation. Plants are grown in a growth chamber with a 16 h photoperiod and temperatures of 24°C day and 16°C night for 16 days. Inoculum is then rubbed on cotyledons. Symptoms can be rated about two weeks later on susceptible lines while there are no symptoms on the resistant plants. Symptomless plants can also be checked by ELISA (Parrella et al. 1997a). The resistance was reported to be of the extreme resistance (ER) type and not a hypersensitive type (Parrella et al. 2004).

Cucumber mosaic virus (CMV). The type member of the Cucumovirus group (Murphy et al. 1995), CMV is an RNA virus with some strains that have satellite RNA that modulates CMV symptoms. This aphid transmitted disease has a very broad host range and many strains. The strains have been divided into subgroups I and II based on nucleic acid or protein sequence composition and on serological studies (see Kaniewski et al. 1999). Strains from both subgroups infect tomato. Symptoms in tomato range from severe stunting, fern leaf, filiform leaves, and necrosis. Resistance or tolerance has been reported in many wild tomato species including *L. pimpinellifolium, L. peruvianum, L hirsutum , L. cheesmanii* var. *minor, L. chilense,* and *Solanum lycopersicoides* (Phills et al. 1977, Gebre et al. 1990, Nitzany 1992, Stoimenova et al. 1992, Stamova 1993, Parella et al. 1997b, Cillo et al. 2005). Inheritance of resistance from these species is largely unknown and no cultivars have yet been developed. *L. hirsutum* PI 247087 expressed resistance most clearly in plants that were 45 days old when inoculated and grown at 20°C. Resistance appeared to be controlled by a single gene (Parella et al. 1997b). More recently, a single dominant resistance gene *Cmr* was introgressed from *L. chilense* accession LA458 and mapped to chromosome 12 (Stamova and Chetelat 2000; Fig. 12.1 bacterial resistance chapter). Seven RFLP markers spanning 24.7 cM were associated with resistance. There was no clear indication of where in this region the resistance gene was and possibly there could even be two linked genes in the region. Nevertheless, CMV resistance was most strongly associated with markers TG68 and CT79 that are 4.3 cM apart. Interestingly, the powdery mildew resistance gene *Lv,* also introgressed from *L. chilense,* is linked to markers in the same general region indicating presence of a resistance gene cluster in this region. Stamova and Chetelat (2000) reported that there was recombination between the two resistances indicating the linkage was not very tight. *Cmr* provided resistance that was not always complete and the virus was isolated from resistant plants indicating immunity was not involved.

Pathogen-derived resistance has shown considerable promise for developing broad spectrum resistance to CMV. Approaches using coat protein genes appear to be the most effective. There is considerable literature in this area that is beyond the scope of this chapter. Two coat proteins from subgroups I and II with a single transgene copy worked well in providing resistance to multiple isolates via mechanical or aphid transmission (Kaniewski et al. 1999). Fuchs et al. (1996) developed lines with one coat protein gene that provided immunity to a range of isolates in field tests.

An inoculation procedure was described by Stamova and Chetelat (2000). The pathogen was maintained on small pumpkin seedlings, at the cotyledon to second leaf stage, by rub inoculation. The leaves were first dusted with Celite abrasive powder. The inoculum was a 1:10 (w/v) ratio of young symptomatic leaves in freshly prepared ice-cold grinding buffer (1:1-0.03 M potassium phosphate buffer, pH 7 and 0.1% sodium sulfite). Seedlings of tomato were inoculated at the first to second true leaf stage, followed by a second inoculation a week later to reduce the chance of escapes. Disease symptoms were scored 2-3 weeks after the last inoculation.

Curly top virus (CTV). This is a member of the Geminiviridae genus and has also been known as beet curly top virus (BCTV). It is vectored by leafhoppers (*Circulifer tenellus* Baker). If tomato plants are infected early they are severely stunted, epinastic, and often die so no fruits are produced. This disease has been a problem in semiarid regions from Mexico to Canada and in the Mediterranean basin (Zitter 1991). Resistance work started in 1930 in Utah with Loran Blood and then was carried on by his successor Oscar Cannon (Martin et al. 1971). Thereafter, Mark Martin of the USDA at Washington State University pursued this work. The resistance that was developed is based on an ability to escape infection from the leafhopper as opposed to a tolerance since once plants become infected the symptoms in resistant lines are as severe as in susceptible lines. The resistance was derived from crosses involving several wild species including *L. peruvianum* var. *dentatum* PI 128660, *L. peruvianum* var. *humifusum* PI 127829, *L. hirsutum* PI 127826 and *L. pimpinellifolium* Utah 45 (Martin 1963, Martin and Clark 1966, Martin et al. 1971). Breeding lines CVF4 and C5 from this program were released in the late 1960's (Martin 1966, Martin and Thomas 1969) and factors involved in the utility of this resistance were discussed in a later paper (Martin and Thomas 1986). The ability to escape infection is multigenically inherited and closely linked with undesirable characters (see Martin and Thomas 1986). Younger plants are more susceptible than older plants. Useful resistance for commercial production can be overwhelmed by severe seedling tests that have greater exposure than that normally seen in the field. Thus, field testing was suggested with an early and late rating of disease incidence. Once infected, plants generally die and thus the need for the early rating. Since the mechanism is escape rather than tolerance, rating disease severity is not useful. Repeating plantings at different times in a season was suggested to elucidate useful, moderate levels of resistance.

It was suggested that a range of resistance levels from the cultigens C5, C27, C193, CVF4, C22, 'VF145', and 'VR Moscow' would provide references for measuring resistances of unknown lines (Martin and Thomas 1986).

There has been no recent work on resistance to this disease since Mark Martin retired. CTV resistant lines were tested in our initial screening for resistance to tomato mottle virus (ToMoV) and they were all very susceptible (Scott and Schuster 1991). Resistance from *L. chilense* accessions have been shown to have resistance to several other tomato viruses as indicated in this writing. It would be interesting to test *L. chilense* accessions and/or early generation lines derived from them for CTV resistance, either of the escape type or for actual tolerance. If tolerance could be found, combining it with the ability to escape infection might provide superior resistance most. Also Martin (1962) reported resistance in *Solanum pennellii* (later=*L. pennellii*) and this may well relate to the insect resistance of that species. Thus, tomato germplasm derived from *L. pennellii* for resistance to other insects (Liedl et al. 1995) might also have some resistance to CTV as would insect resistant lines derived from *L. hirsutum* (Muigai et al. 2003). Martin (1962) also reported that *S. lycopersicoides* Dun. had a high level of CTV resistance and thus it would be worth testing the introgression lines now available at the Tomato Genetic Resource Center for resistance (Chetelat and Meglic 2000).

Pepino mosaic virus (PepMV). This member of the *Potexvirus* genus in the family Potexviridae was first reported on pepino (*Solanum muricatum*) in Peru in 1980 (Jones et al. 1980). It was first reported on greenhouse tomatoes in the Netherlands and the U.K. in 1999 (Van der Vlugt et al. 2000). Since then it has rapidly spread to greenhouse tomato crops in much of Europe (Verhoeven et al. 2003) and, North and South America (French et al. 2001, Soler et al. 2002). The rapid spread is likely due to the ease of mechanical transmission and the ability of the virus to survive in soil and on planting tools for extended periods. There is a high sequence similarity (96-97%) between the Peruvian pepino strain and European isolates, but there is greater sequence variability in tomato isolates from the United States (Maroon-Lango et al. 2003, Ling, unpublished data). Symptoms on tomato vary depending on the age of infection, the cultivar, and environmental conditions. Often symptoms do not appear until up to six weeks after infection and could be most severe on plants that are under some type of growth stress. Infected plants grown under optimal conditions often become symptomless after showing symptoms earlier. Generally young plants develop yellow mosaic that may lead to bright yellow spots or patches with severe leaf distortion or blistering on older leaves. Sometimes the infected leaflets become narrower and pointed so the growing point resembles a nettlehead. Mature fruits may show uneven ripening that resembles marbling. Yield losses of 15% have been reported in the U.K. PepMV is also associated with the collapse syndrome recently reported by Soler-Aleixandre et al. (2005a).

There has been little published information on resistance to pepino mosaic virus. Picó et al. (2002) indicated partial resistance to Spanish isolates existed in *L. chilense* accession LA1963 and *L. peruvianum* accessions PI 126944 and LA1708. Little is known about the total number and type of accessions that have been tested and of the importance of strain effects on resistance. Recently, Soler-Aleixandre et al. (2005b) reported some resistance in several accessions with the best resistance in *L. chilense* accession LA470. Undoubtedly more information will emerge on resistance to this disease since there is much current concern over it. In 2000, it was put under quarantine in the European Community and seedling transplants must be tested by ELISA before shipment. In 2003, it was put on the European and Mediterranean Plant Protection Organization Alert list which monitors diseases considered a serious risk to crop production.

For the mechanical inoculation procedure (Ling, personnel communication), the virus isolate is maintained and propagated on susceptible tomato plants or *N. benthamiana*. Virus inoculum is prepared by grinding the leaf tissue 1:10 (w/v) in phosphate buffered saline. Inoculum is rubbed gently on carborundum-dusted cotyledons and the first true leaf with a cotton swab. Plants are then grown in a greenhouse at 25-30°C and fertilized weekly (20:20:20) to promote symptoms for up to a six week period. Plants are rated on a scale where 0= no symptoms; 1= light yellow mosaic symptoms; 2= strong yellowing patches with leaf curling; and 3= leaf curling and puckering, plant stunting. ELISA tests with PepMV specific antiserum can be performed on symptomless plants.

Potato virus X (PVX). This *Potexvirus* is a disease that is often latent in potato where it causes mosaic symptoms. In tomato it causes mosaic symptoms and slight stunting. When cultivars are infected with both ToMV and PVX whereby a more serious disease than either alone called double streak results (Watterson 1986, Zitter 1991). There has been very little reported on PVX resistance. It has been a troublesome disease when tomatoes and potatoes are grown at the same time or sequentially in a region. Such is the case in Pakistan where cultivars were tested and three: Tobol, Turquesa, and Parana had moderate resistance (Rashid et al. 1989). Lower virus titers were found in plants of these cultivars than in plants of susceptible cultivars.

A high level of resistance to a transformed PVX was shown in some intriguing experiments reported by Tobias et al. (1999). The bacterial speck (*Pseudomonas syringae* pv. *tomato*) resistance gene *Pto* provides resistance to bacterial strains (see chapter 12) that have the avirulence gene *avrPto*. The *avrPto* gene was cloned to a PVX vector and it was shown that when this derivative was inoculated onto tomato plants only those without the *Pto* gene were infected. The recognition of the avirulence gene by plants with *Pto* prevented the PVX virus from being expressed. This work is of interest from a pathogen recognition/plant defense standpoint but will not be helpful for all the wild PVX strains that do not have the *avrPto* gene. In some related work, gene *Prf*, which recognizes a bacterial pathogen, provides resistance to both bacterial and viral pathogens (Oldroyd and Staskawicz 1998).

Inoculation is done mechanically. Rashid et al. (1989) maintained the pathogen on *Datura stramonium* plants. Leaves were mixed in a 0.02 M phosphate buffer, with pH 7.0, in a 1:1 w/v ratio. Plants were inoculated at the 4 to 5 leaf stage by rubbing one leaf with the inoculum. Symptoms were expressed 5-6 days later. Tobias et al. (1999) used a phosphate buffer with other ingredients and inoculated plants by gently rubbing the upper surface of cotyledons with a gloved finger.

Potato virus Y (PVY). This *Potyvirus* is also known as vein banding mosaic. It causes problems primarily in tropical and subtropical regions, has a narrow host range, and is transmitted in a non-persistent fashion by aphids (Watterson 1986, Zitter 1991). Various sources of resistance have been reported and many appear to be strain specific (Legnani et al. 1997). Nagai et al. (1992) incorporated resistance from PI 126410 into 'Santa Cruz' and developed the resistant cultivar 'Angela' that was widely grown in Brazil. They reported resistance was conferred by two recessive genes and that these were the same genes as in *L. pimpinellifolium* accession NAV 1062. PI 128887 which was resistant in Florida was susceptible to the Brazilian isolate tested. Thomas (1981), using greenhouse and field inoculations, tested 14 accessions with reported resistance to potyviruses (PVY, Peru tomato virus, tobacco etch virus) and found uniform resistance only in *L. hirsutum* accession PI 247087. Later, Thomas and McGrath (1988) tested PI 247087 against 10 Australian isolates and it was resistant to all of them. In an interspecific inheritance study it appeared the resistance was controlled by a single recessive gene, but there were fewer resistant plants than expected in the backcross to the resistant parent. This could be related to disturbance in gametes due to the interspecific nature of the study. In another study, Legnani et al (1995) determined that PVY resistance from PI 247087 was conferred by one or two recessive genes depending on the strain used. Parrella et al. (2002) used AFLP markers and a set of *L. hirsutum* introgression lines (Monforte and Tanksley 2000) to map the recessive gene *pot-1* from PI 247807 to a 10cM interval on the short arm of chromosome 3 near the corky root rot resistance locus *py-1*. The *pot-1* gene was flanked by RFLP markers TG135 and CT31 at distances of 2.1 and 4.6 cM, respectively. Green and Hanson (1996) reported PI 247087 was resistant to Taiwan and Australian isolates but was susceptible to isolates from Hawaii and Thailand. In addition they screened 169 accessions from several *Lycopersicon* species and found that *L. hirsutum* L3683 (PI 365904) was resistant (no disease symptoms and ELISA negative). Furthermore, L3683 was resistant to isolates from Australia, Hawaii, California, and Thailand. The F_1 was susceptible suggesting resistance was recessive. Inoculation is the same as that described for tobacco etch virus.

Tobacco etch virus (TEV). This aphid transmitted Potyvirus causes losses in tobacco, pepper, and tomato. It has been found on tomato in Florida, South America, Cuba, the Philippines, Taiwan, Thailand, and Turkey (see Legnani et al. 1996). Both leaves and fruits are mottled and plants are stunted. There are no reports of cultivars with resistance although several reports have identified sources of resistance. These include PI 183692 (Walter 1956); PI 166989 and *L. hirsutum* accessions PI 134417, and PI 127827 (Alexander and Hoover 1955), and *L. hirsutum* PI 247087, *L. pimpinellifolium* LA 1478, and *L. pennellii* LA 716 (Legnani et al. 1996). The latter report indicated that after inoculation the virus was undetectable by ELISA in PI 247087 but virus was detectable in LA 1478 and LA 716. They also reported PI 134417 was susceptible possibly meaning that the reported resistance from this PI is strain specific. PI 247087 was resistant to all four strains that were tested. Even with PI 247087, virus was recovered by back inoculations to tobacco plants. Thus, the resistance mechanism does not provide immunity but impairs virus multiplication and/or virus migration from cell to cell and therefore prevents systemic spread of the virus. Inheritance of resistance from PI 247087 was conferred by a single recessive gene in an intraspecific *L. hirsutum* population (Legnani et al. 1996). PI 247807 also had resistance to PVY and CMV as discussed in those sections of this chapter. Susequently, Parrella et al. (2002) found the *pot-1* gene from PI 247807 conferred resistance to both PVY and TEV. Further information on the *pot-1* gene is presented in the section on PVY. The author is not aware of work indicating if *pot-1* confers resistance to CMV.

The inoculation procedure was described by Legnani et al. (1996). Virus isolate(s) are inoculated to *Datura stramonium* 15 days before inoculation of tomato plants. One gram of young leaves of *Datura* were ground in 4 ml of 0.03 M Na_2HPO_4 containing 0.2% sodium diethyldithiocarbamate. Carborundum and charcoal each at 75 mg/ml were added to the solution before inoculation. The inoculum was rubbed on the first three leaves of 21 day old plants. After inoculation the plants were grown at 25°C with 16 h days. Symptoms were evident before 15 days and correlated well with ELISA data at 45 days after inoculation. Thirty days after inoculation may give similar results to 45 days.

Tomato mosaic virus (ToMV). This virus is closely related to tobacco mosaic virus (TMV) and it used to be commonly referred to as the latter. These Tobamovirus species can be distinguished from each other by differences in their serological affinities and protein compositions. ToMV has been shown to be the predominant virus in tomato crops around the world (Brunt 1986). Historically ToMV has caused considerable crop damage especially under the intensive culture systems used in greenhouse tomato production where it can easily be spread mechanically by pruning and training operations. The disease can also be spread by infected seed transmission where the virus is on seed coats. The virus can be removed by treating seed with tri-sodium phosphate during the seed extraction process. Resistance to ToMV or TMV was first reported in 1939 (Porte 1939) and the early work was reviewed by Pelham (1966). Since then ToMV resistance work has been summarized by Lukyanenko (1991) and the reader is referred to these publications as it is not the intent of this writing to repeat much of the information reported in these publications. Numerous cultivars with ToMV resistance have been released including all modern greenhouse cultivars and a number of field and home garden cultivars.

Three dominant resistance genes have been named and used in the development of resistant cultivars, *Tm-1*, *Tm2*, and *Tm2²* (*Tm2ª*). The *Tm-1* gene was named in 1960 and was made available in PI 235673, an *L. esculentum* accession. The source of resistance from the breeding program at Hawaii was not clear as the pedigree involved several wild species (Pelham 1966). Later evidence strongly supported *L. hirsutum* as the source since the ribosomal DNA sequence of an iso-line with *Tm-1* was similar to that of two *L. hirsutum* accessions (Levesque et al. 1990). Tanksley et al. (1992) mapped *Tm-1* to chromosome 2 (Fig. 12.1 bacterial resistance chapter). Japanese researchers later developed two co-dominant SCAR markers linked to *Tm-1* that would be useful for marker assisted breeding (Ohmori et al. 1996). The Hawaiian program of Frasier was also the source of the *Tm-2* gene that was named by Clayberg et al. (1960). This gene was tightly linked to the deleterious netted virescent (*nv*) gene that causes stunting and yellowing when homozygous. Breeding work with this gene was not successful because the linkage could not be broken. However, Laterrot and Pecaut (1969) developed a line called Perou 2, derived from *L. peruvianum* PI 126926, that had the *Tm-2* allele without *nv*. This is the source of many cultivars that possess the *Tm-2* gene. The *Tm-2* gene was also incorporated with the Fusarium crown rot resistance gene *Frl* that originated from IRB-301 (Vakalounakis et al. 1997; fusarium crown rot section of Chapter 13). In 1963, another resistance gene was discovered that was introgressed from *L. peruvianum* PI 128650 (Alexander 1963). Pecaut (1965) determined that this gene was allelic to *Tm-2* and named it *Tm2²* although Cirulli and Alexander (1969) later suggested *Tm2ᵃ* was more appropriate because it might be a different gene linked to *Tm2* and not an allele. They cite its more broad resistance against pathogen strains as another reason for *Tm2ᵃ* designation. Young et al. (1988) used isogenic lines with and without *Tm2ᵃ* to map the locus in a 0.4 cM region of chromosome 9 near RFLP markers TG101 and TG79.

McRitchie and Alexander (1963) reported there were four strains of the virus, designated I, II, III, and IV, based on their symptoms on host genotypes. Pelham (1969) later indicated they had four races designated 0, 1, 2, and 1.2 based on reactions to the resistance genes. Merging the two systems, I and II = 0, III = 1, IV = 2 while 1.2 is a new race. Cirulli and Alexander (1969) reported a fifth strain (V) in Ohio that was later determined to be the same as strain 1.2 (Pelham 1972). The Pelham strain designations refer to the resistance genes they are virulent on. Thus strain 1 isolates are virulent on *Tm-1*, strain 2 isolates are virulent on *Tm-2*, and strain 1.2 isolates are virulent on *Tm-1* and *Tm-2*.

Systemic necrosis is a deleterious resistance reaction that causes more severe symptoms than does a susceptible reaction. There is considerable literature on the systemic necrosis response of tomato genotypes and the reader is referred to the paper of Hall (1980) for more information on this topic. Reactions of tomato lines with various resistance genes against the strains of the virus are summarized in Table 14.1. It can be seen that genotypes heterozygous for *Tm-2* or *Tm-2²* can exhibit systemic necrosis for some virus strains. The occurrence of systemic necrosis is contingent on several factors; temperature, virus isolate and strain, plant age, resistance genotype including cytoplasmic factors, and the number and combinations of resistance alleles (Hall 1980). Systemic necrosis occurs more commonly under high temperatures (>26°C) and with genotypes heterozygous for *Tm-2* or *Tm-2²*. Kopeliovich et al. (1978) found heterozygotes with resistance on the seed parent side of the cross had more necrosis than their reciprocals. Thus, if cultivars are heterozygous for the above two genes the susceptible parent should be used as the seed parent and the resistant parent as the pollen parent. However, several researchers advise against using heterozygous resistant cultivars because of the potential problem. Many cultivars today are homozygous for resistance at the *Tm-2* locus as years of breeding has eliminated the undesirable traits that were linked to the resistance genes early on (Alexander 1971, Laterrot 1971). It should also be noted that systemic necrosis can occur in homozygous resistant genotypes but this occurred with less frequency and under more severe conditions than with heterozygotes (Hall 1980). To insure against this it is suggested that resistance genes be combined. Thus, when cultivars are homozygous at

Table 14.1 Disease reactions[z] for tomato genotypes against various tomato strains. (Modified from Hall and Bowes 1980: as presented by Lukyanenko 1991)

ToMV strain[y]	Resistance genotypes						
	Tm-1/+	Tm-2/+	Tm-2/Tm-2	Tm2[z]/+	Tm2[z]/Tm2[z]	Tm1/+Tm2/+	Tm1/+Tm2/Tm2[z]
0	T	R*	R	R*	R	R	R
1	S	R*	R	R*	R	R*	R
2	T	S	S	R*[x]	R	R	R
1.2	S	S	S	R*	R	S	R*

[y]Strains 0, 1, 2, and 1.2 are equivalent to Ohio strains 1 & II, III, IV, and V, respectively, see text for references.
[z]S = susceptible – severe mosaic symptoms and stunting
T = tolerance – mild mosaic symptoms, little effect on growth.
R = resistant, no symptoms.
R* = A hypersensitive, deleterious, systemic necrosis reaction can occur under high temperatures.
[x]Cirulli and Alexander (1969) did not find systemic necrosis for this genotype with this strain.

the *Tm-2* locus adding *Tm-1/+* will prevent systemic necrosis from being caused by any isolates controlled by *Tm-1* (Hall 1980).

The *Tm-2²* gene has proven to be a very durable resistance gene effective against all pathogen strains. It has been widely deployed in cultivars since 1970 and there are no published reports of virus strains that have become virulent on it and persisted in the environment. Hall (1980) theorizes that if *Tm-2²* mutated from the *Tm-2* allele, it would take at least two steps toward virulence for the pathogen to overcome it. Another option for resistance breeding is via development of transgenic tomatoes expressing the ToMV coat protein gene (Sanders et al. 1992).

Mechanical inoculation is used to select for resistance in breeding programs using strains controlled by the gene(s) being used. Tobacco plants *Nicotiana tabacum* n n (systemic) are inoculated with the pathogen when they have 5-6 true leaves. Once the plants begin flowering and all leaves are infected, leaves are removed, placed in plastic bags, and frozen until needed. Five to 10 leaves are blended in a Waring blender with enough dH_2O to make 50 ml. The mixture is strained through cheesecloth, diluted 1:5 with dH_2O and a pinch of carborundum (600 mesh) is added. If not used right away the inoculum can be stored in a refrigerator. Tomatoes to be inoculated are at the first true leaf stage. A cotton swab is dipped in the inoculum and rubbed over the cotyledons or the true leaves. Symptoms can be evaluated in three weeks. To distinguish systemic necrosis of heterozygous plants the screen should be done under high temperatures (>30°C) in a growth chamber.

Tomato Spotted wilt virus (TSWV). This is the type species of the *Tospovirus* genus in the Bunaviridae family and it has a broad host range attacking at least 1090 plant species from over 92 botanical families from both the monocots and dicots (Parrella 2003). It is transmitted by at least seven thrips species, *Thrips tabaci* (Pittman 1927), *T. palmi* (Iwaki et al. 1984), *T. setosus* (Kobatake et al. 1984), *Scirtothrips dorsalis* (Amin et al. 1981), *Frankliniella schultzei* (Samuel et al. 1930), *F. fusca* (Sakimura 1963), and *F. occidentalis* (Gardner et al. 1935). Various strains have been identified and are categorized by their symptoms and severity. The most stable are TB (tip blight), N (necrotic), R (ringspot), M (mild), VM (very mild) (Dixon 1981). Today pathogens are categorized as serial groups-species based on DNA sequences-and there are now 13-15 species. Some of the early resistance work was reviewed by Stevens et al. (1992) and the genes were derived from *L. pimpinellifolium*. Finlay (1953) tested resistance sources against 10 strains in Australia and identified two dominant resistance alleles at the *Sw-1* locus (*Sw-1*^{a}, *Sw-1*^{b}) and three recessive resistance genes: *sw-2*, *sw-3*, and *sw-4*. According to the gene list published in 1959 the cultivar 'Pearl Harbor' has *Sw*^{a} and 'Rey de los Tempranos' has *Sw-1*^{b} and the three recessive genes (Clayberg et al. 1959). These genes were strain specific in their effects. Paterson et al. (1989) tested the resistant sources of Finlay and found them to be susceptible to isolates in Arkansas. In 1986 resistance was reported in the South African cultivar 'Stevens' with resistance being introgressed from *L. peruvianum* (van Zijl et al. 1986). The resistant accession used in 'Stevens' is not known for sure but it may be PI 128654 (see Stevens et al. 1992). Stevens et al. (1992) determined that the resistance from 'Stevens' was conferred by a single dominant gene, named it *Sw-5*, and found it not to be race specific. Boiteux and Giordano (1993) found *Sw-5* also provided resistance to the tospoviruses chlorotic spot virus (TCSV) and groundnut ring spot virus (GRSV). The *Sw-5* gene has been mapped to the long arm of chromosome 9 between CT71 and CT220 (Stevens et al. 1995, Brommonschenkel and Tanksley 1997, Folkertsma et al. 1999). A co-dominant SCAR marker linked tightly to the *Sw-5* locus was developed from RAPD primer UBC#421 (Stevens et al. 1996). It is presently in wide use by tomato breeders around the world and *Sw-5* has now been deployed in a number of cultivars. However, under high disease pressure the resistance from *Sw-5* can be partially overcome (Roselló et al. 1997, 1998). Furthermore, strains virulent on *Sw-5* cultivars have been reported in Hawaii (Cho et al. 1996), South Africa (Thompson and van Zijl 1996), and Australia (Latham and Jones 1998). If such strains appear in tomato production regions the resistance in cultivars based on *Sw-5* could be rendered useless. Thus, it is important to have other sources of resistance available. Roselló et al. (1999) used mechanical and thrips inoculation to screen *L. peruvianum* and *L. chilense* accessions and found useful resistance in *L. peruvianum* accessions PI 126935, PI 126944, CIAPAN 16, PE-18, and CIAPAN 17. Earlier, Roselló et al. (1998) reported resistance gene *Sw-6* was introgressed from *L. peruvianum* accession PE-18. The level of resistance of *Sw-6* was not as high as that provided by *Sw-5*. In later work, the *Sw-6* gene present in line UPV 1 was determined to be allelic to *Sw-5*, and the former in heterozygous condition had a higher resistance level than the latter indicating an advantage in developing hybrid cultures. (Roselló et al. 2001). Other sources of resistance have been reported including a recovery type of resistance derived from *L. chilense* accession LA 1938 (Canady et al. 2001). Furthermore, this resistance hold up against the Hawaiian strain that overcomes *Sw-5* (Stevens, personal communication). Recently, advanced breeding lines derived from this source have been obtained that are homozygous resistant to a strain controlled by *Sw-5*. (Scott et al. 2005). Resistance appeared to be controlled by 1 or 2 dominant genes. The inheritance of this resistance to strains that do and do not overcome *Sw-5* needs to be elucidated and the development of molecular markers linked to the resistance gene(s) will greatly facilitate breeding progress with this material.

A genetically engineered possibility for spotted wilt resistance was demonstrated in tomato and tobacco plants transformed with the nucleocapsid protein (*N*) gene from the virus inserted either in sense or antisense orientations (Kim et al. 1994). Another approach to provide broad spectrum resistance was to combine transgenic resistance with *Sw-5* resistance (Gubba et al. 2002).

As mentioned a SCAR marker is the method of choice for the incorporation of the *Sw-5* gene into elite germplasm due in part to difficulties encountered with inoculation procedures. Resistance is verified by field tests or artificial inoculation methods. The two artificial methods used are mechanical and thrips inoculation and these two methods were compared for an array of germplasm (Kumar et al. 1993). Thrips inoculation is cumbersome involving maintainance of viruliferous thrips colonies and infection of susceptible controls is often less than 100% due to erratic behavior of thrips. For mechanical inoculation repeated mechanical passages can result in defective isolates (Ie 1982). Mechanical inoculation does not always reflect field resistance and with some germplasm field resistance could be deemed susceptible if only mechanical inoculation is used. If the resistance mechanism was based on vector resistance it would not be detected by mechanical inoculation for instance. Escapes can be a problem with mechanical inoculation methods as well. To detect the presence of the virus in plants that may be symptomless after inoculation, an ELISA method has been developed (Gonsalves and Trujillo 1986) and a kit is now available (Agdia Inc. Mishawaka, Indiana, USA). Resistance conferred by *Sw-5* is detectable by mechanical inoculation. The method of Stevens et al. (1992) will be described. TSWV isolates are maintained in young (four-leaf) *N. rustica* plants by rubbing 600 mesh carborundum-dusted leaves with sterile pads dipped in cold buffer (0.1 M phosphate buffer, pH 7.4, containing 0.01 M sodium sulfite) containing homogenized symptomatic tissue. This was done every two weeks. Tomato plants 10-15 cm tall were inoculated with a touch-up paint sprayer at 3.56×10^5 N/m^2. Infected *N. rustica* leaves were homogenized in a blender in cold inoculation buffer (10% w/v) followed by filtration through sterile cheesecloth. Carborundum (600 mesh) was added at 1% w/v. The sprayer nozzle was held 3-5 cm from the apices of the plants with inoculation done with 0.4-0.8 second bursts of the sprayer. Inoculations are repeated 6-8 days later to prevent escapes. Symptoms can be evaluated in two weeks and ELISA can be done in that time frame as well. Temperatures above 30°C should be avoided. The thrips inoculation method will not be described here, the reader is referred to the paper of Kumar et al. (1993).

Tomato yellow leaf curl virus (TYLCV). This virus of the family Geminiviridae is transmitted by the tobacco or sweetpotato whitefly *Bemisia tabaci* Genn. biotype B, which has also been classified as the silverleaf whitefly *B. argentifolii* Bellows and Perring. The type species of this whitefly-transmitted group of viruses is bean golden mosaic virus and thus they are members of the *Begomovirus* genus. Over 100 begomoviruses are transmitted to over 20 plant species of economic importance and the list of viruses is expanding rather dramatically. TYLCV is a monopartite virus that originated in the old world and has since spread to the Western hemisphere in tropical and subtropical zones. It is one of the most serious diseases presently affecting tomato crops in these regions (Picó et al. 1996). TYLCV is a heterogeneous complex of geminiviruses (Czosnek and Laterrot 1997) and eight species were proposed based on their country of origin (Fauquet et al. 2000). All have the TYLCV acronym followed by a hyphen with a code for the country of origin as follows; -CH (China), -IS (Israel), -NG (Nigeria), -Sar (Sardinia)[also called -SR], -SA (Saudi Arabia), -TZ (Tanzania), -TH (Thailand), and –YE (Yemen). Other species have evolved and have now been added to this list. Accotto et al. (2000) reported that TYLCV-IS was identified in Portugal and Spain while TYLCV-Sar was found in Italy and Spain.

No resistance was found in early testing of *L. esculentum* accessions. Resistance was found in several accessions of *L. pimpinellifolium*. Pilowsky and Cohen (1974) reported resistance from LA 121 was controlled by a single incompletely dominant gene. In India, Banerjee and Kalloo (1987) also found control by a single incompletely dominant gene from 'A1921'. However, Hassan et al. (1984) found resistance from LA 121 and LA 373 was controlled quantitatively with partially recessive gene action. Kasrawi (1989) studied resistance from *L. pimpinellifolium* accessions hirsute-INRA and LA 1478 and both were controlled by a single dominant gene. The gene symbol *Tylc* was proposed for this gene although complementation tests to prove that two sources had the same gene were not carried out. Later, a resistance gene, referred to as a major QTL that accounted for 27.7% of the resistance from hirsute-INRA was mapped to chromosome 6 between RFLP markers TG153 and CT83 (Chague et al. 1997 see Fig. 12.1). In Sudan, it was reported that resistance from LA 1478 and LA 1582 was each conferred by a single dominant gene (Geneif 1984). More recently, Hassan and Abdel-Ati (1999) reported resistance from three *L. pimpinellifolium* accessions was each conferred by single genes with dominance in PI 407543 and PI 407544 and partial dominance in PI 407555. Laterrot (1995) used only two *L. pimpinellifolium* accessions, hirsute INRA and LA 1478, in the development of his pimpertylc populations that were based on crosses from selections made in different countries.

Because breeding lines derived from LA 121 had reduced plant vigor and yield in the field, Pilowsky and Cohen (1990) started a new breeding program in 1977 based on resistance from *L. peruvianum* accession PI 126935. The tolerant cultivar TY-20 was released from this program with tolerance conferred by five recessive genes. Young TY-20 plants had to be protected from viruliferous whiteflies for best results. In later work at the Volcani Center in Israel, highly resistant lines TY 172 (and TY 197) were derived from *L. peruvianum* (Friedmann et al. 1998). The exact source of resistance was not indicated but PI 126906, PI 126930, PI 390681, and LA441 were used in the crossing. Plants of these lines carry a low level of viral DNA after being inoculated, but do not suffer the yield reductions of some commercial cultivars derived from other sources of resistance (Lapidot et al. 1997, 2001). Thus, they show promise as resistance sources. Hassan et al. (1984) showed two plants of *L. cheesmanii* LA 1401 were resistant and they recovered only one of 118 F_2 plants that was resistant with no symptoms and no virus graft transmission. The same authors found *L. hirsutum* accession LA 386 and the F_1 with a susceptible parent were resistant, but low plant numbers precluded an in depth genetic analysis. Later, Kasrawi et al. (1988) reported a single plant of LA 386 was free of disease symptoms but did harbor the virus after inoculation. Resistance was also reported in *L. hirsutum* accession LA 1777 (Ioannou 1985, Fargette et al. 1996). Vidavski and Czosnek (1998) crossed LA 386 and LA 1777 and developed an inbred, 902, that had good resistance to TYLCV and to a mixture of bipartite begomoviruses in Guatemala (Mejia et al. 2002). Picó et al. (2000) reported that the resistance from LA 386 was superior to that of LA 1777. Furthermore, Momotaz and Scott (2005) were unable to find a good level of resistance to TYLCV or tomato mottle virus (ToMoV) in any of the recombinant inbred lines that were derived from LA 1777 (Monforte and Tanksley 2000). Thus, although breeding work has been done with LA 1777, it does not appear to be a good source of resistance to TYLCV. Pico et al. (2000) also report on *L. pimpinellifolium* and *L. hirsutum* accessions that have improved levels of resistance after agroinoculation than many sources reported above. In India, Banerjee and Kalloo (1987) found resistance from *L. hirsutum* f. *glabratum* 'B6013' was conditioned by two epistatic genes. Later, Kalloo and Banerjee (1990) developed a line, H24, from this accession. H24 showed good resistance in Taiwan which led to the mapping of a resistance gene on chromosome 11 in a 14.6 cM region between RFLP markers TG36 and TG393 (Hanson et al. 2000). A second gene was not detected and might have been lost in the development of H24. Plants of H24 were free of virus in Taiwan and this gene appeared to be completely to partially dominant to TYLCV-TW. In India, there were some virus symptoms in H24 where the virus isolates were different and this could account for the difference in the plant responses between the two locations.

L. chilense derived resistance is also being used in several breeding programs around the world. Zakay et al. (1991) reported *L. chilense* accession LA 1969 had higher level of resistance than that of several other accessions from other species. Later, Zamir et al. (1994) mapped a partially dominant resistance gene from LA 1969, *Ty-1*, to chromosome 6 in a 4 cM region between RFLP markers TG97 and TG297 (bacterial resistance chapter, Fig. 12.1). The *Ty-1* gene has been widely used in breeding programs and has in some regions provided good resistance but in others virulent strains have overcome the resistance with Spain being one example (Barten, personnel communication). A larger introgression on chromosome 6 from LA 1969 is being used by Seminis Seed Company in their cultivars and it may contain two linked genes with *Ty-1* being one of the two genes. This resistance has held up better than *Ty-1* alone in some regions (Mercier, personnel communication). Scott et al. (1995) found six other *L. chilense* accessions bred for ToMoV had resistance to TYLCV in the Dominican Republic. This will be discussed further in the section on ToMoV. Resistance to TYLCV-SR was demonstrated in the lines of Scott et al. (1995) but lines selected specifically for this virus strain were made to obtain higher levels of resistance (Picó et al. 1999). Transgenic approaches have also been used to develop TYLCV resistance. Readers interested in this work are referred to the paper of Yang et al. (2004) where the replication associated protein (*Rep*) sequence from the pathogen has been altered and inserted into tomato to provide resistance and no detectable viral DNA.

Owing to the devastating nature of this disease in a number of important tomato production regions, there have been intensive breeding efforts to develop cultivars with TYLCV resistance. A number of cultivars have been developed using various resistance sources and some of these have been inoculated with TYLCV and evaluated for virus accumulation, symptoms, and yield reductions (Lapidot et al. 1997, Picó et al. 1998, Vidavsky et al. 1998). To date conventionally bred cultivars often have reduced virus titers compared to susceptible cultivars, but none are immune to TYLCV. Private companies have done much of the breeding, but information as to how the resistance being used is generally not published. The heterogeneous nature of the pathogen requires testing of resistant germplasm in the regions of interest to determine the utility of the resistance. Often a genotype that appears resistant in one location may not be as resistant in another region. Combining resistance genes is an approach to provide adequate and more broad-based resistance. However, it is often difficult to obtain resistant cultivars with horticultural characteristics comparable to existing susceptible cultivars and this problem is compounded as the number of resistance genes increases. The insecticide imidocloprid has been effective in many regions in limiting TYLCV damage and growers in such regions are reluctant to grow resistant cultivars due to their horticultural limitations. Significant challenges remain for tomato breeders to develop TYLCV resistant cultivars for some production regions.

Whitefly mediated inoculation is the primary course to screen for resistance. Researchers use various methods for screening disease resistance and some comparisons have been made (see Picó et al. 1998). Some details of a typical method will be described in the next section on ToMoV. Other methods use an inoculation time of 48 hours and in such situations high densities of whiteflies (20-50/plant) are used to avoid escapes. Agroinoculation procedures (Navot et al. 1991) have also been used, but these procedures resulted in symptom development in LA 1969 and LA 1777, genotypes that were essentially virus free when whitefly mediated virus inoculation was used (Kheyr-Pour et al. 1994). This demonstrated that agroinoculation circumvented natural resistance mechanisms which prevent the replication, spread and expression of symptoms.

Tomato mottle virus (ToMoV) and other bipartite begomoviruses. The new world geminiviruses have bipartite genomes as opposed to the monopartite genome of TYLCV. There are numerous bipartite begomoviruses that have been reported in the Western hemisphere with new ones still being documented. These viruses can cause devastating crop losses and some regions have mixtures of several begomoviruses. For instance, six have been reported in Guatemala (Nakhla 2005). Although losses from early infections can be total, the bipartite disease symptoms are generally not as severe as the symptoms of TYLCV. There has been far less resistance work on these viruses since they have not been around as long as TYLCV and none of them are as widespread. ToMoV was the first begomovirus to attack tomato crops in Florida in 1989 and a breeding program was started in 1990. An initial assay of germplasm, including many accessions with reported resistance to other viruses, revealed six accessions of *L. pimpinellifolium* and one each of *L. hirsutum* and *L. peruvianum* had significantly less disease severity than the susceptible control (Scott and Schuster 1991). No symptoms were present in many *L. chilense* accessions and 12 were selected for breeding based on their resistance and larger leaf size. Introgression with this material was described and tomato lines with ToMoV resistance were also found resistant to TYLCV in the Dominican Republic (Scott et al. 1995). Griffiths (1998) used RAPD markers and found three resistance regions on chromosome 6 using lines derived from accessions LA 1938, LA 2779, and LA 1932. Breeding lines from the former two accessions had *L. chilense* introgression in a region that overlapped the *Ty-1* region. A second region present in LA1938 and LA1932 derived lines was closely linked to the self pruning (*sp*) locus that confers plant habit. The third region present in LA 1932 derived lines was about 30 cM from the second region (Griffiths and Scott 2001). The marker work from LA1932 derived lines was supported by a genetic study which estimated two genes conferred resistance. The data fit an additive-dominance model with a high degree of additive gene action. Thus, to provide acceptable resistance to ToMoV requires resistance in both parents of a hybrid. Ji and Scott (2005) have assayed the RAPD markers from the work of Griffiths (1998) and are in the process of developing co-dominant SCAR markers for the resistance genes. In addition to the three genes on chromosome 6 there appeared to be another gene in lines derived from LA 2779 that was previously not detected. These lines were susceptible to gray leafspot (*Stemphyllium* sp.) and it was thought that a begomovirus resistance gene was linked in repulsion to *Sm*, a gene present in the recurrent parents being used. However, preliminary tests did not detect a *L. chilense* introgression in this chromosome 11 region (Ji, unpublished data). Once SCAR markers are found for all the resistance genes, lines with all combinations of the genes will be synthesized and tested for various begomoviruses around the world to discern what gene combinations are effective against the various viruses and strains thereof. Incorporation of the genes for breeding can then be done using marker-assisted selection.

As mentioned, ToMoV resistant lines had resistance to TYLCV in the Dominican Republic. Hundreds of breeding lines with resistance from LA 1932, LA 2779, and LA 1938 combined with Tyking have been screened for both ToMoV and TYLCV since 1997 when TYLCV was found in Florida. Lines with high levels of resistance to ToMoV also have had high levels of resistance to TYLCV (Scott, unpublished data). However, most lines bred for TYLCV resistance have been susceptible to ToMoV and these include Ty-52 (*Ty-1* gene), Fiona, H24, Gempride, Lignon, Avinash 2, Tycoon, Ty 20, Ty 202, pimpertylc-1 to -5, 8 chepertylc populations, 4 chiltylc populations, hirseptylc -1 to -3 populations, pertylc -1 and -2 populations, 10 Cln lines from AVRDC, Peto 150535, HA 3057, and 'Tygress' (Scott, unpublished data). TYLCV resistant lines that have shown some resistance to ToMoV are Tyking, Ty 197, Ty 34, and chiltylc -2 & -3 where the latter three of the five segregated. Several lines with *L. chilense* derived resistance have been tested for TYLCV and bipartite geminiviruses around the world. Some of the lines have had resistance in every test but no one genotype has emerged that is universally resistant, so it is apparent that there is no simple solution to begomovirus resistance breeding. However, ToMoV resistant lines have had good field resistance in Guatemala where there are mixed infections of up to six different viruses as mentioned. Line 902 (Vidavski and Czosnek 1998) has also shown good resistance in Guatemala and Ty 197 has shown fair resistance. It does appear that resistance genes selected for ToMoV and some genes selected for TYLCV resistance confer resistance to a broad array of these viruses and this does at least make the breeding for resistance feasible. If the resistance genes were specific for one, or only a few viruses, breeding of resistant cultivars would be impossible.

Little breeding work has been reported for bipartite viruses other than ToMoV. Piven et al. (1995) reported resistance to tomato yellow mosaic virus (ToYMV) in *L. chilense* accessions LA 1963 and LA 1969, *L. hirsutum* accession LA 1353, and *L. peruvianum* var. *glandulosum* accession LA1292.

Inoculation is whitefly-mediated and has been described (Griffiths and Scott 2001). ToMoV viruliferous colonies are maintained of the dwarf tomato 'Florida Lanai' in a controlled temperature room with a 14-h photoperiod under cool white fluorescent lights with irradiance of 50 mmol·m^{-2}·s^{-1} at 25°C. Six weeks before inoculation, viruliferous whitefly numbers are increased by adding additional non-infected 'Florida Lanai' plants (with 7 to 10 leaves growing in hanging baskets) to screened cages with the infected plants. Plants to be inoculated are grown in styrofoam trays (cell size =3.8 cm^2) for approximately 20 days past the cotyledon stage when they are at the third leaf stage. They are moved to benches in whitefly proof greenhouses along with the infected 'Florida Lanai' plants that are hung above them so that there are approximately 512 seedlings (four trays) for each infected source plant. Source plants are shaken periodically to distribute whiteflies and a dowl is rubbed over the plants to disperse the whiteflies for more uniform distribution. Inoculation is for 14 d or 21 d if there do not seem to be many whiteflies.

Tomato yellow top (TYTV) and Potato leaf roll virus (PLRV). These Luteoviruses are closely related and are persistently transmitted by several species of aphids. They have narrow host ranges, mostly Solanaceae species. The former causes more severe symptoms than the latter. A search for TYTV resistance was conducted using a number of accessions from several *Lycopersicon* species. This led Hassan and Thomas (1988) to inoculate *L. peruvianum* PI 128655, and its hybrids, with 9 isolates of TYTV and one isolate of PLRV. They found very strong resistance, with little virus recovery, in some plants with some isolates. However, there has been virtually no breeding work done with these diseases.

SUMMARY

The difficulties in developing horticulturally acceptable disease resistant cultivars have been recently discussed (Scott 2005). Much earlier, in a review well worth the reading by anyone interested in tomato disease resistance breeding, Walter (1967) quotes a 1953 paper by Andrus who concluded that "if a new disease-resistant variety is to be acceptable to farmers, it must yield fully as well as the old, susceptible variety in all features that influence the net value of the crop even when disease is absent". Molecular markers tightly linked to resistance genes would greatly facilitate combining resistance genes in new cultivars. However, few "breeder friendly" markers are presently available that, despite impressive advancement in identifying and locating resistance genes, many challenges remain for future researchers.

REFERENCES

Accotto, G.P., J. Navas-Castillo, E. Noris, E. Moriones, and D. Louro. 2000. Typing of tomato yellow leaf curl viruses in Europe. Eur J of Plant Pathol 106(2):179-186.

Alexander, L.J. 1971. Host-pathogen dynamics of tobacco mosaic virus in tomato. Phytopathology 61:611-617.

Alexander, L.J. 1963. Transfer of a dominant type of resistance to the four known Ohio pathogenic stains of tobacco mosaic virus (TMV) from *Lycopersicon peruvianum* to *L. esculentum*. Phytopathology 53:896.

Alexander, L.J. and M.M. Hoover. 1955. Disease resistance in wild species of tomato. Ohio Agric Exp Stn Res Bull 752.

Amin, P.W., D.V.R. Reddy, and A.M. Ghanekar. 1981. Transmission of tomato spotted wilt virus, the causal agent of bud necrosis of peanut, by *Scirtothrips dorsalis* and *Frankliniella schultzei*. Plant Dis 65:663-665.

Banerjee, M.K. and G. Kalloo. 1987. Sources and inheritance of resistance to leaf curl virus in *Lycopersicon*. Theor Appl Genet 73:707-710.

Boiteux, L.S. and L. de B. Giordano. 1993. Genetic basis of resistance against two Tospovirus species in tomato (*Lycopersicon esculentum*). Euphytica 71:151-154.

Brommonschenkel, S.H. and S.D. Tanksley. 1997. Map-based cloning of the tomato genomic region that spans the *Sw-5* tospovirus resistance gene in tomato. Mol Gen Genet 256:121-126.

Brunt, A. A. 1986. Tomato mosaic virus. In: M. H. V. Van Regenmortel and H. Fraenkel-Conrat, (eds.), The Plant Viruses, Vol. 2. Plenum Press, New York, pp. 181-204.

Canady, M.A., M.R. Stevens, M.S. Barineau, and J.W. Scott. 2001. Tomato spotted wilt virus (TSWV) resistance in tomato derived from *Lycopersicon chilense* Dun. LA 1938. Euphytica 117:19-25.

Chagué, V., J.C. Mercier, M. Guénard, A. de Courcel, and F. Vedel. 1997. Identification of RAPD markers linked to a locus involved in quantitative resistance to TYLCV in tomato by bulked segregant analysis. Theor Appl Genet 95:671-677.

Chetelat, R.T. and V. Meglic, 2000. Molecular mapping of chromosome segments introgressed from *Solanum lycopersicoides* into cultivated tomato (*Lycopersicon esculentum*). Theor Appl Genet 100:232-241.

Cho, J.J., D.M. Custer, S.H. Brommonschenkel, and S.D. Tanksley. 1996. Conventional breeding: host- plant resistance and the use of molecular markers to develop resistance to tomato spot wilt virus in vegetables. Acta Hort 431: 367-378.

Cillo, F., M.M. Finatti- Sialej, C. De Giovaurni, L. Recciaidi, and D. Gallitelli 2005. Molecular and genetic bases involved in the susceptibility of *Lycopersicon* spp. to *Cucumber mosasic virus* and its satellite RNA, Proc. XV Evorpin Tomato.

Cirulli, M. and L. J. Alexander. 1969. Influence of temperature and strain of tobacco mosaic virus on resistance in a tomato breeding line derived from *Lycopersicon peruvianum*. Phytopathology 59:1287-1297.

Clayberg, C. D., L. Butler, P.A. Young, and C.M. Rick. 1959. List of genes as of January 1959. Rept Tomato Genet Coop 9:6-18.

Clayberg, C. D., L. Butler, C.M. Rick, and P. A. Young. 1960. Second list of known genes in the tomato. J Hered 51:167-174.

Czosnek, H. and H. Laterrot. 1997. A worldwide survey of tomato yellow leaf curl viruses. Arch Virol 142:1391-1406.

Dixon, G.R. 1981. Vegetable Crop Diseases. AVI Publishing Co., Westport, CN.

Fargette, D., M. Leslie, and B.D. Harrison. 1996. Serological studies on the accumulation and localization of three tomato leaf curl geminiviruses in resistant and susceptible *Lycopersicon* species and tomato cultivars. Ann Appl Biol 128:317-328.

Fauquet, C.M., D.P. Maxwell, B. Gronenborn, and J. Stanley. 2000. Revised proposal for naming geminiviruses. Arch Virol 145(8):1743-1761.

Finlay, K.W. 1953. Inheritance of spotted wilt resistance in tomatoes; five genes controlling spotted wilt resistance in four tomatoes. Aust J Biol Sci 6:153-163.

Folkertsma, R.T., M.I. Spassova, M. Prins, M.R. Stevens, J. Hille, and R. W. Goldbach. 1999. Construction of a bacterial artificial chromosome (BAC) library of *Lycopersicon esculentum* cv. Stevens and its application to physically map the *Sw-5* locus. Mol Breed 5:197-207.

French, C.J., M. Bouthiller, M. Bernardy, G.Ferguson, M.Sabourin, R.C. Johnson, C. Masters, S. Godkin, and R. Mumford. 2001. First report of Pepino virus in Canada and the United States. Plant Dis 85:1121.

Friedmann, M., M. Lapidot, S. Cohen, and M. Pilowsky. 1998. A novel source of resistance to tomato yellow leaf curl virus exhibiting a symptomless reaction to viral infection. J Amer Soc Hort Sci 123(6):1004-1007.

Fuchs, M., R. Provvidenti, J.L. Slightom, and D. Gonsalves. 1996. Evaluation of transgenic tomato plants expressing the coat protein gene of cucumber mosaic virus strain WL under field conditions. Plant Dis 80:270-275.

Gardner, M.W., C.M. Tomkins, and O.C. Whipple. 1935. Spotted wilt of truck crops and ornamental plants. Phytopathology 25:17.

Gebre, S.K., H. Laterrot, G. Marshoux, G. Ragozzino, and A. Saccardo. 1990. Resistance to potato virus Y and cucumber mosaic virus in *Lycopersicon hirsutum*. Rept Tomato Genet Coop 40:12-13.

Geneif, A.A. 1984. Breeding for resistance to tomato leaf curl virus in tomatoes in the Sudan. Acta Hort 143:469-484.

Gonsalves, D. and E.E. Trujillo. 1986. Tomato spotted wilt virus in papaya and detection of the virus by ELISA. Plant Dis 70:501-506.

Green, S.K. and P. Hanson. 1996. A new source of PVY resistance. Rept Tomato Genet Coop 46:9.

Griffiths, P.D. and J.W. Scott. 2001. Inheritance and linkage of tomato mottle virus resistance gene derived from *Lycopersicon chilense* accession LA 1932. J Amer Soc Hort Sci 126(4):462-467.

Griffiths, P.D. 1998. Inheritance and linkage of geminivirus resistance genes derived from *Lycopersicon chilense* Dunal in tomato (*Lycopersicon esculentum* Mill.). Ph.D. diss., Univ. of Florida, Gainesville.

Gubba, A., C. Gonsalves, M.R. Stevens, D.M. Tricoli, and D. Gonsalves. 2002. Combining transgenic and natural resistance to obtain broad resistance to tospovirus infection in tomato (*Lycopersicon esculentum* Mill.). Mol Breed 9:13-23.

Hall, T. J. 1980. Resistance at the TM-2 locus in the tomato to tomato mosaic virus. Euphytica 29:189-197.

Hanson, P.M., D. Bernacchi, S. Green, S.D. Tanksley, V. Muniyappa, A.S. Padmajo, H. Chen, G. Kuo, D. Fang, and J. Chen. 2000. Mapping a wild tomato introgression associated with tomato yellow leaf curl virus resistance in a cultivated tomato line. J Amer Soc Hort Sci 125(1):15-20.

Hassan, A.A. and K.E.A. Abdel-Ati. 1999. Genetics of tomato yellow leaf curl virus tolerance derived from *Lycopersicon pimpinellifolium* and *Lycopersicon pennellii*. Egypt J Hort 26:323-338.

Hassan, A.A., H.M. Mazyad, S.E. Moustafa, S.H. Nassar, W.L. Sims, and M.K. Nakhla. 1984. Genetics and heritability of tomato yellow leaf curl virus tolerance derived from L. *pimpinellifolium*. Eur Assoc Res Plant Breed Tomato Working Group, Wageningen, The Netherlands pp. 298.

Hassan, S. and P.E. Thomas. 1988. Extreme resistance to tomato yellow top virus and potato leaf roll virus in *Lycopersicon peruvianum* and some of its tomato hybrids. Phytopathology 78:1164-1167.

Le, T.S. 1982. A sap transmissible, defective form of tomato spotted wilt virus. J Gen Virol 59:387-391.

Ioannou, N. 1985. Yellow leaf curl and other diseases of tomato in Cyprus. Plant Pathol 345:428-434.

Iwaki, M., Y. Honda, K. Hanada, H. Tochihara, T. Yonaha, K. Hokama, and T. Yokoyama. 1984. Silver mottle disease of watermelon caused by tomato spotted wilt virus. Plant Dis 68:1006-1008.

Ji ,Yuanfu and J.W. Scott. 2005. Finding RAPD markers linked to *Lycopersicon chilense* derived geminivirus resistance genes on chromosome 6 of tomato. Acta Hort 695 (ISHS): 407-416.

Jones, R.A.C., R. Koenig, and D.E. Lesemann. 1980. Pepino mosaic virus, a new potexvirus from pepino *Solanum muricatum*. Ann Appl Biol 94:61-68.

Kalloo, G. and M.K. Banerjee. 1990. Transfer of tomato leaf curl virus resistance from *Lycopersicon hirsutum* f. *glabratum* to *L. esculentum*. Plant Breed 105:156-159.

Kaniewski, W., V. Ilardi, L. Tomassoli, T. Mitsky, J. Layton, and M. Barba. 1999. Extreme resistance to cucumber mosaic virus (CMV) in transgenic tomato expressing one or two viral coat proteins. Mol Breed 5:111-119.

Kasrawi, M.A. 1989. Inheritance of resistance to tomato yellow leaf curl (TYLCV) in *Lycopersicon pimpinellifolium*. Plant Dis 73:435-437.

Kasrawi, M.A., M.A. Suwwan, and A. Mansour. 1988. Sources of resistance to tomato-yellow-leaf-curl-virus (TYLCV) in *Lycopersicon* species. Euphytica 37:61-64.

Kheyr-Pour, A., B. Gronenborn, and H. Czosnek. 1994. Agroinoculation of tomato yellow leaf curl virus (TYLCV) overcomes the virus resistance of wild *Lycopersicon* species. Plant Breed 112:228-233.

Kim, J.W., S.S.M. Sun, and T.L. German. 1994. Diseases resistance in tobacco and tomato plants transformed with the tomato spotted wilt virus nuclesocapsid gene. Plant Dis 6:615-621.

Kobatake, H., T. Osaki, and T. Inouye. 1984. The vector and reservoirs of tomato spotted wilt virus in *Nara prefecture*. Ann Phytopathol Soc Japan 50:541-544.

Kopeliovich, E., N. Kedar, and N. Retig. 1978. Genotypic and environmental effects on heat-necrosis of heterozygous TMV resistant lines. Rept Tomato Genet Coop 28:6-7.

Kumar, N.K. K., D.E. Ullman, and J.J. Cho. 1993. Evaluation of *Lycopersicon* germplasm for tomato spotted wilt tospovirus resistance by mechanical and thrips transmission. Plant Dis 77:938-941.

Latham, L.J. and R.A.C. Jones. 1998. Selection of resistance breaking strains of tomato spotted wilt tospovirus. Ann Appl Biol 133:385-402.

Lapidot, M., M. Friedmann, M. Pilowsky, R. Ben-Joseph, and S. Cohen. 2001. Effect of host plant resistance to tomato yellow leaf curl virus (TYLCV) on virus acquisition and transmission by its whitefly vector. Phytopathology 12:1209-1213.

Lapidot, M., M. Friedmann, O. Lachman, A. Yehezkel, S. Nahon, S. Cohen, and M. Pilowsky. 1997. Comparison of resistance level to tomato yellow leaf curl virus among commercial cultivars and breeding lines. Plant Dis 81:1425-1428.

Laterrot, H. 1995. Breeding network to create tomato varieties resistant to tomato yellow leaf curl virus (TYLCV). Fruits 50:439-444.

Laterrot, H. 1971. Pollen deficiency linked with *Tm*-2a. Rept Tomato Genet Coop 21:21-22.

Laterrot, H. and P. Pecaut. 1969. *Tm*-2: new source. Rept Tomato Genet Coop 19:13-14.

Latham, L. J. and R.A.C. Jones. 1998. Selection of resistance breaking strains of tomato spotted wilt tospovirus. Ann Appl Biol 133:385-402.

Legnani, R., K. Gebre-Selassie, G. Marchoux, and H. Laterrot. 1997. Interactions between PVY pathogens and tomato lines. Rept Tomato Genet Coop 47:13-15.

Legnani, R., P. Gognalons, K.G. Selassie, G. Marchoux, A. Moretti, and H. Laterrot. 1996. Identification and characterization of resistance to tobacco etch virus in *Lycopersicon* species. Plant Dis 80:306-309.

Legnani, R., K.G. Selassie, R.N. Womdim, P. Gognalons, A. Moretti, H. Laterrot, and G. Marchoux. 1995. Evaluation and inheritance of the *Lycopersicon hirsutum* resistance against potato virus Y. Euphytica 86:219-226.

Levesque, H., F. Vedel, C. Mathieu, and A.G.L. de Courcel. 1990. Identification of a short rDNA spacer sequence highly specific of a tomato line containing Tm-1 gene introgressed from *Lycoperiscon hirsutum*. Theor Appl Genet 80:602-608.

Liedl, B.E., D.M. Lawson, K.K. White, J.A. Shapiro, D.E. Cohen, W.G. Carson, and M.A. Mutchler. 1995. Acylsugars of wild tomato *Lycopersicon pennellii* alters settling and reduces oviposition of *Bemisia argentifolii* (Homoptera: Aleyrodidae). J Econ Entomol 88:742-748.

Lukyanenko, A. N. 1991. Disease resistance in tomato. *In*: G. Kalloo. (ed.), Genetic Improvement of Tomato. Monographs on Theoretical and Applied Genetics 14, Springer-Verlag, Berlin Heidelberg, Germany, pp. 99-119.

Maroon-Lango, C., M.A. Guaragna, R.L. Jordan, M. Bandla, and S. Marquardt. 2003. Detection and characterization of a US isolate of Pepino mosaic virus. Phytopathology 93 (6 Supplement):S57

Martin, M.W. and P.E. Thomas. 1986. Levels, dependability, and usefulness of resistance to tomato curly top disease. Plant Dis 70(2):136-141.

Martin, M.W., O.S. Cannon, and W.G. Dewey. 1971. Pedigree history of curly-top resistant tomato lines released by USDA. Rept Tomato Genet Coop 21:23-24.

Martin, M.W. and P.E. Thomas. 1969. C5, a new tomato breeding line resistant to curly top virus. Phytopathology 59:1754-1755.

Martin, M.W. and R.L. Clark. 1966. Increasing levels of curly top resistance by transgressive segregation. Rept Tomato Genet Coop 16:20-21.

Martin, M.W. 1963. Responses of curly-top resistant *Lycopersicon* species to curly top exposure in different areas of the west. Plant Dis Repot 47(2):121-125.

Martin, M.W. 1962. *Solanum pennellii*, a possible source of tomato curly-top resistance. Phytopathology 52(11):1230-1231.

McRitchie, J. J. and L. J. Alexander. 1963. Host specific *Lycopersicon* strains of tobacco mosaic virus. Phytopathology 53:394-398.

Mejia, L., R.E. Teni, H. Czosnek, F. Vidavski, A. Bettilyon, M.K. Nakhla, and D.P. Maxwell. 2002. Field evaluation of tomato experimental lines and hybrids for resistance to begomoviruses in Guatemala. Phytopathology 92:S54.

Monforte, A.J. and S.D. Tanksley. 2000. Development of a set of near isogenic and backcross recombinant inbred lines containing most of the *Lycopersicon hirsutum* genome in a *L. esculentum* genetic background: A tool for gene mapping and gene discovery. Genome 43: 803-813.

Momotaz, A., J.W. Scott, and D.J. Schuster. 2005. Searching for silverleaf whitefly and geminivirus resistance genes from *Lycopersicon hirsutum* Accession LA1777. Acta Hort 695 (ISHA): 417-422.

Muigai, S.G., M.J. Bassett, D.J. Schuster, and J.W. Scott. 2003. Greenhouse and field screening of wild *Lycopersicon* germplasm for resistance to whitefly *Bemisia argentifolii*. Phytoparasitica 31(1):27-38.

Murphy, F.A., C.M. Fauquet, D.H.L. Bishop, S.A. Ghabrial, A.W. Jarvis, G.P. Martelli, M.A. Mayo, and M.D. Summers (eds). 1995. Virus Taxonomy. Springer-Verlag, Wie/New York.

Nagai, H., W.J. Siqueira, and A.L. Lourencão. 1992. Tomato breeding for resistance to diseases and pests in Brazil. Acta Hort 301:91-97.

Nakhla, M.K., A. Bettilyon, D.P. Maxwell, L Mejia, P. Ramirez, and J. P. Karkashian. 2005. Molecular characterization and development of DNA-based detection methods for tomato-infecting Begomoviruses in Central America. Acta Hort 695 (ISHS): 277-288.

Navot, N., E. Picherski, M. Zeidan, D. Zamir, and H. Czosnek. 1991. Tomato yellow leaf curl virus: a whitefly-transmitted geminivirus with a single genomic component. Virology 185:151-161.

Nitzany, F.E. 1992. Cucumber mosaic virus in Israel. Phytopathol Med 14:16-20.

Ohmori, T., M. Murata, and F. Motoyoshi. 1996. Molecular characterization of RAPD and SCAR markers linked to the Tm-1 locus in tomato. Theor Appl Genet 92:151-156.

Oldroyd, G.E.D. and B.J. Staskawicz. 1998. Genetically engineered broad-spectrum disease resistance in tomato. Proc Natl Acad Sci USA 95:10300-10305.

Parrella, G., A. Moretti, P. Gognalons, M. Lesage, G. Marchoux, K. Gebre-Selassie, and C. Caranta. 2004. The Am gene controlling resistance to Alfalfa mosaic virus in tomato is located in the cluster of dominant resistance genes on chromosome 6. Phytopathology 94(4):345-350.

Parrella, G., P. Gognalons, K. Gebre-Selassie, C. Vovlas, and G. Marchoux. 2003. An update of the host range of tomato spotted wilt virus. J Plant Pathol 85 (4): 227-264.

Parrella, G., S. Ruffel, A. Moretti, C. Morel, A. Palloix, and C. Caranta. 2002. Recessive resistance genes against potyviruses are localized in colinear genomic regions of the tomato (Lycopersicon spp.) and pepper (Capsicum spp.) genomes. Theor Appl Genet 105:855-861.

Parrella, G., C. Lanave, G. Marchoux, M.M. Finetti-Sialer, A. Di Franco, and D. Gallitelli. 2000. Evidence for two distinct subgroups of Alfalfa mosaic virus (AMV) from France and Italy and their relationships with other AMV strains. Arch Virol 145:2659-2667.

Parrella, G., H. Laterrot, K. Gebre Selassie, and G. Marchoux. 1998. Inheritance of resistance to alfalfa mosaic virus in *Lycopersicon hirsutum* f. *glabratum* PI 134417. J Plant Pathol 80(3):241-243.

Parrella, G., H. Laterrot, G. Marchoux, and K. Gebre Selassie. 1997a. Screening *Lycopersicon* accessions for resistance to alfalfa mosaic virus. J Genet Breed 51:75-78.

Parrella, G., H. Laterrot, R. Legnani, S.K. Gebre, G. Marchoux, M. Ercolano, and L.M. Monti. 1997b. Factors affecting the expression of the resistance to CMV in two accessions of *L. hirsutum*: role of the temperature and plant age. In: Proc Eucarpia XIII, Jerusalem, Israel, p. 48.

Paterson, R.G., S.J. Scott, and R.C. Gergerich. 1989. Resistance in two *Lycopersicon* species to an Arkansas isolate of tomato spotted wilt virus. Euphytica 43:173-178.

Pecaut, P. 1965. Tomate: Resistance au virus de la mosaique du tabac (T.M.V.). Report of Station de'Amelioration des Plantes Maraicheres. INRA 1964: 50-53.

Pelham, J. 1972. Strain-genotype interaction of tobacco mosaic virus in tomato. Ann Appl Biol 71:219-228.

Pelham, J. 1969. Isogenic lines to identify physiologic strains of TMV. Rept Tomato Genet Coop 19:18-19.

Pelham, J. 1966. Resistance in tomato to tobacco mosaic virus. Euphytica 15:258-267.

Phills, B.R., R. Provvidenti, and R.W. Robinson. 1977. Reaction of *Solanum lycopersicoides* to viral diseases of the tomato. Rep Tomato Genet Coop 27:18.

Picó, B., J. Herraiz, J. J. Ruiz, and F. Nuez. 2002. Widening the genetic basis of virus resistance in tomato. Sci Hort 94:73-89.

Picó, B., A. Sifres, M. Elia, M.J. Diez, and F. Nuez. 2000. Searching for new resistance sources to tomato yellow leaf curl virus within a highly variable wild *Lycopersicon* genetic pool. Acta Physiol Plant 22:344-350.

Picó, B., M. Ferriol, M.J. Díez, and F. Nuez. 1999 . Developing tomato breeding lines resistant to tomato yellow leaf curl virus. Plant Breed 118:537-542.

Picó, B., M.J. Díez, and F. Nuez. 1998. Evaluation of whitefly-mediated inoculation techniques to screen *Lycopersicon esculentum* and wild relatives for resistance to tomato yellow leaf curl virus. Euphytica 101:259-271.

Picó, B. M.J. Díez, and F. Nuez. 1996. Viral diseases causing the greatest economic losses to the tomato crop. II The tomato yellow leaf curl virus-a review. Sci Hort 67:151-196.

Pilowsky, M. and S. Cohen. 1974. Inheritance of resistance to tomato yellow leaf curl virus in tomatoes. Phytopathology 64:632-635.

Pilowsky, M. and S. Cohen. 1990. Tolerance to tomato yellow leaf curl virus derived from *Lycopersicon peruvianum*. Plant Dis 74:248-250.

Pittman, H.A. 1927. Spotted wilt of tomatoes. J Aust Coun Sci Ind Res 1:74-77.

Piven, N.M., R.C. de Uzcategui, and H.D. Infante. 1995. Resistance to tomato yellow mosaic virus in species of *Lycopersicon*. Plant Dis 79:590-594.

Porte, W.S., S.P. Doolittle, and F.L. Wellman. 1939. Hybridization of a mosaic-tolerant, wilt-resistant *Lycopersicon hirsutum* with *Lycopersicon esculentum*. Phytopathology 29:757-759.

Rashid, F., S. Khalid, I. Ahmad, and S.M. Mughal. 1989. Potato virus X (PVX) resistance in tomato cultivars. Tropical Pest Management 35(4):357-358.

Roselló, S., B. Ricarte, M. Jose Diez, and F. Nuez. 2001. Resistance to tomato spotted wilt virus introgressed from *Lycopersicon peruvianum* in line UPV 1 may be allelic to *Sw*-5 and can be used to enhance the resistance of hybrids cultivars. Euphytica. 3:357-367.

Roselló, S., S. Soler, M.J. Díez, J.L. Rambla, C. Richarte, and F. Nuez. 1999. New sources for high resistance of tomato to the tomato spotted wilt virus from *Lycopersicon peruvianum*. Plant Breed 118:425-429.

Roselló, S., M. J. Diez, A. Lacasa, C. Jordá, and F. Nuez. 1997. Testing resistance to TSWV introgressed from *Lycopersicon peruvianum* by artificial transmission techniques. Euphytica 98:93-98.

Rosello, S., J. Diez, and F. Nuez. 1998. Genetics of tomato spotted wilt virus resistance coming from *Lycopersicon peruvianum*. Eur J Plant Pathol 104:499-509.

Sakimura, K. 1963. *Frankliniella fusca*, an additional vector for the tomato spotted wilt virus, with notes on *Thrips tabaci*, another vector. Phytopathology 53:412-415.

Samuel, G., J.G. Bald, and H.A. Pittman. 1930. Investigations on "spotted wilt" of tomatoes. Aust Coun Sci Ind Res 44.

Sanders, P.R., B. Sammons, W. Kaniewski, L. Haley, J. Layton,, B.J. LaVallee, X. Delannay, and N.E. Turner. 1992. Field resistance of transgenic tomatoes expressing the tobacco mosaic virus of tomato mosaic virus cout protein genes. Phytopathology 82(6):683-689.

Scott, J.W. and D.J. Schuster. 1991. Screening of accession for resistance to the Florida tomato geminivirus. Rept Tomato Genet Coop 41:48-50.

Scott, J.W., M.R. Stevens, J.H. Barten, C.H. Thome, J.E. Polston, D.J. Schuster, and C.A. Serra. 1995. Introgression of resistance to whitefly-transmitted geminiviruses from *Lycopersicon chilense* to tomato. In: D. Gerling and R.T. Mayer (eds), Bemisia: Taxomomy, Biology, Damage, Control, and Management. Intercept, Andover, United Kingdom, pp. 357-367.

Scott, J. W. 2005. Perspectives on tomato disease resistance breeding; past, present, and future. Acta Hort 695 (ISHS): 217-224.

Scott, J. W., M.R. Stevens, and S.M. Olson. 2005. An alternative source of resistance to tomato spotted witt virus. Rept Tomato Genet Coop 55: 40-42.

Soler, S., J. Prohens, M.J. Díez, and F. Nuez. 2002. Natural occurrence of Pepino mosaic virus in *Lycopersicon* species in Central and Southern Peru. J Phytopathol 150:49-53.

Soler-Aleixandre, S. C., López, M.J. Díez, A. Pepiz De-Costro, and F. Nuez, 2005a. Association of papino mosaic virus with tomato collapse. J Phytopathol 153: 1-6.

Soler-Aleixandre S., J. Cebolla-cornejo, and F. Nuez. 2005b. Sources of resistance to pepino mosaic virus (PepMV) in tomato. Rept Tomato Genet Coop 55:43-45.

Stamova, B.S. and R.T. Chetelat. 2000. Inheritance and genetic mapping of cucumber mosaic virus resistance introgressed from *Lycopersicon chilense* into tomato. Theor Appl Genet 101:527-537.

Stamova, L. 1993. Scope of the breeding tomatoes for disease resistance in Bulgaria. Proc. XII Eucarpia Symp. Plovdiv 11-19.

Stevens, M.R., D.K. Heiny, P.D. Griffiths, J.W. Scott, and D.D. Rhoads. 1996. Identification of co-dominant RAPD markers tightly linked to the tomato spotted wilt virus (TSWV) resistance gene Sw-5 Rept. Tomato Genet Coop 46:27.

Stevens, M.R., E.M. Lamb, and D.D. Rhoads. 1995. Mapping the Sw-5 locus for tomato spotted wilt virus resistance in tomatoes using RAPD and RFLP analyses. Theor Appl Genet 90:451-456.

Stevens, M. R., S.J. Scott, and R.C. Gergerich. 1992. Inheritance of a gene for resistance to tomato spotted wilt virus (TSWV) from *Lycopersicon peruvianum* Mill. Euphytica 59:9-17.

Stoimenova, E., V. Sotirova, and Z. Valkova. 1992. Sources of resistance to the cucumber mosaic virus in the genus *Lycopersicon*. CR Acad Sci Bulg 8:107-109.

Tanksley, S. D., M.W. Ganal, J.P. Prince, M.C. de Vincente, M.W. Bonierbale, P. Broun, T. M. Fulton, J. J. Giovannoni, S. Grandillo, G.B. Martin, R. Messeguer, J. C. Miller, L. Miller, A.H. Paterson, O. Pineda, M.S. Roder, R. A. Wing, W.Wu, and N. D. Young. 1992. High-density molecular linkage maps of the tomato and potato genomes. Genetics 132:1141-1160.

Thomas, J.E. and D.J. McGrath. 1988. Inheritance of resistance to potato virus Y in tomato. Aust J of Agri Res 3:475-479.

Thomas, J.E. 1981. Resistance to potato virus Y in *Lycopersicon* species. Austr Plant Pathol 10:61-68.

Thompson, G.J. and J.J.B. van Zijl. 1996. Control of tomato spotted wilt virus in tomatoes in South Africa. Acta Hort 431:379-384.

Tigchelaar, E.C. 1986. Tomato breeding. In: M. Bassett (ed), Breeding Vegetable Crops. AVI Publishing Co., Westport, CN, pp. 135-171.

Tobias, C.M., G.E.D. Oldroyd, J.H. Chang, and B.J. Staskawicz. 1999. Plants expressing the Pto disease resistance gene confer resistance to recombinant PVX containing the avirulence gene AvrPto. The Plant J 17:41-50.

Vakalounakis, D. J. H. Laterrot, A. Moretti, E. K. Ligoxigakis, and K. Smardas. 1997. Linkage between Fr1 (*Fusarium oxysporum* f. sp. radicis-lycopersici resistance) and Tm-2 (tobacco mosaic virus resistance-2) loci in tomato (*Lycopersicon esculentum*). Ann Appl Biol 130:319-323.

van der Vlugt, R.A.A., C.C.M.M. Stijger, J.Th.J. Verhoeven, and D.E. Lesemann. 2000. First report of pepino mosaic virus on tomato. Plant Dis 84:103.

van Zijl, J.J.B., S.E. Bosch and C.P.J. Coetzee. 1986. Breeding tomatoes for processing in South Africa. Acta Hort 194:69-75.

Verhoeven, J.Th.J., R.A.A. van der Vlugt, and J.W. Roenhorst. 2003. High similarity between tomato isolates of pepino mosaic virus suggests a common origin. Eur J Plant Path 109:419-425.

Vidavsky, F. and H. Czosnek. 1998. Tomato breeding lines resistant and tolerant to tomato yellow leaf curl virus issued from *Lycopersicon hirsutum*. Phytopathology 88:910-914.

Vidavsky, F., S. Leviatov, J. Milo, H.D. Rabinowitch, N. Kedar, and H. Czosnek. 1998. Response of tolerant breeding lines of tomato, *Lycopersicon esculentum*, originating from three different sources (*L. peruvianum*, *L. pimpinellifolium*, and *L. chilense*) to early controlled inoculation by tomato yellow leaf curl virus (TYLCV). Plant Breed 117:165-169.

Walter, J.M. 1956. Combination of resistances to tobacco-etch and tobacco-mosaic viruses in tomato breeding stocks. Phytopathology 46:517-519.

Walter, J.M. 1967. Hereditary resistance to disease in tomato. Ann Rev Phytopath 5:131-162.

Watterson, J. C. 1986. Diseases. In: J. G. Atherton and J. Rudich (eds.), The Tomato Crop: a Scientific Basis for Improvement. Chapman & Hall, London, pp. 443-484.

Yang, Y., T.A. Sherwood, C.P. Patte, E. Hiebert, and J.E. Polston. 2004. Use of tomato yellow leaf curl virus (TYLCV) rep gene sequences to engineer TYLCV resistance in tomato. Phytopathology 94:490-496.

Young, N. D., D. Zamir, M. W. Ganal, and S. D. Tanksley. 1988. Use of isogenic lines and simultaneous probing to identify DNA markers tightly linked to the *Tm-2a* gene in tomato. Genetics 120:579-585.

Zakay, Y., N. Navot, M. Zeidan,, N. Kedar, H. Rabinowitch, H. Czosnek, and D. Zamir. 1991. Screening *Lycopersicon* accessions for resistance to tomato yellow leaf curl virus: presence of viral DNA and symptom development. Plant Dis 75:279-281.

Zamir, D., I. Ekstein-Michelson, Y. Zakay, N. Navot, M. Zeidan, M. Sarfatti, Y. Eshed, E. Harel, T. Pleban, H. van-Oss, N. Kedar, H.D. Rabinowitch, andH. Czosnek. 1994. Mapping and introgression of a tomato yellow leaf curl virus tolerance gene, TY-1. Theor Appl Genet 88:141-146.

Zitter, T.A., J.B. Jones, J.P. Jones, and R.E. Stall (eds). 1991. Diseases caused by viruses. In: Compendium of Tomato Diseases. APS Press, St. Paul, MN, pp. 31-34.

15

Resistance in Tomato and Other *Lycopersicon* Species to Insect and Mite Pests

GEORGE G. KENNEDY

*Department of Entomology, Box 7630, North Carolina State University,
Raleigh, NC 27695-7630 USA
email: george_kennedy@ncsu.edu*

INTRODUCTION

The cultivated tomato, *Lycopersicon esculentum* (L.) Miller, is attacked by a number of arthropod pests, which are capable of causing devastating losses (Lange and Bronson 1981). The pest complex includes species that feed almost exclusively on foliage (e.g. spider mites, dipterous leafminers), species that feed on both foliage and fruit [such as the lepidopterans *Helicoverpa zea* (Boddie) and *Spodoptera exigua* (Boddie)] and species such as aphids and whiteflies that feed on plant sap. Because tomato is a high-value crop, which must meet rigorous market standards that preclude even minimal damage to the harvested fruit, populations of pests that attack the fruit must be maintained at very low levels. Consequently, management of insect pests on tomato has relied heavily on chemical control measures.

Arthropod resistant cultivars and hybrids play an important role in pest management on many crops, including some vegetable crops, and their use holds considerable promise in tomato. During the last three decades, considerable research has been directed towards identifying and developing a mechanistic understanding of traits within the genus *Lycopersicon* that confer resistance to arthropod pests. It is noteworthy that research has resulted in the identification of numerous resistance traits that have potential utility in the development of pest resistant tomato cultivars. It has also revealed levels of detail and complexity in plant-arthropod interactions that are unique among well-studied systems involving crop plants and their close relatives. This chapter provides an overview of the occurrence, mechanisms, and genetics of arthropod resistance traits in the genus *Lycopersicon*. Additional, detailed information on resistance mechanisms in *Lycopersicon* can be found in reviews by Duffey (1986), Duffey and Bloem (1986), Kennedy (1986), Kennedy et al. (1987), Farrar and Kennedy (1991a), and Kennedy (2003)

In practice, arthropod resistance or susceptibility in crops is typically defined in relation to a commonly grown cultivar. Resistance traits, which are genetically controlled, result in lower levels of attack, lower pest populations, and/or lower levels of damage to the crop than observed on plants lacking the resistance trait(s) under conditions of comparable pest pressure (Painter 1951). Arthropod resistance that is of value in crop protection can be categorized as: antibiosis (pest development delayed and/or survival reduced), antixenosis (=non-preference - normal pest behavior is adversely affected in a way that interferes with its ability to utilize the resistant plant), or tolerance (reduced level of damage under a given level of pest infestation), although these categories are not mutually exclusive (Painter 1951, Kogan and Ortman 1978). Most resistance traits identified in *Lycopersicon* involve antibiosis.

The genus *Lycopersicon* is characterized by great diversity within and among its nine species (Miller and Tanksley 1990, Kalloo 1991). Although all species have been examined for resistance to at least some arthropod pests, the highest levels of resistance to the greatest number of arthropod species are found in *L. hirsutum* f. *typicum (=hirsutum)* Humb & Bonpl., *L. hirsutum* f. *glabratum* C.H. Mull and *L. pennellii* (Corr.) D'Arcy. There is extensive variation among accessions of these species in the spectrum and level of arthropod resistance. A diverse array of traits, including the physical and chemical properties of glandular trichomes as well as constitutively expressed and wound-induced chemical defenses present in the leaf lamella, have been associated with resistance.

OCCURRENCE AND MECHANISMS OF RESISTANCE

Aphids

The potato aphid, *Macrosiphum euphorbiae* (Thomas), and to a lesser extent the green peach aphid, *Myzus persicae* Sulzer, are major pests of tomato. Dense populations developing on the young foliage can cause severe yield reductions (Walgenbach 1997). Resistance to *M. euphorbiae* has been reported from *L. esculentum*, *L. esculentum var. cerasiforme*, *L. hirsutum f. typicum*, *L. hirsutum f. glabratum*, and *L. pennellii* (Gentile and

Stoner 1968a, Stoner et al 1968b, Clayberg and Kring 1974, Clayberg 1975, Quiros et al. 1977).

Type VI trichomes, which have a 4-cell glandular head, have been implicated in the entrapment of aphids and other small arthropods on several *Lycopersicon* species, including *L. esculentum, L. hirsutum f. typicum, L. hirsutum f. glabratum* and *L. pennellii* (Duffey 1986, McKinney 1938, Kennedy unpub.). In most cases entrapment accounts for only a low level of resistance to aphids. The chemical mechanisms responsible for entrapment are well documented in *Lycopersicon* and *Solanum berthaultii* Hawkes. Phenolic substrates and the enzymes polyphenol oxidase and peroxidase are compartmentalized within the trichome tip until an insect discharges the trichome tip and the contents are mixed. The ensuing enzymatic reaction (known as the browning reaction) oxidizes the phenolic substrates to quinones, which polymerize. The product of the browning reaction entangles small arthropods or collects on their appendages and mouthparts, inhibiting their ability to move, cling to the plant and feed (Duffey 1986, Duffey and Isman 1981, Gregory et al. 1986). There is extensive variation both within and among *Lycopersicon* in the level of browning reaction associated with type VI trichomes. This variation reflects differences in polyphenol oxidase and peroxidase activity as well as density of type VI trichomes (Duffey 1986).

In *L. esculentum,* and *L. esculentum var. cerasiforme,* resistance to the pink biotype of *M. euphorbiae* has been associated with the density of non-glandular trichomes, the presence of quinic acid in the foliage, and the poor nutritional quality of the foliage as a source of food for the aphids (Quiros et al. 1977). Definitive proof of the specific mechanisms underlying this resistance is lacking.

Resistance to the pink biotype of *M. euphorbiae* is conditioned by the gene Mi, which also confers resistance to root knot nematodes. Mi was introgressed into *L. esculentum* from *L. peruvianum* to confer nematode resistance and has been widely used in tomato breeding to produce nematode resistant cultivars. Its role in aphid resistance was only recently documented (Rossi et al. 1998, Vos et al. 1998). Mi is simply inherited and the aphid resistant phenotype is dominant under field conditions but incompletely dominant under greenhouse conditions (Kaloshian et al. 1995). The precise mechanism conditioning aphid resistance is not known, nor is it known whether the same mechanism conditions both aphid and nematode resistance (Rossi et al. 1998).

Resistance in *L. hirsutum f. glabratum* to the aphids *M. euphorbiae* and *Myzus persicae* (Sulzer) has been reported from accessions PI134417 and

PI126449 (Musetti and Neal 1997, Leite et al. 1999). The high level of resistance to *M. euphorbiae* is manifested as altered feeding behavior (reduced number and duration of probing and a decrease in total time spent probing plant tissue with mouthparts), elevated mortality and a high rate of dispersal by apterous aphids from resistant foliage (Musetti and Neal 1997). This resistance has been associated with the presence of 2-tridecanone in the tips of type VI glandular trichomes (Musetti and Neal 1997a and b, Kennedy and Yamamoto 1979). 2-Tridecanone is toxic to a number of insect species including *M. euphorbiae* and *M. persicae* but is both repellent and toxic to *M. euphorbiae* at concentrations associated with resistant foliage. The repellent and deterrent effects may play a more important role than toxicity in the resistance to *M. euphorbiae* (Williams et al. 1980, Kennedy 1986, Musetti and Neal 1997 a, b, Kennedy 2003).

Aphid resistance has been studied most intensively in *L. pennellii* accession LA716, which exhibits a high and stable level of resistance to *M. euphorbiae* (Goffreda and Mutschler 1989). On LA716, the normal probing and feeding behavior of *M. euphorbiae* and *M. persicae* is disrupted such that fewer aphids feed, and probing and feeding times of aphids are significantly less than on *L. esculentum* (Goffreda et al. 1988, Rodriguez et al. 1999). This difference is related to the presence of acylsugars in the tips of type IV trichomes of LA716, which deter both settling and feeding by aphids, although other factors appear to contribute to aphid resistance as well (Goffreda et al. 1990, Hartman and St. Clair 1999a). Acylsugars, primarily 2,3,4-tri-O-acylated glucose esters possessing C_4 to C_{12} fatty acids, constitute approximately 90% of the exudate of type IV trichomes of LA716 (Forbes et al. 1985). Goffreda et al. (1990) reported a negative relationship between acylsugar levels on the foliage and abundance of *M. euphorbiae* in segregating F_2 populations from crosses between *L. esculentum* and *L. pennellii*. *L. esculentum* does not possess type IV trichomes and does not accumulate acylsugars (Goffreda et al. 1990).

The inheritance of aphid resistance is complex and involves epistatic effects (Goffreda and Mutschler 1989). The occurrence of type IV trichomes is simply inherited and controlled by two dominant genes in crosses between *L. esculentum* and *L. pennellii*. In F_1 hybrids, the presence of either gene confers the presence of type IV trichomes (Lemke and Mutschler 1984). The genetic control of acylsugar synthesis and accumulation in type IV trichome tips is more complex. *L. esculentum* X *L. pennellii* hybrids produce both glucose and sucrose esters with a different fatty acid composition than the glucose esters of *L. pennellii*, which produces primarily glucose esters (Goffreda et al. 1990). An RFLP/QTL analysis revealed 5 genomic regions on 4 chromosomes associated with acylsugar production; 2 regions on chromosome 2 and one region on each of chromosomes 3, 4, and 11. The regions on chromosome 2 and the region on chromosome 3 account for 11-16% and 7-12% of the total variation in acylsucroses and total acylsugars, respectively. The alleles from *L. esculentum* are partially dominant to the alleles from *L. pennellii* in both regions of chromosome 2, whereas the *L. pennellii* allele is at least partially dominant to the *L. esculentum* allele on chromosome 3. The region on chromosome 4 is associated with increases in acylglucoses and accounts for 7 to 9% of the variability in levels of acylglucoses. The *L. esculentum* allele on chromosome 4 is partially dominant or co-dominant to the allele from *L. pennellii*. The region on chromosome 11 accounts for 17.7 to 22.2% of the variation in molar percentage of acylglucose and likely affects the levels of acylglucose directly. The allele from *L. pennellii* is dominant to that from *L. esculentum* (Mutschler et al. 1996). RAPD markers for 3 of the 5 QTLs have been identified and a genomic map consisting of 111 RAPD and 8 acylglucose transferase-related markers has been added to an existing framework of 150 RFLP markers (McNally and Mutschler 1997). RFLP/PCR-based marker assisted selection through 3 backcross generations for the 5 QTL regions associated with acylsugar accumulation successfully transferred acylsugar accumulation traits from *L. pennellii* to *L. esculentum* but indicated that additional QTLs are likely involved in the accumulation of higher levels of acylsugar accumulation (Lawson et al. 1997).

A QTL analysis by Blauth et al. (1998) of an F_2 population generated by an intraspecific cross between *L. pennellii* accessions exhibiting high (LA716) and low (LA1912) acylsugar accumulation revealed that the relative proportion of acylglucoses and acylsucroses are largely controlled by a one QTL on chromosome 3. A QTL on chromosome 10 was associated with total acylsugar levels and a QTL on chromosome 4 was associated with leaf area. Leaf area is important in resistance because acylsugars are confined to the tips of type IV trichomes and there is a significant association between acylsugar levels and type IV density. Despite this association, the genetic control of acylsugar accumulation is such that merely selecting for increased density of type IV trichomes would likely be ineffective in raising acylsugar levels (Blauth et al. 1998).

Whitefly

Research on resistance to whiteflies has emphasized two important pest species, *Trialeurodes vaporariorum* (Westn.), the greenhouse whitefly, and

Bemisia argentifolii Bellows and Perring, the silverleaf whitefly (=strain B of *B. tabaci*). Both species have piercing sucking mouthparts and feed on phloem sap. *T. vaporariorum* is primarily a pest on glasshouse-grown tomatoes. On susceptible cultivars, it is capable of developing large populations, which reduce yields and contaminate the fruits and plants with honeydew. *B. argentifollii* is primarily of concern in the field because it is a vector of geminiviruses that affect tomato and because heavy infestations can cause irregular ripening of tomato fruits.

Resistance to *T. vaporariorum* has been reported from *L. hirsutum f. typicum, L. hirsutum f. glabratum* and *L pennellii.* Entrapment of whitefly adults in the tips of glandular trichomes has been implicated in resistance (Gentile et al. 1968, Clayberg and Kring 1974, Ponti et al. 1975, 1983). Breeding lines developed by Ponti and Stienhuis (1984) apparently express different mechanisms of resistance, although both reduce survival and reproductive rate of *T. vaporariorum*. On breeding line 82207 whitefly contact the phloem more frequently as they feed but spend less time ingesting phloem sap and more time salivating in the phloem than on the susceptible cultivar 'Moneymaker.' No such differences in feeding behavior are observed on another resistant breeding line, 82216, which also reduces survival and reproductive rate of *T. vaporariorum* (Lei et al 1999). A QTL analysis of *T. vaporariorum* resistance and glandular trichome densities in F_2 population of *L. esculentum* x *L. hirsutum f. glabratum* provided no evidence for an association between whitefly resistance and density of type IV glandular trichomes. Two QTL's, which mapped to chromosomes 1 and 12, were associated with reduced oviposition rate. Two QTL's affecting type IV glandular trichomes were identified and mapped to chromosomes 5 and 9. One QTL affecting type VI trichome density was also identified, and mapped to chromosome 1 (Maliepaard et al 1995).

Resistance of *L. pennellii* to *T. vaporariorum* has been reported to involve entrapment of adults in the exudate of foliar glandular trichomes (Gentile et al 1968, Clayberg and Kring 1974, Ponti et al. 1975). Other traits may be involved, however, and resistance has been assumed to be polygenically inherited (Berlinger et al. 1991). Because *Encarsia formosa*, the primary parasitoid used in biological control of *T. vaporariorum* in glasshouse tomato production, is also entrapped by the foliar glandular trichomes, Ponti et al. (1983) considered the resistance incompatible with biological control and abandoned efforts to incorporate the glandular trichome mediated resistance of *L. pennellii* into tomato varieties adapted for glasshouse production.

Resistance to *B. argentifolii* has been reported from *L. pennellii, L. peruvianum* (L.) Mill., and *L. hirsutum f. typicum*. Resistance in *L. pennellii* accession LA716 greatly reduces oviposition. A major portion of the genetic basis for this resistance is associated with chromosome 6 although additional minor components appear to be associated with chromosomes 2, 3, 8, and 11 (Heinz and Zalom 1995). The resistance appears to be independent of trichome density but may be conditioned at least in part by high concentrations of acylsugars in the glandular tips of type IV trichomes of *L. pennellii* (Liedl et al. 1995, Nombela et al. 2000).

Channarayappa et al. (1992) reported resistance to *B. tabaci* (presumably *argentifolii*) in *L peruvianum*, which they attributed to factors other than foliar trichomes. Some preliminary results suggest that the Mi gene from *L. peruvianum* may contribute to this resistance (Nombela et al. 2000); however, additional studies are needed to confirm this.

High levels of resistance to *B. argentifolii* (=*tabaci*) reported from *L. hirsutum f. typicum* are conditioned, in part by high densities of type IV trichomes, which are associated with reduced attractancy to adults and reduced oviposition (Snyder et al. 1998). Type VI trichomes, which entrap adults, may also play a role in the resistance (Channarayappa et al. 1992). This resistance has been associated with a reduction in the incidence of leaf curl virus transmitted by *B. argentifolii* (Channarayappa et al. 1992). The photophase under which *L. hirsutum f. typicum* is grown influences the densities of type IV and type VI trichomes and the expression of resistance to *B. argentifolii*. Type IV trichome densities and resistance to *B. argentifolii* are higher and type VI trichome densities are lower when plants are grown under a photophase of 8 h than 16 h (Snyder et al. 1998). This sensitivity to photophase may limit the utility of *L. hirsutum f. typicum* as a source of resistance.

Thrips

The western flower thrips, *Frankliniella occidentalis* Pergande, is of concern as a pest of tomato primarily because it transmits tomato spotted wilt virus (TSWV), although feeding in blossoms by high populations of thrips can result in external scarring of fruit. Kumar et al. (1995a) reported high levels of resistance to foliage feeding by nymphs in *L. hirsutum f. glabratum* (accession PI 134417), and *L. pennellii* (accession LA716), and a moderate level of resistance in *L chilense* (accession LA1782). They also reported a high level of resistance to adult feeding on *L. hirsutum f. typicum* (accession LA1353). They characterized this resistance as antixenosis and demonstrated that feeding by adults on antixenotic lines was limited to the epidermal cells, whereas feeding on susceptible foliage caused severe damage to epidermal, palisade and spongy parenchyma cells. In a related study, Kumar et al (1995b) reported significant differences among *Lycopersicon* accessions in flower infestations of *F. occidentalis*. After accounting for flower number, and flowering period, *L. pennellii* (LA716) and the *L. esculentum* cultivar 'Ray de los Tempranos' supported the lowest floral infestations, but there was also significant variation in floral infestations among *L. esculentum* cultivars. *L. pennellii* shows resistance to thrips transmission of TSWV (Kumar et al 1993). The value of this resistance in breeding thrips resistant cultivars to reduce the incidence of TSWV has yet to be demonstrated.

Diptera

Resistance has been reported to the fruit fly or vinegar fly, *Drosophila melanogaster* Meigen, and the agromyzid leafminers *Liriomyza trifolii* (Burgess) and *L. sativae* Blanchard (=*L. munda* Frick). *D. malanogaster* oviposit on fruit and larvae develop within damaged and ripe tomato fruits. Potentially useful levels of variation in resistance to *D. melanogaster* have been reported among cultivars and breeding lines of *L. esculentum* but the genetics and underlying mechanisms have not been elucidated (Stoner et al. 1969).

Larvae of the agromyzid leafminers *L. sativae and L. trifolii* feed on mesophyll tissue and produce serpentine leaf mines. Adults oviposit in leaf tissue and feed on sap exuding from leaf cells punctured by their ovipositor. High levels of resistance to *L. trifolii* and *L. sativae* have been reported from *Lycopersicon pennellii* (accessions LA1735 and LA716), *L. cheesmanii* (accession LA1401) and *L. hirsutum f. glabratum* (accession PI126449) (Webb et al. 1971, Laterrot et al. 1987, Erb et al. 1993). Resistance in these *L. pennellii* and *L. cheesmanii* accessions is apparently inherited with a high degree of dominance in crosses with *L. esculentum* because F_1 progeny from crosses with *L. esculentum* express high levels of resistance. This resistance involves larval antibiosis, (apparently associated with the mesophyll tissue), adult antibiosis, and antixenosis, which is due in part to glandular trichomes (Erb et al. 1993). The resistance of *L. pennellii* to oviposition and feeding by adult *L. trifolii* has been attributed to the presence of acylsugar esters in the type IV trichome tips (Hawthorne et al. 1992). Eigenbrode et al. (1993) reported that *Lycopersicon* accessions with higher densities of nonglandular trichomes on the foliage were generally less damaged by *Liriomyza spp.* than those with low densities.

Beetles

Resistance in tomato has been reported two beetle pests, tobacco flea beetle, *Epitrix hirtipennis* (Melsheimer) and the Colorado potato beetle, *Leptinotarsa decemlineata* (Say). *E. hirtipennis* adults feed on the older foliage of plants causing a "shot hole" type of injury and are typically most severe early in the season, when the plants are young. Resistance to *E. hirtipennis* has been reported from several accessions of *Lycopersicon hirsutum f. typicum* and *L. hirsutum f. glabratum*. This resistance has been associated with repellant(s) produced by glandular trichomes on the foliage (Gentile and Stoner 1968b).

L. decemlineata is a severe pest of tomato in some production areas of the eastern United States. Geographically separated populations of *L. decemlineata* differ in their ability to use *L. esculentum* as a host (Kennedy and Farrar 1987, Lu et al. 1997, Lu et al. 2001). Populations adapted to *L. esculentum* are capable of causing severe damage to tomato crops through defoliation as well as direct feeding on fruit (Schalk and Stoner 1979, Kennedy et al. 1983). In general, infestations tend to be most severe on young plants. Resistance to tomato adapted populations of Colorado potato beetle has been reported from *L. hirsutum f. typicum* (accession PI126445 (Carter et al. 1989a)), *L. hirsutum f. glabratum* (PI134417 and 134418; Kennedy et al. 1985, Schalk and Stoner 1976) and *L. pennellii* (Carter and Schurig, cited in Farrar and Kennedy 1991a).

The expression of resistance in *L. hirsutum f. typicum* accession PI126445 as measured by survival of Colorado potato beetle varies seasonally in response to photoperiod and temperature. When plants are grown in spring, potato beetle survival is comparable on PI 126445 and *L. esculentum* but when plants are grown in autumn, survival on PI126445 is very low. This resistance has been associated with the acute toxicity of the sesquiterpene zingiberene, which is present in the tips of type VI glandular trichomes (Carter et al. 1989 a, b). The accumulation of zingiberene in the tips of type VI trichomes is influenced by temperature and photoperiod, such that the quantities zingiberene associated with the foliage of PI126445 plants are acutely toxic to Colorado potato beetle larvae during autumn but not during spring and summer (Gianfagna et al. 1992). The presence of zingiberene is controlled at a single locus. An allele from *L. hirsutum f. typicum*, which controls expression of zingiberene, is recessive to an allele from *L. hirsutum f. glabratum* but dominant to an allele from *L. esculentum* (Rahimi and Carter 1993). Although inherited simply, zingiberene-mediated resistance to Colorado potato beetle is likely to be of limited utility unless the effects of temperature and photoperiod on expression of resistance can be mitigated (Raimi and Carter 1993).

Two genetically distinct mechanisms of resistance to Colorado potato beetle have been documented from *L. hirsutum f. glabratum* accession PI134417 (Kennedy et al. 1985, Kennedy and Sorenson 1985, Sorenson et al. 1989). One mechanism is associated with type VI glandular trichomes, which abound on the foliage and stems. The other is associated with the leaf lamella. The glandular trichome-mediated resistance results from the presence of high concentrations of the methyl ketone, 2-tridecanone, in the trichome tips. 2-Tridecanone is acutely toxic to young larvae at concentrations associated with the foliage of PI134417. When young larvae move about on the foliage, they discharge the trichome tips, contact lethal quantities of 2-tridecanone, and die within 72 hours.

Because 2-tridecanone is contained in the tips of type VI glandular trichomes, the level of resistance is related to both the average amount of 2-tridecanone per trichome tip and the density of type VI trichomes.. Consequently, there is substantial variation in resistance expression. Both 2-tridecanone level and type VI trichome density are under separate genetic control but there are epistatic effects (Fery and Kennedy 1987, Nienhuis et al. 1987). Expression of both traits is strongly influenced by environmental conditions, including daylength, light intensity, and plant nutrient status (Kennedy et al. 1981, Snyder and Hyatt. 1984, Barbour et al. 1991). High levels of 2-tridecanone are conditioned by at least three major genes inherited in a recessive manner (Fery and Kennedy 1987). Restriction fragment length polymorphism (RFLP) analyses of QTLs associated with 2-tridecanone levels in F_2 progeny of crosses between *L. esculentum* and PI134417 identified three different linkage groups associated with the expression of 2-tridecanone. One of the RFLP loci having the highest correlation with 2-tridecanone levels is primarily associated with expression of type VI trichome density (Nienhuis et al. 1987). Zamir et al. (1984) reported a pleiotropic effect of the gene for determinant plant growth, which caused low levels of 2-tridecanone expression in plants having a determinant growth form. As indicated elsewhere in this chapter, the presence of high levels of 2-tridecanone contributes to the resistance of *L. hirsutum f. glabratum* to a number of arthropod species.

F_1 progeny from crosses between *L. hirsutum f. glabratum* accession PI134417 and *L. esculentum* do not express the trichome mediated resistance attributable to 2-tridecanone, but express resistance that causes high levels of mortality during the later larval stages. Factors responsible for this resistance are inherited in a dominant or semi-dominant fashion mainly associated with the leaf lamella rather than the trichomes (Kennedy 1986, Sorenson and Kennedy 1989).

The glycoalkaloid α-tomatine is common among species of *Lycopersicon* acting as a potent feeding deterrent and growth inhibitor for *Leptinotersa decemlineata* larvae and adults (Sturkow and Low 1961, Schreiber 1968, Hsiao and Frankel 1968b, Roddick 1974, Sinden et al. 1978, Juvik et al. 1982a, Mitchell and Harrison 1985, Hare 1987). Sinden et al. (1978) demonstrated that α-tomatine content of foliage of *L. esculentum* and *L. hirsutum f. glabratum* (PI134417) varied with plant age and the daylength under which the plants were grown. They observed significant negative correlations between feeding by adults and foliar concentration of α-tomatine for both species. The correlation was stronger for *L. esculentum* (r=-0.897, P=0.01) than for *L. hirsutum f. glabratum* (r=-0.613, P=0.05). However for plants containing comparably low levels of α-tomatine, foliage consumption was less on *L. hirsutum f. glabratum* than on *L. esculentum*, suggesting other factors contributed to the reduced feeding on *L. hirsutum f. glabratum*. Other in-depth studies have failed to detect any evidence that α-tomatine levels contribute to the lamella-based resistance of PI134417 to *L. decemlineata* (Barbour and Kennedy 1991). It remains possible that α-tomatine contributes to the relatively poor suitability of *L. esculentum* as host for many populations of *L. decemlineata*. Its value as a resistance mechanism appropriate for use in the development of tomato cultivars resistant to *L. decemlineata* is likely to be limited because its activity appears to depend on concurrent levels of phytosterols and proteins in the foliage (Hare 1987).

Lepidoptera

Resistance in *Lycopersicon* has been reported for more species of Lepidoptera than any other taxon of arthropods (Farrar and Kennedy 1991a). The high level of interest in screening germplasm for resistance to Lepidoptera reflects their importance as fruit-feeding, defoliating and leaf mining pests of tomato. Resistance to one or more of the following species of the family Noctuidae has been reported from at least one accession of all *Lycopersicon* species: *Helicoverpa armigera* (Hubner), *H. zea* (Boddie), *Spodoptera exigua* (Hubner), *S. eridania* (Cramer), *S. littoralis* (Boisduval), and *Plusia chalcites* (Esp.) (Kashyap et al. 1990, Farrar and Kennedy 1991a, Eigenbrode and Trumble 1993, Eigenbrode et al. 1993). In addition, resistance to the foliage and fruit feeding noctuid *Trichoplusia ni* (Hubner) has been reported in *L. hirsutum f. typicum* and *L. hirsutum f. glabratum* (Sinha and McLarin 1989). Resistance has also been reported to sphingids, *Manduca sexta* (L.) and *M. quinquemaculata* (Haw.), and to gelechiids, *Keiferia lycopersicella* (Wals.), *Phthorimaea operculella* (Zell.), and *Tuta (=Scrobipalpula) absoluta* (Meyerich) (Kennedy and Henderson 1978, Juvik et al. 1982b, Lin et al. 1987, Farrar, et al. 1994, Maluf et al. 1997, Leite et al. 1999). In many cases the levels of resistance are modest and the underlying mechanisms and genetic basis are not known, but some resistances are of a very high level and their mechanisms have been well studied. Some resistance mechanisms are shared by more than one *Lycopersicon* species and affect more than one insect pest, whereas others are more limited in scope and spectrum of activity.

Although no extremely high levels of resistance to lepidopteran pests of tomato have been reported from *L. esculentum*, there is significant variation in the suitability of *L. esculentum* accessions as hosts of some important species, including *H. zea, H. armigera, S. exigua, K. lycopersicella, T. absoluta,* and *S. eridania* (Eigenbrode et al. 1993, see also Farrar and Kennedy 1991a). Resistance factors are present in the tips of type VI glandular trichomes, in the leaf lamellae, and in green fruit. In general, these factors act to slow larval growth either by deterring larval feeding, intoxicating the larvae or by interfering with the utilization of nutrients. Detailed information about these mechanisms is presented in several reviews (Duffey 1986, Duffey and Bloem 1986, Kennedy, 1986, Kennedy et al. 1991, Kennedy 2003).

Glandular trichomes are important contributors to Lepidoptera resistance in a number of *Lycopersicon* species. Removal of glandular trichomes (primarily type VI) from *L. esculentum* foliage increased its suitability as a food source for *H. zea* larvae. Larvae reared on foliage from which trichomes were removed grew faster and suffered 10 to 17% less mortality than those reared on foliage containing the normal complement of glandular trichomes (Duffey 1986, Farrar and Kennedy 1987). These effects were associated with a 27% reduction in the relative consumption rate (= dry wt. of foliage ingested per initial dry wt. of larva) in the 2nd instar and a 22% reduction in the efficiency of conversion of ingested foliage (= dry wt. gained per dry wt. of foliage ingested) in the 5th instar (Farrar and Kennedy 1987). The specific factors causing these effects have not been determined.

The catecholic phenolics, rutin and chlorogenic acid, present both in the leaf lamella and the tips of type VI trichomes of *L. esculentum*, upon incorporation into artificial diet cause a dose-dependent growth inhibition of *H. zea* and *S. exigua* larvae (Elliger et al. 1981, Isman and Duffey 1982a, Duffey and Bloem 1986,). However, no corrleation was found between total phenolic content of foliage and growth rate of *H. zea* larvae on *L. esculentum* cultivars, despite consistent differences among cultivars in larval growth rate over 2 years (Isman and Duffey 1982, Duffey 1986). An explanation for this lack of correlation lies in the fact that the growth inhibitory effects of phenolics are dependent on protein levels in the diet (Duffey and Bloem 1986). The abundance of both phenolics and proteins in tomato foliage are known to be influenced by nitrogen fertilization, water stress, and light intensity (English-Loeb et al. 1997, Stout et al. 1996, Wilkens et al. 1996). The levels of phenolics and protein also vary greatly and independently among plants of the same and different *L. esculentum* cultivars (Stamp and Horwath 1992).

The phenolic rutin has also been shown to prolong development of *M. sexta* larvae at cool but not at warm temperatures, in studies involving artificial diet, apparently by interfering with physiological processes involved in the initiation of molting (Stamp 1990, Horwath and Stamp 1992, 1993). The effects of rutin on *M. sexta* have not been documented for larvae fed *Lycopersicon* foliage and the effect may be of little consequence in the context of host plant resistance for crop protection.

The glycoalkaloid α-tomatine, present in the leaf lamella and green fruit, but not the tips of glandular trichomes, has been hypothesized as another growth inhibiting compound for *H. zea* and *S. exigua* in *Lycopersicon* (Elliger et al. 1981, Duffey and Bloem 1986). α-Tomatine is also a potent inhibitor of larval growth in both species when incorporated into artificial diets at concentrations commonly found in *L. esculentum* cultivars (Elliger et al. 1981, Duffey 1986, Bloem et al. 1989). Foliar levels of α-tomatine vary within and among *Lycopersicon* species, with the highest levels found in *L. esculentum var. cerasiforme* and *L. pimpinellifolium* (Juvik et al. 1982a). α-tomatine levels of progeny from crosses between *L. esculentum var. cerasiforme* and *L. pimpinellifolium* express high and low levels of this glycoalkaloid, its expression being controlled by segregation of 2 co-dominant alleles at a single locus (Juvik and Stevens 1982a). In *L. esculentum* fruit α-Tomatine content was positively correlated to development time and mortality of *H. zea* larvae fed on different aged fruit and negatively correlated with larval growth rate and adult weight. However, for *S. exigua*, there were no correlations between α-tomatine content of the fruit and growth or survival rate of larvae but there was a significant correlation between cuticular toughness of the fruit and larval mortality in *S. exigua* (Juvik and Stevens 1982b). Like the phenolics, the growth inhibitory effects of α-tomatine on *H. zea* and *S. exigua* are dependent on the composition of the dietary milieu in which the larvae experience them. The toxicity of α-tomatine to these species is dependent on the relative concentrations of 3-β-hydroxy-sterols present in the diet. α-Tomatine is non-toxic to *H. zea* and minimally toxic to young *S. exigua* larvae in the presence of equimolar concentrations of dietary sterols (Bloem et al. 1989).

It is likely that both phenolics and α-tomatine provide significant but variable levels of background resistance to lepidopterous species in *L. esculentum* as well as other *Lycopersicon* species. However, as is the case with α-tomatine and Colorado potato beetle, the potential to select for high and stable resistance to Lepidoptera based on phenolic or α-tomatine concentrations in the foliage or fruit appears to be very limited.

The impact of growth inhibiting resistance mechanisms such as tomatine and phenolics under field conditions has not been well documented and definitive field studies are needed. It is difficult or impossible to predict population level consequences for a pest species or the potential utility in crop protection of resistance traits that reduce larval growth rate in a laboratory assay. Farrar and Kennedy (1990) found that the commonly used procedure of measuring inhibition of larval growth by plant foliage or plant chemicals exaggerated the effects. Because of the sigmoid shape of larval growth curves, a 50% reduction in growth after a fixed time interval does not necessarily translate into a 50% change in either developmental time or final pupal weight. For example, in studies involving α-tomatine, a concentration that caused a 47% reduction in larval size at 10 days caused only a 4.6% reduction in pupal weight and a 15% increase in development time.

In addition to the constitutively expressed resistance mechanisms already discussed, *L. esculentum* and very likely other *Lycopersicon* species, possess an array of inducible defenses that act to significantly reduce the suitability of induced plants as hosts for insect pests and pathogens. These induced defenses are important in reducing the damage potential of many pests. They provide a critical base level of resistance in tomato. Induced defenses of *L. esculentum* affecting *H. zea*, *S. exigua*, and *M. sexta* have been intensively studied. Minor injury to *L. esculentum* foliage by insects, pathogens, or mechanical wounding elicits a systemic response that leads to synthesis of numerous defense-related proteins and alters the suitability of foliage for some insects and plant pathogens at the wound site or elsewhere on the plant (Ryan 2000). The inducible defenses of *L. esculentum* consist of an array of sets of defensive compounds that are differentially inducible by different insects and pathogens acting individually or in combination (Stout et al. 1999). Injury to a plant by different agents may cause different patterns of resistance to any given array of insects and pathogens because individual insect and pathogen species vary in their sensitivity to each set of defensive compounds.

Feeding on *L. esculentum* by caterpillars induces proteinase inhibitors, peroxidase, polyphenol oxidase and lipoxygenase. The resulting elevated levels of proteinase inhibitors and to a lesser extent polyphenol oxidase result in reduced growth and delayed development in *H. zea, M. sexta* and *S. exigua,* and increased mortality of *S. exigua* feeding on the induced foliage (Broadway et al 1986, Stout et al. 1998a, b, Stout and Duffey 1996). The induction of proteinase inhibitors and polyphenol oxidase is systemic, occurs within hours and lasts for at least 21 days. Induction is not uniform throughout the plant. Consequently, some portions of the plant manifest higher levels of resistance than others (Stout et al. 1998a, Stout and Duffey 1996). Levels of induction of proteinase inhibitors and polyphenol oxidase in response to feeding vary among *L. esculentum* cultivars and generally decline with plant age (Stout et al. 1998a, Cipollini and Redman 1999). The tomato genes coding for proteinase inhibitors I and II have been sequenced and cloned into tobacco (Johnson et al. 1989). Growth of *M. sexta* larvae was reduced by 50 to 67% over controls on transgenic tobacco plants expressing tomato proteinase inhibitor II. In contrast, larval growth was only slightly inhibited on transgenic plants expressing tomato proteinase inhibitor I.

The production of jasmonic acid via the octadecanoid pathway in response to caterpillar feeding serves as the signal for expression of inducible proteinase inhibitors, polyphenol oxidase and peroxidase in tomato foliage (Ryan 2000, Thaler et al. 1999, Fidantsef et al. 1999). Mutants that block jasmonic acid biosynthesis and or inhibit the response to jasmonic acid have been identified in tomato. The resistance of plants possessing either of these mutants is compromised (Bergey et al. 1996, Howe et al. 1996, Li et al. 2002, Howe and Schilmiller 2002). Given the level of understanding that exists for induced defenses at the biochemical and genetic level, there would seem to be tremendous potential to enhance and stabilize the expression of these defenses throughout the plant to achieve higher levels of resistance to key insect pests.

Resistance to *H. zea, S. exigua* and *S. littoralis* has been reported from various accessions of *L. pennellii.* These species typically oviposit on the foliage and the larvae feed initially on the foliage before boring into fruit in later instars. Because young larvae feed on foliage, resistance mechanisms in the foliage but not in the fruit can be effective.

L. pennellii accessions LA1277, Atico 716 and Sisicaya 751, possess high levels of resistance to *Spodotera littoralis* - the cotton leafworm (Berlinger et al. 1997). Numerous first instar *S. littoralis* become entrapped in the exudate of foliar glandular trichomes on these accessions and die. Entrapment on all accessions involves physical entanglement but not toxicity. Both Sisicaya 751 and LA1277 also possess additional resistance mechanisms associated with the leaf lamella, which cause reduced larval survival, increased larval development time and an elevated incidence of deformed adults. In addition, the adult males that develop on resistant foliage exhibit anomalous sperm development. *S. littoralis* larvae fed on artificial diet containing foliar extracts from the resistant accessions exhibited similar effects. Based on the solubility characteristics of the extracted compounds, Berlinger et al. (1997) hypothesize at least three compounds contribute to the lamella-based resistance to *S. littoralis*.

Resistance of *L. pennellii* accession LA716 to *H. zea* and *S. exigua* is due at least in part to the acylglucoses produced by the type IV glandular trichomes (Juvik et al. 1994). Application of acylglucose to *L. esculentum* foliage fed to *H. zea* and *S. exigua* larvae caused a reduction in growth rate and survival in both species. When given a choice of acylglucose-treated or non-treated *L. esculentum* foliage, both species preferred the non-treated foliage. When fed synthetic diet containing acylglucose, weight gain by *H. zea* larvae was significantly reduced but survival was unaffected. In contrast, both survival and growth of *S. exigua* larvae were significantly reduced. Interestingly, acylglucoses appear to stimulate egg laying by *H. zea* moths (Juvik et al. 1994). The genetic control of acylsugar production has been described previously in the context of resistance to aphids.

Factors in addition to acylsugars apparently contribute to resistance to *H. zea* and *S. exigua* because inbred backcross populations from crosses between *L esculentum* and *L. pennellii*, which produced acylsugar levels comparable to *L. esculentum*, expressed significant levels of resistance to both species. Further, an additional inbred backcross line expressing acylsugars did not exhibit higher levels of resistance to *H. zea* and *S. exigua*. Resistance to both species was associated with smaller fruit size, delayed maturity and lower fruit yield (Hartman and St. Clair 1998). A multiple trait selection procedure enabled the selection of an inbred backcross line that did not exhibit a negative association between *S. exigua* resistance and fruit size or yield. However, there were differences in general combining ability among elite inbreds for percent *S. exigua* damaged fruit (Hartman and St. Clair 1999b). In these studies, resistance was measured as the proportion of insect damaged fruits, so it is difficult to compare results to those of studies reporting putative resistance mechanisms on the insects.

Resistance in *L. hirsutum f. typicum* accessions LA1777, LA2329 and PI126445 to *S. exigua* causes reduced larval growth and survival, relative to *L. esculentum*. Toxins contained in the tips of type VI glandular trichomes are primarily responsible for the resistance of LA1777 and LA2329, but contribute less to the resistance of PI126445. Factors associated with the leaf lamella are primarily responsible for the resistance of PI126445 but also contribute to the resistance of LA1777 and L2329 (Eigenbrode and Trumble 1993, Eigenbrode et al. 1994, Eigenbrode et al. 1996). The specific lamella-based factors have not been identified. The sesquiterpene carboxylic acids (+)-(E)-α-santalen-12-oic, (-)-(E)-endo-α-bergamoten-12-oic, and (+)-(E)-endo-β-bergamoten-12-oic acid, produced in the type VI trichomes of LA1777, have been implicated as the causal mechanism for resistance to *S. exigua* and *H. zea,* and to other Lepidoptera as well (Frelichowski et al. 2001).

Resistance to Lepidoptera in *L. hirsutum f. glabratum* has been studied extensively and reviewed in detail elsewhere (Kennedy et al. 1991, Kennedy 2003). *L. hirsutum f. glabratum* accessions possess high levels of resistance to *M. sexta, H. zea, S. littoralis, S. exigua, H. armigera, T. ni, T. absoluta, P. operculella,* and *K. lycopersicella* (Kashyap and Verma 1987, Sinha and McLarin 1989, Farrar and Kennedy 1991, Maluf et al. 1997). Resistance to *M. sexta,* which is manifested by mortality of young larvae on the foliage, is due to the presence of lethal concentrations of the methyl ketone 2-tridecanone in tips of type VI glandular trichomes (Kennedy and Yamamoto 1979, Williams et al. 1980). 2-tridecanone contributes to the resistance to numerous other species. It is toxic to a broad range of insect pests and acts as a feeding and oviposition deterrent for some species such as *T. absoluta* (Maluf et al. 1997). Resistance to *S. littoralis, T. absoluta* and *P. operculella* was correlated with 2-tridecanone concentrations in segregating F_2 populations (Ben-David 1983 cited in Zamir et al. 1984, Maluf et al. 1997). 2-Tridecanone plays only a minor role in the resistance to *H. zea* because exposure of eggs and neonates to sub-lethal concentrations of 2-tridecanone vapors surrounding resistant foliage induces the cytochrome P450 detoxification system in the insect, which then detoxifies the methyl ketone (Kennedy et al. 1987). Subsequent feeding on the foliage by induced larvae exposes them to unknown resistance factors present in the leaf lamellae, which reduce growth and survival of the larvae. Only a small percentage of larvae survive to pupation and most of those die as pupae due to exposure to a second methyl ketone, 2-undecanone, present in the tips of type VI trichomes (Farrar and Kennedy 1988). The methyl ketone based resistance is controlled by at least 3 recessive genes although the lamella-based resistance to *H. zea* is inherited as a dominant or semi-dominant trait (Fery and Kennedy 1987, Farrar and Kennedy 1991b). Resistance in *L. hirsutum f. glabratum* to *H. armigera*, which significantly reduces the incidence of fruit damage in the field, is inherited in an additive manner and may involve the same mechanisms as resistance to its close relative *H. zea* (Kalloo et al. 1989). The lamella-based factors responsible for resistance to *H. zea* are inherited independently from the lamella-based factors conditioning resistance to Colorado potato beetle, thus indicating that separate mechanisms contribute to the lamella-based resistances to *H. zea* and *L. decemlineata* (Farrar and Kennedy 1991b).

Although 2-tridecanone and 2-undecanone are acutely toxic to *K. lycopersicella* and *S. exigua,* in contact toxicity assays, other factors are primarily responsible for the resistance. Resistance of *L. hirsutum f. glabratum* to *S. exigua* is not correlated with concentration of either methyl ketone (Eigenbrode and Trumble 1993). *K. lycopersicella* larvae avoid lethal exposure to the methyl ketones contained in tips of type VI trichomes because they are leafminers (Lin et al. 1987). The mechanisms responsible for resistance to these species are not presently known. 2-Tridecanone has also been implicated in resistance of *L. hirsutum f. glabratum* to *T. absoluta* (Maluf et al. 1997). However, it is likely that other factors are involved because production of 2-tridecanone is inherited as a recessive trait (Fery and Kennedy 1987), yet F_1 offspring from *L. esculentum x L. hirsutum f. glabratum* crosses appeared resistant to *T. absoluta,* on the basis of damage assessments (Maluf et al. 1997).

Mites

Spider mites (Acari: Tetranychidae), primarily the twospotted spider mite, *Tetranychus urticae* Koch, and the carmine spider mite, *T. cinnabarinus* (Boisd.), are important pests of tomato, especially in glasshouse production and under hot, dry conditions in the field. Resistance to either or both of these closely related species has been reported in *L. esculentum, L. esculentum var. cerasiforme, L. hirsutum f. typicum, L. hirsutum f. glabratum, L. pennellii* and *L. pimpinellifolium* (Gilbert et al. 1966, Stoner and Stringfellow 1967, Stoner and Gentile 1968, Stoner et al. 1968a, Gentile et al. 1969, Aina et al. 1972, Rodriguez et al. 1972, Cantelo et al. 1974, Gilbert et al. 1974, Patterson et al. 1975a,b, Chiavegato and Mischan 1981,

Weston et al. 1989, Fernandez-Munoz et al. 2000). Resistance to another related spider mite, *T. marianae* McGregor, has been reported from *L. hirsutum f. typicum* and *L. peruvianum*. Resistance levels in *L. esculentum* are generally low to moderate and generally lower than levels in *L. hirsutum f. typicum, L. hirsutum f. glabratum* and *L. pennellii*. In most cases, glandular trichomes contribute significantly to the resistance through entrapment of mites and through the repellent or toxic qualities of their tip contents (e.g. Gentile et al. 1969, Knavel et al. 1972, Snyder et al. 1993, Goncalves et al. 1998).

Resistance in *L. hirsutum f. typicum* accessions to *T. urticae* causes reduced mite survival and is associated with high densities of type IV trichomes (Snyder and Carter 1984, Carter and Snyder 1986). Foliage of resistant accessions is also repellent to *T. urticae* due to the presence of 2,3, dihydrofarnesoic acid in the type IV tips (Guo et al. 1993, Snyder et al. 1993). Resistance in *L. hirsutum f. glabratum* involves both toxicity and repellency of the foliage due in part to the presence of 2-tridecanone in the type VI trichome tips, but other factors also appear to be involved (Chatzivasileiadis and Sabelis 1998, Goncalves et al. 1998).

L. pimpinellifolium accession T0-937 is highly resistant to *T. urticae*. The resistance is inherited as a completely dominant trait reportedly controlled by 2 to 4 genes (Fernandez Munoz et al. 2000).

TRANSGENIC RESISTANCE

Expression in tomato plants of one or more modified insecticidal protein genes from *Bacillus thuringiensis* Berliner (Bt) has resulted in high levels of resistance to several lepidopterous pests including *M. sexta, H. virescens, H. armigera,* and *S. exigua* (Fischoff et al. 1987, Delannay et al. 1989, Reynaerts and Jansens 1994, van der Salm et al. 1994, Mandaokar et al. 2000). Transgenic maize and cotton, and to a much less extent potato, expressing Bt insecticidal crystal protein toxins have been widely and successfully grown for crop protection. Consumer resistance to transgenic crops in the USA and Western Europe has severely limited the use of transgenic potato cultivars expressing *Bt* genes but had a lesser impact on the use of transgenic maize and cotton cultivars (Shelton et al. 2002).

Although *Bt* genes offer tremendous potential for the production of tomato cultivars resistant to most if not all lepidopterous pests and Colorado potato beetle, such cultivars have not been commercialized due largely to concerns over consumer resistance to transgenic food crops. Transgenic technology offers tremendous potential to readily achieve high levels of resistance to major insect pests in elite plant types thereby greatly simplifying insect management and reducing the quantities of synthetic insecticides required to produce high quality fruit. It is highly likely that consumer resistance to this technology will wane over time and transgenic insect resistance based on Bt insecticidal crystal toxins and other novel resistance mechanisms (e.g. Dowd et al. 1998, Gatehouse et al. 1997, Gatehouse et al. 1999) will become widely used in tomato.

NON-TARGET EFFECTS OF RESISTANCE

A number of resistance mechanisms in tomato exert adverse effects on parasitoids and predators of tomato insect and mite pests. These effects have been reviewed recently by Kennedy (2003). Trichome mediated resistance traits have been most extensively implicated in negative effects on parasitoids and predators in both the field and laboratory. Effects range from reduced searching efficiency, resulting from physical interference with movement, to increased mortality rates due to entrapment and contact with lethal contents of glandular trichome tips. 2-tridecanone-mediated resistance can profoundly affect some species of parasitoids and predators (Kauffman and Kennedy 1989a, b, Kashyap et al. 1991a, Kashyap et al. 1991b, Farrar and Kennedy 1991c, Barbour et al. 1993, Farrar and Kennedy 1993). In the field, mortality rates of *H. zea* due to egg parasitoids and predators is significantly reduced on plants expressing 2-tridecanone mediated resistance to *M. sexta, L. decemlineata, and M. euphorbiae* (Barbour et al. 1993, Barbour et al. 1997, Farrar et al. 1994).

The effects of any given plant resistance trait on parasitoids and predators depend on the specific attributes of the trait and the details of the interaction between the natural enemy, its host or prey, and the plant. Effects can be mediated directly by contact between a parasitoid/predator and the plant or indirectly through the host or prey of the natural enemy. The consequences of any such effects on biological control and pest population levels may be positive, negative or neutral and are difficult to predict based on observed effects on individual parasitoids or predators (Hare 1992). The diversity that exists in parasitoid and predator biology with respect to the details of these interactions makes it difficult to predict such effects and their implications for crop protection. It seems clear, however, that glandular trichome-mediated resistance mechanisms are more likely than non-trichome-mediated resistance traits to exert general adverse effects on parasitoids and predators of tomato pests (Kennedy 2003). However, the simple demonstration of a negative effect on parasitoid or predator, either in the field or in the laboratory, should not automatically be construed to mean that the effect will have significant negative effects on biological control that mitigate the benefit of moderate levels of resistance in protecting the crop from damage. Field experiments specifically designed to characterize the net effects of resistance and biological control should be conducted before conclusions are reached regarding the potential value of a resistance trait for crop protection.

VARIATION IN THE EXPRESSION OF RESISTANCE

The expression of many potentially valuable resistance traits in *Lycopersicon* is influenced by plant age, leaf age, and even leaf position on a plant (e.g. Sinden et al. 1978, Kennedy et al. 1985, Lin et al. 1987, Stamp and Horwath 1992). In addition, environmental conditions, including daylength, temperature, light intensity, moisture stress and plant nutrient status affect resistance traits in complex ways that alter the expression of resistance to numerous species including *B. argentifolii, L. decemlineata, T. urticae*, and several lepidopteran pests (Kennedy et al. 1981, Snyder and Hyatt 1984, Kennedy et al. 1985, Lin et al. 1987, Weston et al. 1989, Barbour et al. 1991, Gianfagna et al. 1992, Wilkens et al. 1996, English-Loeb et al. 1997, Snyder et al. 1998). Environmental effects on glandular trichomes of *Lycopersicon* can influence tritrophic interactions and biological control (Nihoul 1993). Variation in the expression, whether associated with plant development or environmentally induced, is an important determinant of the general utility of individual resistance traits in pest management. Such variation is common among resistance traits in *Lycopersicon* and should be fully understood before any commitment is made to breeding for a particular resistance trait.

IDENTIFYING RESISTANCE TRAITS

Identifying potentially useful pest resistance traits requires a knowledge of the biology of the target pest and its relation to crop damage. For pests, such as whitefly and spider mites, which do not feed on fruit and pass many generations within the crop, low pest populations do not cause reductions in yield and quality. For such species low to moderate levels of resistance (that reduce population growth rate by prolonging development time, reducing fecundity and/or increasing mortality) can cause dramatic reductions in population levels in the crop and have great value in managing the pest. In contrast, for species such as fruit-feeding Lepidoptera, such as *H. zea*, *H. armigera* and *S. exigua*, much higher levels of resistance are needed to protect the crop. For such pests, there has generally been little incentive to develop tomato cultivars with only moderate levels of resistance because the resistance would not dramatically alter insect management practices and significantly reduce insecticide use. However, even moderate levels of resistance to such pests can complement and thereby enhance the effectiveness of other pest management measures, including application of conventional insecticides and, in the absence of adverse tritrophic effects, biological control.

Identifying useful sources of resistance presents a significant challenge, especially in wild relatives of tomato that have distinctly different plant characteristics that may affect pest resistance but which are incompatible with commercially acceptable plant types. Such characteristics as vine size, growth habit, and number and size of fruit may affect incidence of insect damaged fruit (Fery and Cuthbert 1973, Zamir et al. 1984, Eigenbrode et al. 1993,.Hartman and St. Clair 1999b). Comparisons among plant lines based on differences in incidence of damaged fruit without accounting for potential effects of such characteristics and without additional information on insect populations have a high potential for missing potentially valuable resistant plants and for misclassifying otherwise susceptible plants as resistant.

Other characteristics also may obscure the existence of valuable resistance traits. For example, Juvik et al. (1988) documented significant variation among *Lycopersicon* accessions in the occurrence of phytochemical oviposition stimulants for *H. zea*. They found extracts of *L. hirsutum f. typicum* accession LA1777 with much more stimulants than extracts of *L. esculentum (UC82-1-8)*. The oviposition stimulants in LA1777 were identified as three structurally similar sesquiterpene carboxylic acids produced by the trichomes (Coates et al. 1988). The occurrence of phytochemically mediated variation in attractiveness for oviposition by *H. zea* has the potential to mask the occurrence of meaningful levels of larval antibiosis. It also invites further study to determine the feasibility of selecting for resistance due to reduced levels or absence of oviposition stimulants for key lepidopterous pests.

In the field, egg populations of *H. zea* averaged 3.2 times greater on resistant *L. hirsutum f. glabratum* accession PI134417 than on tomato *(L. esculentum)*. Additionally because of adverse effects of high densities of 2-tridecanone-containing type VI trichomes, egg parasitism by

Trichogramma on PI134415 averaged less than 1% compared to 43% on *L. esculentum*. Despite these differences, the populations of large larvae of *H. zea* were similar on PI134415 and *L. esculentum* plants because the effects of higher oviposition and reduced parasitism on PI134417 were offset by higher larval mortality in the resistant PI134417 (Farrar et al. 1994). In this case, a simple comparison of larval populations would not reveal the potent resistance of PI134417 to *H. zea*. The full intensity of the resistance is only revealed by detailed life table studies or field observations guided by the results of well-designed laboratory bioassays. When working with exotic plant types and novel resistance traits, a combination of laboratory and field experiments is recommended to ensure that valuable traits are not overlooked. Complete documentation of carefully designed laboratory or greenhouse experiments on the effects of resistance traits and the life history traits of the target pest can prove extremely valuable in designing optimal selection protocols for use in resistance breeding efforts (Romanow et al. 1991).

SUMMARY AND CONCLUSIONS

Resistance to insect and mite pests of tomato is common but highly variable both within and among *Lycopersicon* species. Resistance to numerous important pests has been reported from *L. esculentum* Mill., but the levels of resistance are generally low. High levels of resistance to major pest species are most common in *L. hirsutum f. typicum* Humb and Bonpl., *L. hirsutum f. glabratum* C.H.Mull and *L. pennellii* (Corr.) D'Arcy, but significant levels of resistance to insect pests have been documented from *L. cheesmanii* Riley, *L. chmielewskii* Dun., *L. peruvianum* (L.) Mill., and *L. pimpinellifolium* (Jusl.) Mill.

There is considerable variation among accessions within each of these species in the levels of resistance to individual pest species and the array of pest species resisted. The majority of research on insect resistance in *Lycopersicon* has focused on identifying sources of resistance to selected pest species and elucidating the mechanisms responsible for resistance. In only a few instances has the genetic basis for resistance been investigated. In even fewer cases have the effects of resistance traits on pest populations and crop damage been systematically characterized under field conditions to assess potential utility in crop protection. There is a clear need for systematic field studies of candidate resistance traits to characterize these effects.

The chemical/physical mechanisms responsible for insect resistance are, in most instances, complex. In many cases, multiple mechanisms are involved such as physical entrapment by foliar trichomes, presence of chemical toxins, feeding deterrants, and growth inhibiting compounds. Foliar glandular trichomes are important contributors to resistance to a number of important pest species. In some cases resistance traits associated with the leaf lamella are expressed constitutively by the plants whereas in others resistant expression is induced by insect feeding or mechanical damage. Even in instances where the mechanisms are relatively simple, as in the case of 2-tridecanone/type VI trichome-mediated resistance of *L. hirsutum f. glabratum*, the biological interactions involving the resistance characters can be extremely complex affecting both pests and their natural enemies. In most instances, expression of resistance is conditioned by multiple genes, and in a number of cases, is subject to significant environmental influence. Nonetheless, there are a number of traits that condition resistance to multiple insect pests, such as those found in *L. pennellii* and *L. hirsutum*, which hold considerable promise for incorporation into commercial tomato cultivars using conventional plant breeding techniques. The mechanics and genetic basis for resistance in *Lycoperssion* and tomato to aphids, white fly, thrips, diptera, beetles, caterpillars, and mites are summarised, including non-target effects of resistant traits.

The relative ease with which tomato can be genetically transformed and the success with transgenic insect resistant cotton and maize suggest that, with time, transgenic insect resistant tomato cultivars will become commonplace.

REFERENCES

Aina, O.L., J.G. Rodriguez, and D.E. Knavel. 1972. Characterizing resistance to *Tetranychus urticae* in tomato. J Econ Entomol 65: 641-643.

Barbour, J.D., R.R. Farrar, Jr., and G.G. Kennedy. 1991. Interaction of fertilizer regime with host plant resistance in tomato. Entomol Exp Appl 60: 289-300.

Barbour, J.D., R.R. Farrar, Jr., and G.G. Kennedy. 1993. Interaction of *Manduca sexta* resistance in tomato with insect predators of *Helicoverpa* zea. Entomol Exp Appl 68: 143-155.

Barbour, J.D., R.R. Farrar, Jr, and G.G. Kennedy. 1997. Populations of predaceous natural enemies developing on insect-resistant and susceptible tomato in North Carolina. Biol Control 9: 173-184.

Barbour, J.D. and G.G. Kennedy. 1991. Role of steroidal glycoalkaloid a-tomatine in host plant resistance of tomato to Colorado potato beetle. J Chem Ecol 17: 989-1005.

Bergey, D., G.A. Howe, and C.A. Ryan. 1996. Polypeptide signaling for plant defensive genes exhibits analogies to defensive signaling in animals. Proc Natl Acad Sci USA 93: 12053-12058.

Berlinger, M.J., S. Mordechi, and M. Pilowsky. 1991. Tomato resistance to the tobacco whitefly. Gan Ssadeh VaMeshek 9: 56-59.

Berlinger, M., M. Tamim, M. Tal, and A.R. Miller. 1997. Resistance mechanisms of *Lycopersicon pennellii* accessions to *Spodoptera littoralis* (Boisduval) (Lepidoptera: Noctuidae). J Econ Entomol 90: 1690-1696.

Blauth, S.L., G.A. Churchill, and M.A. Mutschler. 1998. Identification of quantitative trait loci associated with acylsugar accumulation using intraspecific populations of the wild tomato *Lycopersicon pennellii*. Theor Appl Genet 96: 458-467.

Bloem, K., K.C. Kelley, and S.S. Duffey. 1989. Differential effect of tomatine and its alleviation by cholesterol on larval growth and efficiency of food utilization in *Heliothis zea* and *Spodoptera exigua*. J Chem Ecol 15: 387-398.

Broadway, R.M., S.S. Duffey, G. Pearce, and C.A. Ryan. 1986. Plant proteinase inhibitors: a defense against herbivorous insects? Entomol Exp Appl 41: 33-38.

Cantelo, W.W., A.L. Boswell, and R.J. Argauer. 1974. Tetranych mite repellent in tomato. Environ Entomol 3: 128-130.

Carter, C.D. and J.C. Snyder. 1986. Mite responses and trichome characters in a full-sib F2 family of *Lycopersicon esculentum* x *L. hirsutum*. J Amer Soc Hort Sci 111: 130-133.

Carter, C.D., J.N. Sacalis, and T.J. Gianfagna. 1989a. Zingiberene and resistance to Colorado potato beetles in *Lycopersicon hirsutum* f. *hirsutum*. J Agric Food Chem 37: 306-210.

Carter, C.D., T.J. Gianfagna, and J.N. Sacalis. 1989b. Sesquiterpenes in glandular trichomes of a wild tomato species and toxicity to the Colorado potato beetle. J Agric Food Chem 37: 1425-1428.

Channarayappa, G., G. Shrivashankar, V. Muniyappa, and R.H. Frist. 1992. Resistance of *Lycopersicon* species to *Bemisia tabaci*, a tomato leaf curl virus vector. Can J Bot 70: 2184-2192.

Chatzivasileiadis, E. and M.W. Sabelis. 1998. Variability in susceptibility among cucumber and tomato strains of *Tetranychus urticae* Koch to 2-tridecanone from tomato trichomes: effects of host plant shifts. Expt Appl Acarol 22: 455-466.

Chiavegato, L.G. and M.M. Mischan. 1981. Resistance of varieties of tomato (*Lycopersicon esculentum* Mill) to the mite *Tetranychus urticae* Koch 1836) Boudreaux & Dosse, 1963 (Acari: Tetranychidae) under laboratory conditions. Clientifica 9: 267-271.

Clayberg, C.D. 1975. Insect resistance in a graft-induced periclinal chimera of tomato. HortScience 10: 13-15.

Clayberg, C.D. and J.B. Kring. 1974. Breeding tomatoes resistant to potato aphid and white fly. HortScience. 9: 297.

Coates, R.M., J.F. Dennison, J.A. Juvik, and B.A. Babka. 1988. Identification of an α-santalenoic and endo-α-bergamotenoic acids as moth oviposition stimulants from wild tomato leaves. J Org Chem 53: 2186.

Delannay, X., B.J. LaVallee, R.K. Proksch, R.L. Fuchs, S.R. Augustine, J.G. Layton, and D.A. Fischhoff. 1980. Field performance of transgenic tomato plants expressing the *Bacillus thuringiensis* var. Kurstaki insect control protein. Bio/Technol 7:1265-1269.

Dowd, P.F., L.M. Lagrimini, and T.C. Nelson. 1998. Relative resistance of transgenic tomato tissues expressing high levels of tobacco anionic peroxidase to different insect species. Nat Toxins 6: 241-249.

Duffey, S.S. 1986. Plant glandular trichomes: their potential role in defense against insects. *In*: B. Juniper and R. Southwood [eds.], Insects and the Plant Surface. Arnold. London, UK, pp. 151-172.

Duffey, S.S. and K.A. Bloem. 1986. Plant defense-herbivore-parasite interactions and biological control. In: M. Kogan (ed.), Ecological Theory and Integrated Pest Management Practice. Wiley, New York, USA, pp. 135-183.

Duffey, S.S. and M.B. Isman. 1981. Inhibition of insect larval growth by phenolics in glandular trichomes of tomato leaves. Experientia 37: 574-576.

Eigenbrode, S.D. and J.T. Trumble. 1993. Antibiosis to beet armyworm (*Spodoptera exigua*) in *Lycopersicon* accessions. HortScience 28: 932-934.

Eigenbrode, S.D., J.T. Trumble, and R.A. Jones. 1993. Resistance to beet armyworm, hemipterans, and Liriomyza spp. in *Lycopersicon* accessions. J Amer Soc Hort Sci 118: 525-530.

Eigenbrode, S.D., J.T. Trumble, J.G. Millar, and K.K. White. 1994. Topical toxicity of tomato sesquiterpenes to the beet armyworm and the role of these compounds in resistances derived from an accession of *Lycopersicon hirsutum* f. *typicum*. J Agric Food Chem 42: 807-810.

Eigenbrode, S.D., J.T. Trumble, and K.K. White. 1996. Trichome exudates and resistance to beet armyworm (Lepidoptera: Noctuidae) in *Lycopersicon hirsutum* f. *typicum* accessions. Environ Entomol 25: 90-95.

Elliger, C.A., Y. Wong, B.G. Chan, and A.C. Waiss, Jr. 1981. Growth inhibitors in tomato (*Lycopersicon*) to tomato fruitworm (*Heliothis zea*). J Chem Ecol 7: 753-758.

English-Loeb, G., M.J. Stout, and S.S. Duffey. 1997. Drought stress in tomatoes: changes in plant chemistry and potential nonlinear consequences for insect herbivores. Oikos 79: 456-468.

Erb, W.A., R.K. Lindquist, N.J. Flickinger, and M.L. Casey. 1993. Resistance of selected interspecific *Lycopersicon* hybrids to *Liriomyza trifolii*. J Econ Entomol 86: 100-109.

Farrar, R.R., Jr. and G.G. Kennedy. 1987. Growth, food consumption and mortality of *Heliothis zea* larvae on foliage of the wild tomato *Lycopersicon hirsutum* f. *glabratum* and the cultivated tomato *L. esculentum*. Entomol Exp Appl 44: 213-219.

Farrar, R.R., Jr. and G.G. Kennedy. 1988. 2-Undecanone, a pupal mortality factor in *Heliothis zea*: sensitive larval stage and in planta activity in *Lycopersicon hirsutum* f. *glabratum*. Entomol Exp Appl 47: 205-210.

Farrar, R.R., Jr. and G.G. Kennedy. 1990. Growth inhibitors in host plant resistance to insects: examples from a wild tomato with *Heliothis zea* (Lepidoptera: Noctuidae). J Entomol Sci 25: 46-56.

Farrar. R.R., Jr. and G. G. Kennedy. 1991a. Insect and mite resistance in tomato. *In*: G. Kalloo [ed.], Genetic Improvement of Tomato. Monographs on Theoretical and Applied Genetics 14, Springer-Verlag, New York, USA, pp. 121-142.

Farrar, R.R. and G.G. Kennedy. 1991b. Relationship of leaf lemallar-based resistance to *Leptinotarsa decemlineata* and *Heliothis zea* in a wild tomato *Lycopersicon hirsutum* f. *glabratum*. Entomol Exp Appl 58: 61-67.

Farrar, R.R. and G.G. Kennedy. 1991c. Inhibition of *Telenomus sphingis* an egg parasitoid of *Manduca* spp. by trichome/2-tridecanone-based host plant resistance in tomato. Entomol Exp Appl 60: 157-166.

Farrar, R.R., Jr. and G.G. Kennedy. 1993. Field cage performance of two tachinid parasitoids of the tomato fruitworm on insect resistant and susceptible tomato lines. Entomol Exp Appl 67: 73-78.

Farrar, R.R., Jr., J.D. Barbour, and G.G. Kennedy. 1994. Field evaluation of insect reisistance in a wild tomato and its effect on insect parasitoids. Entomol Exp Appl 43: 17-23.

Fernandez-Munoz, R., E. Dominguez, and J. Cuartero. 2000. A novel source of resistance to the two-spotted spider mite in *Lycopersicon pimpinellifolium* (Jusl.) Mill: its genetics as affected by interplot interference. Euphytica 111: 169-173.

Fery, R.L. and F.P. Cuthbert, Jr. 1973. Factors affecting evaluation of fruitworm resistance in the tomato. J Amer Soc Hort Sci 98: 457-459.

Fery, R.L. and G.G. Kennedy. 1987. Genetic analysis of 2-tridecanone concentration, leaf trichome characteristics, and tobacco hornworm resistance in tomato. J Amer Soc Hort Sci 112: 886-891.

Fischoff, D.A., K.S. Bowdish, F.J. Perlak, P.G. Marrone, S.M. McCormick, J.G. Niedermeyer, D.A. Dean, K. Kusano-Kretzmer, E.J. Mayer, D.E. Rochester, S.G. Rogers, and R.T. Fraley. 1987. Insect tolerant transgenic tomato plants. Bio/Technol 5: 807-813.

Fidantsef, A.L., M.J. Stout, J. Thaler, S.S. Duffey, and R.M. Bostock. 1999. Signal interactions in pathogen and insect attack: expression of liposygenase, proteinase inhibitor II, and pathogenesis-related protein P4 in the tomato, *Lycopersicon esculentum*. Physiol Mol Plant Pathol 54: 97-114.

Forbes, J.R., J. Mudd, and M. Marsden. 1985. Epicuticular lipid accumulation on the leaves of *L. pennellii* (Corr.) D'Arcy and *L. esculentum* Mill. Plant Physiol 77: 567-570.

Frelichowski, J.E., Jr. and J.A. Juvik. 2001. Sesquiterpene carboxylic acids from a wild tomato species affect larval feeding behavior and survival of *Helicoverpa zea* and *Spodoptera exigua* (Lepidoptera: Noctuidae). J Econ Entomol 94: 1249-1259.

Gatehouse, A.M.R., G.M. Davison, C.A. Newell, A. Merryweather, W.D.O. Hamilton, E.P.J. Burgess, R.J.C. Gilbert, and J.A. Gatehouse. 1997. Transgenic potato plants with enhanced resistance to the tomato moth, *Lacanobia oleracea*, growth room trials. Mol Breed 3: 49-63.

Gatehouse, A.M.R., G.M. Davison, J.N. Stewart, L.N. Gatehouse, A. Kumar, I.E. Geoghegan, A.N.E. Birch, and J.A. Gatehouse. 1999. Concanavalin A inhibits development of tomato moth (*Lacanobia oleracea*) and peach-potato aphid (*Myzus persicae*) when expressed in transgenic potato plants. Mol Breed 5: 153-165.

Gentile, A.G. and A.K. Stoner. 1968a. Resistance in *Lycopersicon* and *Solanum* species to the potato aphid. J Econ Entomol 61: 1152-1154.

Gentile, A.G. and A.K. Stoner. 1968b. Resistance in *Lycopersicon* spp. to the tobacco flea beetle. J Econ Entomol 61: 1347-1349.

Gentile, A.G., R.E. Webb, and A.K. Stoner. 1968. Resistance in *Lycopersicon* and *Solanum* to greenhouse whiteflies. J Econ Entomol 61: 1355-1357.

Gentile, A.G., R.E. Webb, and A.K. Stoner. 1969. Resistance in *Lycopersicon* and *Solanum* to carmine and two spotted spider mite. J Econ Entomol 62: 834-836.

Gianfagna, T.J., J.N. Sacalis, and C.D. Carter. 1992. Temperature and photoperiod influence trichome density and sesquiterpene content. Plant Physiol 100: 1403-1405.

Gilbert, J.C., J.T. Chinn, and J.S. Tanaka. 1966. Spider mite tolerance in multiple disease resistant tomatoes. Proc Amer Soc Hort Sci 89: 559-562.

Gilbert, J.C., J.S. Tanaka, and K.Y. Takeda. 1974. 'Kewalo' tomato. HortScience 9: 481-482.

Goffreda, J.C., M.A. Mutschler, and W.M. Tingey. 1988. Feeding behavior of potato aphid affected by glandular trichomes of wilt tomato. Entomol Exp Appl 48: 101-107.

Goffreda, J.C. and M.A. Mutschler. 1989. Inheritance of potato aphid resistance in hybrids between *Lycopersicon esculentum* and *L. pennellii*. Theor Appl Genet 78: 210-216.

Goffreda, J.C., J.C. Steffens, and M.A. Mutschler. 1990. Association of epicuticular sugars with aphid resistance in hybrids with wild tomato. J Amer Soc Hort Sci 115: 161-165.

Goncalves, M.I.F., W.R. Maluf, L.A.A. Gomes, and L.V. Barbosa. 1998. Variation of 2-tridecanone level in tomato plant leaflets and resistance to two mite species (Tetranychus sp.). Euphytica 104: 33-38.

Gregory, P., D.A. Ave, P.Y. Bouthyette, and W.M. Tingey. 1986. Insect-defensive chemistry of potato glandular trichomes. *In*: B. Juniper and R. Southwood [eds.], Insects and the Plant Surface. Arnold, London, UK, pp. 173-183.

Guo, Z., P.A. Weston, and J.C. Snyder. 1993. Repellency to two spotted spidermite, *Tetranyuchus urticae* Koch, as related to leaf surface chemistry of *Lycopersicon hirsutum* accessions. J Chem Ecol 19: 2965-2979.

Hare, J.D. 1987. Growth of *Leptinotarsa decemlineata* larvae in response to simultaneous variation in protein and glycoalkaloid concentration. J Chem Ecol 13: 39-46.

Hare, J.D. 1992. Effects of plant variation on herbivore-natural enemy interactions. *In*: R.S. Fritz and E.L. Simms [eds], Plant Resistance to Herbivores and Pathogens; Ecology Evolution, and Genetics. Univ Chicago Press, Chicago, pp. 278-298.

Hartman, J.B. and D.A. St. Clair. 1998. Variation for insect resistance and horticultural traits in tomato inbred backcross populations derived from *Lycopersicon pennellii*. Crop Sci. 38: 1501-1508.

Hartman, J.B. and D.A. St. Clair. 1999a. Variation for aphid resistance and insecticidal acylsugar expression among and within *Lycopersicon pennellii*-derived inbred backcross lines of tomato and their F_1 progeny. Plant Breed 118: 531-536.

Hartman, J.B. and D.A. St. Clair. 1999b. Combining ability for beet armyworm, *Spodoptera exigua*, resistance and horticultural traits of selected *Lycopersicon pennellii*-derived inbred backcross lines of tomato. Plant Breed 118: 523-530.

Hawthorne, D.J., J.A. Shapiro, W.M. Tingey, and M.A. Mutschler. 1992. Trichome-borne and artificially applied acylsugars of wild tomato deter feeding and oviposition of the leafminer *Liriomyza trifolii*. Entomol Exp Appl 65: 65-73.

Heinz K.M. and F.G. Zalom. 1995. Variation in trichome-based resistance to *Bemisia argentifolii* (Homoptera: Aleyrodidae) oviposition on tomato. J Econ Entomol 88:1494-1502.

Horwath, K.L. and N.E. Stamp. 1993. Use of dietary rutin to study molt initiation in *Manduca sexta* larvae. J Insect Physiol 39: 987-1000.

Howe, G.A., J. Lightner, J. Browse, and C.A. Ryan. 1996. An octadecanoid pathway mutant (JL5) of tomato is compromised in signaling for defense against insect attack. Plant Cell 8: 2067-2077.

Howe, G.A. and L. Schilmiller. 2002. Oxylipon metabolism in response to stress. Curr Opin Plant Biol 5: 230-236.

Hsiao, T. and G. Frankel. 1968. The role of secondary plant substances in the food specificity of Colorado potato beetle. Ann Entomol Soc Amer 61: 485-493.

Isman, M.B. and S.S. Duffey. 1982a. Toxicity of tomato phenolic compounds to the fruitworm, *Heliothis zea*. Entomol Exp Appl 31: 370-376.

Isman, M.B. and S.S. Duffey. 1982b. Phenolic compounds in the foliage of tomato cultivars as growth inhibitors to the fruitworm *Heliothis zea*. J Amer soc Hort Sci 107: 167-170.

Johnson, R., J. Narvaez, G. An, and C.A. Ryan. 1989. Expression of proteinase inhibitor I and II in transgenic tobacco plants: effects on natural defense against *Manduca sexta* larvae. Proc Natl Acad Sci USA 86: 9871-9875.

Juvik, J.A. and M.A. Stevens. 1982a. Inheritance of foliar α-tomatine content in tomato. J Amer Soc Hort Sci 107: 1061-1065.

Juvik, J.A. and M.A. Stevens. 1982b. Physiological mechanisms of host-plant resistance in the genus *Lycopersicon* to *Heliothis zea* and *Spodoptera exigua*, two insect pests of the cultivated tomato. J Amer Soc Hort Sci 107: 1065-1069.

Juvik, J.A., M.A. Stevens, and C.M. Rick. 1982a. Survey of the genus *Lycopersicon* for variability in á-tomatine content. HortScience 17: 764-766.

Juvik, J.A., B.A. Babka, and E.A. Timmermann. 1988. Influence of trichome exudates from species of *Lycopersicon* on oviposition behavior of *Heliothis zea* (Boddie). J Chem Ecol 12: 1261-1278.

Juvik, J.A., J.A. Shapiro, T.E. Young, and M.A. Mutschler. 1994. Acylglucoses from wild tomatoes alter behavior and reduce growth and survival of *Helicoverpa zea* and *Spodoptera exigua* (Lepidoptera: Noctuidae). J Econ Entomol 87: 482-492.

Kalloo, G. [Ed.], 1991. Genetic improvement of tomato. Monographs on Theoretical and Applied Genetics 14. Springer-Verlag. New York, USA.

Kallo, G., M.K. Banerjee, R.K. Kashyap, and A.K. Yaday. 1989. Genetics of resistance to fruit borer, *Heliothis armigera* (Hubner), in *Lycopersicon*. Plant Breed 102: 173-175.

Kaloshian, I., W.H. Lange, and V.M. Williamson. 1995. An aphid-resistance locus is tightly linked to the nematode-resistance gene, Mi, in tomato. Proc Natl Acad Sci USA 92: 622-625.

Kashyap, R.K. and A.N. Verma. 1987. Development and survival of fruit borer, *Heliothis armigera* (Hubner), on resistant and susceptible tomato genotypes. Z Pflkrankh Pflschutz 94: 14-21.

Kashyap, R.K., M.K. Banerjee, G. Kalloo, and A.N. Verma. 1990. Survival and development of fruitborer, *Heliothis armigera* (Hubner), (Lepidoptera: Noctuidae) on *Lycopersicon* spp. Insect Sci Appl 11: 877-881.

Kashyap, R.K., G.G. Kennedy, and R.R. Farrar, Jr. 1991a. Behavioral response of *Trichogramma pretiosum* Riley and *Telenomus sphingis* (Ashmead) to trichome/methyl ketone mediated resistance in tomato. J Chem Ecol 17: 543-556.

Kashyap, R.K., G.G. Kennedy, and R.R. Farrar, Jr. 1991b. Mortality and inhibition of *Helioverpa zea* egg parasitism rates by Trichogramma in relation to trichome/methyl ketone-mediated insect resistance of *Lycopersicon hirsutum* f. *glabratum* accession PI134417. J Chem Ecol 17: 2381-2395.

Kauffman, W.C. and G.G. Kennedy. 1989a. Inhibition of *Campoletis sonorensis* parasitism of *Heliothis zea* and of parasitoid development by 2-tridecanone mediated insect resistance of a wild tomato. J Chem Ecol 15: 1919-1930.

Kauffman, W.C. and G.G. Kennedy. 1989b. Toxicity of allelochemics from the wild insect resistant tomato *Lycopersicon hirsutum* f. *glabratum* to Campoletis sonorensis, a parasitoid of *Helicoverpa zea*. J Chem Ecol 15: 2051-2060.

Kennedy, G.G. 1986. Consequences of modifying biochemically mediated insect resistance in *Lycopersicon* species. *In*: M.B. Green, and P.A. Hedin [eds.], Natural Resistance of Plants to Pests: Roles of Allelochemicals. ACS Symp. Ser. 296. American Chemical Society, Washington, DC, USA, pp. 130-141.

Kennedy, G.G. 2003. Tomato, pests, parasitoids and predators: tritrophic interactions involving the genus *Lycopersicon*. Annu Rev Entomol 48: 51-72.

Kennedy, G.G. and R.R. Farrar, Jr. 1987. Response of insecticide resistant and susceptible Colorado potato beetles, *Leptinotarsa decemlineata*, to 2-tridecanone and resistant tomato foliage: the absence of cross resistance. Entomol Exp Appl 45: 187-192.

Kennedy, G.G., R.R. Farrar, Jr., and R.K. Kashyap. 1991. 2-Tridecanone-glandular trichome-mediated insect resistance in tomato: effect on parasitoids and predators of *Heliothis zea*. *In*: P.A. Hedin [ed.], Naturally occurring pest bioregulators. ACS Symp. Ser. 449. Amer Chem Soc Wash DC. USA, pp. 150-165.

Kennedy, G.G., R.R. Farrar, Jr., and M.R. Riskallah. 1987. Induced tolerance in *Heliothis zea* neonates to host plant allelochemicals and carbaryl following incubation of eggs on foliage of *Lycopersicon hirsutum* f. *glabratum*. Oecologia 73: 615-620.

Kennedy, G.G. and W.A. Henderson. 1978. A laboratory assay for resistance to the tobacco hornworm in *Lycopersicon* and *Solanum* spp. J Amer Soc Hort Sci 103: 334-336.

Kennedy, G.G., L.R. Romanow, S.F. Jenkins, and D.C. Sanders. 1983. Insects and diseases damaging tomato fruits in the Coastal Plain of North Carolina. J Econ Entomol 76: 168-173.

Kennedy, G.G. and C.E. Sorenson. 1985. Role of glandular trichomes in the resistance of *Lycopersicon hirsutum* f. *glabratum* to Colorado potato beetle Coleoptera: Chrysomelidae. J Econ Entomol 78: 547-551.

Kennedy, G.G., C.E. Sorenson, and R.L. Fery. 1985. Mechanisms of resistance to Colorado potato beetle in tomato. Mass Agric Expt Stn Res Bull No 704, pp. 107-116.

Kennedy, G.G. and R.T. Yamamoto. 1979. Resistance in a wild tomato to the tobacco hornworm: the presence of a toxic factor. Entomol: Exp Appl 26: 121-126.

Kennedy, G.G., R.T. Yamamoto, M.B. Dimock, W.G. Williams, and J. Bordner. 1981. Effect of daylength and light intensity on 2-tridecanone levels and resistance in *Lycopersicon hirsutum* f. *glabratum* to *Manduca sexta*. J Chem Ecol 7: 707-716.

Kogan, M. and E.F. Ortman. 1978. Antixenosis – a new term proposed to define Painter's "nonpreference" modality of resistance. Bull Entomol Soc Amer 24: 175-176.

Kumar, N.K.K., D.E. Ullman, and J.J. Cho. 1993. Evaluation of *Lycopersicon* germplasm for tomato spotted wilt tospovirus resistance by mechanical and thrips transmission. Plant Dis 77: 938-941.

Kumar, N.K.K., D.E. Ullman, and J.J. Cho. 1995a. Resistance among *Lycopersicon* species to *Frankliniella occidentalis* (Thysanoptera: Thripidae). J Econ Entomol 88: 1057-1065.

Kumar, N.K.K., D.E. Ullman, and J.J. Cho. 1995b. *Frankliniella occidentalis* (Thysanoptera: Thripidae) landing and resistance to tomato spotted wilt tospovirus among *Lycopersicon* accessions with additional comments on *Thrips tabaci* (Thysanoptera: Thripidae) and *Trialeurodes vaporariorum* (Homoptera: Aleyrodidae). Environ Entomol 24: 513-520.

Lange, W.H. and L. Bronson. 1981. Insect pests of tomato. Annu Rev Entomol 26: 345-371.

Laterrot, H., M. Bordat, M. Renand, and A. Moretti. 1987. Various levels of leafminer resistance in *Lycopersicon* genus. Tomato Genet Coop 37: 47-49.

Lawson, D.M., C.F. Lunde, and M.A. Mutschler. 1997. Marker-assisted transfer of acylsugar-mediated pest resistance from the wild tomato *Lycopersicon pennellii* to the cultivated tomato *Lycopersicon esculentum*. Mol Breed 3: 307-317.

Lei, H., J.C. van Lenteren, and W.F. Tjallingii. 1999. Analyses of resistance in tomato and sweet pepper against the greenhouse whitefly using electronically monitored and visually observed probing and feeding behavior. Entomol Exp Appl 92: 299-309.

Leidl, B.E., D.M. Lawson, K.K. White, J.A. Shapiro, D.E. Cohen, W.G. Carson, J.T. Trumble, and M.A. Mutschler. 1995. Acylsugars of wild tomato *Lycopersicon pennellii* alters settling and reduces oviposition of *Bemisia argentifolii* (Homoptera: Aleyrodidae). J Econ Entomol 88: 742-748.

Leite, G.L.D., M. Picanco, R.N.C. Guedes, and L. Skowronski. 1999. Effect of fertilization levels, age and canopy height of *Lycopersicon hirsutum* on the resistance to *Myzus persicae*. Entomol Exp Appl 91: 267-273.

Lemke, C.A. and M.A. Mutschler. 1984. Inheritance of glandular trichomes in crosses between *Lycopersicon esculentum* and *L. pennellii*. J Amer Soc Hort Sci 109: 592-596.

Li, L., C. Li, G.I. Lee, and G.A. Howe. 2002. Distinct roles for jasmonate synthesis and action in systemic wound response of tomato. Proc Natl Acad Sci USA 99: 6416-6421.

Lin, S.Y.H., J.T. Trumble, and J. Kumamoto. 1987. Activity of volatile compounds in glandular trichomes of *Lycopersicon* species against two insect herbivores. J Chem Ecol 13: 837-850.

Lu, W., G.G. Kennedy, and F. Gould. 1997. Genetic variation in larval survival and growth and response to selection by Colorado potato beetle (Coleoptera: Chrysomelidae) on tomato. Environ Entomol 26:67-75.

Lu, W., G.G. Kennedy, and F. Gould. 2001. Genetic analysis of larval growth of two populations of *Leptinotarsa decemlineata* (Coleoptera: Chrysomelidae) on tomato. Entomol Exp Appl 99: 143-155.

Maliepard, C., N. JeBas, S. van Huesden, J. Kos, G. Pet, R. Verkerk, R. Vrielink, P. Zabel, and P. Lindhout. 1995. Mapping of QTL's for glandular trichome densities and *Trialeurodes vaporariorum* (greenhouse whitefly) resistance in an F_2 from *Lycopersicon* x *Lycopersicon hirsutum* f. *glabratum*. Heredity 75: 425-433.

Maluf, W.R., L.V. Barbosa, and L.V. Costa Santa-Cecilia. 1997. 2-tridecanone-mediated mechanism of resistance to South American tomato pinworm *Scrobipalpuloides absoluta* (Meyrick, 1917) (Lepidoptera - Gelechiidae) in *Lycopersicon* spp. Euphytica 93: 189-194.

Mandaokar, A.D., R.K. Goyal, A. Shukla, S. Bisaria, R. Bhalla, V.S. Reddy, A. Chaurasia, R.P. Sharma, I. Altosaar, and P.A. Kumar. 2000. Transgenic tomato plants resistant to fruit borer (*Helicoverpa armigera* Hubner). Crop Prot 19: 307-312.

McKinney, K.B. 1938. Physical characteristics on the foliage of beans and tomatoes that tend to control some small insect pests. J Econ Entomol 31: 630-631.

McNally, K.L. and M.A. Mutschler. 1997. Use of introgression lines and zonal mapping to identify RAPD markers linked to QTL. Mol Breed 3: 202-212.

Miller, J.C. and S.D. Tanksley. 1990. RFLP analysis of phylogenetic relationships and genetic variation in the genus *Lycopersicon*. Theor Appl Genet 80: 437-448.

Mitchell, B.K. and G.D. Harrison. 1985. Effects of *Solanum glycoalkaloids* on chemosensilla in the Colorado potato beetle, a mechanism of feeding deterrence? J Chem Ecol 11: 73-83.

Musetti, L. and J.J. Neal. 1997a. Resistance to the pink potato aphid,*Macrosiphum euphorbiae*, in two accessions of *Lycopersicon hirsutum* f. *glabratum*. Entomol Exp Appl 84: 137-146.

Musetti, L. and J.J. Neal. 1997b. Toxicological effects of *Lycopersicon hirsutum* f. *glabratum* and behavioral response of *Macrosiphum euphorbiae*. J Chem Ecol 23: 1321-1332.

Mutschler, M.A., R.W. Doerge, S.C. Lin, J.P. Kuai, B.E. Leidl, and J.A. Shapiro. 1996. QTL analysis of pest resistance in the wild tomato *Lycopersicon pennellii*: QTLs controlling acylsugar level and composition. Theor Appl Genet 92: 709-718.

Nienhuis, J., T. Helentjaris, M. Slocum, B. Ruggero, and A. Schaefer. 1987. Restriction fragment length polymorphism analysis of loci associated with insect resistance in tomato. Crop Sci 27: 797-803.

Nihoul, P. 1993. Do light intensity, temperature and photoperiod affect the entrapment of mites on glandular hairs of cultivated tomatoes? Exp Appl Acarol 17: 709-718.

Nombela, G., F. Beitia, and M. Muniz. 2000. Variation in tomato response to *Bemisia tabaci* (Homoptera: Aleyrodidae) in relation to acylsugar content and presence of the nematode and potato aphid resistance gene Mi. Bull Entomol Res 90: 161-167.

Painter, R.H. 1951. Insect resistance in crop plants. Macmillan, NY.

Patterson, C.G., D.E. Knavel, T.R. Kemp, and J.G. Rodriguez. 1975a. Chemical basis for resistance to *Tetranychus urticae* in tomatoes. Environ Entomol 4: 670-674.

Patterson , C.G., D.E. Knavel, T.R. Kemp, and J.G. Rodriguez. 1975b. Tomato resistance to mites. Annu Rep Agric Exp Stn Univ Ky 87:76.

Ponti, O.M.B. de and M.M. Steenhuis. 1984. Prospects of resistance to whiteflies from *Lycopersicon hirsutum* glabratum. *In*: Synopsis of the IXth Eucarpia tomato working group meeting, Wageningen, The Netherlands, 22-24 May 1984, pp. 103-106.

Ponti, O.M.B. de, G. Pet, and N.G. Hogenboom. 1975. Resistance to the glasshouse whitefly (*Trialeurodes vaporariorum* Westw.) in tomato (*Lycopersicon esculentum* Mill.) and related species. Euphytica 24: 645-649.

Ponti, O.M.B. de, M.M. Steenhuis, and P. Elzinga. 1983. Partial resistance of tomato to the greenhouse whitefly (*Trialeurodes vaporariorum* Westw.) to promote its biological control Med Fac Landbouww Rijksuniv Gent 48: 195-198.

Quiros, C.F., M.A. stevens, C.M. Rick, M.L. Kok-Yokomi. 1977. Resistance in the tomato to the pink form of the potato aphid (*Macrosiphum euphorbiae* Thomas): the role of anatomy, epidermal hairs, and foliage composition. J Amer Soc Hort Sci 102: 166-171.

Rahimi, F.R. and C.D. Carter. 1993. Inheritance of zingiberene in *Lycopersicon*. Theor Appl Genet 87: 593-597.

Reynaerts, A. and S. Jansens. 1994. Engineered insect resistance in tomato. Acta Hort 376: 347-352.

Roddick, J.G. 1974. The steroidal glycoalkaloid a-tomatine. Phytochemistry. 13: 9-25.

Rodriguez, J.G., D.E. Knavel, and O.J. Aina. 1972. Studies in the resistance of tomatoes to mites. J Econ Entomol 65: 50-53.

Rodriguez, A.E., W.M. Tingey, and M.A. Mutschler. 1993. Acylsugars of *Lycopersicon pennellii* deter settling and feeding of the green peach aphid (Homoptera: Aphididae). J Econ Entomol 86: 34-39.

Romanow, L.R., O.M.B. de Ponti, and C.Mollema. 1991. Resistance in tomato to the greenhouse whitefly: analysis of population dynamics. Entomol Exp Appl 60: 247-259.

Rossi, M., F.L. Goggin, S.B. Milligan, I. Kaloshian, D.E. Ullman, and V. M. Williamson. 1998. The nematode resistance gene MI of tomato confers resistance against the potato aphid. Proc Natl Acad Sci USA 95: 9750-9754.

Ryan, C.A. 2000. The systemic signaling pathway: differential activation of plant defensive genes. Biochem et Biophys Acta 1477: 112-121.

Schalk, J.M. and A.K. stoner. 1976. A bioassay differentiates resistance to the Colorado potato beetle on tomatoes. J Amer Soc Hort Sci 101: 74-76.

Schalk, J.M. and A.K. Stoner. 1979. Tomato production in Maryland: effect of different densities of larvae and adults of the Colorado potato beetle. J Econ Entomol 70: 434-436.

Schreiber, K. 1968. Steroid alkaloids: the *Solanum* group. *In*: R.H.F. Manske [ed.], The Alkaloids: Chemistry and Physiology, Vol 10. Academic Press, New York, USA, pp. 1-192.

Shelton, A.M., J.Z. Zhao, and R.T. Roush. 2002. Economic, ecological, food safety, and social consequences of the deployment of Bt transgenic plants. Ann Rev Entomol 47: 845-881.

Sinden, S.L., J.M. Schalk, and A.K. Stoner. 1978. Effects of daylength and maturity of tomato plants on tomatine content and resistance to the Colorado potato beetle. J Amer Soc Hort Sci 103: 596-600.

Singa, N.K. and D.G. McLaren. 1989. Screening for resistance to tomato fruitworm and cabbage looper among tomato accessions. Crop Sci 29: 861-868.

Snyder, J.C. and C.D. Carter. 1984. Leaf trichomes and resistance of *Lycopersicon hirsutum* and *Lycopersicon esculentum* to spider mites. J Amer Soc Hort Sci 109: 837-843.

Snyder, J.C. and J.P. Hyatt. 1984. Influence of daylength on trichome densities and leaf volatiles of *Lycopersicon* species. Plant Sci Lett 37: 177-181.

Snyder, J.C., Z. Guo, R. Thacker, J.P. Goodman, and J. Pyrek. 1993. 2,3-Dihydrofarnesoic acid, a unique terpene from trichomes of *Lycopersicon hirsutum* repels spider mites. J Chem Ecol 19: 2981-2997.

Snyder, J.C., A.M. Simmons, and R.R. Thacker. 1998. Attractancy and ovipositional rsponse of adult *Bemisia argentifolii* (Homoptera: Aleyrodidae) to type IV trichome density on leaves of *Lycopersicon hirsutum* grown in three day-length regimes. J Entomol Sci 33: 270-281.

Sorenson, C.E., R.L. Fery, and G.G. Kennedy. 1989. Relationship between Colorado potato beetle (Coleoptera: Chrysomelidae) and tobacco hornworm (Lepidoptera: Sphingidae) resistance in *Lycopersicon hirsutum* f. *glabratum*. J Econ Entomol 82: 1743-1748.

Stamp, N.E. 1990. Growth versus molting time of caterpillars as a function of temperature, nutrient concentration and the phenolic rutin. Oecologia 82: 107-113.

Stamp, N.E. and K.L. Horwath. 1992. Interactive effects of temperature and concentration of the flavinol rutin on growth and food utilization of *Manduca sexta* caterpillars. Entomol Exp Appl 64: 135-150.

Stoner, A.K. and T. Stringfellow. 1967. Resistance of tomato varieties to spider mites. Proc Amer Soc Hort Sci 90: 324-329.

Stoner, A.K. and A.G. Gentile. 1968. Resistance of *Lycopersicon* species to the carmine spider mite. USDA Prod Res Rep, pp. 102.

Stoner, A.K., L.L. Mason, and R.E. Webb. 1969. Reaction of tomato breeding lines to *Drosophila melanogaster*. HortScience 4: 292-293.

Stoner, A.K., J.A. Frank, and A.G. Gentile. 1968a. The relationship of glandular hairs on tomatoes to spider mite resistance. Proc Amer Soc Hort Sci 532-533.

Stoner, A.K., R.E. Webb, and A.G. Gentile. 1968b. Reaction of tomato varieties and breeding lines to aphids. HortScience 3: 77.

Stout, J.J. and S.S. Duffey. 1996. Characterization of induced resistance in tomato plants. Entomol Exp Apl 79: 273-283.

Stout, M.J., Brevont, R.A., and S.S. Duffey. 1998a. Effect of nitrogen availability on expression of constitutive and inducible chemical defenses in tomato. J Chem Ecol 24: 945-963.

Stout, M.J., K.V. Workman, R.M. Bostock, and S.S. Duffey. 1998b. Stimulation and attenuation of induced resistance by elicitors and inhibitors of chemical induction in tomato (*Lycopersicon esculentum*) foliage. Entomol Exp Appl 86: 267-279.

Stout, M.J., A.L. Fidantsef, S.S. Duffey, and R.M. Bostock. 1999. Signal interaction in pathogen and insect attack: systemic plant-mediated interactions between pathogens and herbivores of the tomato *Lycopersicon esculentum*. Physiol Mol Plant Pathol 54: 115-130.

Sturkow, B. and I. Low. 1961. The effects of some *Solanum* glycoalkaloids on the potato beetle. Entomol Exp Appl 4: 133-142.

Thaler, J.S., A.L. Fidantsef, S.S. Duffey, and R.M. Bostock. 1999. Trade-offs in plant defense against pathogens and herbivores: a field demonstration of chemical elicitors of induced resistance. J Chem Ecol 25: 1597-1609.

van der Salm, T., D. Bosch, G. Honee, L. Feng, E. Munsterman, P. Bakker, W.J. Stickema, and B. Visser. 1994. Insect resistance of transgenic plants that express modified *Bacillus thuringiensis* cry1A(b) and cry 1C genes: a resistance management strategy. Plant Mol Biol 26: 51-59.

Vos, P. G. Simons, T. Jesse, J. Wijbrandi, L. Heinen, R. Hogers, A. Frijters, J. Groenendijk, P. Diergaarde, M. Reijans, J. Fierens-Onstenk, M. deBoth, J. Peleman, T. Liharska, J. Hontelez, and M. Zabeau. 1998. The tomato Mi-1 gene confers resistance to both root-knot nematodes and potato aphids. Nature Biotech 16: 1365-1369.

Walgenbach, J.F. 1997. Effect of potato aphid (Homoptera: Aphididae) on yield, quality, and economics of staked-tomato production. J Econ Entomol 90: 996-1004.

Webb, R.E., A.K. Stoner, and A.G. Gentile. 1971. Resistance to leafminers in *Lycopersicon* accessions. J Amer Soc Hort Sci 96: 65-67.

Weston, P.A., D.A. Johnson, H.T. Burton, and J.C. Snyder. 1989. Trichome secretion composition, trichome densities, and spider mite resistance of ten accessions of *Lycopersicon hirsutum*. J Amer Soc Hort Sci 114: 492-498.

Wilkens, R.T., J.M. Spoerke, and N.E. Stamp. 1996. Differential responses of growth and two soluble phenolics of tomato to resource availability. Ecology 77: 247-258.

Williams, W.G., G.G. Kennedy, R.T. Yamamoto, J.D. Thacker, and J. Bordner. 1980. 2-Tridecanone: a naturally occurring insecticide from the wild tomato species *Lycopersicon hirsutum* f. *glabratum*. Science 207: 888-889.

Zamir, D., T.S. Ben-David, J. Rudich, and J.A. Juvik. 1984. Frequency distributions and linkage relationships of 2-tridecanone in interspecific segregating generations in tomato. Euphytica 33: 481-488.

16

Tolerance to Abiotic Stresses

Majid R. Foolad
Department of Horticulture, 217 Tyson Building, The Pennsylvania State University
University Park, PA 16802 USA
email: mrf5@psu.edu

INTRODUCTION

Current world agricultural production is largely limited by environmental stresses (Boyer 1982). It is estimated that only 10% of the world's arable land may be free of stress (Dudal 1976). Extreme temperatures, drought, soil salinity, and nutrient imbalances (including mineral toxicities and deficiencies) are among the major environmental constraints to crop production worldwide. Most plant breeding programs have focused on the development of cultivars with high yield potential in favorable (i.e., nonstress) environments. Such efforts have been extremely successful in improving the efficiency of crop production per unit area thus resulting in significant increase in total agricultural production (Duvick 1986). However, with the rapid increase in human population there is a greater need for food production, and with the increasing diminution in natural resources, and arable lands, greater efforts must be made to increasing crop productivity under stressful agricultural environments and bringing marginal lands under cultivation. Marginal agricultural environments could be enriched by growing stress tolerant plants to increase yields, which is an important step toward solving the problem of feeding the world's growing population.

The genetic improvement of plant for stress tolerance can be an economically viable solution for crop production under stressful environments (Blum 1988). The progress in breeding for stress tolerance depends on an understanding of the physiological mechanisms and genetic basis of stress tolerance at the whole plant and cellular levels. Considerable information is presently available regarding the physiological and metabolic aspects of plant stress tolerance as the subject has been extensively reviewed (Fischer and Turner 1978, Lyons et al. 1979, Turner and Kramer 1980, Greenway and Munns 1980, Levitt 1980a, b, Lange et al. 1981, Paleg and Aspinall 1981, Graham and Patterson 1982, Tylor et al. 1983, Gorham et al. 1985, Maas 1990, Levitt 1990, Munns 1993, Bohnert et al. 1995, Volkmar et al. 1998, Flowers 1999, Zhu 2001, Munns 2005). However, research on genetic characterization and breeding for plant stress tolerance has been rather limited although efforts in this direction have been intensified lately (Shannon 1985, Blum 1988, Dvorak et al. 1992, Foolad 1996a, 1997, Winicov 1998, Zhang et al. 2001). Accumulating evidence suggests that plant response to abiotic stress is generally complex; it is often controlled by more than one gene (Blum 1988, Levitt 1980a, Zhu et al. 1997, Subudhi et al. 2000, Zhang et al. 2001) and highly influenced by environmental variation (Ceccarelli and Grando 1996, Richards 1996). The quantification of stress tolerance often poses serious difficulties. Direct selection under field conditions is generally difficult because uncontrollable environmental factors adversely affect the precision and repeatability of such trials (Richards 1996). In addition, stress tolerance appears to be a developmentally-regulated, stage-specific phenomenon; tolerance at one stage of plant development is generally not correlated with tolerance at other developmental stages (Greenway and Munns 1980, Shannon 1985, Maas 1986, Lauchli and Epstein 1990, Johnson et al. 1992, Foolad and Lin 1997a, Foolad 1999). Specific ontogenetic stages, such as seed germination and emergence, seedling survival and growth, and vegetative growth and reproduction should be evaluated separately for assessment of tolerance and the identification, characterization, and utilization of useful genetic complements. The knowledge gained from developmental and physiological aspects of stress tolerance will positively facilitate a better understanding of its genetic basis and the development of stress-tolerant plants.

Recent advances in molecular genetic techniques, including genetic transformation, analysis of gene expression, marker mapping, and quantitative trait loci (QTLs) analysis have contributed greatly to better understanding of the genetic and biochemical basis of plant stress-tolerance and, in some cases, also led to the development of plants with enhanced tolerance to abiotic stress. For example, significant progress has been made in the identification of genes, enzymes or compounds with significant effects on plant stress tolerance at the cellular or organismal level (Morgan 1991, Bohnert and Jensen 1996, Shinozaki and Yamaguchi-Shinozaki 1997, Shen et al. 1997, Winicov 1998, Allakhverdiev et al. 1999, Apse et al. 1999, Bohnert and Shen 1999, Grover et al. 1999, Seki et al. 2003, Wang 2003, Yardanov et al. 2004.). Furthermore, manipulation of the expression or production of the identified genes, proteins, enzymes, or compounds through transgenic approaches have resulted in the development of plants with enhanced stress tolerance in different plant species (Apse et al. 1999, Lilius et al. 1996, Rathinasabapathi 2000,

Rontein et al. 2002, Serrano et al. 1999, Thomas et al. 1995, Xu et al. 1996, Blumwald 2001, Zhang and Blumwald 2001, Zhang et al. 2001, Lea et al. 2004, Yamaguchi and Blumwald 2005). Molecular marker technology also has allowed the identification and genetic characterization of QTLs with significant effects on plant stress tolerance during different developmental stages (Ellis et al. 1997, Foolad and Chen 1999, Foolad et al. 1998a, Foolad et al. 2001, Forster et al. 1997, Mano and Takeda 1997) and has facilitated the determination of the genetic relationships among tolerance to different stresses and tolerance at different developmental stages (Foolad 1999). Such advancements are expected to contribute significantly to the development of plants with tolerance to different stresses in near future (Flowers 2004).

The cultivated tomato, *Lycopersicon esculentum* Mill., is moderately or highly sensitive to many abiotic stresses, including salinity, drought, extreme temperatures, excessive moisture, nutrient imbalances, and environmental pollution. However, within the genus *Lycopersicon*, there are several wild species that represent a rich source of useful genetic variation (Rick 1976, 1979). Such variation has been extensively utilized in tomato breeding programs for improving desirable agricultural characteristics, especially disease resistance. More recently, this variation has also been used to characterize the physiological and genetic basis of abiotic stress tolerance in tomato and to develop stress-tolerant plants. In this chapter, the available variation in relation to abiotic stress tolerance and recent advancements in genetics and breeding of tomato for stress tolerance are reviewed. The potential and limitations of different approaches for developing commercial tomato cultivars with improved stress tolerance are also discussed.

TOLERANCE TO SALT STRESS

Background Information

Salinity is an increasingly important environmental constraint to crop production mostly in the arid and semi-arid regions of the world (Boyer 1982, Tanji 1990, Rowley 1993). Regardless of the suspected physiological cause (ion toxicity, water deficit, and/or nutritional imbalance), high salinity in the root zone severely impedes normal plant growth and development, resulting in reduced crop productivity or total crop failure. Of the total 14 billion ha of land available on earth, 6.5 billion ha are estimated to be arid and semi-arid and about 1 billion ha are natural saline soils. It is estimated that worldwide about 20% of cultivated lands and 33% of irrigated agricultural lands are afflicted by high salinity (Epstein et al. 1980, Flowers et al. 1986, Francois and Maas 1994, Tanji 1990, Szabolcs 1992). These estimates do not account for the lands that are considered not arable due to the very high concentrations of salts, including areas along the seashores in temperate regions of the world and millions of hectares of desert lands of Africa, the Middle East, Asia, and North America. Such marginal lands could be agriculturally productive if more salt tolerant species or cultivars were available. Furthermore, the salinized areas are increasing at a rate of 10% annually. Low precipitation, high surface evaporation, weathering of native rocks, irrigation with saline water, and poor cultural practices are among the major contributors to the increasing salinity (Kalaji and Pietkiewica 1993, Syverstein et al. 1989, Tanji 1990, Szabolcs 1994). In some coastal areas, such as those of the United States, the gradual intrusion of seawater into the fresh-water aquifers threatens water supplies for agricultural, industrial, and municipal users (Cole 1993).

The presence of saline lands and saline agricultural water has been apparent to farmers for centuries. However, it has only been since the beginning of the 20th century that, because of the rapidly growing population and the need for greater food production, cultivation under saline conditions has been considered. Two major approaches have been proposed to minimize the deleterious effects of high soil/water salinity (Epstein et al. 1980) and these must be considered simultaneously to achieve a sustainable crop production. Implementing large engineering schemes for reclamation, drainage and irrigation with high quality water have been one of the first technological approaches used effectively. Although this approach has been successful in some areas, the associated costs are high and it often provides only a temporary solution to the problem. The second approach entails biological strategies focused on the exploitation or development of plants capable of tolerating high soil/water salinity. This approach includes: (a) diversifying cropping systems to include crops that are known to be salt tolerant (e.g., crop substitution), (b) exploiting wild or feral species that are adapted to saline environments (e.g., domestication), and (c) genetically modifying domesticated crops by breeding and selection for improved salt tolerance (ST). Crop substitution has been largely practiced since the beginning of agriculture, and is one of the most practical strategies to deal with salinity when growing field crops (Shannon 1996). However, most vegetable crops, including tomatoes, are generally more sensitive to salinity than field crops. Thus, the idea of developing salt-tolerant vegetable crops, which can produce economic yields under salinity stress, has drawn more attention (Shannon 1979, Rush et al. 1981).

Salt tolerance (ST) is generally defined as the inherent ability of the plant to withstand effects of high salts in the root zone or in the plant's leaves without significant adverse effect on plant productivity. During the past several decades considerable research has been done on the physiological and metabolic aspects of plant ST at the whole plant and cellular levels and the subject has been reviewed extensively (Levitt 1980b, Cheeseman 1988, Munns 1993, Jacoby 1994, Zhu et al. 1997, Bohnert and Shen 1999, Flowers 1999, Hasegawa et al. 2000, Zhu 2001, Ashraf and Mc Neilly 2004). However, efforts devoted to the genetic characterization of plant ST at the cellular or organismal level, and the integration of whole plant response in a developmental context, are more recent and rather incomplete (Blum 1988, Ashraf 1994, Foolad 1997, Jain and Selvaraj 1997, Jaiwal et al. 1997, Shannon 1997, Winicov 1998, Grover et al. 1999, Borsani et al. 2001). Accumulating evidence suggests that plants respond to salt stress (SS) through numerous quantitative traits involving many genes or proteins whose expression is influenced by external environmental variation and plant developmental stage (Levitt 1980a, Blum 1988, Chaubey and Senadhira 1994, Richards 1996). The quantification of plant ST poses serious difficulties. Direct selection in the field is often difficult because of the spatial and temporal variation in soil salinity (Hajrasuha et al. 1980, Richards and Dennett 1980), and the influence of other environmental factors, including variations in rainfall, nutrient availability, and temperature (Richards 1996). There is no reliable field screening technique that could be used year after year or generation after generation. Thus, breeders have been in search of more effective approaches for screening and improving crop ST.

Commercial cultivars of tomato are sensitive to salinity at all stages of plant development, which include seed germination, vegetative stage, and reproduction. Consequently, the growth and economic yield is substantially reduced under SS (Maas 1986, Jones et al. 1988, Bolarin et al. 1993). However, genetic resources for ST have been identified largely within the related wild species of tomato. Attempts to identify sources of genes for ST in tomato were first made by Lyon (Lyon 1941), who suggested that ST of the cultivated tomato might be improved by introgression of genes from *L. pimpinellifolium* (Jusl.) Mill., the most closely related wild species of tomato. Later investigations resulted in the identification of a few other salt-tolerant accessions within *L. esculentum* and the related wild species, including *L. pimpinellifolium*, *L. peruvianum* (L.) Mill., *L. cheesmanii* Riley, *L. hirsutum* Humb. and Bonpl., and *L. pennellii* (Corr.) D'Arcy (Foolad and Lin 1997b, Jones 1986a, Phills et al. 1979, Rush and Epstein 1976, Sarg et al. 1993, Tal 1971, Tal and Shannon 1983). It is expected, however, that more salt-tolerant accessions can be identified within the wild species of tomato with a more thorough evaluation of the available germplasm.

By using the identified salt-tolerant germplasm and various genetic techniques, notable progress has been made during the past few decades in characterizing the genetic basis of ST in tomato. A significant finding has been that ST at each stage of plant development in tomato is genetically not correlated with tolerance at other developmental stages (Asins et al. 1993a, Foolad 1999, Foolad and Lin 1997a, Jones and Qualset 1984), similar to findings in other plant species (Ashraf and McNeilly 1988, Johnson et al. 1992). Furthermore, it has been demonstrated that, in tomato, ST generally increases with plant age (Bolarin et al. 1993), again similar to the findings in other plant species such as barley (*Hordeum* spp.), corn (*Zea mays* L.), rice (*Oryza sativa* L.) and wheat (*Triticum* spp.) (Maas 1986). Therefore, it has been suggested that ST at each developmental stage must be evaluated separately for the assessment of tolerance and for the identification, characterization, and utilization of useful genetic components. In this section, the recent advancements on tomato ST during different developmental stages are presented and the potential for developing salt-tolerant cultivars using different genetic approaches discussed.

Salt Tolerance during Seed Germination and Seedling Emergence

The commercial cultivars of tomato are most vulnerable to salinity stress at the seed germination (SG) and early seedling growth stages (Cook 1979, Foolad and Jones 1991, Jones 1986b, Maas 1986). At these stages, tomato exhibits sensitivity even to low concentrations of salts (~75 mM NaCl) (Cuartero and Fernandez-Munoz 1999, Foolad and Lin 1997b, Jones 1986a). However, surface soils may have salinities several fold that of the subsoil, presenting a serious problem at the germination stage. High salinity delays the onset, reduces the rate and final percentage germination, and increases the dispersion of germination events. This sensitivity has important biological and applied significance. The costly operation of greenhouse seedling production and transplantation into the field has encouraged many tomato producers to grow direct-seeded crops (Liptay and Schopfer 1983). Furthermore, the dependence upon mechanization in modern cultivation systems, and the use of costly hybrid seed, requires rapid, uniform and complete germination. Improving the uniformity and rapidity of SG responses under saline conditions would contribute significantly to the efficiency of stand establishment in tomato.

Genetic resources for ST during SG have been identified within the cultivated and related wild species of tomato, including *L. pennellii*, *L. pimpinellifolium*, and *L. peruvianum* (Cuartero and Fernandez-Munoz 1999, Foolad and Lin 1997b, Jones 1986a). Some of the identified salt-tolerant accessions have been used to investigate the genetic basis of ST during SG in tomato, as briefly described below.

Inheritance
Using parental, filial and backcross populations (total of 10 generations) of a cross between a salt-sensitive tomato breeding line (UCT5) and a salt-tolerant *L. esculentum* plant introduction line PI174263, Foolad and

Jones (1991) determined that the ability of tomato seed to germinate rapidly under SS was genetically controlled with a narrow-sense heritability ($h2$) of 0.75 ± 0.03. Parent-offspring regression analyses, using F_2:F_3 and F_3:F_4 progeny of the same hybrid (UCT5 × PI174263), provided similar estimates of $h2$ for this trait (Foolad and Jones 1992). Both studies indicated that tomato ST during SG was genetically controlled with additivity being the major genetic component and that it could potentially be improved by directional phenotypic selection. In a later study, the effectiveness of phenotypic selection to improve tomato SG under SS was examined using F_2, F_3 and F_4 progeny populations of a cross between UCT5 and PI174263 (Foolad 1996b). Analysis of response to selection indicated that ST during SG in tomato was improved by directional phenotypic selection; estimates of realized $h2$ for rapid SG under SS ranged from 0.67 to 0.76 (Foolad 1996b). The overall results from these studies suggested that ST during SG in tomato was most likely controlled by a few major genes with additive effects, and therefore it could be improved by directional phenotypic selection. A better understanding of the genetic control of ST during SG in tomato, however, was achieved using molecular marker technology as mentioned in below.

QTL Mapping

An approach for a better understanding of the genetic basis of, and improving the selection efficiency for, complex traits, including plant ST, is to discover molecular genetic markers that are associated, either through linkage or pleiotropy, with genes or QTLs that control the trait(s) of interest. Molecular marker technology can facilitate precise determination of the number, chromosomal location, and individual and interactive effects of QTLs that control the trait. Following their identification, useful QTLs can be introgressed into desirable genetic backgrounds by marker-assisted selection (MAS) (Tanksley et al. 1989; *see also Chapter 7*). During the past several years, a few QTL mapping experiments were conducted to determine the number, genomic location and individual effects of QTLs affecting ST during SG in tomato. In one experiment, seed of an F_2 population of a cross between a salt-sensitive tomato breeding line (UCT5) and a salt-tolerant accession (LA716) of *L. pennellii* was subjected to SS and the two extreme tails (salt-tolerant and salt-sensitive) of the response distribution were selected (Foolad and Jones 1993). The selected individuals in each tail were subjected to marker analysis, and marker allele frequency differences between the two tails were determined. A trait-based marker analysis (also known as selective genotyping or distributional extreme analysis), which determines the significance of the marker allele frequency differences between selected tails of a response distribution (Lebowitz et al. 1987, Darvasi and Soller 1992), was used to identify marker-linked QTLs. Five genomic locations were identified on chromosomes 1, 3, 7, 8 and 12 bearing QTLs for ST during SG in this population (Foolad and Jones 1993).

The validity of these QTLs was examined in a few subsequent investigations using various populations of the same (UCT5 × LA716) or different interspecific crosses, including BC_1, BC_1S_1 and recombinant inbred line (RIL) populations of a cross between a L. esculentum breeding line (NC84173) and a salt-tolerant accession (LA722) of L. pimpinellifolium (Foolad et al. 1997, Foolad and Chen 1998, Foolad et al. 1998a, Foolad et al., unpublished data). In all of these studies, larger populations and larger number of molecular markers were used compared to the original study. In addition, different mapping strategies, including trait-based and marker-based (standard interval mapping) analyses, were employed. These studies validated most of the previously identified QTLs and detected a few new QTLs on chromosomes 2 and 9. The combined results supported the previous suggestion that ST during SG in tomato was a quantitative trait controlled by more than one gene. However, in all of these studies it was demonstrated that ST during SG in tomato was controlled by a few QTLs with major effects and several QTLs with small effects (Foolad and Jones 1993, Foolad et al. 1997, Foolad and Chen 1998, Foolad et al. 1998a). Furthermore, these studies indicated the presence of limited or no epistatic interactions among the identified QTLs. The demonstration that a few QTLs with major and independent effects determined most of the variation in ST during SG in tomato was consistent with the previous observation that this trait could be improved by directional phenotypic selection (Foolad 1996b). The overall results suggested that the prospect for improving this trait by MAS was good.

Comparison of QTLs identified for ST during SG in different interspecific populations of tomato, including those derived from L. esculentum × L. pennellii (Foolad et al. 1997, Foolad and Chen 1998) and L. esculentum × L. pimpinellifolium crosses (Foolad et al. 1998a), indicated that some QTLs were conserved across species whereas others were species-specific. However, because in most cases more than one genetic resource is utilized during the life of a breeding project, it may be possible to incorporate ST QTLs from different resources using MAS. Furthermore, comparison of QTLs in different populations (e.g. BC_1S_1 and RILs) of the same L. esculentum × L. pimpinellifolium hybrid indicated that some QTLs were stable across populations/generations while others were population-specific. The combined results suggest that it should be feasible to improve ST during SG in tomato by either phenotypic selection or MAS. However, because tolerance genes are often found within the wild species of tomato, a combination of phenotypic and marker-assisted selections might be more advantageous.

Response to Different Levels of Stress

It has been argued that in many saline soils the concentration of salts varies across the soil horizon, ranging from low to moderate and high (Richards and Dennett 1980). A successful cultivar would be one which exhibits ST at a wide range of SS and whose performance would not decline in the absence of salts. Selection under different levels of SS, however, may not always be practical in a breeding program. Thus, it is essential to examine relationships among responses to different levels of salts and determine whether there is a critical salt concentration at which selections should be conducted. Several investigations have examined relationships among germination responses under different levels of SS in tomato. For example, evaluation of 56 tomato genotypes (commercial lines, plant introductions and wild accessions) for SG at different levels of SS, including 75 mM (low), 150 mM (intermediate) and 200 mM (high) salt concentrations, indicated genotypes that generally germinated rapidly at the low salt concentration also germinated rapidly at the moderate or the high salt concentration (Foolad and Lin 1997b). Linear correlation analysis indicated the presence of a phenotypic relationship ($r = 0.90$, $P < 0.01$) between germination at 75 mM and 150 mM salt. The results suggested that the same genes control the rate of tomato SG under different levels of SS. This suggestion was later confirmed genetically by an analysis of response and correlated response to selection for ST, where selections were made separately under 100 mM (low), 150 mM (intermediate) and 200 mM (high) salt and their progeny responses also examined at all three salt concentrations (Foolad 1996b). In this study, selection for rapid SG at any SS level resulted in progeny with enhanced germination rate at all three (100, 150 and 200 mM) SS levels, regardless of which salt concentration was used during the selection process. A genetic correspondence (genetic correlation) of up to 100% between germination at different SS levels was observed. The results suggested that similar or identical genes with additive effects contributed to rapid SG at different SS levels. This suggestion was also consistent with the finding that in F_2 populations of a cross between a salt-tolerant accession (LA716) of *L. pennellii* and a salt-sensitive tomato cultivar (UCT5) identical QTLs contributed to ST during SG at different SS levels (Foolad and Jones 1993). The combined results indicated that the development of tomato cultivars with improved SG at diverse SS levels could be accomplished at a single SS level. However, because the rate of tomato SG at a moderate level of SS (150 mM salt) was highly correlated with that at both a low (100 mM salt) and a high SS level (200 mM salt), it was suggested that selection and breeding for ST should be performed at an intermediate SS level (Foolad 1996b).

Physiological Genetics

Although QTLs for ST during SG in tomato have been identified and verified, the genetic nature of the QTLs or the physiological mechanism(s) that they modulate has not been determined. However, based on the current genetics and physiological knowledge of ST during SG, some inferences as to the role of these QTLs can be made. ST during SG is a measure of the seed's ability to withstand the effects of high concentrations of salts in the medium. Excessive salt depresses the external water potential, making water less available to the seed. Slower SG under SS compared to nonstress conditions, however, could be due to osmotic and/or ionic effects of the saline germination medium. Physiological investigations to distinguish between these two effects have been limited. However, accumulating evidence in different crop species suggests that low water potential of the external medium, rather than ion toxicity effects, is the major limiting factor to germination under SS (Kaufman 1969, Ungar 1978, Bliss et al. 1986, Haigh and Barlow 1987, Bradford 1995), although a few reports have indicated otherwise (Choudhuri 1968, Younis and Hatata 1971, Redmann 1974). In a recent investigation, germination responses of eight tomato genotypes, including salt-tolerant and salt-sensitive accessions of *L. esculentum* and *L. pimpinellifolium*, were evaluated in iso-osmotic (water potential ~ -700 kPa or ~15 dSm^{-1}) media containing either NaCl, $MgCl_2$, KCl, $CaCl_2$, sorbitol, sucrose, or mannitol (J.R. Hyman and M.R. Foolad, unpublished data). Comparison of germination rates in the SS treatments (NaCl, $MgCl_2$, KCl, and $CaCl_2$) with those in osmotic-stress treatments (mannitol and sorbitol) indicated that all genotypes responded similarly to these two types of stresses. Furthermore, comparison of germination rates among the various SS treatments indicated that the different types of salt generally affected germination similarly. The overall results suggest that ST during SG in tomato is an adaptation to germinate rapidly under low water-potential and that there is little or no ionic toxicity effects.

Anatomically, the tomato seed comprises a seed coat (testa) that encloses a curved filiform embryo surrounded by an endosperm, which practically fills the lumen of the seed not occupied by the embryo (Esau 1953). For germination to occur, the hydraulic extension force of the embryo must exceed the opposing force of the seed coat and the living endosperm tissues at the placental end of the seed (Hegarty 1978, Bradford, Liptay and Schopfer 1983, 1986, Groot and Karssen 1987). Thus, embryo was suggested to play a major role in determining the time to germination of tomato seed (Liptay and Schopfer 1983). According to this hypothesis, differences in salt sensitivity of tomato seeds during germination reside either in the osmotic potential or pressure potential of the germinating embryo. However, external osmotic stress can also negatively affect seed imbibition, and thus retard (or prevent) weakening of the restrictive forces of the endosperm and testa, resulting in reduced rate (or inhibition) of germination (Liptay and Schopfer 1983, Groot and Karssen 1987, Dahal et al. 1990). Hence, the rate of SG in tomato may be influenced by the physical, chemical, and thus, the genetic composition of the embryo, endosperm and/or the testa. The identified QTLs for ST during SG in tomato could therefore affect germination rate by affecting the vigor of the germinating embryo, the variation in the thickness of the endosperm, the physical and permeability properties of the endosperm cell walls, the time of onset or rate of activity of enzymes which modify the properties of the endosperm cell wall, the release of gibberellin by the embryo (which is necessary for endosperm weakening), the base water potential required for seed germination, the hydrotime constant (Bradford 1995), the rate of metabolic activities in the embryo or endosperm under osmotic stress, osmoregulation during germination, or any other physiological or metabolic processes which are essential for the initiation of germination. However, isolation, characterization and comparison of functional genes which facilitate rapid SG under SS would be necessary to determine the actual role(s) of the identified QTLs. Nonetheless, irrespective of the physiological mechanism(s) of ST during SG in tomato, the identified QTLs could potentially be useful for improving tomato ST using MAS and breeding.

Salt Tolerance during Vegetative Growth and Reproduction

For tomato production under saline conditions, ST during the vegetative stage (VS) is more important than ST during SG because most tomato crops are established by seedling transplantation rather than direct seeding in the field. Furthermore, ST during VS may also be more important than ST during reproduction (i.e., flowering and fruit set) because tomato tolerance of salinity generally increases with plant age so that plants are usually most tolerant at maturity (Bolarin et al. 1993). For example, during flowering and fruiting stages, tomato plants can withstand salt concentrations that can kill them at the seedling stage. Furthermore, there is a positive correlation between tomato yield and plant size under SS (Pasternak et al. 1979, Bolarin et al. 1993), indicating the importance of ST during the VS.

Most commercial cultivars of tomato are moderately sensitive to SS during VS (Tal and Shannon 1983, Maas 1986, Foolad and Lin 1997b). At low concentrations of salt (electrical conductivity (EC) = 3-5 dSm^{-1}), tomato growth is mainly restricted by nutritional imbalances, as nutrients become the limiting factor under such conditions (Cuartero and Fernandez-Munoz 1999). At moderate to high levels of salt (EC = 6 dSm^{-1}), in addition to nutrient imbalances, osmotic effects and ion toxicity contribute to reductions in growth. Phenotypic variation for ST during VS in tomato has been identified within the cultivated (Curatero et al. 1992, Sarg et al. 1993, Foolad 1997) and related wild species, including *L. peruvianum* (Tal and Gavish 1973), *L. pennellii* (Dehan and Tal 1978, Saranga et al. 1991, Perez-Alfocea et al. 1994, Cano et al. 1998), *L. cheesmanii* (Rush and Epstein 1976, Asins et al. 1993a), and *L. pimpinellifolium* (Bolarin et al. 1991, Curatero et al. 1992, Asins et al. 1993a, Foolad and Chen 1999). Although much of this phenotypic variation has not been genetically verified, it can be potentially useful for developing tomato cultivars with enhanced ST.

Comparatively, less research has been conducted on tomato ST during reproduction than earlier stages. Limited effort has been devoted to determine tomato pollen viability or stigma receptivity, and/or the ability of the plant to produce flowers and set fruit under SS. This may be due in part to a higher ST generally observed during reproduction than earlier stages in tomato. Adams and Ho (1992) reported that increasing salinity to 10-dSm^{-1} did not significantly affect fruit set in tomato, which was reduced only at 15-dSm^{-1}. Grunberg et al. (1995) reported that salinity does not affect tomato pollen viability, however, the number of pollen grains per flower decreases with the duration of salinity, ca. 30% of the normal in 70 days after the beginning of salinization. The suggestion of a higher ST during reproduction, however, is in contrast with a report that several tomato genotypes that grew adequately under saline conditions failed to produce any fruit (Asins et al. 1993a). In a recent study, 13 accessions from three different tomato species were grown under both saline (300 mM NaCl + 30 mM CaCl$_2$; equivalent to ~Δ Sm^{-1}) and control (no salt) conditions, and their pollen production and *in vitro* pollen germination examined (S. Prakash and M.R. Foolad, unpublished data). For most accessions, there was no significant reduction in pollen production (per flower) in response to SS. Pollen from both salt-grown and control-grown plants was cultured at different salt concentrations, including 0, 0.2, 0.4 and 0.8% NaCl and evaluated for percentage germination after 4 or 8 h of incubation. In all accessions, pollen germinability was decreased under salt compared to the control (no salt) treatment, and the reduction was greater at higher than lower salt concentrations. However, in most accessions, *in vitro* pollen germinability for salt-grown plants was generally higher than that for the control-grown plants at high salt concentrations (0.4 and 0.8% NaCl), suggesting that pollen ST in tomato was increased by acclimating plants under SS.

The cultivated tomato is considered "moderately sensitive" to salinity considering the reduction in fruit yield under SS compared to normal conditions (Maas 1990). Fruit yield of tomato cultivars starts decreasing when the EC of the saturated soil extract exceed 2.5 Sm^{-1} (Maas 1990,

Saranga et al. 1991), although there are reports of higher thresholds for yield reduction in tomato (Adams 1991). A 10% reduction in fruit yield is expected per additional dSm^{-1} beyond the threshold level (Saranga et al. 1991). The major cause of yield reduction in tomato under low to moderate levels of salinity (EC = 3-9 dSm^{-1}) is the reduction in average fruit size, and not a reduction in fruit number (van-Ieperen 1996). A 10%, 30%, and 50% reduction in fruit size is caused following irrigation with 5-6, 8, and 9 dSm^{-1}, respectively (Cuartero and Fernandez-Munoz 1999). Thus, small-fruited cultivars, including cherry tomatoes, may be better than large-fruited cultivars for growing under low to moderate levels of SS. This is because the relative reduction in fruit size is less in smaller fruits (Caro et al. 1991). However, at higher levels of salinity, or prolonged exposure to salinity, a reduction in the total number of fruits per plant, mainly due to a reduction in the number of trusses per plant, is the major cause of yield reduction in tomato (Cuartero and Fernandez-Munoz 1999, van-Ieperen 1996). Thus, under high SS levels, significant yield reduction is expected for both large-fruited and small-fruited cultivars. Furthermore, because upper inflorescences are more sensitive to salinity (Cuartero and Fernandez-Munoz 1999), when breeding for ST in tomato, it would be better to develop cultivars with a compact plant type and early maturity, in which only 4-6 trusses are produced and harvested.

The potential of wild tomato species as sources of ST during reproduction (i.e., flower and fruit set and fruit yield) has not been assessed critically, mainly because most of the wild accessions are self-incompatible and/or produce very small fruits that cannot be easily compared with fruits of the cultivated species. However, progenies derived from interspecific hybrids of the cultivated and wild tomatoes have been used for such studies.

Inheritance

Most of the earlier research on tomato ST during VS was focused on the physiology of plant response to SS. During the past few decades, however, research has also been conducted on genetics of ST in tomato with the goal of facilitating the development of cultivars with enhanced ST. In 1976, Rush and Epstein (1976) proposed the exploitation of genetic resources within the wild *Lycopersicon* species to increase ST of the cultivated tomato. However, at that time there was little information regarding the genetic control of ST in tomato to warrant such an endeavor. Rush and Epstein (1981b) hybridized a "salt-tolerant" *L. cheesmanii* accession (LA1401) with a salt-sensitive tomato cultivar (Walter) and produced filial (F_1 and F_2) and backcross (BC) progeny. They reported that selections for ST in the F_2 and BC generations resulted in progeny with enhanced ST compared to Walter, suggesting that ST of LA1401 was heritable and could be transferred to the cultivated tomato by hybridization and selection. However, recent studies did not confirm the presence of ST in LA1401 or its interspecific progeny (author's unpublished data). Saranga et al. (1992) developed BC_1 and BC_1S_1 (self-progeny of BC_1) populations of a cross between a salt-sensitive tomato processing line (M82) and a "salt-tolerant" *L. pennellii* accession (LA716) and evaluated them for ST under field conditions. Evaluations were based on total dry matter and total fruit yield under saline conditions as well as total dry matter under salt relative to control (no salt). Estimates of *h2* for these traits were moderate (0.30-0.45), suggesting that ST of the cultivated tomato could be improved by selection for dry matter and yield parameters under salinity using LA716 as a genetic resource. By evaluating F_2 progeny of a cross between a salt-sensitive cultivated tomato and a "salt-tolerant" accession of *L. pimpinellifolium* under SS, Asins, et al. (1993b) concluded that total fruit yield and total fruit number were useful selection criteria for improving tomato ST; estimates of broad-sense *h2*s for these two traits were 0.53 and 0.73, respectively. In a greenhouse hydroponics study, using parental, filial and backcross generations (total of 10 populations) of a cross between a salt-sensitive tomato breeding line (UCT5) and a salt-tolerant primitive cultivar (PI174263), Foolad (1996a) determined that growth under SS relative to control (the most widely used index in physiological investigations of ST in tomato) was under additive genetic control and could be an excellent selection criterion for improving tomato ST. However, in none of these studies was any empirical selection made to verify that ST of the tomato could be improved by directional phenotypic selection under saline conditions. Nonetheless, these and other studies (Bolarin et al. 1991, Foolad 1996a) suggest that shoot growth under salinity relative to control (also known as relative growth under SS) is the best indicator of ST in tomato.

It is generally agreed that direct selection for ST under field conditions is difficult because of the confounding effects of numerous other environmental factors (Richards 1983, Tal 1985, Yeo and Flowers 1990). A suggested approach to improve the efficiency of selection for ST has been the adoption of new selection criteria based on genetic knowledge of physiological processes which limit crop production under saline conditions (Tal 1985). Physiological criteria which have been suggested as potential indicators of ST in tomato include tissue water potential, tissue ion content, K^+/Na^+ ratio, osmoregulation, succulence, and water use efficiency (WUE). Whether these physiological parameters are valid indicators of ST in tomato, or if there is genetic variation within *Lycopersicon* for these responses, must be determined before the question of their genetic controls can be addressed. Several researchers have investigated the relationship between tomato ST and various physiological responses and have commented on their utility as indirect selection criteria (Tal and Gavish 1973, Tal et al. 1979, Asins et al. 1993b, Saranga et al.

1993, Foolad 1997, Cuartero and Fernandez-Munoz 1999). A few of these studies and their findings are reviewed below.

Physiological investigations indicate that most of the salt-tolerant genotypes within the cultivated tomato and the closely-related wild species *L. pimpinellifolium* exhibit a glycophytic response to salinity, that is, exclusion of toxic ions (in particular Na^+) at the root or shoot level and the synthesis and accumulation of compatible organic compounds (e.g., sugars and amino acids) for osmoregulation (Caro et al. 1991, Curatero et al. 1992, Bolarin et al. 1993, Perez-Alfocea et al. 1993, Foolad 1997, Santa-Cruz et al. 1998). In contrast, salt-tolerant accessions within the tomato wild species *L. pennellii*, *L. cheesmanii* and *L. peruvianum* generally exhibit a halophytic response to salinity, in which osmotic adjustment is achieved by uptake of inorganic ions from the soil and compartmentalizing them in the vacuoles of the leaf or other plant organs (Sacher et al. 1983, Tal and Shannon 1983, Bolarin et al. 1991, Perez-Alfocea et al. 1994). However, differential accumulation of ions in the leaf tissue has not always been determined as a major factor affecting tomato ST or salt sensitivity. Analysis of BC_1 and BC_1S_1 populations of a cross between a salt-sensitive *L. esculentum* (M82) and a "salt-tolerant" *L. pennellii* (LA716) indicated that tissue ion content was not likely to provide an efficient selection criterion for ST, as no direct relationship was observed (Saranga et al. 1992). Paradoxically, analysis of the relationship between ST and leaf ion compositions in the cultivated and three wild species of tomato prompted Saranga et al. (1993) to conclude that dry matter production under SS was positively correlated with K^+/Na^+ ratio in the stem and negatively correlated with the Cl^- concentration in leaves and stems. The authors suggested that tissue ion content and ion selectivity were good selection criteria for breeding for ST in tomato. Potassium selectivity over Na^+ was also reported as a good indicator of ST in an investigation of several genotypes of the cultivated and wild species of tomato (Curatero et al. 1992). Studies in wild species of tomato, such as *L. peruvianum* (Tal 1971), *L. cheesmanii* (Rush and Epstein 1981a), and *L. pimpinellifolium*, *L. hirsutum*, and *L. pennellii* (Bolarin et al. 1991), demonstrated elevated concentrations of Na^+ in the leaf tissue in relation to plant ST. In contrast, no relationship was observed between tissue ions content and plant ST (as determined by survival under SS) in BC_1 and BC_1S_1 populations of a cross between a moderately salt-sensitive tomato breeding line (NC84173) and a salt-tolerant accession (LA722) of *L. pimpinellifolium* (Foolad and Chen 1999). Other studies also suggested that the ability to regulate Na^+ concentration in the leaf tissue was more closely correlated with ST than Na^+ concentration *per se* (Sacher et al. 1983) and that the distribution of Na^+ in young and mature leaves was an important part of such regulation (Shannon et al. 1987). The overall conclusion that can be drawn from the various investigations is that tissue ion content *per se* may not be a universal indicator of ST across tomato genotypes.

In salt-tolerant genotypes of tomato exhibiting a glycophytic response to SS, increase in ion concentration beyond a threshold results in failure of the exclusion mechanism at the root and/or shoot level. Consequently, further increases in ion concentration in the root zone leads to declining growth and gradual plant death (Perez-Alfocea et al. 1993, Foolad 1997). Thus, salt-tolerant tomato genotypes bearing an exclusion mechanism may be useful for cultivation only in regions with low to moderate levels of salinity. At higher salinity levels, genotypes that exhibit a halophytic type response to SS may be more advantageous. Unfortunately, many salt-tolerant wild accessions of tomato exhibiting a halophytic response to SS often have the undesirable characteristic of slow growth (Foolad 1996a, Tal 1997). Such accessions may survive high levels of salinity, but they often grow extremely slowly—a trait that is highly undesirable for tomato cultivation. Whether this association is due to pleiotropic effects of the same genes or linkage between different genes affecting growth rate and ST is unknown. It has not been determined whether this association can be eliminated in segregating populations derived from crosses between slow-growing salt-tolerant wild accessions and fast-growing salt-sensitive cultivated types of tomato. If possible, the salt-tolerant wild accessions would be potentially useful for ST breeding in tomato. Several studies in tomato and in other plant species, however, have suggested that genes contributing to plant vigor are different from those conferring ST (Foolad 1996a, Forster et al. 1990). When breeding tomatoes for efficient production under saline conditions, genes for both plant vigor and ST are important.

Research to determine the genetic basis of ion accumulation/exclusion in tomato plants grown under SS has been very limited. In one study, analysis of the parental, filial and backcross generations of a cross between a salt-sensitive tomato breeding line (UCT5) and a salt-tolerant primitive cultivar (PI174263) indicated that growth under SS was positively correlated with leaf Ca^{2+} content and negatively correlated with leaf Na^+ content (Foolad 1997). Generation Means Analysis (GMA) of these populations indicated that, under SS, accumulation of both Na^+ and Ca^{2+} in the leaf was genetically controlled with additivity being the major genetic component. Thus, tissue ion concentration was suggested as a useful selection criterion when breeding for improved ST of tomato using PI174263 as a genetic resource (Foolad 1997).

In spite of these and other studies, there is no consensus on what might be the best physiological or morphological trait(s) used as indirect selection criteria for ST breeding in tomato. Most likely a combination of different traits must be considered if salt-tolerant genotypes with commercial value are anticipated. This, by itself, indicates the complexity of ST and the need for identifying better approaches for characterizing genetic basis of tolerance components and facilitating breeding for ST. QTL mapping, MAS, and genetic transformation could be some promising approaches.

QTL Mapping

Significant progress has been made in mapping QTLs for ST during the vegetative and reproductive stages in tomato. In one study, a BC_1S_1 population of a cross between a moderately salt-sensitive tomato breeding line (NC84173) and a salt-tolerant accession of L. pimpinellifolium (LA722) was evaluated for survival under SS using an aerated hydroponics system (Foolad and Chen 1999). The two parents were distinctly different in ST: while 80% of LA722 survived two weeks after the final salt concentration (700 mM NaCl + 70 mM $CaCl_2$; equivalent to ~64 dSm^{-1}) was reached, only 25% of NC84173 remained alive within that period. The BC_1S_1 population (consisting of 119 families) exhibited continuous variation for ST, with a survival rate ranging from 9% to 94% across families and a mean of 51%. Interval mapping identified five putative QTLs for ST on tomato chromosomes 1, 3, 5 and 9, with individual effects ranging between 5.7% and 17.7% and with combined effects of 46% of the total phenotypic variation under SS (Foolad and Chen 1999). All QTLs had the positive QTL alleles from the salt-tolerant L. pimpinellifolium parent. The results supported the hypothesis that ST during VS in tomato was controlled by more than one gene (Foolad 1996a, 1997). However, the involvement of only a few major QTLs, which accounted for a large portion of the total phenotypic variation, suggested the utility of MAS for improving tomato ST using LA722 as a genetic resource. Analyses of leaf ion content (including Na^+, K^+, Mg^{2+}, Ca^{2+}, Cl^-, $NO3^-$, $SO4^{2-}$ and $PO4^{3-}$) in the BC_1S_1 population indicated the absence of a correlation between ST and tissue ion content in this population. Furthermore, despite the presence of significant variation among BC_1S_1 families in concentration of the various ions, no major QTL was identified for tissue ion content under SS.

Using a different BC_1 population of the same L. esculentum × L. pimpinellifolium cross, a selective genotyping approach was employed to verify the previously identified QTLs and possibly identify new QTLs for ST during VS (Foolad et al. 2001). From a population of 792 BC_1 plants, 37 (4.7% of the total) exhibiting the highest ST were selected, grown to maturity and self-pollinated to produce BC_1S_1 seeds. The 37 selected BC_1S_1 families and 119 nonselected (random) BC_1S_1 families were evaluated for ST and their performances compared. A realized h^2 of 0.46 was obtained for ST during VS, consistent with a previous estimate of

h^2 for this trait obtained from an intraspecific cross of tomato (Foolad 1996a). The 37 selected BC_1 plants and the 119 nonselected BC_1 plants were subjected to RFLP analysis using 115 markers, and marker allele frequencies determined. A comparison of marker allele frequency differences between the selected and nonselected populations detected five genomic regions on chromosomes 1, 3, 5, 6 and 11 bearing QTLs for ST (Foolad et al. 2001). Except for one, all QTLs had positive alleles contributed by the salt-tolerant *L. pimpinellifolium* parent. Three of the five QTLs were at the same locations as those identified in the previous study (Foolad and Chen 1999). Only one of the major QTLs that was identified in the previous study was not detected in this study (Foolad et al. 2001). The high level of consistency of the results of the two studies indicated the genuine nature of the detected QTLs and their potential usefulness for ST breeding using MAS. In each of the two studies described, a few BC_1S_1 families were identified with most or all of the identified QTLs and with a ST comparable to that of the salt-tolerant *L. pimpinellifolium* accession.

In a more recent study, 145 F_9 RILs of the same *L. esculentum* × *L. pimpinellifolium* cross were evaluated in replicated trials for ST during VS (plant survival under SS). The RILs were genotyped for 129 RFLP and 62 resistance gene analog (RGA) markers. Interval analysis identified 7 QTLs for ST during VS on tomato chromosomes 3, 4, 5, 7, 8, 9 and 12 (M.R. Foolad et al., unpublished data). The QTLs detected on chromosomes 3, 5, and 9 were the same as those identified in the previous studies (Foolad and Chen 1999, Foolad et al. 2001) and exhibited larger effects than the newly identified QTLs on chromosomes 4, 7, 8 and 12. The overall results from these three studies indicate that the stable QTLs on chromosomes 3, 5 and 9 should be useful for introgression into the cultivated tomato via MAS to improve tomato ST during VS.

Very limited research has been conducted to identify QTLs for ST during reproduction in tomato. Using an F_2 population of a cross between a salt-sensitive *L. esculentum* cultivar and a salt-tolerant accession of *L. pimpinellifolium*, and by using only 14 genetic markers, Breto et al. (1994) identified a few QTLs which appeared to be associated with fruit yield, fruit number and/or fruit size under SS. However, because of the extreme difference in fruit size between the parents of this F_2 population, QTL identification was most likely confounded by the effects of genes controlling fruit size, and thus, the identified QTLs should be considered with caution and should be verified in advanced generations before use for MAS. Similar studies were also conducted using F_2 populations of different crosses between *L. esculentum* and either *L. pimpinellifolium* or *L. cheesmanii*. Several other QTLs were identified for the same fruit-related traits under SS (Monforte et al. 1996, 1997, 1999). However, large morphological and physiological differences between parental lines of these populations, including differences in flowering habits, maturity times, fruit size and number, and total fruit yield, could have adversely affected the power of experiments in detecting genuine QTLs for ST. These QTLs also should be validated in advanced populations where such confounding effects are eliminated or minimized.

Transgenic Approaches

Plant's response to SS involves the functions of many genes that lead to a wide variety of biochemical and physiological changes. These include expression of genes that facilitate compartmentalization of toxic ions in the vacuole, activation of detoxification enzymes, synthesis of late-embryogenesis-abundant (LEA) proteins, and accumulation of low-molecular weight organic compounds (collectively known as compatible solutes or osmolytes). During the past several years, genetic engineering approaches have been employed to produce transgenic plants with enhanced tolerance to various abiotic stresses, including SS, by overexpression of genes controlling different tolerance-related physiological mechanisms (Bajaj et al. 1999, Serrano et al. 1999, Apse and Blumwald 2002, Rontein et al. 2002). For example, plants have been engineered with genes encoding enzymes that enhance the synthesis of compatible solutes, such as mannitol (Thomas et al. 1995), glycine betaine (Lilius et al. 1996), proline (Zhu et al. 1997), and polyamines (Galston et al. 1997), that contribute to osmotic adjustment and improving plant stress tolerance (Rathinasabapathi 2000, Rontein et al. 2002). Compatible solutes may also contribute to stress tolerance through other functions such as protection of enzyme and membrane structure and scavenging of radical oxygen species (Shen et al. 1997, Bohnert and Shen 1999, Rathinasabapathi 2000). Transgenic plants also have been produced with overexpression of different vacuolar antiports, which facilitate exclusion of toxic ions from the cell cytosol (Apse et al. 1999, Serrano et al. 1999, Zhang and Blumwald 2001, Zhang et al. 2001). Furthermore, transgenic plants have been developed with increased expression of detoxification enzymes, which reduce oxidative stress (Tanaka et al. 1999). Although in almost all cases plant growth and stress treatments were conducted in controlled conditions, and in many cases the increased tolerance has been rather marginal, the transgenic approach has provided opportunities for a better understanding of mechanisms leading to stress tolerance. The preliminary results have been particularly encouraging for enabling scientists to better understand the effects of single-gene transfers to plants.

Notwithstanding the efforts to develop tomatoes with enhanced ST using transgenic approaches have been rather limited, a significant advancement was recently reported by Zhang and Blumwald (2001) who developed transgenic tomato plants overexpressing *AtNHX1*, a single-gene controlling vacuolar Na^+/H^+ antiport protein, introduced from *Arabidopsis thaliana*. The overexpression of this gene was previously shown to increase ST in *Arabidopsis* (Apse et al. 1999). Transgenic tomato plants overexpressing this gene were able to grow, produce flower and set fruit in the presence of 200 mM NaCl in greenhouse hydroponics, whereas the nontransgenic (control) plants did not survive the saline conditions (Zhang and Blumwald 2001). The transgenic plants acquired a halophytic response to SS, that is, accumulated salts in the cell and sequestered them in the vacuole. As indicated earlier, the normal response of the cultivated tomato to salinity is of a glycophytic type, that is, excluding salt from the cell at the plasma membrane in the root and/or in the shoot. This was the first reported example of a single-gene transformation in any crop species that resulted in such a significant enhancement in plant ST. According to this report, under high salinity conditions, transgenic tomato plants accumulated very high concentrations of Na^+ and Cl^- in the leaves. The overproduction of the vacuolar Na^+/H^+ antiport protein enhanced the ability of the transgenic plants to sequester Na^+ in their vacuole, averting its toxic effects in the cell cytosol. At the same time, Na^+ was used to maintain an osmotic balance to drive water into the cells and thus use saline water for cell expansion and growth. Furthermore, there were only minimal increases in concentrations of Na^+ and Cl^- in the fruit, a great horticultural advantage for commercial production of such transgenic plants. The low Na^+ content of the fruit was attributed to the ability to maintain a high cytosolic K^+/Na^+ concentration ratio along the symplastic pathway in the transgenic plants. The results indicated that the enhanced accumulation of Na^+, mediated by the vacuolar Na^+/H^+ antiport, allowed the transgenic plants to ameliorate the toxic effects of Na^+. More recently, the transfer and overexpression of the same *AtNHX1* gene into canola (rape seed), *Brassica napus*, resulted in salt-tolerant transgenic plants that were able to grow, flower, and produce seeds in the presence of 200 mM NaCl (Zhang et al. 2001). Although the transgenic tomato and canola plants are yet to be evaluated for ST under field conditions, these findings suggest the potential for producing salt-tolerant plants using the transgenic approach.

It should be noted that transformation technology for improving plant stress tolerance has just begun. There is no report to date of any field studies testing the performance of transgenic plants under SS conditions. Much more work on transgenics is needed to gain a better understanding of the genetics, biochemical and physiological basis of plant ST. Future knowledge of tolerance components along with identification and cloning of responsible genes may allow transformation of plants with multiple genes and production of highly stress-tolerant transgenic plants.

It is possible that transferred multiple genes may act synergistically and additively to improve plant stress tolerance.

Comparison of Salt Tolerance during Different Stages of Plant Development

Knowledge of the genetic relationship between ST at different plant stages is necessary to expedite breeding efforts to develop cultivars with enhanced tolerance throughout the ontogeny of the plant. Early studies in different plant species had suggested the absence of phenotypic relationships among different developmental stages with regard to plant ST (Abel and Mackenzie 1963, Greenway and Munns 1980, Johnson et al. 1992). Recently, however, systematic approaches were taken to examine the phenotypic and genetic relationships between ST during SG and later stages in tomato. In one study, an F_4 population of a cross between a tomato breeding line (UCT5), with salt sensitivity during both SG and VS, and a primitive cultivar (PI174263), with ST during both of these stages, was evaluated for ST independently during SG and VS. Although there were significant variation among F_4 families in ST during both stages, there was no significant correlation ($r = -0.10$, $P > 0.05$) between the ability of the seed to germinate rapidly and the ability of the plant to grow under SS (Foolad and Lin 1997a). In a second approach, to examine the genetic correlation between ST during SG and VS, selection was made for rapid SG under SS in an F_2 population of the same cross, and the selected F_3 progeny were evaluated for ST separately during SG and VS. Selection for ST during SG significantly improved germination ST of the F_3 progeny, indicating that the selection was effective. However, the selection for ST during SG did not affect ST of the F_3 progeny during VS, suggesting that the genetic and physiological mechanisms that contributed to ST during SG in these genetic materials were different from those conferring ST during VS (Foolad and Lin 1997a).

In a more robust approach to determine the genetic relationship between ST during SG and VS, QTLs for ST at these two developmental stages were identified in a BC_1S_1 population of a cross between a tomato breeding line (NC84173, salt sensitive during both SG and VS) and a *L. pimpinellifolium* accession (LA722, salt tolerant during both stages). Comparison of QTLs indicated that in most cases the locations of QTLs for ST during SG were different from the locations of QTLs for ST during VS (Foolad 1999). This study clearly indicated the involvement of different genes (QTLs) controlling ST during SG and VS in these genetic materials. A similar study was recently conducted using 145 F_9 RILs of a cross between NC84173 and LA722 (M.R. Foolad et al., unpublished data). The RILs were evaluated in replicated trials for ST during SG and vegetative growth (plant survival under SS). There was no significant phenotypic correlation between the rate of SG and the plant survival under SS across the RILs. The RILs were genotyped for 129 RFLP and 62 RGA markers, covering 1,505 cM of tomato genome with an average marker distance of 7.9 cM. Marker analysis identified QTLs for ST during SG on chromosomes 2, 3, 4, 8, 9 and 12 and QTLs for ST during vegetable stage on chromosomes 3, 4, 5, 7, 8, 9 and 12. Different QTLs were detected for ST during these two stages of plant development, suggesting the involvement of different genes controlling ST during SG and VS in this population.

The overall results from these investigations indicate that, in tomato, ST during SG is generally independent of ST during VS. This conclusion is consistent with earlier reports that ST of young tomato plants did not correlate with that of mature plants (Shannon et al. 1987) and that ranking of salt-tolerant tomato genotypes based on vegetative characteristics in mature plants differed from the ranking based on fruit yield (Caro et al. 1991). Absence of genetic relationships among different developmental stages with regard to ST have also been reported in many other plant species, including alfalfa (*Medicago sativa* L.) (Johnson et al. 1992), wheat (*Triticum aestivum* L.) (Ashraf and McNeilly 1988), triticale (*Triticale hexaploide* Lart.) (Norlyn and Epstein 1984), and slender wheatgrass [*Elymus trachycalus* spp. *Trachycalus* (Link) Malte] (Pearen et al. 1997). These findings indicate that when breeding for improved ST, each stage of plant development may have to be evaluated separately for the assessment of tolerance and the identification, characterization and utilization of useful genetic components. However, the identification of QTLs for ST at different developmental stages would facilitate simultaneous or sequential introgression of QTLs for tolerance and the development of cultivars with improved tolerance at all desirable stages. Furthermore, the finding that in tomato only a few major QTLs account for a large portion of the total phenotypic variation for ST at each plant stage suggests that MAS for ST should be feasible, providing the opportunity to develop germplasm with enhanced ST at various developmental stages.

Future Prospects for Developing Tomato Cultivars with Enhanced Salt Tolerance

Although no commercial cultivar of tomato with proven field tolerance to salinity has yet been released through the use of either the conventional protocols of plant genetics and breeding, MAS or transgenic approaches, the prospect for such development in the near future is very good. Limited progress in developing salt-tolerant tomatoes during the past few decades has been in part due to the complexity of the trait, complex interactions of ST with other agronomically important traits, insufficient understanding of the basic physiological mechanisms as well as genetic controls of tolerance-related traits, lack of efficient selection criteria, and limited effort devoted to the identification and characterization of genetic resources that could be used for breeding for ST. With the current knowledge of the physiological and genetic basis of ST, the availability of molecular markers, QTLs, MAS, and genetic transformation technologies applied in developing salt-tolerant genotypes, it would not be unexpected to witness tomato cultivars with field ST in near future. It is anticipated that, similar to that in several other major crop plants, the importance of breeding tomatoes for ST will increase in near future and no bably a few research groups in the United States and worldwide are currently attempting to produce salt-tolerant tomato cultivars.

TOLERANCE TO DROUGHT STRESS

Background Information

Drought, defined as the occurrence of a substantial water deficit in the soil or in the atmosphere, is an increasingly important constraint to crop productivity and yield stability worldwide (Ceccarelli and Grando 1996). It is by far the leading environmental stress in agriculture. The worldwide losses in yield due to drought probably exceed the losses from all other causes combined (Kramer 1980, Blum 1988, Schonfeld et al. 1988). In the U.S., for example, up to 45% of the land surface is subject to continuous or frequent water stress (Boyer 1982, Tanji 1990) and a drought occurs somewhere in the country every year, costing billions of dollars in damage to crops and businesses (Ross and Lott 2000).

Most crop plants, including tomato, are sensitive to drought stress (DS) throughout the ontogeny of the plant, from SG to harvest (Hsiao 1973). Plant response to DS can be generally classified into three categories, drought escape (or avoidance), dehydration avoidance (or postponement), and dehydration tolerance (Kramer 1983, Blum 1988). Drought escape includes situations where plants with short growth cycle and early maturity avoid experiencing the drought. Breeding for drought escape should therefore be directed toward developing cultivars with early maturity so that by the time drought occurs the plant has already completed its life cycle. Advantages and disadvantages of plants with a short growth cycle under conditions of water stress and under normal conditions have been reviewed elsewhere (Reitz 1974, Jordan et al. 1983, Saeed and Francis 1983, Blum 1988). Extremely late maturing cultivars may also escape drought damage.

Dehydration avoidance is defined as the ability of the plant to retain a relatively higher level of "hydration" during the period of water stress (Blum 1988). In this situation, the plant protects its various growth related physiological, biochemical, and metabolic processes from the external water stress. A common measure of dehydration avoidance is the maintenance of a higher tissue water or turgor potential under conditions of water stress. Thus, osmotic adjustment, as a means for retaining a higher turgor at a given tissue water potential, is an example of dehydration avoidance at the cell level. Osmotic adjustment is usually obtained by the production and accumulation of compatible organic solutes, such as amino acids, glycine betaine, sugars, proline and ectoine in the cytoplasm. Various mechanisms employed by different species to avoid dehydration have been described elsewhere (Levitt 1980a).

When the tissue is not protected by any of the avoidance mechanisms, cells lose turgor and dehydrate, resulting in various cellular physicochemical injuries (Hsiao and Bradford 1983). Complete loss of free water will result in desiccation or dehydration. While our understanding on the effects of dehydration at the cellular level and various processes at higher levels of plant organization (e.g. photosynthesis, respiration, etc.) has been improving (in consideration of the application of new molecular techniques), it is far from complete, and there are many unanswered questions (Blum 1988, McCue and Hanson 1990, Bohnert et al. 1995, Bohnert and Jensen 1996, Richards 1996, Shinozaki and Yamaguchi-Shinozaki 1996, 1997, Bray 1997, Zhu et al. 1997, Lutfor-Rahman 1998). What is known, however, is that different genotypes exhibit different responses to cellular and whole plant stresses caused by dehydration, and that there are different levels of dehydration tolerance. It should also be noted that characteristics of the three categories of plant response to DS (drought escape, dehydration avoidance, and dehydration tolerance) are not generally independent of each other, and some plants may exhibit characteristics for more than one category (Blum 1988).

A complementary approach in agricultural methods currently followed is to minimize losses incurred by water stress and develop, via genetic means, "drought tolerant" cultivars with the ability to escape, avoid and/ or tolerate effects of water stress. Development of such plants would have a lasting economic impact on crop production worldwide. Despite many decades of research on drought tolerance (DT), so far drought stress continues to be a major challenge to plant breeders. This is in part due to the complexity of the trait. Accumulating evidence suggests that plant response to DS is controlled by the function of many genes and physiological mechanisms (Zhu et al. 1997, Blum 1988, Subudhi et al. 2000, Zhang et al. 2001) and varies depending on the influence of other environmental factors (Ceccarelli and Grando 1996, Richards 1996). Selection and breeding for DT is also difficult because tolerance appears to be a developmentally-regulated, stage-specific phenomenon (Blum 1988, Ludlow and Muchow 1990, Richards 1996, Mitchell et al. 1998). Each stage may be considered as a separate trait and may require a different evaluation method.

Furthermore, no reliable evaluation procedure is known that can effectively and efficiently be used to identify drought-tolerant plants at different developmental stages. These and other complexities have led to a limited success in developing drought-tolerant plants or improving crop yields in dry environments.

Most commercial cultivars of tomato are sensitive to DS during different stages of plant development, yet genotypic variation for DT exist within the cultivated tomato (Wudiri and Henderson 1985) and related wild species, including *L. cheesmanii, L. chilense, L. pennellii, L. pimpinellifolium,* and *L. esculentum* var. *cerasiforme* (Yu 1972, Richards and Phills 1979, Martin et al. 1989, Pillay and Beyl 1990, Rick 1973, 1978, 1979, 1982). The latter species, being native of coastal deserts of western South America (including Galapagos), witness rainless long periods but for the occasional El Niño episodes when warm ocean water generates heavy rains. The species grow at habitats where condensation of dew at night and fog drip are the main source of moisture. They are also remarkably capable of overcoming brief wilting. Rana and Kalloo (1990) evaluated 150 lines of cultivated and wild species of tomato under water-deficit conditions and identified a few selections within *L. esculentum,* and a few accessions of *L. pimpinellifolium* and *L. chilense,* with various DT attributes. However, only limited effort has been devoted to the characterization of the physiology or genetics of this variation (Kahn et al. 1993, Martin et al. 1999, Pillay and Beyl 1990) to warrant its use in breeding programs to develop drought-tolerant tomato cultivars. This is unlike the extensive research that has been conducted on DT in many other crop species, including rice, *Oryza sativa* L. (Nguyen et al. 1997, Zhang et al. 2001), maize, *Zea mays* L. (Ribaut et al. 1997), sorghum, *Sorghum bicolor* L. Moench (Subudhi et al. 2000) and lettuce, *Lactuca sativa* L. (Johnson et al. 2000). Comparatively, also less research has been done on tomato DT than its tolerance to other abiotic stresses such as salinity and extreme temperatures. The available information on genetics of DT in tomato and the prospect for developing drought-tolerant tomatoes is provided in the following sections.

Drought Tolerance during Seed Germination and Seedling Emergence

The ability of the tomato seed to germinate rapidly and uniformly under DS is a desirable trait for direct seeding in the field. Large areas of land are established for tomato production by sowing seed directly into the field instead of using transplants (Liptay and Schopfer 1983). Successful establishment of direct-seeded crops depends on successful seed germination and seedling emergence. Most commercial cultivars of tomato are sensitive to DS during SG, however, sources of tolerance have been identified within the related wild species of tomato, including *L. pennellii* and *L. pimpinellifolium* (M.R. Foolad et al, unpublished data). Recently, using one accession of *L. pimpinellifolium* in crosses with *L. esculentum*, the genetic basis of DT during SG in tomato was determined in our investigations.

Inheritance

A BC_1 population (N = 1000) from a cross between a drought-tolerant *L. pimpinellifolium* accession (LA722) and a drought-sensitive tomato breeding line (NC84173; maternal and recurrent parent) was evaluated for SG under DS (14% PEG, ψ_w ~ –680 kPa), and the most rapidly germinating seeds (first 3% germinated) were selected. The 30 selected BC_1 individuals were grown to maturity and self-pollinated to produce BC_1S_1 progeny seeds. Twenty of the 30 "selected BC_1S_1" progeny families were evaluated for germination under DS and their average performance was compared with that of a "nonselected" BC_1S_1 population of the same cross. Results indicated that selection for rapid SG under DS was effective and significantly improved progeny SG rate under DS (selection gain = 19.6%). A realized *h2* of 0.41 was obtained for DT during SG in this population. The results indicated that DT during SG in tomato was genetically controlled and could be improved by directional phenotypic selection.

QTL Mapping

Two independent studies were recently conducted to identify QTLs for DT during SG in tomato. In one study, a trait-based marker analysis, using BC_1 individuals of a cross between drought-sensitive tomato breeding line NC84173 and drought-tolerant *L. pimpinellifolium* accession LA722, detected four QTLs on chromosomes 1, 4, 8, 9, and 12 for DT (M.R. Foolad et al., unpublished data). The results indicated that DT during SG in tomato was a quantitative trait, controlled by more than one gene. A few BC_1S_1 families were identified with most or all of the QTLs and with a DT comparable to that of LA722. These families should be useful for developing germination drought-tolerant tomato lines using MAS. In a second study, 145 F_9 RILs of the same cross were evaluated for germination rate under DS and, by using composite interval mapping analysis, several QTLs for DT during SG were identified on tomato chromosomes 1, 2, 3, 4, 8, 9, and 12 (M.R. Foolad et al., unpublished data). The results of this study were highly consistent with those of the previous study (M.R. Foolad et al., unpublished data). These results indicated the presence of stable QTLs for DT during SG across tomato populations derived from the NC84173 × LA722 cross, suggesting the usefulness of these QTLs for improving tomato SG under DS by MAS.

Drought Tolerance during Vegetative Growth and Reproduction

Potential sources of genes for DT during vegetative growth and later stages in tomato have been identified within the related wild species *L. chilense* and *L. pennellii*, mostly among accessions native to dry habitats. (Rick 1973, 1978, 1979, 1982). Different tolerance indices (TIs) have been employed to characterize the physiological and/or genetic basis of DT in tomato, including dry weight (DW) of shoot and root, root length, root morphology, leaf rolling, flower and fruit set, fruit weight, yield, water-use efficiency (WUE), recovery after re-watering, stomatal resistance, plant survival, leaf water potential, leaf osmotic potential, osmoregulation, oxidative damages, transpiration rate, photosynthetic rate, enzymatic (e.g. superoxide dismutase and Rubisco) activities, and pollen viability (Richards and Phills 1979, Blum 1988, Martin and Thorstenson 1988, Rana and Kalloo 1989, Pillay and Beyl 1990, Cohen et al. 1991, Kalloo 1991, Lutfor-Rahman 1998). For example, in a germplasm evaluation study, tomato cultivar Saladette sustained severe water stress as determined by a smaller reduction in fruit set compared to other cultivars, which was attributed to its ability to rolling up leaves under a high evaporative demand and thereby maintaining a high leaf water potential (Wudiri and Henderson 1985). Rick (1978) suggested that the physiological basis for DT in *L. chilense* might be related to its deep vigorous root system. Cultivar Red Rock performed better than other cultivars under drought conditions and its DR was similarly attributed to its deep and more vigorous root system (Stoner 1972). DT in the *L. pimpinellifolium* was also attributed to root length (Rana and Kalloo 1989). Conversely, the drought-tolerant *L. pennellii* accession LA716 has a limited root system, and the basis for its DT is most likely due to its ability to conserve moisture during periods of limited rainfall. LA716 has been characterized as having a greater WUE under DS than *L. esculentum*, as measured by g DW produced per Kg of water consumed (Martin and Thorstenson 1988). A higher WUE in this accession than in *L. esculentum* was attributed to different characteristics of this accession, including smaller leaf conductance due to fewer and smaller stomata, longer trichomes, lower chlorophyll content and Rubisco activity per unit leaf area, and larger mesophyll cell surface exposed to intercellular air space (Martin et al. 1999). Although WUE may be a good indicator of DT in tomato, its measurement under field condition is extremely difficult. Thus, attempts have been made to determine the relationship between WUE and stable carbon isotope discrimination (Δ), a measure for proportion of ^{13}C relative to ^{12}C in plant organic matter, easier to measure when dealing with large number of plants (Martin et al. 1999). A recent study suggested that WUE in progeny of crosses between *L. esculentum* and *L. pennellii* LA716 could be increased by selecting for low Δ, however, this could lead to the selection of small plants, an agriculturally undesirable characteristic (Martin et al. 1999). The authors suggest that the small plant size could be corrected by conventional breeding following selection for DT, but no such effort has been reported.

Inheritance and QTL Mapping
Limited research has been conducted to characterize the genetic control of or to develop tomatoes with improved DT. To facilitate selection for low Δ, three QTLs associated with this trait were previously identified using F_3 and BC_1S_1 progeny of a cross between *L. esculentum* UC82B and *L. pennellii* LA716 (Martin et al. 1989). The results were inconclusive in determining whether selection for these QTLs could increase WUE in tomato. Other related research studies on genetics of tomato DT include the identification of several genes or mRNAs whose expressions are elevated in response to DS. For example, four tomato drought induced genes, *le4*, *le16*, *le25* and *le20*, were identified and characterized in *L. esculentum* (Cohen et al. 1991, Plant et al. 1991, Kahn et al. 1993). It was further determined that the increase in the expression of these genes occurred after a longer period of water deficit in *L. pennellii* than in *L. esculentum*, although these genes did not appear to be responsible for DT in *L. pennellii* (Kahn et al. 1993).

Future Prospects for Developing Tomato Cultivars with Enhanced Drought Tolerance

From the preceding discussion it is evident that currently there is limited physiological and/or genetic information on DT in tomato to warrant development of drought-tolerant cultivars through plant breeding. In essence, research must be extended to identify additional genetic resources for DT in tomato. Collections from torrid areas should be tested for DT at different developmental stages, such as SG, VS, flowering and fruit set, to identify useful resources for basic physiological and genetic studies as well as for breeding purposes. Considering the normal climatic conditions for growing tomatoes, where short periods of drought may occur intermittently throughout the growing season, it seems that the ability of the plant to survive transient periods of water stress and to recover rapidly upon re-availability of water is far more important than the ability to survive long-term water stress. Thus, during germplasm evaluation and breeding process, the focus should be on dehydration avoidance.

From a practical point of view, the most reliable criteria for breeding tomatoes for DT are agronomic characteristics such as yield, and absolute and relative plant growth under stress and nonstress environments. Such criteria, however, may not be efficient or feasible to apply because in most initial germplasm evaluation and/or breeding projects often a large number of individuals, families or populations are used. Alternative criteria based on physiological characteristics such as photosynthetic rates, stomatal resistance and leaf water potential might be more efficient. These characteristics are easier to measure, compared to yield, and generally show rather strong correlations with agronomic characteristics. Other alternatives are the identification of biochemical characteristics such as enzyme activities and protein contents. These methods, however, often lack a strong correlation with agronomic characteristics and are expensive. Like other abiotic stresses, the identification and utilization of molecular markers associated with different tolerance-related physiological, morphological, or agronomic criteria might be an efficient way to improve DT in tomato. Furthermore, transgenic approaches have been employed in other plant species to increase DT (Grover et al. 1999, Serrano et al. 1999), and an increasing number of DT-related genes or proteins are being discovered. Transgenic approaches should be successful in developing tomatoes with improved DT.

TOLERANCE TO COLD STRESS

Background Information

Although the cultivated tomato originated and was domesticated in the tropical and subtropical regions of South and Central America (Rick 1975), this crop is also commercially grown in many temperate regions of the world where it often experiences low temperatures (LTs) during at least part of the growing season. Temperatures in the range of 0 to 15 °C (chilling temperatures) injure many crops of tropical origin, including tomatoes. Most commercial cultivars of tomato are sensitive to LTs equally as to other abiotic stresses during all stages of plant development, SG, VS and reproduction (Patterson et al. 1978, Lyons et al. 1979, Graham and Patterson 1982, Patterson et al. 1987). This sensitivity limits the geographic distribution of tomato and puts constraints on cultivation time for annual planting. For example, the spring soil temperatures in temperate regions, which are often below 15 °C, may restrict direct seeding of tomato crops (Liptay et al. 1982). LT sensitivity also affects greenhouse production of tomato, since cold-sensitive cultivars would require expensive greenhouse heating throughout the life of the plant. One approach to minimizing deleterious effects of cold stress (CS) is to develop cold-tolerant tomato cultivars.

When grown under sub-optimal temperatures, cold-tolerant tomato geno-types could exhibit improved earliness, adaptability, water use, and yield of high-quality fruits. Cold-tolerant genotypes also could be planted earlier in the season and harvested earlier when the crop may have higher economic values. Furthermore, the crop could be grown in the field for longer periods of time (i.e., extended growing season), thus the final yield may be higher compared to cold-sensitive genotypes. Likewise, the production may improve because early plantings avoid high temperatures-typically observed during the mid-summer in many climates which can reduce fruit set. Moreover, cold-tolerant genotypes may have lower demand for water in areas with a Mediterranean climate, because early plant growth would make better use of early-season rains and available water in the root zone.

The search for genetic diversity for temperature adaptation logically leads one to examine cultivars or ecotypes adapted to extreme environments. The potential for this geographic approach in identifying tomato genotypes tolerant to LTs was examined by Thompson (1970) and Patterson et al. (1978), later reviewed by Vallejos (1979). Genetic resources for cold tolerance (CT) have been identified both in cultivated and related wild species of tomato (Patterson et al. 1978, Vallejos 1979, Scott and Jones 1982, Patterson and Payne 1983, Wolf et al. 1986, Patterson 1988, Foolad and Lin 2000, Henk-Venema et al. 2000). These resources have been employed to investigate the physiology and genetics of LT tolerance in tomato. The physiological aspects of CT in tomato and other plant species have been extensively studied and reviewed (Lyons 1973, Lyons et al. 1979, Graham and Patterson 1982, Vallejos et al. 1983, Parkin et al. 1989, Guy 1990, Alberdi and Corduera 1991, Leviatov et al. 1994, 1995, Bohnert et al. 1995, Bradford 1995, Keller and Steffen 1995). Comparatively, however, less research has been done on genetics of CT in tomato. Recent findings regarding genetic characterization of tomato for CT during different developmental stages are presented below and the prospects assessed for developing cold-tolerant cultivars through various genetic approaches.

Cold Tolerance during Seed Germination and Seedling Emergence

Germination rate of tomato seed decreases progressively as the temperature in the germination medium is reduced from 25 to 10 °C, and is inhibited below 10 °C (Scott and Jones 1982). Low temperatures (10-15 °C) delay the onset, reduce the rate, and increase the dispersion of germination events. As a result, under such conditions, many seeds either do not germinate or germinate so sporadically that plants grow differentially, thereby delaying plant establishment and leading to variability in crop maturation. Most commercial cultivars of tomato are sensitive to LTs during SG, with considerable genetic variation within the cultivated and related wild species. Genetic variability for CT during SG in *Lycopersicon* was first reported by Smith and Millet (1964) and has since been the subject of numerous studies. For example, Scott and Jones (1982) evaluated a total of 37 accessions of the cultivated and wild species of tomato and identified one accession of *L. chilense* that germinated better at 10 °C than PI120256, the fastest germinating *L. esculentum* accession known at that time. Additional accessions exhibiting rapid SG at 10 °C were also identified within L. *peruvianum* and *L. hirsutum*. In a later study, Scott and Jones (1985b) evaluated SG of three fast germinating *L. esculentum* accessions, PI120256, PI174263 and PI341988, and a slow germinating tomato breeding line, T3, at temperatures 6-20 °C. They concluded that rapid germination of these PIs at LTs might not be due to CT, but to seed characteristics that promoted rapid SG. Furthermore, early seedling emergence of the same three PIs under LTs was attributed to their rapid SG rather than rapid hypocotyl development. More recently, Foolad and Lin (2000) evaluated germination CT of 31 tomato accessions (cultivars, breeding lines and plant introductions) representing six *Lycopersicon* species and identified significant phenotypic variation in the ability of the seed to germinate rapidly under LT. Several accessions were identified within *L. pimpinellifolium* and *L. hirsutum* with high CT during SG. Potential genetic resources for CT during SG have also been identified in *S. lycopersicoides* (Rick 1988, Wolf et al. 1986).

Generally, accessions germinating rapidly at LTs also exhibit fast seedling growth under such conditions (Scott and Jones 1985b). A few accessions within *L. chilense* (LA460) and *L. peruvianum* (PI126435 and PI127832) were identified with the ability of fast hypocotyl growth under LTs (Scott and Jones 1986). The identified phenotypic variation in these and other studies should be useful for improving LT germination and hypocotyl growth of commercial cultivars of tomato. To facilitate breeding for CT during SG in tomato, several researchers have examined the genetic basis of this seed-related characteristic using a number of LT-tolerant accessions of tomato, which is summarised below. The physiological basis of LT seed germination in tomato has been described elsewhere (Simon 1979, Maluf and Tigchelaar 1982, Dahal et al. 1990, Leviatov et al. 1994, 1995).

Inheritance
Studies on CT during SG in tomato attributed the genetic control of this trait to either one recessive gene (Cannon et al. 1973), additive genes (Whittington and Fierlinger 1972), or three to five genes (Ng and Tigchelaar 1973, DeVos et al. 1981) using different germplasms. Broad-sense h^2 was estimated at 0.97 and narrow-sense $h2$ at 0.67-0.69 (Ng and Tigchelaar 1973). These estimates were independently confirmed by (DeVos et al. 1981). In a recent study, an LT fast germinating *L. esculentum* accession (PI120256)

and an LT slow germinating tomato cultivar (UCT5) and their reciprocal F_2, F_3 and BC_1 progeny (total of 10 populations) were evaluated for germination at a low (11 ± 0.5 °C) and normal (20 ± 0.5 °C) temperature regimes (Foolad and Lin 1998). Weighted least square regression analysis indicated that under CS most of the variation in germination time was due to additive genetic effects; dominance and epistatic interactions were not significant. Partitioning of the total genetic variance into those attributable to the effects of embryo, endosperm, testa, and the cytoplasm indicated that additive effects of endosperm could account for 80% of the total genetic variance for germination under CS. These results supported the previous physiological studies suggesting the significant role of the endosperm in determining the rate of germination in tomato seed (Groot and Karssen 1987, Haigh and Barlow 1987).

While examining the response to selection for improved CT during SG in tomato, an F_2 population of a cross between UCT5 and PI120256 was evaluated under a LT (11 ± 0.5 °C) regime and the fastest germinating seeds (the first 5% that germinated) were selected (Foolad and Lin 1998). The selected F_2 progeny were grown to maturity and self-pollinated to produce F_3 seed. The selected F_3 progeny were evaluated for SG under the LT and their mean performance was compared to that of a nonselected F_3 population of the same cross. The results indicated that selection was effective with significant improvement in germination performance; a realized $h2$ of 0.74 was obtained for CT during SG (Foolad and Lin 1998). This study demonstrated that the rate of tomato SG under CS was genetically controlled, with additive effect, and that it could be improved by directional phenotypic selection. A recent similar study, using interspecific BC_1 and BC_1S_1 progeny of a cross between a LT fast-germinating *L. pimpinellifolium* accession (LA722) and a LT slow-germinating tomato cultivar (NC84173), suggested a more moderate realized heritability ($h2 = 0.43$) for CT during SG in tomato (Foolad et al. 2002).

To improve LT seed germination of processing tomatoes, breeders have used *L. esculentum* lines screened at 10 °C, including PI120256, PI174263, and derived lines PI341985 and PI341988 (Cannon et al. 1973, Ng and Tigchelaar 1973, DeVos et al. 1981, Scott and Jones 1985a, b) or wild accessions from high altitudes screened for high germination at 6-7 °C, including *L. hirsutum* accessions LA1363 and LA1777 and *L. chilense* accession LA460 (Patterson et al. 1978, Patterson and Payne 1983). Several processing breeding lines were released in Europe, and field tests conducted with early direct seeding confirmed an advantage of 3 to 5 days in terms of emergence with lower variability compared to standard cultivars (Damidaux and Martinez 1992). In conclusion, it seems that, due to its moderate-to-high heritability, LT tolerance during SG in tomato can be relatively easily transferred to commercial cultivars by hybridization and phenotypic selection. Molecular mapping techniques have also been employed to examine its genetic basis and elucidate the prospect for improving this trait through MAS.

QTL mapping

The number, chromosomal locations and genetic effects of QTLs contributing to CT during SG in tomato have been determined in different interspecific populations of tomato. A BC_1S_1 population of a cross between breeding line NC84173 (cold sensitive) and *L. pimpinellifolium* accession LA722 (cold tolerant during SG) was evaluated for the rate of SG under a LT regime. Interval mapping analysis detected QTLs on tomato chromosomes 1 and 4 with significant effects on CT during SG (Foolad et al. 1998b). The *L. pimpinellifolium* accession had favorable QTLs on chromosomes 1, and NC84173 had favorable QTLs on chromosome 4. The percentage of phenotypic variation explained (PVE) by individual QTLs ranged from 11.9% to 33.4%. Multilocus analysis indicated that the cumulative action of all significant QTLs accounted for 43.8% of the total phenotypic variance. Digenic epistatic interactions were evident only between two QTL-linked markers and two QTL-unlinked markers. Transgressive phenotypes were observed in the direction of cold sensitivity.

The validity of the identified QTLs was examined in two different populations of the same interspecific cross. First, 1000 seeds of BC_1 progeny were evaluated for rate of germination under CS, and the most rapidly germinating seeds (the first 3% that germinated) were selected as the cold-tolerant individuals. These selected BC_1 plants were subjected to molecular marker analysis, using 119 RFLP markers spanned over 12 tomato chromosomes with an average distance of 9.7 cM between markers. A distributional extreme analysis resulted in the identification of QTLs for CT on chromosomes 1, 4, 8, 9 and 12 (Foolad et al. 2003). In a different study, seeds of 145 F_9 RILs of the NC84173 × LA722 cross were evaluated for germination under CS and the time to 50% germination (T50) for each RIL was determined. The RILs were also subjected to marker analysis using 129 RFLP and 62 RGAs and a molecular linkage map was constructed. Interval mapping analysis detected QTLs for CT on tomato chromosomes 1, 2, 3, 7, 8, 9, and 12 (M.R. Foolad et al., unpublished data). The latter two studies validated all of the QTLs identified in the first study (Foolad et al. 1998b) and further detected a few additional QTLs. The combined results supported the suggestion that CT during SG in tomato was a quantitative trait controlled by more than one gene. Comparison of QTLs in different populations (e.g. BC_1, BC_1S_1 and RILs) of the same *L. esculentum* × *L. pimpinellifolium* cross indicated that most QTLs were stable across populations/generations whereas a few QTLs were population-specific. Although no attempt has yet been made to improve CT during SG in tomato via

MAS, the consistency of QTLs across populations suggests that such an attempt might be fruitful.

Cold Tolerance during Vegetative Growth and Reproduction

Chilling tolerance during VS in tomato has been defined as the ability of the plant to resist damage below ≈ 10 °C but above the freezing temperature (Lyons 1973). Went (1957) determined that the cultivated tomato preferred a growth temperature Ca. 7 °C higher than the cultivated potato, (*Solanum tuberosum* L.), which, although closely related to tomato, is much less liable to chilling injury. Although both species originated from the Andean region of South America (Jenkins 1948, Correll 1962), the cultivated tomato originated from the lower altitudes whereas the cultivated potato came from the cooler higher elevations. Plants that regularly experience temperatures below 10 °C in their native habitat would be expected to be comparatively chilling resistant. Near the Equator, night temperatures regularly are below 10 °C only at altitudes of 2000 m or more. High-altitude wild species of tomato have not contributed to the ancestry of the cultivated tomato, which may explain why the cultivated tomato is less tolerant to chilling than the potato.

A green-fruited related wild species of tomato, *L. hirsutum*, grows naturally over a wide range of altitudes, from sea level in Ecuador to 3300 m in Peru (Rick 1973). Patterson et al. (1978) and Patterson and Payne (1983) evaluated several accessions of this wild species and reported that chilling tolerance was greatest in those which originated in the higher altitudes (e.g. LA1363 and LA1777) and reduced in accessions from low altitudes (e.g. LA407). In these studies, the authors used either a survival test (survival of seedlings after chilling at 0 °C for 4-7 days) or a night chill test (growing seedlings at a d/n temperatures of 20/0 °C for several days) to evaluate chilling tolerance of tomatoes. In addition, Wolf et al. (1986) evaluated CT of several high-altitude accessions of wild tomato species including *L. hirsutum* (LA1363, LA1777), *L. chilense* Dun. (LA1969, LA1971) and *Solanum lycopersicoides* Dun. (LA1964) during vegetative growth and compared that with the CT of a *L. esculentum* breeding line, UC82B. These researchers used chlorophyll fluorescence (Walker et al. 1990, Bruggemann and Linger 1994), electrolyte leakage (Van-De-Dijk et al. 1985), and plastochron index (Coleman and Greyson 1976) as evaluation criteria and reported that high-altitude wild accessions were more tolerant to LT than UC82B. Recently, by producing a cybrid between *L. hirsutum* LA1777 (source of the cytoplasm) and *L. esculentum* cultivar 'Large Red Cherry' (source of nucleus), Henk-Venema et al. (2000) demonstrated that the LT tolerance of LA1777 was not due to genes in the cytoplasm. The cybrid was as cold sensitive as the *L. esculentum* parent as determined by various growth- and photosynthesis-related characteristics under suboptimal (d/n 16/14 °C) temperatures. Vallejos et al. (1983) reported a high-altitude accession of *L. hirsutum* having less growth reduction under a low-temperature (d/n 12/5 °C), whereas at high-temperature regime (25/18 °C) the growth was comparable to a low-altitude *L. hirsutum* or a *L. esculentum* breeding line. The high-altitude *L. hirsutum* accession was capable of growing faster and acquiring greater biomass than the low-altitude *L. esculentum* and the low-altitude *L. hirsutum* accessions. In contrast, Raison and Brown (1989) evaluated CT of a high (LA1777, 3200 m), mid (LA1625, 1500 m) and low (LA1361, 50 m) altitudinal ecotypes of *L. hirsutum* based on inhibition of photosynthesis at LTs (d/n 15/5 °C) and observed that the three ecotypes had similarity for chilling-induced photoinhibition (at critical temperature between 10 and 15 °C) and the rate of their response to chilling stress. Three ecotypes thus were similarly sensitive to chilling. Photosynthesis, however, is only one component of response to CS. Miltan et al. (1986) reported that dry weight (DW) accumulation and leaf area increase in the wild species *L. hirsutum* and *L. chilense* were less adversely affected by LTs *as compared to L. esculentum.*

Root growth is an important attribute for selecting genotypes adapted to LTs. Scott and Jones (1986) evaluated hypocotyl and root growth elongation of four wild accessions of *L. chilense* and *L. peruvianum* and two control genotypes of *L. esculentum* under low (10 °C) and control (20 °C) temperatures. Hypocotyl growth rates of the wild accessions were less inhibited at the 10 °C relative to 20 °C than were either of the cultivated accessions, suggesting that these wild accessions had greater CT than the cultivated genotypes for early seedling growth. In a more recent study, 31 accessions of cultivated and related wild species of tomato were evaluated for CT during VS to identify genetic resources that could be potentially useful for improving chilling tolerance of modern tomato cultivars (Foolad and Lin 2000). In this study, plant vegetative growth was evaluated under two d/n temperature regimes of 12/5 °C (cold stress) and 25/18 °C (control) with 12 h photoperiod and photon flux of 350 $\mu mol.m^{-2}.s^{-1}$ in growth chambers. CT during VS was defined as the ratio of shoot DW under CS to shoot DW under control conditions, and referred to as vegetative stage TI. Across accessions, vegetative stage TI ranged from 0.12 to 0.39 indicating the presence of genotypic variation for CT. CT during VS was independent of plant vigor, as judged by the absence of a significant correlation ($r =$ 0.14, $P > 0.05$) between vegetative stage TI and DW under control. Several accessions were identified that exhibited considerable CT, including *L. hirsutum* PI127826, LA1777 and LA386, and *L. esculentum* PI174263 and PI120256. Several of the identified accessions have been used to examine the genetic basis of CT during VS in tomato, as briefly reviewed in the next section. In addition to *Lycopersicon* wild species, potential sources of LT tolerance for tomato breeding have also been identified in *Solanum* species, in particular *S. lycopersicoides* (Wolf et al. 1986, Rick 1988, Walker et al. 1990)

The cultivated tomato sets fruit poorly at or below 10 °C (Charles and Harris 1972) and this has been attributed mainly to: 1) failure of pollen production, 2) reduced pollen viability, 3) limited anther dehiscence, and 4) failure of pollination under LTs (Maisonneuve 1982, Picken 1984). Pollen tube growth also may be very slow to affect fertilization below 10 °C (Dempsey 1970). Other important processes involved in fruit set of tomato, such as the position of stigma in the anther cone, the formation of fertile ovules, or the early development of the embryo, are not adversely affected by chilling temperatures (Fernandez-Munoz and Cuartero 1991, Fernandez-Munoz et al. 1995b). Exposure of immature flowers of *L. esculentum* to repeated night temperatures below 10 °C resulted in sterile pollen grains (Patterson et al. 1987). However, high-altitude *L. hirsutum* accessions LA1363, LA1393 and LA1777 produced functional pollen under such conditions (Maisonneuve 1983, Patterson et al. 1987). Pollen of *L. hirsutum* accession LA1366 was able to germinate at 6 °C (Patterson et al. 1987). Similarly, pollen germination of *L. hirsutum* accession LA1777 was less inhibited at 5 °C than pollen from a cultivated tomato (Zamir et al. 1981). Furthermore, the frequency of *L. hirsutum* gametes contributing to hybrid zygote formation was doubled when controlled fertilizations with pollen mixtures of *L. esculentum* and *L. hirsutum* occurred at d/n temperatures of 12/6 °C as compared to crosses with the same mixtures at 24/19 °C (Zamir et al. 1981). The results of the latter study suggested competitive ability of high-altitude *L. hirsutum* pollens at LTs. Furthermore, this ability was shown to be under genetic control and heritable. In another study, evaluation of pollen fertility and anther desiscence of 170 accessions from 8 different *Lycopersicon* species indicated that, unlike for *L. esculentum*, accessions from *L. hirsutum, L. peruvianum* and *L. pennellii* generally produced fertile pollen below 10 °C and released pollen satisfactorily (Fernandez-Munoz et al. 1995a). Lyakh (1992) demonstrated that pollen of some accessions of *L. pennellii* and *L. hirsutum* fertilized more efficiently than that of *L. esculentum* at 10 °C. In other germplasm evaluation studies, several *L. esculentum* lines were identified with the ability to produce flowers and fruit set at LTs, including Early North, PI205040, PI280597, Cold Set, Precoce, Apedice, Montfavet and Supermarmande (Kemp 1968, Philouze and Maisonneuve 1979, Gautam et al. 1981). The aforementioned studies indicate the presence of potential genetic resources within the cultivated and related wild species of tomato for improving CT of commercial cultivars. The utility of such genetic resources for improving tomato CT has been examined in a few studies, as briefly described below.

Inheritance and QTL Mapping

A high-altitude ecotype of *L. hirsutum*, which was able to develop the first true leaf when grown at a night temperature of 0 °C, was hybridized with a cold-sensitive *L. esculentum* cultivar (pistillate and recurrent parent) and F_1 and BC_1 progeny were produced (Patterson and Payne 1983). Subsequent selections for survival under a night temperature of 0 °C in the BC_1 and its selfed progeny indicated that the CT attributes of the *L. hirsutum* accession could be transferred to *L. esculentum* through hybridization and selection. Using BC_1 population of an interspecific cross between a cold-sensitive *L. esculentum* line and a cold-tolerant *L. hirsutum* accession, Vallejos and Tanksley (1983) identified three QTLs responsible for growth at LTs. In a more recent study, parental and reciprocal F_1, F_2, F_3 and BC_1 progeny (total of 12 populations) of a cross between a cold-sensitive tomato breeding line (UCT5) and a cold-tolerant primitive cultivar (PI120256) were grown under two d/n temperature regimes of 15/10 °C (cold stress) and 25/15 °C (control) (Foolad and Lin 2001a). Plants were evaluated for shoot DW under CS and for tolerance index (TI), measured as the ratio of DW under CS to DW under control conditions. Shoot DW was reduced in all genotypes in response to CS, however, PI120256 exhibited the highest tolerance (TI = 90.5%) and UCT5 the lowest (TI = 38.9%). The TIs of the filial and backcross progeny were intermediate to the parents, suggesting that CT of PI120256 genetically transmitted in the progeny. Across generations, there was a positive correlation ($r = 0.76$, $P < 0.01$) between DW under CS and DW under control conditions, suggesting that growth under CS was influenced by plant vigor. However, the absence of a significant correlation between DW under control conditions and the TI ($r = 0.47$, $P > 0.05$) and, in contrast, the presence of a significant positive correlation ($r = 0.92$, $P < 0.01$) between DW under CS and the TI suggested that plant vigor was not a determining factor in the expression of CT in PI120256 and its progeny. Generation means analyses of DW under CS and TI indicated that the variation among generations was genetically controlled, with additive effects accounting for most of the variation ($\approx 90\%$ for TI). There were no significant dominance effects, and epistatic effects were minor and involved only additive × additive interactions. The results suggested that the inherent CT of PI120256 would be useful for improving LT tolerance of commercial cultivars of tomato.

The inheritance of CT during reproduction has been characterized to vary in different tomato germplasms. Using parental and F_3 progeny of a cross between *L. esculentum* and a high-altitude cold-tolerant accession of *L. hirsutum*, Patterson et al. (1987) demonstrated that some F_3 plants were able to produce pollen of normal appearance at LT, similar to the *L. hirsutum* parent. Fernandez-Munoz et al. (1995b) investigated LT pollen viability of parental, F_1, F_2 and BC progeny of a *L. esculentum* × *L. pennellii* cross and concluded that this trait was polygenic and heritable. In segregating progeny derived from crosses between *L. esculentum* and *L. pimpinellifolium*, the ability to set fruit under LTs was determined to be controlled by recessive factors (Kalloo and Banerjee 1990). In contrast, the CT of hybrid between the sensitive *L. esculentum* cv Sub Arctic Maxi and tolerant *S. lycopersicoides* was suggested to be due to dominant nuclear genes (Kamps et al. 1987). However, the available information in the literature suggests that sources of genetic variation for CT exist within the cultivated and related wild species of tomato and there is potential for improving tomato CT by hybridization and selection. Further research is needed to determine the number and genomic locations of QTLs or major genes in order to facilitate the development of cold-tolerant tomatoes via MAS or genetic transformation.

Comparison of Cold Tolerance during Different Stages of Plant Development

Low temperature tolerance is required at all stages of plant development, including SG, VS, and reproduction, and during post-harvest storage of fruits. When searching for genetic resources for CT, it is desirable to identify genotypes that exhibit CT throughout the life cycle of the plant. However, CT at one stage of plant development may not be correlated with tolerance at other development stages, and there might not be accessions bearing CT at all critical stages (Herner and Kemps 1983). Patterson et al. (1978) studied altitudinal ecotypes of *L. hirsutum* and reported variable tolerance to LTs at the whole plant level. However, equivalent germination rates of *L. hirsutum* (native to 3100 m altitude in Peru) occurred at 3 °C lower temperature than *L. esculentum* cultivar Rutgers. This marginal performance suggests that CT during VS is not necessarily a valid indication of CT during other stages, including SG. In other studies, comparison of different accessions of *L. hirsutum* indicated that the ability of the seed to germinate quickly at LTs was not related with the ability of the pollen to germinate at LTs (Patterson et al. 1979, Patterson 1988). Similarly, evaluation of *L. hirsutum* accessions for CT indicated that tolerance at the seedling stage was not necessarily correlated with tolerance during reproduction (pollen development and germination) (Patterson et al. 1987), and pollen selection at LTs was not effective for improving tomato CT during plant development (Maisonneuve et al. 1986). Lack of positive relationships between CT during SG and reproduction and between seedling growth and reproduction were also reported in other studies (Patterson et al. 1978, Maisonneuve and Den-Jijs 1984, Maisonneuve et al. 1986, Patterson et al. 1987). Although most studies indicate the absence of relationships among CT during different stages of plant development, a few studies suggest otherwise. For example, Zamir et al. (1981) reported that *L. hirsutum* accessions that were cold tolerant during seedling stage also exhibited greater CT during pollen germination and pollen tube growth. Furthermor, the inhibition of root elongation at LTs was less in plants of the crosses in which pollen was used from LTs-adapted *L. hirsutum* plants than from plants of normal temperatures (Zamir and Gadish 1987).

Two recent studies examined in more detail the phenotypic and genetic relationships between CT during SG and VS in tomato. In one study, the phenotypic relationship was investigated by evaluating 31 tomato accessions (cultivars, breeding lines, and plant introductions), representing six *Lycopersicon* sp., for CT during both stages (Foolad and Lin 2000). SG was evaluated under two temperature regimes of 11 ± 0.5 °C (cold stress) and 20 ± 0.5 °C (control) in darkness. CT during SG was defined as the inverse of the ratio of germination time under CS to germination time under control conditions, and it was called germination TI. Across accessions, germination TI ranged from 0.15 to 0.48, indicating the presence of substantial genotypic variation for CT during SG. Vegetative growth was evaluated in growth chambers with d/n temperatures of 12/5 °C (cold stress) and 25/18 °C (control) and a 12-h photoperiod of 350 $\mu mol.m^{-2}.s^{-1}$ (photosynthetic photon flux). CT during VS was defined as the ratio of shoot DW under CS to shoot DW under control conditions and referred to as vegetative stage TI. Across accessions, vegetative TI ranged from 0.12 to 0.39 indicating the presence of notable genotypic variation for CT during VS. CT during VS was independent of plant vigor, as judged by the absence of a significant correlation ($r = 0.14$, $P > 0.05$) between vegetative stage TI and DW under control. Furthermore, CT during VS was independent of CT during SG, as suggested by the absence of a significant rank correlation ($rR = 0.14$, $P > 0.05$) between vegetative stage TI and germination TI. However, a few accessions were identified with CT during both SG and VS (Foolad and Lin 2000). The results indicate that for CT breeding in tomato, each stage of plant development would have to be evaluated and selected independently.

To determine the genetic relationship between CT during SG and VS, an F_2 population of a cross between *L. esculentum* PI120256 (cold tolerant during both SG and VS) and UCT5 (cold sensitive during both stages) was evaluated for germination under CS and the most cold-tolerant individuals (the first 5% that germinated) were selected (Foolad and Lin 2001b). Selected F_2 individuals were grown to maturity and self-pollinated to produce F_3 families (referred to as the selected F_3 population). The selected F_3 population was evaluated for CT separately during SG and VS and its performance was compared with that of a nonselected F_3 population of the same cross. The results indicated that selection for CT during SG significantly improved

CT of the progeny during SG, and a realized $h2$ of 0.75 was obtained for this trait. However, selection for CT during SG did not affect plant CT during the VS. There was no significant difference between the selected and nonselected F_3 populations in either absolute CT (defined as shoot FW under cold stress) or relative CT (defined as shoot FW under cold as a percentage of control). The results indicated that, in PI120256, CT during SG was genetically independent of CT during VS (Foolad and Lin 2001b).

The overall evidence indicates that there is very little or no relationships among CT in different development stages in tomato, including SG, VS and reproduction. Thus, to develop tomato cultivars with improved CT throughout the ontogeny of the plant, selection protocols that include all relevant developmental stages are necessary. In this regard, the use of molecular markers and MAS may be desirable for achieving this goal.

Future Prospects for Developing Tomato Cultivars with Enhanced Cold Tolerance

Genetic resources for CT during different stages of plant development, including SG, VS and RS, have been identified within the cultivated and related wild species of tomato, in particular *L. hirsutum*. Significant progress has been made in characterizing the physiological mechanisms and metabolic aspects of CT in tomato. We have some knowledge of the genetic controls of CT during different stages of plant development in tomato. However, progress in developing tomatoes with improved CT has been very limited, in part due to an insufficient understanding of the genetic basis of CT and the limited breeding effort that has been devoted to this goal. Significant progress has been made in molecular genetic basis of CT in many other plant species, in particular *A. thaliana*. Several genes, proteins, enzymes and other compounds have been identified with direct or indirect effects on LT tolerance. Utilization of such genes or compounds in tomato using the available transgenic approaches may become a useful method for improving tomato CT. This will require a better understanding of the genetic basis of CT in tomato at the cellular and molecular levels and transferring of tolerance components from CT accessions of *L. hirsutum* into the cultivated tomato. Research to identify QTLs with major effects on CT may also be useful for improving tomato LT tolerance.

TOLERANCE TO HEAT STRESS

Background Information

Although tomato plants can grow in a wide range of climatic conditions, their vegetative and reproductive growth are severely impaired at high temperatures, resulting in reduced yield and fruit quality (Abdul-Baki 1991, Dane et al. 1991, Wessel-Beaver and Scott 1992, Scott 1993). Generally, when the ambient temperature exceeds 35 °C, tomato SG, seedling and vegetative growth, flowering and fruit set, and fruit ripening are adversely affected (Kalloo 1991). This high-temperature sensitivity is particularly important in areas with tropical or subtropical climates. In such environments, heat stress (HS) may become a major limiting factor for field production of tomatoes. Although tomato plants are sensitive to high temperatures during all stages of plant development, flowering and fruit set are the most sensitive stages. Fruit set is somewhat affected at d/n temperatures above 26/20 °C and is severely affected above 35/26 °C (Rudich et al. 1977, George et al. 1984, Stevens and Rudich 1987, Berry and Rafique-Uddin 1988). For example, a 4-h exposure to a day temperature of 40 °C during the reproductive stage prevents fruit set in most cultivars of tomato (Charles and Harris 1972, Rudich et al. 1977). Villareal (1978) defined heat tolerance (HT) in tomato as the ability to set fruits under night temperatures not lower than 21 °C. Other researchers, however, have argued that day and night temperatures may not affect tomato fruit set independently, and that diurnal mean temperature is a better predictor of plant response to high temperature, with day temperature having a secondary role (Peet and Bartholemew 1996, Peet et al. 1997, Peet and Willits 1998, Peet et al. 1998). Accordingly, the sensitivity to high temperature should be determined as the diurnal mean temperature above which tomato fruit set is reduced. By examining different d/n temperature regimes, Peet et al. (1997, 1998) demonstrated that percent fruit set in tomato decreased as mean diurnal temperature rose above 25 °C and was severely impaired at 29 °C.

Reproductive processes adversely affected by high temperature in tomato include meiosis in the microspore and megaspore mother cells (Kinet and Peet 1997), amount of pollen produced (El-Ahmadi and Stevens 1979b, Stevens and Rudich 1987), anther dehiscence and pollen release (El-Ahmadi and Stevens 1979b), pollination (Charles and Harris 1972, Shelby et al. 1978), pollen germination and pollen tube growth (Weaver and Timm 1989), ovule viability (Kinet and Peet 1997), stigmatic and stylar position (Charles and Harris 1972, El-Ahmadi and Stevens 1979b), number of pollen grains retained by the stigma, fertilization as well as post-fertilization processes, growth of the endosperm, proembryo and fertilized embryo (Kinet and Peet 1997, Peet et al. 1998). However, the most noticeable morphological effect of high temperature is the production of an exserted style (when stigma is elongated beyond the anther cone), which may prevent self-pollination (Rick and Dempsey 1969). Poor fruit set at high temperature has also been associated with low levels of carbohydrates and growth regulators released in plant sink tissues (Kinet and Peet 1997). Furthermore, high-temperature effects on fruit set depend on the stage of floral development (Iwahori 1965) and the genotype (Rudich et al. 1977, Shelby et al. 1978, El-Ahmadi and Stevens 1979a, b, Shen and Li 1982). Growth chamber and greenhouse studies suggest that high temperature is most deleterious when flowers are first visible and sensitivity continues for 10-15 days (Calvert 1969). Reproductive phases most sensitive to high temperature are gametogenesis (8-9 days before anthesis) and fertilization (1-3 days after anthesis) (Iwahori 1966). Both male and female gametophytes are sensitive to high temperature and response varies with genotype (Shelby et al. 1978, El-Ahmadi and Stevens 1979a, b), ovules are generally less heat sensitive than pollen (Charles and Harris 1972, Rudich et al. 1977, Peet et al. 1998).

A strong positive correlation has been observed between fruit set and yield under high temperature (El-Ahmadi and Stevens 1979a, Abdul-Baki 1991, Wessel-Beaver and Scott 1992). Therefore, evaluation of tomato germplasm to identify sources of HT has regularly been accomplished by screening for fruit set under high temperature (El-Ahmadi and Stevens 1979a, Berry and Rafique-Uddin 1988, Abdul-Baki 1991). Since poor fruit set at high temperature can not be attributed to a single factor, decreases in pollen germination and/or pollen tube growth are among the most commonly reported factors. Thus, pollen viability has been suggested as an indirect selection criterion for HT (Weaver and Timm 1989). Levi et al. (1978) reported that heat-tolerant cultivars often exhibited higher pollen viability than heat-sensitive cultivars under high temperature. Production of viable seed also is often reduced under high temperature and thus, high seed set has been arguably reported as an indication of HT (Charles and Harris 1972, El-Ahmadi and Stevens 1979a, Berry and Rafique-Uddin 1988, Abdul-Baki 1991). High-temperature conditions also induce the incidence of fruit disorders, e.g., cracks, blossom-end rot, deterioration of fruit color, watery tissue, and small immature fruit. All these cause reductions in marketable yield (Charles and Harris 1972, Scott et al. 1986, Abdul-Baki 1991, Johjima 1995). The non-reproductive processes in tomato which are affected by HS include photosynthetic efficiency (Bar-Tsur et al. 1985), assimilate translocation (Tanaka et al. 1974), mesophyll resistance (Stevens and Rudich 1987), and disorganization of cellular membranes (Chen et al. 1982). Overall the criteria most often used for evaluating tomatoes for HT are yield, fruit set, fruit quality, and seed production.

Genotypic variation has been observed in the cultivated and related wild species of tomato for the effect of high temperature on pollen and ovule production and viability, anther dehiscence, pollination effectiveness (Shelby et al. 1978), style elongation (Rudich et al. 1977, El-Ahmadi and Stevens 1979a), splitting of the antheridial cone, stigma exsertion (Levi et al. 1978), and the ability to set fruit (Rudich et al. 1977, Stoner and Otto

1975, Berry and Rafique-Uddin 1988, Abdul-Baki 1991, Dane et al. 1991, Abdul-Baki and Stommel 1995). However, a causal relationship between each of these characteristics and HT has not always been demonstrated (Charles and Harris 1972, El-Ahmadi and Stevens 1979a, Lohar and Peat 1998). Rudich et al. (1977) reported that in heat-tolerant lines generally stigma position was near the staminal cone whereas in heat-sensitive lines the style was elongated and exserted. However, Lohar and Peat (1998) suggested that the use of stigma exsertion as a criterion for selecting against fruit set at high temperature might be misleading and should be avoided. El-Ahmadi and Stevens (1979a) reported that for an optimal HT response, a cultivar must exhibit a combination of essential characteristics under high temperature, including high number of flowers per plant, absence of stigma exsertion, high pollen production, ovule viability, and substantial fruit and seed set. However, none of the heat-tolerant cultivars studied by these researchers possessed all of the model traits.

Two common and undesirable characteristics generally observed in heat-tolerant tomato genotypes are production of small fruit and restricted foliar canopy (Rudich et al. 1977, Shelby et al. 1978, Villareal 1978, El-Ahmadi and Stevens 1979a, Hanna and Hernandez 1982, Scott et al. 1986, Dane et al. 1991, Wessel-Beaver and Scott 1992, Scott 1993, Scott et al. 1997). The production of small fruit is most likely due to adverse effects of high temperature on the production of auxins in the fruit, and the poor canopy is due to the highly reproductive nature of the heat-tolerant genotypes. Dane et al. (1991) evaluated fruit set ability of 47 tomato genotypes under high temperatures in the field and/or in the greenhouse and concluded that small-fruited genotypes were generally less affected by heat than large-fruited cultivars, consistent with many other reports on the negative correlations between HT and fruit size. However, despite all the complexities of HT and the difficulties encountered during the transfer of tolerance, heat-tolerant inbred lines and hybrid cultivars with commercial acceptability have been developed and released (see Section: Inheritance and Breeding).

Seed germination in tomato is greatly affected by high temperature. Tomato seed germinate best between 20 °C and 25 °C. The germination is reduced at 30 °C, highly inhibited at 35 °C (Jaworski and Valli 1965, Thompson 1974, Lorenz and Maynard 1988, Coons et al. 1989), and almost nil at 39-39.5 °C (Thompson et al. 1977). There is significant genotypic variation in response to germination under high temperature, total germination ranging from 0% to 95% at or above 35 °C in some genotypes (Berry 1969, Thompson 1974, El-Hassan 1978, Taylor et al. 1982, Coons et al. 1989). The ability to germinate under high-temperature conditions is important for the fall production in tropical and subtropical environments where seeding is required during late summer and soil temperatures often exceed 35 °C (Coons et al. 1989).

Inheritance and Breeding

Development of tomato cultivars for improved fruit set and yield under high temperature is desirable for tomato production in regions where temperatures during the growing season reach 35 °C (Scott et al. 1986, Stevens and Rudich 1987, Scott 1993). However, breeding for HT in tomato is a difficult task, in part due to the complexity of the trait and its low to moderate $h2$ (Charles and Harris 1972, Rudich et al. 1977, Levi et al. 1978, Aung 1979, Villareal and Lai 1979, Kuo et al. 1979, El-Ahmadi and Stevens 1979b, Scott et al. 1986, Dane et al. 1991). Other drawbacks being that heat-tolerant lines tend to produce smaller fruit than is commercially acceptable (Rudich et al. 1977, Shelby et al. 1978, Villareal 1978, Villareal and Lai 1979, El-Ahmadi and Stevens 1979a, Hanna and Hernandez 1982, Dane et al. 1991, Wessel-Beaver and Scott 1992, Scott 1993, Scott et al. 1997) and poor foliar canopy (Scott 1993, Scott et al. 1997). To overcome these impediments and for developing widely adapted lines with acceptable horticultural characteristics, it is recommended to utilize in a breeding program genotypes capable of fruit set at high temperatures in combinations with genotypes with large fruit and vigorous vine growth in different environmental conditions (Berry and Rafique-Uddin 1988, Wessel-Beaver and Scott 1992, Scott et al. 1997).

To estimate $h2$ for HT and produce heat-tolerant tomato lines, Wessel-Beaver and Scott (1992) developed a synthetic population of tomato from polycrosses among seven selected genotypes, and evaluated the population for HT in two different locations (Florida and Puerto Rico) during one summer. Single-location $h2$s were high for percent fruit set (0.74-0.77), yield (0.65-0.81), and fruit weight (0.89-0.97). Across-location $h2$ was low for yield (0.14), intermediate for fruit set (0.60), and high for fruit weight (0.92). Genotype × environment interaction was most important for yield and least important for fruit weight. Large genetic correlations ($r = 0.71$-0.74) were observed between yield and fruit set under high temperatures at both locations (Wessel-Beaver and Scott 1992). El-Ahmadi and Stevens (1979b) reported that inheritance of fruit set under high temperature of heat-tolerant genotypes was due to additive and dominant gene actions with moderate $h2$. Hanna et al. (1982) concluded that additive gene action was more important than nonadditive gene action for percent fruit set, percent flower drop, and percent undeveloped ovaries under high field temperatures. In a diallel analysis using several heat-tolerant and heat-sensitive tomato genotypes, Dane et al. (1991) determined that pollen fertility and fruit set under high field temperatures were primarily under additive genetic controls. However, evaluating parental and hybrid progeny of crosses between several genotypes with various levels of fruit-set ability under high temperature, Scott et al. (1986) reported the presence of dominance for high fruit set under high temperature and recommended the development and use of hybrid cultivars with improved HT. A more recent study, including parental and hybrid progeny of crosses among several heat-tolerant and heat-sensitive genotypes, confirmed the presence of dominance for fruit-set ability under high temperature (Scott et al. 1997). The advantage of the dominant nature of high temperature fruit-setting ability is that commercial heat-tolerant fresh-market tomato cultivars can be feasibly developed by producing F_1 hybrids between heat-tolerant and heat-sensitive parents. In this approach, defects of the heat-tolerant parent, such as disease susceptibility, small fruit and poor vine coverage, could be at least partially ameliorated, if the proper heat-sensitive parent is used for the hybrid production.

A review of the literature indicates that, despite various difficulties in transferring HT characteristics in tomato, several breeding lines and hybrid cultivars with improved HT and rather acceptable horticultural characteristics (medium to large size fruit, vigorous vine and good leaf coverage) have been developed and released (Scott et al. 1989, Hanna et al. 1992, Scott et al. 1995, Scott et al. 1997, Scott 2000). The problem of small fruit size has largely been overcome by crossing small-fruited heat-tolerant lines with extremely large-fruited heat-sensitive lines and producing heat-tolerant F_1 hybrids. The use of such hybrid cultivars has already resulted in dramatic increases in tomato yield in areas with high day or night temperatures during the growing season, such as that in Florida (Scott et al. 1997, Scott 2000). Most of the heat-tolerant varieties developed to date have jointed pedicel, however, Scott and his colleagues were able to develop heat-tolerant, large-fruited, jointless inbred lines. Such genotypes, however, do not perform as good as the equivalent genotypes with jointed pedicel (Scott et al. 1997, Scott 2000). The available heat-tolerant genotypes are currently used extensively to develop commercially acceptable tomato cultivars with enhanced HT.

Future Prospects for Developing Tomato Cultivars with Enhanced Heat Tolerance

Germplasm resources for HT have been widely identified within *Lycopersicon* species. The physiological and morphological components of heat tolerance/sensitivity have been resolved and the genetic basis of HT at the whole plant level determined. During the past several years, many breeding lines and cultivars have been developed with improved fruit-setting ability under high temperature. Many of these lines exhibit some undesirable horticultural characteristics, in particular small fruit size, low yield, poor vine coverage, and jointed pedicel. However, many years of tomato research and breeding efforts at the University of Florida have resulted in the development of new high-yielding breeding lines and hybrid cultivars with medium to large size fruit, high fruit setting ability, jointless pedicel, and other HT attributes. Currently, further efforts are being devoted to the development of heat-tolerant tomato cultivars with larger fruit size, higher yield, and more vigorous vines for production in tropical and subtropical environments. Judging from the recent advances in this area, we expect to witness improved cultivars, in particular F_1 hybrids, with excellent fruit setting ability under high temperature and acceptable horticultural characteristics in the near future. However, limited research has been conducted on molecular genetic basis of HT in tomato. This line of research as well as research on the relationship between tolerances to high and low temperatures during different stages of plant development should be strengthened.

RELATIONSHIPS OF TOLERANCES TO DIFFERENT ABIOTIC STRESSES

Tolerance to Different Stresses during Seed Germination

It has been hypothesized that similar or identical genes and physiological mechanisms might control the rate of tomato SG under different stresses, (SS, DS and CS). Excessive salt depresses the water potential of the germination medium, making water less available to the seed, and thus lowers the rate of, or completely inhibits, germination. Most studies suggest that under SS low water potential of the external medium, rather than ion toxicity effects, is the major limiting factor to germination in different crop species, including tomato (Kaufman 1969, Ungar 1978, Bliss et al. 1986, Haigh and Barlow 1987, Bradford 1995). Drought Stress also causes low rate of SG due to reduced water potential of the germination medium (Bradford 1995). Therefore, it is expected that seeds that germinate rapidly under SS would also germinate rapidly under DS, and *vice versa*. Low temperatures (cold stress) also affects the water status of the cell and thus, it could delay SG by causing water stress (Liptay and Schopfer 1983). The genetic and physiological processes that impart rapid SG under different stress conditions are exactly unknown. It is important, both for scientific and practical purposes, to determine whether the same genes contribute to rapid SG under different stress conditions. It is also equally important to determine whether genotypes with the ability to germinate rapidly under various environmental conditions can be identified. Such information may contribute to the development of cultivars with superior germination performance under a wide range of environmental conditions. Recently, three different undermentioned approaches were taken to determine the relationship in the rate of tomato SG to different stress conditions, including SS, CS and DS.

(i) Germplasm Evaluation

Three independent investigations were followed to evaluate germsplasm by finding phenotypic relationships in the rate of tomato SG under different abiotic stresses. In the first study, germination responses of 30 tomato accessions representing six *Lycopersicon* species were examined under CS and SS (Foolad and Lin 1999). In the second study, germination responses of approximately 70 *L. pimpinellifolium* accessions were examined under CS, SS and DS (M.R. Foolad et al., unpublished data). In the third study, germination responses of 145 RILs, derived from a cross between a slow germinating *L. esculentum* breeding line (NC84173) and a fast germinating *L. pimpinellifolium* accession (LA722) were examined under CS, SS and DS. In these studies, the pair-wise phenotypic correlation coefficients between germination responses under the three stress conditions ranged between 0.62 and 0.80, and all were statistically significant at the 0.01 probability level. The overall results indicated the presence of significant phenotypic relationship between germination ability under three stress conditions (CS, SS and DS). However, these results must be supported on the genetic basis of these relationships.

(ii) Analysis of Response, and Correlated Response, to Selection

Two independent studies were conducted to examine genetic relationships among tomato SG responses under different stress conditions. In the first study, seeds of F_2 progeny of a cross between a slow germinating (UCT5) and a fast germinating tomato line (PI120256) were evaluated separately for germination under CS and SS, and in each treatment the most rapidly (first 5%) germinating seeds were selected. Selected seedling were grown to maturity and self-pollinated to produce F_3 progeny. The F_3 progeny from each selection experiment were evaluated separately for germination under CS and SS, and their performances compared with germination rate of non-selected F_3 progeny of the same cross. The results indicated that selection under CS or SS significantly improved progeny germination rate under each of cold- or salt-stress conditions (Foolad, et al. 1999b). The genetic correlation between the rate of SG under CS and SS was estimated as $rG = 1.00$, as determined by $rG = (CRxCRY/RxRY)1/2$, where, CRx is the correlated response under treatment X due to selection under treatment Y, and Rx is the direct response under treatment X due to selection under the same treatment, X. The results supported the suggestion that same genes might contribute to rapid SG under both CS and SS.

In the second study, seeds of BC_1 progeny of an interspecific hybrid between a slow germinating tomato breeding line (NC84173; maternal and recurrent parent) and a fast germinating *L. pimpinellifolium* accession (LA722) were evaluated separately for germination rate under CS, SS and DS, and in each treatment the most rapidly (first 2%) germinating seeds were selected. Selected individuals were grown to maturity and self-pollinated to produce BC_1S_1 progeny. The selected BC_1S_1 progeny from each experiment were evaluated for germination rate under each of CS, SS and DS, and their performances under each stress were compared with those of a non-selected BC_1S_1 population of the same hybrid. Results indicated that selection for rapid SG in each of the three stress treatments was effective and significantly improved progeny germination rate under all three stress conditions (Foolad et al. 2002). The observation of large genetic correlations (rG) between CS and SS (0.94), CS and DS (0.81), and SS and DS (1.00) suggested that the same genes or physiological mechanisms might control the rate of SG under these three stress conditions. Furthermore, these studies suggest the presence of some common stress related genes that facilitate rapid SG under different stress conditions. In practice, therefore, selection for rapid SG under a single stress environment might result in progeny with improved SG under a wide range of stress conditions.

(iii) QTL Mapping
To further examine the genetic relationships of germination under different stress conditions, three QTL comparison investigations were conducted. In all three studies, different populations of the same interspecific hybrid between a slow germinating breeding line, NC84173, and a fast germinating *L. pimpinellifolium* accession, LA722, were used to compare QTLs contributing to rapid SG under CS, SS and DS. In the first experiment, QTLs for CT and ST during SG were identified in BC_1S_1 populations using an interval mapping approach (Foolad et al. 1999a). In the second experiment, a selective genotyping approach was employed to identify QTLs for CS, SS and DS using different BC_1 populations of the same cross (M.R. Foolad et al., unpublished data). In the third experiment, a RIL population was evaluated for SG under CS, SS and DS and QTLs were identified for rapid SG under the three stress conditions (M.R. Foolad et al., unpublished data). The combined results indicated the presence of two types of QTLs. Some QTLs were identified that contributed to rapid SG under two or three stress conditions. These QTLs were referred to as germination-related, stress-nonspecific QTLs. Some QTLs were identified that contributed to rapid SG only under one stress condition. Such QTLs were referred to as germination-related, stress-specific QTLs.

The general conclusions from the above investigations suggest the presence of genes or physiological mechanisms in some accessions of tomato that facilitate rapid SG under different stress conditions, including CS, SS and DS. Apparently such genetic or physiological mechanisms can function under different stress conditions and stimulate rapid SG. The results also suggest the presence of other genes that might be expressed only under specific stress conditions and which may stimulate rapid SG under such conditions. However, isolation, characterization and comparison of functional genes which facilitate rapid SG under different stress conditions would be necessary to validate these suggestions. The results of these studies, however, indicate that to develop tomato cultivars with rapid SG under diverse environmental conditions, it may be sufficient to conduct selection and breeding only under a single stress treatment.

Germination Responses under Stress and Nonstress Conditions

The ability of the seed to germinate rapidly and uniformly under different stress and nonstress (NS) conditions is a desirable trait for many crop species, including tomato. Germplasm evaluation studies have indicated that tomato seeds that germinate rapidly under NS conditions may also tend to germinate rapidly under stress conditions for some germplasm. For example, there was a significant phenotypic correlation ($rP = 0.75$, $P < 0.01$) between the rate of SG in NS and CS treatments among 36 accessions of the cultivated and wild species of tomato (Scott and Jones 1982). Similarly, a significant phenotypic correlation ($rP = 0.89$, $P < 0.01$) between the rate of SG under NS and CS was reported among 30 tomato accessions from six different *Lycopersicon* species (Foolad and Lin 1999). A significant, but smaller, phenotypic correlation ($rP = 0.54$, $P < 0.01$) between the rate of SG under NS and CS was also observed among 145 F_9 RILs of a cross between a slow germinating tomato breeding line (NC84173) and a fast germinating *L. pimpinellifolium* accession (LA722) (M.R. Foolad et al., unpublished data). Furthermore, significant phenotypic correlations between the rate of SG under NS and SS were reported in different studies, including among 45 cultivated and wild accessions of tomato ($rP = 0.62$, $P < 0.01$) (Foolad and Lin 1997b), among 30 accessions from different *Lycopersicon* species ($rP = 0.63$, $P < 0.01$) (Foolad and Lin 1999), and among 145 F_9 RILs of the cross between NC84173 and LA722 ($rP = 0.58$, $P < 0.01$) (M.R. Foolad et al., unpublished data). Moreover, a significant phenotypic correlation ($rP = 0.57$, $P < 0.01$) between the rate of SG under NS and DS was observed among 145 F_9 RILs of the hybrid between NC84173 and LA722 (M.R. Foolad et al., unpublished data). These correlations are consistent with the suggestion that similar physiological mechanisms might contribute to rapid SG under both NS and stress conditions (Bradford 1995). Similarly, they may indicate that genetic factors facilitating rapid SG under stress conditions have no undesirable effects on performance in the absence of stress. Notably, however, in most of these studies, the magnitudes of the correlation coefficients for germination under NS and stress conditions were moderate or small. A common observation in these studies was that some accessions or lines that germinated rapidly under NS exhibited poor germination under stress conditions, and *vice versa*. This observation supports the presence of genes which might be stress-specific and contribute to rapid SG only under specific stress conditions.

Several investigations were conducted to determine the genetic relationship between the rate of tomato SG under stress and NS conditions. In one study, an F_2 population of a cross between a slow germinating (UCT5) and a fast germinating (PI120256) tomato lines was evaluated separately for germination under stress (SS and CS) and NS conditions, and in each treatment, selection was made for rapid SG. Evaluation of response and correlated response to selection in the F_3 progeny indicated that selection for rapid SG under either SS or CS resulted in progeny with improved germination under both stress and NS conditions (Foolad et al. 1999b). However, selection for rapid SG under NS conditions did not significantly improve progeny SG under any of the three conditions. This lack of response to selection could be due to a limited expression of genetic variation in the F_2 population under NS conditions. Large genetic correlations were revealed between germination under NS and either CS ($rG = 0.81$) or SS ($rG = 0.66$). In a different study, selection for rapid SG under SS, CS or DS in a BC_1 population of a hybrid between NC84173 and *L. pimpinellifolium* accession LA722 resulted in moderate improvements (significant only at $P < 0.05$) in progeny SG under NS conditions (Foolad et al. 2002). No selection was made under NS condition because there was little variation in germination rate in the NS treatment. Furthermore, in this study, moderate phenotypic correlations were observed between SG under NS and either CS ($rP = 0.53$, $P < 0.05$), SS ($rP = 0.50$, $P < 0.05$), or DS ($rP = 0.46$, $P < 0.05$). The results suggested the presence of some genetic relationships between germination under NS and stress (salt, cold or drought) conditions. However, these studies indicated that, for practical purposes of improving tomato SG under NS conditions, it is better to conduct selections under a stress treatment. This is because of the higher variance for germination and higher selection efficiency under stress conditions. Higher genetic variance in stress environments is one of the more favorable situations for plant breeders (Rosielle and Hamblin 1981) although it does not appear to be a common occurrence (Daday et al. 1973).

To further examine the genetic relationship between rate of tomato SG under stress and NS conditions, QTLs contributing to these traits were compared in three different studies M. R. Foolad et al., unpublished data). First, QTLs were identified for germination under NS, CS and SS in BC_1S_1 populations of a cross between NC84173 and *L. pimpinellifolium* accession

LA722 and compared for co-localization (Foolad et al. 1999a). In the second study, QTLs for the rate of tomato SG under NS, CS, SS and DS were determined in different BC_1 populations of a NC84173 × LA722 hybrid using selective genotyping. In the third study, QTLs for the same four traits were identified and compared using a RIL population of a cross between NC84173 and LA722. Results of the these investigations were comparable and indicated the presence of a genetic relationship between the ability of tomato seed to germinate rapidly under NS and stress conditions. A few QTLs were detected which contributed to rapid SG under both NS and stress conditions; these QTLs were referred to as germination-related, stress-nonspecific QTLs. Several other QTLs were identified which affected germination under one or more of the stress conditions but not under NS conditions; these QTLs were referred to as germination-related, stress-specific QTLs.

The detection of stress-nonspecific QTLs indicated the presence of a genetic relationship between the ability of the tomato seed to germinate rapidly under NS and stress conditions. However, whether such a genetic relationship was due to pleiotropic effects of the same genes, physical linkage of different genes, or a combination of the two, could not be determined in these studies. Isolation and characterization of functional genes affecting germination rate under different stress and NS conditions may be necessary to determine the nature of the genetic relationship. The detection of stress-nonspecific QTLs, however, is consistent with the significant phenotypic and genetic correlations observed between germination rate under NS, CS, SS and DS conditions. The results are in agreement with the suggestion that genetic parameters or physiological mechanisms that facilitate rapid tomato SG under NS conditions also contribute to improved germination rate under stress conditions (Jones 1986a, Foolad and Jones 1991, Bradford 1995). In this regard, the stress-nonspecific QTLs could affect germination rate by controlling the vigor of the germinating embryo, the variation in the thickness of the endosperm, the physical and permeability properties of the endosperm cell walls, the time of onset or rate of activity of enzymes that modify the properties of the endosperm cell wall, the release of gibberellins by the embryo (which is necessary for endosperm weakening), or other unknown factors essential for the initiation of germination. Theoretically, such effects should contribute to rapid SG under both NS and stress conditions.

Conversely, the detection of stress-specific QTLs indicates the presence of genes that facilitate rapid germination only under specific stress conditions. These QTLs may affect germination-related physiological processes that are triggered by a specific stress, and thus contribute to rapid germination under such conditions. For example, the QTLs for rapid germination under CS might affect the rate of germination-related metabolic activities in the embryo or endosperm under LTs, the thermal time requirements for germination (Bradford 1995) or the rate of embryo growth and its ability to overcome the mechanical restraint imposed by the surrounding endosperm. Similarly, the QTLs for rapid germination under SS or DS might affect the base water potential required for SG, the hydrotime constant (Bradford 1995) or the rate of metabolic activities in the embryo or endosperm under osmotic stress. Moreover, such QTLs may contribute to a better osmoregulation in rapidly germinating seeds. Finally, the finding that some QTLs affect germination only under specific stress conditions may explain the less-than-perfect correlations observed between germination under NS and stress conditions or between germination under different stress conditions in different studies (Foolad 1999a, Foolad, et al. 1999b). The QTL results are consistent with the previous report that selection for rapid SG under NS was less effective than selection under stress conditions (Foolad, et al. 1999a). For practical breeding purposes, however, the stress-nonspecific QTLs should be more useful as they relate to rapid germination under a wide range of environmental conditions.

The identification of germination-related, stress-specific and stress-non-specific QTLs indicates that marker-assisted selection for such QTLs may results in the development of germplasm with improved germination under both NS and stress conditions. Between 5 and 6 QTLs have been identified in the wild accession LA722 for germination under each germination condition. Introgression of this rather small number of QTLs by MAS is feasible, providing opportunities to rapidly develop cultivars with enhanced germination under different conditions. However, further genetic and physiological investigations are needed to examine the nature of the common and/or stress-specific genes (or physiological mechanisms) which affect germination under different conditions.

Tolerance to Different Stresses during Vegetative Growth and Reproduction

Different abiotic stresses, salinity, drought and extreme temperatures, often adversely affect the same growth-related physiological, biochemical, and metabolic aspects of the plant and same protective mechanisms are often activated to these stresses, albeit sometimes through different signaling pathways (Holmberg and Bulow 1998, Shinozaki and Winicov 1998, Zhu 2001, Yamaguchi 2005). Research in other plant species, including A. *thaliana*, has demonstrated that the expression of some common genes is increased in response to different stresses (Kasuga et al. 1999). In fact, single-gene transgenic A. *thaliana* plants have been developed with enhanced salt, cold and freezing tolerance (Holmberg and Bulow 1998, Kasuga et al. 1999). However, in tomato very little research has been done to determine relationships among tolerance to different stresses during VS or reproduction although some studies have been made in our laboratory. The overall results from our investigations indicated the absence of a relationship between ST and CT across the accessions studied. Sometimes tomato genotypes that are capable of fruit setting at low temperatures also have the ability for high temperature fruit setting (Charles and Harris 1972). Further research is therefore needed to determine tolerance to different stresses during vegetative growth and reproduction in tomato.

CURRENT STATUS AND FUTURE PERSPECTIVES

Abiotic stresses are major constraints to crop production worldwide. Most commercial cultivars of tomato are sensitive to abiotic stresses during all stages of plant development, and, thus, tomato production is limited in stressful environments. The cultivated species of tomato has a very narrow germplasm base due to several genetic bottlenecks during its domestication and evolution. Consequently, genetic resources having desirable agricultural characteristics, such as abiotic stress tolerance, are not found in the cultivated species. Fortunately, however, the eight related wild species within *Lycopersicon* are a rich source of desirable genes for tomato crop improvement. Although thus far only a superficial assessment of the extent of the genetic variation for abiotic stress tolerance within *Lycopersicon* has been made, some germplasms with significant tolerance to different abiotic stresses, including salinity, drought and extreme temperatures, have been identified. Such resources are extensively used to characterize the physiological mechanisms and cellular bases of stress tolerance in tomato. Comparatively, little progress has been made on genetics and breeding for stress tolerance in tomato. With the availability of advanced tools of plant molecular biology the focus has largely been shifted to the genetic basis of stress tolerance in tomato. Some notable progress has been achieved. Tolerance components have been defined and their genetic controls characterized, and many controlling genes or QTLs with major effects have been identified and/or cloned. Several breeding lines, cultivars, or germplasms with improved tolerance to different abiotic stresses, in particular salt stress, cold stress and heat stress, have been developed. The new biotechnological approaches of gene transfer have provided opportunities to engineer tomatoes with enhanced stress tolerance. Although the transgenic plants have only been subjected to artificial laboratory tests of stress tolerance, the prospect for engineering tomato plants with field stress tolerance is improving. The recent achievements, however, should be considered only a beginning and it is predicted that the importance of breeding for stress tolerance in tomato will increase substantially in the future. To accelerate the development of tomato cultivars with improved tolerance to different abiotic stresses, various research activities deemed necessary are listed in Appendix 16.1.

SUMMARY

Abiotic stresses are among the most important challenges to agricultural production worldwide. The cultivated tomato, L. esculentum Mill., is sensitive to most abiotic stresses, including salinity, drought, and extreme temperatures. Considerable genotypic variation for abiotic-stress tolerance, however, exists within the related wild species of tomato. During the past few decades, significant progress has been made in characterizing the physiological mechanisms and whole plant response to abiotic stress in tomato. Progress also has been made in discerning the genetic basis of tolerance to different stresses. Most physiological and genetic investigations indicate that tolerance to abiotic stress is a complex trait, controlled by more than one gene, and highly influenced by environmental variation. Furthermore, abiotic stress tolerance in tomato is a developmentally regulated, stage-specific phenomenon; tolerance at one stage of plant development is often not correlated with tolerance at other developmental stages. Thus, specific ontogenic stages have often been evaluated separately for the assessment of tolerance and for the identification, characterization and genetic manipulation of tolerance components. Partitioning of the tolerance into its developmental and physiological components has provided a better understanding of the plant's response to abiotic stress. The use of traditional plant genetics and breeding protocols as well as contemporary molecular biological techniques, such as molecular marker maps, quantitative trait locus (QTL) mapping, marker-assisted selection (MAS), and genetic transformation have resulted in genetic characterization and/or development of tomato genotypes with improved stress tolerance. The application of QTL mapping in particular has contributed to a better understanding of the genetic relationship among tolerances to different stresses. Furthermore, some tomato germplasms, comprising breeding lines and cultivars, have been developed with improved stress tolerance, such as heat tolerance, cold tolerance, and salt tolerance. With the recent advancements in molecular genetic techniques, and the isolation, cloning, and characterization of new stress-related genes and proteins, the prospect for developing commercial cultivars of tomato with enhanced stress tolerance is improving.

Acknowledgements

I thank my colleagues Professors Richard Craig and Dennis Decoteau for reviewing this chapter before submission and for their useful comments and criticisms. I also would like to thank Dr. L.P. Zhang and Mr. G.Y. Lin for their technical support in conducting various experiments in my laboratory. The financial support through research grants from funding agencies, such as the National Research Initiative Competitive Grants Program, U.S. Department of Agriculture (#96-35300-3647), the Agricultural Research Funds administered by The Pennsylvania Department of Agriculture (#ME447275), The Pennsylvania Vegetable Marketing and Research Program, and the College of Agricultural Sciences, The Pennsylvania State University, are highly appreciated. This is contribution no. 427 of the Department of Horticulture, the Pennsylvania State University.

REFERENCES

Abdul-Baki, A.A. 1991. Tolerance of tomato cultivars and selected germplasm to heat stress. J Am Soc Hort Sci 116: 1113-1116.

Abdul-Baki, A.A., and J.r. Stommel. 1995. Pollen viabililty and fruit set of tomato genotypes under optimum- and high-temperature regimes. HortScience 30: 115-117.

Abel, G.H. and A.J. Mackenzie. 1963. Salt tolerance of soybean varieties (*Glycine max* L. Merrill) during germination and later growth. Crop Sci 3: 159-161.

Adams, P. 1991. Effects of increasing the salinity of the nutrient solution with major nutrients or sodium chloride on the yield, quality and composition of tomatoes grown in rockwool. J Hort Sci 66: 201-207.

Adams, P. and L.C. Ho. 1992. The susceptibility of modern tomato cultivars to blossom-end rot in relation to salinity. J Hort Sci 67: 827-839.

Alberdi, M. and L.J. Corduera. 1991. Cold acclimation in plants. Phytochemistry 30: 3177-3184.

Allakhverdiev, S.I., Y. Nishiyama, I. Suzuki, Y. Tasaka, and N. Murata. 1999. Genetic engineering of the unsaturation of fatty acids in membrane lipids alters the tolerance of *Synechocystis* to salt stress. Proc Natl Acad Sci USA 96: 5862-5867.

Apse, M. P., and E. Blumwald. 2002. Engineering salt tolerance in plants. Curr Opin Biotech 13: 146-150.

Apse, M.P., G.S. Aharon, W.A. Snedden, and E. Blumwald. 1999. Salt tolerance conferred by overexpression of a vaculolar Na/H anitort in *Arabidopsis*. Science 285: 1256-1258.

Ashraf, M. 1994. Breeding for salinity tolerance in plants. Critical Rev Plant Sci 13: 17-42.

Ashraf, M. and T. McNeilly. 1988. Variability in salt tolerance of nine spring wheat cultivars. J Agron Crop Sci 160: 14-21.

Ashraf, M., and T. McNeilly. 2004. Salinity tolerance in Brassica oilseeds. Crit Rev Plant Sci 23: 157-174.

Asins, M.J., M.P. Breto, M. Cambra, and E.A. Carbonell. 1993a. Salt tolerance in *Lycopersicon* species. I. Character definition and changes in gene expression. Theor Appl Genet 86: 737-743.

Asins, M.J., M.P. Breto, and E.A. Carbonell. 1993b. Salt tolerance in *Lycopersicon* species. II. Genetic effects and a search for associated traits. Theor Appl Genet 86: 769-774.

Aung, L.H. 1979. Temperature regulation of growth and development of tomato during ontogeny. Proceedings of the 1st International Symposium on Tropical Tomato. Asian Vegetable Ressearch and Development Center, Taiwan, pp. 79-93.

Bajaj, S., J. Targolli, L.F. Liu, T.H.D. Ho, and R. Wu. 1999. Transgenic approaches to increase dehydration-stress tolerance in plants. Mol Breed 5: 493-503.

Bar-Tsur, A., J. Rudich, and b. Bravdo. 1985. High temperature effects on CO_2 gas exchange in heat-tolerant and sensitive tomatoes. J Am Soc Hort Sci 110: 582-586.

Berry, S.Z. 1969. Germinating response of the tomato at high temperature. HortScience 4: 218-219.

Berry, S.Z. and M. Rafique-Uddin. 1988. Effect of high temperature on fruit set in tomato cultivars and selected germplasm. HortScience 23: 606-608.

Bliss, R.D., K.A. Platt-Aloia, and W.W. Thomson. 1986. Osmotic sensitivity in relation to salt sensitivity in germinating barley seeds. Plant Cell Environ 9: 721-725.

Blum, A. 1988. Plant Breeding for Stress Environment. CRC Press, Boca Raton.

Bohnert, H.J. and B. Shen. 1999. Transformation and compatible solutes. Scientia Hort 78: 237-260.

Bohnert, H.J. and R.G. Jensen. 1996. Strategies for engineering water-stress tolerance in plants. Trends Biotech 14: 89-97.

Bohnert, H.J., D.E. Nelson, and R.G. Jensen. 1995. Adaptation to environmental stresses. The Plant Cell 7: 1099-1111.

Bolarin, M.C., F. Perez-Alfocea, E.A. Cano, M.T. Estan, and M. Caro. 1993. Growth, fruit yield, and ion concentration in tomato genotypes after pre- and post-emergence salt treatments. J Am Soc Hort Sci 118: 655-660.

Bolarin, M.C., F.G. Fernandez, V. Cruz, and J. Cuartero. 1991. Salinity tolerance in four wild tomato species using vegetative yield-salinity response curves. J Am Soc Hort Sci 116: 286-290.

Borsani, O., J. Cuartero, J.A. Fernandez, V. Valpuesta, and M.A. Botella. 2001. Identification of two loci in tomato reveals distinct mechanisms for salt tolerance. The Plant Cell 13: 873-887.

Boyer, J.S. 1982. Plant Productivity and environment. Science 218: 443-448.

Bradford, K.J. 1986. Manipulation of seed water relations via osmotic priming to improve germination under stress conditions. HortScience 21: 1105-1112.

Bradford, K.J. 1995. Water relations in seed germination, In: J. Kigel and G. Galili [eds.], Seed Development and Germination. Marcel Dekker, Inc., New York, pp. 351-396.

Bray, E.A. 1997. Plant responses to water deficit. Trends Plant Sci 2: 48-54.

Breto, M.P., M.J. Asins, and E.A. Carbonell. 1994. Salt tolerance in *Lycopersicon* species: III. detection of quantitative trait loci by means of molecular markers. Theor Appl Genet 88: 395-401.

Bruggemann, W. and P. Linger. 1994. Long-term chilling of young tomato plants under low light. IV. Differntial responses of chlorophyll fluorescence quenching coefficients in *Lycopersicon* species of different chilling sensitivity. Plant Cell Physiol 35: 585-591.

Calvert, A. 1969. Flower initiation and development in the tomato. Natl Agric Adv Serv Quart Rev 70: 79-88.

Cannon, O.S., D.M. Gatherum, and W.G. Miles. 1973. Heritability of low temperature seed germination in tomato. HortScience 8: 404-405.

Cano, E.A., F. Perez-Alfocea, V. Moreno, M. Caro, and M.C. Bolarin. 1998. Evaluation of salt tolerance in cultivated and wild tomato species through *in vitro* shoot apex culture. Plant Cell, Tissue and Organ Cult 53: 19-26.

Caro, M., V. Cruz, J. Cuartero, M.T. Estan, and M.C. Bolarin. 1991. Salinity tolerance of normal-fruited and cherry tomato cultivars. Plant and Soil 136: 249-255.

Ceccarelli, S. and S. Grando. 1996. Drought as a challenge for the plant breeder. Plant Growth Regulation 20: 149-155.

Charles, W.B. and R.E. Harris. 1972. Tomato fruit-set at high and low temperatures. Can J Plant Sci. 52: 497-506.

Chaubey, C.N. and D. Senadhira. 1994. Conventional plant breeding for tolererance to problem soils. In: A.R. Yeo and T.J. Flowers [eds.], Soil mineral stresses approaches to crop improvement. Springer-Verlag, Berlin, pp. 11-36.

Cheeseman, J.M. 1988. Mechanisms of salinity tolerance in plants. Plant Physiol 87: 547-550.

Chen, H., Z.Y. Shen, and P.H. Li. 1982. Adaptability of crop plants to high temperature stress. Crop Sci 22: 719-725.

Choudhuri, G.N. 1968. Effect of soil salinity on germination and survival of some steppe plants in Washington. Ecology 49: 465-471.

Cohen, A., A.L. Plant, M.S. Moses, and E.A. Bray. 1991. Organ-specific and environmentally regulated expression of two abscisic acid-induced genes of tomato. Plant Physiol 97: 1367-1374.

Cole, J. 1993. Plenty of water but less to use. The Packer 100: A1.

Coleman, W.K. and R.I. Greyson. 1976. The growth and development of the leaf in tomato (Lycopersicon esculentum). I. The plastochron index, a suitable basis for description. Can J Bot 54: 2421-2428.

Cook, R.E. 1979. Patterns of juvenile morbidity and recruitment in plants. In: O.T. Solbrig, S. Jain, G.B. Johnson, and P.H. Raven [eds.]. Topics in plant population biology. Columbia University Press, Los Angeles, pp. 207-301.

Coons, J.M., R.O. Kuehl, N.F. Oebker, and N.R. Simons. 1989. Germination of eleven tomato phenotypes at constant or alternating high temperatures. HortScience 24: 927-930.

Correll, D.S. 1962. The Potato and its Wild Relatives. Texas Research Foundation, Renner.

Cuartero, J. and R. Fernandez-Munoz. 1999. Tomato and salinity. Scientia Hort 78: 83-125.

Curatero, J., A.R. Yeo, and T.J. Flowers. 1992. Selection of donors for salt-tolerance in tomato using physiological traits. New Phytol 121: 63-69.

Daday, H., F.E. Binet, A. Grassia, and J.W. Peak. 1973. The effect of environment on heritability and predicted selection response in Medicago sativa. Heredity 31: 293-308.

Dahal, P., K.J. Bradford, and R.A. Jones. 1990. Effects of priming and endosperm integrity on seed germination rates of tomato genotypes: I. Germination at suboptimal temperature. J Expt Bot 41: 1431-1440.

Damidaux, R. and J. Martinez. 1992. Tomato cold resistance: Present status and future trends. Acta Hort 301: 73-86.

Dane, F., A.G. Hunter, and O.L. Chambliss. 1991. Fruit set, pollen fertility, and combining ability of selected tomato genotypes under high-temperature field conditions. J Am Soc Hort Sci 116: 906-910.

Darvasi, A. and M. Soller. 1992. Selective genotyping for determination of linkage between a marker locus and a quantitative trait locus. Theor Appl Genet 85: 353-359.

Dehan, K. and M. Tal. 1978. Salt tolerance in the wild relatives of the cultivated tomato: Responses of Solanum pennellii to high salinity. Irrig Sci 1: 71-76.

Dempsey, W.H. 1970. Effect of temperature on pollen germination and tube growth. Rep Tomato Genet Coop 20: 15-16.

DeVos, D.A., J. R. R. Hill, R.W. Helper, and D.L. Garwood. 1981. Inheritance of low temperature sprouting ability in F$_1$ tomato crosses. J Am Soc Hort Sci 106: 352-355.

Dudal, R. [ed.]. 1976. Inventory of major soils of the world with special reference to mineral stress. Cornell Univ Agric Expt Stn., Ithaca.

Duvick, D. 1986. Plant breeding: Past achievement and expectations for the future. Eco Bot 40: 289-294.

Dvorak, J., E. Epstein, A. Galvez, P. Gulick, and J.A. Omielan. 1992. Genetic basis of plant tolerance of soil toxicity. In: H.T. Stalker and J.P. Murphy [eds.], Plant Breeding in the 1990s. C. A. B. International, North Carolina State University, Raleigh, pp. 201-217.

El-Ahmadi, A.B. and M.A. Stevens. 1979a. Reproductive responses of heat-tolerant tomatoes to high temperatures. J Am Soc Hort Sci 104: 686-691.

El-Ahmadi, A.B. and M.A. Stevens. 1979b. Genetics of high-temperature fruit set in the tomato. J Am Soc Hort Sci 104: 691-696.

El-Hassan, G.W. 1978. Inheritance studies of high temperature germination of tomato: Egypt. J Hort 5: 29-41.

Ellis, R.P., B.P. Forster, R. Waugh, L.L. Handley, D. Robinson, D.C. Gordon, and W. Powell. 1997. Mapping physiological traits in barley. New Phytol 137: 149-157.

Epstein, E., J.D. Norlyn, D.W. Rush, R.W. Kingsbury, D.B. Kelly, G.A. Gunningham, and A.F. Wrona. 1980. Saline culture of crops: A genetic approach. Science 210: 399-404.

Esau, K. 1953. Plant Anatomy. John Wiley, New York.

Fernandez-Munoz, R. and J. Cuartero. 1991. Effects of temperature and irradiance on stigma exsertion, ovule viability and embryo development in tomato. J Hort Sci 66: 395-401.

Fernandez-Munoz, R., J.J. Gonzalez-Fernandez, and J. Cuartero. 1995a. Viability of pollen tolerance to low temperatures in tomato and related wild species. J Hort Sci 70: 41-49.

Fernandez-Munoz, R., J.J. Gonzalez-Fernandez, and J. Cuartero. 1995b. Genetics of the viability of pollen grain produced at low temperatures in *Lycopersicon* Mill. Euphytica 84: 139-144.

Fischer, R.A. and N.C. Turner. 1978. Plant productivity in the arid and semiarid zones. Ann Rev Plant Physiol 29: 277-.

Flowers, T. J. 2004. Improving crop salt tolerance. J Exp Bot 55: 307-319.

Flowers, T.J. [ed.] 1999. Salinity and Horticulture. Elsevier, Oxford.

Flowers, T.J., M.A. Hajibagheri, and N.C.W. Clipson. 1986. Halophytes. Quart. Rev. Biol. 61: 313-337.

Foolad, M.R. 1996a. Genetic analysis of salt tolerance during vegetative growth in tomato, *Lycopersicon esculentum* Mill. Plant Breed 115: 245-250.

Foolad, M.R. 1996b. Response to selection for salt tolerance during germination in tomato seed derived from P.I. 174263. J Am Soc Hort Sci 121: 1006-1011.

Foolad, M.R. 1997. Genetic basis of physiological traits related to salt tolerance in tomato, *Lycopersicon esculentum* Mill. Plant Breed 116: 53-58.

Foolad, M.R. 1999. Comparison of salt tolerance during seed germination and vegetative growth in tomato by QTL mapping. Genome 42: 727-734.

Foolad, M.R. and F.Q. Chen. 1998. RAPD markers associated with salt tolerance in an interspecific cross of tomato (*Lycopersicon esculentum X L. pennellii*). Plant Cell Rep 17: 306-312.

Foolad, M.R. and F.Q. Chen. 1999. RFLP mapping of QTLs conferring salt tolerance during vegetative stage in tomato. Theor Appl Genet 99: 235-243.

Foolad, M.R. and G.Y. Lin. 1997a. Absence of a relationship between salt tolerance during germination and vegetative growth in tomato. Plant Breed 116: 363-367.

Foolad, M.R. and G.Y. Lin. 1997b. Genetic potential for salt tolerance during germination in *Lycopersicon* species. HortScience 32: 296-300.

Foolad, M.R. and G.Y. Lin. 1998. Genetic analysis of low temperature tolerance during germination in tomato, *Lycopersicon esculentum* Mill. Plant Breed 117: 171-176.

Foolad, M.R. and G.Y. Lin. 1999. Relationships between cold- and salt-tolerance during seed germination in tomato: Germplasm evaluation. Plant Breed 118: 45-48.

Foolad, M.R. and G.Y. Lin. 2000. Relationship between cold tolerance during seed germination and vegetative growth in tomato, Germplasm evaluation. J Am Soc Hort Sci 125: 679-683.

Foolad, M.R. and G.Y. Lin. 2001a. Genetic analysis of cold tolerance during vegetative growth in tomato, *Lycopersicon esculentum* Mill. Euphytica 122: 105-111.

Foolad, M.R. and G.Y. Lin. 2001b. Relationship between cold tolerance during seed germination and vegetative growth in tomato. Analysis of response and correlated response to selection. J Amer Soc Hort Sci 126: 216-220.

Foolad, M.R. and R.A. Jones. 1991. Genetic analysis of salt tolerance during germination in *Lycopersicon*. Theor Appl Genet 81: 321-326.

Foolad, M.R. and R.A. Jones. 1992. Parent-offspring regression estimates of heritability for salt tolerance during germination in tomato. Crop Sci 32: 439-442.

Foolad, M.R. and R.A. Jones. 1993. Mapping salt-tolerance genes in tomato (*Lycopersicon esculentum*) using trait-based marker analysis. Theor Appl Genet 87: 184-192.

Foolad, M.R., F.Q. Chen, and G.Y. Lin. 1998a. RFLP mapping of QTLs conferring salt tolerance during germination in an interspecific cross of tomato. Theor Appl Genet 97: 1133-1144.

Foolad, M.R., F.Q. Chen, and G.Y. Lin. 1998b. RFLP mapping of QTLs conferring cold tolerance during seed germination in an interspecific cross of tomato. Mol Breed 4: 519-529.

Foolad, M.R., G.Y. Lin, and F.Q. Chen. 1999a. Comparison of QTLs for seed germination under non-stress, cold stress and salt stress in tomato. Plant Breed 118: 167-173.

Foolad, M.R., J.R. Hyman, and G.Y. Lin. 1999b. Relationships between cold- and salt-tolerance during seed germination in tomato: Analysis of response and correlated response to selection. Plant Breed 118: 49-52.

Foolad, M.R., L.P. Zhang, and G.Y. Lin. 2001. Identification and validation of QTLs for salt tolerance during vegetative growth in tomato by selective genotyping. Genome 44: 444-454.

Foolad, M.R., L.P. Zhang, and P. Subbiah. 2003. Relationships among cold, salt and drought tolerance during seed germination in tomato: Inheritance and QTL mapping. Acta Hort 618:47-57.

Foolad, M.R., P. Subbiah, C. Kramer, G. Hargrave, L.P. Zhang, and G.Y. Lin. 2002. Genetic relationships between cold, salt and drought tolerance during seed germination in an interspecific cross of tomato. Euphytica (accepted).

Foolad, M.R., T. Stoltz, C. Dervinis, R.L. Rodriguez, and R.A. Jones. 1997. Mapping QTLs conferring salt tolerance during germination in tomato by selective genotyping. Mol Breed. 3: 269-277.

Forster, B.P., J.R. Russell, R.P. Ellis, L.L. Handley, D. Robinson, C.A. Hackett, E. Nevo, R. Waugh, D.C. Gordon, R. Keith, and W. Powell. 1997. Locating genotype and genes for abiotic stress tolerance in barley: a strategy using maps, markers and the wild species. New Phytol 137: 141-147.

Forster, B.P., M.S. Phillips, T.E. Miller, E. Baird, and W. Powell. 1990. Chromosome location of genes controlling tolerance to salt (NaCl) and vigor in *Hordeum vulgare* and *H. chilense*. Heredity 65: 99-107.

Francois, L.E. and E.V. Maas. 1994. Crop response and management on salt-affected soils. In: M. Pessarakli [ed.], Handbook of Plant Crop Stress. Marcel Dekker, Inc., New York, pp. 149-181.

Galston, A.W., R. Kaur-Sawhney, T. Altabella, and A.F. Tiburcio. 1997. Plant polyamines in reproductive activity and response to abiotic stress. Bot Acta 110: 197-207.

Gautam, R.R., G. Kalloo, and B.S. Dhankhar. 1981. Evaluation of tomato genotypes for fruit set under low temperature condition. Haryana J Hortic Sci 10: 81-85.

George, W.L.J., J.W. Scott, and W.E. Splittstoesser. 1984. *In:* J. Janick [ed.], Parthenocarpy in tomato, pp. 65-84. Hort. Rev. AVI Publishing Co., Westport, CT.

Gorham, J., R.G.W. Jones, and E. McDonnell. 1985. Some mechanisms of salt tolerance in crop plants. Plant and Soil 89: 15-40.

Graham, D. and B.R. Patterson. 1982. Responses of plants to low, nonfreezing temperatures: Proteins, metabolism, and acclimation. Ann Rev Plant Physiol 33: 347-372.

Greenway, H. and R. Munns. 1980. Mechanism of salt tolerance in nonhalophytes. Ann Rev Plant Physiol 31: 149-190.

Groot, S.P.C. and C.M. Karssen. 1987. Gibberellins regulate seed germination in tomato by endosperm weakening: A study with gibberellin-deficient mutant. Planta 171: 525-531.

Grover, A., C. Sahi, N. Sanan, and A. Grover. 1999. Taming abiotic stresses in plants through genetic engineering: current strategies and perspective. Plant Sci 143: 101-111.

Grunberg, K., R. Fernandez-Muñoz, and J. Cuartero. 1995. Growth, flowering, and quality and quantity of pollen of tomato plants grow under saline conditions. Acta Hort 412: 484-489.

Guy, C.L. 1990. Cold acclimation and freezing stress tolerance: Role of protein metabolism. Plant Mol Biol 41: 187-223.

Haigh, A.H. and E.W.R. Barlow. 1987. Water relations of tomato seed germination. Austral J Plant Physiol 14: 485-492.

Hajrasuha, S.N., J. Baniabbassi, J. Metthey, and D.R. Nielson. 1980. Spatial variabililty in soil sampling for salinity studies in southwest Iran. Irrig Sci 1: 197-208.

Hanna, H.Y. and T.P. Hernandez. 1982. Response of six tomato genotypes under summer and spring weather conditions in Louisiana. HortScience 17: 758-759.

Hanna, H.Y., A.J. Adams, and L.L. Black. 1992. LHT24 heat-tolerant tomato breeding line. HortScience 27: 1337.

Hanna, H.Y., T.P. Hernandez, and K.L. Koonce. 1982. Combining ability for fruit set, flower drop, and underdeveloped ovaries in some heat-tolerant tomatoes. HortScience 17: 760-761.

Hasegawa, P.M., R.A. Bressan, and J.K. Zhu. 2000. Plant cellular and molecular responses to high salinity. Ann Rev Plant Physiol Plant Mol Biol 51: 463-499.

Hegarty, T.W. 1978. The physiology of seed hydration and dehydration, and the relation between water stress and the control of germination: A review. Plant, Cell Environ 1: 101-119.

Henk-Venema, J., M. Eekhof, and P.R. Van-Hasselt. 2000. Analysis of low-temperature tolerance of a tomato (*Lycopersicon esculentum*) cybrid with chloroplasts from a more chilling-tolerant *L. hirsutum* accession. Ann Botany 85: 799-807.

Herner, R.C. and T. Kemps. 1983. Chilling tolerance of wild tomato species, 4th Tomato Quality Workshop, University of Florida, pp. 26-37.

Holmberg, N. and L. Bulow. 1998. Improving stress tolerance in plant by gene transfer. Trends Plant Sci 3: 61-66.

Hsiao, T.C. 1973. Plant responses to water stress. Ann Rev Plant Physiol 24: 519-570.

Hsiao, T.C. and K.J. Bradford. 1983. Physiological consequences of cellualar water deficits, In: H.M. Taylor, W.R. Jordan, and T.R. Sinclair [eds.], Limitations to efficient water used in crop production. American Society of Agronomy, Madison, Wis, pp. 227.

Iwahori, S. 1965. High temperature injuries in tomato: IV. Development of normal flower buds and morphological abnormalities of flowers treated with high temperature. J Jpn Soc Hort Sci 34: 33-41.

Iwahori, S. 1966. High temperature injuries in tomato. V. Fertilization and development of embryo with special reference to abnormalities caused by high temperature. J Jpn Soc Hort Sci 35: 55-62.

Jacoby, B. 1994. Mechanisms involved in salt tolerance by plants. In: M. Pessarakli [ed.], Handbook of Plant and Crop stress. Marcel Dekker, New York, pp. 97-123.

Jain, R.K. and G. Selvaraj. 1997. Molecular genetic improvement of salt tolerance in plants. Biotech Ann Rev 3: 245-267.

Jaiwal, P.K., R.P. Singh, and A. Gulati [eds.]. 1997. Strategies for Improving Salt Tolerance in Higher Plants. Science Publishers, Inc., USA.

Jaworski, C.A. and V.J. Valli. 1965. Tomato seed germination and plant growth in relation to soil temperatures and phosphorus levels. Proc Fla State Hort Soc 77: 177-183.

Jenkins, J.A. 1948. The origin of the cultivated tomato. Eco Bot 21: 379-392.

Johjima, T. 1995. Inheritance of heat tolerance of fruit coloring in tomato. Acta Hort 412: 64-70.

Johnson, D.W., S.E. Smith, and A.K. Dobrenz. 1992. Genetic and phenotypic relationships in response to NaCl at different developmental stages in alfalfa. Theor Appl Genet 83: 833-838.

Johnson, W.C., L.E. Jackson, O. Ochoa, R.V. Wijik, J. Peleman, D.A.S. Clair, and R.W. Michelmore. 2000. Lettuce, a shallow-rooted crop, and *Lactuca serriola*, its wild progenitor, differ at QTL determining root architechture and deep soil water exploitation. Theor Appl Genet 101: 1066-1073.

Jones, R.A. 1986a. High salt-tolerance potential in *Lycopersicon* species during germination. Euphytica 35: 576-582.

Jones, R.A. 1986b. The development of salt-tolerant tomatoes: breeding strategies. Acta Hort 190: 101-114.

Jones, R.A. and C.O. Qualset. 1984. Breeding crops for environmental stress tolerance. In: G.B. Collins and J.F. Petolino [eds.], Application of Genetic Engineering to Crop Improvement. Nijihoff/Junk, The Hague, pp. 305-340.

Jones, R.A., M. Hashim, and A.S. El-Beltagy. 1988. Developmental responsiveness of salt-tolerant and salt-sensitive genotypes of *Lycopersicon*. In: E. Whitehead, F. Hutchison, B. Timmeman, and R. Varazy [eds.], Arid Lands: Today and Tomorrow. Westview Press, Boulder, pp. 765-772.

Jordan, W.R., W.A.J. Dougas, and P.J. Shouse. 1983. Strategies for crop improvement for drought prone regions. Agric Water Manage 7: 281-.

Kahn, T.L., S.E. Fender, E.A. Bray, and M.A. O'Connell. 1993. Characterization of Expression of Drought- and Abscisic Acid-Regulated Tomato Genes in the Drought-Resistant Species *Lycopersicon Pennellii*. Plant Physiol 103: 597-605.

Kalaji, M.H. and S. Pietkiewica. 1993. Salinity effects on plant growth and other physiological processes. Acta Physiol Plant 15: 89-124.

Kalloo, G. 1991. Breeding for environmental resistance in tomato, In: G. Kalloo [ed.], Genetic Improvement of Tomato. Springer-Verlag, Berlin, Heidelberg, pp. 153-165.

Kalloo, G. and M.K. Banerjee. 1990. Low temperature fruit set attribute in cultivated variety of tomato derived from *Lycopersicon pimpinellifolium*, Eucarpia Tomato 90: 11th Meet Eucarpia Tomato Working Group, Malaga, Spain, pp. 99-103.

Kamps, T.L., T.G. Isleib, R.C. Herner, and K.C. Sink. 1987. Evaluation of techniques to measure chilling injury in tomato. HortScience. 22: 1309-1312.

Kasuga, M., W. Liu, S. Miura, K. Yamaguchi-Shinozaki, and K. Shinozaki. 1999. Improving plant drought, salt, and freezing tolerance by gene transfer of a single stress-inducible transcription factor. Nature America Inc. 17: 287-291.

Kaufman, M.R. 1969. Effects of water potential on germination of lettuce, sunflower, and citrus seeds. Can J Bot 47: 1761-1764.

Keller, E. and K.L. Steffen. 1995. Increased chilling tolerance and altered carbon metabolism in tomato leaves following application of mechanical stress. Physiol Plant 93: 519-525.

Kemp, G.A. 1968. Low-temperature growth response of the tomato. Can J Plant Sci 48: 281-286.

Kinet, J.M. and M.M. Peet. 1997. Tomato. In: H.C. Wien [ed.], The Physiology of Vegetable Crops. CAB International, Wallingford, U.K, pp. 207-258.

Kramer, P.J. 1980. Water Relations of Plants. Academic Press, New York.

Kramer, P.J. 1983. Water Relations of Plants. Academic Press, New York.

Kuo, C.G., B.W. Chon, M.H. Chou, C.L. Tsai, and T.S. Tsay. 1979. Tomato fruit-set at high temperature, Proceedings of the 1st International Symposium in Tropical Tomato. Asian Vegetable Research and Development Center, Taiwan, pp. 94-109.

Lange, O.L., P.S. Nobel, C.B. Osmond, and H. Ziegler [eds.]. 1981. Physiological plant ecology. I. Responses to physical environment. Springer-Verlag, Berlin.

Lauchli, A. and E. Epstein. 1990. Plant responses to saline and sodic conditions, In: K.K. Tanji [ed.]. Agricultural salinity assessment and management. Am Soc Civil Engrs, New York, pp. 113-137.

Lea, P. J., M. A. Parry, and H. Medrano. 2004. Improving resistance to drought and salinity in plants. Ann Appl Biol 144: 249-250.

Lebowitz, R.J., M. Soller, and M. Beckmann. 1987. Trait-based analysis for the detection of linkage between marker loci and quantitative loci in crosses between inbred lines. Theor Appl Genet 73: 556-562.

Levi, A., H.D. Rabinowitch, and N. Kedar. 1978. Morphological and physiological characters affecting flower drop and fruit set of tomatoes at high temperatures. Euphytica 27: 211-218.

Leviatov, S., O. Shoseyov, and S. Wolf. 1994. Roles of different seed components in controlling tomato seed germination at low temperature. Scientia Hort 56: 197-206.

Leviatov, S., O. Shoseyov, and S. Wolf. 1995. Involvement of endomannanase in the control of tomato seed germination under low-temperature conditions. Ann Bot 76: 1-6.

Levitt, J. 1980a. Responses of plants to environmental stresses: Water, salt and other stresses. Academic Press, New York.

Levitt, J. 1990. Stress interactions - back to the future. HortScience 25: 1363-1365.

Levitt, L. 1980b. Responses of plants to environmental stresses: Chilling, freezing and high temperature stresses. Academic Press, New York.

Lilius, G., N. Holmberg, and L. Bulow. 1996. Enhanced NaCl stress tolerance in trangenic tobacco expressing bacterial choline dehydrogenase. Biotechnology 14: 177-180.

Liptay, A. and P. Schopfer. 1983. Effect of water stress, seed coat restraint, and abscisic acid upon different germination capabilities of two tomato lines at low temperature. Plant Physiol 73: 935-938.

Liptay, A., E.F. Bolton, and V.A. Dirks. 1982. A comparison of field-seeded and transplanted tomatoes grown on a clay soil. Can J Plant Sci 62: 483-487.

Lohar, D.P. and W.E. Peat. 1998. Floral characteristics of heat-tolerant and heat-sensitive tomato (*Lycopersicon esculentum* Mill.) cultivars at high temperature. Scientia Hort 73: 53-60.

Lorenz, O.A. and D.N. Maynard. 1988. Knott's handbook for vegetable growers. Wiley, New York.

Ludlow, M.M. and R.C. Muchow. 1990. A critical evaluation of traits for improveing crop yields in water-limited environments. Advances in Agron. 43: 107-153.

Lutfor-Rahman, S.M. 1998. Eco-physiological study on tomato drought tolerance. Division of Environmental Science and Technology. Kyoto University, Kyoto, Japan, pp. 80.

Lyakh, V.A. 1992. Competence of pollen of wild species and cultivar of tomato to affect fertilization at low temperature. Sex Plant Reprod 5: 128-30.

Lyon, C.B. 1941. Responses of two species of tomatoes and the F_1 generation to sodium sulphate in the nutrient medium. Bot Gaz 103: 107-122.

Lyons, J.M. 1973. Chilling injury in plants. Ann Rev Plant Physiol 24: 445-466.

Lyons, J.M., D. Graham, and J.K. Raison [eds.]. 1979. Low temperature stress in crop plants. The Role of the Membrane. Academic Press, New York.

Maas, E.V. 1986. Salt tolerance of plants. Appl Agric Res 1: 12-26.

Maas, E.V. 1990. Crop salt tolerance, *In*: K.K. Tanji [ed.], Agricultural Salinity Assessment and Management. ASCE Mannuals and Reoprts on Engineering No. 71, New York, pp. 262-304.

Maisonneuve, B. 1982. Effect of low night temperature on a collection of varieties of tomato (*L. esculentum*). I. Study of fruit production and fertilizing quality of the pollen. Agronomie 2: 443-451.

Maisonneuve, B. 1983. Cold resistance of *Lycopersicon hirsutum* pollen. Rep. Tomato Genet Coop 33: 8.

Maisonneuve, B. and A.P.M. Den-Jijs. 1984. In vitro pollen germination and tube growth of tomato (*Lycopersicon esculentum* Mill.) and its relation with plant growth. Euphytica 33: 833-840.

Maisonneuve, B., N.G. Hogenboom, and A.P.M. Den-Jijs. 1986. Pollen selection in breeding tomato (*Lycopersicon esculentum* Mill.) for adaptaton to low temperature. Euphytica 35: 983-992.

Maluf, W.R. and E.C. Tigchelaar. 1982. Relationship between fatty acid composition and low-temperature seed germination in tomato. J Am Soc Hort Sci 107: 620-623.

Mano, Y. and K. Takeda. 1997. Mapping quantitative trait loci for salt tolerance at germination and the seedling stage in barley (*Hordeum vulgare* L.). Euphytica 94: 263-272.

Martin, B. and Y.R. Thorstenson. 1988. Stable carbon isotope composition (delta ^{13}C), water use efficiency and biomass productivity of *Lycopersicon esculentum*, *Lycopersicon pennellii*, and the F_1 hybrid. Plant Physiol 88: 213-217.

Martin, B., C.G. Tauer, and R.K. Lin. 1999. Carbon isotope discrimination as a tool to improve water-use efficiency in tomato. Crop Sci 39: 1775-1783.

Martin, B., J. Nienhuis, and G. King. 1989. Restriction fragment length polymorphisms associated with water use efficiency in tomato. Science 243: 1725-1728.

McCue, K.F. and A.D. Hanson. 1990. Drought and salt tolerance: towards understanding and application. TIBTECH 8: 358-362.

Miltan, O., D. Zamir, and J. Rudich. 1986. Growth rates of *Lycopersicon* species at low temperatures. Z Pflanzenzucht 96: 193-199.

Mitchell, J.H., D. Siamhan, M.H. Wamala, J.B. Risimeri, E. Chinyamakobvu, S.A. Henderson, and S. Fukai. 1998. The use of seedling leaf death score for evaluation of drought resistance of rice. Field Crops Res 55: 129-139.

Monforte, A.J., M.J. Asins, and E.A. Carbonell. 1996. Salt tolerance in *Lycopersicon* species. IV. Efficiency of marker-assisted selection for salt tolerance improvement. Theor Appl Genet 93: 765-772.

Monforte, A.J., M.J. Asins, and E.A. Carbonell. 1997. Salt tolerance in *Lycopersicon* species. V. Does genetic variability at quantitative trait loci affect their analysis? Theor Appl Genet 95: 284-293.

Monforte, A.J., M.J. Asins, and E.A. Carbonell. 1999. Salt tolerance in *Lycopersicon* spp. VII. Pleiotropic action of genes controlling earliness on fruit yield. Theor Appl Genet 98: 593-601.

Morgan, J.M. 1991. A gene controlling difference in osmoregulation in wheat. Aust J Plant Physiol 18: 249-257.

Munns, R. 1993. Physiological processes limiting plant growth in saline soils: some dogmas and hypotheses. Plant, Cell Environ 16: 15-24.

Munns, R. 2005. Genes and salt tolerance: bringing them together. New Phytol 167: 645-663.

Ng, T.J. and E.C. Tigchelaar. 1973. Inheritance of low temperature seed sprouting in tomato. J Amer Soc Hort Sci 98: 314-316.

Nguyen, H.T., R.C. Babu, and A. Blum. 1997. Breeding for drought resistance in rice: Physiology and molecular genetics considerations. Crop Sci. 37: 1426-1434.

Norlyn, J.D. and E. Epstein. 1984. Variablility in salt tolrance of four Triticale line at germination and emergence. Crop Sci 24: 1090-1092.

Paleg, L.G. and D. Aspinall [eds.]. 1981. The physiology and biochemistry of drought resistance in plants. Academic Press, New York.

Parkin, K.L., A. Marangoni, R.L. Jackman, R.Y. Yada, and D.W. Stanley. 1989. Chilling injury. A review of possible mechanisms. J Food Biochem 13: 127-153.

Patterson, B.D. 1988. Genes for cold resistance from wild tomatoes. HortScience 23: 795-795.

Patterson, B.D. and L.A. Payne. 1983. Screening for chilling resistance in tomato seedlings. HortScience 18: 340-341.

Patterson, B.D., D. Graham, and R. Paull. 1979. Adaptation to chilliing: Survival, germination, respiration and protoplasmic dynamics, *In*: J.M. Lyons, D. Graham, and J.K. Raison [eds.], Low Temperature Stress In Crop Plants. The Role of the Membrane. Academic Press, New York, pp. 25-35.

Patterson, B.D., L. Mutton, R.E. Paull, and V.Q. Nguyen. 1987. Tomato pollen development: Stages sensitive to chilling and a natural environment for the selection of resistant genotypes. Plant Cell Environ 10: 363-368.

Patterson, B.D., R. Paull, and R.M. Smillie. 1978. Chilling resistance in *Lycopersicon hirsutum* Humb. & Bonpl., a wild tomato with a wide altitudinal distribution. Austral J Plant Physiol 5: 609-617.

Patterson, B.D., R.E. Paull, and D. Graham. 1987. Adaptation to chilling: survival, germination, respiration and protopolasmic dynamics, In: J.J. Lyons, D. Graham, and J.K. Raison [eds.], Low temperature stress in crops plants: the role of membrane. Academic Prss, New York, pp. 25-35.

Pearen, J.R., M.D. Pahl, M.S. Wolynetz, and R. Hermesh. 1997. Association of salt tolerance at seedling emergence with adult plant performance in slender wheatgrass. Can J Plant Sci 77: 81-89.

Peet, M.M. and D.H. Willits. 1998. The effect of night temperature on greenhouse grown tomato yields in warm climate. Agric Forest Meteorol 92: 191-202.

Peet, M.M. and M. Bartholemew. 1996. Effect of night temperature on pollen characteristics, growth, and fruit set in tomato. J Am Soc Hort Sci 121: 514-519.

Peet, M.M., D.H. Willits, and R.G. Gardner. 1997. Response of ovule development and post-pollen production processes in mail-sterile tomatoes to chronic, sub-acute high temperatures stress. J Expt Bot 48: 101-111.

Peet, M.M., S. Sato, and R.G. Gardner. 1998. Comparing heat stress effects on male-fertile and male-sterile tomatoes. Plant, Cell and Environ 21: 225-231.

Perez-Alfocea, F., G. Guerrier, M.T. Estan, and M.C. Bolarin. 1994. Comparative salt responses at cell and whole-plant levels of cultivated and wild tomato and their hybrid. J Hort Sci 69: 639-644.

Perez-Alfocea, F., M.T. Estan, M. Caro, and M.C. Bolarin. 1993. Response of tomato cultivars to salinity. Plant and Soil 150: 203-211.

Phills, B.R., N.H. Peck, G.E. McDonald, and R.W. Robinson. 1979. Differential responses of *Lycopersicon* and *Solanum* species to salinity. J Am Soc Hortic Sci 104: 349-352.

Philouze, J. and B. Maisonneuve. 1979. Breeding tomatoes for their ability to set fruit at low temperatures. Rev Plant Pathol 58: 54-64.

Picken, A.J.F. 1984. A review of pollination and fruit set in the tomato (*Lycopersicon esculentum* Mill.). J Hort Sci 59: 1-13.

Pillay, I. and C. Beyl. 1990. Early responses of drought-resistant and -susceptible tomato plants subjected to water stress. J Plant Growth Regul 9: 213-219.

Plant, A.L., A. Cohen, M.S. Moses, and E.A. Bray. 1991. Nucleotide sequence and spatial expression pattern of drought- and ABA-induced gene for tomato. Plant Physiol 97: 900-906.

Raison, J.K. and M.A. Brown. 1989. Sensitivity of altitudinal ecotypes of the wild tomato *Lycopersicon hirsutum* to chilling injury. Plant Physiol 91: 1471-1475.

Rana, M.K. and G. Kalloo. 1989. Morphological attributes associated with the adaptation under water deficit conditions in tomato (*L. esculentum* Mill.), 12th Eucarpia Congress 1989, Vortrage Pflanzenzucht, pp. 23-27.

Rana, M.K. and G. Kalloo. 1990. Evaluation of tomato genotypes under drought conditions (Abstr.). 23rd International Horticultre Congress, Firenze, Italy.

Rathinasabapathi, B. 2000. Metabolic engineering for stress tolerance: Installing osmoprotectant synthesis pathways. Ann Bot 86: 709-716.

Redmann, R.E. 1974. Osmotic and specific ion effects on the germination of alfalfa. Can J Bot 52: 803-808.

Reitz, L.P. 1974. Breeding for more efficient water-use - is it real or a mirage. Agric Meteorol 14: 3-10.

Ribaut, J.M., C. Jiang, D. Gonzalez-de-Leon, G.O. Edmeades, and D.A. Hoisington. 1997. Identification of quantitative trait loci under drought conditions in tropical maize. 2. Yield components and marker-assisted selection strategies. Theor Appl Genet 94: 887-896.

Richards, M.A. and B.R. Phills. 1979. Evaluation of *Lycopersicon* species for drought tolerance (Abstr.). HortScience 14: 121.

Richards, R.A. 1983. Should selection for yield in saline regions be made on saline or non-saline soils? Euphytica 32: 431-438.

Richards, R.A. 1996. Defining selection criteria to improve yield under drought. Plant Growth Regulation 20: 157-166.

Richards, R.A. and C.W. Dennett. 1980. Variation in salt concentration in a wheat field. Soil and Water 44: 8-9.

Rick, C.M. 1973. Potential genetic resources in tomato species: clues from observation in native habitats. In: A.M. Srb [ed.], Genes, Enzymes, and Populations. Plenum, New York, pp. 255-269.

Rick, C.M. 1975. The tomato, *In:* R.C. King [ed.], Handbook of Genetics. Plenum Press, New York, pp. 247-280.

Rick, C.M. 1976. Natural variability in wild species of *Lycopersicon* and its bearing on tomato breeding. Genet Agraria 30: 249-259.

Rick, C.M. 1978. The Tomato. Sci Amer 23: 76-87.

Rick, C.M. 1979. Potential improvement of tomatoes by controlled introgression of genes from wild speies, Proc Conf Broadening Genetic Base of Crops. Pudoc Wageningen, pp. 167-173.

Rick, C.M. 1982. The potential of exotic germplasm for tomato improvement, *In:* I.K. Vasil, W.R. Scowcroft, and K.J. Freys [eds.], Plant Improvement and Somatic Cell Genetics. Academic Press, New York, pp. 1-28.

Rick, C.M. 1988. Tomato-like nightshades: affinities, autoecology and breeders' opportunities. Eco Bot 42: 145-154.

Rick, C.M. and W.H. Dempsey. 1969. Position of the stigma in relation to fruit setting of the tomato. Bot Gaz 130: 180-186.

Rontein, D., G. Basset, and A.D. Hanson. 2002. Metabolic engineering of osmoprotectant accumualation in plants. Metabolic Engineering 4: 49-56.

Rosielle, A.A. and J. Hamblin. 1981. Theoretical aspects of selection for yield in stress and non-stress environments. Crop Sci 21: 943-946.

Ross, R. and N. Lott. 2000. A climatology of recent extreme weather and climate events. U. S. Department of Commerce. NOAA/NESDIS, National Climatic Data Center.

Rowley, G. 1993. Multinational and national competition for water in the Middle East: Towards the deepening crisis. J Environ Mgt 39: 187-197.

Rudich, J., E. Zamski, and Y. Regev. 1977. Genotypic variation for sensitivity to high temperature in tomato: pollination and fruit set. Bot Gaz 138: 448-452.

Rush, D.W. and E. Epstein. 1976. Genotypic responses to salinity: differences between salt-sensitive and salt-tolerant genotypes of the tomato. Plant Physiol 57: 162-166.

Rush, D.W. and E. Epstein. 1981a. Comparative studies on the sodium, potassium, and chloride relations of a wild halophytic and domestic salt-sensitive tomato species. Plant Physiol 68: 1308-1313.

Rush, D.W. and E. Epstein. 1981b. Breeding and selection for salt tolerance by the incorporation of wild germplasm into a domestic tomato. J Am Soc Hort Sci 106: 699-704.

Rush, D.W., D.B. Kelly, R. Richards, J.D. Norlyn, Q.W. Kingsbury, and G.A. Cunningham. 1981. Salt-Tolerance Crops Solution to a complex problem. Crop and soil magazine Oct.: 12-16.

Sacher, R.F., R.C. Staples, and R.W. Robinson. 1983. Ion regulation and response of tomato to sodium chloride: A homeostatic system. J Am Soc Hort Sci 108: 566-569.

Saeed, M. and C.A. Francis. 1983. Yield ability in relation to maturity in grain sorghum. Crop Sci 23: 683-.

Salt-affected soils and their management. FAO Soils Bull, FAO, Rome.

Santa-Cruz, A., F. Perez-Alfocea, M. Caro, and M. Acosta. 1998. Polyamines as short-term salt tolerance traits in tomato. Plant Sci 138: 9-16.

Saranga, Y., A. Cahaner, D. Zamir, A. Marani, and J. Rudich. 1992. Breeding tomatoes for salt tolerance: Inheritance of salt tolerance and related traits in interspecific populations. Theor Appl Genet 84: 390-396.

Saranga, Y., D. Zamir, A. Marani, and J. Rudich. 1991. Breeding tomatoes for salt tolerance: Field evaluation of *Lycopersicon* germplasm for yield and dry matter production. J Am Soc Hort Sci 116: 1067-1071.

Saranga, Y., D. Zamir, A. Marani, and J. Rudich. 1993. Breeding tomatoes for salt tolerance: variation in ion concentration associated with response to salinity. J Am Soc Hort Sci 118: 405-408.

Sarg, S.M.H., R.G. Wyn-Jones, and F.A. Omar. 1993. Salt tolerance in the Edkawy tomato. *In*: H. Lieh and A. Al-Masoom [eds.]. Towards the rational use of high salinity tolerant plants. Kluwer Academin Publishers, The Netherlands, pp. 177-184.

Schonfeld, M.A., J. R. C., B.D. Carver, and D.W. Mornhinweg. 1988. Water relations in wheat as drought resistance indicators. Crop Sci 28: 526-531.

Scott, J.W. 1993. Breeding tomatoes for resistance to high temperatures, biotic and abiotic diseases. Hort Bras 11: 167-170.

Scott, J.W. 2000. Fla. 7771, a medium-large, heat tolerant, jointless-pedicel tomato. HortScience 35: 968-969.

Scott, J.W., H.H. Bryan, and L.J. Ramos. 1997. High temperature fruit setting ability of large-fruited, jointless pedicel tomato hybrids with various combinations of heat-tolerance. Proc Fla State Hort Soc 110: 281-284.

Scott, J.W., R.B. Volin, H.H. Bryan, and S.M. Olson. 1986. Use of hybrids to develop heat tolerant tomato cultivars. Proc Fla State Hort Soc 99: 311-315.

Scott, J.W., S.M. Olson, H.H. Bryan, T.K. Howe, P.J. Stoffella, and J.A. Bartz. 1989. Solar Set - A heat tolerant, fresh market tomato hybrid. Florida Agr Expt Sta Circ S:359.

Scott, J.W., S.M. Olson, T.K. Howe, P.J. Stoffella, J.A. Bartz, and H.H. Bryan. 1995. 'Equinox' heat-telerant hybrid tomato. HortScience 30: 647-648.

Scott, S.J. and R.A. Jones. 1982. Low temperature seed germination of *Lycopersicon* species evaluated by survival analysis. Euphytica 31: 869-883.

Scott, S.J. and R.A. Jones. 1985a. Quantifying seed germination responses to low temperatures: Variation among *Lycopersicon* spp. Environ Expt Bot 25: 129-137.

Scott, S.J. and R.A. Jones. 1985b. Cold tolerance in tomato. I. Seed germination and early seedling growth of *Lycoperscion esculentum*. Physiol Plant 65: 487-492.

Scott, S.J. and R.A. Jones. 1986. Cold tolerance in tomato: II. Early seedling growth of *Lycopersicon sp.* Physiol Plant 66: 659-663.

Seki, M., A. Kamei, K. Yamaguchi-Shinozakiz, and K. Shinozaki. 2003. Molecular responses to drought, salinity and frost: Common and different paths for plant protection. Curr Opin Biotech 14: 194–199.

Serrano, R., F.A. Culiañz-Macia, and V. Moreno. 1999. Genetic engineering of salt and drought tolerance with yeast regulatory genes. Scientia Hort 78: 261-269.

Shannon, M.C. 1979. In quest of rapid screening techniques for plant salt tolerance. HortScience 14: 587-589.

Shannon, M.C. 1985. Principles and strategies in breeding for higher salt tolerance. Plant and Soil 89: 227-241.

Shannon, M.C. 1996. New insights in plant breeding efforts for improved salt tolerance. HortTech 6: 96-98.

Shannon, M.C. 1997. Genetics of salt tolerance in higher plants: In: P.K. Jaiwal, R.P. Singh, and A. Gulati [eds.], Strategies for Improving Salt Tolerance in Higher Plants. Science Publishers, Inc., USA.

Shannon, M.C., J.W. Gronwald, and M. Tal. 1987. Effects of salinity on growth and accumulation of organic and inorganic ions in cultivated and wild tomato species. J Am Soc Hort Sci 112: 416-423.

Shelby, R.A., W.H. Greenleaf, and C.M. Paterson. 1978. Comparative floral fertility in heat tolerant and heat sensitive tomatoes. J Am Soc Hort Sci 103: 778-780.

Shen, B., R.G. Jensen, and J.J. Bohnert. 1997. Mannitol protects against oxidation by hydroxyl radicals. Plant Physiol 115: 527-532.

Shen, Z.Y. and P.H. Li. 1982. Heat adaptability of the tomato. HortScience 17: 924-925.

Shinozaki, K. and K. Yamaguchi-Shinozaki. 1996. Molecular responses to drought and cold stress. Curr Opin Biotechnol 7: 161-167.

Shinozaki, K. and K. Yamaguchi-Shinozaki. 1997. Gene expression and signal transduction in water-stress response. Plant Physiol 115: 327-334.

Simon, W.W. 1979. Seed germination under low temperatures. In: J.M. Lyons, D. Graham, and J.K. Raison [eds.], Low Temperature Stress in Crop Plants. The Role of the Membrane. Academic Press, New York, pp. 37-45.

Smith, P.G. and R.H. Millet. 1964. Germinating and sprouting responses of the tomato at low temperatures. Proc Am Soc Hort Sci 84: 480-484.

Stevens, M.A. and J. Rudich. 1987. Genetic potential for overcoming physiological limitations on adaptability, yield, and quality of the tomato. HortScience 13: 673-679.

Stoner, A.K. 1972. Merit, Red Rock and Potomac-tomato varieties adapted to mechanical harvesting. USDA Prod. Res. Rep.

Stoner, A.K. and D.E. Otto. 1975. A greenhouse method to evaluate high temperature setting ability in the tomato. HortScience 10: 264-265.

Subudhi, P.K., D.T. Rosenow, and H.T. Nguyen. 2000. Quantitative trati loci for the stay green trait in sorghum (Sorghum bicolor L. Moench): Consistency across genetic backgrounds and environments. Theor Appl Genet 101: 733-741.

Syverstein, J.P., B. Boman, and D.P.H. Tucker. 1989. Salinity in Florida citrus production. Proc Fl State Hort Soc 102: 61-64.

Szabolcs, I. 1992. Salinization of soils and water and its relation to deserification. Desertification Control Bull 21: 32-37.

Szabolcs, I. 1994. Soil salinization. In: M. Pessarakli [ed.], Handbook of plant crop stress. Marcel Dekker, New York, pp. 3-11.

Tal, M. 1971. Salt tolerance in the wild relatives of the cultivated tomato: Responses of Lycopersicon esculentum, L. peruvianum, and L. esculentum minor to sodium chloride solution. Aust J Agric Res 22: 631-638.

Tal, M. 1985. Genetics of salt tolerance in higher plants: Theoretical and practical considerations. Plant and Soil 89: 199-226.

Tal, M. 1997. Wild germplasm for salt tolerance in plants. *In*: P.K. Jaiwal, R.P. Singh, and A. Gulati [eds.], Strategies for Improving Salt Tolerance in Higher Plants. Science Publishers, Inc., USA, pp. 291-320.

Tal, M. and M.C. Shannon. 1983. Salt tolerance in the wild relatives of the cultivated tomato: Responses of *Lycopersicon esculentum*, *L. cheesmanii*, *L. peruvianum*, *Solanum pennellii* and F$_1$ hybrids to high salinity. Aust J Plant Physiol 10: 109-117.

Tal, M. and U. Gavish. 1973. Salt tolerance in the wild relatives of the cultivated tomato: Water balance and abscisic acid in *Lycopersicon esculentum* and *L. peruvianum* under low and high salinity. Aust J Agric Res 24: 353-361.

Tal, M., A. Katz, H. Heikin, and K. Dehan. 1979. Salt tolerance in the wild relatives of the cultivated tomato: proline accumulation in *Lycopersicon esculentum* Mill., *L. peruvianum* Mill., and *Solanum pennellii* Cor. treated with NaCl and polyethylene glycol. New Phytol 82: 349-355.

Tanaka, A., K. Fujita, and K. Kikuchi. 1974. Nutrio-physiological studies on the tomato plant: Photosynthetic rates of individual leaves in relation to the dry matter production in plants. Soil Sci Plant Nutr 20: 173-183.

Tanaka, Y., T. Hibino, Y. Hayashi, A. Tanaka, S. Kishitani, T. Takabe, S. Yokota, and T. Takabe. 1999. Salt tolerance of trangenic rice overexpressing yeast mitochondrial Mn-SOD in chloroplasts. Plant Sci 148: 131-138.

Tanji, K.K. 1990. Nature and extent of agricultural salinity. In: K.K. Tangi [ed.], Agricultural Salinity Assessment and Mangement. American Society of Civil Engineers, New York, pp. 1-13.

Tanksley, S.D., N.D. Young, A.H. Paterson, and M.W. Bonierbale. 1989. RFLP mapping in plant breeding: new tools for an old science. Bio/Technol 7: 257-264.

Taylor, A.G., J.E. Motes, and M.B. Kirkham. 1982. Germination and seedling growth characteristics of three tomato species affected by water deficit. J Am Soc Hort Sci 107: 282-285.

Thomas, J.C., M. Sepahi, B. Arndall, and H.J. Bohnert. 1995. Enhancement of seed germination in high salinity by engineering mannitol expression in *Arabidopsis thaliana*. Plant Cell Environ 18: 801-806.

Thompson, K., J.P. Grime, and G. Mason. 1977. Seed germination in response to diurnal fluctuations of temperature. Nature (London) 267: 147-149.

Thompson, P.A. 1970. Characterization of the germination response to temperature of species and ecotypes. Nature 225: 827-831.

Thompson, P.A. 1974. Characterization of the germination reponses to temperatures of vegetable crops. I. Tomatoes Scient Hort 2: 35-54.

Turner, N.C. and P.J. Kramer [eds.]. 1980. Adaptation of plants to water and high temperature stress. Wiley-Interscience, New York.

Tylor, H.M., W.R. Jordan, and T.R. Sinclair [eds.]. 1983. Limitation to Efficient Water Use in Crop Production. American Society of Agronomy, Madison, Wis. USA.

Ungar, I.A. 1978. Halophyte seed germination. The Bot Rev 44: 233-264.

Vallejos, C.E. 1979. Genetic diversity of plants for response to low temperature and its potential use in crop plants. *In*: J.M. Lyons, D. Graham, and J.K. Raison [eds.]. Low-temperature stress in crop plants. Academic Press, New York, pp. 473-489.

Vallejos, C.E. and S.D. Tanksley. 1983. Segregation of isozyme markers and cold tolerance in an interspecific backcross of tomato. Theor Appl Genet 66: 241-247.

Vallejos, E., J.M. Lyons, R.W. Breidenbach, and M.F. Miller. 1983. Characterization of a differential low-temperature growth response two species of *Lycopersicon*: the plastochron as a tool. Planta 159: 487-496.

Van-De-Dijk, S.J., J.A. Maris, and P.R.V. Hasselt. 1985. Genotypic variation in chillingi induced leakage of electrolytes from leaf tissue of tomato (*Lycopersicon esculentum* Mill.) in relation to growth under low-energy conditions. J Plant Physiol 120: 39-45.

van-Ieperen, W. 1996. Effects of different day and night salinity levels on vegetative growth, yield and quality of tomato. J Hort Sci 71: 99-111.

Villareal, R.L. 1978. Screening for heat tolerance in the genus *Lycopersicon*. HortScience 13: 479-481.

Villareal, R.L. and S.H. Lai. 1979. Development of heat-tolerant tomato varieties in the tropics, Proceedings of the 1st International Symposium in Tropical Tomao. Asian Vegetable Research and Development Center, Taiwan, pp. 188-200.

Volkmar, K.M., Y. Hu, and H. Steppuhn. 1998. Physiological responses of plants to salinity: A review. Can J Plant Sci 78: 19-27.

Walker, M.A., D.M. Smith, K.P. Pauls, and B.D. McKersie. 1990. A chlorophyl fluorescence screening test to evaluate chilling tolerance in tomato. HortScience 25: 334-339.

Wang, W.-X., B. Vinocur, and A. Altman. 2003. Plant responses to drought, salinity and extreme temperatures: towards genetic engineering for stress tolerance. Planta 218: 1-14.

Weaver, M.L. and H. Timm. 1989. Screening tomato for high-temperature tolerance through pollen viability tests. HortScience 24: 493-495.

Went, F.W. 1957. The Experimental Control of Plant Growth. Ronald Press Co., New York.

Wessel-Beaver, L. and J.W. Scott. 1992. Genetic variability of fruit set, fruit weight, and yield in a tomato population grown in two high-temperature environments. J Am Soc Hort Sci 117: 867-870.

Whittington, W.J. and P. Fierlinger. 1972. The genetic control of time to germination in tomato. Ann Bot 36: 873-880.

Winicov, I. 1998. New molecular approaches to improving salt tolerance in crop plants. Ann Bot 82: 703-710.

Wolf, S., D. Yakir, M.A. Stevens, and J. Rudich. 1986. Cold temperature tolerance of wild tomato species. J Am Soc Hort Sci 111: 960-964.

Wudiri, B.B. and D.W. Henderson. 1985. Effects of water stress on flowering and fruit set in processing tomatoes. Sci Hortic 27: 189-198.

Xu, D., X. Duan, B. Wang, B. Hong, T.H.D. Ho, and R. Wu. 1996. Expression of a late embryogenesis abundant protein gene, HVA1, from barley confers tolerance to water deficit and salt stress in transgenic rice. Plant Physiol 110: 249-257.

Yamaguchi, T. and E. Blumwald. 2005. Developing salt-tolerant crop plants: Challenges and opportunities. Trends Plant Sci. 10: 615-620.

Yeo, A.R. and T.J. Flowers. 1990. Screening of rice (*Oryza sativa* L.) genotypes for physiological characters contributing to salinity resistance, and their relationship to overall performance. Theor Appl Genet 79: 377-384.

Yordanov, I., V. Velikova, and T. Tsonev. 2004. Plant responses to drought and stress tolerance. Bulg J Plant Physiol Special Issue: 187-206.

Younis, A.F. and M.A. Hatata. 1971. Studies on the effects of certain salts on germination, on growth of root, and on metabolism. I. Effects of chlorides and sulphates of sodium, potassium, and magnesium on germination of wheat grains. Plant and Soil 13: 183-200.

Yu, T.T. 1972. The genetics and physiology of water usage in *Solanum pennellii* Corr. and its hybrids with *Lycopersicon esculentum* Mill. University of Cal, Davis, USA.

Zamir, D. and I. Gadish. 1987. Pollen selection for low temperature adaptation in tomato. Theor Appl Genet 74: 545-548.

Zamir, D., S.D. Tanksley, and R.A. Jones. 1981. Low temperature efffect on selective fertilization by pollen mixtures of wild and cultivated tomato species. Theor Appl Genet 59: 235-238.

Zhang, H.X. and E. Blumwald. 2001. Transgenic salt-tolerant tomato plants accumulate salt in foliage but not in fruit. Nature Biotech 19: 765-768.

Zhang, H.X., J.N. Hodson, J.P. Williams, and E. Blumwald. 2001. Engineering salt-tolerant
    Brassica plants: Characterization of yield and seed oil quality in transgenic plants with
    increased vacuolar sodium accumulation. Proc Natl Acad Sci USA 98: 12832-12836.

Zhang, J., H.G. Zheng, A. Aarti, G. Pantuwan, T.T. Nguyen, J.N. Tripathy, A.K. Sarial, S.
    Robin, R.C. Babu, B.D. Nguyen, S. Sarkarung, A. Blum, and H.T. Nguyen. 2001.
    Locating genomic regions associated with components of drought resistance in rice:
    comarative mapping within and a cross species. Theor Appl Genet 103: 19-29.

Zhu, J.K. 2001. Plant salt tolerance. Trends Plant Sci 2: 66-71.

Zhu, J.K., P.M. Hasegawa, and R.A. Bressan. 1997. Molecular aspects of osmotic stress in
    plants. Crit Rev Plant Sci 16: 253-277.

APPENDIX 16.1

FUTURE PROSPECTS OF PRODUCING TOMATO TOLERANT TO ABIOTIC STRESSES

1. Large screening experiments must be conducted in search of additional genetic resources for tolerance to different abiotic stresses, in particular drought tolerance. An almost inexhaustible supply of unexplored diversity exists within the wild species of *Lycopersicon*. A few species within *Solanum* have also been shown to be rich sources of genes for abiotic stress tolerance and which could be used for tomato breeding (Rick 1988).

2. For each abiotic stress, the major components of tolerance have to be identified and characterized. Often it is not only one physiological mechanism or genetic factor that contributes to plant stress tolerance. Furthermore, different physiological and genetic mechanisms may be involved in stress tolerance in different genotypes. The identification and characterization of major tolerance components in different germplasms would facilitate the transfer and possibly pyramiding of the tolerance contributing factors in desirable genetic backgrounds.

3. Research is needed to extend our knowledge of the genetic controls of tolerance related components in different germplasms and during different developmental stages. This includes studies on the inheritance of tolerance traits, the identification and characterization of controlling genes (or QTLs), and assessing the methods for transferring genes individually or together across genotypes.

4. The search for potential tolerance components, including genes and proteins, must go beyond the limits of species within *Lycopersicon*, and it should include other genera such as model plants and microbial organisms.

5. A close collaboration is highly needed among physiologists, geneticists, breeders and molecular biologists interested in plant stress tolerance. Successful development of commercial cultivars with proven field stress tolerance is beyond the capabilities of one individual scientist.

Author Index

Subject Index

Color Plate Section

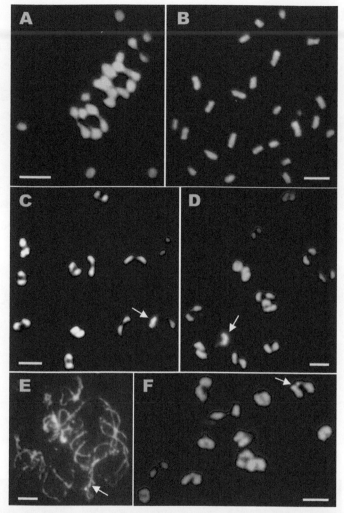

Fig. 3.2 Examples of the use of genomic in situ hybridization (GISH) for genome analysis of tomato interspecific/intergeneric hybrids, and monosomic addition (MA), substitution (SL) and introgression (IL) lines. (A) disrupted pairing at metaphase I of F$_1$ *L. esculentum* x *S. lycopersicoides*; (B) partial differentiation at mitotic metaphase of homologous chromosomes in F$_1$ *S. lycopersicoides* x *S. sitiens*; (C) *S. lycopersicoides* SL-8 at diakinesis with pairing between homeologous chromosomes (bivalent; arrow); (D) *S. sitiens* MA-8 at diakinesis showing pairing between homeologues (bivalent; arrow) and an unpaired *L. esculentum* chromosome (univalent; arrowhead); (E) *S. lycopersicoides* SL-7 at pachytene showing the unpaired *S. lycopersicoides* chromosome (arrow); (F) heterozygous introgression line containing a segment from *S. lycopersicoides* chromosome 7 of ~42 cM (TG499 - TG128) in length (arrow). (A, C, E-F) Red = *S. lycopersicoides*, Blue = *L. esculentum*; (B) Blue = *S. lycopersicoides*, Red = *S. sitiens*; (D) Blue = *L. esculentum*, Red = *S. sitiens*. Bars represent 5 μm.

Chapter 4

Fig. 4.1 Sporogenous (*ms 10*) sterility in tomato. Flowers with exserted and non-exserted stigma

Fig. 4.2 Structural (*sl*) sterility in tomato.

Fig. 4.3 Functional (*ps*) sterility in tomato.

Fig. 4.4 Functional (*ps 2*) sterility in tomato. Longitudinal section of tomato anther cones; on the left: *ps 2*-indehiscent anthers; on the right: fertile anthers

Fig. 4.5 Functional (*ex*) sterility in tomato. On the left: fertile flower (P₂); on the right: *exserted stigma* sterile flower (P₁); in the middle: flower with exserted stigma (F₁)

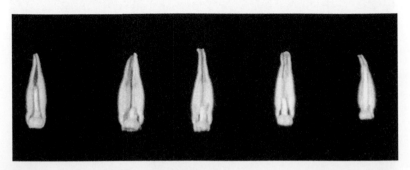

Fig. 4.6 Longitudinal section of tomato flowers with normal and short style.

Fig. 4.7 Emasculation in tomato *ps 2-* line possessing relatively low level stigma without using forceps

Chapter 5

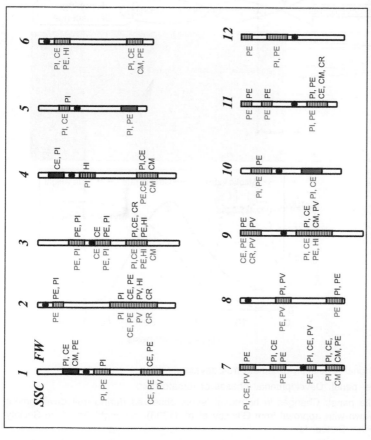

Fig. 5.3 Genetic map of the tomato genome showing the regions where QTLs were detected in at least two different progeny for fruit weight (right of the chromosome) or soluble solid content (left). The map is based on data involving *L. pimpinellifolium* **P I** (Grandillo and Tanksley 1996a, Tanksley et al. 1996, Chen et al. 1999) *L. cheesmanii* **C E** (Paterson et al. 1991, Goldman et al. 1995), *L. chmielewskii* **C M** (Paterson et al. 1988), *L. hirsutum* **H I** (Bernacchi et al. 1998a), *L. peruvianum* **P V** (Fulton et al. 1997), *L. pennellii* **P E** (Eshed and Zamir 1996), *L. esculentum var cerasiforme* **C R** (Saliba-Colombani et al. 2001).

Chapter 10

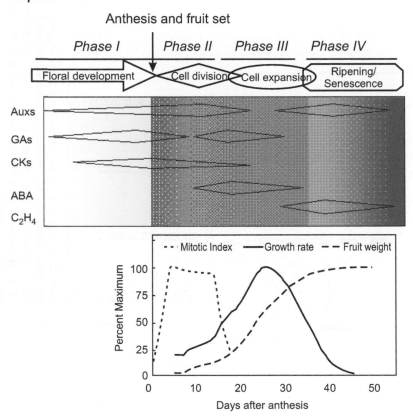

Fig. 10.1 Hormonal regulation of tomato fruit development;

Upper panel: Developmental phases of tomato fruit.

Middle panel: Changes in hormonal levels observed during fruit development (redrawn with approval from Gillaspy et al. (1993), copyright American Society for Plant Biologists).

Lower panel: Cell division as mitotic index is redrawn with approval from Cong et al. (2002). Copyright (2002) National Academy of Sciences, U.S.A.; fruit weight gain and growth rate (weight gained/day) are redrawn and calculated from Abdel-Rahman (1977) with approval. All data are normalized as percent maximum observed during fruit development.

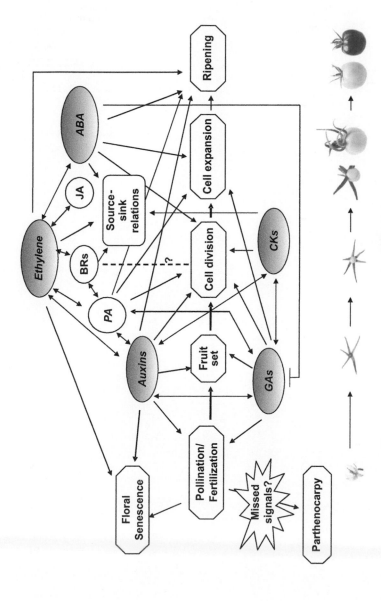

Fig. 10.2 Hormonal cross-talk during tomato fruit development.

Printed and bound by CPI Group (UK) Ltd, Croydon, CR0 4YY

23/10/2024

01778227-0007